Springer Textbooks in Earth Sciences, Geography and Environment

The Springer Textbooks series publishes a broad portfolio of textbooks on Earth Sciences, Geography and Environmental Science. Springer textbooks provide comprehensive introductions as well as in-depth knowledge for advanced studies. A clear, reader-friendly layout and features such as end-of-chapter summaries, work examples, exercises, and glossaries help the reader to access the subject. Springer textbooks are essential for students, researchers and applied scientists.

More information about this series at http://www.springer.com/series/15201

Marco G. Malusà • Paul G. Fitzgerald
Editors

Fission-Track Thermochronology and its Application to Geology

 Springer

Editors
Marco G. Malusà
Department of Earth and
 Environmental Sciences
University of Milano-Bicocca
Milan, Italy

Paul G. Fitzgerald
Department of Earth Sciences
Syracuse University
Syracuse, NY, USA

ISSN 2510-1307   ISSN 2510-1315 (electronic)
Springer Textbooks in Earth Sciences, Geography and Environment
ISBN 978-3-030-07766-2   ISBN 978-3-319-89421-8 (eBook)
https://doi.org/10.1007/978-3-319-89421-8

Denali (formerly Mt. McKinley), America's highest mountains at 6,190 m. This photo is taken from the north, along
the Muldrow glacier. Apatite fission-track data from Denali are discussed in Chapter 9. Photo by Jeff Benowitz
(University of Alaska, Fairbanks).

Printed on acid-free paper

This Springer imprint is published by the registered company Springer International Publishing AG
part of Springer Nature
The registered company address is: Gewerbestrasse 11, 6330 Cham, Switzerland

This book is dedicated to the memory of Chuck Naeser (1940–2016) and Giulio Bigazzi (1941–2018)

Preface

This book *Fission-Track Thermochronology and its Application to Geology* edited by Marco G. Malusà and Paul G. Fitzgerald is the result of the efforts of many recognised scientists, including those who have had long and successful careers and those who are closer to the beginning of their careers. All authors have made contributions in improving our understanding of the application of fission-track analysis, plus other thermochronologic techniques to geologic problems. As a scientific community, we are fortunate that over the years there have been a number of books published on fission tracks, from *Nuclear Tracks in Solids* by Robert Fleischer, Buford Price and Robert Walker in 1975 to *Fission-Track Dating* by Günther Wagner and Peter van den Haute in 1992. In the last decade or so, there have been a number of "thermochronology-themed" books, for example the 2004 Geological Society of America Special Paper 378 *Detrital Thermochronology—Provenance Analysis, Exhumation and Landscape Evolution of Mountain Belts* edited by Matthias Bernet and Cornelia Spiegel, the 2005 MSA Volume 58 on *Low Temperature Thermochronology, Techniques, Interpretations and Applications* edited by Peter Reiners and Todd Ehlers and the 2006 book on *Quantitative Thermochronology. Numerical Methods for the Interpretation of Thermochronological Data* written by Jean Braun, Peter van der Beek and Geoff Batt. There have also been a number of excellent review papers such as Reiners and Brandon (2006) *Using Thermochronology to Understand Orogenic Erosion* in Annual Reviews of Earth and Planetary Sciences Volume 34.

We have no wish to repeat the information in these earlier publications. This book focuses on the basics of applying thermochronology to geologic and tectonic problems, with the emphasis on fission-track thermochronology. Because of the complementary nature of different thermochronologic techniques, many other techniques are also discussed and concepts and approaches are often similar for all techniques, as comes out in the various chapters. The objective of this book was to aim for relatively concise review-type chapters that emphasise the basics, the concepts and the various approaches of applying thermochronology within a geologic framework. In the course of writing and editing this book, certain themes have emerged in the chapters: the importance of sampling strategy and the frame of reference to which data will be referenced; the complementary nature of different thermochronology techniques and the benefits of applying multiple techniques to samples or single detrital grains; the benefits of using techniques with kinetic parameters; that all ages cannot be interpreted as closure temperature ages—especially relevant for detrital studies; and the creativeness in which these thermochronologic methods are applied to a wide variety of geologic problems. The relevance of low-temperature thermochronology is exemplified by the increasing numbers of papers that often include some element of thermochronology in order to better understand geologic processes and temporally constrain the geologic history.

In this book, we deliberately avoided incorporating chapters on modeling, instead concentrating to a certain extent on emphasising the importance of data collection and interpretation of data before modeling is used to test the interpretation or evaluate competing hypotheses. Modeling programs are evolving continuously, and the creators and user groups of most of the commonly used programs (e.g. HeFTy—Rich Ketcham; QTQt—Kerry

Gallagher; and PeCube—Jean Braun) do an excellent job of updating the programs, incorporating new approaches with greater latitude or variables when undertaking both forward and inverse modelling. One of the aims of this book was to make it a volume for relatively new practitioners to thermochronology, as well as scientists experienced in the various methods. In essence, a resource that scientists would pick up before planning a thermochronology study or proceeding with data interpretation, in order to have an overview of the potentials and limitations of the various approaches as illustrated with a few clear examples and citations to other relevant studies.

This book is structured in two parts (Part I and Part II). *Part I* is devoted to an historical perspective on fission-track thermochronology (Chap. 1), to the basic principles of the fission-track method (Chaps. 2–4) and to its integration with other geochronologic methods on single crystals (Chap. 5). Part I also includes two chapters concerning the basic principles of statistics for fission-track dating (Chap. 6) and sedimentology applied to detrital thermochronology (Chap. 7). *Part II* is devoted to the geologic interpretation of the thermochronologic record. Chapters 8–10 analyse the basic principles, conceptual ideas and different approaches for the interpretation of fission-track data within a geologic framework. Chapters 11–21 present concepts and selected case histories to illustrate the potential of fission-track thermochronology from the specific perspective of different Earth's science communities (e.g. tectonics, petrology, stratigraphy, hydrocarbon exploration, geomorphology), with chapters on the application to basement rocks in orogens, passive continental margins and cratonic interiors as well as detrital thermochronology.

Part I, entitled *Basic Principles*, starts with an historical perspective on the topics developed in this book. In Chap. 1, Tony Hurford reviews the background, beginnings and early development of the fission-track method, starting from the discovery of spontaneous fission of uranium in the '40s. He also illustrates the derivation of the fission-track age equation, the progressive advances towards the calibration of the dating method and the development of the external detector method and zeta comparative approach that are in use today.

Chapter 2, written by Barry Kohn, Ling Chung and Andy Gleadow, is a state-of-the-art, handbook style chapter on using apatite, zircon and titanite for fission-track thermochronology. This is based on the University of Melbourne approach which is similar to many laboratories around the world, and this chapter can be used by new practitioners and trained scientists as a reference for practical details for all stages of the method. Topics illustrated in this chapter start with field sampling, mineral separation and preparation for analysis, and move onto the measurement of essential parameters that are required in fission-track thermochronology according to the external detector and LA-ICP-MS methods.

Chapter 3, written by Rich Ketcham, reviews the evolving state of knowledge concerning fission-track annealing based on theory, experiments and geologic observations. This chapter illustrates experiments in which spontaneous or induced tracks are annealed and then etched and measured, and the results and predictions compared against geologic benchmarks. It also describes insights into track structure, formation and evolution based on transmission electron microscopy, small-angle X-ray scattering, atomic force microscopy and molecular dynamics computer modeling, and discusses the implications for thermal history modeling. Areas for future research to enhance our understanding and improve the models are outlined.

Chapter 4, written by Andy Gleadow, Barry Kohn and Christian Seiler, provides ideas about the future directions and research of fission-track thermochronology given the trends of the past and the development of new technologies. These technologies, including LA-ICP-MS analysis for ^{238}U concentrations that eliminates the need for neutron irradiation, new motorised digital microscopes and new software systems for microscope control and image analysis, allow for new image-based and highly automated approaches to fission-track thermochronology.

Chapter 5, written by Martin Danišík, illustrates how fission-track thermochronology can be integrated with U-Pb and (U-Th)/He dating on single crystals, and describes a triple-dating approach that combines fission-track analysis by LA-ICP-MS and *in situ* (U-Th)/He dating, whereby the U-Pb age is obtained as a by-product of U-Th analysis by LA-ICP-MS. This procedure allows simultaneous collection of U-Pb, trace element and REE data that can be used as annealing kinetics parameters or as provenance and petrogenetic indicators.

In Chap. 6, Pieter Vermeesch describes statistical tools to extract geologically meaningful information from fission-track data that were collected using both the external detector and LA-ICP-MS methods. He describes cumulative age distributions, kernel density estimates and radial plots that can be used to visually assess multigrain fission-track data sets. Chapter 6 also describes the concepts of pooled age, central age and overdispersion, describes the application of finite and continuous mixing models to detrital fission-track data and provides the statistical foundations for fission-track dating based on LA-ICP-MS.

In Chap. 7, Marco Malusà and Eduardo Garzanti describe the basic sedimentology principles applied in detrital thermochronology studies in order to fully exploit the potential of the single-mineral approach. They illustrate the fundamentals of hydraulic sorting, how these can be used to improve procedures of mineral separation in the laboratory, and discuss the impact on the detrital thermochronology record of a range of potential sources of bias in the source-to-sink environment, as well as simple strategies for bias minimisation.

Part II of this book, entitled *The Geologic Interpretation of the Thermochronologic Record*, starts with three chapters jointly written by the editors. These three chapters outline the conceptual framework for the geologic interpretation of thermochronologic data and lay the foundations for the subsequent more specific chapters that conclude the second part of this book.

Chapter 8, written by Marco Malusà and Paul Fitzgerald, reviews the nomenclature and basic relationships related to cooling, uplift and exhumation. This chapter examines the characteristics of the thermal reference frame that is used for the interpretation of thermochronologic data, and illustrates strategies to constrain the paleogeothermal gradient at the time of exhumation. It also discusses situations where ages recorded by low-temperature thermochronometers are not related to exhumation, but either reflect transient changes in the regional thermal structure of the crust, episodes of magmatic crystallisation or more localised transient thermal changes due to hydrothermal fluid circulation, frictional heating or wildfire.

In Chap. 9, Paul Fitzgerald and Marco Malusà discuss the approach of collecting samples over a significant elevation range to constrain the timing and rate of exhumation. They also illustrate the concept of the exhumed partial annealing (or retention) zone and use a number of well-known examples from Alaska, the Transantarctic Mountains and the Gold Butte Block to illustrate sampling strategies, common mistakes, factors and assumptions that must be considered when interpreting thermochronologic data from age–elevation profiles.

In Chap. 10, Marco Malusà and Paul Fitzgerald provide an overview of different approaches and sampling strategies for bedrock and detrital thermochronologic studies. They describe bedrock approaches based on multiple thermochronologic methods from the same sample, single methods on multiple samples collected over significant relief or across a geographic region, and multiple methods on multiple samples. They also illustrate different approaches to detrital thermochronology using modern sediments and sedimentary rocks, and discuss the underlying assumptions and the potential impact on data interpretation.

Three subsequent chapters provide insights into the application of bedrock thermochronology within geologic frameworks of increasing complexity.

In Chap. 11, aimed at structural geology and tectonics, Dave Foster reviews the application of fission-track thermochronology in extensional tectonic settings. He presents examples showing how fission-track data can be used to constrain displacements of normal faults, slip rates, paleogeothermal gradients and the original dip of low-angle normal faults.

In Chap. 12, Takahiro Tagami discusses the application of fission-track thermochronology to fault zones. He describes factors controlling the geothermal structure of fault zones, with emphasis on frictional heating and hot fluid flow. He also illustrates how the timing and thermal effects of fault motion can be constrained by fission-track thermochronology and other thermochronologic analyses of fault rocks.

In Chap. 13, Suzanne Baldwin, Paul Fitzgerald and Marco Malusà summarise the ways that the thermal evolution of plutonic and metamorphic rocks in the upper crust may be revealed, and the role that fission-track thermochronology plays in conjunction with other methods such as metamorphic petrology. A simple interpretation of thermochronologic ages in terms of monotonic cooling may be precluded when assumptions based on bulk closure temperatures are violated, for example in cases involving fluid flow and recrystallisation below the closure temperature. They show that geologically well-constrained sampling strategies and application of multiple thermochronologic methods provide reliable constraints on the timing, rates and mechanisms of crustal processes. They present case studies on exhumation of (U)HP metamorphic terranes, an extensional orogeny, a compressional orogeny and a transpressional plate boundary zone.

Chapters 14–18 illustrate the application of detrital thermochronology with examples and strategies of increasing complexity, aimed at inferring realistic geologic scenarios based on the analysis of the thermochronology record preserved in a stratigraphic succession.

Chapter 14, written by Andy Carter, reviews the development and applications of thermochronology to address stratigraphic and provenance problems. This chapter also describes approaches to provenance discrimination based on double- and triple-dating strategies, and potential improvements by integration with mineral trace element data.

In Chap. 15, Matthias Bernet describes the application of fission-track dating to the analysis of the long-term exhumation history of convergent mountain belts. He demonstrates how fission-track ages from sediments and sedimentary rocks of known depositional age can be transferred into average exhumation or erosion rates using the lag-time concept, and how double dating of single grains can be used to detect volcanically derived grains that may obscure the exhumation signal.

Chapter 16, written by Marco Malusà, provides a guide for interpreting complex detrital age patterns collected from stratigraphic sequences. This chapter illustrates how different geologic processes produce different trends of thermochronologic ages in detritus and how these basic age trends are variously combined in the stratigraphic record. This chapter also discusses the potential sources of bias that may affect the thermochronologic signal inherited from the eroded bedrock, and how this bias can be considered and minimised during sampling, laboratory processing and data interpretation.

Chapter 17, written by Paul Fitzgerald, Marco Malusà and Josep-Anton Muñoz, illustrates the benefits of detrital thermochronology analysis using cobbles, either from modern sediments or from basin stratigraphy, to constrain the exhumation history of the adjacent orogen or hinterland. Cobbles are useful because all grains share a common thermal history; thus, inverse thermal modeling of fission-track data in combination with the lag-time concept approach can constrain the timing as well as rate of cooling/exhumation in the hinterland. This chapter also shows the potential of this approach when multiple techniques are applied to each cobble.

Chapter 18, written by David Schneider and Dale Issler, reviews the basics of low-temperature thermochronology when applied to constraining the thermal evolution of a hydrocarbon-bearing sedimentary basin. This chapter is presented in the context of project workflow, from sampling to modeling. Within this framework, the authors illustrate the application of multi-kinetic apatite fission-track dating and the usefulness of the r_{mr0} parameter for interpreting complex apatite age populations that are often present in sedimentary rocks.

The last three chapters of this book are devoted to the application of fission-track and other low-temperature thermochronology methods to investigate the geologic and geomorphologic evolution of orogenic systems, passive continental margins and cratons.

In Chap. 19, Taylor Schildgen and Peter van der Beek discuss the application of low-temperature thermochronology to the geomorphologic evolution of orogenic systems. They review recent studies aimed at quantifying relief development and modification associated with river incision, glacial modifications of landscapes and shifts in the position of range divides. Selected examples point out how interpretations of some data sets are non-unique, and underline the importance of understanding the full range of processes that may influence landscape morphology, and how these processes may affect the spatial patterns of thermochronologic ages.

Chapter 20, written by Mark Wildman, Nathan Cogné and Romain Beucher, reviews the application of fission-track thermochronology to decipher the long-term development of passive continental margins and resolve the spatial and temporal relationships between continental erosion and sediment accumulation in adjacent offshore basins. Examples provided in this chapter suggest that these margins may have experienced significant post-rift activity and that several kilometres of material may be removed from the onshore margin following rifting.

Chapter 21, written by Barry Kohn and Andy Gleadow, is devoted to the application of low-temperature thermochronology to long-term craton evolution. They discuss processes involved in cratonic heating and cooling, with emphasis on the impact of low-conductivity blanketing sediments, the importance of linking the cooling history inferred from thermochronology to onshore and offshore geologic evidence, the impact of dynamic topography and the effects of far-field plate tectonic forces.

Case histories presented in Part II of this book include examples from the East African rift (Chap. 11), the Basin and Range Province of North America (Chaps. 9, 11), the conjugate passive continental margins of the North and South Atlantic (Chap. 20), the Arctic continental margin of north-western Canada (Chap. 18), the cratonic areas of Fennoscandia, Western Australia, southern Africa and Western Canada (Chap. 21), the rift-flank Transantarctic Mountains (Chaps. 9, 13), the transpressional plate boundary zone of New Zealand (Chap. 13), (ultra) high-pressure terranes such as eastern Papua New Guinea and the Western Alps (Chap. 13), compressional orogens and associated foreland basins with examples from the Pyrenees (Chaps. 13, 17), the European Alps (Chaps. 15, 17), the Himalaya (Chap. 15) and the central Alaska Range of North America (Chap. 9), studies from major seismogenic faults in Japan (Chap. 12) and more specific analyses of canyon incision in the Andes, eastern Tibet and North America (Chap. 19). In the light of the wide range of geodynamic settings and topics covered, and of the worldwide distribution of selected examples, Part II of this book may thus be seen as a collection of self-contained chapters that provide an overview of continental tectonics from a thermochronologic perspective.

The chapters benefited from careful peer reviews by Owen Anfinson (Sonoma State University), Phil Armstrong (California State University, Fullerton), Suzanne Baldwin (Syracuse University), Maria Laura Balestrieri (CNR-IGG, Florence), Mauricio Bermúdez (Universidad Central de Venezuela), Ann Blythe (Occidental College), Stéphanie Brichau (Université de Toulouse), Barbara Carrapa (University of Arizona), Andy Carter (Birkbeck, University of London), David Chew (Trinity College, Dublin), Martin Danišík (Curtin University), Alison Duvall (University of Washington), Eva Enkelmann (University of Cincinnati), Rex Galbraith (University College London), Ulrich Glasmacher (Universität Heidelberg), Andy Gleadow (University of Melbourne), Noriko Hasebe (Kanazawa University), Raymond Jonckheere (TU Bergakademie Freiberg), Shari Kelley (New Mexico Tech), Scott Miller (University of Utah), Paul O'Sullivan (GeoSep Services), Jeffrey Rahl (Washington and Lee University), Meinert Rahn (Universität Freiburg), Alberto Resentini (Università di Milano-Bicocca), Diane Seward (Victoria University of Wellington), Kurt Stüwe (Universität Graz), Takahiro Tagami (Kyoto University), Peter van der Beek (Université Grenoble Alpes), Pieter Vermeesch (University College London), Mark Wildman (Université de Rennes 1) and Massimiliano Zattin (Università di Padova). Chapter authors and the reviewers are gratefully acknowledged for their contributions and for their insightful comments and suggestions that allowed us to improve the clarity and completeness of this book.

We are confident that the concepts and ideas summarised in this book will help provide a baseline for future thermochronology-based investigations and insightful interpretations of the intrinsic complexities of the geologic record.

Milan, Italy   Marco G. Malusà
Syracuse, USA   Paul G. Fitzgerald

Contents

About the Editors

Marco G. Malusà is a geologist at the University of Milano-Bicocca whose main research emphasis is the tectonic evolution and exhumation processes of orogenic belts and associated detrital fluxes to sedimentary basins. He obtained his MSc and PhD at the University of Torino, and began his research career contributing to extensive geologic mapping projects in the Western Alps with the National Research Council of Italy. His research integrates bedrock and detrital thermochronology with field geology (sedimentology, stratigraphy, structural geology) and geophysics. Study areas include orogenic belts and sedimentary basins of the Mediterranean and North Africa.

Paul G. Fitzgerald is a Professor of Earth Sciences at Syracuse University in New York. He obtained his BSc and BSc (Hons) at Victoria University of Wellington in New Zealand and his PhD at the University of Melbourne in Australia. He was a post doctoral researcher at Arizona State University and then a research scientist at the University of Arizona. His research involves the application of low-temperature thermochronology to geologic and tectonic problems, mainly associated with the formation of orogens and understanding geologic processes. He has worked extensively in Antarctica, Alaska, the Basin and Range Province of South-western USA, Papua New Guinea and the Pyrenees.

Part I
Basic Principles

An Historical Perspective on Fission-Track Thermochronology

Anthony J. Hurford

1

Abstract

This chapter reviews the background, beginnings and early development of fission-track (FT) thermochronology. In the 1930s, it was discovered that uranium would break into two lighter products when bombarded with neutrons and, subsequently, that uranium was capable of natural, spontaneous fission. The fission process produced damage tracks in solid-state detectors, which could be revealed by chemical etching and observed by electron and, later, by optical microscopy. Fleischer, Price and Walker at the General Electric R&D laboratories developed diverse track-etching procedures, estimates of track registration and stability in different materials, track formation models, uranium determination in terrestrial, lunar and meteorite samples, neutron dosimetry and mineral dating using ^{238}U spontaneous fission. Application to dating of natural and man-made glass was frustrated by low-uranium content and relative ease of track fading (annealing). In the 1970s–1980s, most FT analyses used apatite, zircon and titanite (sphene) to date tephra and acid intrusive rocks with the recognition of differing sensitivities of track annealing in each mineral. Studies in the Alps showed apatite with its greater susceptibility to annealing could provide estimates of the timing and rate of exhumation. The landmark 1980 Pisa FT Workshop highlighted problems with FT system calibration and emphasised the value of annealing in apatite to reveal thermal history. System calibration eventually reached a consensus agreement in 1988 at the Besançon FT Workshop with the majority of analysts adopting the zeta comparative approach. Multiple laboratory and borehole studies have determined the conditions for track annealing in apatite leading to widespread applications in exhumation, sedimentary basin, hydrocarbon exploration and other areas.

1.1 FT Thermochronology: The Fundamentals

If a crystal of apatite ($Ca_5(PO_4)_3(F, Cl, OH)$) is immersed in dilute nitric acid at room temperature for about 20 s, minute etch figures are revealed which can be observed and counted under an optical microscope at high magnification (Fig. 1.1a). Similar etch figures are found if zircon ($ZrSiO_4$) and titanite (sphene) ($CaTiSiO_5$) are immersed in different chemical etchants for appropriate times and temperatures (Fig. 1.1b, c) —see Chap. 2, (Kohn et al. 2018) for specific conditions. These etched tracks have accumulated over time and result from the natural, spontaneous fission of ^{238}U atoms in the crystal lattice of the minerals.[1] Each track results from the fission of a single atom. The number of these spontaneous tracks also depends upon the amount of uranium in the apatite, which is determined conventionally by irradiating the sample with low-energy neutrons. Irradiation induces a proportion of the less abundant ^{235}U isotope to fission, giving a second generation of induced tracks that are revealed and counted in the apatite itself or, more usually, in a detector held against the mineral. The ratio of spontaneous-to-induced tracks provides a measure of the atoms decayed by fission to the total uranium remaining. Allowance has to be made in calculation for the additional decay of uranium by alpha emission. This spontaneous-to-induced track ratio, when used with the rate of spontaneous fission decay, a physical constant, yields the time during which fission tracks have been accumulating in the mineral which in some circumstances may equate to the age of the sample. This method is known as fission-track (FT) dating.

---

[1]Spontaneous fission also occurs in ^{234}U, ^{235}U and ^{232}Th, but their spontaneous fission half-lives are too long and/or abundances too low to produce significant numbers of natural tracks compared to ^{238}U.

A. J. Hurford (✉)
32 Tempest Ave, Potters Bar, Hertfordshire EN65JX, UK
e-mail: tony.hurford@btinternet.com

© Springer International Publishing AG, part of Springer Nature 2019
M. G. Malusà and P. G. Fitzgerald (eds.), *Fission-Track Thermochronology and its Application to Geology*,
Springer Textbooks in Earth Sciences, Geography and Environment, https://doi.org/10.1007/978-3-319-89421-8_1

Fig. 1.1 Fission tracks etched in commonly used materials (scale bars approx. 50 μm).
a Spontaneous ^{238}U tracks in apatite from Tioga ash bed, Pennsylvania, USA (Roden et al. 1990). Etched 5 M HNO$_3$ at 20 °C for 20 s. **b** Spontaneous ^{238}U tracks in zircon from Fish Canyon Tuff. Etched in 100 N NaOH at 220 °C for 6 h (these samples predate use of the NaOH–KOH eutectic etchant). **c** Spontaneous ^{238}U tracks in titanite (sphene) from the Thorr Granite in Donegal, Ireland. Etched in 1HF:2HCl:3HNO$_3$:6H$_2$O acid mixture at 20 °C for 6 min. **d** Induced ^{235}U tracks in a nineteenth century man-made U-bearing glass from Robert Brill at Corning (Brill et al. 1964). Etched in 40% HF at 20 °C for 5 s. Images courtesy of Andy Gleadow and Andy Carter

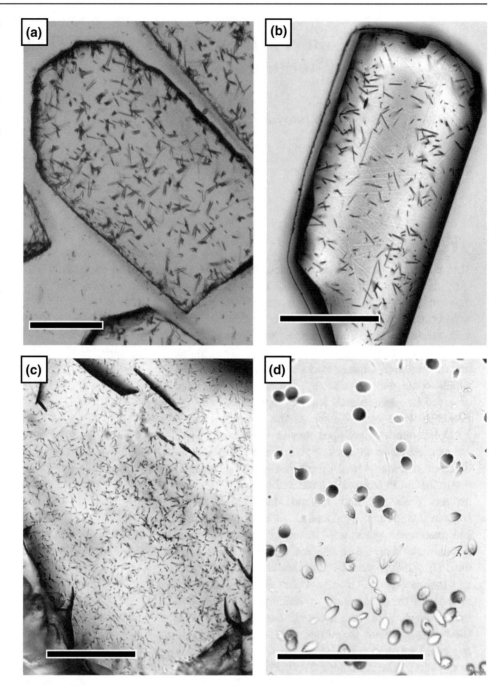

Spontaneous tracks heal or fade naturally at temperatures above ambient values, a process known as annealing. Early applications to natural and man-made glasses found annealing of tracks (along with low-uranium contents) to be major problems. Apatite is more resistant to track fading than glass, but more susceptible than zircon or titanite, especially at temperatures typically found in the upper few kilometres of the Earth's crust. Thus, calculated FT ages may not relate to the age of formation of a sample but to its subsequent history; measured ages should be considered alongside all available geological information.

Understanding the annealing parameters of fission tracks has provided the means for interpreting sample thermal history and interpreting the measured FT age. The particular sensitivity of apatite to annealing has made it especially useful in reconstructing thermal history, in particular by relating the reduction of the lengths of spontaneous tracks (specifically confined track lengths—see below) to temperature and time. Since each track is formed at a different time in the life of a sample, it is exposed to a different portion of that sample's thermal history. The overall track-length distribution thus preserves the integrated temperature record of that sample. Such FT thermochronology has proved especially valuable in sedimentary basin analysis, in hydrocarbon prospection and in understanding the exhumation of crustal rocks. These applications form the basis of many of the chapters that follow.

1.2 The Fission-Track Story Begins

In the 1930s, Enrico Fermi (1934) and his collaborators bombarded uranium with neutrons, concluding that the resulting particles were new elements, lighter than uranium but heavier than lead. Ida Noddack (1934) criticised Fermi's work in that he failed to analyse for elements lighter than lead, suggesting that:

> "It is conceivable that the (uranium) nucleus breaks up into several large fragments, which would, of course, be isotopes of known elements but would not be neighbors of the irradiated element."

Although her paper went largely unnoticed, Noddack effectively predicted nuclear fission. In 1939, Otto Hahn and Fritz Strassman in attempting to reproduce Fermi's results clearly identified the reaction as a break-up of uranium into two lighter products with a range of atomic numbers. Lise Meitner and her nephew Otto Frisch (1939) confirmed these findings coining the term "fission" for the first time because of the similarity of the process to biological cell division. Having fled Germany in July 1938 first to the Netherlands and then to Stockholm (Meitner) and Copenhagen (Frisch), these authors gave the first qualitative discussion of the fission process, drawing the analogy with the liquid-droplet model and contrasting the disrupting effects of Coulomb repulsion with the stabilising influence of surface tension. On 1 September 1939, a propitious month in world history, Niels Bohr and John Wheeler presented a theoretical description which placed the recent developments firmly within the context of the liquid-droplet model. One year later, Flerov and Petrjak (1940) reported the first evidence of the natural spontaneous fission of a nuclide, namely ^{238}U, eliminating any possible cosmic-ray-induced fission by working in an underground laboratory at the Dinamo Station of the Moscow Metro. The formal certificate of discovery stated:

> "The new type of radioactivity with mother nucleus decays into two nuclei, that have kinetic energy of about 160 meV."

The historical events of the 1940s dictated the next steps: the military implications of fission were seized upon, and the remainder of the decade saw frantic applied and basic experimentation. It was not until the 1950s and 60s that advances were made in basic theory largely due to the development of nuclear irradiation facilities.

The first positive recognition of uranium fission tracks can be attributed to D. A. Young in 1958, working at the Atomic Energy Research Establishment, Harwell, UK. Young found that if a lithium fluoride crystal was sandwiched against an uranium foil, irradiated with thermal (low energy) neutrons and then chemically etched, a series of etch pits would be produced. The number of etch pits showed close agreement with the calculated number of fission fragments originating from the uranium foil, and thus, it appeared that each etch pit related to some solid-state damage associated with the passage of the fission fragment. One year later, Silk and Barnes (1959), also working at Harwell, published the first direct transmission electron micrographs of damage tracks in mica created by the passage of ^{235}U fission fragments. These discoveries started a chain reaction of electron microscopic observations of heavy charged-particle tracks in thin films of various materials, although the appearance of the tracks depended critically upon the structure and thickness of the film.

1.3 Fleischer, Price and Walker at GEC Schenectady

In 1961, P. Buford Price and R. M. Walker were working on nuclear tracks at the General Electric Research Laboratory at Schenectady, New York, being joined by R. L. Fleischer the following year. As reported in Fleischer (1998), together the trio were:

> "Charged with doing science – with the long range expectation that some fraction of qualitatively new advances would be of practical use to society, and (crucially) to General Electric."

They resolved to explore the use of solid-state track detectors in nuclear research and, subordinately, hoped to find fossil tracks of cosmic-ray-induced events in meteorites. Price and Walker (1962a) using an electron microscope to observe mica with zircon and apatite inclusions first identified fossil fission tracks which had resulted from the spontaneous fission of ^{238}U. Unaware at this stage of Young's experiments, Price and Walker (1962b) similarly found that tracks in mica could be "developed and fixed" by chemical etching giving permanent channels visible under an optical microscope, thus avoiding track fading problems common when viewing samples by electron microscopy. Anomalous etch figures which had long puzzled crystallographers (e.g. Baumhauer 1894; Honess 1927) were now identified as etched spontaneous fission tracks (Fleischer et al. 1964a). Tracks are revealed because of an accelerated etch rate (V_T) along the damaged track zone relative to the general (or bulk) etch rate (V_G) for the material. In minerals, $V_T \gg V_G$ for some crystallographic orientations (e.g. prismatic c-axis parallel sections of apatite and zircon) giving a linear etch

(a)

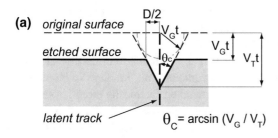

(b) $\theta > \theta_c$

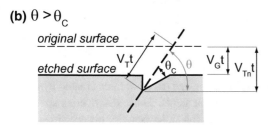

(c) $\theta < \theta_c$

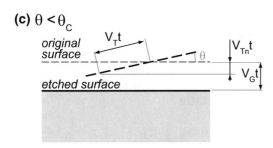

Fig. 1.2 Fundamental track registration and etching geometry. **a** Latent fission tracks are revealed after time t by an accelerated etch rate (V_T) along the damaged track zone relative to the general (or bulk) etch rate (V_G) of the material. Track shape is determined by the difference in the two etch rates. Where $V_T \gg V_G$ as in many minerals, the track is linear (Fig. 1.1a–c). Where V_T is only slightly greater than V_G as in glasses, the track is an elliptical pit (Fig. 1.1d). **b** Tracks lying at an angle of incidence θ greater than a critical value θ_c are registered on the etched and viewed surface. **c** For tracks lying less than the critical angle θ_c, the surface of the material is removed faster during etching than the normal component of V_T and thus the track is totally removed. V_T/V_G defines a material's etching efficiency, that is fraction of tracks intersecting a given surface that are etched on that surface under specified conditions (after Fleischer et al. 1975)

figure (Fig. 1.2). In glasses, V_T approximates to V_G forming conical etch pits with circular to elliptical outlines (see Fig. 1.1d). For comprehensive descriptions of etching geometry, see Chap. 2 of each of Fleischer et al. (1975) and Wagner and van den Haute (1992).

Fleischer, Price and Walker subsequently applied the nuclear track technique to a wide variety of other materials, including minerals, glasses and plastics, finding great variation in the sensitivity of materials to particles of different mass and charge. A profusion of papers from these authors and their associates during the mid- and late 1960s included track-etching recipes, estimates of the stabilities of tracks in different media, registration characteristics of different detectors, models of track formation, applications in uranium

determination of terrestrial, meteoritic and lunar samples, and neutron dosimetry. A comprehensive review of their work was provided by the three scientists in their seminal volume "*Nuclear Tracks in Solids*" (Fleischer et al. 1975).

Pertinent to our story, Price and Walker (1963) first suggested that the spontaneous fission decay of ^{238}U could be usefully employed as a radiometric rock and mineral dating method. FT ages of 3.5×10^8 yr for muscovite from Renfrew, Canada, and of 5×10^8 yr for phlogopite mica from Madagascar were taken as substantiating the concept, and suggestions were proffered for the use of other minerals. FT dating had been born!

1.4 Deriving a FT Age

As with other isotopic dating methods, a FT age depends on the decay of a naturally radioactive parent isotope to a stable daughter product—in our case a fission damage track. From Rutherford and Soddy's law of radioactive decay, the rate of decay is proportional to the number of unstable parent atoms N remaining after any time t:

$$dN/dt = -\lambda N \qquad (1.1)$$

where λ is a constant of proportionality, the decay constant, with a specific value for each radionuclide, expressed in units of inverse time and representing the probability that a nucleus will decay within a given period of time. Integrating Eq. 1.1 with respect to time and accepting that at time $t = 0$ none of the original parent atoms have decayed gives:

$$N = N_o \exp(-\lambda t) \qquad (1.2)$$

which is the general formula for describing the number N of radioactive nuclei remaining after time t, where N_o is the initial number of atoms present. The number of decay events D_t which have occurred since time t is given by:

$$D_t = N_o - N \qquad (1.3)$$

Since it is easier to deal with N, the measurable number of atoms of a radionuclide still remaining in a system, rather than estimating N_o, the original number of atoms present, Eq. 1.2 can be arranged to give an expression for N_o:

$$N_o = N (\exp \lambda t)$$

which, substituted into Eq. 1.3, gives:

$$D_t = N (\exp \lambda t - 1) \qquad (1.4)$$

^{238}U decays both by α-emission, eventually forming ^{206}Pb, and by spontaneous nuclear fission. Each process has a specific decay constant which can be summed to give λ_D, the total constant for the decay of ^{238}U by both mechanisms.

λ_D is essentially just the α-decay constant because this is more than six orders of magnitude greater than the fission decay constant λ_f. Thus for the total decay of ^{238}U, Eq. 1.4 reads:

$$D_t = N \left(\exp \lambda_D t - 1\right) \tag{1.5}$$

D_t includes both α-decay and spontaneous fission decay events, the small fraction due to spontaneous fission, D_s, being given by:

$$D_s = \lambda_f / \lambda_D N \left(\exp \lambda_D t - 1\right) \tag{1.6}$$

Equation 1.6 gives the number of fission events which have taken place over time t in a material now containing N atoms of ^{238}U. Provided the material contained no fission tracks at the beginning of time t and that it has lost no fission tracks through annealing, then it will now contain D_s spontaneous fission tracks, and t will be its FT age. Equation 1.6 can be solved for t provided that D_s can be evaluated and expressed as a proportion of N.

The number of fission events, D_s, in a unit volume can be measured from the area density of etched fission tracks ρ on an internal surface of the uranium-bearing material. Only a fraction of the fission events within one fission-track range of the surface will actually intersect it to give an etchable track. This fraction can be expressed in terms of an effective distance R from the etched surface, within which every fission event can reach the surface although not all may be registered as an etchable track. R is obtained by integrating the contribution of etched tracks from all distances up to R_o, the maximum fission-fragment range. This integral can be evaluated to show that R is equivalent to the etchable range of *one fission fragment*, that is half a track range on either side of an internal surface. The number of fission tracks crossing an internal surface is therefore given by $D_s R$.

The track density ρ actually observed on any surface will also be determined by the etching efficiency η that is the fraction of tracks crossing the surface which are revealed by etching. Thus for the etched FT density on any internal surface, we can write:

$$\rho = D_s R \eta \tag{1.7}$$

For the ^{238}U spontaneous FT density, ρ_s, resulting from D_s fissions per unit volume, Eqs. 1.6 and 1.7 combine to give:

$$\rho_s = \lambda_f / \lambda_D [^{238}U] \left(\exp \lambda_D t - 1\right) R \eta \tag{1.8}$$

where N has been replaced by $[^{238}U]$, the number of atoms of ^{238}U per unit volume remaining in the uranium-bearing material. Conventional evaluation of uranium uses neutron activation, but direct determination of $[^{238}U]$ requires higher energy neutrons which also induce fission in other uranium and thorium isotopes. Lower-energy, thermal neutron irradiation causes ^{235}U alone to fission, creating a second, induced track density in the host material. The number of thermal neutron-induced ^{235}U fissions per unit volume, D_i, can be expressed as:

$$D_i = [^{235}U] \sigma \phi \tag{1.9}$$

where $[^{235}U]$ is the number of ^{235}U atoms per unit volume; σ the ^{235}U nuclear cross section for thermal neutron-induced fission; and ϕ the neutron fluence received by the sample, in neutrons per cm^2.

These induced fission tracks can be etched on an internal surface of the host material in exactly the same way as for the spontaneous tracks. Substituting D_i for D in Eq. 1.7 gives the induced track density, ρ_i, observed on an internal surface:

$$\rho_i = [^{235}U] \sigma R \phi \eta \tag{1.10}$$

on the reasonable assumption that R and η are essentially identical for ^{238}U and ^{235}U fission-fragment tracks. By taking the ratio of the two track densities from Eqs. 1.8 and 1.10, these range and etching efficiency terms cancel, giving:

$$\rho_s / \rho_i = \left(\lambda_f [^{238}U] \exp \lambda_D t - 1\right) / \left(\lambda_D [^{235}U] \sigma \phi\right) \tag{1.11}$$

Assuming the atomic ratio $[^{235}U]/[^{238}U]$ for natural uranium is constant in nature, it may be represented by the isotopic abundance ratio I. Solving Eq. 1.11 for t gives the general form of the FT age equation (Price and Walker 1963; Naeser 1967):

$$t = 1/\lambda_D \ln \left[1 + (\lambda_D \sigma I \phi \rho_s)/(\lambda_f \rho_i)\right] \tag{1.12}$$

Equation 1.12 assumes that the spontaneous and induced track densities, ρ_s and ρ_i are measured on surfaces with similar geometry or else modified by a suitable factor. A useful approximation can be made by noting that $\ln(1+x) = x$, where x is much smaller than 1. Where FT ages are less than a few hundred million years, this approximation can be applied to Eq. 1.12 which then simplifies to:

$$t = (\sigma I \phi \rho_s)/(\lambda_f \rho_i) \tag{1.13}$$

This approximation simply means that over this period of time, which is short compared to the half-life of ^{238}U, there is so little reduction in the total amount of ^{238}U by radioactive decay that effectively it can be regarded as constant. The total decay constant, λ_D, therefore cancels out in the simplification.

1.5 Aspirations for the Infant FT Method

Price and Walker's (1963) paper triggered a series of studies to date mica, ubiquitous in many igneous rocks (e.g. Fleischer et al. 1964a; Bigazzi 1967; Miller 1968). Much activity was also invested in determining recipes for etching tracks in a wide variety of minerals including tourmaline, epidote and garnet, the tracks being derived by implantation of heavy ions (see Fleischer et al. 1975; Wagner and van den Haute 1992 and references therein).

Volcanic and other glasses also received considerable attention in the infancy of the method. In *ocean floor basalts* correlation of K–Ar ages with preserved magnetic reversal signatures offered support for the then-new concept of sea-floor spreading. This prompted attempts to measure FT data on the glassy skins of pillow basalt lavas erupted from the Mid-Atlantic Ridge spreading centre (e.g. Aumento 1969; Fleischer et al. 1971). The dating of *glass shards from acidic volcanic tephra* was seen as a direct means of determining the age of volcanic eruption with data published from studies in Tanzania (Fleischer et al. 1965a), New Zealand (Seward 1974, 1975) and the Yukon (Briggs and Westgate 1978). FT dating was used to supplement K–Ar ages determined on *tektites*, those enigmatic glassy objects probably resulting from meteorite impacts (see discussion in Wagner and van den Haute 1992). In *archaeometry,* the infant FT dating technique prompted excitement with studies undertaken on the dating of artifacts heated in fires, glazes on ceramic bowls and man-made glasses (see Wagner 1978; Wagner and van den Haute 1992; Bigazzi et al. 1993).

Glasses have low-uranium contents and require long analysis times to search for just a few tracks, yielding very poor statistics and high analytical errors; they are also especially susceptible to track annealing at temperatures $\ll 100$ °C. Most FT studies on glasses were in the early years of the method, and this application is now generally considered to be inaccurate if used for direct dating. Similarly virtually no dating analyses have been reported since the 1970s for micas or other low-uraniferous minerals whose etching characteristics had been investigated. In contrast, in uranium ores the crystal structure becomes metamict (effectively destroyed) as a result of α-recoil damage from uranium and thorium decay, and no fission tracks can be discerned. To take the infant FT method, forward attention turned to using accessory minerals.

1.6 Accessorise: Apatite, Zircon and Titanite

The accessory minerals apatite, zircon and titanite (sphene) have trace amounts of uranium in the ppm range giving resolvable numbers of spontaneous tracks. They have proved the ideal material for routine FT analysis. Credit must go to Charles (Chuck) Naeser and to Günther Wagner for pioneering the use of accessory minerals to solve geological problems. Naeser, working in the Southern Methodist University, Dallas, Texas, and subsequently the US Geological Survey, Denver, established the fundamental preparation and analytical techniques and readily passed on his knowledge and expertise to the next generations of FT exponents. Initially, other workers analysed only apatite and titanite because of the simpler handling procedures (see Chap. 2, Kohn et al. 2018). Naeser (1969) had determined FT ages on zircons, but his sample handling techniques were complicated and not readily replicated by others. A revised technical approach (Gleadow et al. 1976) based on Naeser's early experiments opened up FT dating of zircon to all workers, with zircon ages frequently equating with those of titanite. Wagner was pursuing very similar studies on apatite at the University of Pennsylvania and then at the Max-Planck Institut für Kernphysik, Heidelberg (Wagner 1968, 1969—and see references below).

The philosophy in the 1960s and 1970s was to use the FT technique as a dating tool analogous to K–Ar and Rb–Sr methods to determine the age of a sample—by implication the crystallisation age. Investigations concentrated on the dating of acid volcanic and intrusive rocks. A major study dating Tertiary ash-flow tuffs from Central Nevada (Naeser and McKee 1970) furthered the application of FT dating to tephrochronology, with many studies providing time-marker horizons in stratigraphic sequences which have served in Phanerozoic timescale calibration, e.g. zircons and apatites from tuffs and bentonites in UK Ordovician and Silurian stratotypes (Ross et al. 1982). Early studies on acid intrusive rocks yielded apatite FT (AFT) and titanite FT (TFT) ages which equated with K–Ar and Rb–Sr ages, and similarly represented the timing of crystallisation (see, e.g., Naeser 1967; Christopher 1969). However, the recognition of the different annealing characteristics of apatite and titanite (Wagner 1968; Naeser and Faul 1969) produced a tool for the detection of thermal events whereby discordant ages from pairs of co-existing minerals gave a clear indication of minor heating. Calk and Naeser (1973) studied the 10 Ma-old intrusion of a 100-m-diameter basaltic plug into the 80 Ma-old Cathedral Park quartz monzonite in Yosemite National Park, California. Their AFT ages in the country rock were totally or partially annealed up to 150 m from the contact, whilst TFT ages were unaffected just 10 m from basalt.

Clear differences between AFT and K–Ar biotite and hornblende ages were reported by Wagner (1968) for the same granite, granodiorite and gabbro samples from the Odenwald basement, Germany: apatite ages ranged between 69 and 105 Ma, whilst K–Ar spanned 315–340 Ma. Since no post-Hercynian thermal event could be presumed to have affected these rocks, Wagner argued that the AFT ages represented the approximate time of cooling through ~ 100 °C, thereby revealing a difference in uplift rate between the northern and southern Odenwald. This probably represents the first approach towards interpreting FT data in terms of exhumation.

Church and Bickford (1971) related AFT ages of 45–50 Ma from a suite of igneous rocks in the Sawatch Range of Colorado (Rb–Sr whole-rock isochron age of 1650 ± 35 Ma) together with other similar AFT and TFT ages reported by Naeser and Faul (1969) to Laramide igneous activity. Stuckless and Naeser (1972) reported three time–temperature events in the evolution of the Precambrian plutonic basement around the Superstition volcanic field in Arizona. The intrusion of granite at ~ 1390 Ma recorded by Rb–Sr whole-rock isochron and TFT systems, completely reset the titanite and disturbed the Rb–Sr systems in a quartz diorite intruded earlier (Rb–Sr whole-rock age 1540 ± 45 Ma; TFT age 1390 ± 60 Ma). A third episode at ~ 50 Ma recorded by the AFT age in the granite represented cooling due to uplift and erosion subsequent to the Laramide orogeny.

In the Tatra mountains, Poland, AFT ages from 10 to 36 Ma in Hercynian granitic and metamorphic basement significantly post-dated the main nappe-style folding at ~ 80 Ma, leading Burchart (1972) to conclude that the AFT ages related to Miocene post-orogenic uplift and erosion.

AFT was a chronometer capable of recording "events" at temperatures much less than those detected by other mineral and isotopic systems. Publication of the seminal work "*The tectonic interpretation of fission track ages*" by Wagner and Reimer (1972) represented a major advance in the understanding and use of AFT ages in active mountain belts. The study showed the distribution of AFT ages in the Central and Southern European Alps to be concordant with regional tectonic elements when considered with elevation: AFT age generally increased with elevation. This was interpreted as the earlier cooling of the upper sample below the temperature at which fission tracks are retained; thus, for a limited lateral distance, the difference in sample elevation divided by the difference in AFT age may provide a direct measure of uplift. Wagner and Reimer reported an uplift rate of 0.4 mm/yr for the Monte Rosa nappe area and that the fastest uplift around the Simplon Pass had occurred in recent times. This study (and the less readily available Wagner et al. (1977)) represent the foundation of many subsequent tectonic uplift and exhumation studies in active mountain belts worldwide, although in modern thinking the trend of increasing age with elevation may often reflect an uplifted partial annealing zone (see Chap. 8, Malusà and Fitzgerald 2018, and Chap. 9, Fitzgerald and Malusà 2018).

1.7 Problems and Renaissance: Pisa 1980

Despite this promising beginning and the flurry of activity to use the fledgling method to determine ages, FT dating faced serious criticism in the 1970s. Some established geochronologists who measured isotope abundances using mass spectrometry underlined the low analytical precision of the FT method, with the results of a relatively small number of fissioned uranium atoms being observed and counted manually with the inherent potential for misidentification. This was contrasted with other isotopic methods using mass spectrometric automated counting of many ions of precisely known mass. The high level of sensitivity of fission tracks to annealing (especially in apatite) was viewed as rendering the method of little, if any value in determining age—remember the thinking was still dominantly of formation or crystallisation age. A third criticism was the lack of a single system of calibrating the FT system—in particular with no agreed value for the decay constant λ_f—giving little confidence to the comparison of results from different workers and laboratories. Together, these perceived failings caused one pre-eminent British isotope geologist to dismiss fission tracks as the "*Cinderella of dating methods*". By the late 1970s, the earth sciences community had heard of fission tracks and they wondered if the method might provide something interesting to solve geological problems, but was suspicious as to whether the technique was fundamentally flawed.

Under this cloud of uncertainty, a group of FT workers attending the 4th ICOG[2] in 1978 met to discuss shared problems in methodology, in fundamental principles and in interpretation. They determined to hold a workshop specifically to deal with FT matters and hopefully resolve some of the outstanding problems. Thus in September 1980, 44 FT practitioners from 15 countries took up this proposal, meeting in Pisa,[3] Italy, at the Domus Galilaeana, founded in 1942 to commemorate the tercentenary of the death of Galileo Galilei. The Pisa Workshop was a critical week in the history of FT thermochronometry and truly represented the renaissance of the method with four main issues being debated:

---

[2]4th International Conference on Geochronology, Cosmochronology and Isotope Geology (ICOG) in August 1978 at Snowmass-at-Aspen, Colorado, USA.

[3]Note the Pisa workshop is sometimes regarded as the first FT Dating Workshop and is so titled in the Proceedings. However, strictly it was the second such workshop, the first being within the 4th ICOG in 1978, which initiated the Pisa Workshop. Subsequent FT Workshops have followed this latter numbering convention.

- Methodology: what experimental strategies are used by different laboratories?
- Calibration: how is the FT dating system calibrated?
- Statistics: what are the uncertainties of the method and how do different laboratories compare?
- Interpretation: what does track annealing mean and can lowered ages be corrected?

Each of these issues is considered below. The formal confrontation of these problems at Pisa determined the FT community to seek effective solutions. Possibly most significant was the recognition that the understanding of annealing provides a unique means to reveal thermal history (Gleadow and Duddy 1981; Green 1981a) rather than being the hitherto perceived burden. Here lies the foundation of the diversity of geoscience applications we see today including application to hydrocarbon prospection (Gleadow et al. 1983). On FT calibration, the issue was subsequently taken up by the IUGS Subcommission on Geochronology with widespread consultation leading to a recommendation and agreement on procedure (see Sect. 1.9).

1.8 Experimental Strategies

Ideally, the spontaneous and induced track densities measured in FT dating should be registered, revealed and counted under identical conditions over areas having exactly the same uranium content. At the Pisa Workshop Andrew Gleadow underlined that a wide variety of strategies of differing reliability were in use by FT analysts and that the sample may also impose a compromise from the ideal. Variation between strategies derives essentially from differences in registration geometry and etching efficiency of the surfaces on which the tracks are counted. Two distinct groups of procedures can be distinguished (Fig. 1.3), (Gleadow 1981; Hurford and Green 1982).

Multi-grain Methods use separate sample aliquots to determine spontaneous and induced tracks, the track counts from each being summed over the total grains counted (see, e.g. Price and Walker 1963; Naeser 1967, 1979). Typically, areas usually of equal size are measured in 100 grains for each track count. The *population method* has been widely used and requires one aliquot to be heated to anneal fully the spontaneous tracks before irradiation to induce tracks in a portion of ^{235}U atoms. The two aliquots are then processed identically and counted sequentially, one aliquot containing only spontaneous tracks the other only induced tracks. The *subtraction method* omits the heating step so that the irradiated portion contains both induced *and* spontaneous tracks, the induced track density being calculated by subtracting the spontaneous density, inherently less precise since two uncertainties are involved in determining the induced track density.

With multi-grain methods both track counts are made on internal surfaces of the same material with identical registration and etching characteristics. However, statistically equivalent uranium concentrations are assumed in the two aliquots, which supposes a relatively homogeneous uranium distribution both within and between the sample grains, frequently not true. Multi-grain approaches also assume a single age population, probably true for a rapidly cooled volcanics but invalid for samples with multiple age populations, e.g. a contaminated volcaniclastic. Further, laboratory annealing can remove accumulated α-recoil radiation damage in the crystal lattice resulting from uranium and thorium decay. This appears to have no effect on apatite (possibly because α-recoil damage is not preserved) but for zircon and titanite the etching characteristics may be substantially altered, invalidating one of the principal advantages of the multi-grain strategy.

Single-grain Methods allow ages to be measured for each suitable grain even at relatively small sizes (~ 100 μm) and are essential for samples containing multiple populations. Historically 6–10 grains were analysed for each sample. Today, when the age structure of a sample is used to reveal source and thermal history, many more grains are examined —see Chap. 2, (Kohn et al. 2018). The *re-polish and re-etch* strategies are variations with a sample re-polished, re-etched and recounted after irradiation to determine the induced track density (e.g. Fleischer and Price 1963); these approaches compound uncertainties and are no longer used. In contrast, the *external detector method (EDM)* (Fleischer et al. 1964b; Naeser and Dodge 1969) has been used almost exclusively as the FT strategy of choice for several decades. In the EDM ^{238}U spontaneous fission tracks are etched in the grain and ^{235}U induced tracks recorded and etched in a detector of low-uranium muscovite held against the grain mount during irradiation. Analysis examines the etched grain and its mirror image recorded by ^{235}U induced tracks in the detector allowing exact comparison of the identical uranium contents. The EDM records tracks with different track registration geometries. Spontaneous tracks have a 4π geometry with tracks originating from fissioned uranium atoms in the grain lying below the viewed surface, and from above in the part of the grain removed during polishing. In contrast, the detector has 2π geometry with tracks contributed only from below. A correction factor must therefore be applied to the spontaneous/induced ratio which, in the ideal circumstance, is 0.5. Deviation from this value would arise if the bulk etch rate (see Fig. 1.2) was high in either grain and/or detector, thereby possibly removing tracks lying at low angles to the etched and viewed surface. Fortunately,

Fig. 1.3 Experimental strategies for FT dating (modified from Hurford and Green 1982). Amongst the methods described in this chapter, only the external detector (and to a much lesser extent population) methods are nowadays employed. Some laboratories now use LA-ICP-MS to determine uranium content (Hasebe et al. 2004), the equations for which are given in Chap. 6

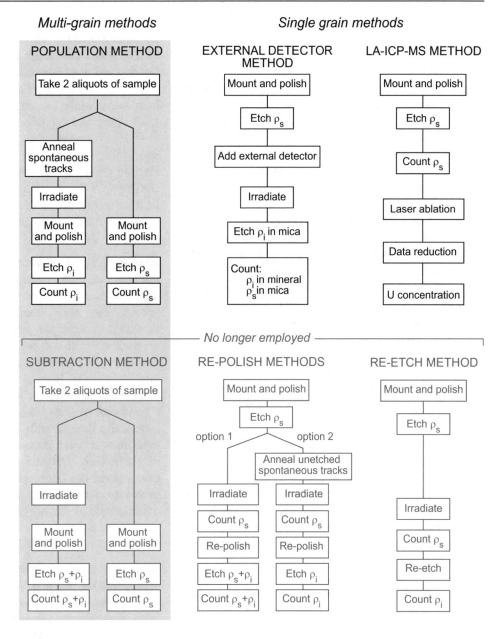

the minerals in common use, apatite, zircon and titanite all have high etching efficiencies for prismatic (c-axis parallel) faces, as does muscovite mica mainly used as detector, and thus, the ideal geometry of 0.5 is appropriate (Gleadow and Lovering 1977; Green and Durrani 1978). Note that etch rate is dependent upon crystal orientation and other faces may exhibit different and anisotropic etching. Some early studies used polycarbonate plastic as a detector. The track registration mechanism may be quite different for such plastics (see Fleischer et al. 1975), and a geometry factor of 0.5 may also be inappropriate. Accordingly, low-uranium muscovite is preferred as a detector by most analysts.

1.9 Calibration of the FT Dating System

A major problem confronting the FT method from its inception until the early 1980s was the fundamental question of system calibration and hinged, in part, upon the value of the ^{238}U spontaneous fission decay constant λ_f. Some 45 experiments revealed a spectrum of results which centred around either 7×10^{-17} yr^{-1} or 8.5×10^{-17} yr^{-1}, $\sim 20\%$ difference (see, e.g., Wagner et al. 1975; Thiel and Herr 1976; Bigazzi 1981; van den Haute et al. 1998). Four groups of experimental procedures have been used to determine λ_f:

- direct determination of fission events using ion chambers and other particle systems
- radiochemical or mass spectrometric measurement of uranium and fission products
- detection and counting of fission tracks in solid-state track detectors (SSTDs) or in photographic emulsions
- comparison of measured FT ages with independently known ages.

Most λ_f values determined using SSTDs and by comparison with minerals from rocks of known age supported a lower value. These experiments require neutron irradiation, and intuitively such methods might appear more appropriate to FT dating with calibration and sample determinations using similar procedures. However, evaluation of neutron fluence can vary according to the dosimetry system used. Thus when a λ_f value determined using specific neutron dosimetry is used to evaluate a sample, similar irradiation and dosimetry protocols should be followed. Those λ_f determinations using direct measurements and radiochemical methods tend to support the higher value. The independence of these values from any neutron dosimetry, track registration or track-etching process argues that they may be more robust when seeking an absolute calibration approach to FT analysis.

Neutron Fluence Measurement[4] can use either metal activation monitors or uranium-doped glass dosimeters. With a *metal activation technique* the neutron-induced gamma activity in foils of gold, cobalt or copper is measured using a scintillation counter and, using the mass of isotope present in the foil, the reacting neutron fluence is calculated. However, neutron beams are not mono-energetic and the response of an isotope varies according to the energy of the bombarding neutron. Different isotopes respond to neutrons of different energies in different ways: the response of one monitor compared with another, and with the ^{235}U in a dating sample, cannot be assumed to be equivalent.

Fortunately for thermal neutrons (energies of 0–0.25 eV with the highest probability of 0.0253 eV at 20.4 °C), the responses of ^{235}U, ^{197}Au and ^{59}Co are virtually parallel with a 0.5 slope denoting a 1/V behaviour (see Wagner and van den Haute 1992, their Fig. 3.3). Thus, in principle, for reactors with well-thermalised fluxes the fluences reacting with these isotopes should be equivalent. However, at higher energy levels (in the epithermal neutron energy range) this

relationship breaks down with large resonance peaks in the Au capture cross section. Use of a poorly thermalised reactor facility would thus result in deviation from this 1/V relationship and would complicate the comparison of the ^{235}U, ^{197}Au and ^{59}Co responses necessitating complex corrections. In addition, whilst thermal neutrons induce fission only in ^{235}U, epithermal neutrons induce fission in both ^{235}U and ^{238}U. Fast neutrons induce fission in both uranium isotopes and ^{232}Th, the resulting tracks being indistinguishable. Neutron dosimetry has been treated in detail by Green and Hurford (1984) who also recommend minimum specifications for thermalisation of reactor facilities used for FT work expressed in terms of cadmium ratios of 3 for Au, 24 for Co and 48 for Cu.[5]

Uranium-Doped Glasses can be used to monitor neutron-induced ^{235}U fission events by counting the resulting tracks in the glass or, more usually, in a detector held against the glass during irradiation. Detectors of low-uranium mica are preferable in that they preserve a permanent record of the irradiation whilst leaving the dosimeter glass undamaged and reusable for multiple irradiations (see, e.g., Hurford and Green 1983). As a dosimetry method it is appealing in that neutron-induced ^{235}U fission is used in both standard glass and unknown sample. However, quantitative determination of fluence is complicated by the need for evaluation of the detection and etching efficiencies of the system. Nevertheless, if identical dosimeter and experimental conditions are always used, the fluence ϕ is related to the measured induced track density ρ_d in the dosimeter (or its detector) by:

$$\phi = B\rho_d \qquad (1.14)$$

where B is a constant specific to the dosimeter (Fleischer et al. 1975; Hurford and Green 1981a).

To avoid the difficult empirical evaluation of fission-fragment ranges and etching efficiencies (see, e.g., Wagner and van den Haute 1992), some workers have used metal activation monitors to evaluate B, often over multiple irradiations: Hurford and Green (1983) showed values of B for two dosimeter glasses measured over a 5-year period. B may then be used to derive fluence values in subsequent irradiations which, in turn, are substituted into the age equation to calculate sample ages.

A comprehensive listing of uranium-doped glasses used as neutron dosimeters is given in Hurford (1998, Table 1). The two sets of glasses used most extensively are the NIST

---

[4]Neutron irradiation induces fission in a proportion of ^{235}U in a sample as a measure of its uranium content and requires determination of the total number of neutrons to which the sample is exposed, known as the neutron fluence. Fluence is the neutron flux (or dose) integrated with respect to the irradiation time and is expressed in neutrons cm^{-2}.

[5]Cadmium absorbs thermal neutrons ($<\sim 0.4$ eV) whilst permitting higher energy neutrons passage through. A cadmium ratio gives the activities of a bare monitor/ a Cd-shielded monitor, thus recording neutrons of all energies/neutrons with energies >0.5 eV; the higher the cadmium ratio, the better thermalised the reactor facility.

(NBS) SRM 600 series and six glasses subsequently denoted CN1–CN6 prepared by Jan Schreurs at the Corning Glass Works. A subset of SRM 600, numbered SRM 961–964 was issued as pre-irradiated glasses with neutron fluences measured at NBS by activation monitors (Carpenter and Reimer 1974). These pre-irradiated glasses offered a common baseline for all workers, whereby a virgin SRM glass wafer of the same series could be irradiated in an unknown reactor facility and then etched simultaneously with the equivalent NBS-irradiated glass. After counting the track densities in the two glasses, the fluence in the unknown reactor ϕ_{UNK} could be calculated directly by comparison with the cited NBS fluence value ϕ_{NBS}:

$$\phi_{UNK} = \phi_{NBS}\, \rho_{d2}/\rho_{d1} \qquad (1.15)$$

where ρ_{d1} is the ^{235}U induced track density measured in the NBS pre-irradiated glass, and ρ_{d2} is the ^{235}U induced track density measured in the equivalent glass irradiated in the unknown reactor.

Despite the conceptual excellence of this calibration experiment, its effectiveness was marred by differences of up to 11% in the fluences measured by gold and copper activation monitors for the same NBS irradiation. Unlike the SRM 600 series, the Corning glasses were prepared:

- using an uranium salt with natural uranium isotopic ratios —the salt was prepared before 1939, that is before depletion of ^{235}U was initiated
- without the addition of other trace elements which can produce unwanted additional activity after irradiation and/or disturb the neutron fluence, e.g. Gd, can absorb neutrons.

Repeated analysis has shown an homogeneous distribution of uranium both within and between the Corning glass wafers. Thus for most analysts, CN1, CN2 and CN5 have become the dosimeter glasses of choice with, respectively 39.81 ± 0.69, 36.5 ± 1.4 and 12.17 ± 0.62 uranium ppm by weight (see Hurford 1990; Bellemans et al. 1995).

An additional series of unirradiated and irradiated uranium-doped glasses IRMM-540 has been produced at the Institute for Reference Materials and Measurements of the European Commission, Geel, Belgium, with fluence measurement using both ^{97}Au and ^{59}Co monitors (De Corte et al. 1998; Roebben et al. 2006).

Interdependence of λ_f and ϕ In the 1970s and early 1980s, analysts were publishing FT ages using different λ_f values and neutron dosimetry schemes. In the age equation (Eq. 1.13), a change in either λ_f or ϕ could be compensated for by a change in the other. Such changes were not a matter of artifice but resulted from the selection by the analyst of a λ_f value and neutron irradiation scheme which gave the right

ages on standard material (see below). This interrelation of λ_f and neutron dosimetry has been discussed by Khan and Durrani (1973), Wagner et al. (1975), Thiel and Herr (1976), Hurford and Green (1981a, 1982 and 1983). It was the ratio ϕ/λ_f which was effective in giving these concordant results, indicating that simple agreement with an independent age cannot be taken to validate *individually* either the neutron dosimetry or the value of λ_f. Correct calculation of a sample age cannot be made by casual selection of λ_f and a neutron fluence dosimetry scheme, but Eq. 1.13 can be used to discern which combination of parameters gives the right answers on age standards of known age.

Standards are critical for all analytical procedures, providing baselines to assess the reproducibility of measurement, comparison of analysts, equipment and laboratories, and to compare different methodological approaches. *Age standards* have long been used in geochronology to provide confidence in the data to the geological end-user and should meet strict criteria:

- come from an accessible, geologically well-documented horizon
- be sufficient to fulfil immediate and future needs
- be homogeneous in age; if a mineral is separated it should be a single generation and free from older, derived crystals
- have unambiguous, precise independent calibrating ages (e.g. ^{40}Ar–^{39}Ar, K–Ar, U–Pb and Rb–Sr), preferably determined in more than one laboratory and compatible with known stratigraphy
- require no correction to the FT density based on track-size measurement (see G. A. Wagner in Hurford and Green 1981b, 1983).

Although a prospective standard may conform to these criteria, if the FT and independent ages were inequal, a systematic error would be introduced into the FT calibration. Such a difference could exist if a sample had been heated sufficiently to cause partial track loss, but insufficiently to affect the independent (e.g. argon) system: such an FT calibration would subsequently yield sample ages which were too old. Conversely, inclusion of older derived crystals in an FT calibration analysis would give an FT value disproportionately high when compared with the calibrating age, resulting in the subsequent calculation of sample ages that are too young. The self-consistency of results obtained on **a series of putative age standards** from wide geographical, temporal and uranium concentration ranges safeguards against the introduction of such a systematic error since it is inconceivable that such an error could be of constant proportion for each of a widespread group of samples. Ideal age standards are sub-aerial volcanic rocks or minor high-level intrusives,

Table 1.1 Age standards used most frequently in FT dating

Fish Canyon Tuff Apatite and Zircon (27.8 ± 0.5 Ma)
Description: Crystal-rich, welded ash-flow unit, up to 1000 m in thickness in San Juan mountains of southern California. K–Ar, ^{40}Ar–^{39}Ar phenocrysts of plagioclase, sanidine, biotite and hornblende give concordant ages; U–Pb zircon ages (Steven et al. 1967; Hurford and Hammerschmidt 1985; Cebula et al. 1986; Lanphere and Baadsgaard 2001; Philips and Matchan 2013)

Durango Apatite (31.4 ± 0.5 Ma)
Readily available euhedral lemon-yellow, gem quality apatites up to 3 cm long, gangue minerals in iron ore deposits, Cerro de Mercado mine, Durango City, Mexico (Paulick and Newesely 1968; Young et al. 1969). K–Ar and ^{40}Ar–^{39}Ar sanidine-anorthoclase ages given by underlying Aguila Formation tuffs and volcanic breccia, and overlying Santuario crystal-vitric ash-flow tuff (Swanson et al. 1978; McDowell and Keizer 1977; McDowell et al. 2005). Readily identifiable crystallographic faces have proved useful in FT annealing and track-length studies

Buluk Member Tuff (FTBM) Zircon (16.4 ± 0.2 Ma)
Miocene Bakate Formation of volcanics, volcaniclastic sediments, fluviatile clays, silts, sands and conglomerates 45 km east of Lake Turkana, northern Kenya (Key and Watkins 1988). K–Ar high temperature alkali-feldspar and plagioclase ages internally consistent from Buluk Member and overlying Gum Dura Member (McDougall and Watkins 1985). Zircon separated from 5 rapidly reworked and redeposited air-fall tuffs free of inherited zircons (Hurford and Watkins 1987)

Mt Dromedary Igneous Complex Titanite (Zircon and Apatite) (98.7 ± 0.6 Ma)
South of Narooma, NSW, Australia; 6 km diameter quartz monzonite and quartz syenite pluton, surrounded by a monzonitic rim, probably emplaced at high level and rapidly cooled. Weighted mean of 98.7 ± 0.6 Ma (Green 1985) from K–Ar hornblende and biotite and Rb–Sr biotite ages. (McDougall and Wellman 1976; McDougall and Roksandic 1974; Williams et al. 1982). Contains abundant titanite, zircon and apatite

Note that zircon from the Tardree rhyolite (Fitch and Hurford 1977) was included in the IUGS recommendation (Hurford 1990). Many zircon phenocrysts in this Tertiary (Danian) rhyolite from Northern Ireland show inhomogeneous uranium distribution with pronounced zoning (Tagami et al. 2003), and for this reason, the Tardree zircon is not ideal as an age standard. Moldavite glass with an age of 15.1 ± 0.7 Ma (Gentner et al. 1963) was also included in the IUGS recommendation

which have cooled quickly and which show no evidence of post-formational disturbance. Few samples completely fulfil the conditions demanded of an age standard. Table 1.1 lists those samples most widely used in FT analysis.

Zeta and the Besançon Agreement An alternative approach to FT calibration which circumvented the need for explicit evaluation of λ_f and an absolute determination of thermal neutron fluence was proposed by Fleischer and Hart in 1972—and then ignored. They suggested that the simplified age equation (Eq. 1.13) may be rewritten in terms of three track densities and a proportionality factor zeta ζ:

$$t = \zeta(\rho_s/\rho_i)\,\rho_d \qquad (1.16)$$

Note that neutron fluence ϕ is represented by ρ_d the track density in a specific dosimeter glass or adjacent mica detector. Zeta for the dosimeter glass may be evaluated against a mineral standard whose age t_{std} is known (or reasonably inferred) from independent age determinations:

$$\zeta = (t_{std}\,\rho_i)/(\rho_s\,\rho_d) \qquad (1.17)$$

Substituting ζ into the full FT age Eq. 1.12, and introducing an appropriate geometry factor G, gives:

$$t = 1/\lambda_D \ln\left(1 + G\,\lambda_D\zeta\,\rho_d\,\rho_s/\rho_i\right) \qquad (1.18)$$

where ζ for a given glass dosimeter is evaluated from age standards according to:

$$\zeta = [\exp(\lambda_D t_{std}) - 1]/[\lambda_D(\rho_s/\rho_i)_{std}\,\rho_d] \qquad (1.19)$$

and where $G = 0.5$ for the external detector method, and $G = 1$ for the population method.

The first zeta values were published by Paul Green and myself for dosimeter glasses SRM 612, CN1 and CN2 measured using zircons from four sources over a seven-year period (Hurford and Green 1983). Green (1985) extended this initial zeta study to a larger number of zircon samples and, for the first time, to apatite and titanite. Three dosimeter glasses were used in his study although the results were recalculated to a common baseline for glass SRM 612 and showed overall weighted mean zetas of 381.8 ± 10.3 for zircon, 353.5 ± 7.8 for apatite and 320.0 ± 12.4 for titanite. Green notes that the study illustrates the need for all FT age determinations to be derived from a calibration scale based on the use of age standards and that consistent FT counting must be achieved in controlled experiments before reliable ages may be determined on unknown samples. In contrast to Green's results, Takahiro Tagami (1987) found no difference within analytical uncertainties between zetas which he measured for glass SRM 612 using zircon, apatite and titanite age standards: 348.4 ± 8.3 (2σ) zircon; 330.1 ± 15.2 apatite; and 335.7 ± 11.5 titanite.

During the 1980s, the FT community recognised that the zeta approach circumvented the uncertainties of FT calibration, with age standards providing a common baseline for all analysts. More zeta values were published and the long-forgotten proposal of Fleischer and Hart (1972) took root. Under the auspices of the IUGS Subcommission on Geochronology, all known FT workers were questioned about their approach to calibration, following the IUGS precedent of agreeing decay constants for other geochronological methods (Steiger and Jäger 1977). Long discussions and extensive consultations culminated in 1988 at the 6th International FT Workshop in Besançon, France, in the consensus recommendation of two approaches to calibration (Hurford 1986a, 1990):

- the zeta approach was recommended for all minerals and techniques using the age standards listed in Table 1.1.
- the absolute approach with selection of λ_f and measurement of neutron fluence was recommended for population method analyses of apatite only.

Since 1988, FT workers have almost exclusively followed the IUGS recommendation with the vast majority opting for the zeta calibration approach publishing their zeta values or referencing their data in doctoral theses. The FT method has burgeoned in the subsequent decades and problems that beset the method during the 1970s have been overcome, securing widespread acceptance of the method throughout the geoscience community. Zeta has enabled the FT community to produce results which are comparable between analysts and which relate directly to other radiometric ages. To counter the 1970s comment from the eminent British isotope geologist: *Cinderella had gone to the ball!*

However, zeta combines physical constants with empirically determinable factors. As Peter van den Haute and his colleagues have eloquently pointed out, precision neutron metrology is today practicable and an exact determination of λ_f achievable (van den Haute et al. 1998). Perhaps future deconvolution of zeta might be possible with an agreed λ_f value and a defined activation monitor protocol. However, for the present the zeta approach remains the preferred calibration method for most laboratories.

1.10 Uncertainties and Data Reporting

All analytical methods need to define experimental uncertainty to provide the user with the limitations placed on the measured result. FT thermochronometry is dependent in part upon personal technique and experience, and so reproducibility of analysis needs to be demonstrated; the Besançon agreement recommended that analysts make a minimum of 5 measurements per mineral phase over at least 3 irradiations to establish their calibration (Hurford 1990). In most laboratories, new workers are more likely to analyse at least 30 age standard mounts to establish their mean zeta value, whilst age standard mounts are regularly included in irradiation packages as a control of irradiation and experimental conditions.

Consistency in reporting of results and data enables other workers to review the techniques and parameters used, and to permit comparison of data from different laboratories (see Hurford 1990 after Naeser et al. 1979). It was also recommended that data for the calibration analyses be included or referenced and that some measure of the dispersion of the count data should be reported. It is normally assumed that FT counts can be described by a Poissonian distribution, an assumption supported by the work of Gold et al. (1968) who counted 4053 tracks in over 345 mica detector strips in contact with a uranium foil source. The standard deviation, σ, of a track count, N, is approximated by:

$$\sigma(N) = N^{0.5} \qquad (1.20)$$

which can be applied to both the spontaneous and induced track densities. As most analysts monitor neutron fluence using a uranium dosimeter glass, Eq. 1.20 can also be applied to the dosimeter track density. For practical purposes, an error for the calculated age, t, can be written as:

$$\sigma(t) = t(1/N_s + 1/N_i + 1/N_d)^{0.5} \qquad (1.21)$$

where N_s is the sum of spontaneous tracks counted in all grains, N_i is the sum of all induced tracks counted in all grains, and N_d sum of all tracks counted in the dosimeter glass detector (or glass).

This error estimation (frequently termed the conventional error analysis after the terminology of Green 1981b) has been that used by most workers for EDM analyses. However, extra-Poissonian error can be present in an analysis resulting from a multiplicity of sources, e.g. track misidentification, use of a mica detector with significant uranium content, inappropriate etching, microscope or irradiation preparation technique, or reactor flux gradients (Burchart 1981, Green 1981b). The analysis of different populations of FT ages within sediment can provide important information about sediment provenance. The use of probability density curves (Clarke and Carter 1987) and peak fitting programs such as BinonomFit (Brandon 1996) have enabled the discrimination of different age modes (e.g. Hurford et al. 1984).

In 1981, Rex Galbraith suggested using a χ^2 test to detect the presence of uncertainty additional to that allowed for by Poissonian variation in track counts, and this has become standard in reporting results. An additional measure of data dispersion results from construction of a radial plot, a graphical means of displaying single-grain age estimates, taking into account the different standard error of each age (Galbraith 1988, 1990). Alongside the radial plot, Galbraith and Laslett (1993) developed the central age estimate for the population of single-grain ages measured for a sample, each age being weighted according to the numbers of track counted.

1.11 FT Annealing and Steps Towards Data Interpretation

Annealing of tracks presents a problem in the recovery of formation age since it violates the requirement, common to all isotopic dating systems, for retention of the daughter product. Annealing is essentially a function of time and temperature although other mechanisms have been considered.[6] Accumulation of the radiogenic daughter product below a certain temperature range is the concept of closure temperature, above which temperature the system is open and radiogenic products are lost (after Dodson 1973). Such closure cannot be instantaneous, and there exists a transitional temperature range, the partial stability zone, in which daughter products are partially lost and partially retained. Wagner (1972) proposed four schematic thermal histories applicable equally to all mineral-isotopic systems where samples pass variously between temperature zones of varying stability (Fig. 1.4a). The setting of time and temperature boundaries for partial track stability (usually now referred to as the partial annealing zone) represents the beginning of a numerical description of the annealing process which may be used to interpret measured FT ages in terms of a sample's thermal history.

Laboratory Annealing Experiments artificially heat different aliquots of sample at comparatively high temperatures over short annealing times. The measured FT parameters are compared with those in an unannealed sample to determine the percentage reduction at each temperature step, the laboratory data being plotted on Arrhenius-type plots (Fig. 1.5). For a single sample, points denoting similar annealing levels and thus similar percentages of track loss define straight lines forming a fanning array, with the increase in slope indicating increase in track loss. The earliest laboratory experiments by Fleischer et al. (1965b) examined annealing of tracks in zircon, olivine, micas and glass, whilst Wagner

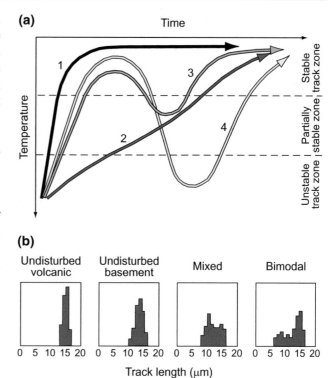

Fig. 1.4 **a** Schematic cooling curves for rock and mineral systems considered in terms of fission-track accumulation (redrawn after Wagner 1972). Curve 1 rapid cooling (e.g. Fish Canyon Tuff): all tracks stable and measured age approximates to formation age. Curve 2 slow cooling (e.g. an Alpine gneiss): stability reached some time after formation; tracks formed during residence in partial stability (or partial annealing) zone are partially annealed. Curves 1 and 2 only consider differences in initial accumulation of tracks. Curves 3 and 4 show reheating of already cooled samples. Curve 3 gives a mixed age: earlier-formed tracks partially annealed to a level dependent upon time and temperature experienced in partial stability zone; second family of tracks are stored on re-entry into stability zone. Intrusion of a dyke into granite could cause such partial resetting in granite samples near the contact. Curve 4 indicates a reset age: initial cooling into stability field is followed by reheating back into unstable zone. FT age would relate only to the time of re-emergence into the track stability zone. **b** Comparison of horizontal confined track-length distributions in apatite (after Gleadow et al. 1986). Undisturbed volcanic distributions characterise rapid cooling (Curve 1) with a dominance of long lengths, a narrow distribution and a mean track length ~14 μm. Undisturbed basement distributions are broader, with mean track lengths ~12–13 μm and would be found in samples following cooling curve 2. Partially reset samples (cooling curve 3) will also show a broad track-length mixed distribution, sometimes resolving into bimodality. Total annealing (cooling curve 4) may show volcanic or basement-type distributions depending on the rate of cooling in the stability zone

(1968) and Naeser and Faul (1969) reported the first annealing data for apatite. Listings of other early annealing studies can be found in Green (1980, Fig. 3) and Wagner and van den Haute (1992, Appendix B). Apatite has been the focus of most annealing studies (e.g. Green et al. 1986; Carlson et al. 1999; Barbarand et al. 2003), and this is considered in detail in Chap. 3 Ketcham 2018.

---

[6]Hydrostatic pressure, static shear stress, fluids, irradiation with non-track forming particles and weathering appear to have no, or minimal affect on annealing of tracks in crystalline materials although may cause some modification in glasses. Chemical composition may have significant changes in track annealing rates, in particular variation of the F/Cl ratios of apatite—see Chap. 3, Ketcham 2018.

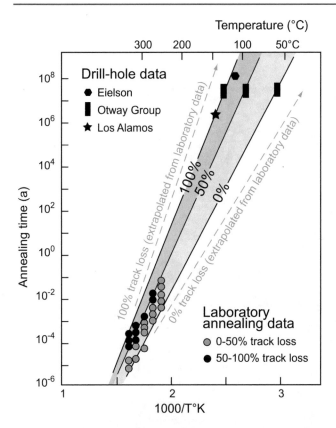

Fig. 1.5 Arrhenius plot of AFT annealing data, which combines the results of the laboratory studies of Naeser and Faul (1969) with data from the Eielson and Los Alamos boreholes (Naeser 1979) and the Otway Group (after Gleadow and Duddy 1981). The dashed lines represent the 0 and 100% track loss lines extrapolated from the laboratory data. Good agreement is found for the 50% annealing curves

The linearity of empirical annealing data enables their extrapolation to geological time, albeit over many orders of magnitude. For apatite extrapolation a "closure temperature" of between 70 and 125 °C for cooling rates of between 1 and 100 Ma was proposed by Wagner and Reimer (1972); Haack (1977); Gleadow and Lovering (1978). Note that whilst offering a first approximation for the stability of tracks over time, the closure temperature concept is of little value in many samples that have experienced complex thermal histories: multiple periods of burial and residence in the upper crust, magmatic activity, contact metamorphism or the passage of hot fluids all may occur at temperatures where tracks in apatite are partially stable.

Tracks in zircon and titanite are more resistant to annealing. Extrapolation of more limited laboratory studies has estimated the zircon partial annealing zone as (rather broadly) 390–170 °C (Yamada et al. 1995) or ∼ 300–200 °C (Tagami 2005)—assuming heating for 10^6 years or more.

Borehole Samples As tracks in apatite are affected by very low geological temperatures, significant annealing effects should be observable in deep boreholes. Naeser (1979) reported apatite ages from a 2.9 km drill hole (and associated wells) in a metamorphic complex at the Eielson US Airforce Base, Alaska, uplifted slowly since the Mesozoic. Extrapolation of the decreasing age trend showed a zero age at a present-day temperature of ∼ 105 °C, giving an estimate of the 100% apatite annealing temperature for a heating period of ∼ 10^8 years.

Geothermal test wells GT1 and GT2 at Los Alamos, New Mexico, lie on the flank of a Pleistocene volcanic centre, with the lower 2.2 km of the 2.9 km deep wells penetrating a Precambrian crystalline complex. A zero AFT age was reached where the present-day temperature is 135 °C, although TFT ages were unaffected until the temperature exceeded 177 °C. Samples here have been heated recently, and thus, 135 °C represents an estimate for 100% track loss temperature in apatite for a heating period of just 10^6 years.

Andrew Gleadow and Ian Duddy (1981) reported a study of outcrop and borehole samples in the Otway Basin, southern Victoria, Australia, one of a series of graben formed during early continental rifting between Australia and Antarctica. Early Cretaceous sedimentation produced 3 km of non-marine sandstones, with the Otway Group, a prominent formation of lithic sandstones, having abundant volcanic rock fragments, fresh phenocryst phases and some glassy clasts. The Otway Group outcrops in two broad areas of the basin, elsewhere being covered by flat-lying late Cretaceous and early Tertiary marine deposits. Outcrop samples yielded indistinguishable apatite, titanite and zircon FT ages ∼ 120 Ma, indicating a cooling history similar to curve 1 of Fig. 1.4a, ∼ 120 Ma representing the age of volcanism immediately prior to deposition. The similarity of FT ages from different minerals also indicated that outcrop samples had not been significantly heated since deposition. Similar ages ∼ 120 Ma were found for titanites from borehole samples at all depths down to 3.4 km (124 °C), demonstrating the Otway Group sandstones had an homogeneous FT age prior to burial in the deeper parts of the basin. AFT ages of ∼ 120 Ma were found for near-surface borehole samples, but age reduction began at about 60 °C, with a zero age determined where the temperature was 120 °C. The decrease in apparent age between 60 and 120 °C delineates the apatite annealing zone (roughly equating with Wagner's partial stability zone of Fig. 1.4a) resulting from gradual temperature increase over 120 Ma due to progressive burial. Stratigraphic evidence suggests that sediments reached their maximum burial depths ∼ 30 Ma ago and have essentially remained at constant depth since that time: the effective duration of heating of these samples is thus of the order of 10–40 Ma. Figure 1.5 (after Gleadow and Duddy 1981) combines the results of the Otway Group,

Fig. 1.6 a Schematic diagram comparing semi- and confined tracks in an apatite crystal. Projected lengths of semi-tracks (or surface tracks) do not reflect their true length and are subject to complex bias. Measurement of horizontal confined tracks (parallel or near parallel to the viewed surface) gives an accurate measurement of length whose bias can be described. TINCLE is a confined track which intersects a crack or cleavage; TINT is a confined track which intersects a surface track. **b** Photomicrograph of an apatite showing a near horizontal confined track (a TINT) amongst surface tracks. The surface tracks are counted to determine FT age; horizontal confined tracks are measured to determine track length. Photomicrograph courtesy of Paul Green

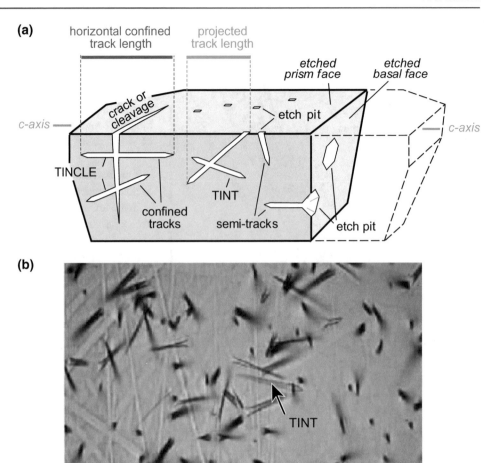

Eielson and Los Alamos borehole studies with the laboratory annealing results of Naeser and Faul (1969).

Borehole studies of annealing in zircon have also been attempted but temperatures sufficient to cause complete annealing of tracks (>300 °C) are seldom encountered, unless tectonic inversion has uplifted a fossil partial annealing zone; some zircon results have been reported from boreholes in the Vienna Basin (Tagami et al. 1996) and KTB Bavaria (Coyle and Wagner 1996). Rahn et al. (2004) have considered the affect of accumulated α-damage from uranium and thorium decay on the stability of fission tracks. Comparison of measured FT ages with ages from other mineral-isotopic systems in exhuming orogens has also provided constraints for zircon track stability (e.g. Harrison et al. 1979; Hurford 1986b).

FT Size Thus far we have mainly considered *track density*: the number of tracks stored, revealed and counted for a sample as measure of time. However, early workers recognised that a fission track may shrink in size with temperature,

giving a second parameter for detecting annealing. Bigazzi (1967) and Mehta and Rama (1969) showed that spontaneous tracks were shorter than induced tracks in a range of muscovite samples. Similar size reductions were found for apatites (Wagner and Storzer 1970; Bhandari et al. 1971; Märk et al. 1973) and for natural glasses (Storzer and Wagner 1969; Durrani and Khan 1970).

What is meant by track size? We need to differentiate between two types of tracks (Fig. 1.6a). *Surface tracks* outcrop on the viewed plane and are those tracks counted in determining an age. These are sometimes called semi-tracks because the upper part of the track has been removed during polishing and etching. Length is measured either from the 2D image projected onto the viewed surface (projected length) or by including the z-dimension by careful focusing of the microscope and use of a calibrated micrometer. In contrast, *confined tracks* occur totally within the mineral grain and are revealed because they cross a surface track or a crack or cleavage plane which allows a pathway for the etchant. Bhandari et al. (1971) termed these "TINTs" (track-in-track—Fig. 1.6b) or "TINCLEs" (track-in-cleavage). Unlike surface tracks, confined tracks are full length and provide a more complete record of any track annealing. Note that all track-length measurements are subject to bias although for horizontal confined tracks the bias can be determined (Laslett et al. 1982). Annealing can also be anisotropic. Green and Durrani (1977) found a marked anisotropy of projected track lengths in apatite, with tracks parallel to the c-crystallographic axis the most resistant to annealing; subsequent measurement of confined tracks in apatite confirmed this finding (Green 1981a) with a more detailed study showing that anisotropy increases directly with annealing level (Green et al. 1986).

Of historical curiosity, two approaches have attempted to use reduction in track size or length to see beyond a "partial overprinting event" (e.g. curve 3 in Fig. 1.4a) and recover the original FT age—although each method is very time-consuming. The *track-size correction* compared spontaneous track size with that of freshly induced tracks, using the degree of size reduction to correct the measured age. Reduction is not a 1:1 relationship and requires construction of empirical calibration curves for each type of material derived from the reduction in track size and track density measured over a series of laboratory annealing experiments. Such curves have been applied to ages measured on glass, mica and apatite (e.g. Durrani and Khan 1970; Selo and Storzer 1981; Wagner and Storzer 1970). Some concordance of corrected FT age with independent age constraints was found for glasses where the heating event was in the recent past but the approach was not able to resolve older "events" or more complex thermal histories. The *plateau correction method* endeavoured to mimic the annealing experienced by spontaneous tracks by artificially heating the induced tracks. Pairs of sample aliquots (one with spontaneous and the other induced tracks) were heated in several steps at increasing temperature or duration. The spontaneous-to-induced track density ratio ρ_s/ρ_i increases until the level of experimental annealing in the induced aliquot equates to that experienced by the partially annealed spontaneous tracks. From this point, the annealing rates are assumed to be equal and a plateau value for ρ_s/ρ_i is reached and used to calculate an age corrected for the partial track loss. Plateau-corrected ages have been reported for glasses (e.g. Storzer and Poupeau 1973; Seward et al. 1980) and apatite (Poupeau et al. 1980). Separate measurement of spontaneous and induced track densities requires a multi-grain strategy which imposes constraints of variation in uranium content and anisotropy of track etching. Both the so-called correction methods are fraught with unresolved problems, and neither approach has any relevance to modern FT analysis.

In direct contrast to projected lengths, *horizontal confined track lengths* in apatite are diagnostic of a sample's thermal history. Each track is formed with approximately identical length—all freshly induced tracks are ~16–16.5 μm—and is then shortened to a length dependent upon the maximum temperature it has experienced. Since the fission of uranium atoms in a sample is a continuing process, each track is of a different age and thus experiences a different record of temperature change, the combined track-length distribution recording the overall sample thermal history. Spontaneous lengths show a definite relationship to geological setting with clear differences between rapidly cooled volcanic samples, slowly exhumed crystalline basement samples and bimodal samples which have been reheated with the earlier-formed tracks partially annealed (Fig. 1.4b after Gleadow et al. 1986). More recent annealing studies in apatite have used confined track length as a quantitative measure of track fading. The quartet of papers from Melbourne (Annealing 1 to 4—Green et al. 1986; Laslett et al. 1987; Duddy et al. 1988; Green et al. 1989) represented a landmark in our understanding of annealing of tracks in apatite, codifying the annealing parameters, underlining the influence of composition on annealing and using the annealing model to simulate track annealing for specific time–temperature pathways to predict age and length parameters and hence thermal history (Ketcham et al. 1999, 2007). Chapter 3 (Ketcham 2018) looks at annealing and modelling in much greater detail.

Acknowledgements This chapter represents my attempt to piece together the story of tracks as I recall it, underlining the landmarks in the development of the method and mentioning some of the people responsible. I hope that this story will serve as a foundation for others, showing them how the FT method arrived where it is today and encouraging them to use and develop fission tracks to help understand all manner of geoscience problems. Remember that no chronometric method provides all the answers and measured data and modelled thermal histories should be evaluated against all other geological information. My apologies go to those whom I have omitted to mention or misrepresented—my sins are not wilful. I thank Andy Carter, Paul Green, Andy Gleadow, Diane Seward and Pieter Vermeesch who offered valuable comments on earlier drafts—some of which I took on board!—and Marco G. Malusà for drafting the figures. The previous volumes of Fleischer et al. (1975) and Wagner and van den Haute (1992) provide much more information on the basics of track formation, registration and etching, and I heartily commend them. I have made many good friendships in the FT community over the past 45 years and owe a debt of gratitude to all my colleagues for their support, discussion, enlightenment and correction. I especially acknowledge my gratitude to three people who have now passed away: Frank Fitch who started me thinking about tracks; and Bob Fleischer and Chuck Naeser who took much time and patience to teach me the trade. I am indebted to Günther Wagner for facilitating my sojourn in Berne, Switzerland; to Andy Gleadow, Paul Green and Andy Carter for their longstanding collaboration and personal friendships; and to my colleagues, students and friends past and present in the laboratory at University College London and Birkbeck, University of London, especially Rex Galbraith. I thank you all.

References

Aumento F (1969) The mid-Atlantic ridge near 45° N.V. Fission track and ferro-manganese chronology. Can J Earth Sci 6:1431–1440

Barbarand J, Carter A, Wood I, Hurford T (2003) Compositional and structural control of fission-track annealing in apatite. Chem Geol 198:107–137

Baumhauer H (1894) Die Resultate der Aetzmethode in der krystallographischen Forschung. Verlag von Wilhelm Engelmann, Leipzig 131 pp

Bellemans F, De Corte F, van den Haute P (1995) Composition of SRM and CN U-doped glasses: significance for their use as thermal neutron fluence monitors in fission-track dating. Radiat Meas 24:153–160

Bhandari N, Bhat SG, Lal D, Rajagopalan G, Tamhane AS, Venkatavaradan VS (1971) Fission fragment tracks in apatite: recordable track lengths. Earth Planet Sci Lett 13:191–199

Bigazzi G (1967) Length of fission tracks and age of muscovite samples. Earth Planet Sci Lett 3:434–438

Bigazzi G (1981) The problem of the decay constant λ_f of ^{238}U. Nucl Tracks 5:35–44

Bigazzi G, Ercan T, Oddone M, Özdoğan M, Yeğingil Z (1993) Application of fission track dating to archaeometry: provenance studies of prehistoric obsidian artifacts. Nucl Tracks Rad Meas 22:757–762

Bohr N, Wheeler JA (1939) The mechanism of nuclear fission. Phys Rev 56:426

Brandon MT (1996) Probability density plot for fission-track grain-age samples. Radiat Meas 26:663–676

Briggs ND, Westgate JA (1978) A contribution to the Pleistocene geochronology of Alaska and the Yukon territory: fission track age of distal tephra units. In: Short papers of 4th International Conference, Geochronology, Cosmochronology, Isotope Geology 1978. US Geology Survey Open-File Rep vol 78, no 701, pp 49–52

Brill RH, Fleischer RL, Price PB, Walker RM (1964) The fission-track dating of man-made glasses. J Glass Studies 6:151–155

Burchart J (1972) Fission track determinations of accessory apatite from the Tatra mountains, Poland. Earth Planet Sci Lett 15:418–422

Burchart J (1981) Evaluation of uncertainties in fission track dating: some statistical and geochemical problems. Nucl Tracks 5:87–92

Calk LC, Naeser CW (1973) The thermal effect of a basalt intrusion on fission tracks in quartz monzonite. J Geol 81:189–198

Carlson WD, Donelick RA, Ketcham RA (1999) Variability of apatite fission-track annealing kinetics: I. experimental results. Am Mineral 84:1213–1223

Carpenter BS, Reimer GM (1974) Calibrated glass standards for fission track use. NBS Special Publication 260–49

Cebula GT, Kunk MJ, Mehnert HH, Naeser CW, Obradovich JD, Sutter JF (1986) The Fish Canyon Tuff, a potential standard of the ^{40}Ar–^{39}Ar and fission-track methods. In: Abstracts, 6th International Conference on Geochronology, Cosmochronology and Isotope Geology, Terra Cognita vol 6, no 2, pp 139–140

Christopher PA (1969) Fission track ages of younger intrusions in southern Maine. Geol Soc Am Bull 80:1809–1814

Church SE, Bickford ME (1971) Spontaneous fission track studies of accessory apatite from granitic rocks of the Sawatch Range, Colorado. Geol Soc Am Bull 82:1727–1734

Clarke ACWV, Carter A (1987) Handling of counting data for fission track dating. Nucl Tracks 13:105–110

Coyle DA, Wagner GA (1996) Fission-track dating of zircon and titanite from the 9101 m deep KTB: observed fundamentals of track stability and thermal history reconstruction. In International Workshop on Fission-Track Dating, Gent 1996, Abstracts, 22

De Corte F, Bellemans F, van den Haute P, Ingelbrecht C, Nicholl C (1998) A new U-doped glass certified by the European Commission for the calibration of fission-track dating. In: van den Haute P, De Corte F (eds) Advances in fission-track geochronology pp 67–78. Kluwer Academic Publishers, Dordrecht

Dodson MH (1973) Closure temperature in cooling geochronological and petrological systems. Contrib Miner Petrol 40:259–274

Duddy IR, Green PF, Laslett GM (1988) Thermal annealing of fission tracks in apatite 3: variable temperature behaviour. Chem Geol (Isot Geosci Sect) 73:25–38

Durrani SA, Khan HA (1970) Annealing of fission tracks in tektites: corrected ages of Bediasites. Earth Planet Sci Lett 9:431–445

Fermi E (1934) Possible production of elements of atomic number higher than 92. Nature 133:898–899

Fitch FJ, Hurford AJ (1977) Fission track dating of the Tardree Rhyolite, Antrim. Proc Geol Assoc 88:261–266

Fitzgerald PG, Malusà MG (2018) Concept of the exhumed partial annealing (retention) zone and age-elevation profiles in thermochronology (Chapter 9). In: Malusà MG, Fitzgerald PG (eds) Fission-track thermochronology and its application to geology. Springer, Berlin

Fleischer RL (1998) Tracks to innovation: nuclear tracks in science and technology. Springer, Berlin, 605 pp

Fleischer RL, Hart HR (1972) Fission track dating: techniques and problems. In: Bishop WW, Miller JA, Cole S (eds) Calibration of hominoid evolution. Scottish Academic Press, Edinburgh, pp 135–170

Fleischer RL, Price PB (1963) Charged particle tracks in glass. J Appl Phys 34:2903–2904

Fleischer RL, Price PB, Symes EM, Miller DS (1964a) Fission track ages and track annealing behavior of some micas. Science 143:349–351

Fleischer RL, Price PB, Walker RM (1964b) Fission track ages of zircons. Geophys Res 69:4885–4888

Fleischer RL, Price PB, Walker RM, Leakey LSB (1965a) Fission track dating of Bed I, Olduvai Gorge. Science 148:72–74

Fleischer RL, Price PB, Walker RM (1965b) Effects of temperature, pressure and ionization on the formation and stability of fission tracks in minerals and glasses. J Geophys Res 70:1497–1502

Fleischer RL, Viertl JRM, Price PB, Aumento F (1971) A chronological test of ocean-bottom spreading in the North Atlantic. Rad Effects 11:193–194

Fleischer RL, Price PB, Walker RM (1975) Nuclear tracks in solids. University of California Press, Berkeley, 605pp

Flerov GN, Petrjak KA (1940) Spontaneous fission of uranium. J Phys 3:275–280

Galbraith RF (1981) On statistical models for fission track counts. Math Geol 13:471–488

Galbraith RF (1988) Graphical display of estimates having differing standard errors. Technometrics 30:271–281

Galbraith RF (1990) The radial plot: graphical assessment of spread in ages. Nucl Tracks Rad Measure 17:207–221

Galbraith RF, Laslett GM (1993) Statistical models for mixed fission track ages. Nucl Tracks Rad Measure 21:459–470

Gentner W, Lippolt H, Schäffer OA (1963) Argonbestimmungen an Kaliummineralien-XI: Die Kalium-Argon-Alter der Gläser des Nordlinger Rieses und der bohmisch-mahrischen Tektite. Geochim Cosmochim Acta 27:191–200

Gleadow AJW (1981) Fission track dating methods: what are the real alternatives? Nucl Tracks 5:1–14

Gleadow AJW, Duddy IR (1981) A natural long-term annealing experiment for apatite. Nucl Tracks 5:169–174

Gleadow AJW, Lovering JF (1977) Geometry factor for external detectors in fission track dating. Nucl Track Detect 1:99–106

Gleadow AJW, Lovering JF (1978) Thermal history of granitic rocks from western Victoria: a fission track dating study. J Geol Soc Aust 25:323–340

Gleadow AJW, Hurford AJ, Quaife RD (1976) Fission track dating of zircon: improved etching techniques. Earth Planet Sci Lett 33:273–276

Gleadow AJW, Duddy IR, Lovering JF (1983) Fission track analysis: a new tool for the evaluation of thermal histories and hydrocarbon potential. Aust Pet Explor Assoc J 23:92–102

Gleadow AJW, Duddy IR, Green PF, Lovering JF (1986) Confined fission track lengths in apatite: a diagnostic tool for thermal history analysis. Contrib Miner Petrol 94:405–415

Gold R, Armani RJ, Roberts JH (1968) Absolute fission rate measurements with solid-state track recorders. Nucl Sci Eng 34:13–32

Green PF (1980) On the cause of shortening of spontaneous fission tracks in certain minerals. Nucl Tracks 4:91–100

Green PF (1981a) Track-in-track length measurements in annealed apatites. Nucl Tracks 5:121–128

Green PF (1981b) A new look at statistics in fission track dating. Nucl Tracks 5:77–86

Green PF (1985) Comparison of zeta calibration baselines for fission track dating of apatite, zircon and sphene. Chem Geol (Isotope Geosci Sect) 58:1–22

Green PF, Durrani SA (1977) Annealing studies of tracks in crystals. Nucl Track Detection 1:33–39

Green PF, Durrani SA (1978) A quantitative assessment of geometry factors for use in fission track studies. Nucl Tracks 2:207–214

Green PF, Hurford AJ (1984) Thermal neutron dosimetry for fission track dating. Nucl Tracks 9:231–241

Green PF, Duddy IR, Gleadow AJW, Tingate PR, Laslett GM (1986) Thermal annealing of fission tracks in apatite: I—a qualitative description. Chem Geol (Isotope Geosci Sect) 59:237–253

Green PF, Duddy IR, Laslett GM, Hegarty KA, Gleadow AJW, Lovering JF (1989) Thermal annealing of fission tracks in apatite 4: quantitative modeling techniques and extension to geological timescales. Chem Geol (Isotope Geosci Sect) 79:155–182

Haack U (1977) The closing temperature for fission track retention in minerals. Am J Sci 277:459–464

Hahn O, Strassmann F (1939) Über den Nachweis und das Verhalten der bei der Bestrahlung des Urans mittels Neutronen entstehenden Erdalkalimetalle. Naturwissenschaften 27:11–15

Harrison TM, Armstrong RL, Naeser CW, Harakal JE (1979) Geochronology and thermal history of the Coast Plutonic complex, near Prince Rupert, B.C. Can J Earth Sci 16:400–410

Hasebe N, Barbarand J, Jarvis K, Carter A, Hurford A (2004) Apatite fission-track chronometry using laser ablation ICP-MS. Chem Geol 207:135–145

Honess AP (1927) The nature, origin and interpretation of the etch figures on crystals. Wiley, New York, p 171

Hurford AJ (1986a) Standardization of fission-track dating calibration: results of questionnaire distributed by International Union of Geological Sciences Subcommission on Geochronology. Nucl Tracks 11:329–333

Hurford AJ (1986b) Cooling and uplift patterns in the Lepontine Alps, South Central Switzerland and an age of vertical movement on the Insubric fault line. Contrib Miner Petrol 92:413–427

Hurford AJ (1990) Standardization of fission-track dating calibration: recommendation by the Fission Track Working Group of the I.U.G. S. Subcommission on Geochronology. Chem Geol (Isotope Geosci Sect) 80:171–178

Hurford AJ (1998) Zeta: the ultimate solution to fission-track analysis calibration or just an interim measure? In: De Corte F, van den Haute P (eds) Advances in fission-track geochronology. Kluwer Academic Publishers, Dordrecht, pp 19–32

Hurford AJ, Green PF (1981a) A reappraisal of neutron dosimetry and ^{238}U λ_f values in fission-track dating. Nucl Tracks 5:53–61

Hurford AJ, Green PF (1981b) Standards, dosimetry and the ^{238}U λ_f decay constant: a discussion. Nucl Tracks 5:73–75

Hurford AJ, Green PF (1982) A users' guide to fission-track dating calibration. Earth Planet Sci Lett 59:343–354

Hurford AJ, Green PF (1983) The zeta age calibration of fission-track dating. Isotope Geosci 1:285–317

Hurford AJ, Hammerschmidt K (1985) ^{40}Ar-^{39}Ar and K-Ar dating of the Bishop and Fish Canyon tuffs: calibration ages for fission-track dating standards. Chem Geol (Isotope Geosci Sect) 58:23–32

Hurford AJ, Watkins RT (1987) Fission-track age of the tuffs of the Buluk Member, Bakate Formation, northern Kenya: a suitable fission-track age standard. Chem Geol (Isotope Geosci Sect) 66:209–216

Hurford AJ, Fitch FJ, Clarke ACV (1984) Resolution of the age structure of the detrital zircon populations of two Lower Cretaceous sandstones from the Weald of England by fission-track dating. Geol Mag 121:269–277

Wagner GA, Reimer GM, Jäger, E (1977) Cooling ages derived by apatite fission-track, mica Rb-Sr and K-Ar dating: the uplift and cooling history of the Central Alps. Memoria degli Isituti di Geologia e Mineralogia dell'Universita di Padova 27 pp

Ketcham R (2018) Fission track annealing: from geologic observations to thermal modeling (Chapter 3). In: Malusà MG, Fitzgerald PG (eds) Fission-track thermochronology and its application to geology. Springer, Berlin

Ketcham RA, Donelick RA, Carlson WD (1999) Variability of apatite fission track annealing kinetics III: Extrapolation to geological time scales. Am Mineral 84:1235–1255

Ketcham RA, Carter A, Donelick R, Barbarand J, Hurford A (2007) Improved modeling of fission-track annealing in apatite. Am Mineral 92:799–810

Key RM, Watkins RT (1988) Geology of the Sabarei area. Mines and Geology Department, Ministry of Environment and Natural Resources, Nairobi, Kenya, Report No. 111, 57 pp

Khan HA, Durrani SA (1973) Measurements of spontaneous fission decay constant of ^{238}U with a mica solid state track detector. Rad Effects 17:133–135

Kohn B, Chung L, Gleadow A (2018) Fission-track analysis: field collection, sample preparation and data acquisition (Chapter 2). In: Malusà MG, Fitzgerald PG (eds) Fission-track thermochronology and its application to geology. Springer, Berlin

Lanphere MA, Baadsgaard H (2001) Precise K–Ar, ^{40}Ar/^{39}Ar, Rb–Sr and U/Pb mineral ages from the 27.5 Ma Fish Canyon Tuff reference standard. Chem Geol 175:653–671

Laslett GM, Kendall WS, Gleadow AJW, Duddy IR (1982) Bias in measurement of fission-track length distributions. Nucl Tracks 6:79–85

Laslett GM, Green PF, Duddy IR, Gleadow AJW (1987) Thermal annealing of fission tracks in apatite 2: a quantitative analysis. Chem Geol (Isotope Geosci Sect) 65:1–15

Malusà MG, Fitzgerald PG (2018) From cooling to exhumation: setting the reference frame for the interpretation of thermocronologic data (Chapter 8). In: Malusà MG, Fitzgerald PG (eds) Fission-track thermochronology and its application to geology. Springer, Berlin

Märk E, Pahl M, Purtsceller F, Märk TD (1973) Thermische Ausheilung von Uran-Spaltspuren in Apatiten alterskorrekturen un Beitrage zur Geothermochronologie. Tscher Miner Petrog Mitt 20:131–154

McDougall I, Roksandic Z (1974) Total fusion ^{40}Ar-^{39}Ar ages using HIFAR reactor. J Geol Soc Aust 21:81–89

McDougall I, Watkins RT (1985) Age of hominoid-bearing sequence at Buluk, Northern Kenya. Nature 318:175–178

McDougall I, Wellman P (1976) Potassium-argon ages for some Australian Mesozoic igneous rocks. J Geol Soc Aust 23:1–9

McDowell FW, Keizer RP (1977) Timing of mid-Tertiary volcanism in the Sierra Madre Occidental between Durango City and Mazatlan, Mexico. Geol Soc Am Bull 88:1479–1487

McDowell FW, McIntosh WC, Farley KA (2005) A precise ^{40}Ar-^{39}Ar reference age for the Durango apatite (U-Th)/He and fission track dating standard. Chem Geol 214:249–263

Mehta PP, Rama (1969) Annealing effects in muscovite and their influence on dating by the fission track method. Earth Planet Sci Lett 7:82–86

Meitner L, Frisch O (1939) Disintegration of Uranium by Neutrons: a new type of nuclear reaction. Nature 143:239–240

Miller DS (1968) Fission track ages on 250 and 2500 m.y. micas. Earth Planet Sci Lett 4:379–383

Naeser CW (1967) The use of apatite and sphene for fission track age determinations. Geol Soc Am Bull 78:1523–1526

Naeser CW (1969) Etching tracks in zircons. Science 165:388–389

Naeser CW (1979) Fission track dating and geologic annealing of fission tracks. In: Jäger E, Hunziker JC (eds) Lectures in isotope geology. Springer, Heidelberg, pp 154–169

Naeser CW, Dodge FCW (1969) Fission-track ages of accessory minerals from granitic rocks of the central Sierra Nevada Batholith, California. Geol Soc Am Bull 80:2201–2212

Naeser CW, Faul H (1969) Fission track annealing in apatite and sphene. J Geophys Res 74:705–710

Naeser CW, McKee EH (1970) Fission track and K/Ar ages of Tertiary ash-flow tuffs, north-central Nevada. Geol Soc Am Bull 81:3375–3384

Naeser CW, Gleadow AJW, Wagner GA (1979) Standardization of fission track data reports. Nucl Tracks 3:133–136

Noddack I (1934) Über das Element 93. Angew Chem 47:653–655

Paulick J, Newesely H (1968) Zur Kenntis der Apatite der Cerro de Mercado, Durango, Mexiko. Neu Jb Mineral, Mh 1(2):224–235

Philips D, Matchan E (2013) Ultra-high precision ^{40}Ar/^{39}Ar ages for Fish Canyon Tuff and Alder Creek Rhyolite sanidine: new dating standards required? Geochim Cosmochim Acta 121:229–239

Poupeau G, Carpena J, Chambaudet A, Romary P (1980) Fission track plateau-age dating. In: Francois H et al (eds) Solid state nuclear track detectors. Pergamon Press, Oxford, pp 965–971

Price PB, Walker RM (1962a) Observation of fossil particle tracks in natural micas. Nature 196:732–734

Price PB, Walker RM (1962b) Chemical etching of charged-particle tracks in solids. J Appl Phys 33:3407–3412

Price PB, Walker RM (1963) Fossil tracks of charged particles in mica and the age of minerals. J Geophys Res 68:4847–4862

Rahn MK, Brandon MT, Batt GE, Garver JI (2004) A zero-damage model for fission-track annealing in zircon. Am Mineral 89:473–484

Roden MK, Parrish RR, Miller DS (1990) The absolute age of the Eifelian Tioga ash bed, Pennsylvania, J Geology 98:282–285

Roebben G, Derbyshire M, Ingelbrecht C, Lamberty A (2006) Certification of uranium mass fraction in IRMM-540R and IRMM-541 uranium-doped oxide glasses. European Community Institute for Reference Materials and Measurements Report EUR 22111 EN, Scientific and Technical Research series, Luxembourg, 27 pp, ISBN 92-79-01630-X

Ross RJ, Naeser CW, Izett GA, Whittington HB, Hughes CP, Rickards RB, Zalasiewicz J, Sheldon PR, Jenkins CJ, Cocks LRM, Bassett MA, Toghill P, Dean WT, Ingham JK (1982) Fission track dating of Lower Palaeozoic bentonites in British stratotypes. Geol Mag 119:135–153

Selo M, Storzer D (1981) Uranium distribution and age pattern of some deep-sea basalts from the Entrecasteaux area, South-Western Pacific: a fission-track analysis. Nucl Tracks 5:137–145

Seward D (1974) Age of New Zealand Pleistocene substages by fission track dating of glass shards from tephra horizons. Earth Planet Sci Lett 24:242–248

Seward D (1975) Fission track ages of some tephras from Cape Kidnappers, Hawke's Bay, New Zealand. N Z J Geol Geophys 18:507–510

Seward D, Wagner GA, Pichler H (1980) Fission track ages of Santorini volcanics (Greece). In: Doumas C, (ed) Thera and the Aegean World II, Papers & Proceedings 2nd International Science Congress, Santorini, Greece, August 1978

Silk ECH, Barnes RS (1959) Examination of fission fragment tracks with an electron microscope. Phil Mag 4:970–971

Steiger RH, Jäger E (1977) Subcommission on geochronology: convention on the use of decay constants in geo- and cosmochronology. Earth Planet Sci Lett 36:359–362

Steven TA, Mehnert HH, Oradovich JD (1967) Age of volcanic activity in the San Juan mountains, Colorado. US Geol Surv Prof Paper 575-D: 47–55

Storzer D, Poupeau G (1973) Ages-plateaux de mineraux et verres par la methode de traces de fission. CR Acad Sci Paris 276(D):137–139

Storzer D, Wagner GA (1969) Correction of thermally lowered fission track ages of tektites. Earth Planet Sci Lett 5:463–468

Stuckless JS, Naeser CW (1972) Rb-Sr and fission track age determinations in the Precambrian plutonic basement around the Superstition Volcanic Field, Arizona. US Geol Surv Prof Paper 800-B: B191–B194

Swanson ER, Keizer RP, Lyons JI, Clabaugh SE (1978) Tertiary volcanism and caldera development near Durango City, Sierra Madre Occidental, Mexico. Geol Soc Am Bull 89:1000–1012

Tagami T (1987) Determination of zeta calibration constant for fission track dating. Int J Rad Appl Instrum D 13:127–130

Tagami T (2005) Zircon fission-track thermochronology and applications to fault studies. Rev Mineral Geochem 58:95–122

Tagami T, Carter A, Hurford AJ (1996) Natural long-term annealing of the zircon fission-track system in Vienna Basin deep borehole samples: constraints upon the partial annealing zone and closure temperature. Chem Geol 130:147–157

Tagami T, Farley KA, Stockli DF (2003) (U-Th)/He geochronology of single zircon grains of known Tertiary eruption age. Earth Planet Sci Lett 207:57–67

Thiel K, Herr W (1976) The ^{238}U spontaneous fission decay constant re-determined by fission tracks. Earth Planet Sci Lett 30:50–56

van den Haute P, De Corte F, Jonckheere R, Bellemans F (1998) The parameters that govern the accuracy of fission-track age determinations: a re-appraisal. In: van den Haute P, De Corte F (eds) Advances in fission-track geochronology, Kluwer Academic Publishers, Dordrecht, pp 33–46

Wagner GA (1968) Fission track dating of apatites. Earth Planet Sci Lett 4:411–415

Wagner GA (1969) Spuren der spontanen Kernspaltung des ^{238}Urans als Mittel zur Datierung von Apatiten und ein Beitrag zur Geochronologie des Odenwaldes. Neues Jb Miner Abh 110:252–286

Wagner GA (1972) The geological interpretation of fission track ages. Trans Am Nucl Soc 15:117

Wagner GA (1978) Archeological applications of fission-track dating. Nucl Track Detect 2:51–63

Wagner GA, Reimer GM (1972) Fission track tectonics: the tectonic interpretation of fission track ages. Earth Planet Sci Lett 14:263–268

Wagner GA, Storzer D (1970) Die Interpretation von Spaltspurenaltern (fission track ages) am Beispiel von naturlichen Gläsern, Apatiten und Zirkonen. Eclogae Geol Helv 63:335–344

Wagner GA, van den Haute P (1992) Fission-track dating. Kluwer Academic Publishers, Dordrecht, Solid Earth Sciences Library vol 6, 285 pp

Wagner GA, Reimer GM, Carpenter BS, Faul H, Van den Linden R, Gijbels R (1975) The spontaneous fission rate of ^{238}U and fission track dating. Geochim Cosmochim Acta 39:1279–1286

Williams IS, Tetley NW, Compston W, McDougall I (1982) A comparison of K/Ar and Rb/Sr ages of rapidly cooled igneous rocks: two points in the Palaeozoic time scale re-evaluated. J Geol Soc London 139:557–568

Yamada R, Tagami T, Nishimura S, Ito H (1995) Annealing kinetics of fission tracks in zircon: an experimental study. Chem Geol (Isotop Geosci Sect) 104:251–259

Young DA (1958) Etching of radiation damage in lithium fluoride. Nature 182:375–377

Young EJ, Myers AT, Munson EL, Conklin NM (1969) Mineralogy and Geochemistry of fluorapatite from Cerro de Mercado, Durango, Mexico. US Geol Surv Prof Paper 650-D: D84–D93

Fission-Track Analysis: Field Collection, Sample Preparation and Data Acquisition

Barry Kohn, Ling Chung and Andrew Gleadow

Abstract

Fission-track (FT) analysis for geological applications involves a range of practical considerations, which are reviewed here. These include field sampling, the separation of the most commonly used minerals (apatite, zircon and titanite), the preparation of these minerals for analysis (including for double or triple-dating of the same grains) and measurement of the essential parameters required. Two main analytical strategies are described, the External Detector Method (EDM) and Laser Ablation-Inductively Coupled Plasma-Mass Spectrometry (LA-ICP-MS). Although the initial steps ranging from sample selection to mineral separation are common to both approaches, the next practical steps vary with the specific dating strategy adopted. The workflow outlined here for sample preparation and aspects of data acquisition follows a widely used standard sequence of steps, but some of the specific details described are those developed over many years by the Melbourne Thermochronology Group. While these protocols may be readily applicable or adaptable, it is recognised that many laboratories may have developed their own particular recipes for different aspects of these methods.

Gleadow et al. (2002), Tagami and O'Sullivan (2005), Donelick et al. (2005), Kohn et al. (2005), Galbraith (2005), Braun et al. (2006) and Lisker et al. (2009). The topics covered here include field sampling and identification of suitable target material, step-by-step mineral separation and sample preparation (including double- or triple-dating of the same grains), and measurement of the key parameters required for FT thermochronology. Some of the methods outlined for sample preparation and aspects of data acquisition are commonly used and follow a sequence of steps, while the Melbourne Thermochronology Group has specifically developed some others over many years. These protocols are regarded as being readily applicable, if required, to most laboratories carrying out routine FT analysis at this time. However, we emphasise that the methods described here are in no way meant to be prescriptive and it is acknowledged that other laboratories will often have developed their own procedures for certain aspects of sample preparation and data acquisition. In general, proper training conducted by the laboratory supervisor involving an induction process covering relevant procedures and familiarity with Occupational, Health and Safety (OH & S) requirements (especially for handling strong chemicals, heavy liquids and radioactive material) should be an essential first-step for all users who wish to work in a FT laboratory.

2.1 Introduction

This chapter focuses on the practical details of FT analysis for geological applications. Several works have been published emphasising other aspects of FT thermochronology including the fundamental principles, interpretation of data, statistics and application to geological problems; these include Fleischer et al. (1975), Wagner and van den Haute (1992), Gallagher et al. (1998), Dumitru (2000),

2.2 FT Dating Strategies

For FT dating, the most common strategy for studying minerals is the External Detector Method (EDM), which involves sending off polished grain mounts for a thermal neutron irradiation in a nuclear reactor (see Chap. 1, Hurford 2018). More recently however, Laser Ablation-Inductively Coupled Plasma-Mass Spectrometry (LA-ICP-MS) has emerged as an alternative method for the direct acquisition of uranium content in target minerals (Hasebe et al. 2004). Only these two techniques will be

B. Kohn (✉) · L. Chung · A. Gleadow
School of Earth Sciences University of Melbourne,
Melbourne, VIC 3010, Australia
e-mail: b.kohn@unimelb.edu.au

© Springer International Publishing AG, part of Springer Nature 2019
M. G. Malusà and P. G. Fitzgerald (eds.), *Fission-Track Thermochronology and its Application to Geology*,
Springer Textbooks in Earth Sciences, Geography and Environment, https://doi.org/10.1007/978-3-319-89421-8_2

considered here, as they remain the best alternatives in most FT dating situations. Fundamental age equations for age calculation related to using these methods have been reviewed by Wagner and van den Haute (1992), Gleadow et al. (2002), Hasebe et al. (2004), Tagami and O'Sullivan (2005) and Donelick et al. (2005) (see Chap. 1, Hurford 2018 and Chap. 6, Vermeesch 2018).

In FT dating by EDM, alternative strategies are possible for measuring the ratio of spontaneous (ρ_s) to induced (ρ_i) track densities, upon which the age depends. Not all of these dating strategies are equally reliable in every case, and care is required to ensure that an appropriate method is selected. In practice, a variety of factors such as the registration geometry of the etched surface, accumulated radiation damage, anisotropic etching and uneven intra-grain uranium distribution must be considered when selecting a suitable method of FT dating for a particular sample. The various FT dating methods differ importantly in the registration geometries of the etched surfaces used to count spontaneous and induced tracks, and the corrections required if these are not equivalent. All of these methods, however, require first that the uranium-bearing mineral grains sought be physically separated from their host rock as outlined in Sect. 2.4 and Fig. 2.1. However, the next practical steps after that may rely on the specific dating strategy adopted.

Previous reports describing different laboratory procedures for FT analysis in some detail include Naeser (1976), Hurford and Green (1982), Gleadow (1984), Wagner and van den Haute (1992), Ravenhurst and Donelick (1992), Dumitru (2000), Gleadow et al. (2002), Donelick et al. (2005), Tagami (2005) and Bernet and Garver (2005). Here, we briefly review some of the main steps required for FT analysis, emphasising some recently developed procedures that have become available over the past decade or so.

For volcanic glass, relatively homogeneous uranium concentrations occur between fragments (shards) or within bulk samples (e.g. obsidian). The separation steps, mounting and etching conditions for volcanic glass are outlined in Sects. 2.4–2.6. In most types of natural glass, fission tracks are not fully stable at ambient temperatures over geological time, therefore different methods are used for age determination than the two mentioned above. These will not be discussed further here, but for more information see Dumitru (2000 and references therein) and for a more recent glass dating protocol using a less complicated age correction procedure in combination with a LA-ICP-MS procedure (i.e. no neutron irradiation required) see Ito and Hasebe (2011).

2.3 Sample Collection—Suitable Geological Materials

2.3.1 Sample Collection

Where possible the freshest and cleanest rock material available should be collected for analysis. It is advantageous to remove as much biological material, soil and weathered surface (the outer few centimetres) as possible and break down samples into fist-sized pieces, while sampling *in the field*. Fire prone areas should be avoided if possible, as heat may affect the ability of some minerals to retain their daughter products. However, if sampling is carried out in such areas, then at least the outer ∼3 cm of a bedrock sample should be removed (e.g. Reiners et al. 2007). If importing samples from overseas, then the steps outlined above will help to alleviate any concerns by local Quarantine and Customs Authorities.

Wear safety glasses whenever hammering rocks. For some highly altered lithologies, you may have to use a percussion hammer to obtain fresh material. Ensure that samples are *representative* of the lithology being examined and that all are in situ from outcrops. For detrital samples, especially recent or loosely consolidated sediment, it may be possible to do some 'gold panning' in the field in order to obtain a first-order heavy mineral separation and reduce the sample size (e.g. Bernet and Garver 2005). It is critical when sampling to accurately record (usually with a GPS) the sample location, i.e. horizontal datums (in latitude and longitude or another coordinate system) and vertical datums (as either elevation or depth). This is particularly important for future 'users' of the data and for databases.

Depending on the rock type, outcrop samples generally ranging in weight from ∼1–3 kg should be collected, but larger samples (∼4–7 kg) are recommended from loose sediments, which haven't been panned in the field (e.g. Bernet and Garver 2005). Some samples, such as from cores or cuttings in drill holes, will invariably be much smaller than 1 kg. If drill hole samples are very small, it might be possible to combine samples from a range of limited depths, i.e. over a few 10's m, to form a single sample. For drill hole cuttings, it is important to ensure that down-hole cavings or contamination (particularly from drilling mud) are not a problem. For dating of tephra, near-source coarser pumice blocks or pumiceous lapilli are less likely to be contaminated, but when working with pumiceous material it is critical to ensure that potential contaminants in vesicles have been removed by ultrasonic cleaning. For more distal,

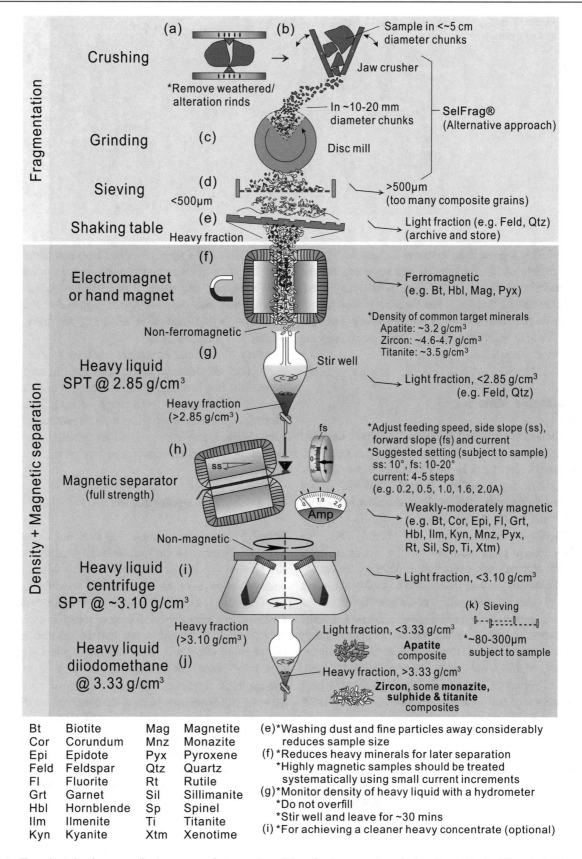

Fig. 2.1 Flow chart showing generalised sequence of steps and conditions for the separation of minerals suitable for FT analysis. See text for further details

Table 2.1 General lithology guide for target minerals sought for FT analysis

Preferred	Less favourable to problematic
• Igneous rocks: silicic to intermediate intrusives (granite, granodiorite, diorite, tonalite) and volcanics (lavas and pyroclastics) and less commonly basic intrusives (gabbro) • Metamorphic rocks: gneisses, granulites, amphibolites, meta-sandstones, some schists • Sedimentary rocks: immature sandstones, red beds, arkoses, some conglomerates and greywackes, occasionally more mature sandstones and quartzites	• Mafic volcanic rocks • Ultramafic rocks • Eclogites • Mafic schists (often contain metamorphic titanite and apatite, but with low U content) • Shales, slates and phyllites • Siltstones and claystones • Mylonites • Evaporites and carbonates (although see dating of detrital minerals in carbonates by Arne et al. 1989) • Highly altered or mineralised rocks (but see Gleadow and Lovering 1974 for mineral suitability in strongly weathered rocks)

thinner and finer tephra, it is important to sample from as deep as possible within the outcrop and to search for the coarsest material available, at the same time taking every precaution to ensure that there has been no potential contamination from overlying beds. Common lithologies containing minerals routinely used for FT thermochronology are summarised in Table 2.1.

2.3.2 Suitable Minerals

Etching studies have revealed fission tracks in more than 100 uranium-bearing minerals and glasses (e.g. Fleischer et al. 1975; Wagner and van den Haute 1992). But factors such as uranium content, track stability characteristics and mineral abundance result in very few of these minerals being routinely analysed for FT thermochronology. The most widely dated minerals are apatite, zircon and titanite. These are commonly present as primary accessory minerals in many igneous and metamorphic rocks and as detrital components in some sedimentary rocks (see Table 2.1). The annealing behaviour of other minerals, which could potentially be used in FT studies, most notably epidote group minerals and some types of garnets and micas are not well understood and their uranium content may be quite low and variable, thus they are not generally suitable for rigorous study. Volcanic, pseudotachylitic, impact and man-made glasses have also been dated occasionally (see Wagner and van den Haute 1992, their Chap. 6.2.11).

2.4 Mineral Separation

Mineral separations for FT dating aim at recovering relatively clean concentrates of suitable uranium-bearing accessory minerals. This is achieved by rock fragmentation, crushing and use of a shaking table (if available) followed by the exploitation of differences in mineral density using heavy liquids and differences in magnetic

susceptibility. The steps described here and summarised in Fig. 2.1 are a general, widely used sequence. While many have been tried and tested, it is recognised that some laboratories may use different specific procedures during the workflow (e.g. Donelick et al. 2005; see also Chap. 7, Malusà and Garzanti 2018).

Many of the finer points are not easy to describe, thus there is no substitute for the actual hands-on acquisition of these skills in a functioning laboratory. It is emphasised that absolute cleanliness at each stage of rock crushing and mineral separation is of paramount importance. The final data obtained are only as good as the attention given to prevent possible contamination by foreign mineral grains.

2.4.1 Rock Fragmentation

Prior to commencing rock fragmentation, remove any remaining weathered rind or surfaces possibly exposed to fire (within ∼3 cm of the outer surface) with a diamond saw, wire brush or rock splitter. If possible, it is most advantageous if these outer layers are removed and samples broken down into fist-sized pieces, while sampling in the field. Hard rock samples must first be reduced into small pieces using either a hammer or a mechanical splitter, followed by processing through a jaw crusher to produce smaller rock fragments. A wide variety of crushing and grinding equipment can be used to further reduce the particle size and disaggregate grains; this will often be a disc pulveriser with a rotating plate grinding mill fitted with hardened steel plates. In this case, it is crucial to properly adjust the mill plates to minimise the yield of mineral composites (mostly in fractions >500 μm), but at the same time prevent over-grinding of grains, which may result in grains being too small and unsuitable for FT analysis. Once the particle size for a sample has been adequately reduced, samples should be sieved to yield a fraction <500 μm, which may then be passed over a shaking table (e.g. Wilfley or Gemini) if available. This procedure can greatly speed up the separation process, allowing large samples (several kilograms) to be processed and a heavy mineral fraction concentrated. However, if samples are fine-grained or very small, it is best to avoid using a shaking table and in the case of the former, they should be thoroughly washed with tap water to decant off the finer material (e.g. Donelick et al. 2005). A general scheme is summarised in Fig. 2.1a–e. The separation process is then continued with the heavy concentrate, which first needs to be dried either by heating at low temperature (~ 50 °C) or rinsing in acetone.

SelFrag® An alternative approach to the mechanical fragmentation of rocks and liberation of minerals as described above, but which requires considerable capital investment in equipment and infrastructure, is through an electrodynamic disaggregation method. This protocol seeks to break grains along their boundaries or internal grain discontinuities. In recent years, this possibility has been achieved through the development of the selFrag Lab®. Studies on apatite and zircon grains, which were liberated using this protocol, have been compared with samples separated by conventional mechanical preparation as described above and indicate no adverse affects for FT thermochronology (Giese et al. 2009; Sperner et al. 2014).

2.4.2 Gravity Separation Using Heavy Liquids and Magnetic Separation

As most target minerals are usually of higher density (>3.2 g/cm^3) and non- or weakly magnetic, combined heavy liquid and magnetic (using a Frantz® Isodynamic Magnetic Separator) separation allows for refinement of the mineral concentrate and removal of any minerals of lighter density not required. Figure 2.1f–j summarises a commonly used sequence of steps for heavy liquid and magnetic separation, although these can vary with sample grain size and mineral composition. Repeated heavy liquid and magnetic separation (using settings with varying degrees of magnetic susceptibility) (see Fig. 2.1h) may provide further purification.

In the past, many laboratories used heavy liquids that are volatile and classed as toxic chemicals such as bromoform (specific gravity 2.89 g/cm^3) and tetrabromoethane (TBE) (specific gravity ~ 2.96 g/cm^3). However, these have now been largely replaced by non-toxic lithium metatungstate (LMT), lithium heteropolytungstate (LST, density ~ 2.9 g/cm^3 at 25 °C) and sodium polytungstate (SPT) (Callahan 1987; Torresan 1987; Chisholm et al. 2014), as well as diiodomethane (DIM), also known as methylene iodide (density ~ 3.31 g/cm^3 at 25 °C).

Two additional methodologies previously reporting the effective use of liquids for carrying out mineral separations are:

- Use of organic liquids for diluting heavy liquids (bromoform and DIM) to create a range of liquid densities, which maintain relatively constant specific gravities for use and storage, as well as an efficient method for recovery of heavy liquids (Ijlst 1973).
- Froth flotation of crushed and sieved sand-size mineral fractions employs chemicals that change the electrical surface properties of specific minerals and make them selectively hydrophobic. When air is blown into a suspension, hydrophobic grains stick to ascending air bubbles and concentrate in a foam on the surface of the flotation cell. Hejl (1998) outlined the practical steps for what is described as a low cost procedure for the successful separation of apatite and zircon from silicate rocks.

For magnetic separation—make sure the Frantz and operating environment is absolutely clean. Use compressed air on the feed hopper, chute and collection buckets and wipe with alcohol. Set the Frantz with a forward slope of 10°–20° (depending on sample) and a side-slope (top towards back) of +10°. These settings can be varied somewhat for special applications when some experience has been gained. Lower side-slopes can be used at a later stage for cleaning up the final mineral fractions—see below.

Use the mechanical vibrator and a moderate feed rate to process the SPT sink heavy mineral fraction through the Frantz in a number of steps, increasing the current at each stage. The exact number of passes depends upon the nature of the sample and can be varied with experience. After each stage, the least magnetically susceptible sample should be reprocessed. Four passes using current (A) settings of 0.4, 0.8, 1.2 A and full-scale (1.6 A) are usually adequate for titanite-bearing samples. Fewer steps can be used if no titanite is present in the 0.4 or 0.8 A fractions. Minerals that typically behave magnetically at various current settings are listed in Fig. 2.1.

Titanite will often separate out in the magnetic fraction at 0.8 and 1.2 A, but may be contaminated with a variety of minerals, e.g. amphibole and pyroxene. In general, playing around with different current settings, slope and tilt, may provide a relatively clean separate. Otherwise handpicking may be used. Another DIM step may also be useful.

Apatite and zircon tend to separate out on the non-magnetic side following the four passes outlined above. To clean up the apatite fraction, reduce the side-slope on the Frantz to +5° and run at full-scale current, then at +2°, but note at slopes less than +2°, some apatites behave magnetically. To clean up zircon, reduce the side-slope on the Frantz to −2° (top towards front) and run at full-scale current. This should be done in gradual steps, i.e. +5° then 0° to −1°, then if still dirty −2°. This procedure can remove sulphides, aluminous titanites (grothite) and metamict zircons, but is not always successful.

After completing routine separations for apatite and zircon, other minerals can still contaminate these fractions. In the apatite fraction, these include:

(a) Barite—often occurs in cuttings samples from oil wells where barite has been used in the drilling mud. Barite contamination can be avoided if cutting chips are washed and are large enough so that they can be sieved to retain the >500 μm fraction. This fraction is then ground and processed in the usual way.

(b) Fluorite—generally occurs in particular granite provinces, e.g. some S-type granites and tin-granites and in sediments derived from such parent sources. Often such an apatite fraction is unworkable as fluorite has almost identical physical properties to apatite and cannot be separated by any of the usual techniques. However, apatite has very low abundance in such rocks.

(c) Sulphide/quartz composite grains—because of their composite properties these may be very difficult to handle. As the composites tend to be larger than the apatite, such separates can sometimes be cleaned up using a small nylon of 200 or 300 μm sieve size. Otherwise handpicking may be the only way to remove such grains.

In the zircon fraction, the contaminants may include non-magnetic sulphides. If sulphide grains are large, then first sieve in the same way as for apatite (see above). Otherwise dissolve sulphide in aqua regia in a small beaker under a heat lamp. This may need to be done several times to remove all the sulphides. The recovered grains will then need to be subjected to a further heavy liquid separation (e.g. SPT) in order to remove the light minerals liberated from the sulphide composite grains.

DIM is typically used in the last stage of the mineral separation treatment to separate apatite (floats) from zircon and titanite (sink) (see Fig. 2.1j).

Volcanic glass is usually separated using a Frantz® Isodynamic Magnetic Separator. Bubble junction glass shards form the best surfaces for counting fission tracks and are weakly magnetic and separate out between 1.2 and 1.6 A with side-slope between 5° and 10°. Pumiceous glass, which is slightly more magnetic, is often vesicular and does not usually provide an ideal surface for counting fission tracks. Note that volcanic zircons often have glass overgrowths that may cause them to float in heavy liquids. To dissolve the glass so that the zircons sink, as would be expected from their density, the heavy mineral concentrate should be soaked in concentrated HF for 1–3 min (care being taken not to inadvertently dissolve other minerals of possible interest).

2.4.3 Further Possible Final Treatment

The last three steps shown in Fig. 2.1 (steps i–k) can be used selectively coupled with handpicking, especially if only a small amount of heavy and non-magnetic fractions remained after step (h).

Step i involves an additional heavy liquid separation method for further purifying and reducing sample size. This can be achieved by centrifuging a mixture of the non-magnetic mineral fraction, which formed the sink fraction in SPT (at a density of 2.85 g/cm^3), in a further solution of SPT made up to a maximum achievable density of ~ 3.10 g/cm^3.

A final sieving step k is often useful for FT grain mount preparation by concentrating an optimal grain size. This involves sieving a small volume of grains using either small brass sieves or disposable sieve cloths (e.g. nylon bolting cloth) secured over small plastic cups. The range of grain sizes present in a mineral separate depends on a variety of factors and is highly variable, but typically the most suitable grains for FT analysis will fall in the range of ~ 80–300 μm. However in any particular separate, the largest grains will generally be the most suitable for providing large clear areas for analysis, so further subdivision into a more restricted size range may be desirable. In addition, the ideal scenario for a grain mount is that all grains are ground and polished to a desired internal surface. Working with mineral separates of similar size allows all grains to attain a common surface/depth during grinding and polishing. However, final sieving should be carried out carefully because for some studies, such as on detrital grains; it may bias the representation of different populations (see Chap. 16, Malusà 2018).

2.5 Sample Mounting and Polishing

Prior to carrying out this step, if (U-Th)/He dating is also planned for the sample, then it is recommended to first handpick the best-quality grains (in terms of size, shape and clarity) from the final mineral concentrate for that purpose, as grain morphology requirements for FT analysis are less stringent.

The aim of the different steps described in this section (see Fig. 2.2) is to establish a flat and well-polished surface, in an appropriate crystallographic orientation, that results from the removal of sufficient external grain material to expose an internal grain surface for analysis (i.e. 4π geometry—see Wagner and van den Haute 1992, and Chap. 1, Hurford 2018).

It is important that care be taken at every stage of sample and mount preparation to produce the best-quality

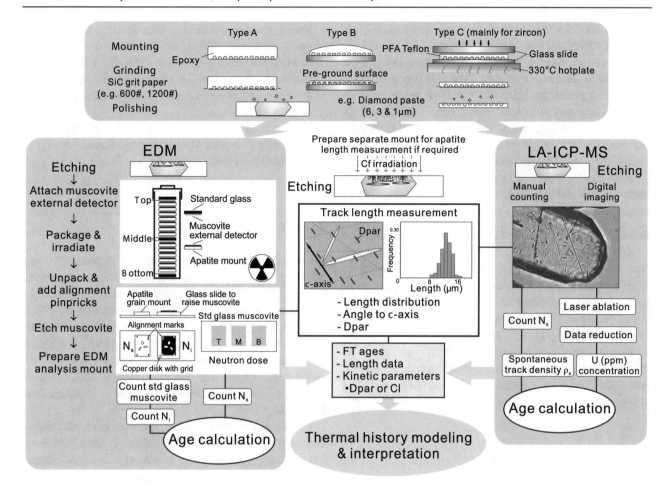

Fig. 2.2 Alternative methods for preparing and collecting data from mineral separates for FT analysis using the External Detector Method (EDM) and LA-ICP-MS protocols, as well as the measurement of track lengths and a kinetic indicator in apatite, either Dpar or Cl. Dpar is measured parallel to the crystallographic *c*-axis. On grain mount

N_s = spontaneous fission tracks counted on grain mount and N_i = induced fission tracks counted on mica detector, ρ_s = the spontaneous fission-track density (tracks/cm^{-2}) calculated from N_s. See text for further information

polished surface. Polishing time is variable for individual samples, so it needs to be monitored periodically. Proper washing between each polishing stage is also crucial for preventing cross-contamination between different grades of diamond paste or any other polishing media used, e.g. alumina powder, colloidal silica or other suspensions. Factors contributing to poor polishing outcomes may include cracking of grains induced by rock crushing, inadequately mixed or cured resins and grain shatter during grinding. Over-grinding and polishing (leading to the production of excessive relief) should be avoided. Grinding the grains too thinly or leaving insufficient thickness of mounting media to secure the grains may result in grains falling out during polishing or etching, and lead to contamination or damage of polishing laps.

2.5.1 Apatite

Mounting A wide variety of epoxy resins suitable for making apatite or volcanic glass mounts are available. It is important to follow the manufacturer's instructions exactly with respect to the resin-to-hardener ratio, curing time and temperature. Improperly set epoxy may result in a soft or gluey texture after curing and lead to cracking of grains and contamination of polishing laps, as well as some grain loss.

There are several different mounting media available. Two suitable resins are EpoFix™ (from Struers) for mounting at room temperature and Petropoxy 154™ (with very low volatility and toxicity, available from Burnham

Petrographics) for mounting at elevated temperature (suggested at 135 °C on a hotplate). Using these resins, one can either make an epoxy-only button (type A in Fig. 2.2) or mount the grains directly on glass (type B in Fig. 2.2). Many laboratories prefer the latter and use cut-down mounts on glass, so they can fit into irradiation cans for neutron irradiation for the EDM protocol. For LA-ICP-MS, samples are mounted directly on either an epoxy-only button or glass slide without the need for modification of the slide. Volcanic glass is usually mounted in an epoxy resin (cold setting) button mount and ground and polished in a similar manner to apatite.

Grinding and Polishing Grains in type A mounts are already exposed at or near the surface, whereas grains in type B mounts are fully enclosed within the epoxy. Both types of mount can be further processed by grinding manually with SiC grit paper (e.g. 1200# or 600#) on a glass plate before polishing on a rotating lap with different grades of diamond paste (e.g. 6, 3 and 1 µm). However, an extra time-saving pre-grinding step for type B mounts is to directly expose target grains prior to SiC grinding by using an automatic cut-off machine (e.g. Struers Accutom™), if available.

2.5.2 Zircon and Titanite

The use of FEP Teflon (type C in Fig. 2.2) for mounting zircons for etching was developed by Gleadow et al. (1976). However, during long periods of etching, as required for zircons with low radiation damage, grains often tend to fall out. Following work by Tagami (1987), many laboratories changed over to PFA Teflon. This Teflon has a higher melting temperature and maintains its transparency even after prolonged etching. For mounting with PFA Teflon, it is recommended to use quartz glass or a release agent, as it is often difficult to remove the Teflon sheet from other types of glass slides. One issue with PFA Teflon, however, is that it is often sold in bulk and may be more difficult to access commercially.

Tagami (2005) described the mounting, grinding and polishing of zircon in some detail. Of special note is the following:

- Teflon should not be allowed to overheat and create bubbles, which will ruin the mount.
- As zircons are exposed at the surface of the Teflon, very little grinding is necessary. Start with SiC grit paper (e.g. #600 or #1200) on a glass plate and only sand about 2–3 times over each before using different grades of diamond paste (e.g. 6, 3 and 1 µm) or some other medium on a rotating lap. There is no actual adhesion between the

Teflon and the zircons, so over-grinding is a common cause of grains falling out of Teflon mounts during etching as it removes the small Teflon lip, which encloses and holds the grains. Removal of all the original shine from the Teflon surface is an indication that grinding is nearing an end.
- Titanite may also be mounted in Teflon, but is most commonly mounted in an epoxy button mount as described for apatite above.

2.6 Chemical Etching

Spontaneous fission tracks are revealed by chemical etching in a track-recording material because the etchant preferentially attacks the highly disordered material in the core of the track (e.g. Fleischer et al. 1975). The bulk-etching rate in minerals is not uniform and varies in different crystallographic directions, so that the tracks take on different shapes and sizes, depending on which crystal surface they are etched (e.g. Wagner and van den Haute, 1992). Common etching recipes for different minerals are outlined in Table 2.2.

Apatite and zircon grains are often prismatic and during mounting often tend to align on prismatic faces, which are approximately parallel to the c-axis. This is the orientation sought for optimal track revelation for FT dating (e.g. Wagner and van den Haute 1992; Gleadow et al. 2002; Donelick et al. 2005; Tagami 2005). In apatite, zircon and titanite etching is anisotropic (i.e. tracks are revealed preferentially parallel to the crystallographic c-axis but also need to be fully revealed perpendicular to that axis for etching to be judged as being optimal), and this is particularly evident in low-radiation damaged grains (e.g. Gleadow 1981). In glass, fission tracks have a circular or elliptical cross-section, even after prolonged etching, because the host material etches isotropically (e.g. Dumitru 2000, see also Chap. 1, Hurford 2018).

Over or under etching may jeopardise the quality of data obtained. The correct etching time in minerals, such as zircon and titanite, is highly variable and will depend on a variety of factors, especially the general radiation damage level, which is reflected in the track density. The particular choice of etchant used is also important in many cases. The anisotropic etching characteristics of both zircon and titanite are significantly greater when using acid etchants shown in Table 2.2 or listed in references cited therein, than for the hydroxide etchants. This can be seen in Fig. 2.3, which shows the same zircon etched in a hydroxide and an acid etchant for comparison. Except where the radiation damage levels are fairly high, therefore, the hydroxide etchants are to be preferred and give better results for both these minerals. It

Table 2.2 Commonly used etching recipes and notes for different minerals and glass

Mineral	Etchant	Conditions	Comments
Apatite	5 N HNO_3 (Gleadow and Lovering 1978)	20 ± 1 °C 20 s	Different HNO_3 strengths have also been reported, e.g. HNO_3 conc for etching time of 10–30 s (Fleischer and Price 1964) and 1.6 M (7 vol.%) at 25 °C (or at room temperature) for 20–40 s (Naeser 1976), but those listed to left are the most commonly used (see also Seward et al. 2000 and Sobel and Seward 2010)
	5.5 N HNO_3 (Carlson et al. 1999, see also Donelick et al. 2005)	21 ± 1 °C 20 ± 0.5 s	Following etching, wash thoroughly under tap, dry with clean paper tissue and leave in air for a few hours to ensure all etchant has dried out from within mount before inspecting under a microscope. If a longer etch is required, wet mount briefly before re-etching. This aids the entry of etchant into partially etched tracks, as normally any extra etching will only be for a matter of seconds
Zircon	Binary eutectic mixture of KOH:NaOH (in proportions by weight: 8.0 g KOH and 11.2 g NaOH—Gleadow et al. 1976) A variant of the eutectic is NaOH:KOH:LiOH (6:14:1), which is reported to increase etching efficiency at lower etching temperatures for comparable etching times (Zaun and Wagner 1985) For other possibilities, see also Garver (2003) and Tagami (2005)	225–230 °C 4–120 h (or more)	Vessel: (ceramic, platinum or Teflon)—place on hotplate (monitor temperature with a thermometer) and cover with inverted beaker to prevent crust from forming around top. Check temperature of etchant solution with thermocouple Place zircon mounts face down in etchant (they will float), leave initially for 4 h and then check etching progress. Note —grain surfaces with highest etching efficiency are those showing the presence of sharp polishing scratches Following etching place the mount in 48% HF in a Teflon dish for 15–30 min in order to clean up grains. This will not affect the quality of the etched grains After etching, mounts will almost always deform slightly. In order to flatten the mount so that a muscovite detector with a good contact can be later applied, more heating should be carried out, but not enough to melt the Teflon or push grains further into the mount *Note* Generally the time required for proper fission track revelation in zircon is inversely proportional to the accumulated radiation damage, which is related to uranium concentration and age of the grain. Hence, etching duration varies over a broad range
Titanite	37% HCl (Naeser and Dodge 1969)	90 °C 15–60 min	Vessel: Stainless steel, Teflon or alkali-resistant ceramic beaker Preparing the acid etchant produces an exothermic reaction, so leave to cool down over night
	$1HF:2HCl:3HNO_3:6H_2O$ (Naeser and McKee 1970)	20 ± 1 °C 10–60 min	Etching times typically vary between 10 and 60 min, depending on the level of radiation damage (see Fig. 2.4). Check track etching rate and characteristics after 10 min. Before viewing under microscope wash mount thoroughly. Return to etchant to complete etching—repeat as often as necessary with varying times
	50 M NaOH solution (40 g NaOH: 20 g H_2O) (Calk and Naeser 1973)	130 °C 10–60 min	*Note* The anisotropic etching characteristics of titanite are significantly greater when using the acid etchant (especially for low radiation damage grains)—so the NaOH etchant is preferred for such grains. This etch is also suitable for track revelation in garnets and epidote (Naeser and Dodge 1969), but etching time required may be >1 h. A 75 M NaOH etch may be required for etching in some garnet compositions (Haack and Gramse 1972). Over long etching times the NaOH solution may dehydrate and become less efficient, so a condenser may be useful when using this etchant
Muscovite	HF (Fleischer and Price 1964); strengths of both 48 and 40% have been reported	20 ± 1 °C Reported: 5–45 min Commonly used: 20–25 min	Vessel: Teflon dish After etching, wash muscovite thoroughly in warm running water. Then place the muscovite on a glass slide on a hot plate at ∼100 °C to drive off any HF absorbed between the cleavage planes, this will prevent any later inadvertent etching of the microscope objective
Glass	HF (Fleischer and Price 1964), different strengths ranging from 48 to 12% have been reported	23 °C 5–90 s depending in part on acid strength	Vessel: Teflon dish After etching, neutralise HF and wash mount thoroughly in warm running water *Note* obsidian usually takes longer to etch than glass shards

Fig. 2.3 Spontaneous fission tracks in zircon from the Mud Tank Carbonatite in Central Australia etched in (**a**) the KOH: NaOH eutectic etchant of Gleadow et al. (1976) and (**b**) in the equivolume HF:H_2SO_4 etchant of Krishnaswami et al. (1974). The eutectic etch has revealed the tracks more isotropically than the acid etchant and is the preferred etchant for zircon. The polished surface is parallel to the (010) cleavage, and the extended faint lines running across each frame are polishing scratches. The scale bar on each frame is approximately 10 μm

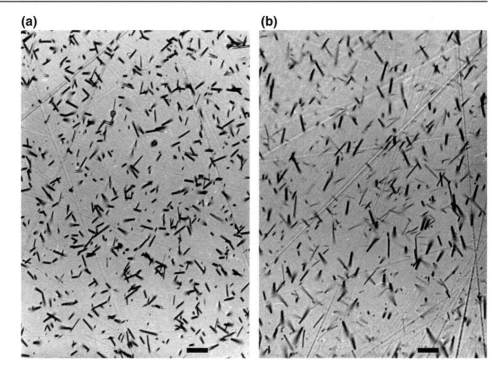

(a)   (b)

is extremely important in these minerals that the degree of etching be judged from the appearance of the tracks and not from the application of some standard etching time (Fig. 2.4). Increased radiation damage results in markedly shorter etching times required to fully reveal tracks for microscope observation (Fig. 2.5). Hence, it is common practice, especially when working with sediments, to prepare more than one mount of a zircon and titanite grain population and etch for different times in order to enable FT analysis to be performed on the entire population (Naeser et al. 1987). For strategies on handling zircons with very-low or very-high fission-track densities, see Appendix in Naeser et al. (2016).

Because radiation damage does not accumulate to any degree in apatite, it is more consistent in its etching behaviour and a standard etching time is commonly used, although exceptions will be found from time to time. However, for most apatites, it is important to adhere to a strict etching protocol, so that key parameters such as track lengths and Dpars (see Sects. 2.11.5 and 2.11.7) can be measured in a consistent fashion and reproduced in other laboratories.

2.7 The External Detector Method (EDM)

The most common protocol for studying minerals with heterogeneous uranium content between different grains is the External Detector Method (EDM). In order to determine their ^{238}U content, the EDM requires that etched grains be

sent to a nuclear reactor for thermal neutron irradiation. The sequence of steps involved in preparing samples for irradiation for a number of minerals has been described in some detail in a number of works cited in Sect. 2.2 (see also figures in Gleadow et al. 2002 and Tagami and O'Sullivan 2005) and is shown in Fig. 2.2.

Briefly, the spontaneous fission tracks are etched on an exposed internal polished surface on the grains and the induced tracks on a muscovite external detector attached to the grain surface during neutron irradiation. After irradiation, the external detector is etched to reveal an induced fission-track mirror image corresponding to grains in the mount (Chap. 1, Hurford 2018).

Even though this results in a second set of tracks being produced within the grains themselves, these tracks will not be detected because they are not etched after the irradiation. Also, because ages can be measured on individual grains, a careful selection of grains can be made to avoid those which may be badly etched or contain dislocations. The external detector is usually a sheet of low-uranium muscovite (commonly Brazil Ruby—with ASTM Visual Quality of V-1, which is designated as clear and free of stains and inclusions, cracks, waves and other defects, with about 5 ppb uranium). Suitable high-quality muscovite external detectors can be bought commercially already cleaved to ∼45–55 μm thickness and pre-cut to a suitable size (typically ∼ 12 × 12 mm). Muscovite is more suitable as a detector in dating applications than plastics, such as Lexan, because its track registration and etching properties are much more like those of the minerals with which it is being compared.

Fig. 2.4 Spontaneous tracks in titanite following various etching times; (**a**)–(**d**) show tracks etched for 5, 10, 15 and 20 min, respectively, in the mixed acid etchant (1HF:2HCl:3HNO$_3$:6H$_2$O). The tracks in (**b**) represent the minimum degree of etching for which a reliable track density can be obtained

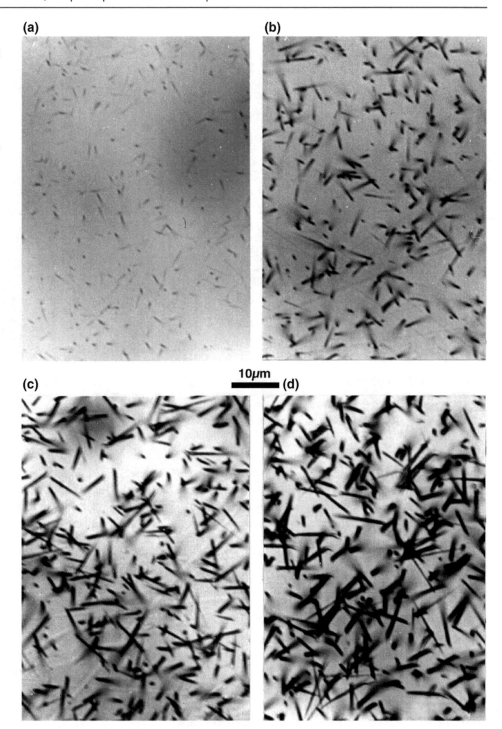

To measure the FT age by the EDM involves determining the spontaneous track density in a selected grain and finding the mirror image area on the muscovite external detector where the induced track density is counted over exactly the same area. Because the geometry of track registration is not the same for the internal surface (4π) on which the spontaneous tracks are measured and the external detector surface (2π) used for induced tracks, a geometry factor must be introduced to correct for this difference (e.g. Wagner and

van den Haute 1992). The geometry factor is $\sim$0.5, but this is not exact because of small differences in detection efficiency of the two surfaces and differences in the range of fission tracks in the two different materials (e.g. Iwano and Danhara 1998; Gleadow et al. 2002). A consequence of anisotropic etching rates in minerals such as zircon or titanite is that some grain surfaces in a mount may have a low etching efficiency (e.g. Fig. 2.4). Comparison of spontaneous tracks on such surfaces with induced tracks in an

(a)   **(b)**   **(c)**

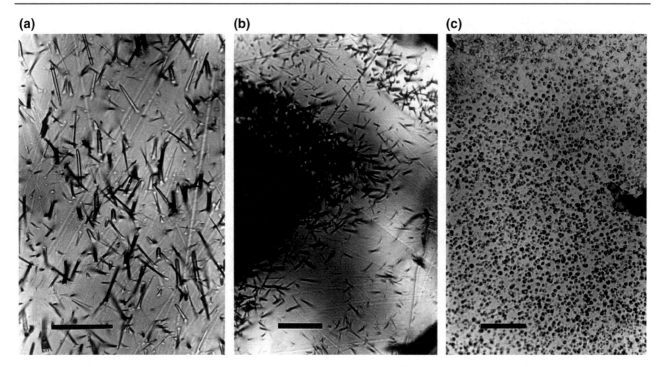

Fig. 2.5 Etching of titanites at different stages. **a** Shows the highly anisotropic etching of induced tracks in annealed titanite, etched 25 min in the 1HF:2HCl:3HNO$_3$:6H$_2$O etch at 20 °C (see Sect. 2.6). **b** Shows the less anisotropic but very variable etching at intermediate radiation damage levels in a zoned titanite, ranging from under etched in low U zones, to strongly over etched and unresolvable in the highest-uranium core zone (etched 12 min). **c** Shows the small rounded, conical etch pits, which resemble tracks in glass, found in titanite with an extreme track density of $\sim 8 \times 10^7$ cm^{-2} and etched for 1 min. The scale bar is 20 μm in each case

adjacent muscovite detector, where the etching efficiency approaches 100%, will clearly give erroneous results. A very careful selection of only the highest etching efficiency surfaces, as identified by sharp polishing scratches (Figs. 2.3, 2.5a) (e.g. Gleadow 1978), is therefore essential for the EDM. Care is also necessary when dealing with low track densities in minerals that etch anisotropically to ensure that etching has been sufficient for even the most weakly etched tracks to be revealed (see Fig. 2.3). In some samples, such as young zircons or titanites, it is extremely difficult to reveal tracks in certain orientations. This effect is moderated by accumulated radiation damage so that most zircons and titanites with track densities between $\sim 10^5$ and 10^7 cm^{-2} can be readily analysed by the EDM (Gleadow 1981, see also Naeser et al. 2016—Appendix 1). Montario and Garver (2009) developed a scanning electron microscope technique, which permits counting of track densities to as high as 2 $\times 10^8$ cm^{-2} allowing for a greater range of high track density zircons to be counted, particularly in populations containing very old (Precambrian) grains, which had previously been considered uncountable.

Also noteworthy is a further method complementary to the EDM for FT dating of moderate-to-high uranium zircons carried out by electron probe microanalysis (EMPA). This method involves determining uranium concentration and imaging of the number of spontaneous fission tracks intersecting the surface, using an electron backscatter detector (Gombosi et al. 2014). Dias et al. (2017) reported an alternative approach to FT dating of zircons, but also using EMPA to measure uranium concentration.

Because of its relative ease of handling, amenability to automation (see Sects. 2.11.4 and 2.12) and its provision for single-grain age information, the EDM is currently the preferred dating method for apatite, zircon and titanite in most FT laboratories.

2.7.1 Preparing Mounts for EDM Age Dating and Neutron Irradiation

2.7.1.1 Wrapping and Packing for Irradiation

- Cut down etched apatite mount size to fit into the irradiation can (e.g. 1 × 1.5 cm) and thoroughly clean with soapy lukewarm water and dry in alcohol.
- Overlay mounts with dust-free pre-cut (if possible) low-uranium muscovite—always handle with tweezers. A clean muscovite surface may be formed by placing it on a piece of sticky tape and lifting off a thin flake of muscovite. To ensure a good contact with grains during irradiation make sure that the muscovite does not extend beyond edges of the mount.

- Prepare heat shrink plastic bags with an opening sufficient to slide the muscovite and grain mount pair in easily.
- Place muscovite–mount pair inside the bag and holding contents firmly with tweezers. Use a kitchen-type bag sealer, to close the opening as close as possible to the muscovite–mount. Trim the edges of the bag to provide passage for escaping air. Heat two clean glass microscope slides on a hotplate at 100 °C. Place bag and its contents on one slide keeping the muscovite face up and cover immediately with the other slide. Press firmly with tweezers to achieve good contact. Some laboratories use low chlorine-content reactor-friendly Scotch 3M® magic tape or Parafilm M to affix the muscovite to the mount.
- Wrap the standard glass–muscovite pair in the same way as described above.
- Place mounts in an irradiation can and record order of samples in the stack. The number of samples that can be fitted into an irradiation stack will depend on local reactor protocols. Use two standard glasses in each irradiation package one at top and bottom of the stack, to monitor any spatial neutron flux gradients. A further glass standard may also be inserted in the middle of the package (Fig. 2.2). If possible, it is also recommended to include an appropriate mineral age standard (e.g. Durango apatite or Fish Canyon Tuff zircon) within the irradiation package (see Chap. 1, Hurford 2018).
- If the irradiation can is sucked pneumatically into the reactor to the irradiation position, it may be necessary to allow for the addition of packing material at each end of the can (e.g. Al foil padding), which acts as a shock absorber preventing glass breakage on impact.
- *Note*: Irradiation results in significant levels of radioactivity due to short-lived isotopes, mainly related to Na content present in conventional soda lime petrographic glass slides, but also from sweat in fingerprints (*so gloves should always be worn when preparing material for irradiation*). In this respect, silica glass slides are less problematic.

2.7.1.2 Standard Glasses

Hurford and Green (1983), Wagner and van den Haute (1992) and Bellemans et al. (1995) provided information on different glass standards and evaluations of their suitability for monitoring neutron fluence and determining ^{238}U content of individual grains. In the early years of FT dating, glasses produced by the National Institute of Standards and Technology (NIST) (Carpenter and Reimer 1974) the SRM-series were commonly used for monitoring. However, these are depleted in ^{235}U and contain a variety of trace elements, and have gradually been replaced by the Corning CN1–CN6 series with a natural ^{235}U/^{238}U ratio and fewer trace elements (see Bellemans et al. 1995). CN1 and CN2 are generally suitable for zircon and titanite and CN5 for apatite. However, worldwide stocks of CN5 have now been exhausted, but uranium-doped oxide glass IRMM-540R (15 ppm uranium) produced by the European Commission's Institute for Reference Materials and Measurements (IRRM) has proven to be an effective substitute (De Corte et al. 1998). Further, uranium oxide-doped glass IRMM-541 (50 ppm uranium) is suitable as an alternative standard, e.g. for zircon and titanite, which require shorter irradiation times due to their generally higher uranium content.

The use of induced tracks in muscovites over standard glasses has been instrumental in acquiring zeta (ζ) calibrations based on the analysis of geological age standards analysed by other geochronological techniques (Hurford and Green 1983; Chap. 1, Hurford 2018). This procedure settled earlier disagreements about the ^{238}U spontaneous fission decay-constant, neutron dosimetry calibrations and ambiguities in corrections for measuring spontaneous and induced fission tracks on different surfaces and in different materials, and was deemed to be a workable solution to some long-standing problems (Hurford 1990). The ζ-method calibration is the most widely used for age determinations in FT laboratories today. But this method also has some drawbacks because it yields calibration factors that are to some extent personal, vary for different mineral species and combine known and unknown factors (e.g. Wagner and van den Haute 1992; Hurford 1998). An alternative procedure to using standard glasses, but not that widely used, is the ϕ-method. This involves an absolute determination of the thermal neutron fluence by measurement of neutron-induced gamma activity in Au and Co metal activation monitors included in irradiation cans together with samples (e.g. van den Haute et al. 1998; Enkelmann et al. 2005).

2.7.1.3 Neutron Irradiation

In a nuclear reactor, the total neutron flux may comprise three neutron components of different energy ranges; fast, epithermal and thermal. When choosing a reactor for sample irradiation, it is crucial that only a well-thermalised neutron facility is used (e.g. Wagner and van den Haute 1992). This is required in order to avoid track production by epithermal neutrons from ^{235}U fission or by fast neutrons from ^{238}U and ^{232}Th fission. Such tracks would be indistinguishable from the thermal neutron-induced ^{235}U fission tracks required. Since the Th/U ratio of material used for FT analysis is highly variable, it is important that the nature of the neutron flux in the irradiation position used is well known. Ratios of thermal/epithermal and thermal/fast neutrons of >100 and >80, respectively, provide some certainty that practically all induced tracks measured in muscovite external detectors originate from ^{235}U fission (see Green and Hurford 1984 and Wagner and van den Haute 1992 for further details).

The integrated neutron dose requirement for different minerals varies and is dictated by the expected uranium content of a particular mineral. Typical nominal doses requested for the most common minerals used for FT studies at the University of Melbourne in past years have been apatite $\sim 1 \times 10^{16}$ n cm^{-2}, zircon $\sim 1 \times 10^{15}$ n cm^{-2}, titanite ~ 4–5×10^{15} n cm^{-2}. These values are not absolute, however, as the fluence requested in relation to that actually received and monitored within the irradiation package may vary from reactor to reactor.

2.7.1.4 Post-irradiation Sample Handling and Slide Preparation

- After the irradiated package is received, place it in properly lead-shielded storage until radiation levels, which should be monitored periodically, are safe and samples can be unpacked.
- Prior to unwrapping samples, use a sharp pin to make holes in each corner of the glass–muscovite pair making sure that one corner has two holes.
- If tape is used to secure muscovite to glass mounts, remove very gently so as not to lift off large amounts of muscovite flakes or alternatively cut around the detector with a scalpel and place in etchant and any remaining tape will fall off.
- Etch muscovites in HF as listed in Table 2.2.

Each apatite, zircon and titanite grain mount and muscovite can be mounted together on a standard petrographic glass slide (typically 26 × 76 × 1.5 mm) so that one is the mirror image of other. These can be mounted using a small amount of Petropoxy® 154, but note that in cases where grains are embedded in Teflon, then these may be mounted using double-sided sticky tape. To minimise focussing requirements during track counting the mica should be glued on a thin glass slide to bring it to the same level as the grain mount. Ensure that the sample number is labelled on the back of the slide. A reference point should be placed in the centre of the mount, between the grain mount and muscovite. The most convenient is a metal (copper) disc with grid (used for locating and referencing specific areas in Transmission and Scanning Electron Microscopy), which is used as a central coordination point for commencing the later alignment procedure keyed to the pin-pricks between specific grains and their mirror image on the muscovite (see Fig. 2.2 in Sect. 2.11.4). Such coordination points can also be used later for locating the exact grains on which track measurements have been carried out for electron microprobe analyses if required (see Sect. 2.11.7).

2.8 The LA-ICP-MS Method

LA-ICP-MS is the first technique that has been able to compete with the traditional neutron irradiation method for determination of ^{238}U content in terms of high spatial resolution and ppm sensitivity. This analytical development has added a new approach for FT analysis, whereby ^{238}U can be determined directly in mineral grains, rather than by using ^{235}U-induced fission tracks as a proxy, as required by the EDM. LA-ICP-MS facilities are now becoming widely available, and this mode of analysis has considerable advantages over the conventional EDM, as it no longer requires neutron irradiations and the long delays in sample processing (typically many weeks) that they require. Other advantages of this approach are that it eliminates the need for handling radioactive materials and as only one track density measurement (the spontaneous FT density) is required, it reduces the overall requirement for FT counting (e.g. Hasebe et al. 2004; Donelick et al. 2005; Vermeesch 2017, and Chap. 4, Gleadow et al. 2018).

Hasebe et al. (2004) carried out the first systematic study using LA-ICP-MS for FT analysis, an approach foreshadowed by Cox et al. (2000) and Košler and Svojtka (2003). More recent studies by Donelick et al. (2005), Hasebe et al. (2009, 2013) and Chew and Donelick (2012) have provided additional experimental details and demonstrated the effectiveness of this approach.

The sequence of steps using LA-ICP-MS is illustrated in Fig. 2.2. The first three steps are the same as for the EDM, in that a grain mount is prepared, polished and then etched to reveal the spontaneous fission tracks. However, prior to mount preparation, it is important to confirm that the mount size can be accommodated by both the microscope stage and laser cell, as each laser cell may have specific sample dimension requirements. The spontaneous tracks are then counted manually or on sets of digital images that are acquired using an automated image capture system referenced to a coordinate system best defined by three metal discs with grids placed around the grain mount (see Sect. 2.12 and Chap. 4, Gleadow et al. 2018). The grain coordinates and the slide are then transferred to the laser ablation cell and analysed by LA-ICP-MS. Most studies conducted have used a single ablation spot of ~ 20–30 μm diameter or a rastered scan centred around the area where the tracks were counted. As all of the tracks, which are etched on a surface in zircon and apatite, originate within ~ 5.5–8.5 μm of the surface, respectively (e.g. Hasebe et al. 2009), it is important to ablate the surface only to about that depth, in case there is zoning of uranium in the vertical dimension.

The ^{238}U concentration is determined relative to suitable external standards of accurately known and uniform uranium abundance (e.g. NIST610 and 612 glasses) and measurement of an appropriate internal standard, e.g. ^{43}Ca (apatite), ^{29}Si (zircon), to correct for variations in ablation volume (e.g. Hasebe et al. 2004, 2009, 2013; Chew and Donelick 2012). These works also list operating conditions, but these are by no means universal and may vary with specific instrumentation used.

LA-ICP-MS for direct determination of ^{238}U is a relatively new approach to FT analysis. It is usually calibrated using a variant of the zeta calibration approach (Hurford and Green 1983; Donelick et al. 2005) or can be calculated as an absolute age using explicit values for all the constants in the age equation (e.g. Hasebe et al. 2004; Gleadow et al. 2018). Efforts are also underway to identify some well-characterised and matrix-matched minerals with relatively homogeneous uranium content to correct for elemental fractionation during laser ablation (e.g. Soares et al. 2014; Chew et al. 2016). Vermeesch (2017) has outlined the statistical treatment of analytical uncertainties arising from different approaches to LA-ICP-MS-derived FT age dating (see also Chap. 6, Vermeesch 2018).

As outlined earlier, the measurement of ^{238}U by LA-ICP-MS is a relatively new approach to FT dating. To date, mainly LA-ICP-MS data on apatite, zircon and volcanic glass have been reported and sample preparation is similar to that described previously for these minerals.

2.9 Double–Triple-Dating

With technological advances in instrumentation and an improved understanding of the behaviour of different geo-thermochronological systems, it has become possible to carry out dating of individual grains from an aliquot using independent isotopic systems, i.e. combinations of FT, U/Pb and (U-Th)/He analyses on either apatite or zircon. This so-called double- or triple-dating approach is a powerful new development in the geo-thermochronology toolbox because the radioactive decay schemes for these systems have different temperature sensitive ranges and can provide more robust constraints for computing time–temperature histories (Chap. 5, Danišík 2018).

Earlier approaches using different combinations of methods on subsets of particular sample aliquots were prepared for analysis using standard procedures (e.g. Carter and Moss 1999; Carrapa et al. 2009). When using FT dating in combination with other techniques on single grains, principally by use of LA-ICP-MS, some modification of practical steps may be required in sample preparation. Chew and Donelick (2012) described the double-dating of apatite using the FT and U-Pb methods on single grains, while Hasebe et al. (2013) also outlined a similar approach for dating both apatite and zircon. As all the data (apart from the counting of spontaneous fission tracks) in both studies were acquired using LA-ICP-MS, the preparation of mounts was essentially similar to that described in Sects. 2.5 and 2.6 above. For triple-dating (U/Pb, FT and U-Th/He) of apatite (Danišík et al. 2010), analyses were carried out on subsets of an exceptionally large apatite aggregate precipitated in cavities and veins in a late Palaeozoic rhyolite; therefore, the standard method was used for the preparation of samples for each dating method. Reiners et al. (2007) reported a combined apatite FT and (U-Th)/He double-dating study. In this case, the apatite FT age was determined by conventional mounting in epoxy and counting of spontaneous FT density (ρ_s) and uranium content determined by LA-ICP-MS on the same grains, following which grains were plucked from the mount and dated by standard (U-Th)/He procedures.

Note: With ongoing technical developments, apatite FT dating might in the future be combined with in situ (U-Th)/He dating, in which case samples will need to be embedded in Teflon, due to the excessive degassing of epoxy mounts preventing the attainment of the ultra-high vacuum required (e.g. Evans et al. 2015). A similar mounting procedure would also be applicable to double- or triple-dating of zircon grains in such an in situ (U-Th)/He dating approach.

2.10 Microscope Requirements

The microscope is the single most important component in a FT dating laboratory, but the choice of equipment is frequently not given the scrutiny it deserves. Any old microscope will not do. What is required is a research-grade microscope of the highest quality fitted with objectives, condenser and illumination to give optimum performance at the highest magnifications (at least 1000×). Polarising equipment and a rotating stage are not necessary for track counting and place significant restrictions on the other equipment that can be used on a microscope. Also, most other microscope stages are more robust and have superior mechanical slide movements than do rotating stages.

For work at high magnification, it is important to have high intensity illumination available and preferably a light source of 50 W or more. In recent years however, illumination for optical microscopy using light-emitting diodes (LEDs) has shown considerable promise as a spatially and temporally stable and cost-effective technology compared to traditional arc lamp illumination sources. It is essential to have both reflected and transmitted light illumination available on the microscope, arranged so that the user can readily switch back and forth between the two. Tracks are usually counted in transmitted light, but reflected light can be useful for resolving complex track overlaps, locating the end of tracks which intersect the surface, and for counting very-high track densities. Reflected light can also be very useful for locating horizontal confined tracks used in length measurement (see Sect. 2.11.5).

In general, it is more important to have flat-field objectives, such as planachromats, than those that have a very high degree of colour correction. However, some lenses rate highly in both characteristics. Objectives should be parfocal with each other, and it is more convenient if they are mounted in a multiple revolving nosepiece rather than on an individual bayonet-type objective carrier.

An important question is whether to use dry or oil-immersion objectives for the highest magnifications. This is partly a matter of personal choice and both systems are in routine use in FT laboratories. Technically, oil-immersion objectives have superior resolution but other effects often outweigh this advantage. Very fine images are obtained under oil immersion in zircon and titanite, which both have high refractive indices. However, in some other minerals, oil has distinct disadvantages because the refractive index of the usual immersion oil, 1.515, is almost the same as that of the mineral. In apatite and muscovite, this results in tracks losing contrast with the surrounding mineral and becoming more difficult to observe. This problem cannot be avoided by using a different immersion medium because the objectives can only be used with an oil of the refractive index for which they were designed. For this reason, it is often preferable to use dry objectives, which must also be corrected for no cover glasses.

A binocular eyepiece tube is regarded as an essential component of any microscope used for observing and counting fission tracks. The most commonly available eyepieces have magnifications of $10\times$, although higher magnifications, e.g. $12.5\times$ or $15\times$ are often preferable for FT analysis.

For counting fission tracks, one of the eyepieces should be a focusing type fitted with a graticule, usually in the form of a 10×10 grid. In the most commonly used graticules, each grid square is 1 mm across on the carrier disc. For track length measurements, it is essential to have an eyepiece fitted with a scale bar or eyepiece micrometre. Calibration of the scale bar and the area of the graticule are carried out by measuring their dimensions against a stage micrometre, which can be obtained with divisions down to 2 μm. If such a calibration slide is not available, then a satisfactory alternative is to use a piece of optical diffraction grating, although this requires observation under incident light illumination, as the metallic coating of the grating is opaque. Diffraction gratings are accurately ruled with very fine lines, at a known spacing of the order of 1 μm.

Modern research microscopes are all equipped with a sub-stage condenser and field and aperture diaphragms required to produce Köhler illumination (see: http://zeiss-campus.magnet.fsu.edu/articles/basics/kohler.html) for providing the most uniform and optimal specimen illumination across the field of view and the maximum optical resolution for a particular objective lens. It is particularly important to set up the optimum illumination conditions for the particular objective to be used, especially for image capture for automation (see Sect. 2.12). This will almost always be with the $100\times$ lens, the highest lens available.

2.11 Data Collection

2.11.1 Identification of Fission Tracks

In order to count fission tracks, they need to be reliably identified and distinguished from etch pits or features of other origins, such as dislocations or inclusions. Etched fission tracks have certain properties (Fleischer and Price 1964), which enable their discrimination from spurious dislocation etch pits (Table 2.3).

Dislocations are most commonly encountered in large numbers in relatively young volcanic apatites, although they are seldom found in all grains mounted from a single sample. Apatites from slowly cooled plutonic rocks usually show few, if any, dislocations and discrimination is fairly

Table 2.3 Properties that distinguish etched fission tracks from dislocations

Fission tracks
Etched fission tracks: • Straight features, as fission-fragments travel essentially in straight lines • Have a limited length, as fission fragments have a limited range of about 5–10 μm (depending on the host material) and the maximum track length is up to the maximum etched range of both fission fragments and varies from ~10–20 μm in different minerals • Randomly oriented, although highly annealed tracks are preferentially aligned parallel to the c-axis in apatite • The distribution of spontaneous tracks must be statistically the same as that of uranium, and hence of induced tracks in a particular material Unetched fission tracks have a limited thermal stability that is characteristic of the registering material and is usually different from that of dislocations or micro-inclusions

Dislocations
• Often bent, branching, curved or wavy • Often occur in swarms, the distribution of which is unrelated to that of uranium (see Gleadow et al. 2002) • Lengths frequently similar to each other and often much greater than those of fission tracks • Often occur with a strongly preferred orientation, either as sub-parallel swarms, or as lines of parallel etch channels • May act as nucleation sites for the precipitation of impurities from the host material, in which case they are readily identified

straightforward. Zircons and titanites tend to have relatively few dislocation etch pits and discrimination is not usually a problem. Zircons may contain minute crystalline inclusions, however that can sometimes be mistaken for tracks, especially where the fossil track density is very low. Often such crystallites show a regular orientation in relation to some crystallographic direction and a range of sizes that can aid in their discrimination.

In general, the problem of discriminating fission tracks from other etch features is not severe, but can become significant when dating grains with very low track densities. It is always an advantage when there are sufficient tracks present so that they can be compared with each other. Experience and an appropriate selection of dating technique are important in handling difficult cases, but it may be wiser to simply go on to another sample.

2.11.2 Counting Techniques

Microscope work for an EDM FT age determination, using the ζ-calibration approach, involves the counting of three track densities: spontaneous (ρ_s), induced (ρ_i) and standard glass (ρ_d). For each of these, the tracks are normally counted at magnifications of $1000\times$ or more, and usually in transmitted light. The depth of focus under these conditions is very limited so that the fine focus of the microscope needs to be moved up and down frequently during counting to follow the three-dimensional nature of each track.

The position of a track in relation to the counting graticule is defined by the intersection of the track with the surface. With experience, this surface-end of the track can be recognised at a glance but at first should be judged from the following characteristics. All the tracks (long and short) will only be in focus together at the surface, so that progressively fewer tracks will be seen as the focus is moved down into the grain. Moving the fine focus up and down can therefore be used to identify the surface-end of the tracks. Also, the intersection of each track with the surface is clearly visible as a dark hole (etch pit) in reflected light (if available). Even in transmitted light, the two ends of a track do not look the same, but can still be identified.

On both the exposed internal grains surfaces and the external detector surfaces, the tracks vary in length from essentially zero up to the maximum for the particular mineral. It is important not to overlook the smallest tracks, or at least to use some consistent criterion as to which of the short tracks will be included in the final track density. Counting the tracks then involves systematically scanning across an appropriate number of eyepiece grid squares, so that each track is included once only. Where the surface intersection of the track lies exactly on a grid line, some consistent convention must be used to assign the track to a particular grid square. For example, a track might be included in a square if it lies along the top or right-hand edge, but not if it is on the bottom or left-hand edge. Further, if no tracks are observed in the grid, then that zero *must* be recorded accordingly and included in the final tally of counted tracks per total surface area analysed. A repositioning technique for achieving more accurate counts of induced tracks in muscovite external detectors in grains with low and/or inhomogeneous uranium concentrations has been described by Jonckheere et al. (2003).

Typically, more than 20 grains should be counted if possible, and the results combined to give an age for the sample. In crystalline samples with complex age spectra or detrital samples, a greater number of grains should be targeted, aiming for about 50–100 grains or even more, as a higher number is especially important for the discrimination of different age populations if present (e.g. Garver et al. 1999; Bernet and Garver 2005; Coutand et al. 2006; see also Chap. 16, Malusà 2018).

2.11.3 Standard Glasses

The standard glasses used for neutron dosimetry are produced with uniform uranium concentrations so that as a result of irradiation, their corresponding muscovite detectors receive a uniform-induced track density, usually over a large area of about 1 cm^2. The tracks in each muscovite detector should be counted on a regular pattern covering the whole available area using the same counting criteria as with counting N_s on the grain surfaces. The simplest method is to move around the muscovite on a regular 1 or 0.5 mm grid and at each location, count the tracks in a predetermined area of the graticule. The number of locations and the number of grid squares counted at each will depend on the track density. Typically tracks are counted at a number of locations to give a total track count of at least 1000. It is good practice to count each glass over different areas at least twice to verify the track density obtained and to increase the precision on the combined measurement. The track density (per cm^2) is determined from the total number of tracks and the total area counted. If a significant and reproducible difference is found in standard glass track density along the irradiation canister, then this indicates a neutron flux gradient and an intermediate value should be interpolated for each mount in the package. Calculation of the uranium concentration for individual mineral grains may be estimated approximately by measuring the ratio of the induced track density over the grain to the induced track density in an external detector over the standard glass multiplied by the known uranium content of the standard glass.

2.11.4 Automation of the EDM

To determine the ratio of spontaneous to induced tracks, identical areas are counted on each mineral grain and its muscovite detector mirror image. A typically used sequence of steps is to select a suitable grain, count the spontaneous tracks, locate its mirror image on the muscovite and count the induced tracks. Before counting an external detector mount, it is recommended to scan the muscovite at low power to check that the detector had remained in close contact with the mount during irradiation. Grains with good contact and sufficient uranium and N_i will have sharp, clearly defined mirror images and grain boundaries in the muscovite. Areas of poor contact, indicated by diffuse, rounded grain image boundaries often with a splayed track pattern should be avoided, as they will give an underestimate for the induced track density.

The ability to select suitable grains for counting only comes after the completion of a significant amount of training to recognise identifiable features associated with each grain. Suitable grains include those having well-etched tracks, sharp polishing scratches, reasonably uniform track density and minimum interference from inclusions, cracks and dislocations. Where spontaneous track densities (N_s) are low, the numbers of tracks in each grain, and the apparent single-grain ages, can vary substantially due to the natural statistical variation of the decay process. It is easy to select only those grains with relatively higher track densities, but this can lead to a seriously biased age. In such cases, it is important to ensure that grains are selected covering the whole range of variation in track densities, even including grains with no tracks if they are present. Having selected a suitable grain, the spontaneous tracks are counted, but the zone within the range of one track of any external grain margin should be avoided.

Locating the corresponding area on the muscovite can be carried out manually, although this is extremely tedious and mistakes are easily made resulting in erroneous ages. However, the time-consuming task of locating matching points on the grain mount and its external detector is now mostly automated using a computer-controlled microscope stage system (e.g. Smith and Leigh-Jones 1985).

Such automated stage systems are now almost universally used for the EDM and provide a much faster and more reliable technique for accurately and repeatedly locating matching points on a muscovite external detector. Another advantage of such automated methods is that they provide a framework for the systematic collection and organisation of the FT data.

An example of the sequence followed using an automated stage system involves the following steps:

- Find and mark the zero reference point (e.g. the copper disc in Fig. 2.2).
- Coarse alignment of the mount and detector using at least two different pinprick positions (alignment marks) preferably between opposite corners of the mount.
- Refine the alignment using mineral grains and their induced track images.
- Select and label suitable grains for counting.
- Count spontaneous and induced tracks over each grain.
- Measure confined track lengths as these are observed.
- Measure Dpars (see Sect. 2.11.7) for each grain from which age or length data are collected—using only etch pits from spontaneous tracks.
- Save all data to a computer file.

Most systems are capable of operating in three axes, so that relative offsets in x, y and z, as well as rotations of the muscovite relative to the mount can be corrected for by automated movements of the stage. Once the alignment procedure is completed, the stage system retains an exact knowledge of the positions of matching points on the mineral mount and their mirror image positions on the muscovite, and can move between them as required.

2.11.5 FT Length Measurements

In order to carry out thermal history modelling of a sample measurement of the distribution of horizontal or close-to-horizontal confined FT lengths (i.e. below the polished surface within the mineral) is a critical parameter required to accompany any FT age (Gleadow et al. 1986). A variety of measurements have been used in FT dating studies to estimate the distribution of track lengths (e.g. Bhandari et al. 1971; Wagner and Storzer 1972; Dakowski 1978). Some of these measurements (e.g. projected lengths) contain little useful information about the true length distribution (Dakowski 1978; Green 1981; Laslett et al. 1982). A much better procedure is to measure the lengths of *confined* tracks that do not intersect the surface and are entirely located within the crystal interior but have been etched from an intersection with either a track at the surface or a crack or cleavage plane emerging at the surface (see Chap. 1, Hurford 2018). These tracks are called TINTs (track-in-track) and TINCLEs (track-in-cleavage) after Lal et al. (1969), who first suggested their use. However, the measurement of TINCLE fission tracks may lead to unreliable data, so they should be avoided (e.g. Donelick et al. 2005). The measurement of semi-tracks (the preserved parts of tracks that have intersected a polished prismatic surface, i.e. the spontaneous tracks counted for age determination) is a further

(a) **(b)** **(c)**

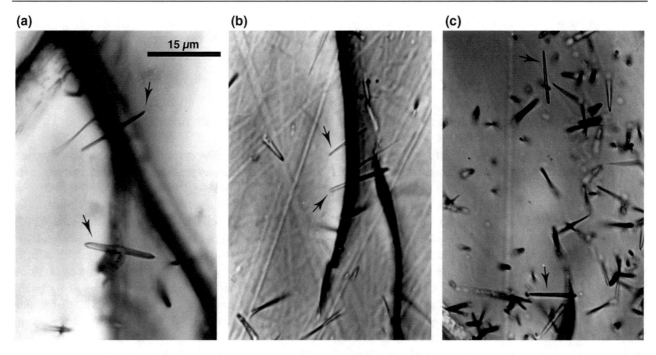

Fig. 2.6 Confined spontaneous fission tracks in apatite suitable for length measurements. TINCLEs are indicated by arrows in frames **(a)** and **(b)**, while frame **(c)** shows one TINT (top) and one TINCLE (bottom). Scale is similar for each frame

possibility (Laslett and Galbraith 1996) that may have some advantages in automated measurement systems (see also Chap. 4, Gleadow et al. 2018). For further information on the measurement of FT lengths in apatite see Gleadow et al. (2002) and Donelick et al. (2005), and zircon see Tagami (2005).

Only tracks with rounded or angular ends should be measured, indicating that the etchant has penetrated right to the end of the track to reveal its full length. Prior to measurement of confined tracks, care should be taken that the mount is clean and dry, as liquids, especially oils from greasy fingerprints, can lodge in the end of a confined track making the tip of the track very difficult to see. Washing the mount in a strong detergent will usually remove any liquid from the confined tracks. Examples of well-etched confined tracks in apatite are shown in Fig. 2.6.

In principle, it is possible to measure both the horizontal and vertical components of the length of confined tracks to give their actual length, regardless of their orientation. In practice on most of the older microscopes however, the vertical distance is not easily measured and reduces the precision of the overall measurement. A simpler and more rigorous procedure is to select only those confined tracks, which are ± 10 °C from the horizontal (Ketcham et al. 2009) and to measure their apparent length directly. Such measurements have the closest relationship to the true length distribution and are less subject to inherent sampling bias than other parameters (Laslett et al. 1982). Horizontal tracks

can be readily identified as those that are in focus along their entire length under a high-power objective. In reflected light, horizontal tracks are very obvious because they have a very bright reflection, without the diffraction bands, which characterise shallowly dipping tracks. Scanning in reflected light (if available) for suitable horizontal confined tracks for measurement can be very useful. In many cases, though certainly not all, horizontal tracks can show up very obviously because they have a bright reflection without the diffraction bands, which often characterise shallow dipping tracks.

Most laboratories carry out confined track length measurements using a drawing tube attachment to a microscope in association with a digitising tablet attached to a computer. The drawing tube superimposes an image of the digitising tablet with its cursor on the usual microscope image. The cursor carries a bright light-emitting diode to mark the measuring point in the optical image. Once a suitable track is located the cursor is simply moved to each end of the image of the track in turn and the positions marked. The raw coordinates for the track ends are translated into a length measurement, and these data are transferred to a computer for storage and statistical analysis. For each length measured, the azimuth direction of the c-axis from the elongation of the track etch pit is usually carried out at the same time as a reference frame for the orientation of the confined tracks. The sample mount should be scanned systematically and the lengths of the confined tracks measured.

Most of the time during a track length analysis is spent locating suitable tracks. In terms of the minimum number of lengths that should be counted, there is no firm rule of thumb, but one should endeavour to collect as much length data as possible. However, samples containing both long and short tracks, which usually reflect a more complicated thermal history, should require more measurements. Mean track length values in apatites generally stabilise after $\sim$50–120 measurements (Barbarand et al. 2003a). However, using c-axis projection to normalise track lengths in relation to crystallographic angle due to differences in annealing characteristics with orientation leads to improved measurement reproducibility and earlier stabilisation of a mean length value compared to non-projected tracks (Ketcham et al. 2007). In detrital zircons, length measurements are not carried out routinely due to complications arising from the variability in etching requirements often encountered between grains, possibly arising from the presence of different provenance populations, but mainly due to variations in α-radiation damage (Bernet and Garver 2005). In detrital apatites, it is important to measure confined track lengths only in those crystals that are dated, so that discrete grain-age populations can be identified and robust geological interpretations made. Several methods for decomposing FT ages using peak fitting programs are available (e.g. Brandon 1992; Galbraith and Laslett 1993; Vermeesch 2009; see also Chap. 6, Vermeesch 2018).

2.11.6 Californium (^{252}Cf) Irradiation

^{252}Cf irradiation is a technique used to enhance the number of measurable confined lengths in apatite (Donelick and Miller 1991) and also to lessen observer bias (Ketcham 2005). Such irradiations, carried out prior to chemical etching, may be performed on a masked area of the grain mount or on a second grain mount prepared for each sample. Alternatively, one could use the same mount made for N_s determination and re-etch for length measurements after counting for N_s (see Donelick et al. 2005). By placing grain mounts under vacuum at a distance of several mm from a planar ^{252}Cf spontaneous fission fragment source ($T_{1/2} = 2.645$ yr), a substantial number of fission particles are created, which penetrate the exposed internal grain surface. The resultant tracks act as 'pathways' for the etchant to reach confined tracks at depth below the exposed grain surface within the apatite, effectively increasing the number of confined FT lengths available for measurement (see Donelick et al. 2005). This method has been widely applied to samples with low uranium concentration and/or those young in age where insufficient track-in-tracks (TINTs) were observed. In zircon, which is a denser mineral, ^{252}Cf irradiation does not work well for implanting long tracks and

other techniques for increasing the number of measurable confined tracks, such as irradiation by heavy nuclides and artificial fracturing have been reviewed by Yamada et al. (1998).

2.11.7 Kinetic Parameters

FT annealing in apatite is a complicated, nonlinear process that is not completely understood, but is known to be dominantly controlled by temperature (markedly so above >60 °C), the duration of heating and to a lesser degree by crystallographic orientation (e.g. Donelick et al. 2005). However, annealing is also related to a complex interplay of anion (Cl, F, OH) and cation substitutions (e.g. REE, Mn, Sr, Fe, Si), with Cl playing a primary role (Green et al. 1985). REE in more F-rich apatites have been suggested to exercise some control on annealing (Barbarand et al. 2003b) as have some other possible chemical factors as outlined by Donelick et al. (2005) and Spiegel et al. (2007). The bulk track-etching rate of apatite has also been proposed as a proxy for bulk chemical composition, and this involves measurement of the Dpar—the arithmetic mean of FT etch pit lengths measured parallel to the crystallographic c-axis (e.g. Donelick 1993; Burtner et al. 1994).

Dpar length and Cl content are the two kinetic parameters most routinely measured and should be collected from every grain analysed for either age or track length data. Along with FT age and length data, the measurement of either parameter is considered essential for carrying out quantitative thermal history modelling of individual apatite grains or populations of grains. Thermal history models in common use can accommodate the input of either parameter. The choice of which measurement should be preferred, however, is still the subject of some debate (e.g. Barbarand et al. 2003b; Green et al. 2005; Hurford et al. 2005; Donelick et al. 2005).

Data for both parameters (as for age and length data) should only be acquired on prismatic sections parallel to the crystallographic c-axis, and this orientation can be readily checked by the presence of sharp polishing scratches and the alignment of etch pit openings under reflected light parallel to the c-axis (see Donelick et al. 2005).

The measurement of Dpar for a conventional microscope set-up is essentially the same as for track length measurement, but in this case the scale of measurement is considerably smaller. With digital imaging (see Sect. 2.12 and Chap 4, Gleadow et al. 2018) Dpars, however, can now be measured automatically with greater resolution down to almost the pixel level, within a matter of seconds and the smaller parameter Dper (the etch pit minor axis perpendicular to the c-axis) can also be resolved, allowing for a more precise and complete characterisation of etch pit geometry.

Measurement of apatite Cl content has now largely become a routine procedure. However, traditionally it involves another analytical step and is perceived as being more expensive and time-consuming, in that grain x–y coordinates recorded from the measurement of both grain ages and track lengths are now required to be transferred to a suitable stage for electron probe microanalysis for grain-by-grain halogen analysis (F and Cl content). Special care should be taken when carrying out such analysis on apatite as the halogen X-ray intensity can fluctuate strongly with operating conditions, grain orientation and bulk F and Cl content (Goldoff et al. 2012; Stock et al. 2015). The use of infrared microspectroscopy for the semi-quantitative determination of apatite anion composition has been described by Siddall and Hurford (1998), but is not used routinely. The measurement of Cl content in apatite using LA-ICP-MS (Chew et al. 2014) is an important development towards collecting kinetic information together with uranium (and REE) content for a more integrated approach towards data acquisition for FT analysis.

2.12 Digital Imaging and Automated FT Analysis

Conventional FT counting is very labour-intensive. However, new analytical approaches are now promising significant improvements in data quality and analytical productivity in FT analysis. Gleadow et al. (2009, 2015) described a method that combines autonomous digital microscopy and automatic image analysis for the recognition and counting of fission tracks in minerals such as apatite, along with new tools for the enhanced measurement of Dpar and FT lengths. This new technique takes full advantage of the capabilities of the new generation of digital microscopes, such as the Zeiss Axio-Imager series. Much of the operator time previously tied to the microscope is now freed up to do other things as the microscope/software system captures and processes the images autonomously, without the need for operator involvement after the first setting up has been completed. Multiple slides can be imaged overnight, and the processed images analysed offline on a computer using the analysis software. Digital coordinates of analysed grains are exported to other computer-operated devices such as a laser ablation stage or an electron probe microanalyser stage, for further analysis. This new approach is described in more detail in Chap. 4, (Gleadow et al. 2018).

Acknowledgements We are grateful to numerous past and present researchers and graduate students in the Thermochronology Group at the University of Melbourne for their contributions towards establishing some of the methodologies described in this work. The National Collaborative Research Infrastructure Strategy AuScope programme supports the University of Melbourne thermochronology facility. Martin Danišík and Paul O'Sullivan provided thoughtful and constructive reviews, which helped to improve the clarity of this work.

References

Arne DC, Green PF, Duddy IR, Gleadow AJW, Lambert IB, Lovering JF (1989) Regional thermal history of the Lennard Shelf, Canning Basin, from apatite fission track analysis: Implications for the formation of Pb-Zn deposits. Aust J Earth Sci 36:495–513

Barbarand J, Carter A, Hurford T (2003a) Variation in apatite fission-track length measurement: implications for thermal history modelling. Chem Geol 198:77–106

Barbarand J, Carter A, Wood I, Hurford T (2003b) Compositional and structural control of fission-track annealing in apatite. Chem Geol 198:107–137

Bellemans F, De Corte F, Van den Haute P (1995) Composition of SRM and CN U-doped glasses: significance for their use as thermal neutron monitors in fission track dating. Radiat Meas 24:153–160

Bernet M, Garver JI (2005) Fission-track analysis of detrital zircon. In: Reiners P, Ehlers T (eds) Low-temperature thermochronology. Rev Min Geochem 58:205–238

Bhandari N, Bhat SC, Lal D, Rajagoplan G, Tamhane AS, Venkatavaradan VS (1971) Fission fragment tracks in apatite: recordable track lengths. Earth Planet Sci Lett 13:191–199

Brandon MT (1992) Decomposition of fission-track grain-age distributions. Am J Sci 292:535–564

Braun J, van der Beek P, Batt G (2006) Quantitative thermochronology. Cambridge University Press, Cambridge

Burtner RL, Nigrini A, Donelick RA (1994) Thermochronology of Lower Cretaceous source rocks in the Idaho-Wyoming Thrust Belt. Bull Am Assoc Petrol Geol 78:1613–1636

Calk LC, Naeser CW (1973) The thermal effect of a basalt intrusion on fission tracks in quartz monzonite. J Geol 81:189–198

Callahan J (1987) A nontoxic heavy liquid and inexpensive filters for separation of mineral grains. J Sed Pet 57:765–766

Carlson WD, Donelick RA, Ketcham RA (1999) Variability of apatite fission-track annealing kinetics: I. Experimental results. Am Min 84:1213–1223

Carpenter SB, Reimer GM (1974) Standard reference materials: calibrated glass standards for fission track use. Nat Bur Stand Spec Pub 260–49

Carrapa B, DeCelles PG, Reiners PW, Gehrels GE, Sudo M (2009) Apatite triple dating and white mica ^{40}Ar/^{39}Ar thermochronology of syntectonic detritus in the Central Andes: a multiphase tectonothermal history. Geology 37:407–410

Carter A, Moss SJ (1999) Combined detrital-zircon fission-track and U-Pb dating: a new approach to understanding hinterland evolution. Geology 27:235–238

Chew DM, Donelick RA (2012) Combined apatite fission track and U-Pb dating by LA-ICP-MS and its application in apatite provenance analysis. In: Sylvester P (ed) Quantitative mineralogy and microanalysis of sediments and sedimentary rocks. Mineralogical Association of Canada Short Course, pp 219–248

Chew DM, Donelick RA, Donelick MB, Kamber B, Stock MJ (2014) Apatite chlorine concentration measurements by LA-ICP-MS. Geostand Geoanalyt Res 38:23–35

Chew DM, Babechuk MG, Cogné N, Mark C, O'Sullivan GJ, Henrichs IA, Doepke D, Mckenna CA (2016) (LA, Q)-ICPMS trace-element analyses of Durango and McClure Mountain apatite and implications for making natural LA-ICPMS mineral standards. Chem Geol 435:35–48

Chisholm E-K, Sircombe K, DiBugnara D (2014) Handbook of geochronology mineral separation laboratory techniques. Geoscience Australia Record 2014/46, 45 p

Coutand I, Carrapa B, Deeken A, Schmitt AK, Sobel ER, Strecker MR (2006) Propagation of orographic barriers along an active range front: insights from sandstone petrography and detrital apatite fission-track thermochronology in the intramontane Angastaco basin, NW Argentina. Basin Research 18:1–26

Cox R, Košler J, Sylvester P, Hodych J (2000) Apatite fission-track (FT) dating by LAM-ICO-MS analysis. Abstracts Goldschmidt 2000. J Conf Abstracts 5(2):322

Dakowski M (1978) Length distributions of fission tracks in thick crystals. Nucl Track Detect 28:181–189

Danišík M, Pfaff K, Evans N, Manoloukos C et al (2010) Tectonothermal history of the Schwarzwald Ore District (Germany): an apatite triple dating approach. Chem Geol 278:58–69

Danišík M (2018) Integration of fission-track thermochronology with other geo-chronologic methods on single crystals (Chapter 5). In: Malusà MG, Fitzgerald PG (eds) Fission-track thermochronology and its application to geology. Springer, Berlin

De Corte F, Bellemans F, van den Haute P, Inglebrecht C, Nicholl C (1998) A new U doped glass certified by the European Commission for calibration of fission-track dating. In: den Haute Van, De Corte F (eds) Advances in fission-track geochronology. Springer, Dordrecht, pp 67–78

Dias ANC, Chemale F Jr, Soares CJ, Guedes S (2017) A new approach for electron microprobe zircon fission track thermochronology. Chem Geol 459:129–136

Donelick R (1993) Apatite etching characteristics versus chemical composition. Nucl Tracks Radiat Meas 21:604

Donelick RA, Miller DS (1991) Enhanced TINT fission track densities in low spontaneous track density apatites using ^{252}Cf-derived fission fragments tracks: a model and experimental observations. Nucl Tracks Radiat Meas 18:301–307

Donelick RA, O'Sullivan PB, Ketcham RA (2005) Apatite fission-track analysis. In: Reiners P, Ehlers T (eds) Low-temperature thermochronology. Rev Min Geochem 58:49–94

Dumitru TA (2000) Fission-track geochronology. In: Noller JS, Sowers JM, Lettis, WR (eds) Quaternary geochronology: methods and applications. Am Geophys Union Ref Shelf 4, Washington, DC, American Geophysical Union, pp 131–155

Enkelmann E, Jonckheere R, Wauschkuhn B (2005) Independent fission-track ages (φ-ages) of proposed and accepted apatite age standards and a comparison of φ-, Z-, ζ- and ζ0-ages: implications for method calibration. Chem Geol 222:232–248

Evans NJ, McInnes BIA, McDonald B, Danišík M, Becker T, Vermeesch P, Shelley M, Marillo-Sialer E, Patterson DB (2015) An in situ technique for (U–Th–Sm)/He and U-Pb double dating. J Anal At Spectrom 30:1636–1645

Fleischer RL, Price PB (1964) Techniques for geological dating of minerals by chemical etching of fission fragment tracks. Geochim et Cosmochim Acta 28:1705–1714

Fleischer RL, Price PB, Walker RL (1975) Nuclear tracks in solids: principles and applications. University of California Press, Berkeley

Galbraith RF (2005) Statistics for fission track analysis. Chapman & Hall, Boca Raton

Galbraith RF, Laslett GM (1993) Statistical models for mixed fission-track ages. Nucl Track Radiat Meas 21:459–470

Gallagher K, Brown R, Johnson C (1998) Fission track analysis and its applications to geological problems. Ann Rev Earth Planet Sci 26:519–572

Garver JI (2003) Etching age standards for fission track analysis. Radiat Meas 37:47–54

Garver JI, Brandon MT, Roden-Tice MK, Kamp PJJ (1999) Exhumation history of orogenic highlands determined by detrital fission track thermochronology. In: Ring U, Brandon MT, Willett SD, Lister GS (eds) Exhumation processes: normal faulting, ductile flow, and erosion. Geol Soc London Spec Pub 154:283–304

Giese J, Seward D, Stuart FM, Wüthrich E et al (2009) Electrodynamic disaggregation: does it affect apatite fission-track and (U-Th)/He analyses? Geostand Geoanal Res 34:39–48

Gleadow AJW (1978) Anisotropic and variable track etching characteristics in natural sphenes. Nuclear Track Detect 2:105–117

Gleadow AJW (1981) Fission track dating methods: what are the real alternatives. Nucl Tracks 5:3–14

Gleadow AJW (1984) Fission track dating methods—II: a manual of principles and techniques. Workshop on fission track analysis: principles and applications. James Cook University, Townsville, Australia, 4–6 September 1984, 35 p

Gleadow AJW, Lovering JF (1974) The effect of weathering on fission track dating. Earth Planet Sci Letts 22:163–168

Gleadow AJW, Lovering JF (1978) Thermal history of granitic rocks from Western Victoria: a fission-track study. J Geol Soc Aust 25:323–340

Gleadow AJW, Hurford AJ, Quaife RD (1976) Fission track dating of zircon: improved etching techniques. Earth Planet Sci Letts 33:273–276

Gleadow AJW, Duddy IR, Green PF, Lovering JF (1986) Confined fission track lengths in apatite: a diagnostic tool for thermal history analysis. Contrib Mineral Petrol 94:405–415

Gleadow AJW, Belton DX, Kohn BP, Brown RW (2002) Fission track dating of phosphate minerals and the thermochronology of apatite. Rev Min Geochem 48:579–630

Gleadow AJW, Gleadow SJ, Belton DX, Kohn BP, Krochmal MS (2009) Coincidence mapping a key strategy for automated counting in fission track dating. In: Ventura B, Lisker F, Glasmacher UA (eds) Thermochronological methods: from palaeotemperature constraints to landscape evolution models. Geol Soc Lond Spec Pub 324, pp 25–36

Gleadow A, Harrison M, Kohn B, Lugo-Zazueta R, Phillips D (2015) The Fish Canyon Tuff: a new look at an old low-temperature thermochronology standard. Earth Planet Sci Letts 424:95–108

Gleadow A, Kohn B, Seiler C (2018) The future of fission-track thermochronology (Chapter 4). In: Malusà MG, Fitzgerald PG (eds) Fission-track thermochronology and its application to geology. Springer, Berlin

Goldoff B, Webster JD, Harlov DE (2012) Characterization of fluor-chloroapatites by electron microprobe analysis with a focus on time-dependent intensity variation of halogens. Am Min 97:1103–1115

Gombosi DJ, Garver JI, Baldwin SL (2014) On the development of electron microprobe zircon fission-track geochronology. Chem Geol 363:312–321

Green PF (1981) "Track-in-track" length measurements in annealed apatites. Nucl Tracks 5:121–128

Green PF, Hurford AJ (1984) Neutron dosimetry in fission track dating: a theoretical background and some practical precautions. Nucl Tracks 9:231–241

Green PF, Duddy IR, Gleadow AJW, Tingate PR, Laslett GM (1985) Fission track annealing in apatite: track length measurements and the form of the Arrhenius plot. Nucl Tracks 10:323–328

Green PF, Duddy IR, Hegarty KA (2005) Comment on "Compositional and structural control of fission track annealing in apatite" by Barbarand J, Carter A, Wood I and Hurford AJ, Chem Geol 198 (2003);107–137; Chem Geol 214:351–358

Haack UH, Gramse M (1972) Survey of garnets for fossil fission tracks. Contrib Mineral Petrol 34:258–260

Hasebe N, Barbarand J, Jarvis K, Carter A, Hurford AJ (2004) Apatite fission-track chronometry using laser ablation ICP-MS. Chem Geol 207:135–145

Hasebe N, Carter A, Hurford AJ, Arai S (2009) The effect of chemical etching on LA-ICP-MS analysis in determining uranium concentration for fission-track chronometry. In: Ventura B, Lisker F, Glasmacher UA (eds) Thermochronological methods: from palaeotemperature constraints to landscape evolution models. Geol Soc Lond Spec Pub 324, pp 37–46

Hasebe N, Tamura A, Arai S (2013) Zeta equivalent fission-track dating using LA-ICP-MS and examples with simultaneous U-Pb dating. Island Arc 22:280–291

Hejl E (1998) The Zeta-Potential of apatite and zircon: its significance for mineral separation. On Track—Newsletter of the International Fission Track Community 8(no 1, Issue 16):7–8

http://zeiss-campus.magnet.fsu.edu/articles/basics/kohler.html. Zeiss resource—Configuring a Microscope for Köhler Illumination In: Education in microscopy and digital imaging. Accessed 5 July 2016

Hurford AJ (1990) Standardization of fission track dating calibration: recommendation by the fission track working group of the I.U.G.S. subcommission on geochronology. Chem Geol (Isot Geosci Sect) 80:171–178

Hurford AJ (1998) Zeta: the ultimate solution to fission-track analysis calibration or just an interim measure? In: van den Haute P, De Corte F (eds) Advances in fission-track geochronology. Kluwer Academic Publishers, Dordrecht, pp 19–32

Hurford AJ (2018) An historical perspective on fission-track thermochronology (Chapter 1) In: Malusà MG, Fitzgerald PG (eds) Fission-track thermochronology and its application to geology. Springer, Berlin

Hurford AJ, Green PF (1982) A user's guide to fission track dating calibration. Earth Planet Sci Lett 59:343–354

Hurford AJ, Green PF (1983) The zeta age calibration of fission-track dating. Chem Geol Isot Geosci 1:285–317

Hurford AJ, Barbarand J, Carter A (2005) Reply to comment on "Compositional and structural control of fission track annealing in apatite" by Barbarand J, Carter A, Wood I, Hurford AJ, Chem Geol 198(2003):107–137; Chem Geol 214:359–361

Ijlst L (1973) New diluents in heavy liquid mineral separation and an improved method for the recovery of liquids. Am Min 58:1084–1087

Ito K, Hasebe N (2011) Fission-track dating of Quaternary volcanic glass by stepwise etching. Radiat Meas 46:76–182

Iwano H, Danhara T (1998) A re-investigation of the geometry factors for fission-track dating of apatite, sphene and zircon. In: van den Haute P, De Corte F (eds) Advances in fission-track geochronology. Kluwer Academic Publishers, Dordrecht, pp 47–66

Jonckheere R, Ratschbacher L, Wagner GA (2003) A repositioning technique for counting induced fission tracks in muscovite external detectors in single-grain dating of minerals with low and inhomogeneous uranium concentrations. Radiat Meas 37:217–219

Ketcham RA (2005) Forward and inverse modeling of low-temperature thermochronometry data. In: Reiners P, Ehlers T (eds) Low-temperature thermochronology. Rev Mineral Geochem 58:275–314

Ketcham RA, Carter A, Donelick RA, Barbarand J, Hurford AJ (2007) Improved measurement of fission-track annealing in apatite using c-axis projection. Am Min 92:789–798

Ketcham RA, Donelick RA, Balestrieri ML, Zattin M (2009) Reproducibility of apatite fission-track length data and thermal history reconstruction. Earth Planet Sci Letts 284:504–515

Kohn BP, Gleadow AJW, Brown RW, Gallagher K, Lorencak M, Noble WP (2005) Visualising thermotectonic and denudation histories using apatite fission track thermochronology. In: Reiners P, Ehlers T (eds) Low-temperature thermochronology, Rev Min Geochem 58:527–565

Košler M, Svojtka M (2003) Present trends and the future of zircon in geochronology: laser ablation ICPMS. In: Manchar JM, Hoslin PWO (eds) Zircon. Rev Min Geochem 54:243–275

Krishnaswami S, Lal D, Prabhu N, MacDougall D (1974) Characteristics of fission tracks in zircon: applications to geochronology and cosmology. Earth Planet Sci Letts 22:51–59

Lal D, Rajan RS, Tamhane AS (1969) Chemical composition of nuclei of Z > 22 in cosmic rays using meteoritic minerals as detectors. Nature 221:33–37

Laslett GM, Galbraith RF (1996) Statistical properties of semi-tracks in fission track analysis. Radiat Meas 26:565–576

Laslett GM, Kendall WS, Gleadow AJWLaslett GM, Kendall WS, Gleadow AJW, Duddy IR (1982) Bias in measurement of fission-track length distributions. Nucl Tracks 6:79–85

Lisker F, Ventura B, Glasmacher UA (2009) Apatite thermochronology in modern geology. In: Ventura B, Lisker F, Glasmacher UA (eds) Thermochronological methods: from palaeotemperature constraints to landscape evolution models. Geol Soc Lond Spec Publ 324, pp 1–23

Malusà MG (2018) A guide for interpreting complex detrital age patterns in stratigraphic sequences (Chapter 16). In: Malusà MG, Fitzgerald PG (eds) Fission-track thermochronology and its application to geology. Springer, Berlin

Malusà MG, Garzanti E (2018) The sedimentology of detrital thermochronology (Chapter 7). In: Malusà MG, Fitzgerald PG (eds) Fission-track thermochronology and its application to geology. Springer, Berlin

Montario MJ, Garver JI (2009) The thermal evolution of the Grenville terrane revealed through U-Pb and fission-track analysis of detrital zircon from Cambro-Ordovician quartz arenites of the Potsdam and Galway formations. J Geol 117:595–614

Naeser CW (1976) Fission track dating. US Geological Survey Open-File Report 76-190 86 p

Naeser CW, Dodge FCW (1969) Fission track ages of accessory minerals from granitic rocks of the Central Sierra Nevada batholith, California. Geol Soc Am Bull 80:2201–2212

Naeser CW, McKee EH (1970) Fission-track and K-Ar ages of Tertiary ash-flow tuffs, north central Nevada. Geol Soc Am Bull 81:3375–3384

Naeser ND, Zeitler PK, Naeser CW, Cerveny PF (1987) Provenance studies by fission track dating of zircon—etching and counting procedures. Nucl Tracks Radiat Meas 13:121–126

Naeser CW, Naeser ND, Newell WL, Southworth S, Edwards LE, Weems RE (2016) Erosional and depositional history of the Atlantic passive margin as recorded in detrital zircon fission-track ages and lithic detritus in Atlantic coastal plain sediments. Am J Sci 316:110–168

Ravenhurst CE, Donelick RA (1992) Fission track thermochronology. In: Zentilli M, Reynolds PM (eds) Short course handbook on low temperature thermochronology. Mineral Assoc Can, Ottawa, pp 21–42

Reiners PW, Thomson SN, McPhilips D, Donelick RA, Roering JJ (2007) Wildfire thermochronology and the fate and transport of apatite in hillslope and fluvial environments. J Geophy Res 112: F04001. https://doi.org/10.1029/2007jf000759

Seward D, Spikings R, Viola G, Kounov A, Ruiz GMH, Naeser N (2000) Etch times and operator variation for spontaneous track length measurements in apatites: an intra-laboratory check. OnTrack 10(21):16–21

Siddall R, Hurford AJ (1998) Semi-quantitative determination of apatite anion composition for fission-track analysis using infrared microspectroscopy. Chem Geol 150:181–190

Smith MJ, Leigh-Jones P (1985) An automated microscope scanning stage for fission-track dating. Nucl Tracks 10:395–400

Soares C, Guedes S, Hadler J, Mertz-Kraus R, Zack T, Iunes P (2014) Novel calibration for LA-ICP-MS-based fission-track thermochronology. Phys Chem Miner 41:65–73

Sobel ER, Seward D (2010) Influence of etching conditions on apatite fission-track etch pit diameter. Chem Geol 271:59–69

Sperner B, Jonckheere R, Pfander J (2014) Testing the influence of high-voltage mineral liberation on grain size, shape and yield and on fission track and $^{40}Ar/^{39}Ar$ dating. Chem Geol 371:83–95

Spiegel C, Kohn BP, Raza A, Rainer T, Gleadow AJW (2007) The effect of long-term low temperature exposure on apatite fission track stability: A natural annealing experiment in the deep ocean. Geochim Cosmochim Acta 71:4512–4537

Stock MJ, Humphreys MCS, Smith VC, Johnson RD et al (2015) New constraints on electron-beam induced halogen migration in apatite. Am Min 100:281–293

Tagami T (1987) Determination of zeta calibration constant for fission track dating. Nucl Tracks Radiat Meas 13:127–130

Tagami T (2005) Zircon fission-track thermochronology and applications to fault studies. In: Reiners P, Ehlers T (eds) Low-temperature thermochronology. Rev Min Geochem 58:95–122

Tagami T, O'Sullivan PB (2005) Fundamentals of fission-track thermochronology. In: Reiners P, Ehlers T (eds) Low-temperature thermochronology. Rev Min Geochem 58:19–47

Torresan M (1987) The use of sodium polytungstate in heavy mineral separations. US Geological Survey of Open-File Report 87-590, 18 p

van den Haute P, De Corte F, Jonckheere R, Bellemans F (1998) The parameters that govern the accuracy of fission-track age determinations: a-reappraisal. In: van den Haute P, De Corte F (eds) Advances in fission-track geochronology. Kluwer Academic Publishers, Dordrecht, pp 33–46

Vermeesch P (2009) Radialplotter: a java application for fission track, luminescence and other radial plots. Radiat Meas 44:409–410

Vermeesch P (2017) Statistics for LA-ICP-MS based fission track dating. Chem Geol 456:19–27

Vermeesch P (2018) Statistics for fission-track thermochronology (Chapter 6). In: Malusà MG, Fitzgerald PG (eds) Fission-track thermochronology and its application to geology. Springer, Berlin

Wagner GA, Storzer D (1972) Fission track length reductions in minerals and the thermal history of rocks. Trans Amer Nucl Soc 15:127–128

Wagner GA, van den Haute P (1992) Fission track dating. Kluwer Academic, Dordrecht

Yamada R, Yoshioka T, Watanabe K, Tagami T, Nakamura H, Hashimoto T, Nishimura S (1998) Comparison of experimental techniques to increase the number of measurable confined fission tracks in zircon. Chem Geol (Isotope Geosci Sect) 149:99–107

Zaun PE, Wagner GA (1985) Fission-track stability in zircons under geological conditions. Nucl Tracks 10:303–307

Fission-Track Annealing: From Geologic Observations to Thermal History Modeling

Richard A. Ketcham

Abstract

This chapter reviews the evolving state of knowledge concerning fission-track (FT) annealing, primarily in apatite and zircon, based on theory, experiments, and geological observations. Multiple insights into track structure, formation, and evolution arise from transmission electron microscopy, small-angle X-ray scattering, atomic force microscopy, and molecular dynamics computer modeling. Our principal knowledge, however, comes from experiments in which spontaneous or induced tracks are annealed, etched, and measured, the results statistically fitted, and their predictions compared against geological benchmarks. This empirical approach has proven effective and resilient, though physical understanding remains an ultimate goal. The precise mechanism by which lattice damage anneals, and how it varies among minerals and damage types, remains unknown. Multiple similarities between apatite and zircon suggest equivalent underlying processes. Both minerals demonstrate annealing anisotropy, and its characterization is crucial for understanding both track shortening and density reduction. The fanning curvilinear equation, featuring curved iso-annealing lines on an Arrhenius-type diagram, has been the most successful for matching data spanning timescales from seconds to hundreds of millions of years. A super-model featuring a single set of iso-annealing lines describes all apatite experimental data to date. Annealing rates vary with both anion and cation substitutions, and more work is required to ascertain how these substitutions interact. Other areas for further research include differences between spontaneous and induced tracks, and possible additional processes affecting length and density evolution, such as seasoning. Thermal history inversion simultaneously leverages and tests our models, and accounting for kinetic variation is key for doing it soundly.

3.1 Introduction

The power of fission tracks for geological investigations derives precisely from their thermal sensitivity. Fission-track (FT) methods are unique among thermochronometers in that upon formation each daughter product becomes a sensitive recorder of the thermal history its host mineral subsequently undergoes. If the temperature information in a large number of tracks that formed over a long time interval can be successfully recovered, that information can be integrated and merged with other geological constraints to ascertain detailed thermal histories (e.g., Green et al. 1989; Gallagher 1995, 2012; Issler 1996; Ketcham 2005).

The ability to derive reliable thermal history information from fission tracks depends critically on how well the annealing process is understood and characterized. Here, we run into difficulty, as the annealing of fission damage, and radiation damage in general, is a complex physical process that is not well understood. There are many reasons for our lack of progress. The mechanism by which energy is deposited into the crystal lattice remains a matter of debate and may vary among minerals and damage types, and the disposition and configuration of displaced atoms are difficult to ascertain and inevitably varied. The fission damage zone is very inconveniently shaped, 5–10 nm in diameter but 12–20 μm long, confounding the ability of nano-analytical methods to observe them across their full extents without disturbing them, and the capacity of computational resources to encompass them, as would be required for the formation or shortening process to be observed or modeled in complete detail. Given these difficulties, the precise mechanism by which damage anneals, and whether it varies among

R. A. Ketcham (✉)
Department of Geological Sciences, Jackson School of
Geosciences, University of Texas at Austin, Austin, TX, USA
e-mail: ketcham@jsg.utexas.edu

© Springer International Publishing AG, part of Springer Nature 2019
M. G. Malusà and P. G. Fitzgerald (eds.), *Fission-Track Thermochronology and its Application to Geology*,
Springer Textbooks in Earth Sciences, Geography and Environment, https://doi.org/10.1007/978-3-319-89421-8_3

minerals or damage types, remains a matter of conjecture. Finally, insofar as we are primarily interested in the behavior of fission tracks over geological timescales, we cannot be sure that our theories are correct, or that our laboratory experiments or molecular models are capturing the processes that operate over these timescales, unless and until we test their implications against geological observations.

As a result, the degree of success that has been achieved in usefully describing and quantifying FT annealing has been due to simultaneous work across multiple fronts: theory, experiment, and geological application. All of these have evolved in parallel over the past fifty years, and their continued progress can be expected to further improve our understanding of FT annealing and our ability to use it to derive thermal history information.

3.2 Radiation Damage

To discuss annealing, we first need to review how heavy ions create radiation damage, and what aspects of this damage are reflected in the measurements we make. For a recent and comprehensive technical review of heavy ion track formation and damage, readers are referred to Wesch and Wedler (2016); here, only lattice damage will be considered. The two primary types of lattice damage in U+Th-bearing materials stem from fission and alpha recoil. Both need to be considered, as the latter is responsible for more lattice damage and is thought to possibly influence both topics of direct interest such as zircon FT etching and annealing rates (Rahn et al. 2004) and allied ones such as helium diffusivity (Flowers et al. 2009; Gautheron et al. 2009; Guenthner et al. 2013).

3.2.1 Fission Versus Alpha Recoil Tracks and Damage

Fission fragments are heavy (mass 80–155 amu) and energetic (~ 160 MeV/decay), and their stopping distances are on the order of 8–11 μm in typical minerals (Jonckheere 2003b). Alpha decay energies from U and Th chains are in the 3–9 MeV range, but most of the energy goes to the alpha particle (^{4}He ion), which is thought to cause relatively little damage, several hundred atomic displacements along a ~ 20 μm path (Weber 1990), due to its small size and charge. About 70–140 keV of decay energy is imparted to the recoiling nucleus, which due to its large size and low velocity interacts more strongly with atoms along its path, and typical stopping distances and thus latent track lengths are on the order of 25 nm and feature thousands of displacements. However, even though the energy deposited by each recoil nucleus is $\sim 10^3$ times less than for a pair of fission fragments, alpha decay in ^{238}U and its daughter products is $\sim 10^7$ times more frequent than spontaneous fission, and thus, alpha recoil tracks (ARTs) are 10^7 more plentiful and are responsible for depositing 10^4 more energy into the lattice than fission tracks.

The mechanisms by which fission particles and alpha recoil nuclei impart energy to the enclosing material and create damage are different (Fleischer et al. 1975). The energy released during fission is high enough that the daughter nuclei leave their electron clouds behind, and they travel through the enclosing crystal as highly charged ions (+30 to +50), leaving a lattice charge in their wake that displaces atoms by electronic repulsion, the "ion spike" model (Fleischer et al. 1965b). Conversely, the comparatively low energy and charge of the recoiling nucleus probably results in kinetic, or hard-sphere (rigid nucleus and electron cloud) "wrecking ball" interactions and damage. The principal alternative to the ion spike model for FT formation is the thermal spike, which posits energy transfer as a thermal process. These processes are not exclusive of each other, and a "compound spike" understanding of ion tracks has been put forward (Chadderton 2003). Although most calculations of track interactions with their host material use or approximate an electronic model (e.g., Rabone et al. 2008; Li et al. 2012), the thermal spike is more conducive to making some predictions such as track radius (Szenes 1995), and has been useful in examining TEM observations of fission tracks in zircon (Li et al. 2014).

Whether damage trails from fission and alpha recoil anneal in the same way, or at the same temperatures, is a subject of great current interest in thermochronology, due to the importance of understanding radiation damage and its annealing for (U−Th)/He dating. On this topic, the data are mixed and sparse. The most direct comparison can be made in micas, as their perfect cleavage makes them among the only minerals in which alpha recoil tracks can be etched and observed directly (Huang and Walker 1967; Gögen and Wagner 2000; Stübner et al. 2015). Annealing experiments by Yuan et al. (2009) indicate that the closure temperature (T_c) for ARTs in phlogopite is on the order of 26 °C for 10 °C/Myr cooling, whereas earlier experiments by Parshad et al. (1978) suggest that fission tracks in phlogopite persist at up to 200 °C on million-year timescales. Previous work on biotite has also suggested that the retention temperature for ARTs is about 100 °C lower than that of fission tracks (Saini and Nagpaul 1979; Hashemi-Nezhad and Durrani 1983).

In zircon, some component of non-fission lattice damage is thought to be responsible for color change which persists to higher temperatures than fission tracks over geological timescales (Garver and Kamp 2002), enhanced etching rate (Gleadow et al. 1976), and possibly lowering resistance to FT annealing (Rahn et al. 2004). At short timescales, however, Braddy et al. (1975) found that a brief pre-annealing step could reduce etching rates enhanced by radiation damage-induced metamictization, while preserving FT density. This may indicate a "kinetic crossover" (Reiners 2009), in which relative annealing rates change as one goes from laboratory to geological timescales. Alternatively, damage recovery as gauged by Raman spectroscopy from a near-metamict state has been shown to be faster than from lower damage magnitudes (Zhang et al. 2000; Geisler et al. 2001), perhaps indicating that the Braddy et al. (1975) result reflects a bulk process rather than a property of individual recoil tracks. However, Tagami et al. (1990) and Yamada et al. (1995b) also report a strong reduction in etching rate in far less-damaged zircons at only the incipient stages of FT annealing at laboratory timescales, which they attribute to removal of alpha recoil damage. Taken together, these observations point to a complex picture, with the bulk of alpha recoil damage annealing well before fission tracks in zircon, but final remnants persisting to higher temperatures.

3.2.2 Revelation by Etching

The stopping distances of fission particles are considerably longer than the track lengths as revealed by etching. For example, the two fission nuclei are estimated to end up 21.2 ± 0.9 μm from each other in apatite (Jonckheere 2003b), whereas the etchable length of the fission track is on the order of 16 μm by standard protocols (e.g., Ketcham et al. 2015), a "length deficit" of 25%. Similarly, in zircon, fission fragments travel 16.7 ± 0.8 μm, but etched lengths are on the order of 11 μm (Tagami et al. 1990; Yamada et al. 1995a), a length deficit of about 34%. TEM observations (Li et al. 2012) verify that atomic-scale damage does exist beyond the etchable length, and thus, there evidently exists a degree of damage below which etchability is not significantly enhanced. A related observation is that, at high degrees of annealing, some tracks form "unetchable gaps" (Green et al. 1986), zones of slow etching separating zones of fast etching along a track.

There is thus a disconnect between the atomic-scale process of annealing and track shortening as we infer it from etched fission tracks. Annealing equations describing length reduction do so in terms of etchable length, and equations describing density reduction do so in terms of the zone of enhanced etchability intersecting a polished internal surface. Thus, they reflect annealing of damage to below some etchability limit, usually defined by the etching protocol, not the complete eradication of damage. Similarly, the formation of slow-etching gaps can make the zone of enhanced etchability appear shorter than it really is, leading to an apparent acceleration of shortening in the final stages of annealing that varies with the crystallographic orientation of the track. In apatite, this latter problem is largely bypassed by

characterizing the annealing of tracks parallel to the crystallographic c-axis, which are the last ones to undergo accelerated annealing (Donelick et al. 1999; Ketcham et al. 2007a).

Another salient point is that length measurements can depend on the etching protocol used. In apatite, the usual procedure is to etch using HNO_3 at a certain strength and temperature for a certain duration (see Chap. 2, Kohn et al. 2018). The most common protocols use 5.0 or 5.5 M HNO_3 at 20 °C for 20 s, linking them to the annealing data sets produced by Barbarand et al. (2003) and Carlson et al. (1999), respectively. However, some laboratory groups have preferred using weaker acids, compensated by longer etching times (e.g., Crowley et al. 1991), and a wide range of protocols remain in use (Ketcham et al. 2015). Recent results also indicate that enhanced etchability declines gradually rather than as a step function, and that standard, single-step etching protocols may omit a zone beyond the end of the etched track in which etching rates are reduced but still greater than for bulk crystal (Jonckheere et al. 2017). Zircon is more difficult to etch than apatite, typically requiring a KOH-NaOH eutectic at 210–250 °C, and etching times from several hours up to days (Gleadow et al. 1976; Garver 2003).

Etching rates can even vary on a grain-by-grain basis. Apatite etching rates increase with increasing Cl and OH content, leading to the suggestion that etching times should be varied to achieve a constant etch figure width (Ravenhurst et al. 2003). This procedure has not been adopted for apatite, as more efficiencies accrue from treating all samples identically. Conversely, in zircon, etching rates vary greatly as a function of radiation damage (Gleadow et al. 1976), and the preference has been to etch until the tracks are a given width, usually 1 ± 0.5 μm (Yamada et al. 1995a). Different zircon grains in a given mount can etch at different rates, necessitating multiple etches with interspersed measurements, or multiple aliquots etched for different durations (Garver et al. 2005). These complications make zircon FT lengths more difficult to interpret, in terms of both discerning thermal histories and studying thermal annealing behavior.

3.3 Annealing Experiments

Most of our knowledge of annealing comes from laboratory experiments in which tracks are subjected to controlled conditions and inspected to reveal the extent to which they have changed from an "unannealed" reference point. The tracks used for these studies have been either natural spontaneous tracks, usually from a sample with a sufficiently low-temperature geological history that there is evidence of no to minimal annealing, or tracks that have been induced in a nuclear reactor after pre-existing damage has been erased by heating. In fact, some degree of annealing occurs in apatite over geological timescales at near-surface temperatures, and even in the seconds to days following irradiation (Donelick et al. 1990), but accounting for this initial component is not straightforward, as discussed below, and is thus usually omitted. From the standpoint of formation, spontaneous and induced tracks can be safely considered as identical, as the energy released and decay products from the fission of ^{238}U and ^{236}U are nearly equivalent.

Induced tracks are generally preferred for annealing experiments because they provide the cleanest interpretation, as all tracks in a grain will have a near-identical history, whereas the ages of natural spontaneous tracks in a sample will vary widely. However, spontaneous tracks may be preferred in some cases if there is concern that the pre-annealing step could alter the mineral in an undesirable way, such as by erasure of background radiation damage from alpha recoil.

3.3.1 Early Work

Initial investigations of the causes for fission damage annealing quickly settled on temperature as the primary factor; Fleischer et al. (1965a) ruled out other possibilities such as pressure and ionizing radiation as being comparatively insignificant based on experiments on zircon, olivine, micas, and glasses. The first thermal annealing study on fission tracks was by Fleischer and Price (1964), in which they measured reduction in FT density (tracks observed per unit area on a polished surface) (see Chap. 1, Hurford 2018). They found that the transition between unannealed and fully annealed fission tracks in a volcanic glass followed a linear trend on an Arrhenius-like plot of ln(*t*) versus 1/*T* (Fig. 3.1a), suggesting the relation:

$$\tau_a = Ae^{E/kT} \qquad (3.1)$$

where τ_a is the time necessary to obtain a certain degree track density reduction at absolute temperature *T*, *A* is a pre-exponential constant, *E* is an activation energy, and k is Boltzmann's constant. Subsequent work on apatite and titanite (e.g., Naeser and Faul 1969; Wagner and Reimer 1972) found that the slope of the line describing a given degree of density reduction steepens as annealing progresses (Fig. 3.1b). This pattern was subsequently observed in several other minerals, such as garnet (Haack and Potts 1972; Lal et al. 1977), epidote (Naeser et al. 1970; Haack 1978), vesuvianite (Haack 1978), and phlogopite (Parshad et al. 1978), as well as volcanic glasses (Storzer 1970).

Equation 3.1 is not strictly an Arrhenius equation, as it describes the time necessary for a certain degree of reaction progress, rather than a reaction rate; plots of the type shown in Fig. 3.1 are thus referred to as "Arrhenius-type" diagrams in this chapter, rather than Arrhenius diagrams. The progression in the activation energy inferred from the shifting slope as annealing progresses is enigmatic; it may imply that the activation energy required for further annealing increases as tracks shorten, or alternatively that different tracks may have different activation energies. The activation energies derived in this way for a mineral typically vary by factor of 2–3 between first onset of annealing and completion, leading to a wide range of thermal sensitivities. For example, results for epidote by Haack (1978) suggest that FT density will fall by 10% after 10 million years at ~204 °C, but 90% reduction over the same time interval requires temperatures of 460 °C.

Fig. 3.1 Arrhenius-type diagrams of FT annealing based on track density. **a** Experimental observations on the effects of time and temperature on tracks in an indochinite volcanic glass (based on Fleischer and Price 1964); filled circles = tracks observed, empty circles = no tracks. **b** 100, 50, 0% track-loss curves for apatite (based on Naeser and Faul 1969)

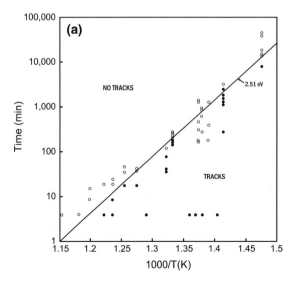
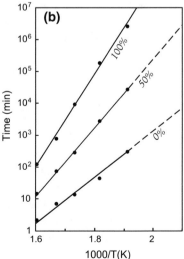

3.3.2 Length-Based Approach

The recognition that lengths of horizontal confined tracks in apatite contain detailed thermal information (Bertagnolli et al. 1983; Gleadow et al. 1986) led to new annealing experiments that focused on length rather than density (Green et al. 1985, 1986; Crowley et al. 1991; Donelick 1991; Carlson et al. 1999; Barbarand et al. 2003; Ravenhurst et al. 2003). Length-based experiments have the advantage of not requiring U homogeneity in the sample material, although homogeneity of other chemical components that may affect annealing rates remains necessary. Overall, the experiments are conducted in a similar fashion, and evaluated in a similar way, to density-based ones.

A key advantage of length-based experiments is the more detailed information they provide concerning the annealing process—for a density measurement, only the presence of a track is recorded, whereas each track length reveals a degree of shortening. Several key observations can be obtained from the length data shown in Fig. 3.2, which summarises a set of experiments on apatite with progressively increasing degrees of annealing, and the effect of track crystallographic orientation (Donelick et al. 1999). At low degrees of annealing, there is continuous, gradual shortening of tracks, indicating that annealing at this stage progresses from the tips inward. Although unannealed track lengths are nearly isotropic (Fig. 3.2a), tracks at higher angles to the apatite crystallographic c-axis anneal more quickly than tracks at low angles (Fig. 3.2b, c), a relationship that can be described well by a polar-plot ellipse (Donelick 1991; Donelick et al. 1999; Ketcham 2003; Ravenhurst et al. 2003). Once individual tracks anneal to about 10.5 μm (Fig. 3.2d–f), their annealing rate accelerates, causing a breakdown of the elliptical relationship. At least part of this acceleration is caused by the formation of slow-etching gaps, which can prevent the full etchable length of the track from being etched in standard protocols. This phenomenon suggests that track-tip annealing is augmented by annealing from the track sides (Carlson 1990).

The only mineral aside from apatite that has had extensive length-based annealing experiments is zircon (Tagami et al. 1990, 1998; Yamada et al. 1995b; Murakami et al. 2006). Even though the etching techniques have been different, as described above, general patterns of zircon FT annealing resemble those for apatite (Fig. 3.3), including anisotropy in both etching and annealing behavior. An interesting difference is that, while individual tracks appear to undergo accelerated annealing, the elliptical relationship does not appear to break down. To counteract over-etching, Yamada et al. (1995a) recommend that only fission tracks with c-axis angles >60° be used for analysis, and most subsequent studies have followed this advice.

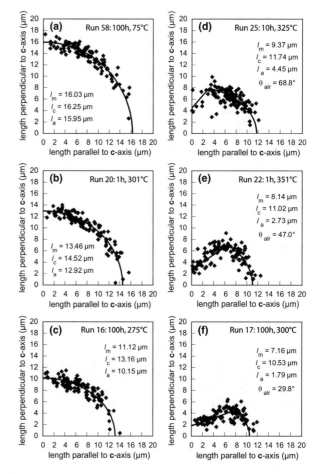

Fig. 3.2 Annealing anisotropy of confined horizontal FT lengths in apatite shown using polar plots (from Donelick et al. 1999). Length distribution remains well described by an ellipse at low levels of annealing; l_c and l_a are fitted lengths parallel to the crystallographic c- and a-axes, respectively, which fall at differing rates as mean length l_m falls. At advanced levels of annealing the elliptical relationship breaks down, with tracks at angles above θ_{alr} showing evidence of accelerated length reduction, which is represented as a line segment connecting the ellipse to the a-axis at intercept I_a

3.3.3 Link Between Length and Density

With the transition from density-based to length-based studies came the need to characterize and understand the relationship between the two. The most definitive study on this topic was by Green (1988), in which both spontaneous and induced tracks were annealed and compared in multiple apatite specimens. The relationship between density reduction and length reduction (Fig. 3.4) is not straightforward, and interesting phenomena appear at both early and advanced stages of annealing.

A key point of the approach recommended by Green (1988), and since adopted by the majority of the FT community, is that comparing spontaneous and induced tracks requires renormalizing both the length and density measurements to a common baseline corresponding to an initial

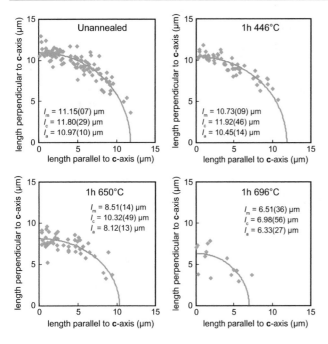

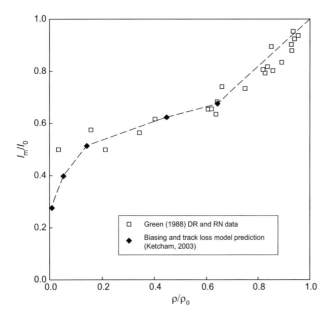

Fig. 3.3 Annealing anisotropy of confined horizontal FT lengths in zircon (data from R. Yamada and T. Tagami). Variables as described in Fig. 3.2 caption; numbers in parentheses are 1σ uncertainties. Patterns are similar as for apatite, except there is no sharp boundary where accelerated length reduction begins to dominate

Fig. 3.4 Relationship between FT length reduction and density reduction in apatite based on induced track data from Green (1988), and prediction from biasing and track-loss model (Ketcham 2003)

condition of induced tracks. Because spontaneous tracks in apatite are invariably 4–11% shorter than induced ones (Green et al. 1986; Carlson et al. 1999; Spiegel et al. 2007), there is an implication that all spontaneous tracks have been

slightly annealed, even at Earth-surface or ocean-bottom conditions (Vrolijk et al. 1992; Spiegel et al. 2007). If this is the case, then according to line segment theory the number of tracks intersecting a unit area of the polished (etched) grain surface (i.e., the surface track density) should also have diminished by a similar amount (Fleischer et al. 1975; Parker and Cowan 1976; Laslett et al. 1982, 1984), and spontaneous track densities should be renormalized to a value corresponding to what they would have been in the absence of this annealing. This assertion is subtly provocative, however, as it essentially posits that absolute apatite FT ages (i.e., obtained without using a zeta calibration) should never reproduce crystallization ages. It has thus remained contentious for those continuing to look in detail at the feasibility of absolute dating (Enkelmann et al. 2005; Jonckheere et al. 2015) and affects the assumptions underlying thermal history inverse modeling as well. This point is revisited later in this chapter.

At low levels of annealing, the Green (1988) renormalization results in a gap in spontaneous track data between reduced length and density values from 0.9 to 1.0 (Fig. 3.4). The induced track data in this region generally follow a near-1:1 relationship, as predicted by line segment theory adapted to account for annealing anisotropy (Laslett et al. 1984; Ketcham 2003).

As annealing progresses, rather than continuing along the ~1:1 trend, there is a sudden deviation at reduced length and density values of about 0.65, in which density reduction seems to accelerate with respect to length reduction. The mechanism behind this was properly but incompletely identified by Green (1988) as a result of annealing anisotropy combined with length biasing (Laslett et al. 1984): Because a shorter confined track is less likely to be intersected by an etchant pathway from the polished surface, it is correspondingly less likely to be observed. Thus, in a track length distribution that has variable lengths due to anisotropy, mean length would remain long (because those tracks are more likely to be seen), while density would fall more quickly. However, in re-examining the annealing data from Carlson et al. (1999), Ketcham (2003) found that length biasing alone is not a sufficient explanation. There is a stark shift in observational frequency between longer, low-angle tracks and shorter, high-angle ones as annealing progresses —the relative chances of measuring a low-angle versus a high-angle track increase by a factor of well over 100. This observation can only be explained if significant proportions of the high-angle tracks have become undetectable, or zero-length for practical purposes. Because zero-length tracks cannot contribute to measurements of mean track length, mean lengths are skewed to larger values that do not reflect the true extent of annealing of the entire track population. This makes mean track length data a very imperfect gauge of annealing progress, even compared to density, unless the effects of anisotropy are accounted for, as discussed later.

3.3.4 Short-Timescale Annealing Experiments

Another implication of apatite FT annealing at Earth-surface conditions is that some annealing of induced tracks may occur during and immediately after irradiation, making the "initial state" used to normalize length and density measurements in fact an already-annealed one. Donelick et al. (1990) performed an important experiment in which they etched irradiated mounts in the minutes and hours immediately after irradiation, while they were still "hot" out of the reactor, essentially conducting a room-temperature annealing experiment. His results on four different apatite varieties indicate that 0.3–0.5 μm of shortening does take place during the time interval from ∼10 min to ∼3 weeks after irradiation.

Another valuable set of experiments at short timescales and high temperatures was conducted by Murakami et al. (2006), in which a high-speed graphite furnace was used to anneal zircon fission tracks at ∼500–900 °C in experiments carefully controlled to last only 4, 10, and 100 s. The importance of such experiments is that they extend observations across several more log units of time in comparison to what is possible with normal furnaces, which typically take a few minutes to heat a sample to the target temperature after it is introduced. At the same time, as shorter experiments require higher temperatures, there is a danger that annealing rate may be affected by operation of some different mechanism. For example, Murakami et al. (2006) report some track length distributions obtained after 4 and 10 s that show unusually high standard deviations, which they attribute to an earlier onset of segmentation compared to tip shortening than observed in longer-time, lower-temperature experiments. Girstmair et al. (1984) conducted similar experiments on Durango apatite, but only measured track density.

3.4 Geological Observations

Necessary clues for understanding FT annealing come from geological benchmarks, or studies in which the thermal history is relatively well established from independent information and thus can serve to test whether annealing models adequately reproduce geological timescale behavior. Generally, the easiest cases to identify and utilize will be where the samples have spent long periods of time at their present-day temperature, whether it be high (e.g., at depth, obtained from a well or borehole) or low (e.g., ocean bottom or Earth surface).

3.4.1 High-Temperature Benchmarks

High-temperature benchmarks are necessary for testing FT behavior near the total annealing point, and down-well studies have been long recognized to be a natural setting for this purpose (e.g., Naeser and Forbes 1976; Naeser 1981). Unfortunately, the requirement that a rock body at depth has maintained a near-constant temperature for tens of millions of years is not an easy condition to fulfill. Geologically, it requires that there has been no significant change in boundary conditions such as basal heat flow and surface temperature, and negligible advective factors such as erosion or deposition, faulting, intrusive activity, and fluid flow (see Chap. 8, Malusà and Fitzgerald 2018). These conditions are difficult to meet over million-year timescales; for example, mean global surface temperature was probably roughly 14 °C higher in the Eocene and has been falling fairly steadily since then (Zachos et al. 2008). Perhaps the most likely settings to be free of subsurface disturbances are deep boreholes on continental shields, but the few examples we have do not provide encouragement. The *Kola superdeep well* on the Scandinavian shield encountered unexpectedly high temperatures and active fluid flow at over 9 km depth (Kozlovsky 1984). The *Kontinentale Tiefbohrung (KTB) continental deep drilling project* in Germany likewise had to cease drilling early due to high temperatures at depth well in excess of predictions (O'Nions et al. 1989), and permeability supporting the possibility of convective flow was found down to 9 km (Clauser et al. 1997).

Two important down-well data sets for study of the FT system are the Otway Basin on the southern Australian passive margin and the KTB well in the Bavarian shield. The *Otway Basin* was initiated in the Late Jurassic through Early Cretaceous by the continental breakup of Australia and Antarctica, with rapid subsidence during rifting leading to deposition of 3–5 km of sediments in the Otway Supergroup between ∼123 and 106 Ma, mostly from contemporaneous volcanic sources. Based on detailed basin analysis, the western Otway Basin is thought to have undergone nearly monotonic burial since deposition, with little activity and thus near-static temperature since 30 Ma (Gleadow and Duddy 1981; Duddy 1997). Green et al. (1985) reported that end-member F-apatites in the *Flaxmans-1 well* become fully annealed at a present-day downhole temperature of 92 °C, while apatites with a Cl content of 0.6 apfu (∼2.2 wt%) retain their depositional age, indicating that Cl content exerts a first-order effect on annealing kinetics.

The KTB deep drilling project consists of two boreholes, reaching depths of 4001 and 9101 m, penetrating the western margin of the Bohemian Massif through Variscan crystalline rocks. These rocks host a range of accessory phases, which has made it a useful proving ground for evaluating and comparing thermochronometers (Wauschkuhn et al. 2015a).

Fig. 3.5 Apatite FT length and age data from the KTB borehole. Length data from Coyle et al. (1997) and Wauschkuhn et al. (2015a, b); analysts EH = E. Heijl, RJ = R. Jonckheere, BW = B. Wauschkuhn. Age data from Wauschkuhn et al. (2015a, b)

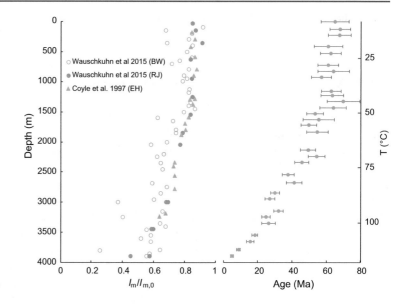

The Cenozoic geological history of the area drilled by the KTB has been an area of dispute, however. Based on apatite FT ages and lengths that show little variation in the top two kilometers of the section (Fig. 3.5), Coyle et al. (1997) posit that Cenozoic reverse faulting thickened this section by up to 1000 m, leading to repetition of ages and lengths. This conclusion is disputed by Wauschkuhn et al. (2015a), who argue that there is no "positive support from independent geological evidence" for such a scenario, and instead favor the assumption that the block has been for all practical purposes quiescent since the end of the Cretaceous, and test various annealing models based on this assumption.

Corrigan (1993) examines apatites from a series of surface and subsurface samples from south Texas and reports down-well samples with fission tracks persisting to as high as 130 °C in the *Frio growth-fault trend*. These data continue to pose an interesting potential challenge to the conventional wisdom on apatite annealing, as these temperatures exceed those at which F-apatite is thought to retain tracks over geological time, and fission tracks will tend to reflect the highest temperature they experience. Thus, recent cooling is more likely to provide a harmfully misleading idea of down-well thermal history than recent heating. However, interpretation of these data is not straightforward. The wells are in active natural gas fields and show extremely variable downhole temperatures from site to site; according to the National Geothermal Data System database (http://geothermal.smu.edu/gtda), wells within several hundred meters of the hottest wells used by Corrigan (1993) are some 10–20 °C cooler. Also, although Corrigan (1993) measured Cl composition to gauge kinetic variability, only 300 random grains were analyzed, rather than matching compositions to specific density and length measurements. As some Cl content variation was observed, there is the

danger of a selection bias, and that deeper, hotter measurements reflect more resistant grains. Finally, as discussed below, other chemical substitution aside from Cl can increase annealing resistance.

A series of studies on fission tracks in zircon has resulted in a variety of high-temperature constraints. By comparing zircon FT ages with apatite FT and mica Rb–Sr and K–Ar systems reflecting a supposedly continuous unroofing history at various localities in the Lepontine Alps, Hurford (1986) derived a widely cited T_c estimate of 240 ± 50 °C (for ∼15 °C/Myr cooling), which was corroborated in subsequent analyses by Brandon et al. (1998) and Bernet (2009) that include additional field studies. Subsequent upward revisions of inferred mica T_c (Villa and Puxeddu 1994; Grove and Harrison 1996; Jenkin et al. 2001; Harrison et al. 2009) suggest that some of these estimates could be increased by some tens of degrees, however. In addition, thermochronological interpretation of some high-temperature isotopic systems in localities of the Lepontine Alps featuring extensive fluid flow and deformation has been challenged (Villa 1998; Challandes et al. 2008).

Well-based studies are more challenging for zircon, and one has to go much deeper and hotter than for apatite to observe any annealing effects. Tagami et al. (1996) found only minor length shortening (mean length 10.5 μm) down a borehole in the Vienna Basin up to temperatures of 197 °C that reached maximum burial ∼5Myr ago. Hasebe et al. (2003) similarly report evidence of only limited zircon FT annealing at up to 200 °C in the *MITI-Nishikubiki* and *MITI-Mishima* boreholes in southern and central Japan over ∼1 Myr timescales. Zircon fission tracks at 205 °C in the MITI-Nishikubiki are slightly more shortened, with a mean length of 9.8 μm, while the MITI-Mishima well features 10.5 µm mean track lengths at 200 °C. In these studies, there may be some inherited, shortened tracks (particularly evident in the MITI-Mishima well), and equating the time at peak temperature with the cessation of fast burial may omit longer timescale re-equilibration. In a core traversing a hydrothermal system in the Valles Caldera, Ito and Tanaka (1995) estimate up to 30% length reduction after inferred heating for 1 Myr at 256 °C and 56% at 294 °C, although these time-temperature estimates hinge on the stability of a system for which instability is to be expected.

3.4.2 Low-Temperature Benchmarks

Geological settings in which samples have spent a long time at low temperatures are important for interrogating apatite annealing behavior in the near-surface environment. The optimal locality for such studies is the ocean floor, due to the stable thermal boundary layer provided by deep Hansen et al. (2013), based on data from Zachos et al. (2008), estimate long-term deep ocean variation of only about 4 °C during the Cenozoic. Studies of apatite FT lengths in ODP samples by Vrolijk et al. (1992) and Spiegel et al. (2007) confirm that spontaneous tracks that accumulate over geological time in low-temperature settings are 4–11% shorter than induced tracks. Although not straightforward to compare because of their various ages, reconstructed thermal histories, and compositions, this variation can be understood to first order in terms of peak temperature and Cl content (Fig. 3.6). The F-apatites (smaller circles) form the lower-temperature part of a consistent trend (dashed line) suggesting that shortening increases from 4 to 11% as peak paleotemperature rises from ∼10 °C to ∼25 °C. The Spiegel et al. (2007) data also feature some apatites with higher Cl content, which is expected to make them more resistant to annealing; the effect in Fig. 3.6 is to push them upwards compared to their expected response if they were F-apatites.

The principal disadvantage of ocean-bottom samples is that material quantity is limited and may include multiple sources and thus may not be straightforward to characterize in terms of unannealed induced length (which can vary by 0.64 µm even among near-end-member F-apatites (Carlson et al. 1999)) or annealing properties. It thus makes sense to complement them by also examining standards such as Durango and Fish Canyon, whose annealing characteristics have been extensively studied and that are likely to have been near the Earth surface since formation, although excursions such as limited burial can be difficult to rule out (e.g., Gleadow et al. 2015). Data from Durango and Fish Canyon reported by Carlson et al. (1999) and Ketcham et al. (1999) are included in Fig. 3.6, and it is apparent that they are generally consistent with the ocean-floor data. A compilation of values from the literature is given in Table 3.1 and

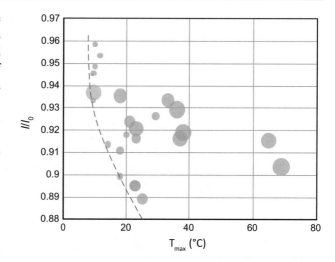

Fig. 3.6 Low-temperature annealing data, from ocean-bottom core samples (blue circles) reported by Spiegel et al. (2007) and Durango (orange) and Fish Canyon (gray) standards reported by Ketcham et al. (1999). Circle size corresponds to Cl content, with largest circle = 1 wt % and smallest 0.03 wt%. Dashed line approximates minimum length reduction for F-apatite at geological timescales

shows that Durango apatite mean spontaneous track lengths average about 11% shorter than induced lengths, while the Fish Canyon apatite spontaneous lengths are only about 6% shorter, which is consistent with their increased annealing resistance as determined experimentally (Ketcham et al. 1999).

The low-temperature sections of deep wells can also be used for this purpose, although they are subject to the same sources of uncertainty as their higher-temperature brethren. Shallow measurements in the KTB borehole reported by Coyle et al. (1997) and Wauschkuhn et al. (2015a) in aggregate suggest ∼15–20% length reduction (Fig. 3.5) at holding temperatures the latter authors interpret to have remained as low as 10–15 °C, which is clearly not consistent with the ocean-floor and standard data shown in Fig. 3.6. For example, Durango apatite has undergone only about 11% length reduction, and the current mean annual surface temperature of the area is 18 °C and was likely higher in the Miocene. Wauschkuhn et al. (2015a) thoroughly analyzed apatite composition in several samples and found no anomalies that would be expected to affect annealing rates. The most likely explanation, to this author, is thus that the shallowest samples in the KTB have seen somewhat higher temperatures than their present-day state and that isothermal stasis throughout the Cenozoic is not a good assumption for the KTB.

For zircon, spontaneous tracks from near-surface settings have lengths that are indistinguishable from induced ones (Kasuya and Naeser 1988; Tagami et al. 1990; Hasebe et al. 1994; Yamada et al. 1995b); example values are given in Table 3.2. The simplest interpretation is that length annealing at Earth-surface temperatures is negligible for zircon.

Table 3.1 Measurements of spontaneous and induced track lengths in apatites with low-temperature histories

Apatite	Spon. (µm)	Source	Ind. (µm)	Source	l/l_0	Analyst	Etch
Durango	14.24 (08)	G88	15.91 (09)	G88	0.895 (7)	PFG	5.0 M, 20 °C, 20 s
Durango	14.47 (06)	K&99	16.21 (07)	C&99	0.893 (5)	RAD	5.5 M, 21 °C, 20 s
Durango	14.58 (13)	G&86	16.24 (09)	G&86	0.898 (9)	AJWG	5.0 M, 20 °C, 20 s
Durango	14.68 (09)	G&86	16.49 (10)	G&86	0.890 (8)	IRD	5.0 M, 20 °C, 20 s
Durango	14.20 (10)	J&07	16.30 (10)	J&07	0.871 (8)	RJ	4.0 M, 25 °C, 25 s
				Mean Durango	*0.889*		
Fish canyon	15.35 (06)	K&99	16.38 (09)	C&99	0.937 (6)	RAD	5.5 M, 21 °C, 20 s
Fish canyon	15.60 (09)	G&86	16.27 (09)	G&86	0.959 (8)	AJWG	5.0 M, 20 °C, 20 s
Fish canyon	15.00 (11)	G&86	16.16 (09)	G&86	0.928 (9)	PFG	5.0 M, 20 °C, 20 s
				Mean Fish Canyon	*0.941*		
Mt. Dromedary	14.57 (86)	G&86	15.89 (09)	G&86	0.917 (8)	PFG	5.0 M, 20 °C, 20 s
Otway	14.58 (11)	G88	16.17 (09)	G88	0.902 (8)	PFG	5.0 M, 20 °C, 20 s

Spon. = mean spontaneous track length. Ind. = mean induced track length. Numbers in parentheses 1 SE. Analysts: RAD = Raymond Donelick; PFG Paul Green; AJWG Andrew Gleadow; IRD Ian Duddy; RJ Raymond Jonckheere. *Source* C&99, Carlson et al. (1999); G88, Green (1988); G&86, Gleadow et al. (1986); J&07, Jonckheere et al. (2007); K&99, Ketcham et al. (1999). Etch = concentration HNO_3, temperature, and time used for etching

Table 3.2 Measurements of spontaneous and induced track lengths in zircons with low-temperature histories

Zircon	Spon. (µm)	Ind. (µm)	Source	T_{etch} (°C)	Measured
Koto Rhyolite	10.89(14)	10.94(15)	Tagami et al. (1990)	225	all
Nisatai Dacite	11.05(08)	11.03(10)	Yamada et al. (1995b)	248	>60°
Bulk Member Tuff	10.48(14)	10.78(09)	Hasebe et al. (1994)	225	>60°
Fish Canyon	10.67(11)	10.61(10)	Hasebe et al. (1994)	225	>60°
Mt. Dromedary	10.45(18)	10.65(13)	Hasebe et al. (1994)	225	>60°

Spon. = mean spontaneous track length. Ind. = mean induced track length. Numbers in parentheses 1 SE. Measured = whether all tracks or only tracks with *c*-axis angle >60° included in mean

3.5 Other Observation Modalities

In addition to chemical etching, a number of other techniques have been used to investigate the structure of latent fission tracks and the mechanisms of annealing, both experimental and computational.

3.5.1 Transmission Electron Microscopy

Transmission electron microscopy (TEM) data provide the most direct look at latent fission tracks (Paul and Fitzgerald 1992) and ion tracks generated by particle accelerators (Jaskierowicz et al. 2004; Lang et al. 2008; Li et al. 2010, 2011, 2012, 2014), and can provide invaluable information on atomic-scale structure and annealing. Accelerated ions are considered good analogues for fission particles with respect to track formation, with the key parameter being

energy loss per unit path length (Lang et al. 2008), and offer precise control of experimental design parameters such as track density and orientation. TEM data have shown that tracks appear to have sharp boundaries, which in apatite have facets in some crystallographic orientations (Paul and Fitzgerald 1992; Li et al. 2014). Early data by Paul and Fitzgerald (1992) suggested that latent fission tracks are wider in the crystallographic *c*-direction than perpendicular to it, but later analysis by Li et al. (2014) using ion tracks found no such distinction. A particularly important result is that the latent track radius in apatite falls continuously from the origin point ($\sim$4.5 nm, the radius of 80 MeV Xe ion tracks) to its terminus, with a break-in-slope at about the half-way point (Fig. 3.7) that may correspond to the transition between tip shortening and segmentation modes of annealing (Li et al. 2012). Li et al. (2012) also observe dispersion of ions from their original trajectories as predicted by SRIM calculations (Ziegler et al. 2008) by $\pm$ 15°, demonstrating how fission tracks can occasionally be curved

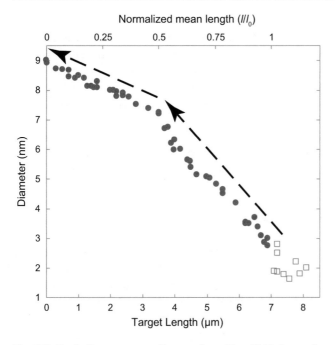

Fig. 3.7 Track diameter versus distance along 80 meV Xe ion tracks through apatite target measured by TEM (based on Li et al. 2012). Red symbols mark where track appears continuous in TEM images; blue symbols denote tracks appearing as discontinuous droplets

features. Thermal annealing in apatite as imaged by TEM appears to take place by segmentation of tracks that is dependent on their radius (Li et al. 2011, 2012).

A puzzling feature of many TEM observations of tracks in apatite is that they appear to be hollow, or "porous," as opposed to zircon tracks which are filled with amorphous material (Li et al. 2010). TEM observations of track annealing in apatite also appear to show significant damage persisting at annealing conditions far above those at which tracks are no longer revealed by etching (e.g., 2 h at 700 °C), as discontinuous chains of droplets (Li et al. 2010), whereas tracks in zircon tracks appear to anneal in TEM data at temperatures comparable to those observed for etched tracks (Li et al. 2011). The apparent dichotomy in track structure and annealing behavior between apatite and zircon led Li et al. (2011, 2012) to conclude that they must anneal by very different mechanisms, with zircon annealing by defect elimination and apatite by some combination of thermal emission of vacancies and mechanisms driven by high surface energy and high diffusivity such as Rayleigh instability, Brownian motion, or preferential motion.

The observations of porous tracks in apatite lead to the natural question of where the atoms formerly occupying the track volume went, to which there is no readily apparent answer; there is no evidence in the TEM imagery of excessive strain or surplus density in the crystal structure immediately surrounding the tracks. Li et al. (2010) postulate that volatile components may pass through planar

defects, but no residuum of nonvolatile elements remains. Although Li et al. (2010) consider many lines of evidence when establishing that the tracks are hollow, it may be possible that these observations can be explained by some combination of their creation and imaging methodologies. Jaskierowicz et al. (2004), studying tracks created by 30 meV C_{60} implantation, observe roughly spherical dark features that they interpret as droplets of ejected material, presumably created during bombardment. Fluorine is known to migrate rapidly in an electron beam (Stormer et al. 1993; Li et al. 2010), even when bound within the apatite crystal structure; in an amorphous, poorly-bound mass, it may be even more mobile, possibly leading to tracks becoming quickly evacuated at TEM operating conditions.

3.5.2 Small-Angle X-Ray Scattering

Another series of interesting experiments on implanted ion tracks has been conducted using small-angle X-ray scattering (SAXS) using synchrotron radiation (Dartyge et al. 1981; Afra et al. 2011, 2012, 2014; Schauries et al. 2013, 2014). X-ray scattering patterns from ion tracks are usually well-fit by a model of a cylindrical track shape, though the radius derived represents an average along its length (Afra et al. 2011). Track radius changes as a function of energy loss rate, but in a fashion that varies on either side of the Bragg peak of maximum energy loss rate in the vicinity of 600–800 MeV. Lower particle energies create wider damage zones at the equivalent energy loss rate because higher-energy particles disperse their damage over a larger region (Afra et al. 2012; Schauries et al. 2014). Schauries et al. (2014) demonstrated this transition in behavior by modeling and measuring the effects of deceleration of 2.2-GeV Au particles, showing how their damage zones widen as they decelerate past the Bragg peak. Fission particle energies are all below the Bragg peak, and their energy loss rate decreases as they decelerate and ultimately stop, corroborating the conical track shape measured by Li et al. (2012) for 80 MeV Xe ions.

Some SAXS results appear inconsistent with the porous-track model based on TEM observations. In addition to measuring track dimensions, analysis of SAXS spectra also provides an estimate of density contrast. Schauries et al. (2013) report that tracks made by 2.2 GeV Au ions in 200-μm-thick slabs of Durango apatite have densities only 0.5% less than the surrounding crystalline material, which seems far from porous. Annealing studies by Schauries et al. (2013) suggest that tracks in apatite anneal at temperatures comparable to those observed using etched tracks and moreover that annealing is marked by a sudden and monotonic drop in track radius. This seems at odds with TEM observations indicating that conical tracks anneal by their ends dissolving into droplets, while their inner sections show little change (Li et al. 2012), which would be expected to increase the mean radius of the track. A possible explanation for this divergence is that the tracks observed by TEM were evacuated, and those measured with SAXS were not.

3.5.3 Atomic Force Microscopy

Atomic force microscopy (AFM), which measures the atomic-scale topography of a surface, has been used to study etched tracks in zircon (Ohishi and Hasebe 2012) and unetched tracks created by ^{129}I ion bombardment of zircon, apatite, and muscovite (Kohlmann et al. 2013). The latter study is notable in that all ion tracks appear as "hillocks" on the mineral surface, with diameters larger than indicated by TEM, ranging from 52 ± 2 to 27 ± 2 nm in apatite and from 64 ± 3 to 28 ± 1 nm in zircon. The authors attribute their observations to amorphization of the track center due to melting by a thermal spike followed by expansion of the material partially out of the polished free surface. These observations appear to contradict those made by Li et al. (2010) using electron energy loss spectroscopy on the TEM that show apatite tracks as holes similar to ones drilled by a focused electron beam, rather than hillcocks, again suggesting that TEM may have evacuated the tracks.

3.5.4 Raman Spectroscopy

While not used to study fission tracks directly, Raman spectra have been used as a measure of radiation damage in zircon (Geisler et al. 2001; Nasdala et al. 2001, 2004; Marsellos and Garver 2010) and are thought to primarily reflect alpha dose. In particular, the $v_3(SiO_4)$ Raman band

at ~ 1000 cm^{-1} widens with increasing effective alpha dose. This information has been used to estimate the extent to which radiation damage may have affected helium diffusivity (Reiners et al. 2002; Guenthner et al. 2013) and could be valuable in investigations of zircon FT annealing rates, which may vary with alpha recoil damage (Rahn et al. 2004; Garver et al. 2005). Separately, Raman spectroscopy has been studied by Zattin et al. (2006) as a means to estimate apatite unit cell parameters, which correlates with FT annealing kinetics in some apatites (Barbarand et al. 2003).

3.5.5 Molecular Dynamics Simulation

Although the complete region disturbed by a fission track contains far too many atoms to be encompassed in molecular dynamics (MD) models, Rabone et al. (2008) were able to simulate the formation of a cylindrical cross section a few unit cells thick by using a periodic boundary condition. The high energetics of track formation, and the ion spike mechanism, are beyond the scope of operations MD was built for, necessitating a number of approximations and simplifications, but overall results seem to be congruent with observations. The modeled energy loss rate of 500 eV Å^{-1}, corresponding to the loss rate at about half the travel distance of a fission ion, and the resulting disrupted region was 5-6 nm in diameter, consistent with subsequent observations by Li et al. (2012). The disrupted region (Fig. 3.8) features an amorphous, depleted center and higher-density rim during formation after 30 ps has elapsed, which is then allowed to equilibrate to a glassy core surrounded by stressed crystal (Rabone et al. 2008). The mean disruption (Å travelled per atom) is highest near the core and falls toward the rim along a sigmoidal trend. An interesting outcome of the re-equilibration step was the formation of fluorite (CaF$_2$)

(a)

Disruption within cylindrical shells around the track

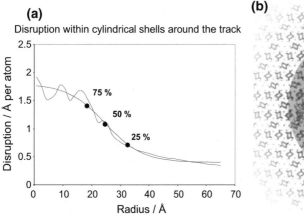

(b)

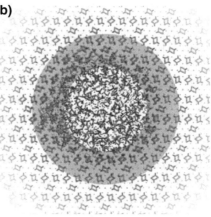

Fig. 3.8 MD simulation of fission damage, from Rabone et al. (2008). **a** Logistic curve (red line) fitted to the trend of atomic disruption (blue line) from the track center. **b** Visualization of track cross section with displacements. The upper and lower quartiles from (a) provide a good definition of the interface region shown in the gray circle

clusters at the center of the track, which is in some ways reminiscent of the formation of subcrystals of ZrO_2 during the annealing of highly damaged zircon (Weber et al. 1994). Unfortunately, the long timescales necessary for annealing pose a severe challenge for the picosecond timescales accessible by MD methods (Rabone and De Leeuw 2007).

3.5.6 Implications for Fission Track Structure

The incompatibility of the TEM data for apatite with data from SAXS and AFM casts into doubt the idea that tracks in apatite are "porous" and thus differ substantially from those in zircon in their structure and annealing mechanisms. Similarly, TEM observation of track annealing in apatite may be compromised by evacuation of the tracks prior to annealing. However, TEM observations that tracks are long, skinny cones, with a break-in-slope at approximately the half-way point along the path of each fission fragment, are likely to be key for understanding track structure and its effect on annealing. Similarly, the TEM observations of very sharp and sometimes euhedral track boundaries are likely to represent the true transition between the track core (and anything in it) and the undisturbed crystal beyond. The gradation in nature and degree of disruption from the track core toward the rim inferred from the MD simulations may also be a key point, providing an intuition for how different parts of the track may anneal at different rates.

3.6 Annealing Mechanisms

Generally speaking, track annealing consists of re-establishing the mineral crystal structure in the disrupted zone. The proper way to conceive of this zone is not straightforward, and the differing conceptions lead to different potential mechanisms for annealing. Carlson (1990) posits the disrupted zone as "laden with defects dominated by vacancies and interstitials," which are then repaired "by short translations of atoms (perhaps on the order of $\sim$5–10 Å or less) from abnormal positions into their proper sites." This implies a partially intact lattice whose extent of damage can be quantified as a defect density, and for which annealing consists of lowering the defect density below some threshold, and can occur throughout the damage zone (Carlson 1990). Alternatively, conceiving of the damage zone as partially or completely amorphous as suggested by MD simulation (Trachenko et al. 2002; Rabone et al. 2008), combined with the sharp boundaries suggested by TEM imaging, may suggest a boundary-controlled mechanism, such as epitaxial recrystallization from the margins of the

zone inwards or snap-off and migration of damaged zones by a surface energy minimization process (Li et al. 2011, 2012).

If one considers annealing as a form of recrystallization, a fission track may constitute a uniquely challenging setting for whatever atomic migrations and rearrangements are necessary, compared to the assumptions commonly made for crystallization reactions. To the extent that disrupted structure is partially amorphous, it should be substantially less dense than the crystal, but any expansion can only come at the expense of local elastic strain in the lattice. This arrangement probably raises the free energy of the disrupted region but may inhibit migration of atoms and molecules to attachment sites compared to grain-boundary transport. Another complicating factor may be that the crystallizing surface is tightly concave in cross section. Yet another complication suggested by MD simulation and annealing experiments is that in sufficiently disordered areas intermediate crystallization products may form, such as fluorite in apatite or ZrO_2 in zircon, possibly increasing the activation energy barrier for re-assimilation. Finally, insofar as some substitutions require certain ordering in the crystal structure, such as alternation of F, Cl, and OH on the halogen site (Hughes et al. 1990), some reattachments may be disadvantageous for further annealing if this ordering is not maintained.

Fortunately, or ambiguously, the rate equations used to portray various mechanisms are in many cases similar. The equation used for the defect elimination rate (dN/dt) used by Carlson (1990) is:

$$\frac{dN}{dt} = -c\left(\frac{kT}{h}\right)\exp\left(\frac{-Q}{RT}\right) \qquad (3.2)$$

where k is Boltzmann's constant, h is Planck's constant, c is an empirical rate constant, Q is the activation energy, R is the universal gas constant, and T is absolute temperature. Equation 3.2 is a simplification of the kinetic equation for atomic motions across a coherent interface (Turnbull 1956) and thus can be used to represent a recrystallization process just as easily as defect elimination. The first temperature term expresses the influence of temperature on the frequency factor, or the rate at which the relevant reaction is attempted, while the second influences the proportion of attempts that are successful. In many applications, the effect of the first temperature term is insignificant compared to the second, and so it is often omitted for the sake of simplification. The single activation energy Q implies that all defects in a fission track are kinetically equivalent. Carlson (1990) omitted the reverse reaction term from Turnbull (1956), which if included could provide an avenue for "undoing" unfavorable intermediate reactions.

3.7 Empirical Annealing Models

The considerable complexity of annealing and etching led Green et al. (1988) to recommend that fitting the experimental data well is preferable to forcing a "physically meaningful" model that aspires to depict a hypothesized underlying process but does not respect the measurements. Though frequently derided as being merely "empirical" (e.g., Wendt et al. 2002; Li et al. 2010), this has proven to be a practical and resilient approach. The charge of empiricism is in some respects misleading, as even putatively physical models are simplified and idealized depictions of complex underlying mechanisms and usually require empirical adjustments to match observations, such as the c parameter in Eq. 3.2. Furthermore, in addition to better respecting the data, to the extent that successful empirical equations are analogous to theoretical ones used to represent physical processes they may provide evidence of which processes are operational or significant.

Accordingly, a mapping between the empirical equations used to date and possible physical models can be obtained by further inspecting the Carlson (1990) model in terms of iso-annealing lines on the Arrhenius-type plot of $\ln(t)$ versus $1/T$ that has been used to interpret annealing data (Fig. 3.9). Carlson (1990) translates the defect elimination rate into an extent of annealing after an isothermal experiment as:

$$l_{as} = l_0 - A\left(\frac{kT}{h}\right)^n \exp\left(\frac{-nQ}{RT}\right) t^n \qquad (3.3)$$

where l_{as} is etchable length after axial (i.e., tip) shortening, l_0 is initial etchable length, t is time, n is a power law exponent

describing the radial defect density distribution, and A is a revised rate constant that incorporates the defect distribution and taper of the disrupted zone. This equation can be rearranged as:

$$\ln(1 - r) = \ln\left(\frac{A}{l_0}\right) + n\ln\left(\frac{kT}{h}\right) + n\left(\ln t - \frac{Q}{R}T^{-1}\right) \qquad (3.4)$$

where r is the reduced track length l/l_0. The temperature in the frequency factor term causes curvature of iso-annealing lines on the standard Arrhenius-type plot, while the one in the activation energy term corresponds to straight lines. A single activation energy implies that iso-annealing lines are parallel, and the fanning of iso-annealing lines observed in density- and length-based annealing studies has been interpreted to mean that there may be a range of activation energies.

Parallel, fanning, and curved iso-annealing lines overlap considerably over the timescales at which laboratory experiments can be conducted, but lead to very different predictions of geological timescale behavior (Fig. 3.9).

3.7.1 Parallel Model

The preliminary annealing model of Green et al. (1985) fitted to Durango apatite annealing data had the form:

$$\ln(1 - r) = c_0 + c_1 \ln t + c_2 T^{-1} \qquad (3.5)$$

simplifying Eq. 3.4 and omitting the temperature effect on the frequency factor. The parameter c_2 corresponds to a single activation energy, leading to parallel linear iso-annealing lines on the Arrhenius-type plot.

Fig. 3.9 Arrhenius-type diagram illustrating parallel, fanning, and fanning curvilinear model forms (from Ketcham 2005)

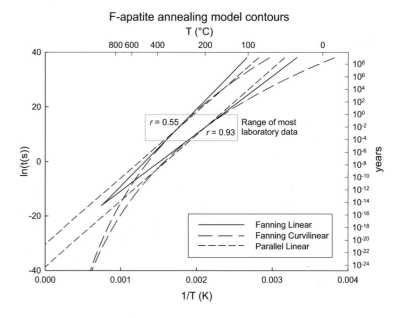

Ultimately, the parallel model was rejected (along with the near-parallel model of Carlson (1990)) because of structured time residuals, suggesting that extrapolation to geological timescales would not be reliable.

3.7.2 Fanning Arrhenius Model

Laslett et al. (1987) fitted an expanded Durango data set with a more sophisticated model with the form:

$$\frac{\left[\left(1 - r^{\beta}\right)/\beta\right]^{\alpha} - 1}{\alpha} = c_0 + c_1 \left[\frac{\ln t - c_2}{(1/T) - c_3}\right] \quad (3.6)$$

In this formulation, lines of constant annealing emanate from points with coordinates (c_2, c_3) on the Arrhenius-type plot. Laslett et al. (1987) found that c_3 could be set to zero when fitting the Green et al. (1986) data, but subsequent data sets by Crowley et al. (1991), Carlson et al. (1999) and Barbarand et al. (2003) required non-zero values for an optimal fit.

The revised left-hand term in Eq. 3.6 is a transform proposed by Box and Cox (1964), which was used by Laslett et al. (1987) as a "super-model" that encapsulates a number of potential forms, including $\ln(1 - r)$ if $\alpha = 0$ and $\beta = 1$. In a practical sense, it added a couple of degrees of freedom to the transformation of variable r, which were also used by Crowley et al. (1991) and Ketcham et al. (1999) to optimize data fit, but later efforts sought to re-simplify the r-transform (Laslett and Galbraith 1996; Ketcham et al. 2007b).

3.7.3 Fanning Curvilinear Model

In examining the multi-apatite annealing data set by Carlson et al. (1999), Ketcham et al. (1999) found that although the best fit to mean track length data was provided by Eq. 3.6, its geological timescale predictions did not match well with benchmarks based on field studies. In particular, the extrapolation of fits to Eq. 3.6 implied that low-temperature annealing at Earth-surface conditions was negligible and ~ 20 °C degree of heating (i.e. ~ 1 km of burial) would be required to explain, for example, the extent of annealing observed in Durango and Fish Canyon apatites, and even ocean-bottom sediments. Likewise, fits to Eq. 3.6 implied a higher T_c than indicated by borehole studies.

Ketcham et al. (1999) thus looked for avenues for fitting the laboratory data while honoring the geological observations, following the original recommendation of Green et al. (1988). Their solution was based on fitting mean c-axis-projected length, essentially removing anisotropy effects, and using a different model form originally proposed but abandoned by Crowley et al. (1991):

$$\frac{\left[\left(1 - r^{\beta}\right)/\beta\right]^{\alpha} - 1}{\alpha} = c_0 + c_1 \left[\frac{\ln t - c_2}{\ln(1/T) - c_3}\right] \quad (3.7)$$

With this equation, the iso-annealing lines have significant curvature on the Arrhenius-type plot (Fig. 3.9). Subsequent work by Ketcham et al. (2007b) found that Eq. 3.7 fitted the annealing data set of Barbarand et al. (2003) as well or slightly better than Eq. 3.6 and maintains good agreement with field observations. Ketcham et al. (2007b) also found that the left-hand expression could be simplified with only minimal degradation of the fits of both data sets by setting β to -1, resulting in:

$$(1/r - 1)^{\alpha} = c_0 + c_1 \left[\frac{\ln t - c_2}{\ln(1/T) - c_3}\right] \quad (3.8)$$

Equation 3.8 underlies the currently most used apatite annealing model. Equation parameters for a theoretical most annealing-resistant apatite, close to F–Cl–OH-apatite B2 from Bamble, Norway (Carlson et al. 1999), are provided in Table 3.3.

Table 3.3 Fitted parameters for annealing models in this chapter

Data	l_0 (µm)	N	Model	c_0	c_1	c_2	c_3	α	χ_r^2
1	various	579	FC	0.39528	0.01073	−65.130	−7.9171	0.04672	3.10
2	11.04	38	FA	−12.216	0.00028	32.125	7.54e−05	−0.02904	1.49
			FC	−91.659	2.09366	−314.94	−14.287	−0.05721	1.15
3A	16.66	75	FA	−16.266	0.00057	−21.856	0.00047	−0.59154	3.47
			FC	−899.07	19.871	−1614.3	−42.712	−0.77735	3.23
3B	17.08	81	FA	−19.422	0.00069	−27.200	0.00027	−0.81075	3.52
			FC	−999.56	22.291	−1828.7	−47.453	−0.77955	3.17

Data 1, 26 apatites from Carlson et al. (1999) and Barbarand et al. (2003), fit by Ketcham et al. (2007b); 2, NST zircon from Yamada et al. (1995b), Tagami et al. (1998), and Murakami et al. (2006); 3, Tioga apatite from Donelick (1991) and Donelick et al. (1990) (A = high T only; B = high T + 23 °C). c_0, c_1, c_2, c_3, α = parameters for Eq. 3.6 (with $\beta = -1$) for FA (fanning Arrhenius) and Eq. 3.8 for FC (fanning curvilinear). N = number of experiments fitted. χ_r^2 = reduced chi-squared value of fit

3.7.4 Multi-kinetics

Green et al. (1985) first reported apatite FT ages in Otway Basin well samples that increased with Cl content, indicating that apatite composition can affect annealing rates. Subsequent annealing studies by Crowley et al. (1991), Ravenhurst et al. (2003), Carlson et al. (1999), and Barbarand et al. (2003) demonstrated that a range of chemical substitutions can affect annealing kinetics. The influence of Cl is probably best considered in the context of mixing on the halogen site, as OH can also affect annealing rates (Ketcham et al. 2007b; Powell et al. 2017), and near-ternary F–Cl–OH is more resistant to annealing than end-member Cl-apatite (Carlson et al. 1999; Gleadow et al. 2002). Cation substitutions, including Sr, Mn, Fe, and Si, have also been documented to affect annealing rates in laboratory experiments (Crowley et al. 1991; Carlson et al. 1999; Ravenhurst et al. 2003; Tello et al. 2006). Apatite solubility, as expressed in the etch figure diameter parallel to the c-axis, D_{par}, is also correlated with annealing kinetics, with high-D_{par} apatites being more resistant to annealing (Burtner et al. 1994; Carlson et al. 1999).

Ketcham et al. (1999) found that laboratory annealing results for any two apatites annealed at the same conditions can be described by the relation:

$$r_{lr} = \left(\frac{r_{mr} - r_{mr0}}{1 - r_{mr0}} \right)^{\kappa} \qquad (3.9)$$

where r_{lr} and r_{mr} are the reduced track lengths of the less-resistant and more resistant apatite variety, respectively, and r_{mr0} and κ are fitted parameters (Fig. 3.10a). The parameter r_{mr0} has a physical meaning as the reduced track length of the more resistant apatite at the conditions where the less-resistant apatite first becomes fully annealed.

Figure 3.10b shows r_{mr0} and κ values for the 26 apatite varieties studied by Carlson et al. (1999) and Barbarand et al. (2003), as fitted by Ketcham et al. (2007b). Generally, as one proceeds from right to left, annealing resistance and T_c rise. The most annealing-resistant apatite studied to date, B2 (Bamble, Norway), is on the $r_{mr0}= 0$ axis and has an estimated T_c of 160 °C assuming 10 °C/Myr cooling. B2 is probably set apart by its near-ternary F–Cl–OH composition in the anion site. The second-most resistant apatite, PC (Portland Connecticut), has an estimated T_c almost as high at 158 °C, but for entirely distinctive reasons: It is an end-member F-apatite but with 7.0 wt% (1.0 apfu) MnO. The third-most resistant, TI (Tioga, Pennsylvania), has an estimated T_c of 149 °C, and no clear-cut culprit; it has minor Cl, but close to an even split between F and OH on the anion site, and 0.1 and 0.2 apfu of MnO and FeO, respectively.

It bears pointing out that the temperature implications of r_{mr0} and κ are very nonlinear; the cluster of apatites on the

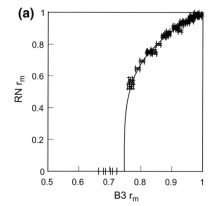

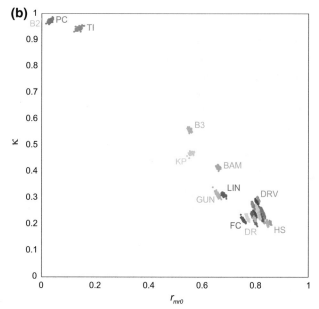

Fig. 3.10 a Apatite–apatite fitting using Eq. 3.9, as demonstrated based on reduced mean lengths for apatites RN and B3 from Ketcham et al. (1999). **b** Relationship between r_{mr0} and κ for 26 apatites (data from Ketcham et al. 2007b). Each apatite is shown as a cluster of values generated by Monte Carlo analysis, to give an idea of their uncertainties. Apatite abbreviations from Carlson et al. (1999) and Barbarand et al. (2003)

lower right in Fig. 3.10b have estimated T_c values ranging from 90 to 127 °C as r_{mr0} falls from 0.857 (HS—Holly Springs OH-apatite) to 0.757 (FC—Fish Canyon Tuff), a T_c span comparable to going from 0.757 to zero. F-apatites have T_c values that vary over a 10 °C range from 95–105 °C, imposing a limit to the precision of thermal information we can hope for, and Durango apatite has a notably higher estimated T_c of 112 °C.

The general alignment of the apatites on Fig. 3.10b indicates a helpful simplification, as the two fitted parameters are strongly correlated, implying that:

$$r_{mr0} + \kappa \approx 1 \qquad (3.10)$$

(Ketcham et al. 1999) or 1.04 (Ketcham et al. 2007b)

allowing κ to be estimated from r_{mr0}. Taken together, these relations allow for the creation of a "super-model" in which the annealing equation is used to characterize the most annealing-resistant apatite, and any less-resistant apatite can be characterized by determining an r_{mr0} value for it. In addition to composition and D_{par}, other suggested proxies include total ionic porosity (Carlson 1990) and unit cell parameter a (Barbarand et al. 2003), but none are able to explain the full range of experimental results gathered to date. Relations put forward by Ketcham et al. (2007b) include:

$$r_{mr0} = 0.83 \left[(Cl^* - 0.13)/0.87 \right]^{0.23} \qquad (3.11a)$$

$$r_{mr0} = 0.84 \left[(4.58 - D_{par})/2.98 \right]^{0.21} \qquad (3.11b)$$

$$r_{mr0} = 0.84 \left[(9.509 - a)/0.162 \right]^{0.175} \qquad (3.11c)$$

$$r_{mr0} = \begin{pmatrix} -0.0495 - 0.0348\,F + 0.3528\,Cl^* + 0.0701\,OH^* \\ -0.8592\,Mn - 1.2252\,Fe - 0.1721\,Others \end{pmatrix}^{0.1433} \qquad (3.11d)$$

where elemental compositions are in atoms per formula unit (apfu) based on $Ca_{10}(PO_4)_6(F,Cl,OH)_2$, and starred elemental compositions are calculated as abs(apfu−1). *Others* are the sum of all other substituting cations other than Mn and Fe, D_{par} is in μm, and a is in Å.

Mathematically, the super-model approach posits that there is only one fanning point and set of iso-annealing contours emanating from it defined by Eq. 3.8 and that Eq. 3.9 through 3.11 assign these contours to specific degrees of annealing for a given apatite. This is likely to be an over-simplification, but Ketcham et al. (2007b) found no evidence that it has a significant effect on geological time-scale predictions compared to individual-apatite models.

At this stage, there is no physical model that explains how composition affects annealing rates. The super-model interpretation of a common set of iso-annealing contours might suggest that composition influences the activation energy for annealing, although the meaning of an activation energy in a curvilinear model is unclear. The way that Cl and OH are incorporated into the r_{mr0} equation form in Eq. 3.11a and 3.11d suggests that mixing of components on a given crystallographic site may be crucial—although Cl is generally known to increase resistance to annealing, the near-ternary F–Cl–OH-apatite B2 from Carlson et al. (1999) is much more annealing-resistant than end-member Cl-apatite B3. When there is mixing in the halogen site in apatite, the different components are not random, but occur as an ordered sequence (Hughes et al. 1989, 1990). This requirement for ordering, probably to minimize local distortions in the crystal structure, may in turn inhibit

annealing, particularly if displaced halogens are re-integrated out of their optimal sequence. Similar reasoning may apply to cation substitutions.

3.8 Annealing Models for Zircon

As previously discussed, zircon FT length measurements are complicated by the variability of etching rates caused by alpha recoil (and perhaps particle) damage. An additional and very relevant question is whether alpha damage affects annealing rates, which in turn may affect how annealing experiments should be conducted and interpreted. Kasuya and Naeser (1988) conducted 1-hour annealing experiments on spontaneous fission tracks in zircon samples with a range of spontaneous FT densities (9e5, 1.5e6, 4e6, 1e7 tracks cm^{-2}) and thus alpha doses, and found no significant differences in annealing rates. However, a number of geological studies have found a strong negative correlation between uranium content and zircon FT age, suggesting that increasing radiation damage lowers annealing resistance (Garver and Kamp 2002; Garver et al. 2005). Rahn et al. (2004) noted a significant change in annealing rate between spontaneous tracks measured by Tagami et al. (1990) and Yamada et al. (1995b) and attributed it to alpha damage (spontaneous FT densities 7e6 and 4e6 tracks cm^{-2}, respectively), although Yamada et al. (1995a) claim that the studies' different etching conditions provides a sufficient explanation for their discrepancy. The two studies also report different temperatures at which the change in etching rate occurred, 450–500 °C in Tagami et al. (1990) and 550–600 °C in the 1-h experiments by Yamada et al. (1995b). Yamada et al. (1995b) did find induced tracks to be more resistant to annealing than spontaneous ones, however. It is also noteworthy that density-based annealing experiments by Carpéna (1992) on pre-annealed zircons with induced tracks found evidence of annealing resistance being inversely correlated with U content exclusive of radiation damage, suggesting a compositional effect linked to trace element composition. Given the available experimental data, it is thus difficult to quantify any radiation damage effect. For that matter, it is also difficult to be assured that laboratory thermal annealing experiments are a robust means for inferring annealing behavior on geological timescales if different damage types anneal at different rates.

Two approaches for quantifying annealing rates have been proposed. Rahn et al. (2004) propose a "zero-damage" model based on induced tracks in pre-annealed zircons, with the idea that it could be modified to include damage effects once both the annealing behavior of alpha recoil damage and its effect on FT annealing are quantified. They fit both parallel and fanning linear models and find that the fanning model fits better. Their model predicts a T_c of 325 °C at 10 °C/Myr cooling, substantially above most estimates based on field data and comparison with other thermochronometers, indicating that the step of including radiation damage will be required for it to be useful for thermal history analysis, except possibly in very young rocks.

Most zircon annealing models (Yamada et al. 1995b, 2007; Tagami et al. 1998; Guedes et al. 2013) have been based on spontaneous track data from the Nisatai Dacite (Yamada et al. 1995b; Tagami et al. 1998; Murakami et al. 2006), which has a K–Ar biotite age of 20.5 ± 0.5 Ma and spontaneous FT density of 4e6 cm^{-2}. As discussed previously, the Murakami et al. (2006) contribution expanded the time coverage of the experimental data in log space by three orders of magnitude compared to apatite studies. As with apatite, annealing in zircon is anisotropic, but the recommended response was essentially the reverse approach to apatite: Rather than projecting track lengths to c-axis parallel, anisotropy is dealt with by only considering high-c-axis angle tracks (>60°; Hasebe et al. 1994; Yamada et al. 1995a).

The Yamada et al. (2007) analysis did not find a satisfactory single model that encompassed data across all experimental timescales, and their preferred solution is a hybrid that fits the low-T and high-T data separately and merges the two fits. The component that extrapolates to geological timescales, which uses a fanning linear form, predicts a high T_c of 317 °C for a 10 °C/Myr cooling rate, again substantially above field-based estimates (Brandon et al. 1998; Bernet 2009). Other attempts using a fanning linear form have had the same issue (Yamada et al. 1995b; Tagami et al. 1998). Yamada et al. (2007) also considered parallel and fanning curvilinear model forms, but were unable to get the latter to converge to a solution, and the former was rejected based on geological comparisons.

In fact, a fanning curvilinear model can be fitted to the complete Nisatai Dacite zircon data set, with results that echo those obtained for apatite annealing. Such a model, fitted using the methods described in Ketcham et al. (1999; Appendix A), was incorporated into the Guenthner et al. (2013) zircon (U−Th)/He diffusivity model that attempts to incorporate radiation damage and annealing effects and is discussed further here. This model predicts a T_c for 10 °C/Myr cooling of 281 °C, broadly consistent with upward revision of earlier studies if higher mica K/Ar and Rb/Sr T_c are used, as discussed previously, but ~40 °C cooler than previous fanning linear models. Figure 3.11a compares this model with a fanning linear one fitted using the same methods (parameters in Table 3.3), based on comparison to some of the geological benchmarks cited previously. In general, the fanning curvilinear form does a better job of matching the more highly annealed benchmarks, and the two models split the low-temperature

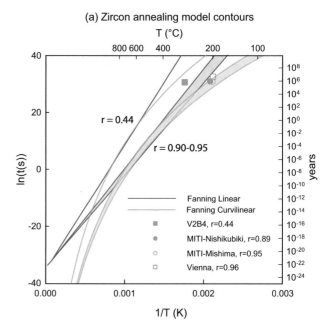

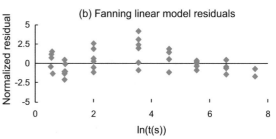

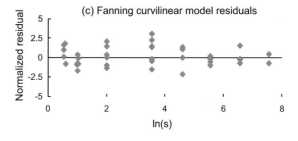

Fig. 3.11 Zircon annealing model. **a** Pseudo-Arrhenius diagram showing fanning linear and fanning curvilinear fits of data from Yamada et al. (1995b), Tagami et al. (1998), and Murakami et al. (2006), and selected geological benchmarks. **b** Normalised time residuals as model-data/standard error for fanning linear model. **c** Normalised time residuals for fanning curvilinear model

benchmarks, with the fanning curvilinear coming closer to the more annealed MITI-Nishikubiki data and the fanning linear being closer to MITI-Mishima and Vienna. The residuals present a more stark contrast (Fig. 3.11b, c), with the fanning linear model featuring very structured residuals indicating poor extrapolation to geological timescales (as also observed by Yamada et al. 2007). In contrast, the fanning curvilinear model exhibits no structure, indicating that it is able to successfully encompass the expanded time range provided by the Murakami et al. (2006) data set. Various fanning curvilinear forms are also fitted and discussed by Guedes et al. (2013).

This model may serve as an adequate placeholder, but clearly more work is called for. In addition to better understanding the possible influence of alpha recoil damage, another opportunity may lie in accounting for track angle in a better way than simply excluding low-angle tracks, by using a *c*-axis-projection scheme or something similar. In particular, again assuming that apatite is a sufficiently analogous case, the exclusive use of tracks >60° may actually be counter-productive, as it mixes annealing modes of tip shortening and segmentation in a way that may over-task the empirical equations with which we try to both fit laboratory-scale observations and predict geological-scale behavior.

3.9 Alternative Annealing Mechanisms

A number of claims have been made that some component of FT annealing is non-thermal or is controlled by an additional process that is not easily observable in the laboratory because it cannot be accelerated by turning up the heat. Each deserves careful scrutiny. The chief candidates are low-temperature ambient annealing (also called "seasoning"), pressure-driven annealing, and radiation effects.

3.9.1 Seasoning

"Seasoning" (Durrani and Bull 1987) is a general and somewhat ambiguous term that (along with "aging" and "hardening") has been used in various contexts to posit a low-temperature mechanism whereby some degree of annealing or track modification takes place that is not reflected in higher-temperature annealing experiments. For example, Durrani and Bull (1987) posit that "hardening" of fission tracks (i.e., an increase in annealing resistance) in lunar materials may be due to the ambient radiation field. Seasoning has most recently been used to refer to a possible process that causes shortening of the etchable length of confined tracks, but that does not result in a (proportional) decrease in track density (Jonckheere et al. 2017).

One possible instance of seasoning is the ~ 0.5 μm of annealing in apatite observed by Donelick et al. (1990) at room temperature in the seconds to hours immediately following irradiation. Subsequent work has suggested some component of annealing in the years following irradiation by examining apatites irradiated over the past few decades (Gleadow 2014). Is this seasoning, or is it an aspect of the same thermal process operating on geological timescales? Laslett and Galbraith (1996) observed that if the Donelick et al. (1990) data are correct, and the process underlying the observed annealing is the same thermal one that operates on geological timescales, then we may be using an inappropriate normalizing value for FT lengths. Rather than a presumed "unannealed" length measured some months after irradiation, during which tracks have undergone some degree of ambient-temperature annealing, they posited that a longer initial length value may be a more physically defensible starting point. Their attempt to use this assumption to achieve a superior fit to data for two apatite varieties from Crowley et al. (1991) by making the true initial track length a fitted value was questionably successful, however. The values they obtained, 16.71 +0.57/−0.45 μm for F-apatite and 18.96 +2.08/−1.12 μm for Sr–F-apatite are substantially more divergent than the actual measurements by Crowley et al. (1991) of 16.34 ± 0.09 and 16.29 ± 0.10 μm, respectively; these two apatites would have had to start at very different lengths and coincidentally converge and become statistically indistinguishable. Ketcham et al. (1999, Appendix B) also attempted to determine whether allowing initial length to vary among the 15 apatites studied by Carlson et al. (1999) resulted in a superior fit, and found the differing degrees of implied initial length adjustment, 4–14%, to be unjustifiable.

Another approach to gauging the effect of low-temperature, short-duration annealing is to incorporate the Donelick et al. (1990) data into a complete annealing model that also utilizes longer-duration experiments. At this time, such an exercise can only be done with data for Tioga apatite published by Donelick (1991), as that is the only other data set obtained using the same etching protocol (5N HNO$_3$ at 21 °C for 25 s). Parameters for fanning linear and curvilinear models fitted with and without the Donelick et al. (1990) data are provided in Table 3.3, and example iso-annealing contours are compared in Fig. 3.12. All models feature a high χ_r^2 value, indicating considerable scatter, but there is no penalty for incorporating the 23 °C data; in fact, the best-fitting model includes it. However, although the fanning points (c_2 and c_3 for both model types) are offset, the geological timescale predictions of the models with and without the 23 °C data are almost indistinguishable, supporting the contention that the room-temperature annealing observed by Donelick et al. (1990) is the same process, and can be described by the same equations, as geological timescale annealing. This finding is somewhat tempered by the likelihood that the high-temperature annealing experiments by Donelick (1991) were affected by a poor furnace temperature calibration, as discussed by Carlson et al. (1999).

Seasoning has also been cited as an explanation for why Durango apatite appears to be amenable to absolute dating (i.e., dating without using age standards with the zeta method). Because Durango spontaneous fission tracks are shortened by $\sim 10\%$ compared to induced tracks, the

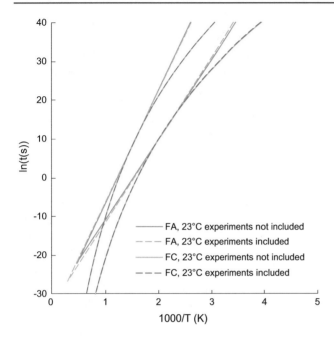

Fig. 3.12 Arrhenius-type diagram comparing Tioga apatite fanning Arrhenius (FA) and curvilinear (FC) models based on data from Donelick (1991) fitted with and without 23 °C annealing data from Donelick et al. (1990). Left iso-annealing contours for each model are $r = 0.41$, right contours are $r = 0.90$

expectation is that the absolute FT age should be $\sim 10\%$ younger due to reduced intersection probabilities of tracks with the polished interior surface (Fleischer et al. 1975). However, such a step is apparently unnecessary if all experimental and biasing factors are meticulously accounted for (Enkelmann et al. 2005; Jonckheere et al. 2015). Another observation corroborating the lack of a length-shortening effect is that the Durango apatite plateau method-determined age of 31.4 Ma (Jonckheere 2003a) closely matches its $^{40}Ar/^{39}Ar$ reference age of 31.44 ± 0.18 Ma (McDowell et al. 2005).

Yet another related observation using Durango apatite is that side-by-side annealing of spontaneous and induced tracks in Durango apatite indicates non-congruent behavior, as observed by Wauschkuhn et al. (2015a) and foreshadowed in earlier work (e.g., Naeser and Fleischer 1975). Overall, induced tracks are shown to anneal slightly more quickly than spontaneous ones, but to remain longer due to their longer starting point, until conditions for total annealing are approached. If instead the induced tracks are pre-annealed to match the spontaneous track initial mean length, their subsequent annealing is slower than spontaneous tracks. These results imply that induced tracks are slightly more resistant to annealing than spontaneous ones, reminiscent of observations in zircon (Yamada et al. 1995b).

Contrasting with these observations is that the fanning curvilinear models (e.g., Ketcham et al. 2007b) based on

induced tracks successfully reproduce the observed shortening of confined horizontal tracks in Durango and Fish Canyon apatites, as well as numerous ocean-floor samples, as discussed previously (Fig. 3.6). It is unlikely that this degree of fit is coincidental.

These inconsistencies may instead indicate that the connection between confined horizontal track length and the length applicable for predicting FT density is not as straightforward as originally considered. One recent and possibly promising observation is that the tips of spontaneous and induced tracks etch at different rates (Jonckheere et al. 2017) and moreover that the model of track etching based on only two end-member etching rates for track and bulk mineral is oversimplified, a point made long ago for zircon (Yamada et al. 1993). This leads to the possibility that the effective length of a spontaneous track for etching during density determination is somewhat longer than its horizontal confined length (Wauschkuhn et al. 2015b).

3.9.2 Effects of Alpha Decay and Damage

Hendricks and Redfield (2005) suggest that alpha particles may cause appreciable annealing of radiation damage in apatite. Studying apatites from the Fennoscandian shield, they note a negative correlation between FT age and uranium concentration and go on to posit that the culprit is the higher flux of alpha particles that fission tracks in higher-U apatites will experience. Rutherford backscattering spectrometry data have shown that alpha particles in sufficient quantities will anneal some component of radiation damage (Ouchani et al. 1997; Soulet et al. 2001). Hendricks and Redfield also find that weighted mean apatite (U−Th)/He ages in their samples are frequently older than their AFT ages.

Kohn et al. (2009) dispute these conclusions. In similarly old rocks in the southern Canadian and Western Australian shields, they find the relationship between age and uranium to be inconsistent, with different samples showing weak positive and negative correlations, or no significant correlation at all. They additionally find that including the alpha dose contribution of thorium in the southern Finland data significantly weakens the correlations observed by Hendricks and Redfield (2005) and suggest instead that Cl content may be responsible for the age variation observed.

The link between the laboratory observations of alpha-induced annealing and field conditions can be made by considering the relative fluxes of alpha particles. To enable comparison to the experimental data, we can calculate the flux of naturally generated alpha particles that cross an internal plane (i.e. particles/cm²). Considering an interior 1 cm × 1 cm plane, the probability of a decay-generated He atom striking that plane is 0.5 for immediately adjacent decay events and falls with increasing distance from the plane, reaching zero at about 20 μm, the alpha particle stopping distance. At distance d from the plane, the proportion p crossing it corresponds to the fraction of the surface area of a 20-μm sphere which crosses that plane:

$$p = \frac{2\pi Rh}{4\pi R^2} = \frac{h}{2R} \tag{3.12}$$

where R is the stopping distance (20 μm) and h is equal to $R - d$. The integrated total of alpha particle crossings over the interval from 0 to 20 μm from the plane then corresponds to a thickness of:

$$\int_0^{20} \frac{h}{40}\,dh = \frac{h^2}{80}\bigg|_{20} = 5 \tag{3.13}$$

Thus, assuming uniform U+Th concentration, the total number of alpha particles crossing the plane from one side corresponds to the total number generated within 5 μm of it, and so the equivalent total dose will equal the number of alpha particles generated within a 1 cm × 1 cm × 10 μm volume of mineral. Accounting for the densities of apatite and uranium, this wedge generates a dose of about 1.1×10^{11} α/cm²/ppm U in 1 Ga. This estimate can be doubled to include Th decay, assuming a typical Th/U ratio of 3.8. Assuming a U content of 20 ppm, in the midrange of the Fennoscandian samples, a total dose of $\sim 4.4 \times 10^{12}$ α/cm² can be estimated for a 1 Ga history.

Conversely, Ouchani et al. (1997) and Soulet et al. (2001) used doses of 0.25–15×10^{15} and 3.2×10^{15} He/cm², respectively, to study alpha particle-induced annealing; even the lowest-dose experiment in these studies is $\sim 100 \times$ higher than the naturally occurring internal alpha dose. Combined with the fact that later-forming spontaneous tracks will experience even lower doses, it is evident that, while alpha-induced annealing is certainty a real process, the existing experimental data do not support it being a geologically important one.

However, it bears pointing out that the negative age-U correlations noted by Hendricks and Redfield (2005) mimic those reported by Garver et al. (2005) for the zircon system, and thus their observations may be interpreted in a different way. Rather than the alpha particles causing annealing, heavily damaged apatite may have a lower annealing resistance than less-damaged apatite, as has been proposed for zircon. To bring about a negative age-U correlation, thermal histories would have to feature the appropriate degree of reheating to cause differential annealing, providing an explanation for why such correlations may sometimes be present and sometimes not.

3.9.3 Pressure Effects

Wendt et al. (2002) reported evidence for a geologically relevant pressure effect on FT annealing in apatite, contradicting earlier experiments by Fleischer et al. (1965a) and in particular Naeser and Faul (1969). Their approach was severely criticized by Kohn et al. (2003), who cited a lack of evidence for such an effect in both previous experimental work and multiple borehole studies. Perhaps the most significant issue is that Wendt et al. (2002) used different furnaces for annealing experiments at different pressures, increasing the potential harm of any problem in calibration. Subsequent efforts to reproduce these results (e.g., Donelick et al. 2003) were unsuccessful, and instead supported the earlier findings indicating that, although there may be a pressure effect on annealing, it is not significant in a geological context; the temperature increase associated with any reasonable geothermal gradient will thermally anneal fission tracks before sufficient pressures are encountered.

3.10 Thermal History Modeling

The ultimate test of our annealing models comes when they are applied in geological settings. The approach for doing so was laid out by Green et al. (1989): For any possible thermal history, our models allow us to predict the age and length distribution that would be measured. This prediction requires an annealing equation, a length-density conversion, and proper accounting for observational biases. Another important component is the equivalent time hypothesis, which states that the annealing behavior of a track depends solely on its current length, and not the particular history by which it achieved that length (Duddy et al. 1988). When we combine this set of assumptions into a computer program, we can then pose a series of geologically reasonable thermal histories and see which ones match our measurements (e.g., Issler 1996; Gallagher 2012; Ketcham et al. 2016). This approach has been used in hundreds of studies to help extract the maximum amount of information from FT measurements, and its wide use attests to its underlying soundness, or at least resilience.

However, while most practitioners use FT thermochronology and modeling to solve geological problems, they are also testing this soundness and resilience. Another "feature" of FT thermochronology is the large uncertainties inherent in the method—we are essentially trying to work out geological histories spanning millions of years by studying the aftermaths of a handful of spontaneous fission decays. Large uncertainties in ages stem from the unforgiving Poissonian statistics of track counts in the hundreds to thousands, and the equivalent time hypothesis combined with some tens up to a hundred track lengths generally ensures that many time–temperature paths are consistent with the data. If there are few other quantitative constraints on the geology, it is generally possible to find many thermal histories that are consistent with our measurements. The flip side to luxuriating within generous uncertainty bounds is that we may get incorrect answers, notwithstanding the ingenuity of our theoretical, computational, and statistical approaches. Successfully fitting a model can lead to complacency, and within the statistical noise lie both freedom and danger.

The largest pitfall in modeling using apatite fission tracks lies in their variable annealing kinetics, which are probably

the rule rather than the exception in sedimentary rocks and occur in igneous rocks as well, even in individual plutons (O'Sullivan and Parrish 1995). Just going from end-member F-apatite to Durango to Fish Canyon spans a T_c range of about 30 °C. The large uncertainties can make it tempting to lump kinetically diverse grains into a single population for modeling, and the software will often still find reasonable-looking fits. However, modeling a kinetically diverse population using a single set of kinetics has been compared to putting one's hand in the oven and foot in the freezer and claiming to feel about right on average.

Powell et al. (2017) provide a nice example of the potential upside of thoroughly addressing variable annealing

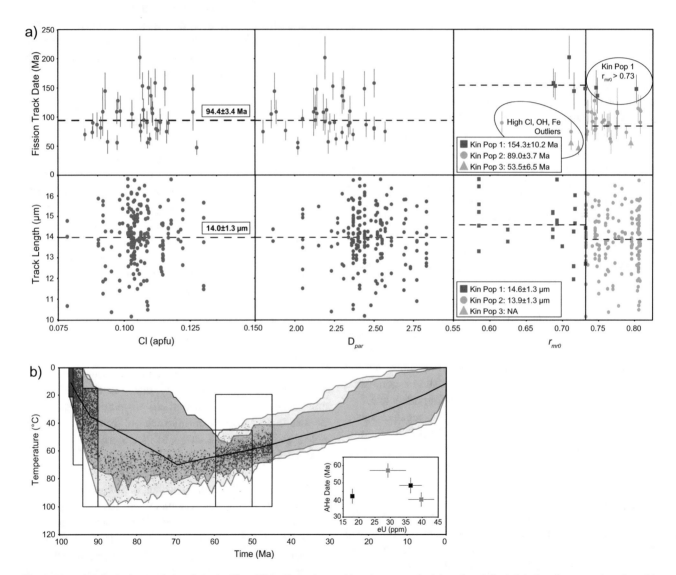

Fig. 3.13 **a** Apatite fission-track data from the Slater River Formation, Mackenzie Plain, Northwest Territory, Canada, from Powell et al. (2017), combining their Figs. 3 and S1. **b** HeFTy thermal history inverse model showing broad regions of possible thermal histories (pink and green envelopes, encompassing good and acceptable histories, respectively) and peak burial (red and green points from 90 to 60 Ma). Inset shows associated apatite (U−Th)/He dates and their relation to effective uranium content; points in black included in fitting (from Powell et al. 2017, Fig. 7)

kinetics (Fig. 3.13). Single-grain ages of apatites from a bentonite layer in the Upper Cretaceous Slater River Formation in the Mackenzie Plain of the Canadian Northwest Territories fail the chi-squared test, and neither Cl content nor etch figures differentiate them. However, a multi-compositional analysis using r_{mr0} (Eq. 3.11d) provided a means to separate their measurements into two statistically significant kinetic populations with distinct track length distributions (Fig. 3.13a), with the main culprit being F–OH mixing. This separation allowed them to estimate the timing and extent of maximum burial with a reasonable degree of confidence (Fig. 3.13b) using HeFTy (Ketcham 2005). That being said, the separation is not perfect and still features outliers, likely due to the need for further progress in understanding apatite annealing kinetics.

The associated apatite (U–Th)/He (AHe) grain data shown in the Fig. 3.13b inset provide an instructive illustration of the vagaries of working with FT and helium systems for thermal history modeling. Because the error bars on the AHe single-grain ages are small, it is impossible to have a single model that encompasses them all to within their respective uncertainties, based on our current knowledge of how that thermochronometric system works; there is evidently some process or factor we do not yet understand and is not included in our modeling, that is responsible for this excess variation. Thus, the choices are to leave the AHe data out entirely, to choose which data points to fit and provide a reason for the choice (as the authors did), or to fit all of the data and tolerate the misfit. There are no really satisfying options, and the third, although often chosen, is arguably the worst one, as it abdicates thinking and lets the statistics interpret ignorance as excess noise. However, if those had been apatite FT single-grain ages, the uncertainties in the ages would have been substantially larger, and we may not have even been aware of a problem.

3.11 Future Directions

This chapter has documented a great deal of progress in our understanding of FT annealing and tried to point out both nagging questions and prospective avenues for research in the near and medium term. In apatite, continued work on the "seasoning" effect and on kinetics will help to put our thermal history modeling on an ever-firmer foundation and likely provide new means for extracting even more information from our data. Zircon has received detailed experimental and analytical attention but relatively sparse application; the recent incorporation of length-based modeling into thermal history inversion software should help address this and may help clarify abiding issues in that system. The most pressing of these are probably the effects of alpha decay and recoil and how the various damage types evolve and interact.

An interesting, and probably underappreciated, insight that arises from the research described in this chapter is how similar FT annealing seems to be in apatite and zircon, from the effects of anisotropy to the success of a fanning curvilinear model form in describing and extrapolating from laboratory data to the differences between induced and spontaneous tracks. These similarities provide another argument against there being fundamental differences in their structure or annealing mechanisms. In fact, they probably point to insufficiently tapped reservoirs within the distinct communities specializing in each mineral for informing and helping each other, as similar problems in such areas as etching and anisotropy have led to very different, but all carefully considered, solutions.

Acknowledgements I thank M. Tamer for help with data transcription for drafting figures, and R. Yamada for providing the zircon FT length data. Thorough and thoughtful reviews by the editors, T. Tagami, and particularly R. Jonckheere, helped improve the manuscript.

References

Afra B, Lang M, Bierschenk T, Rodriguez MD, Weber WJ, Trautmann C, Ewing RC, Kirby N, Kluth P (2014) Annealing behaviour of ion tracks in olivine, apatite and britholite. Nucl Instr Meth Phys Res B 326:126–130

Afra B, Lang M, Rodriguez MD, Zhang F, Giulian R, Kirby N, Ewing RC, Trautmann C, Toulemonde M, Kluth P (2011) Annealing kinetics of latent particle tracks in Durango apatite. Phys Rev B 83:064116

Afra B, Rodriguez MD, Lang M, Ewing RC, Kirby N, Trautmann C, Kluth P (2012) SAXS study of ion tracks in San Carlos olivine and Durango apatite. Nucl Instr Meth Phys Res B 286:243–246

Barbarand J, Carter A, Wood I, Hurford AJ (2003) Compositional and structural control of fission-track annealing in apatite. Chem Geol 198:107–137

Bernet M (2009) A field-based estimate of the zircon fission-track closure temperature. Chem Geol 259:181–189

Bertagnolli E, Keil R, Pahl M (1983) Thermal history and length distribution of fission tracks in apatite: part I. Nucl Tracks 7:163–177

Box GEP, Cox DR (1964) An analysis of transformations. J Royal Statistical Soc B 26:211–252

Braddy D, Hutcheon ID, Price PB (1975) Crystal chemistry of Pu and U and concordant fission track ages of lunar zircons and whitlockites. In: Merrill RB, Hubbard NJ, Mendell WW, Williams RJ (eds) Proceedings of the 6th lunar science conference. Pergamon Press, New York, United States, pp 3587–3600

Brandon MT, Roden-Tice MK, Garver JI (1998) Late Cenozoic exhumation of the Cascadia accretionary wedge in the Olympic Mountains, northwest Washington State. Geol Soc Am Bull 110:985–1009

Burtner RL, Nigrini A, Donelick RA (1994) Thermochronology of lower Cretaceous source rocks in the Idaho-Wyoming thrust belt. Am Assoc Petrol Geol Bull 78:1613–1636

Carlson WD (1990) Mechanisms and kinetics of apatite fission-track annealing. Am Mineral 75:1120–1139

Carlson WD, Donelick RA, Ketcham RA (1999) Variability of apatite fission-track annealing kinetics I: experimental results. Am Mineral 84:1213–1223

Carpéna J (1992) Fission track dating of zircon: zircons from mont blanc granite (French-Italian Alps). J Geol 100:411–421

Chadderton LT (2003) Nuclear tracks in solids: registration physics and the compound spike. Rad Meas 36:13–34

Challandes N, Marquer D, Villa IM (2008) P-T-t modelling, fluid circulation, and ^{39}Ar-^{40}Ar and Rb-Sr mica ages in the Aar Massif shear zones (Swiss Alps). Swiss J Geosci 101:269–288

Clauser C, Giese P, Huenges E, Kohl T, Lehmann H, Rybach L, Šafanda J, Wilhelm H, Windloff K, Zoth G (1997) The thermal regime of the crystalline continental crust: Implications from the KTB. J Geophys Res 102:18417–18441

Corrigan JD (1993) Apatite fission-track analysis of Oligocene strata in South Texas, U.S.A.: testing annealing models. Chem Geol 104:227–249

Coyle DA, Wagner GA, Hejl E, Brown RW, van den Haute P (1997) The Cretaceous and younger thermal history of the KTB site (Germany): apatite fission-track data from the Vorbohrung. Geol Rundsch 86:203–209

Crowley KD, Cameron M, Schaefer RL (1991) Experimental studies of annealing etched fission tracks in fluorapatite. Geochim Cosmochim Acta 55:1449–1465

Dartyge E, Duraud JP, Langevin Y, Maurette M (1981) New model of nuclear particle tracks in dielectric materials. Phys Rev B 23:5213–5229

Donelick RA (1991) Crystallographic orientation dependence of mean etchable fission track length in apatite: an empirical model and experimental observations. Am Mineral 76:83–91

Donelick RA, Farley KA, Asimow P, O'Sullivan PB (2003) Pressure dependence of He diffusion and fission-track annealing kinetics in apatite?: experimental results. Geochim Cosmochim Acta 67:A82

Donelick RA, Ketcham RA, Carlson WD (1999) Variability of apatite fission-track annealing kinetics II: crystallographic orientation effects. Am Mineral 84:1224–1234

Donelick RA, Roden MK, Mooers JD, Carpenter BS, Miller DS (1990) Etchable length reduction of induced fission tracks in apatite at room temperature ($\sim$23 °C): Crystallographic orientation effects and "initial" mean lengths. Nucl Tracks 17:261–265

Duddy IR (1997) Focussing exploration in the Otway basin: understanding timing of source rock maturation. Aust Pet Prod Explor Assoc J 37:178–191

Duddy IR, Green PF, Laslett GM (1988) Thermal annealing of fission tracks in apatite 3. Variable temperture behaviour. Chem Geol 73:25–38

Durrani SA, Bull RK (1987) Solid state nuclear track detection. Pergamon, Oxford

Enkelmann E, Jonckheere R, Wauschkuhn B (2005) Independent fission-track ages (f-ages) of proposed and accepted apatite age standards and a comparison of f-, Z-, z-, and z_0- ages: Implications for method calibration. Chem Geol 222:232–248

Fleischer RL, Price PB (1964) Glass dating by fission fragment tracks. J Geophys Res 69:331–339

Fleischer RL, Price PB, Walker JD (1965a) Effects of temperature, pressure, and ionization of the formation and stability of fission tracks in minerals and glasses. J Geophys Res 70:1497–1502

Fleischer RL, Price PB, Walker RM (1965b) Ion explosion spike mechanism for formation of charged-particle tracks in solids. J Appl Phys 36:3645–3652

Fleischer RL, Price PB, Walker RM (1975) Nuclear tracks in solids; principles and applications. University of California Press, Berkeley, California, United States

Flowers RM, Ketcham RA, Shuster DL, Farley KA (2009) Apatite (U-Th)/He thermochronometry using a radiation damage accumulation and annealing model. Geochim Cosmochim Acta 73:2347–2365

Gallagher K (1995) Evolving temperature histories from apatite fission-track data. Earth Planet Sci Lett 136:421–435

Gallagher K (2012) Transdimensional inverse thermal history modeling for quantitative thermochronology. J Geophys Res 117:B02408

Garver JI (2003) Etching zircon age standards for fission-track analysis. Rad Meas 37:47–53

Garver JI, Kamp PJJ (2002) Integration of zircon color and zircon fission-track zonation patterns in orogenic belts: application to the Southern Alps, New England. Tectonophysics 349:203–219

Garver JI, Reiners PW, Walker LJ, Ramage JM, Perry SE (2005) Implications for timing of andean uplift from thermal resetting of radiation-damaged zircon in the Cordillera Huayhuash, Northern Peru. J Geol 113:117–138

Gautheron C, Tassan-Got L, Barbarand J, Pagel M (2009) Effect of alpha-damage annealing on apatite (U-Th)/He thermochronology. Chem Geol 266:157–170

Geisler T, Pidgeon RT, Van Bronwijk W, Pleysier R (2001) Kinetics of thermal recovery and recrystallization of partially metamict zircon: a Raman spectroscopic study. Eur J Mineral 13:1163–1176

Girstmair A, Ritter W, Märk E, Märk TD (1984) High temperature fission track annealing in natural fluorapatite. Nucl Tracks 8:381–384

Gleadow AJW (2014) Thermochronology of the future. In: 14th International conference on thermochronology, Chamonix-Mont Blanc, pp 3–4

Gleadow AJW, Belton DX, Kohn BP, Brown RW (2002) Fission track dating of phosphate minerals and the thermochronology of apatite. Rev Mineral Geochem 48:579–630

Gleadow AJW, Duddy IR (1981) A natural long-term track annealing experiment for apatite. Nucl Tracks 5:169–174

Gleadow AJW, Duddy IR, Green PF, Lovering JF (1986) Confined fission track lengths in apatite: a diagnoastic tool for thermal history analysis. Contrib Mineral Petrol 94:405–415

Gleadow AJW, Harrison TM, Kohn BL, Lugo-Zazueta R, Phillips D (2015) The fish canyon tuff: a new look at an old low-temperature thermochronology standard. Earth Planet Sci Lett 424:95–108

Gleadow AJW, Hurford AJ, Quaife RD (1976) Fission track dating of zircon: improved etching techniques. Earth Planet Sci Lett 33:273–276

Gögen K, Wagner GA (2000) Alpha-recoil track dating of quaternary volcanics. Chem Geol 166:127–137

Green PF (1988) The relationship between track shortening and fission track age reduction in apatite: combined influences of inherent instability, annealing anisotropy, length bias and system calibration. Earth Planet Sci Lett 89:335–352

Green PF, Duddy IR, Gleadow AJW, Tingate PR, Laslett GM (1985) Fission-track annealing in apatite: track length measurements and the form of the Arrhenius plot. Nucl Tracks 10:323–328

Green PF, Duddy IR, Gleadow AJW, Tingate PR, Laslett GM (1986) Thermal annealing of fission tracks in apatite 1. A qualitative description. Chem Geol 59:237–253

Green PF, Duddy IR, Laslett GM (1988) Can fission track annealing in apatite be described by first-order kinetics? Earth Planet Sci Lett 87:216–228

Green PF, Duddy IR, Laslett GM, Hegarty KA, Gleadow AJW, Lovering JF (1989) Thermal annealing of fission tracks in apatite 4. Quantitative modeling techniques and extension to geological time scales. Chem Geol 79:155–182

Grove M, Harrison TM (1996) ^{40}Ar* diffusion in Fe-rich biotite. Am Mineral 81:940–951

Guedes S, Moreira PAFP, Devanathan R, Weber WJ, Hadler JC (2013) Improved zircon fission-track annealing based on reevaluation of annealing data. Phys Chem Min 40:93–106

Guenthner WR, Reiners PW, Ketcham RA, Nasdala L, Giester G (2013) Helium diffusion in natural zircon: radiation damage, anisotropy, and the interpretation of zircon (U-Th)/He thermochronology. Am J Sci 313:145–198

Haack U (1978) The stability of fission tracks in epidote and vesuvianite. Earth Planet Sci Lett 30:129–134

Haack UK, Potts MJ (1972) Fission track annealing in garnet. Contrib Mineral Petrol 34:343–345

Hansen J, Sato M, Russell G, Kharecha P (2013) Climate sensitivity, sea level and atmospheric carbon dioxide. Phil Trans R Soc London A 371:20120294

Harrison TM, Célérier J, Aikman AB, Hermann J, Heizler MT (2009) Diffusion of ^{40}Ar in muscovite. Geochim Cosmochim Acta 73:1039–1051

Hasebe N, Mori S, Tagami T, Matsui R (2003) Geological partial annealing zone of zircon fission-track system: additional constrains from the deep drilling MITI-Nishikubiki and MITI-Mishima. Chem Geol 199:45–52

Hasebe N, Tagami T, Nishimura S (1994) Towards zircon fission-track thermochronology: Reference framework for confined track length measurements. Chem Geol 112:169–178

Hashemi-Nezhad SR, Durrani SA (1983) Annealing behaviour of alpha-recoil tracks in biotite mica: implications for alpha-recoil dating method. Nuclear Tracks 7:141–146

Hendricks BWH, Redfield TF (2005) Apatite fission track and (U-Th)/He data from Fennoscandia: An example of underestimation of fission track annealing in apatite. Earth Planet Sci Lett 236:443–458

Huang WH, Walker RM (1967) Fossil alpha-particle recoil tracks: a new method of age determination. Science 155:1103–1106

Hughes JM, Cameron M, Crowley KD (1989) Structural variations in natural F, OH, and Cl apatites. Am Mineral 74:870–876

Hughes JM, Cameron M, Crowley KD (1990) Crystal structures of natural ternary apatite: solid solution in the $Ca_5(PO_4)_3X$ (X = F, OH, Cl) system. Am Mineral 75:295–304

Hurford AJ (1986) Cooling and uplift patterns in the Lepontine Alps South Central Switzerland and an age of vertical movement on the Insubric fault line. Contrib Mineral Petrol 92:413–427

Hurford AJ (2018) An historical perspective on fission-track thermochronology (Chapter 1). In: Malusà MG, Fitzgerald PG (eds) Fission-track thermochronology and its application to geology. Springer

Issler DR (1996) An inverse model for extracting thermal histories from apatite fission track data: instructions and software for the Windows 95 environment. Geolgical Survey Canada, p 84

Ito H, Tanaka K (1995) Insights on the thermal history of the Valles caldera, New Mexico: evidence from zircon fission-track analysis. J Volcan Geotherm Res 67:153–160

Jaskierowicz G, Dunlop A, Jonckheere R (2004) Track formation in fluorapatite irradiated with energetic cluster ions. Nucl Instr Meth Phys Res B 222:213–227

Jenkin GRT, Ellam RM, Rogers G, Stuart FM (2001) An investigation of closure temperature of the biotite Rb-Sr system: The importance of cation exchange. Geochim Cosmochim Acta 65:1141–1160

Jonckheere R (2003a) On methodical problems in estimating geological temperature and time from measurements of fission tracks in apatite. Rad Meas 36:43–55

Jonckheere R (2003b) On the densities of etchable fission tracks in a mineral and co-irradiated external detector with reference to fission-track dating of minerals. Chem Geol 200:41–58

Jonckheere R, Tamer MT, Wauschkuhn B, Wauschkuhn F, Ratschbacher L (2017) Single-track length measurements of step-etched fission tracks in Durango apatite: "Vorsprung durch Technik". Am Mineral 102

Jonckheere R, van den Haute P, Ratschbacher L (2015) Standardless fission-track dating of the Durango apatite age standard. Chem Geol 417:44–57

Kasuya M, Naeser CW (1988) The effect of α-damage on fission-track annealing in zircon. Nucl Tracks 14:477–480

Ketcham RA (2003) Observations on the relationship between crystallographic orientation and biasing in apatite fission-track measurements. Am Mineral 88:817–829

Ketcham RA (2005) Forward and inverse modeling of low-temperature thermochronometry data. Rev Mineral Geochem 58(1):275–314

Ketcham RA, Carter A, Hurford AJ (2015) Inter-laboratory comparison of fission track confined length and etch figure measurements in apatite. Am Mineral 100:1452–1468

Ketcham RA, Carter AC, Donelick RA, Barbarand J, Hurford AJ (2007a) Improved measurement of fission-track annealing in apatite using c-axis projection. Am Mineral 92:789–798

Ketcham RA, Carter AC, Donelick RA, Barbarand J, Hurford AJ (2007b) Improved modeling of fission-track annealing in apatite. Am Mineral 92:799–810

Ketcham RA, Donelick RA, Carlson WD (1999) Variability of apatite fission-track annealing kinetics III: extrapolation to geological time scales. Am Mineral 84:1235–1255

Ketcham RA, Mora A, Parra M (2016) Deciphering exhumation and burial history with multi-sample down-well thermochronometric inverse modelling. Basin Res (early view)

Kohlmann F, Kohn BL, Gleadow AJW, Siegle R (2013) Scanning force microscopy of ^{129}Iodine surface impact structures in muscovite, zircon and apatite as proxies for damage of simulated fission fragments in solids. Rad Meas 51–52:83–91

Kohn BP, Belton DX, Brown RW, Gleadow AJW, Green PF, Lovering JF (2003) Comment on: "Experimental evidence for teh pressure dependence of fission track annealing in apatite" by A.S. Wendt et al. [Earth Planet. Sci. Lett. 201 (2002) 593–607]. Earth Planet Sci Lett 215:299–306

Kohn BP, Lorencak M, Gleadow AJW, Kohlmann F, Raza A, Osadetz KG, Sorjonen-Ward P (2009) A reappraisal of low-temperature thermochronology of the eastern Fennoscandia shield and radiation-enhanced apatite fission-track annealing. Geol Soc Spec Publ 324:193–216

Kohn B, Chung L, Gleadow A (2018) Fission-track analysis: field collection, sample preparation and data acquisition (Chapter 2). In: Malusà MG, Fitzgerald PG (eds) Fission-track thermochronology and its application to geology. Springer

Kozlovsky YA (1984) The superdeep well of the Kola Peninsula. Springer-Verlag

Lal N, Parshad R, Nagpaul KK (1977) Fission track annealing characteristics of garnet. Lithos 10:129–132

Lang M, Lian J, Zhang F, Hendricks BWH, Trautmann C, Neumann R, Ewing RC (2008) Fission tracks simulated by swift heavy ions at crustal pressures and temperatures. Earth Planet Sci Lett 274:355–358

Laslett GM, Galbraith RF (1996) Statistical modelling of thermal annealing of fission tracks in apatite. Geochim Cosmochim Acta 60:5117–5131

Laslett GM, Gleadow AJW, Duddy IR (1984) The relationship between fission track length and track density in apatite. Nucl Tracks 9:29–38

Laslett GM, Green PF, Duddy IR, Gleadow AJW (1987) Thermal annealing of fission tracks in apatite 2. A quantitative analysis. Chem Geol 65:1–13

Laslett GM, Kendall WS, Gleadow AJW, Duddy IR (1982) Bias in measurement of fission-track length distributions. Nucl Tracks 6:79–85

Li N, Wang L, Sun K, Lang M, Trautmann C, Ewing RC (2010) Porous fission fragment tracks in fluorapatite. Phys Rev B 82:144109

Li W, Kluth P, Schauries D, Rodriguez MD, Zhang F, Zdorvets MV, Trautmann C, Ewing RC (2014) Effect of orientation on ion track formation in apatite and zircon. Am Mineral 99:1127–1132

Li W, Lang M, Gleadow AJW, Zdorvets MV, Ewing RC (2012) Thermal annealing of unetched fission tracks in apatite. Earth Planet Sci Lett 321–322:121–127

Li W, Wang L, Lang M, Trautmann C, Ewing RC (2011) Thermal annealing mechanisms of latent fission tracks: Apatite vs. zircon. Earth Planet Sci Lett 302:227–235

Malusà MG, Fitzgerald PG (2018) From cooling to exhumation: setting the reference frame for the interpretation of thermocronologic data (Chapter 8). In: Malusà MG, Fitzgerald PG (eds) Fission-track thermochronology and its application to geology. Springer

Marsellos AE, Garver JI (2010) Radiation damage and uranium concentration in zircon as assessed by Raman spectroscopy and neutron irradiation. Am Mineral 95:1192–1201

McDowell FW, McIntosh WC, Farley KA (2005) A precise ^{40}Ar-^{39}Ar reference age for Durango apatite (U-Th)/He and fission-track dating standard. Chem Geol 214:249–263

Murakami M, Yamada R, Tagami T (2006) Short-term annealing characteristics of spontaneous fission tracks in zircon: a qualitative description. Chem Geol 227:214–222

Naeser CW (1981) The fading of fission tracks in the geologic environment—data from deep drill holes. Nucl Tracks 5:248–250

Naeser CW, Engels JC, Dodge FC (1970) Fission track annealing and age determination of epidote minerals. J Geophys Res 75:1579–1584

Naeser CW, Faul H (1969) Fission track annealing in apatite and sphene. J Geophys Res 74:705–710

Naeser CW, Fleischer RL (1975) Age of the apatite at Cerro de Mercado, Mexico: a problem for fission-track annealing corrections. Geophys Res Lett 2:67–70

Naeser CW, Forbes RL (1976) Variation of fission track ages with depth in two deep drill holes. Eos 57:363

Nasdala L, Reiners PW, Garver JI, Kennedy AK, Stern RA, Balan E, Wirth R (2004) Incomplete retention of radiation damage in zircon from Sri Lanka. Am Mineral 89:219–231

Nasdala L, Wenzel M, Vavra G, Irmer G, Wenzel T, Kober B (2001) Metamictization of natural zircon: accumulation versus thermal annealing of radioactivity-induced damage. Contrib Mineral Petrol 141:125–144

O'Nions RK, Griesshaber E, Oxburgh ER (1989) Rocks that are too hot to handle. Nature 341:391

O'Sullivan PB, Parrish RR (1995) The importance of apatite composition and single-grain ages when interpreting fission track data from plutonic rocks: a case study from the Coast Ranges, British Columbia. Earth Planet Sci Lett 132:213–224

Ohishi S, Hasebe N (2012) Observations of fission-tracks in zircons by atomic force microscope. Rad Meas 47:548–556

Ouchani S, Dran JC, Chaumont J (1997) Evidence of ionization annealing upon helium-ion irradiation of pre-damaged apatite. Nucl Instr Meth Phys Res B 132:447–451

Parker P, Cowan R (1976) Some properties of line segment processes. J Appl Prob 13:255–266

Parshad R, Saini HS, Nagpaul KK (1978) Fission track etching and annealing phenomenon in phlogopite and their applications. Can J Earth Sci 15:1924–1929

Paul TA, Fitzgerald PG (1992) Transmission electron microscopic investigation of fission tracks in fluorapatite. Am Mineral 77:336–344

Powell JW, Schneider DA, Issler DR (2017) Application of multi-kinetic apatite fission track and (U-Th)/He thermochronology to source rock thermal history: a case study from the Mackenzie Plain, NWT, Canada. Basin Res (early view)

Rabone JAL, Carter A, Hurford AJ, De Leeuw NH (2008) Modelling the formation of fission tracks in apatite minerals using molecular dynamics simulations. Phys Chem Min 35:583–596

Rabone JAL, De Leeuw NH (2007) Molecular dynamics simulations of fission track annealing in apatite. Geochim Cosmochim Acta 71: A816

Rahn MK, Brandon MT, Batt GE, Garver JI (2004) A zero-damage model for fission-track annealing in zircon. Am Mineral 89:473–484

Ravenhurst CE, Roden-Tice MK, Miller DS (2003) Thermal annealing of fission tracks in fluorapatite, chlorapatite, manganoapatite, and Durango apatite: experimental results. Can J Earth Sci 40:995–1007

Reiners PW (2009) Nonmonotonic thermal histories and contrasting kinetics of multiple thermochronometers. Geochim Cosmochim Acta 73:3612–3629

Reiners PW, Farley KA, Hickes HJ (2002) He diffusion and (U-Th)/He thermochronometry of zircon: initial results from Fish Canyon Tuff and Gold Butte. Tectonophysics 349:297–308

Saini HS, Nagpaul KK (1979) Annealing characteristics of fission tracks in minerals and their applications to earth sciences. Int J Appl Rad Isotop 30:213–231

Schauries D, Afra B, Bierschenk T, Lang M, Rodriguez MD, Trautmann C, Li W, Ewing RC, Kluth P (2014) The shape of ion tracks in natural apatite. Nucl Instrum Methods Phys Res, Sect B 326:117–120

Schauries D, Lang M, Pakarinen OH, Botis S, Afra B, Rodriguez MD, Djurabekova F, Nordlund K, Severin D, Bender M, Li WX, Trautmann C, Ewing RC, Kirby N, Klutha P (2013) Temperature dependence of ion track formation in quartz and apatite. Appl Crystall 46:1558–1563

Soulet S, Carpena J, Chaumont J, Kaitasov O, Ruault MO, Krupa JC (2001) Simulation of the α-annealing effect in apatitic structures by He-ion irradiation: influence of the silicate/phosphate ratio and of the OH^-/F^- substitution. Nucl Instr Meth Phys Res B 184:383–390

Spiegel C, Kohn BL, Raza A, Rainer T, Gleadow AJW (2007) The effect of long-term low-temperature exposure on apatite fission track stability: a natural annealing experiment in the deep ocean. Geochim Cosmochim Acta 71:4512–4537

Stormer JCJ, Pierson ML, Tacker RC (1993) Variation of F and Cl X-ray intensity due to anisotropic diffusion in apatite during electron microprobe analysis. Am Mineral 78:641–648

Storzer D (1970) Fission track dating of volcanic glasses and the thermal history of rocks. Earth Planet Sci Lett 8:55–60

Stübner K, Jonckheere R, Ratschbacher L (2015) The densities and dimensions of recoil-track etch pits in mica. Chem Geol 404:52–61

Szenes G (1995) General features of latent track formation in magnetic insulators irradiated with swift heavy ions. Phys Rev B 51:8026–8029

Tagami T, Carter A, Hurford AJ (1996) Natural long-term annealing of the zircon fission-track system in Vienna Basin deep borehole samples: constraints upon the partial annealing zone and closure temperature. Chem Geol 130

Tagami T, Galbraith RF, Yamada R, Laslett GM (1998) Revised annealing kinetics of fission tracks in zircon and geological implications. In: van den Haute P, De Corte F (eds) Advances in fission-track geochronology. Kluwer Academic Publishers, Netherlands, pp 99–112

Tagami T, Ito H, Nishimura S (1990) Thermal annealing characteristics of spontaneous fission tracks in zircon. Chem Geol 80:159–169

Tello CA, Palissari R, Hadler JC, Iunes PJ, Guedes S, Curvo EAC, Paulo SR (2006) Annealing experiments on induced fission tracks in apatite: Measurements of horizontal-confined track lengths and track densities in basal sections and randomly oriented grains. Am Mineral 91:252–260

Trachenko K, Dove MT, Salje EKH (2002) Structural changes in zircon under a-decay irradiation. Phys Rev B 65:180101–180103

Turnbull D (1956) Phase Changes. In: Seitz F, Turnbull D (eds) Solid state physics. Academic Press, New York, pp 226–309

Villa IM (1998) Isotopic closure. Terra Nova 10:42–47

Villa IM, Puxeddu M (1994) Geochronology of the Larderello geothermal field: new data and the "closure temperature" issue. Contrib Mineral Petrol 115:415–426

Vrolijk P, Donelick RA, Queng J, Cloos M (1992) Testing models of fission track annealing in apatite in a simple thermal setting: site 800, leg 129. In: Larson RL, Lancelot Y (eds) Proceedings of the ocean drilling program, scientific results. Ocean Drilling Program, College Station, TX, pp 169–176

Wagner GA, Reimer GM (1972) Fission track tectonics: the tectonic interpretation of fission track apatite ages. Earth Planet Sci Lett 14:263–268

Wauschkuhn B, Jonckheere R, Ratschbacher L (2015a) The KTB apatite fission-track profiles: Building on a firm foundation? Geochim Cosmochim Acta 167:27–62

Wauschkuhn B, Jonckheere R, Ratschbacher L (2015b) Xe- and U-tracks in apatite and muscovite near the etching threshold. Nucl Instr Meth Phys Res B 343:146–152

Weber WJ (1990) Radiation-induced defects and amorphization in zircon. J Mater Res 5:2687–2697

Weber WJ, Ewing RC, Wang LM (1994) The radiation-induced crystalline-to-amorphous transition in zircon. J Mater Res 9:688–698

Wendt AS, Vidal O, Chadderton LT (2002) Experimental evidence for the pressure dependence of fission track annealing in apatite. Earth Planet Sci Lett 201:593–607

Wesch W, Wendler E (2016) Ion beam modification of solids; ion-solid interaction and radiation damage. In: Car R, Ertl G, Freund HJ, Lüth H, Rocca MA (eds) Springer series in surface sciences. Springer, Switzerland, p 534

Yamada R, Murakami M, Tagami T (2007) Statistical modeling of annealing kinetics of fission tracks in zircon; reassessment of laboratory experiments. Chem Geol 236:75–91

Yamada R, Tagami T, Nishimura S (1993) Assessment of overetching factor for confined fission-track length measurement in zircon. Chem Geol 104:251–259

Yamada R, Tagami T, Nishimura S (1995a) Confined fission-track length measurement of zircon: assessment of factors affecting the paleotemperature estimate. Chem Geol 119:293–306

Yamada R, Tagami T, Nishimura S, Ito H (1995b) Annealing kinetics of fission tracks in zircon. Chem Geol 122:249–258

Yuan W, Ketcham RA, Gao S, Dong J, Bao Z, Deng J (2009) Annealing behavior of alpha recoil tracks in phlogopite. Chem Geol 266:352–358

Zachos JC, Dickens GR, Zeebe RE (2008) An early Cenozoic perspective on greenhouse warming and carbon-cycle dynamics. Nature 415:279–283

Zattin M, Bersani D, Carter A (2006) Raman microspectroscopy: a nondestructive tool for routine calibration of apatite composition for fission-track analyses. In: European conference on thermochronology, Bremen, Germany

Zhang M, Salje EKH, Capitani GC, Leroux H, Clark AM, Schlüter J, Ewing RC (2000) Annealing of α-decay damage in zircon: a Raman spectroscopic study. J Phys: Condens Matter 12:3131–3148

Ziegler JF, Biersack JP, Ziegler MD (2008) SRIM the stopping and range of ions in matter, v05 edn. SRIM Co., Chester, Maryland

The Future of Fission-Track Thermochronology

Andrew Gleadow, Barry Kohn and Christian Seiler

Abstract

The methods of fission-track (FT) thermochronology, based on a combination of the external detector method, zeta calibration against independent age standards and measurements of horizontal confined track lengths, have undergone relatively little change over the last 25 years. This conventional approach has been highly successful and the foundation for important thermal history inversion methods, supporting an expanding range of geological applications. Several important new technologies have emerged in recent years, however, that are likely to have a disruptive effect on this relatively stable approach, including LA-ICP-MS analysis for ^{238}U concentrations, new motorised digital microscopes and new software systems for microscope control, digital imaging and image analysis. These technologies allow for new image-based and highly automated approaches to FT dating and eliminate the need for neutron irradiations. Together they are likely to have a major influence on the future of FT analysis and gradually replace the older, highly laborious manual methods. Automation will facilitate the acquisition of larger and more comprehensive data sets than was previously possible, assist with standardisation and have important implications for training and distributed analysis based on image sharing. Track length measurements have been more difficult to automate, but 3D measurements and automated semi-track length measurements are likely to become part of future FT methods. Other important trends suggest that FT analysis will increasingly be combined with other isotopic dating methods on the same grains, and multi-system methods on coexisting minerals, to give much more comprehensive accounts of the thermal evolution of rocks. There are still a range of important fundamental issues in FT analysis that are poorly

understood, such as a full understanding of the effects of composition and radiation damage on the annealing properties of different minerals, which are likely to be fruitful areas for future research in this field.

4.1 Introduction

A variety of experimental methods emerged during the early development of FT dating to measure the essential ratio of the ^{238}U spontaneous fission track density to the parent uranium concentration. Mostly the uranium concentration could not be measured explicitly, but rather the track density of induced ^{235}U fission tracks resulting from a thermal neutron irradiation was used as a proxy (Fleischer et al. 1975). By the end of the 1980s, just one of these experimental methods, the external detector method (EDM) (e.g. Naeser and Dodge 1969; Gleadow 1981; see Chap. 2, Kohn et al. 2018) had come to dominate the practice of FT analysis. As discussed below, there were several reasons for the success of this method, but by far the most significant was that it provided age information on individual ~ 100–$200\ \mu m$ mineral grains, the first of any geochronological technique to routinely do so.

The EDM has become central to what we may now describe as conventional FT analysis, a combination of procedures that has progressively become the standard, and almost universally adopted, approach to FT dating in laboratories around the world. In this approach, FT ages are determined by the EDM using the empirical 'zeta' calibration against a set of independent age standards (Fleischer and Hart 1972; Hurford and Green 1983), combined with measurements of horizontal confined track lengths (Gleadow et al. 1986) usually measured using a microscope drawing tube and digitising tablet. In essence, these two complementary data sets comprising the FT grain-age distribution and the FT length distribution, contain, respectively, the time information and the temperature information that are

A. Gleadow (✉) · B. Kohn · C. Seiler
School of Earth Sciences University of Melbourne, Victoria, 3010, Australia
e-mail: gleadow@unimelb.edu.au

required by forward modelling schemes to reconstruct the thermal history of a particular sample (Fig. 4.1). This approach has been widely applied to the common uranium-bearing accessory minerals apatite, titanite and zircon, of which apatite has been by far the most important, because of its common occurrence, its simple and consistent etching behaviour and its well-documented annealing properties relative to other minerals. Apatite will therefore be the principal focus in the rest of this chapter, although most of the discussion would apply equally to other minerals.

There has been little change in this conventional approach to FT thermochronology over a period now approaching 25 years, although important advances have been made in

understanding FT annealing and in the development of multi-compositional annealing models over this time, accompanied by a comprehensive description of the statistics of FT analysis (Galbraith 2005; see Chap. 6, Vermeesch 2018). Together, these advances are applied in the inverse modelling techniques used for thermal history reconstruction from FT data (Ketcham 2005; Gallagher 2012). The recognition that apatite FT annealing properties vary with composition (e.g. Green et al. 1985; Barbarand et al. 2003) means that measurement of some specific 'kinetic parameter' for each grain is now also required. The two kinetic parameters in common use are the average diameter of track etch pits parallel to the c-axis, Dpar (Donelick 1993;

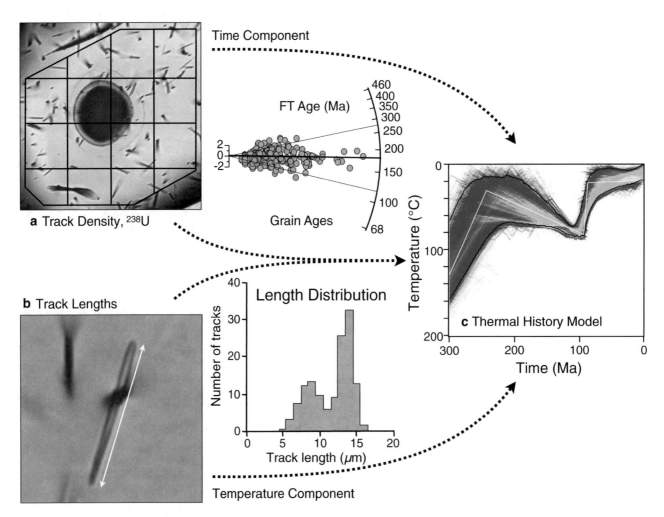

Fig. 4.1 Analytical sequence required for the reconstruction of thermal histories from primary measurements of **a**, the number, and **b**, the lengths of fission tracks in a mineral sample, together with the ^{238}U concentration, here indicated by a circular laser ablation pit in (**a**). Image **a** includes a region of interest outlining the area to be counted and an internal grid to assist in counting. Image **b** shows a confined fission track and its measured length. In the centre, a radial plot shows the distribution all of grain ages, and a histogram the distribution of all lengths measured in a sample. These results can be thought of as primarily reflecting the time component, and the temperature

component of the underlying thermal history, respectively. The thermal history reconstructed from such data is shown in the temperature–time plot, **c**, on the right. In practice, this FT analytical sequence is currently fully developed only for apatite and is the basis of the conventional approach described here. Future developments in FT thermochronology, however, are likely to follow a similar sequence, but will include new technologies for collecting the data, greater use of automation, richer data sets and routine integration of the results with other thermochronometers

Donelick et al. 2005) and/or the chlorine concentration of the apatite, usually measured by electron probe micro-analysis (EPMA, Green et al. 1985). With these additional refinements, the conventional approach has provided a stable platform that has underpinned major growth in the field, and a widening diversity of applications.

Following the general adoption of this standardised approach, the dominant emphasis of most FT studies over the past decade or more has been, appropriately, on geological applications, rather than on ongoing development of the methods themselves, or probing remaining areas of uncertainty. There is a danger in this state of affairs, however, that FT analysis may be applied in a formulaic way with little understanding of some of the more fundamental assumptions and unresolved issues that underlie the technique. To continue in this mode would be to invite stagnation, but, fortunately, important new trends have been emerging over the last decade, which are likely to have a disruptive influence on FT analysis as it is now practiced. Ultimately, these changes have the potential to improve the quality and consistency of FT data and thereby provide new insights into the thermochronology of the continental crust.

The early part of the twenty-first century has seen the emergence of several important new technologies including (i) laser ablation inductively coupled plasma mass spectrometry (LA-ICP-MS), (ii) fully motorised digital microscopes that allow all microscopy functions to be placed under computer control and (iii) practical image analysis techniques based on captured and stored digital images obtained using (ii). Underlying all of these, of course, is the steady advance of desktop computing power and, crucially, the availability of abundant, cheap disc storage for captured digital images. These developments are likely to have a major impact, and indeed come to dominate, the foreseeable future of FT analysis, and are the central subject of this chapter. A number of other important trends, in the way that FT data are used and combined with other thermochronometers, will also be discussed below.

4.2 The Current State of FT Analysis

The stability and practical convenience of the conventional approach to FT analysis have had some important advantages, particularly in providing single grain-age information and the ability to statistically test the coherence of the sampled age populations (Galbraith 1990). The induced tracks could be measured over exactly the same area counted for spontaneous tracks, and the zeta calibration largely settled, or at least avoided, earlier disputes about the ^{238}U spontaneous fission decay constant, and neutron dosimetry calibrations. Horizontal confined track lengths were shown by Laslett et al. (1982) to give the closest approximation to

the true underlying distribution of unetched fission-track lengths. The widespread adoption of this approach has also aided consistency in data production, although the degree to which this has been universally achieved is doubtful.

There are also a number of disadvantages to the conventional approach, however. The most significant of these include the very long sample turnaround times, typically months, necessitated by neutron irradiation, and the accompanying laboratory safety issues arising from handling radioactive materials. Moreover, the procedure is highly laborious, with long hours needed at the microscope to measure the three track densities and confined track lengths required for each sample. Even then, the number of counts and lengths are relatively limited, thereby reducing the precision of the measurements. The method also requires a major commitment to operator training, typically over a period of months, in order to develop the appropriate levels of expertise. In addition, the quality of measurements is significantly dependent on operator experience and requires an individual calibration. Finally, the zeta calibration means the technique is dependent on other methods and therefore is not an absolute dating method in its own right.

Reference to widely available age standards has brought a measure of consistency to FT age determinations between individual researchers, but recent inter-laboratory comparison experiments have shown an alarming lack of similar consistency when it comes to track length measurements (Ketcham et al. 2015). There are a variety of factors contributing to this situation, but one crucial issue is the lack of any comparable 'length standards', and indeed it is currently hard to see how an independent reference value for FT lengths could be obtained. A resolution is urgently needed to this problem so that track length measurements, and the thermal history models based on them, can be accepted with confidence. This must be a major short-term goal for the future of FT analysis.

4.3 LA-ICP-MS Analysis of Uranium Concentrations

The progressive refinement and decreasing costs of access to LA-ICP-MS technologies during the 1990s has led to their increasingly widespread availability to researchers in the twenty-first century. This technology provides an alternative to neutron irradiation for analysing the uranium concentrations of mineral grains with ppm sensitivity and a spatial resolution comparable to fission-track dimensions. The first significant adaptations of this technology to FT analysis (Hasebe et al. 2004, 2013) showed this to be a practical and highly effective approach, and, as a result, it is now being progressively adopted by an increasing number of FT groups. The major advantages of this method include the speed of analysis, direct analysis of parent ^{238}U, rather than using ^{235}U as a proxy, at least comparable precision and accuracy to the conventional approach, and elimination of the need to handle irradiated materials. There can be no doubt that LA-ICP-MS will be an important part of the future of FT thermochronology, a change being accelerated by the increasing difficulty of access to suitable neutron irradiation facilities, following closure of many research reactors around the world.

One of the most important features of using the LA-ICP-MS method is that only a single track density measurement (spontaneous track density, ρ_s) is required, compared to three for the EDM, which, given the relatively large uncertainties associated with most track density measurements, has the potential to improve the precision of measurements using the new approach. So far, it appears that comparable precision is being obtained using either method (e.g. Seiler et al. 2014), although this has not yet been rigorously tested and further developments in LA-ICP-MS technology are likely to see the precision from this method exceeding that from conventional analyses.

An important issue raised by the adoption of LA-ICP-MS in FT analysis is that it requires rethinking of how the FT age equation is applied (Hasebe et al. 2004; Donelick et al. 2005; Gleadow et al. 2015; see also Chap. 2, Kohn et al. 2018). As with most new technologies the initial adoption of this approach has tended to follow earlier practices and most reported measurements have used a 'modified zeta' calibration based on reference to age standards (Hasebe et al. 2013; Gleadow and Seiler 2015). However, from the beginning it has been clear that the emergence of this new technology gives an important opportunity to revisit the question of using an absolute calibration, based on explicit use of the constituent constants making up this 'modified zeta' (e.g. Hurford 1998; Soares et al. 2014; Gleadow et al. 2015).

Many of these constituent constants, such as Avogadro's number, or the density of apatite are well known, but other, such as the FT detection efficiency in different materials, and indeed, the spontaneous ^{238}U decay constant, are not. The near future of FT analysis is therefore likely to include renewed experimental efforts to ascertain precise values for these constants (e.g. Jonckheere and van den Haute 2002; Soares et al. 2014). Another important consideration that will be essential for the future evolution of this method will be the development of a series of widely accessible, matrix-matched compositional standards for the analysis of apatite and other FT minerals.

Interestingly, this version of the FT age equation (see Chap. 6, Vermeesch 2018, Eq. 6.25) also explicitly requires the average etchable range of a single fission fragment, which can be approximated by half the confined track length. This is the key geometric parameter required to relate the planar, 2D,

spontaneous fission track density to the 3D volume concentration of ^{238}U atoms in the mineral. We might then ask which value should be used for this length factor. If one is using a modified zeta calibration, then this factor is implicitly assumed to be the characteristic track length in the age standards used, typically around 14.5–15.0 µm. Used this way the results obtained are directly comparable to those from the conventional, zeta calibrated EDM, which also makes the same implicit assumption about the track length component.

The question then arises about what is the significance of the FT age calculated in the conventional way when applied to samples that have a mean track length that is quite different to the age standards used for the zeta calibration. In such cases, the FT age calculated with the conventional approach is in fact a kind of 'model age', rather than the 'true' FT age anticipated in the derivation of the FT age equation. This model age is the FT age that would be obtained for a sample, if it had the same length distribution as the age standards used. In the great majority of samples, however, this assumption is invalid as the average track length is mostly shorter, sometimes significantly so, than in the age standards. In almost all samples, therefore, if the actual etchable fission fragment range was used in the equation, a significantly older age will result because this length factor occurs on the denominator of the age equation. In the conventional, or the 'modified zeta' LA-ICP-MS approach, the importance of the length component is obscured by the zeta calibration, and the calculated 'model' age will be nearly always younger than the 'true' FT ages based on the actual mean track length.

This is an important question and one that is likely to be debated in ongoing discussions about FT thermochronology, especially as fission track results are increasingly integrated with other thermochronometers, where comparable assumptions and 'model age' calculations do not apply. At one level, this is not a major issue for reconstructed thermal histories, because these histories are derived from the actual length distributions. However, there can be no doubt this 'model age' calculation is also responsible for many of the apparent inversions observed in the expected relative sequence of apatite FT and apatite (U–Th)/He ages, because the two ages are not being calculated in the same way. One of the benefits that may come from more widespread adoption of the LA-ICP-MS approach will be a clearer understanding of what we actually mean by a 'FT age'.

The adoption of LA-ICP-MS, of course, raises a range of new practical considerations and calibration issues of its own, which are still being worked through. As was the case with the conventional approach, the development of standard analytical protocols and appropriate reference materials for U analysis will take time and is likely to be a fertile research field in the near future. At present, this is largely being pursued on a laboratory-by-laboratory basis, but there are important opportunities for greater inter-laboratory standardisation and adoption of a more unified approach. Such issues as the optimum ablation parameters (e.g. power levels, pulse rate and ablation pattern), potential heterogeneous ablation, controlling ablation pit depth, and detecting and monitoring compositional zoning, have not yet been fully explored for their potential influence on FT ages derived by this method compared to other applications of LA-ICP-MS. Similarly, the statistics of FT dating using LA-ICP-MS has lagged behind that of the EDM but is now receiving much needed attention (Chap. 6, Vermeesch 2018).

The number of direct comparisons between the conventional and LA-ICP-MS approaches is still quite limited although the concordance between the two is generally excellent (e.g. Hasebe et al. 2004; Seiler et al. 2014; Gleadow et al. 2015). LA-ICP-MS can analyse for many elements simultaneously, so that this new method could progressively build an analytical database on apatite compositions and etching properties that could be of significant value to future annealing models for apatite and other minerals. The issue of well-calibrated, homogeneous and matrix-matched standards for LA-ICP-MS is another area that is currently limiting the more effective utilisation of this approach.

One exciting possibility might be to measure the concentration of chlorine and other elements that might influence annealing kinetics in apatites simultaneously with the measurement of ^{238}U. Chlorine has an extremely high ionisation potential and therefore has a much higher detection limit in apatite (Chew and Donelick 2012), but early experiments (Chew et al. 2014) show great promise for this to become a standard procedure in future. This would have the great advantage for FT analysis of eliminating an entire analytical procedure, electron probe micro-analysis (EPMA) that is currently required to measure the halogen content in apatite. EPMA for fluorine and, to a lesser extent, chlorine can be problematic, and very careful analytical protocols are required to produce satisfactory results (e.g. Goldoff et al. 2012). It is not hard to imagine this cumbersome additional analytical step being replaced by LA-ICP-MS in the near future. With either method, well-calibrated compositional standards are required to produce satisfactory results, but the method still lacks suitable international reference materials (e.g. Chew et al. 2016).

4.4 Computer-Controlled Digital Microscopy

The emergence of a new generation of fully motor-driven research microscopes over the past decade is a major advance that has the potential to put all aspects of FT microscopy under computer control. For the first time, the most advanced research microscopes can assist the operator with virtually all routine microscope tasks. Coupled with high-quality digital imaging and appropriate control software, these instruments can provide fully autonomous microscopy and multiple slide-handling capabilities that largely eliminate the long periods at the microscope, previously required for the conventional approach. Such motorised microscopes are also allowing new kinds of track measurements to be made that were not practical, or even possible, using earlier manual systems (e.g. Jonckheere et al. 2017).

This trend towards automated microscopy actually began with the development by Gleadow et al. (1982) of a computer-controlled three-axis stage system that substantially improved the efficiency of mirror-image matching between grain mounts and muscovite track detectors in the external detector method. Alternative manual solutions to the mirror-image matching problem in the EDM have also been found, including laying out grains in a grid pattern prior to mounting, and the precise relocation of the etched detector over the grain mount and counting the two sets of tracks at two different focal levels—the so-called sandwich technique (Jonckheere et al. 2003). Both require considerable manual dexterity and patience by the analyst, and neither has been widely adopted. In most laboratories, various versions and generations of either two-, or three-axis motorised stage systems have become part of the conventional EDM approach (Smith and Leigh-Jones 1985; Dumitru 1993).

The capabilities of modern digital microscopes, however, provide a fundamental change in the opportunities for precise computer control, and image-based analysis that can transform the way FT microscopy is conducted. These capabilities include motor-driven objective nosepieces, reflector turrets, diaphragms, filter wheels, illumination shutters, condenser components, stage movements and focusing. The focus motors are particularly important as they now give control on the vertical or z-axis with a precision of a few tens of nanometres, enabling the capture of vertical stacks of images (z-stacks) spaced at a fraction of a micrometre. In this way, it is possible to image the full 3D structure of etched fission tracks in transmitted light and of the etched surface in reflected light at very high magnification. Essentially all of the information available to the operator in conventional microscopy can now be permanently captured in digital form for display on a high-resolution monitor, or subsequent digital image analysis.

Precise control of the vertical movement of the microscope stage is of fundamental importance for automatic focusing and for digital imaging of fission tracks. The ability to capture images in multiple image planes at sub-micron spacing of typically 100–300 nm means that the full 3D character of etched tracks can be captured conveniently and quickly at high magnification for the first time. In this, it is vital that the microscope is stable against both vibrations of the bench and thermal expansion of the microscope body. In some locations, it may be necessary to isolate the microscope from building vibrations using some form of damping table or tablet, although this is mostly not required.

Thermal expansion of the microscope body can be a significant problem for precise, sub-micron, focus control. Large research microscopes, especially as more and more electronics and powerful incandescent light sources have been added to the basic frame, can show substantial thermal expansion over the temperature range typically encountered in the operating environment. If this factor is not controlled, then the distance between the specimen on a stage and the objective can change by up to 20 μm over time in larger format microscopes, simply due to expansion of the microscope frame, as illustrated in Fig. 4.2. The most advanced microscopes now compensate for this thermal expansion and remain stable during an imaging session, but analysts need to be aware that not all motorised microscopes have successfully controlled this problem. High-intensity LED lamps are now appearing that greatly reduce the amount of heat generated close to the microscope and also have the advantage of exhibiting a constant colour temperature over their entire power range, giving a constant white-balance in the captured images.

In addition to motorised control of the microscope itself, there have been key improvements in digital cameras for microscope use over the last few years. Most microscope cameras over the past decade have used CCD sensors that produced images of excellent quality and resolution, but had very slow frame rates of only a few frames per second. This has been an important limitation on the speed at which functions such as autofocus and digital imaging could be carried out, and has been the rate-limiting step in autonomous image capture for FT analysis. In the last few years, however, this has changed significantly with the emergence of a new generation of CMOS sensors with similar image quality and resolution, but very high frame rates of tens to even hundreds of frames per second, and making use of faster data transfer protocols, such as USB3 and GigE. This technological advance has greatly reduced the time required for digital image acquisition in FT analysis.

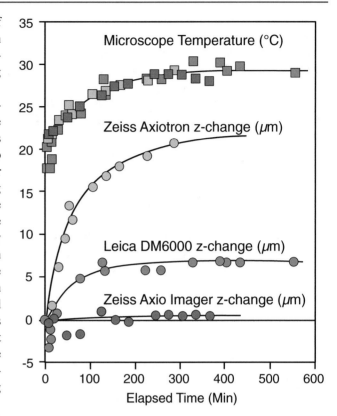

Fig. 4.2 Thermal expansion in three motor-driven, large-format research microscopes that have been used for FT analysis. The upper group of square symbols shows how the temperature of the microscope stand, measured using an infrared non-contact thermometer, is observed to change as a function of elapsed time since the microscope was switched on. Perhaps surprisingly, all three microscopes follow a very similar temperature increase from a room temperature of ∼20 °C to eventually reach a stable operating temperature of nearly 30 °C after about 4 h. For each of the three microscopes, a specimen was set in focus at the start and then the z-axis adjustment in μm required to bring the specimen back into focus was recorded at different times up to 300–550 min. The lower three curves show the amount of adjustment required due to this thermal expansion of the microscope stand. The early 1990s Axiotron microscope shows an expansion of over 20 μm, consistent with the thermal expansion expected in the aluminium microscope body over this temperature range. In contrast, the two modern motorised microscopes (DM6000 and Axio Imager) show only a small, or negligible degree of expansion as temperature increases, although both have some initial instability, before stabilizing. This, along with vibrational stability, is an important consideration for reliably capturing image stacks at sub-micron spacing for automated FT analysis

4.5 Automation in FT Analysis

The goal of automating FT analysis has had an enduring appeal, and the highly labour-intensive and operator-dependent nature of the required microscopy makes the automation of routine tasks particularly desirable. The potential benefits of an automated system, in addition to relieving the operator of the need for long periods at the microscope, include advantages in providing more consistent data collection, in reducing the dependence on operator experience, and accelerated training for new practitioners. Another factor driving the need for change is that microscope drawing tubes, a key hardware component for conventional track length measurements, are now obsolete and no longer available. Ultimately, the goal of any automated system must be to significantly increase the efficiency of analysis and improve the quality of the data obtained. Many thermochronology studies are also demanding larger quantities of data than can easily be met with traditional analytical approaches.

Until recently, however, the goal of comprehensive automation of FT analysis has remained difficult and impractical, not least because the necessary hardware was not yet available. Early attempts at developing automated systems therefore required specially constructed hardware for the task (e.g. Wadatsumi et al. 1988; Wadatsumi and Masumoto 1990) and principally for this reason have had very little subsequent impact on the field. This situation has now changed with new digital microscopes and computing resources, leading to significant progress in the development of automated analysis systems over the past decade. Important conceptual and software advances are now also being made (e.g. Gleadow et al. 2009a, 2015) that allow these hardware developments to be fully utilised. Importantly, both hardware and software continue to develop rapidly and it is almost certain that automated FT analysis will play a major role in the future of the field.

Several research groups are now pursuing the goal of FT automation including automatic track counting (Gleadow et al. 2009a, 2015; Kumar 2015), automatic or assisted location of suitable grains for counting (Gleadow et al. 2009b; Booth et al. 2015), determination of crystallographic orientation of grains (Peternell et al. 2009; Gleadow et al. 2009b, 2015) and the characterisation of fission track etch pits (Reed et al. 2014; Gleadow et al. 2015). A novel alternative to these automated methods (Vermeesch and He 2016) seeks to use crowdsourcing for counting tracks in captured digital images to average interpersonal differences in fission track counts, while also generating larger data sets than would be possible by a single analyst alone. Many of these approaches are still at an early, experimental stage of development, and the only one now fully operational as an integrated FT analytical system is that developed at the University of Melbourne, which will be detailed here as a case study.

The approach developed in Melbourne is divided into two components. The first is a microscope control system (*TrackWorks*) that autonomously captures a comprehensive set of digital images on previously marked mineral grains, and the second is an image analysis and review system (*FastTracks*) that processes the captured and stored images offline on a computer. The operator is required mainly to supervise a brief setting up routine at the microscope, and finally to review the results obtained from the image analysis on a computer and make any corrections or adjustments required.

In this system, the automated track counting is achieved by an image analysis routine called '*coincidence mapping*', described in detail by Gleadow et al. (2009a), based on a digital superimposition of transmitted and reflected light images. After applying a fast Fourier transform bandpass filter to flatten background variations, the two images are segmented by applying an automatic threshold value to produce a binary image separating 'features' from 'background'. These images include both track and non-track features, but the 'coincidence' between the transmitted and reflected images, derived by extracting only those elements that are common to both, is almost entirely fission tracks. Spurious non-track features, such as dust particles, polishing scratches and inclusions, are rarely observed equally in both reflected and transmitted light images, and the few that may be erroneously counted can easily be removed during a rapid manual review of the results.

The best results are obtained with this image analysis technique when careful attention is given to consistent sample preparation. The most suitable mounts have flat polished surfaces with little grain relief, obtained by using diamond, rather than alumina, polishing compounds. A more significant factor is that polished apatite grains commonly show strong internal reflections when observed in incident light, due to their very low surface reflectivity, which degrades the reflected light image required for coincidence mapping. Applying a thin metal coating, typically by depositing a thin (~ 10 nm) layer of gold onto the etched surface, eliminates these internal reflections. This coating step is most important for apatite, but reflected light imaging of zircon, titanite and muscovite external detectors, are all improved by a reflective coating. Applying the metal coating is easily implemented using either a sputter coater or vacuum-evaporation unit and greatly improves automatic track counting without being dependent on operator skill or experience.

The current Melbourne system is designed to work with either the conventional EDM or the LA-ICP-MS methods, but has particular advantages when coupled with laser ablation because it produces a permanent digital replica of the etched grain surfaces, which are otherwise partially destroyed at the ablation step. The optical information in both transmitted and reflected light, previously accessible only in real time at the microscope, is still available for later observation and analysis through the stored images. The locations of coordination markers and analysed grains can also be exported to a laser ablation or electron microprobe stage, so that further geochemical analysis is streamlined.

As was typical in earlier microscope stage systems (Gleadow et al. 1982; Smith and Leigh-Jones 1985; Dumitru 1993), grain coordinates need to be related to an internal reference frame defined by coordination markers attached to or engraved in the surface of each slide or grain mount. In the Melbourne system, three Cu electron microscope grids are usually attached to each slide for this purpose. The process of locating and accurately centring these markers can itself be automated, so that multiple slides can be loaded on the microscope stage and coordinated automatically (Fig. 4.3).

The task of locating optimally oriented grains (*c*-axis parallel) of a suitable size for FT analysis can be automated using digital imaging of grains under circular polarised light

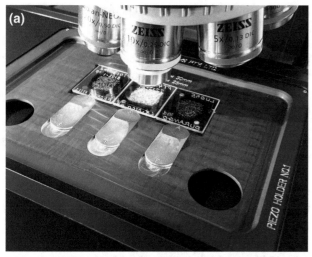

Fig. 4.3 Upper image **a** shows a microscope stage with multiple slide insert from *Autoscan Systems* to carry three fission track mounts for automatic slide coordination, automatic grain detection and autonomous grain imaging. This insert is fitted to a piezo-motor stage on a Zeiss Axio Imager microscope. **b** is an enlarged view of the same three apatite grain mounts showing the three Cu electron microscope grids attached to each that are used as coordination markers. The grey band across each is a thin (~ 10 nm) gold coating deposited on the surface using a sputter coater to enhance the surface reflectivity, while still allowing for observation and imaging in transmitted light (Gleadow et al. 2009a). These custom microscope slides have dimensions of 25 × 30 mm, a size that better allows for multiple slide handling and imaging on a normal sized stage. Later versions of this stage insert will carry six slides of this size at one time

(Gleadow et al. 2009b) where the brightness of similarly sized grains is controlled by their orientation. In this observation mode, the brightest grains will be those with their *c*-axes parallel to the polished surface. A particular advantage of automatic grain detection is that it provides an unbiased selection of grains for analysis, independent of where tracks are located, or whether a particular grain contains any tracks at all. This factor is especially important for low track densities where inappropriate grain selection can significantly bias an age determination. Once a suitable set of grains have been selected for analysis, an automatic image capture sequence can be run autonomously over perhaps 40–80 grains with no further need for operator supervision. The digital image sets captured over a batch of several slides can then be then archived to disc for later retrieval and image analysis.

4.5.1 Automatic FT Counting

An advantage of any automated image analysis system working on captured images is that the processing can be done independently of the microscope on a different computer. In the Melbourne system, image analysis is typically an interactive process that starts with running the automatic *coincidence mapping* routine for track counts, adjusting measurement parameters and filter settings as necessary, and reviewing the results. A typical workflow involved in using the *FastTracks* system for a complete analysis of all parameters is shown schematically in Fig. 4.4, and representative results are shown

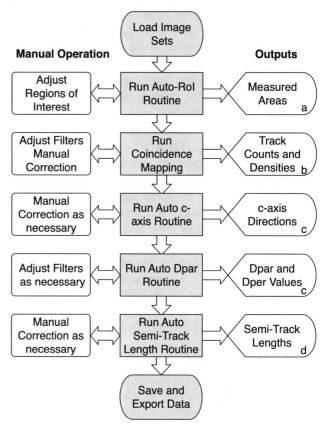

Fig. 4.4 Flow chart showing the sequence of steps involved in conducting a typical automatic FT analysis session using the *FastTracks* image processing system. The system operates on a set of images captured autonomously using a motorised digital microscope in transmitted and reflected light using the *TrackWorks* control package. Even though most of the analysis is made automatically, some input or review from the operator is usually required at various steps, as indicated in the left side column. Outputs are shown on the right, and **a–d** correspond to results in the image overlays in Fig. 4.5. The procedure is essentially the same if the EDM is used, with the addition that tracks are also counted over the paired mica external detector images. Confined FT lengths can be measured at any time though this sequence using 3D measurement tools but are usually measured on a separate set of captured images, either with or without implantation of ^{252}Cf fission tracks to enhance the yield

overlaid on some transmitted light images for a Durango apatite sample in Fig. 4.5.

In this sequence, the first step would involve setting a region of interest (ROI) on each grain (Fig. 4.5a), to define the area for analysis and exclude areas that are unsuitable for counting such as regions outside the grain boundary, areas that are poorly illuminated, or with fractures, inclusions, etc. In a departure from the simple rectangular shapes dictated by eyepiece grids used in conventional microscopy, an automated system can use a ROI of any arbitrary size or shape, or even multiple ROIs within a particular field of view to avoid interference from features such as fractures or inclusions.

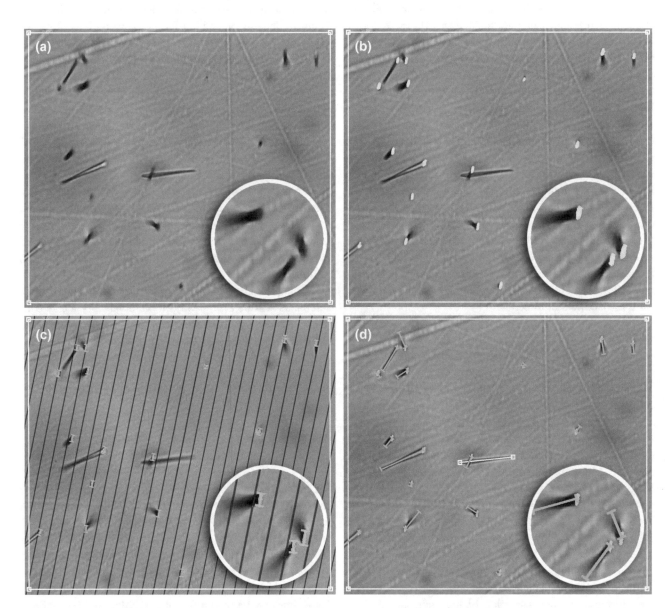

Fig. 4.5 Typical sequence of images with overlaid results obtained using the automated steps in Fig. 4.4. Each image shows a different step in the procedure and the enlarged circle in each frame shows more detail of the lower right area. **a** shows etched spontaneous fission tracks in a surface transmitted light image in a sample of Durango apatite. The white box is the region of interest (ROI) that defines the area within which the analysis is carried out. **b** shows the automatically counted tracks from the *coincidence mapping* algorithm with their surface intersections highlighted. **c** shows the automatically detected c-axis azimuth direction highlighted by the dark blue parallel lines, and a small bar on each track etch pit shows the automatically determined Dpar values. **d** shows the results of a prototype automated semi-track length measurement tool, the overlaid lines being the surface projection of the semi-track lengths that are determined in 3D. For each of these semi-tracks, the system determines the true length from the centroid of the surface etch pit to the end of the track, the true angle to the c-axis, the projected length, the depth to the end and the true dip angle below the surface (corrected for refractive index). This image also shows the length of a single horizontal confined track near the centre that has been manually measured by clicking on each end of the track at high magnification. None of the automated measurements in **b–d** have had any manual correction, although this is often required for a small fraction of the tracks

The area of the ROI can be simply determined from the number of pixels contained so that an accurate track density can be calculated for each grain.

Once image analysis for automatic track counting is complete (Fig. 4.5b), typically taking less than one minute for 30–40 grains using *FastTracks*, the results may be reviewed and corrected if necessary by the operator for any misidentified features or uncounted tracks (Gleadow et al. 2009a). Such features typically make up only a few per cent of the total, and the manual review can be carried out quite quickly. A worthwhile focus for future research will be to develop additional image analysis routines to correctly identify the small proportion of tracks, mostly very shallow dipping tracks, which may be overlooked or misidentified. Machine learning techniques (Kumar 2015, Donelick and Donelick 2015) may be useful in this regard. After an automatic count has been completed, further automatic image analysis routines can be used to determine other parameters from the stored image sets, such as the *c*-axis direction and Dpar values (Fig. 4.5c). This secondary analysis typically takes only seconds once the initial image processing has been carried out, and, since the *c*-axis direction is known within the selected prismatic grain surfaces, the orientation of all other features can be automatically referred to that direction.

4.5.2 Confined FT Length Measurements

The most time-consuming part of measuring the lengths of confined tracks is not actually the length measurement as such, but the location of suitable horizontal confined tracks. It is this location task that makes automation of length measurements significantly more difficult than automatic counting. Measuring the positions of track ends is relatively trivial to do manually, but can be carried out much more precisely on greatly enlarged digital images than on a live microscope image using a drawing tube and digitiser. An automated microscopy system can greatly assist in this process, but so far no fully automated system has emerged. One partial solution is for the operator to locate suitable confined tracks for length measurement and simply mark their positions during inspection of the grain mount. It is then straightforward to autonomously capture cropped image stacks locally around these locations for later measurement by manually clicking the cursor on each end of the confined track in the on-screen 3D image stack (Gleadow et al. 2015). The length, angle to the *c*-axis and dip angle can then be automatically recorded (Fig. 4.5d). Although not completely automated, such an image-based system has advantages in that a permanent record is made of the tracks measured, and much greater precision and consistency are possible, relative to a manual measurement under the microscope.

With this system, length measurements can be made accurately over a greater range of dip angles, compared to traditional 'horizontal' confined tracks, by recording both the position and the depth plane of the track ends in a transmitted light image stack. The depth component of the true length may be automatically corrected for the refractive index of apatite (Laslett et al. 1982; Gleadow et al. 2015). Such 3D lengths potentially introduce an additional source of sampling bias that still needs further investigation, because dipping tracks are more likely to intersect the surface, and therefore become disqualified as confined tracks if they are long than if they are short. However, the likely increase in sample size from 3D measurements makes understanding this potential bias a worthwhile objective. Even restricting the true dip angles to the usual 'horizontal' criterion of <10–15° (Ketcham et al. 2009; Laslett et al. 1982) will make the measurements more precise by correcting for that dip, thereby improving data quality and consistency. Indeed, because the actual dips of conventional 'horizontal' track lengths are poorly constrained, and greater than they appear due to the refractive index, the projected lengths recorded without reference to dip may include errors of up to several per cent, or even more.

An important goal for the future of track length measurements will be to find ways to automatically locate confined tracks for measurement, and to develop fully automated measurement tools. This is certainly the most challenging image analysis problem encountered in the automation of FT analysis so far, but there are no fundamental reasons why a solution should not be found.

4.5.3 Automated Semi-track Length Measurements

Recently, a possible alternative pathway to the automated collection of track length data has emerged, building on the theoretical understanding of the true lengths of 'semi-tracks' developed by Laslett and Galbraith (1996) and Galbraith (2005). Semi-tracks are those tracks intersecting the etched surface, the same features that are counted for the spontaneous track density measurement (Chap. 1, Hurford 2018). These tracks have the great advantage that, unlike confined tracks, their locations are already known from the automatic counting result, and the etch pit marking the counted feature also precisely defines the surface end of the track. The termination of the track can then be located by digital analysis of the captured image stack starting from the surface etch pit and tracing the feature until the end of the track is reached. Current attempts at automating this procedure are extremely promising (Fig. 4.5d). The number of semi-tracks available for measurement vastly exceeds the number of the inherently rare confined tracks by a factor of about 50–100, so that much larger length data sets could be collected automatically. Another advantage of semi-tracks is that they are all etched to the same degree, unlike confined tracks that usually show significant variation in the degree of etching because of the variable geometry of their etching access pathways.

Laslett and Galbraith (1996) have shown that, although semi-track lengths are less useful than confined track lengths, they nonetheless show distinctly different length distributions for different thermal history styles and are significantly more informative than 'projected' track lengths. This new source of track length information could therefore potentially be used for thermal history modelling. The statistical assessment of Laslett and Galbraith (1996) and Galbraith (2005) allows the semi-track length distribution corresponding to any particular confined length distribution to be calculated, opening the possibility that the inverse procedure can also be achieved by a Monte Carlo modelling approach. The ability to model the underlying confined track length distribution would also mean that, in principle, the results of both confined and semi-track length measurements could be combined in reconstructing a thermal history.

4.5.4 Other Benefits of Automated FT Analysis

So far, this discussion has been concerned with automation of measurements that have previously been made manually, albeit in a much less convenient way. Other measurements could be made by an automated system, however, that go beyond what has been done before. One such possibility would be the routine assessment of the widths and other dimensions of confined tracks and the semi-tracks from which they are etched. In principle, this might allow for a correction to be applied for the degree of etching of each individual track, and then calculating the etchable range of the latent track before etching began. Such 'unetched' lengths of confined tracks might give a more precise representation of the true underlying track length distribution and remove the effects of variable degrees of etching that currently blur the detail of confined track length distributions, and by implication the thermal histories derived from them.

Other significant benefits flow from a digital image-based system, most importantly in the area of training. Previously processed image sets provide a novel and highly efficient way for new analysts to be exposed to the detailed results from an experienced observer. The latest release of the *FastTracks* system, for example, includes a 'Training Extension' that allows a new analyst to import and overlay the results of a previous analysis of the same image set. This system makes a digital comparison with the previous results and highlights any discrepancies for consideration, correction or discussion. Experience in the Melbourne laboratory

has shown that this image-based automated system substantially reduces the training time for new students and other novice analysts to a matter of just weeks. Previously, using manual methods, it took many months for a new analyst to achieve a similar level of competence. Widespread sharing of such processed digital image sets between laboratories could therefore play a key role in training a new generation of FT analysts and, importantly, in standardisation between individuals and groups. Indeed, common measurements on the same track length images could provide widely distributed standard reference materials for track lengths that have so far been so conspicuously lacking in FT measurements (Ketcham et al. 2015).

The ability to rapidly transfer digital FT image sets over the Internet also means that completely new modes of distributed analysis can be imagined. In this way, a single centralised digital microscopic imaging facility could support a number of different analysts dispersed at remote locations. One well-equipped laboratory could thus potentially support a number of individual researchers and groups around the world who would only need an Internet connection and a computer with the appropriate software tools to undertake FT analysis on the downloaded image sets. The size of the comprehensive digital image packages currently in use is quite large, typically 5–10 GB per sample, which has implications for how easily these can be transferred. However, files of comparable size are already routinely downloaded today, so such transfers are unlikely to be a significant problem. Some image compression techniques are available to reduce the file sizes, and future implementation of high efficiency 'lossless' image compression techniques might further reduce the size of files for transfer. The so-called lossy compression formats, such as jpeg, are best avoided as they can introduce image artefacts and loss of detail during image processing. Of course, continuing improvements in network bandwidth will also assist in the process of efficient transfer of large files.

4.6 Future Technological Developments

New approaches in FT thermochronology such as those described here have often followed or coincided with the arrival of significant new technologies. Important ongoing developments can be anticipated in all the hardware components relevant to FT analysis, including digital microscopy, imaging systems, computational power, communications, and mass spectrometry for trace element and isotopic analysis. Such ongoing technological advances are likely to significantly influence the future evolution of FT analysis.

Continuing developments in automated digital microscopy are moving rapidly at the present time and likely to lead to a new generation of powerful autonomous instruments that will require less and less supervision by a human operator. One important trend already underway is the emergence of a new generation of 'headless' microscopes that are designed for fully automated scanning of large numbers of slides for later offline inspection and analysis. Several major microscope manufacturers are already producing digital microscopy systems that are entirely directed at on-screen examination and image capture. Some are designed to automatically scan up to 100 slides at a time and capture high-resolution digital images for biomedical and even petrographic applications. These systems are designed to operate without human intervention and therefore make no provision for a binocular head and eyepieces, greatly simplifying the design. The microscope systems of the future are likely to look very different to those now in use, and one can imagine a dedicated digital microscopic imaging system constructed specifically for the requirements of FT analysis. Such a system would need many fewer components and options compared to the general purpose instruments of today, potentially reducing costs and increasing the imaging capacity.

An intermediate step with current generation digital microscopes is the ability to automatically scan and image apatite grains over multiple slides, with substantial savings in operator time. Once set up, the present *TrackWorks* microscope control system can automatically locate slide coordination markers, detect suitable grains and autonomously capture image sets over three FT slides at a time. Increasing camera speeds, however, have meant that this capture time is now reduced from several hours, previously run overnight, to less than one hour for such a batch. A new generation of larger format motorised stages mean it should soon be possible to increase this autonomous capture capacity to six slides at a time, and further increases can be foreseen beyond that.

One other area where new developments in microscopy could have an impact on FT analysis is the emergence of various so-called super-resolution imaging techniques. These techniques are able to exceed the normal diffraction-limited resolution of conventional optics and could potentially provide more sharply resolved images for FT analysis. There are several such systems currently available, directed almost exclusively at biomedical applications and particularly fluorescence microscopy. None of these appear at present to be suitable for imaging fission tracks, but this is a very active area of innovation at present, and it is possible that a suitable technology might appear in future.

Developments in computing power and display technologies continue at a rapid pace so that even greater computational capacity will be available to replace current desktop and laptop computers, allowing ever more powerful image analysis procedures. Other opportunities for advanced image analysis could arise by utilising the massively parallel processing power of graphics processor cards to greatly accelerate performance, and also from unconventional computing architectures, such as neural networks and even quantum computing. The point here is not so much that greater speed of analysis is urgently needed, as it is already very fast, but these new technologies are likely to substantially expand the kinds and quality of data that can be collected and processed automatically. For example, higher-resolution imagery and more closely spaced image stacks could be captured and analysed. This has significant implications for the necessary image storage, but the capacity of computer storage is also continually increasing and becoming cheaper—essentially following 'Moore's Law'. Similarly, the capacity and speed of global communications capable of transferring very large data sets between laboratories will continue to increase. These developments will accelerate the opportunities for inter-laboratory collaboration, standardisation, training and distributed analysis.

The accuracy of uranium concentration measurements with current generation laser ablation ICP-MS instruments is typically limited to a few per cent. However, looking at the rapid development of this technology over the last 20 years gives confidence that it, and potentially other kinds of mass spectrometry systems, such as ICP time-of-flight (TOF) mass spectrometers, will continue to evolve, leading to increased precision and sensitivity for future applications in FT analysis. Such methods for the direct, rapid measurement of ^{238}U for FT analysis is also likely to be accompanied by greater levels of automation, faster analysis, better software integration, simultaneous multi-element capabilities and multi-system dating, all with many fewer demands on operator time.

One final comment here is that all of these developments at the most advanced level of FT analysis still depend on the basic requirement of crushing rock samples and separating their constituent minerals. These are still mostly carried out using quite primitive mechanical crushing, heavy liquid and magnetic separation techniques that are dirty, slow and potentially dangerous. Despite some developments in the emergence of non-toxic heavy liquids, these methods for the most part have changed little over the last century. One important new technology is the electrodynamic disaggregation of rocks with the *Selfrag* devices (Giese et al. 2010), which are starting to appear in some larger groups, but at a cost that puts them well out of reach for most laboratories. It is possible that this kind of disaggregation device could find a more widespread application as part of a centralised FT sample preparation and digital imaging facility as discussed above. Even so, the ongoing necessity of large-scale mineral separation suggests that a substantial research effort to develop new, innovative and highly automated approaches to this laborious task would surely be of great benefit to the analytical community.

4.7 Other Trends in FT Analysis

A number of other important trends can be discerned in the wider field of FT thermochronology that go beyond the technical data acquisition advances that have been the main emphasis here. An example is the increasing integration of FT analysis of apatite and zircon in thermochronology studies with other systems such as $^{40}Ar/^{39}Ar$ dating of co-genetic mineral phases (e.g. Carrapa et al. 2009; see Chap. 5, Danišík 2018). Another is the simultaneous modelling of multiple FT samples that have a known spatial relationship to each other, such as in a borehole or in a vertical sampling profile across topographic relief. The thermal history inversion will be much more powerfully constrained in such cases, and this approach is now being facilitated by the addition of multi-sample capabilities to modelling codes (e.g. Gallagher 2012). Beyond this, there are many exciting applications for thermochronology emerging through the integration of thermal history modelling with thermokinematic and surface process models through software such as PECUBE (Braun et al. 2012) to provide better-constrained and mutually consistent solutions to geological problems.

Similarly, FT analysis can be combined with other U-decay schemes applied to the same mineral grains in the so-called double dating (e.g. FT and U–Pb, or (U–Th)/He and U–Pb) or 'triple dating' (e.g. FT, (U–Th)/He and U–Pb) (see Chap. 5, Danišík 2018). This is another important trend that is likely to accelerate in coming years with the wider adoption of LA-ICP-MS techniques, which allows the measurement of multiple isotopic systems simultaneously with the measurement of ^{238}U for FT analysis (e.g. Carrapa et al. 2009; Shen et al. 2012). With advances in mass spectrometry, it is even possible to imagine a system that could simultaneously, or sequentially make in situ measurements of FT, (U–Th)/He, U–Th–Pb and even U–Xe ages in a single grain.

Within FT thermochronology itself, another trend is towards analysis of significantly larger numbers of grains in a sample, particularly for detrital applications (~ 100 grains, Bernet and Garver 2005; Carrapa et al. 2009). For this trend to continue requires greater analytical efficiency and sample throughput which would be greatly facilitated by the application of automation. Automatic grain detection, in particular, might help to ensure representative sampling of different age populations in a detrital suite and reduce the potential for operator bias in grain selection. The adoption of LA-ICP-MS for uranium measurements also gives the opportunity to analyse for a range of additional trace elements simultaneously, such as the REEs, which could help in fingerprinting different detrital sources and characterising their petrogenetic origins. Another possible trend, after several decades where FT analysis has been dominated by apatite, is an renewed interest in multi-mineral FT studies, including titanite, zircon and other previously studied minerals, and exploration of the potential of additional minerals, such as monazite which has potential for ultra-low temperature thermochronometry (Gleadow et al. 2002, 2005).

The development of new approaches and new technologies gives an opportunity for many fundamental issues in FT analysis to be revisited, new calibrations undertaken and new modelling strategies developed to take advantage of alternative kinds of data. New measurements of the track detection efficiency of apatite and other minerals are needed if FT analysis is to move away from a purely empirical calibration and become an independent absolute dating method (e.g. Jonckheere and van den Haute 2002; Soares et al. 2014). New statistical studies are needed to fully understand the implications of 3D track length measurements. If automated measurements of semi-track lengths become a routine option, as is highly likely, then new FT annealing models based directly on such measurements need to be developed, or alternatively, robust inversion methods to derive confined track length distributions from semi-track data. The influence of radiation damage on the annealing properties of zircon, titanite and other minerals is another area that requires further study. While this factor is increasingly being recognised as a significant influence in thermochronology, it is nonetheless very poorly understood.

There are also many aspects of the effects of mineral composition on FT annealing that have received almost no attention, and are worthy of further exploration. These include the potential influence of OH substitution in fluorapatites. The difficulty in measuring this anion has meant that it has been largely ignored to date, and the annealing properties of hydroxyapatites are almost unknown. One important observation, however, is that etching rates in diagenetic hydroxyapatites are comparable to those in chlorapatites, so it may be that some OH substitution could be responsible for the observed tendency for Dpar values to increase towards the fluorapatite end of the compositional spectrum (Green et al. 2005; Spiegel et al. 2007). Although anion substitutions, especially Cl, in apatite clearly have a dominant control on the annealing properties, various cation substitutions might also have an influence, such as REE, Sr and Mn (Carlson et al. 1999; Barbarand et al. 2003), which need to be investigated further. In other minerals, such as zircon and titanite we have scarcely begun to assess the influence of compositional factors on their annealing properties, and how these might interact with accumulating radiation damage. Some studies (Haack 1972; Dahl 1997; Carlson et al. 1999) have provided evidence of a systematic relationship between annealing properties and the underlying crystal chemistry of FT minerals reflected in such parameters as the ionic porosity. In principle, such relationships, if adequately understood, could allow the annealing properties of an unknown mineral to be predicted from its composition, which would be of great benefit. Current knowledge of the underlying mechanisms falls far short of such a capability, but further research into some of these fundamental questions is clearly warranted and could significantly influence the future of FT studies.

Another area that remains poorly understood is the relationship between the fundamental mechanisms of FT annealing at the atomic scale and the empirically observed behaviour of etched tracks. Several recent investigations are providing important new insights into the annealing behaviour of unetched fission tracks (Li et al. 2011, 2012; Afra et al. 2011) that are now starting to close this gap in our understanding. A complete convergence between studies of the properties of latent fission tracks and the properties of etched tracks is still some way off, but must surely be an important research goal for the future of FT thermochronology.

4.8 Conclusions

After a long period of relative stability in the widely used methods of FT thermochronology, a number of disruptive developments are now occurring that will have a formative effect on future practice in this field. Importantly, these have the potential for much more rapid analysis and improved data quality, which should in turn propagate through to more reliable thermal history reconstructions. These changes include the gradual adoption of LA-ICP-MS analysis for the direct determination of ^{238}U concentrations in mineral grains, rather than using induced ^{235}U fission tracks as a proxy, and the adoption of automated analytical methods based on digital images captured with a new generation of computer-controlled, motor-driven microscopes. Image analysis of captured image sets can provide automatic counting of fission tracks and a range of support features including automatic slide coordination, automatic grain detection and batch processing of multiple grain mounts at one time. The c-axis direction in uniaxial minerals such as apatite and zircon can also be determined automatically, as can the track etch pit size parameters Dpar and Dper. 3D confined track length measurements can then be automatically related to crystallographic orientation. New approaches to the automatic measurement of track lengths are now under development that will complete the analytical sequence required for automatic FT analysis. Significant improvements in all these methods can be anticipated, as is the involvement of an increasing number of groups in their development. This trend towards increasingly effective automation is likely to be a major component of future developments in FT analysis.

Importantly, these new approaches are not simply new ways of doing the same thing as before, but include opportunities for obtaining new and much richer data sets that were not previously practical to measure manually. This is likely to lead to more robust, consistent and better-standardised analyses than were previously possible, and much improved training procedures for new analysts. Sharing of digital image sets also means that much greater inter-laboratory standardisation will be possible, as will new distributed modes of analysis where one well-equipped laboratory could provide captured images for a number of other analysts in different places. Working with captured digital images of etched mineral grains has particular advantages when combined with laser ablation analysis, which is destructive of the grain surfaces on a micro-scale. LA-ICP-MS also enables a range of other isotopes to be analysed simultaneously with ^{238}U, including ^{232}Th, Pb isotopes and potentially ^{35}Cl as a kinetic parameter in apatite, which allows not only for more rapid and streamlined analytical methods, but also for multiple dating systems to be applied to single grains at the same time. Thus, the current trend towards so-called double and triple dating in apatite and zircon is likely to become a major part of the future of FT thermochronology.

Continued technical development in instrumentation, computational power and software can be expected to lead to further enhancements and new opportunities across all of these current trends in FT thermochronology. The outcome is expected to be a future FT analysis based on more comprehensive data sets, better standardisation and more robust thermal history inversions, as part of a broader multi-mineral, multi-system approach to thermochronology.

Acknowledgements Many of the ideas presented here have emerged from discussions with members of our research group at the University of Melbourne over many years, and we particularly thank David Belton, Rod Brown, Asaf Raza and Ling Chung for their input at various times. The group has received sustained funding to support its work over many years from the Australian Research Council (ARC) including grant LP0348767 with Autoscan Systems Pty Ltd which supported the initial development of automatic fission track counting. We also thank our software engineers, Stewart Gleadow, Artem Nicolayevski, Josh Torrance, Sumeet Ekbote and Tom Church who have implemented so much of our automated fission track analysis system. The group has also received support from the AuScope Program funded by the National Collaborative Research Infrastructure Strategy and the Education Investment Fund, which has provided dedicated LA-ICP-MS facilities and ongoing maintenance and operational support. Major equipment purchases for microscopes and laser ablation were also provided by ARC grant LE0882818. We thank Pieter Vermeesch and Noriko Hasebe for their very helpful reviews, which have significantly improved the manuscript.

References

Afra B, Lang M, Rodriguez MD, Zhang J, Giulian JR, Kirby N, Ewing RC, Trautmann C, Toulemonde M, Kluth P (2011) Annealing kinetics of latent particle tracks in Durango apatite. Phys Rev B 83:064116

Barbarand J, Carter A, Hurford AJ (2003) Compositional and structural control of fission-track annealing in apatite. Chem Geol 198:107–137

Bernet M, Garver JI (2005) Fission-track analysis of detrital zircon. Rev Mineral Geochem 58:205–238

Booth JT, Jones J, Schaeffer K, Woodall K, Kumar R, Dodds Z, Donelick R (2015) Mapping for microscopes: automating apatite-image handling. Goldschmidt Abstr 2015:341

Braun J, van der Beek P, Valla P, Robert X, Herman F, Glotzbach C, Pedersen V, Perry C, Simon-Labric T, Prigent C (2012) Quantifying rates of landscape evolution and tectonic processes by thermochronology and numerical modeling of crustal heat transport using PECUBE. Tectonophysics 524–525:1–28

Carlson WD, Donelick RA, Ketcham RA (1999) Variability of apatite fission-track annealing kinetics: I. Experimental results. Am Mineral 84:1213–1223

Carrapa B, DeCelles PG, Reiners PW, Gehrels GE, Sudo M (2009) Apatite triple dating and white mica $^{40}Ar/^{39}Ar$ thermochronology of syntectonic detritus in the Central Andes: a multiphase tectonothermal history. Geology 37:407–410

Chew DM, Donelick RA (2012) Combined apatite fission track and U-Pb dating by LA-ICP-MS and its application in apatite provenance studies. Mineral Assoc Canada Short Course 42:219–247 (St Johns, NL)

Chew D, Donelick RA, Donelick MB, Kamber BS, Stock MJ (2014) Apatite chlorine concentration measurements by LA-ICP-MS. Geostand Geoanal Res 38:23–35

Chew D, Babechuk MG, Cogné N, Mark C, O'Sullivan GJ, Henrichs IA, Doepke D, McKenna CA (2016) (LA, Q)-ICPMS trace-element analyses of Durango and McClure Mountain apatite and implications for making natural LA-ICPMS mineral standards. Chem Geol 435:35–48

Dahl PS (1997) A crystal-chemical basis for Pb retention and fission-track annealing systematics in U-bearing minerals, with implications for geochronology. Ear Planet Sci Lett 150:277–290

Danišík M (2018) Chapter 5. Integration of fission-track thermochronology with other geochronologic methods on single crystals. In: Malusà MG, Fitzgerald PG (eds) Fission-track thermochronology and its application to geology. Springer, Berlin

Donelick RA (1993) Apatite etching characteristics versus chemical composition. Nucl Tracks Radiat Meas 21:604

Donelick A, Donelick R (2015) Machine learning applied to finding and characterizing the tips of etched fission tracks. Goldschmidt Abstr 2015:759

Donelick RA, O'Sullivan PB, Ketcham RA (2005) Apatite fission-track analysis. Rev Mineral Geochem 58:49–94

Dumitru TA (1993) A new computer-automated microscope stage system for fission-track analysis. Nucl Tracks Radiat Meas 21:575–580

Fleischer RL, Hart HR (1972) Fission track dating: techniques and problems. In: Bishop WW, Miller DA, Cole S (eds) Calibration of hominid evolution. Scottish Academic Press, Edinburgh, pp 135–170

Fleischer RL, Price PB, Walker RM (1975) Nuclear tracks in solids. University of California Press, Berkeley, p 605

Galbraith RF (1990) The radial plot; graphical assessment of spread in ages. Nucl Tracks Radiat Meas 17:207–214

Galbraith RF (2005) Statistics for fission track analysis. Chapman & Hall, Boca Raton, p 219

Gallagher K (2012) Transdimensional inverse thermal history modeling for quantitative thermochronology. J Geophys Res 117:B02408

Giese J, Seward D, Stuart FM, Wüthrich E, Gnos E, Kurz D, Eggenberger U, Schruers G (2010) Electrodynamic disaggregation: does it affect apatite fission-track and (U-Th)/He analyses? Geostand Geoanal Res 34:39–48

Gleadow AJW (1981) Fission-track dating methods: what are the real alternatives? Nucl Tracks 5:3–14

Gleadow AJW, Seiler C (2015) Fission track dating and thermochronology. In: Rink WJ, Thompson JW (eds) Encyclopedia of scientific dating methods. Springer, Dordrecht, pp 285–296

Gleadow AJW, Leigh-Jones P, Duddy IR, Lovering JF (1982) An automated microscope stage system for fission track dating and particle track mapping. In: Workshop on fission track dating. Fifth international conference on geochronology, cosmochronology and isotope geology, Nikko Japan, Abstract, pp 22–23

Gleadow AJW, Duddy IR, Green PF, Lovering JF (1986) Confined fission track lengths in apatite: a diagnostic tool for thermal history analysis. Contrib Mineral Petrol 94:405–415

Gleadow AJW, Belton DX, Kohn BP, Brown RW (2002) Fission track dating of phosphate minerals and the thermochronology of apatite. Rev Mineral Geochem 48:579–630

Gleadow AJW, Raza A, Kohn BP, Spencer SAS (2005) The potential of monazite for fission-track dating. Geochim Cosmochim Acta 69 (Supp 1):A21

Gleadow AJW, Gleadow SJ, Belton DX, Kohn, BP, Krochmal MS (2009a) Coincidence Mapping, a key strategy for automated counting in fission track dating. In: Ventura B, Lisker F, Glasmacher UA (eds) Thermochronological methods: from palaeotemperature constraints to landscape evolution models, vol 324. Geological Society of London Special Publication, pp 25–36

Gleadow AJW, Gleadow SJ, Frei S, Kohlmann F, Kohn, BP (2009b) Automated analytical techniques for fission track thermochronology. Geochim Cosmochim Acta 73(Suppl):A441

Gleadow A, Harrison M, Kohn B, Lugo-Zazueta R, Phillips D (2015) The Fish Canyon Tuff: a new look at an old low-temperature thermochronology standard. Ear Planet Sci Lett 424:95–108

Goldoff B, Webster JD, Harlov DE (2012) Characterization of fluor-chlorapatites by electron probe microanalysis with a focus on time-dependent intensity variation in halogens. Am Mineral 97:1103–1115

Green PF, Duddy IR, Gleadow AJW, Tingate PR, Laslett GM (1985) Fission track annealing in apatite: track length measurements and the form of the Arrhenius plot. Nucl Tracks 10:323–328

Green PF, Duddy IR, Hegarty KA (2005) Comment on compositional and structural control of fission track annealing in apatite by Barbarand J, Carter A, Wood I, and Hurford AJ. Chem Geol 214:351–358

Haack U (1972) Systematics in the fission track annealing of minerals. Contrib Mineral Petrol 35:303–312

Hasebe N, Barberand J, Jarvis K, Carter A, Hurford AJ (2004) Apatite fission-track chronometry using laser ablation ICP-MS. Chem Geol 207:135–145

Hasebe N, Tamura A, Arai S (2013) Zeta equivalent fission-track dating using LA-ICP-MS and examples with simultaneous U-Pb dating. Island Arc 22:280–291

Hurford AJ (1998) Zeta: the ultimate solution to fission-track analysis calibration or just an interim measure? In: van den Haute P, De Corte F (eds) Advances in fission-track geochronology. Kluwer Academic Publishers, pp 19–32

Hurford AJ (2018) Chapter 1. An historical perspective on fission-track thermochronology. In: Malusà MG, Fitzgerald PG (eds) Fission-track thermochronology and its application to geology. Springer, Berlin

Hurford AJ, Green PF (1983) The zeta age calibration of fission track dating. Chem Geol (Isot Geosci Sect) 1:285–317

Jonckheere R, van den Haute P (2002) On the efficiency of fission-track counts in an internal and external apatite surface and in a muscovite external detector. Radiat Meas 35:29–40

Jonckheere R, Ratschbacher L, Wagner GA (2003) A repositioning technique for counting induced fission tracks in muscovite external detectors in single-grain dating of minerals with low and inhomogeneous uranium concentrations. Radiat Meas 37:217–219

Jonckheere R, Tamer M, Wauschkuhn F, Wauschkuhn B, Ratschbacher L (2017) Single-track length measurements of step-etched fission tracks in Durango apatite: Vorsprung durch Technik. Am Mineral (in press)

Ketcham RA (2005) Forward and inverse modeling of low-temperature thermochronometry data. Rev Mineral Geochem 58:275–314

Ketcham RA, Donelick RA, Balestrieri ML, Zattin M (2009) Reproducibility of apatite fission-track length data and thermal history reconstruction. Ear Planet Sci Lett 284:504–515

Ketcham RA, Carter A, Hurford AJ (2015) Inter-laboratory comparison of fission track confined length and etch figure measurements in apatite. Am Mineral 100:1452–1468

Kohn B, Chung L, Gleadow A (2018) Chapter 2. Fission-track analysis: field collection, sample preparation and data acquisition. In: Malusà MG, Fitzgerald PG (eds) Fission-track thermochronology and its application to geology. Springer, Berlin

Kumar R (2015) Machine learning applied to autonomous identification of fission tracks in apatite. Goldschmidt Abstr 2015:1712

Laslett GM, Galbraith RF (1996) Statistical properties of semi-tracks in fission track analysis. Radiat Meas 26:565–576

Laslett GM, Kendall WS, Gleadow AJW, Duddy IR (1982) Bias in measurement of fission track length distributions. Nucl Tracks 6:79–85

Li W, Wang L, Lang M, Trautmann C, Ewing RC (2011) Thermal annealing mechanisms of latent fission tracks: apatite vs. zircon. Ear Planet Sci Lett 302:227–235

Li W, Lang M, Gleadow AJW, Zdorovets MV, Ewing RC (2012) Thermal annealing of unetched fission tracks in apatite. Ear Planet Sci Lett 321–322:121–127

Naeser C, Dodge FCW (1969) Fission-track ages of accessory minerals from granitic rocks of the central Sierra Nevada Batholith, California. Geol Soc Am Bull 80:2201–2212

Peternell F, Kohlmann F, Wilson CJL, Gleadow AJW (2009) A new approach to crystallographic orientation measurement for apatite fission track analysis: effects of crystal morphology and implications for automation. Chem Geol 265:527–539

Reed L, Vigue K, Kumar R, Ndefo-Dahl A, Dodds Z, Donelick R (2014) Automated fission track and etch figure characterisation in apatite crystals (Abstract). In: 14th International conference on thermochronology, Chamonix, September 2014, p 23

Seiler C, Kohn B, Gleadow A (2014) Apatite fission track analysis by LA-ICP-MS: an evaluation of the absolute dating approach. In: 14th International conference on thermochronology, Chamonix, September 2014, pp 11–12

Shen CB, Donelick RA, O'Sullivan PB, Jonckheere R, Yang Z, She ZB, Miu XL, Ge X (2012) Provenance and hinterland exhumation from LA-ICP-MS zircon U-Pb and fission-track double dating of Cretaceous sediments in the Jianghan Basin, Yangtze block, central China. Sed Geol 281:194–207

Smith MJ, Leigh-Jones P (1985) An automated microscope scanning stage for fission track dating. Nucl Tracks 10:395–400

Soares CJ, Guedes S, Hadler JC, Mertz-Kraus R, Zack T, Iunes PJ (2014) Novel calibration for LA-ICP-MS-based fission-track thermochronology. Phys Chem Mineral 41:65–73

Spiegel C, Kohn B, Raza A, Rainer T, Gleadow A (2007) The effect of long-term low-temperature exposure on apatite fission track stability: a natural annealing experiment in the deep ocean. Geochim Cosmochim Acta 71:4512–4537

Vermeesch P (2017) Statistics for LA-ICP-MS based fission track dating. Chem Geol (in press)

Vermeesch P (2018) Chapter 6. Statistics for fission-track thermochronology. In: Malusà MG, Fitzgerald PG (eds) Fission-track thermochronology and its application to geology. Springer, Berlin

Vermeesch P, He J (2016) geochron@home: a crowdsourcing app for fission track dating. In: 15th International conference on thermochronology, Maresias, Brazil, September 2016, p 2

Wadatsumi K, Masumoto S (1990) Three-dimensional measurement of fission-tracks: principles and an example in zircon from the Fish Canyon Tuff. Nucl Tracks Radiat Meas 17:399–406

Wadatsumi K, Matsumoto S, Suzuki K (1988) Computerised image-processing: system for fission-track dating; system configuration and functions. J Geosci Osaka City Univ 31:19–46

Integration of Fission-Track Thermochronology with Other Geochronologic Methods on Single Crystals

Martin Danišík

Abstract

Fission-track (FT) thermochronology can be integrated with the U–Pb and (U–Th)/He dating methods. All three radiometric dating methods can be applied to single crystals (hereafter referred to as "triple-dating"), allowing more complete and more precise thermal histories to be constrained from single grains. Such an approach is useful across a myriad of geological applications. Triple-dating has been successfully applied to zircon and apatite. However, other U-bearing minerals such as titanite and monazite, which are routinely dated by single methods, are also candidates for this approach. Several analytical procedures can be used to generate U–Pb—FT—(U–Th)/He age triples on single grains. The procedure introduced here combines FT dating by LA-ICPMS and in situ (U–Th)/He dating approach, whereby the U–Pb age is obtained as a by-product of U–Th analysis by LA-ICPMS. In this case, U–Pb, trace element and REE data can be collected simultaneously and used as annealing kinetics parameter or as provenance and petrogenetic indicators. This novel procedure avoids time-consuming irradiation in a nuclear reactor, reduces multiple sample handling steps and allows high sample throughput (predictably on the order of 100 triple-dated crystals in 2 weeks). These attributes and the increasing number of facilities capable of conducting triple-dating indicate that this approach may become more routine in the near future.

5.1 Introduction

The FT method is a powerful dating technique that can be used to constrain the timing and rates of a wide range of geological processes occurring in the uppermost kilometres of Earth's crust in the temperature range of ~ 60–$350\ ^{\circ}C$. The major applications of the method include delineating the timing of rock exhumation (central for understanding the dynamics of orogenic systems and cratonic area), basin studies (revealing provenance of the material and burial/exhumation history), dating of volcanic eruptions, fault activity, genesis and preservation potential of economic mineralisations and many others (see Part II of this book, review books by Wagner and van den Haute (1992), Bernet and Spiegel (2004), Reiners and Ehlers (2005), Lisker et al. (2009), and some classic papers, e.g. (Wagner and Reimer 1972; Gleadow et al. 1983; Hurford 1986; Gleadow and Fitzgerald 1987; Green et al. 1989a, b; Gallagher et al. 1998; Ketcham et al. 1999; Kohn and Green 2002).

The FT method is based on the spontaneous fission of ^{238}U (Price and Walker 1963; Fleischer et al. 1975) in minerals like zircon, apatite and titanite (see Chap. 1 Hurford 2018). Spontaneous fission is only one of several decay mechanisms (e.g. U–Pb, (U–Th)/He, Lu/Hf, and Sm/Nd) that can be applied as geochronometers to these minerals and which provide complementary information on the cooling history. Until the late 1990s, the combined application of FT and other geochronometer(s) to the same crystals was not feasible due to technical limitations, although minerals from the same rock were often analysed using different techniques to constrain a time–temperature history. Thus in the majority of studies, the FT method was applied as a stand-alone technique focused solely on low-temperature geological processes. In the years to follow, technical and methodological advances paved the way for development of so-called in situ *multi-dating*. These include the ability to analyse smaller sample volumes, advances in in situ analytical techniques (e.g. laser ablation inductively coupled plasma mass spectrometry (LA-ICPMS), ion microprobe dating by secondary ion mass spectrometry (SIMS) or sensitive high-resolution ion microprobe (SHRIMP) instruments), the introduction of a complementary FT dating methodology utilising LA-ICPMS (Cox et al. 2000; Svojtka and Košler 2002; Hasebe et al. 2004) and the emergence of

M. Danišík (✉)
John de Laeter Centre, School of Earth and Planetary Sciences, Curtin University, Perth, Australia
e-mail: m.danisik@curtin.edu.au

© Springer International Publishing AG, part of Springer Nature 2019
M. G. Malusà and P. G. Fitzgerald (eds.), *Fission-Track Thermochronology and its Application to Geology*,
Springer Textbooks in Earth Sciences, Geography and Environment, https://doi.org/10.1007/978-3-319-89421-8_5

the (U–Th)/He method as an additional and complementary low-temperature method (Zeitler et al. 1987; Farley 2002). In the in situ multi-dating approach, single minerals are analysed by FT method in combination with U–Pb or (U–Th)/He methods (hereafter termed *double-dating*; Carter and Moss 1999; Carter and Bristow 2003; Donelick et al. 2005; Chew and Donelick 2012) or by all three methods together (hereafter termed *triple-dating*; Reiners et al. 2004a; Carrapa et al. 2009; Danišík et al. 2010a; Zattin et al. 2012).

Multi-dating offers several advantages over the single method approach. For instance, it allows unprecedented, detailed reconstruction of thermal histories on single grains. These histories may cover the full spectrum of geological processes from crystal formation, through metamorphic overprint to final exhumation. This has an enormous application potential in Earth Sciences, in particular for detrital geochronology. However, this multi-dating approach on a single crystal has been only rarely applied and is still relatively new. For this reason, in this chapter a brief review of the history of multi-dating involving the FT method will be provided, along with a description of the rationale and theoretical background, and potential and limitations. Existing triple-dating analytical procedures will be described, and a brief introduction given to a new triple-dating approach that is currently being developed at Curtin University. The chapter will close with some applications and proposals for future directions.

5.2 Historical Perspective

Double-dating (FT and U–Pb methods applied to the same crystal) was first introduced by Carter and Moss (1999). These authors analysed detrital zircons from the Khorat Basin (Thailand) by FT using the external detector method (EDM, e.g. Gleadow 1981) to unravel the low-temperature thermotectonic evolution of the source terrains. Then, the same grains were U–Pb dated using SHRIMP to identify their crystallisation age. In addition to introducing the double-dating concept, this study highlighted that without complementary U–Pb data, the FT data alone would lead to misinterpretation of the source area and other erroneous conclusions (Carter and Moss 1999; Carter and Bristow 2000). Despite the demonstrated potential for provenance studies and exhumation studies (Chap. 14; Carter 2018; Chap. 15; Bernet 2018), the combined SIMS/SHRIMP U–Pb and EDM FT double-dating approach was subsequently used only twice (Carter and Bristow 2003; Bernet et al. 2006), likely because U–Pb dating by ion microprobe is a time-consuming and expensive technique, and more appropriate for other applications. Soon after, zircon U–Pb—(U–Th)/He double-dating (Rahl et al. 2003; Campbell et al. 2005; Reiners et al. 2005; McInnes et al. 2009; Evans et al.

2013) was also introduced, in which the zircon FT method is replaced by the (U–Th)/He thermochronometer that has a similar temperature sensitivity range and offers a less labour-intensive, higher sample throughput approach.

The renewed interest in U–Pb—FT double-dating started in the mid-2000s and was associated with the advent of LA-ICPMS into the field of thermochronology. First, a new methodology of FT dating employing LA-ICPMS to measure ^{238}U directly, replacing the conventional thermal neutron irradiation approach, was introduced (Cox et al. 2000; Svojtka and Košler 2002; Hasebe et al. 2004), dramatically increasing the speed of FT analysis and sample throughput. Soon after, LA-ICPMS methodology for FT dating of zircon was enhanced by adding the capability to determine the U–Pb age for each FT dated zircon grain by default (Donelick et al. 2005). In addition, the relatively recent introduction of new matrix-matched reference materials and new approaches to the common Pb correction allowed combined FT and U–Pb dating by LA-ICPMS to be routinely applied on apatite (Chew et al. 2011; Chew and Donelick 2012; Thomson et al. 2012). LA-ICPMS thus provided a more convenient, faster, less expensive but sufficiently precise and accurate means for routine U–Pb dating (e.g. Košler and Sylvester 2003). Nowadays, in FT studies using the LA-ICPMS approach, the provision of both FT and U–Pb ages on single grains is routine and an increased application of FT—U–Pb double-dating is noticeable in the literature (e.g. Shen et al. 2012; Liu et al. 2014; Moore et al. 2015).

At the time of writing, only one conference abstract and three research papers on triple-dating involving the FT method have been published (Reiners et al. 2004a; Carrapa et al. 2009; Danišík et al. 2010a; Zattin et al. 2012). The concept of zircon triple-dating was first introduced by Reiners et al. (2004a). The authors applied a combination of EDM FT, LA-ICPMS U–Pb and conventional (U–Th)/He dating to detrital zircon grains. Assuming that the measured ages record the time of cooling through the closure temperature (Dodson 1973; cf. Chap. 10; Malusà and Fitzgerald 2018a, b), they reconstructed cooling trajectories constrained by crystallisation ages, and cooling ages marking the passage through the $\sim$240 and $\sim$180 °C isotherms (i.e. nominal closure temperatures for zircon FT and zircon (U–Th)/He systems, respectively; Hurford 1986; Reiners et al. 2004b) for single zircon crystals. This demonstrated the potential of triple-dating to provide more information than double-dating approaches (both U–Pb—FT and U–Pb—(U–Th)/He). A similar concept utilising EDM FT, multi-collector LA-ICPMS (LA-MC-ICPMS) U–Pb and conventional (U–Th)/He dating was applied to detrital apatite by Carrapa et al. (2009) and Zattin et al. (2012). Both studies demonstrated the capability of this approach to obtain ages that were interpreted by the authors to represent cooling through the $\sim$500, $\sim$110 and $\sim$65 °C isotherms (i.e. nominal closure temperatures for

U–Pb, FT and (U–Th)/He systems in apatite, respectively; Cherniak et al. 1991; Wagner and van den Haute 1992; Farley 2000). Using the triple-ages, the authors elucidate the provenance of the grains and their thermal history at higher temporal and spatial resolution than would be possible by using single-dating methods. An alternative approach to, and application of, apatite triple-dating was presented by Danišík et al. (2010a) who applied ID-TIMS U–Pb, EDM FT and conventional (U–Th)/He dating to a hydrothermal apatite aggregate in an attempt to constrain the thermotectonic evolution of basement rocks and investigate apparent discrepancies between FT and (U–Th)/He data obtained by single methods.

Currently, the major obstacle of the triple-dating approaches described above relates to the relatively complicated, time-consuming, labour-intensive analytical procedures with multiple handling steps and to accessibility of the analytical instruments. There are only a few institutions pursuing research directions both in high-temperature and low-temperature geochronology and which are accordingly equipped with a FT laboratory, and LA-ICPMS and noble gas mass spectrometer instruments. However, recent progress in the automation of FT counting and offsite data processing (Gleadow et al. 2009), together with increasing numbers of accessible LA-ICPMS laboratories and the development of fast throughput in situ (U–Th)/He dating techniques (Boyce et al. 2006; van Soest et al. 2011; Vermeesch et al. 2012; Tripathy-Lang et al. 2013; Evans et al. 2015; Horne et al. 2016) may help to overcome at least some of these limitations and holds great promise for triple-dating in the future.

5.3 Rationale of Multi-dating

The rationale of the double- and triple-dating approaches is in the combined application of dating methods with different temperature sensitivities to the same crystals, which enables geoscientists to extract a more complete and more detailed picture of the cooling history. Principles of the radiometric dating techniques complementing FT in double- and triple-dating approaches have been described in several comprehensive review books and papers (e.g. Farley 2002; Reiners 2005 for (U–Th)/He dating; Hanchar and Hoskin 2003; Schaltegger et al. 2015 for U–Pb dating), and therefore are not detailed here. In brief, the FT method is based on the accumulation (and annealing) of linear damage (fission tracks) produced by spontaneous fission of ^{238}U; the U–Pb method is based on the accumulation of Pb produced by a series of alpha and beta decays of U; and the (U–Th)/He method is based on the accumulation of ^{4}He produced by alpha decay of U, Th and Sm.

Temperature sensitivity ranges for the U–Pb, FT and (U–Th)/He systems in most common mineral phases suitable for double- and triple-dating are illustrated in Fig. 5.1. In general, the U–Pb system is sensitive to higher temperatures (i.e. 350–1000 °C) than the FT and (U–Th)/He systems and typically records process occurring in upper-mantle to mid-crustal levels (e.g. Chew et al. 2011; Cochrane et al. 2014; Schaltegger et al. 2015). FT and (U–Th)/He systems, in contrast, are sensitive to lower temperatures (40–350 °C) and typically record upper crustal processes (e.g. Ehlers and Farley 2003; Danišík et al. 2007; Malusà et al. 2016).

It should be noted that the concept of applying a range of methods to different aliquots of the same mineral from the same rock is not new (see, e.g., McInnes et al. 2005; Vermeesch et al. 2006; Siebel et al. 2009). So, the question to be asked is—what is the advantage of combining FT dating with U–Pb and/or (U–Th)/He methods and applying these to a single crystal? To answer this question, it is worthwhile to consider the strengths and limitations of the FT method and appreciate possible ambiguity in some FT data. Even though U–Pb and (U–Th)/He methods have also strengths and limitations to be aware of that discussion is beyond the scope of this chapter.

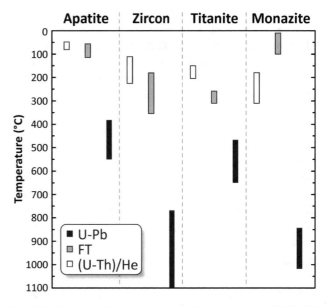

Fig. 5.1 Characteristic temperature sensitivity ranges for U–Pb, FT and (U–Th)/He radiometric systems in most common minerals suitable for triple-dating. Partial annealing/retention zones were calculated with closure software (Brandon et al. 1998) using the data for apatite after Cherniak et al. (1991), Ketcham et al. (1999), Chamberlain and Bowring (2001), Farley (2000); for zircon after Brandon et al. (1998), Cherniak and Watson (2001, 2003), Cherniak (2010), Rahn et al. (2004), Guenthner et al. (2013); for titanite after Cherniak (1993, 2010), Coyle and Wagner (1998), Hawkins and Bowring (1999), Reiners and Farley (1999); for monazite after Cherniak et al. (2004), Gardés et al. (2006), Boyce et al. (2005), Weise et al. (2009)

The major strength of the FT method is in its ability to discriminate not only the timing of cooling, but also the style of cooling within the partial annealing zone (PAZ), as recorded by the track length distribution (e.g. Gleadow et al. 1986a). FT age and track length data in apatite often enable robust reconstructions of best-fit time–temperature envelopes via forward and/or inverse modelling (Ketcham 2005; Gallagher 2012), i.e. whether they record a distinct geologic event or they are the result of a more complex path (see Chap. 3; Ketcham 2018; Chap. 8; Malusà and Fitzgerald 2018a, b). Despite the possibility of Californium irradiation technique that allows etchant to reach confined tracks (Donelick and Miller 1991; Chap. 2; Kohn et al. 2018), measurement of statistically robust track length distributions (typically 100 track length per sample or per significant age population) may not always be feasible, in particular in samples with young (Late Cenozoic) FT ages and/or low uranium concentration. A poorly defined confined track length distribution makes it difficult to interpret non-reset detrital samples that have multiple age populations. Whether track length data can be obtained or not, the addition of U–Pb and/or (U–Th)/He ages provides additional higher- and lower-temperature constraints to cooling trajectories that can greatly improve the understanding and interpretation of FT ages. When track length data are available, the addition of U–Pb and/or (U–Th)/He data can still greatly benefit the interpretation by permitting more accurate reconstruction of the time–temperature history, whereby U–Pb age constrains the high-temperature part of the cooling trajectory and (U–Th)/He data provide additional constraints for low-temperature processes (e.g. Stockli 2005; Green et al. 2006; Emmel et al. 2007).

Because of the relatively low number of fission tracks counted in each grain, the FT method yields relatively low precision on single-grain ages. For example, standard errors in young samples are commonly >20% at 1σ compared with <2 and 2–5% errors for the U–Pb by LA-ICPMS and conventional (U–Th)/He methods, respectively. In addition, dispersion of single-grain FT ages is common, even in apatites from rapidly cooled rocks. For example, a typical range of 25–50 single shard/grain EDM FT ages obtained on age standards regularly measured for zeta-calibration is commonly from ∼17 to ∼55 Ma (on Durango apatite) and from ∼15 to ∼50 Ma (on Fish Canyon zircons) (M. Danišík, unpublished data). Dispersion is due to the relatively few numbers of tracks counted as mentioned above, as well as compositional and structural variation of individual crystals, and thermal evolution in which slow or complex cooling through or into and out of the PAZ causes increased scatter of single-grain ages (e.g. Gleadow et al. 1986b). In the FT applications where single-grain FT ages are expected to form single age population (e.g. quickly cooled igneous rocks or quickly cooled, fully reset sediments), these problems are mitigated by

analysing >20 crystals and calculating a population geometric mean age (a.k.a. *central age*, see Chap. 6; Vermeesch 2018) with corresponding standard error, which is commonly ∼3–5% at 1σ (Galbraith and Laslett 1993). However, in applications such as detrital dating studies, where crystals are derived from multiple sources and commonly produce complex FT age distributions, dispersion of single-grain FT ages will be greater. In detrital studies, confined track length distributions are typically not representative of the collective cooling history, because of different grain populations from different provenances, thus making interpretation of data sets challenging. Complementing FT ages with U–Pb and/or (U–Th)/He ages from the same grains is, therefore, invaluable in this respect, and triple-dating offers several advantages to overcome these issues:

- First, the combination of three ages allows a direct internal data quality check, where ages for apatite, zircon and titanite should follow the general trend of U–Pb age ≥ FT age ≥ (U–Th)/He age (e.g. Hendriks 2003; Lorencak 2003; Belton et al. 2004; Hendriks and Redfield 2005; Green et al. 2006; Ksienzyk et al. 2014), as dictated by the closure temperature concept (Dodson 1973), although there are exceptions. For example in old, slowly cooled terrains, an apatite FT age may be < (U–Th)/He age, as discussed in Chap. 21 (Kohn and Gleadow 2018). The multi-method approach therefore allows identification of analytical outliers (improving data set robustness), and of contaminant, diagenetic or authigenic grains that may present important geological information but that would not be otherwise detected using the FT method alone.

- Second, a lack of geological context can hamper accurate interpretation of single ages whereas multiple ages on the same crystal may mitigate this. For instance, without the prior knowledge of U–Pb and FT data, it is not possible to discriminate between "apparent" (U–Th)/He ages resulting from complex thermal histories causing partial resetting of the (U–Th)/He system and "cooling" (U–Th)/He ages resulting from simple cooling paths where the closure temperature concept applies (e.g. Stockli et al. 2000; Danišík et al. 2015).

- Third, the combination of the three ages may better constrain the cooling trajectories for single crystals—from crystallisation (or high-grade metamorphism) to final cooling. Detailed cooling histories may provide diagnostic fingerprints of the source terrain, allowing a more robust interpretation of detrital data than could be achieved by using one method alone.

Finally, it may be argued that single-grain multi-dating, notably for detrital studies should be pursued because it is technologically possible and markedly more efficient. When compared to traditional FT dating protocols, modern LA-ICPMS U–Pb and (U–Th)/He dating procedures are largely automatised and do not require permanent attendance of an operator. Given recent technological and methodological innovations (e.g. FT dating by using LA-ICPMS providing U–Pb dating as a by-product (Donelick et al. 2005; Chew and Donelick 2012), the automation of FT counting (Gleadow et al. 2012), developments of high-throughput in situ U–Pb and (U–Th)/He dating techniques (Boyce et al. 2006, 2009; van Soest et al. 2011; Vermeesch et al. 2012; Tripathy-Lang et al. 2013; Evans et al. 2015) and the improved accessibility of the required instruments (e.g. growing number of LA-ICPMS, FT and He laboratories, the possibility of remotely controlled analytical measurements, offsite data reduction, etc.), the acquisition of double- and triple-age data on single crystals are now practical and may soon become the norm.

5.4 Analytical Procedures for Combined FT, U–Pb and (U–Th)/He Dating

Analytical protocols for triple-dating using combined U–Pb, FT and (U–Th)/He methods involve measurements of parent nuclides (U, Th, ±Sm) and their daughter products (Pb isotopes, spontaneous fission tracks and He, respectively) in the same crystals. Published and newly proposed workflows for triple-dating are summarised in Fig. 5.2.

Selection of the analytical procedure depends on several factors such as size and quantity of minerals, accessibility of analytical instruments, time available for analytical work and desired data quality with regard to precision and accuracy. Large, ($>\sim 2$ mm) single crystals or crystal aggregate(s) (with identical magmatic and cooling history) can be crushed or disaggregated in a mortar in order to obtain small shards (preferably >50 μm), and these can be dated separately using fully destructive (e.g. ID-TIMS U–Pb, conventional (U–Th)/He dating) or semi-destructive methods (U–Pb by SIMS or LA-(MC)-ICPMS, in situ (U–Th)/He, FT dating by EDM or LA-ICPMS). The main advantage of this approach is in the possibility of obtaining high-precision U–Pb data when using ID-TIMS (Parrish and Noble 2003) and more accurate and precise conventional (U–Th)/He ages by eliminating the need for alpha ejection correction (Farley et al. 1996) when analysing shards from grain interiors or applying mechanical abrasion removing the outer ~20 μm of grain surface (Krogh 1982; Danišík et al. 2008). The primary limitation of this approach is that large U-bearing crystals suitable for triple-dating are rare in the nature. An example of the combined TIMS U–Pb, EDM FT and conventional (U–Th)/ He dating of a cm-size, hydrothermal apatite can be found in Danišík et al. (2010a).

Mineral(s) or mineral concentrate >2 mm

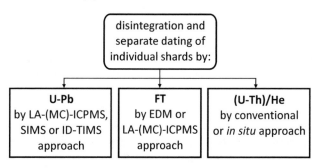

Mineral(s) or mineral concentrate <2 mm

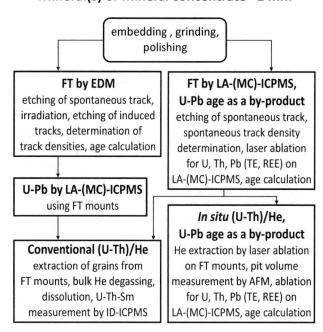

Fig. 5.2 Flow charts summarising feasible methodologies for triple-dating

Accessory heavy minerals that are typically found in the <250 μm size fraction need to be analysed using semi-destructive techniques during intermediate dating stages. The three studies reporting triple-dating of apatite (Carrapa et al. 2009; Zattin et al. 2012) and zircon (Reiners et al. 2004a), followed an almost identical protocol in which crystals were first dated by conventional EDM FT methods involving embedding, grinding, polishing, etching of spontaneous fission tracks, irradiation in a nuclear reactor, etching of induced tracks in the mica external detector and track counting allowing the age calculation. Then, the FT dated grains were dated by U–Pb method using LA-(MC)-ICPMS, and finally, the FT + U–Pb dated grains were extracted from the FT mounts and dated by a conventional (U–Th)/He method that involved determination of bulk He content using noble gas mass spectrometry, dissolution of crystals in acids and U–Th analysis by solution isotope dilution ICPMS

(Reiners et al. 2004a; Carrapa et al. 2009; Zattin et al. 2012). This approach suits detrital crystals of common size (typically >50 µm) and has the added advantage of providing the opportunity to simultaneously collect geochemical data during LA-(MC)-ICPMS analysis (e.g. Cl, F, trace elements, REE or Hf isotopes). These data can be very useful for characterising annealing kinetics in apatite for thermal modelling purposes (Barbarand et al. 2003; Ketcham et al. 2007a, b), as indicators of source rock lithology (e.g. trace elements in apatite; Morton and Yaxley 2007; Malusà et al. 2017), as petrogenetic tracers (e.g. REE in zircon; Schoene et al. 2010; Jennings et al. 2011), or as tracers of host rock origin (Hf in zircon; Kinny and Maas 2003; Flowerdew et al. 2007). The major limitations of this approach relate to sample irradiation in a nuclear reactor, which in some cases can be time-consuming, and in certain countries can sometimes take up to three months. There are also more sample handling steps when dealing with irradiated samples and then there is also a risk of crystal loss during the extraction from the FT mount (crystals are typically plucked out from the FT mount submerged in ethanol by using a sharp needle or tweezers with sharp ends) and loading into micro-tubes prior to He analysis. In addition, correction for the high common Pb content in young, low-U apatites can be difficult when using a quadrupole ICPMS for U–Pb dating (Chew et al. 2011; Thomson et al. 2012).

Some of these issues can be circumvented using a new triple-dating approach that is currently being developed in the GeoHistory Facility at the John de Laeter Centre (Curtin University). This approach combines FT dating by LA-ICPMS (Hasebe et al. 2004; Donelick et al. 2005; Chew and Donelick 2012) and in situ (U–Th)/He dating (Evans et al. 2015), with the LA-(MC)-ICPMS U–Pb age obtained as a by-product of either of the two methods. Preliminary results obtained on Durango apatite (Fig. 5.3) are in excellent agreement with the expected value, which holds a great promise for the future. This new approach can be briefly described as follows:

Crystals of interest (both apatite and zircon) are embedded into Teflon (DuPont PFA, Type 6000LP) as it, unlike epoxy, does not excessively degas and allows the desired pressure in the UHV cell to be obtained. Crystals are then ground to 4π geometry and sequentially polished using 9-, 3-, 1-µm-diamond and 0.3-µm-colloidal silica suspensions. After a thorough cleaning of the mounts, spontaneous fission tracks in polished crystals are revealed by etching using standard etching protocols. Spontaneous track density (i.e. number of tracks per known area), confined track lengths and Dpar values (Burtner et al. 1994) are then measured for selected grains under an optical microscope equipped with a high-resolution camera. The concentration of ^{238}U in the grains is then directly determined by LA-(MC)-ICPMS (Agilent 7700 s or NU Plasma II ICPMS both connected to

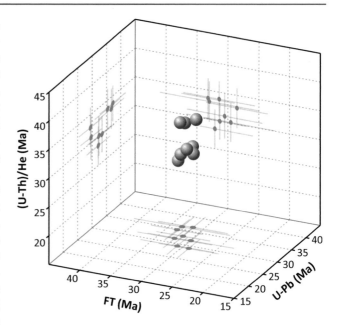

Fig. 5.3 3D scatter plot showing initial results of the newly proposed triple-dating methodology applied to shards of Durango apatite age standard (Ar–Ar reference age: 31.44 ± 0.18 Ma (2σ); McDowell et al. 2005). The procedure employed FT dating by LA-ICPMS (Chew and Donelick 2012) and in situ (U–Th)/He dating (Evans et al. 2015); spontaneous fission tracks were counted under a Zeiss Axioskop 2 microscope; isotopic data were collected on quadrupole noble gas mass spectrometer (Pfeiffer PrismaPlus) and an Agilent 7700 s single quadrupole ICPMS both connected to 193 nm ArF RESOlution COMPexPro 102 excimer laser. U–Pb ages are ^{207}Pb corrected ^{238}U/^{206}Pb ages calculated as Tera–Wasserburg concordia lower-intercept ages anchored through common Pb (after Stacy and Kramer 1975); FT ages were calculated using the scheme presented by Donelick et al. (2005) and Chew and Donelick (2012); in situ (U–Th)/He ages were calculated by using the first principle approach (Boyce et al. 2006). Uncertainties on U–Pb, FT and (U–Th)/He ages are reported at 1σ level

193 nm ArF RESOlution COMPexPro 102 excimer laser with S155 Laurin Technic laser ablation flow-through cell) and the FT age for each grain is determined following the protocols described by Hasebe et al. (2004), Donelick et al. (2005) and Chew and Donelick (2012). In addition to ^{238}U, the LA-(MC)-ICPMS permits measurement of Pb content (as well as a range of trace elements and REE, if desired), permitting calculation of a U–Pb age as a by-product (Donelick et al. 2005; Chew and Donelick 2012). Almost non-destructive FT analysis by LA-ICPMS utilising $\sim$23 to 50 µm circular laser spots preserves enough space on the polished crystal surfaces for in situ (U–Th)/He analysis. Analytical procedures for in situ (U–Th)/He dating follow the protocols of Evans et al. (2015). The Teflon mount with target crystals is loaded into the UHV cell of the RESOchron instrument, and He is extracted from intact polished surfaces by laser ablation (using 33 or 50 µm spots). He content is determined on a quadrupole mass spectrometer (Pfeiffer PrismaPlusTM) by isotope dilution using a known volume of

[3]He spike. The mount is then retrieved from the UHV cell, and the volume of the He ablation pits is measured using an atomic force microscope (although a confocal laser scanning microscope could also be employed; Boyce et al. 2006). This allows the He concentration to be determined which is required for the in situ (U–Th)/He age calculation (Boyce et al. 2006; Vermeesch et al. 2012). The mount is then reloaded into the flow-through cell (S155) for a third ablation to determine U, Th and Sm contents by LA-(MC)-ICPMS, allowing the final calculation of (U–Th)/He age. Similar to the first laser ablation for FT dating, this stage of LA-(MC)-ICPMS analysis permits determination of the U–Pb age as by-product (Boyce et al. 2006; Vermeesch et al. 2012; Evans et al. 2015).

The major advantages of such approach are in the significantly shorter analytical time and higher sample throughput achieved by an overall simplification of sample handling procedures for what is essentially three methods. There is no need to extract dated grains from the FT mounts and load them into micro-tubes for conventional He extraction, most of the analytical steps are automated and samples do not require irradiation for FT analyses. A realistic workflow suggests that ∼100 crystals may be triple-dated in 2 weeks. Further advantages include the ability to date "problematic" crystals, not suitable for conventional (U–Th)/He analysis (e.g. crystals with extreme zonation of parent nuclides, structural inhomogeneities caused by radiation damage, mineral and/or fluid inclusions), and circumvention of an alpha ejection correction both achieved by targeting grain interiors (Boyce et al. 2006), and improved worker safety is that there is no need for grain dissolution for (U–Th)/He dating, thereby avoiding the use of hydrofluoric, nitric or perchloric acids, and there is no need for irradiation and training of workers in use of radioactive samples. However, at least four limitations to this approach should be considered:

- First, although FT dating by LA-ICPMS provides higher precision on relative uranium concentrations compared to the conventional EDM method (Donelick et al. 2006), this approach may not be suitable for strongly zoned crystals (Hasebe et al. 2004; Donelick et al. 2005).
- Second, the precision and accuracy of in situ (U–Th)/He ages may not be as good as conventional (U–Th)/He ages (Horne et al. 2016) due to the simplified assumption of the homogeneity in distribution of parent nuclides in dated minerals, inherited lower analytical precision of LA-ICPMS data in comparison to isotope dilution ICPMS data, and an additional source of uncertainty related to the pit volume measurements.

- Third, as in the previous EDM FT + LA-ICPMS U–Pb + conventional (U–Th)/He approach, triple-dating of young, low-U apatite can be problematic due to the high amount of common Pb and low abundance of radiogenic Pb and He.
- Fourth, currently there are only four laboratories publishing in situ (U–Th)/He data so the accessibility to this methodology is currently limited.

However, the advantages of triple-dating using this approach outweigh the limitations for many applications, such as, a case where large numbers of grains need to be analysed. In the future, it is anticipated that the number of laboratories with similar in situ capabilities will increase; thus, triple-dating is likely to become much more widely used in coming years.

The effect of chemical etching on (U–Th)/He systematics One of the critical requirements for successful triple-dating is that all radiometric decay schemes used are undisturbed during the multiple analytical steps. While sample embedding, grinding, polishing, FT counting and "cold" ablation by excimer laser should not alter the parent–daughter systems, the effect of chemical etching (required to reveal spontaneous fission tracks) on (U–Th)/He systematics may be a concern. The etching of apatite, titanite and monazite is safe in this regard as it is carried out at temperatures well below the temperature sensitivity of the (U–Th)/He system in these minerals. Routine procedures include etching in 5 or 5.5 M HNO_3 solution at 21 °C for 20 s for apatite (e.g. Donelick et al. 1999), HF–HNO_3–HCl–H_2O solution at 23 °C for 6–30 min for titanite (e.g. Gleadow and Lovering 1974) and in boiling (∼50 °C) 37% HCl for 45 min for monazite (Fayon 2011). However, the effect of long and aggressive etching of fission tracks in zircon (10–100 h at 215–230 °C in a eutectic KOH–NaOH mixture; Zaun and Wagner 1985; Garver 2003; Bernet and Garver 2005) on He diffusion may be of concern as the etching temperatures are above the lower limit of the zircon He partial retention zone (∼150 °C; Guenthner et al. 2013). To test this hypothesis, an experiment was conducted in which 15 zircon crystals from the Fish Canyon Tuff were etched in a eutectic KOH–NaOH mixture at 215 °C for 100 h, at which point the spontaneous fission tracks were revealed in all crystals. The etched crystals were then analysed, together with 15 unetched Fish Canyon Tuff zircons by conventional (U–Th)/He method. Comparison of the results (Fig. 5.4) shows no significant difference in (U–Th)/He ages of etched and unetched crystals, suggesting that the long and aggressive etching of spontaneous fission tracks in zircon is not an issue for subsequent (U–Th)/He dating and the triple-dating approach.

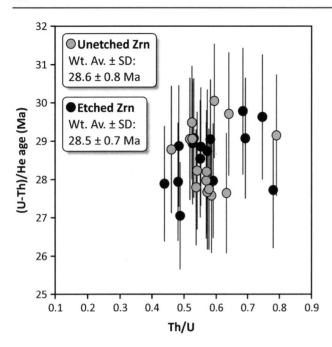

Fig. 5.4 Comparison of conventional (U–Th)/He ages obtained on chemically untreated and etched zircon crystals from Fish Canyon Tuff (reference (U–Th)/He age: 28.3 ± 1.3 Ma; Reiners 2005). Weighted averages of unetched and etched zircon are similar, demonstrating no significant effect of etching on (U–Th)/He system in the investigated sample. Etching conditions uses: NaOH–KOH eutectic melt, 215 °C, 100 h (Zaun and Wagner 1985; Garver 2003; Bernet and Garver 2005)

5.5 Applications

Triple-dating of single crystals yields an unprecedented amount of chronological information and is ideally suited to detrital dating studies (e.g. Carrapa 2010) where a large number of crystals need to be analysed in order to identify statistically significant age components with confidence (e.g. Vermeesch 2004). Depending on the dated mineral, single crystal triple-dating can resolve the magmatic, metamorphic and exhumation history of source terrains, establish maximum depositional ages and detect post-depositional heating events. Hypothetically, the combination of (tectono-)thermal events experienced by detrital grains should result in a unique combination of U–Pb, FT and (U–Th)/He ages for each grain (Fig. 5.5).

As discussed above, triple-dating provides significant advantages over single-dating methods in several aspects. The studies of Carrapa et al. (2009) and Zattin et al. (2012) show that triple-ages obtained on a relatively low number of grains allowed to derive more geological conclusions than those resolvable from 100 grains (the commonly adopted number in detrital studies) dated by a single method. Evaluation of the robustness of triple-dating with regard to identifying age components is beyond the scope of this study and needs to be further tested in future. However, it is

likely that fewer grains will provide at least as much (and likely more) detailed provenance information than single-dating method approaches, applied to a high number of grains.

In addition to its application to zircon and apatite, triple-dating should be applicable to other common detrital minerals such as titanite and monazite that are (with some limitations) datable by U–Pb, FT and (U–Th)/He methods (e.g. Reiners and Farley 1999; Stockli and Farley 2004; Boyce et al. 2005; Siebel et al. 2009; Fayon 2011; Weisheit et al. 2014; Kirkland et al. 2016a, b). In addition to different closure temperatures for given radiometric systems (Fig. 5.1), each of these minerals may be representative of different source lithologies and also have different mechanical and chemical properties that translate to different stability during sedimentary transport (Chap. 7; Malusà and Garzanti 2018). Triple-dating applied to different minerals can therefore potentially enable more reliable identification of source areas and reconstruction of their individual thermal histories and can provide critical information on thermal events during different stages of the mineral recycling processes. For instance, while highly refractory zircons can survive multiple orogenic and sedimentary cycles and transport over extremely long distances, less durable apatites are more likely to represent first cycle detritus and will likely record the thermal history of relatively proximal sources.

In addition to detrital studies aimed at provenance discrimination, multi-dating may be also useful for tectonic studies since it permits more complete and more accurate reconstruction of thermal histories for different minerals and thus provides a potentially powerful tool for exploring the link between deep and shallow processes. The combined application of FT and (U–Th)/He may allow a more robust and more detailed reconstruction of thermal histories within the corresponding partial annealing/retention zones. This is usually achieved using thermal modelling packages like HeFTy or QTQt (Ketcham 2005; Gallagher 2012), which offer a wide range of options and parameters to be defined in order to achieve reliable results. Even though the models have proved viable for reconstructing thermal histories in many studies, caution is recommended when attempting to model a combination of only FT and (U–Th)/He data. In particular, in some situations it is challenging and sometimes even impossible to obtain satisfactory and geologically reasonable results even where the model is constrained by both FT data (age and length) and (U–Th)/He data (i.e. age, size, zonation of parent nuclides, diffusion parameters) (e.g. Danišík et al. 2010b, 2012). While application of inverse thermal modelling often produces geologically reasonable best-fit time–temperature envelopes and paths, there are still challenges and improvements to be made. For example, testing the reliability and reproducibility of modelling results obtained by the combined modelling approach on natural

Fig. 5.5 U–Pb, FT and (U–Th)/He data potentially recovered in detritus derived from erosion of a hypothetical geologic landscape shown in (**a**). Detrital apatite and zircon grains eroded from different subareas (A to F) will yield distinct combinations of U–Pb—FT—(U–Th)/He ages (**b**), that can be plotted on a three-dimensional bubble chart (**c**). AFT (ZFT) and AHe (ZHe) indicate FT and (U–Th)/He ages on apatite (and zircon), respectively; PAZ, partial annealing zone; PRZ, partial retention zone

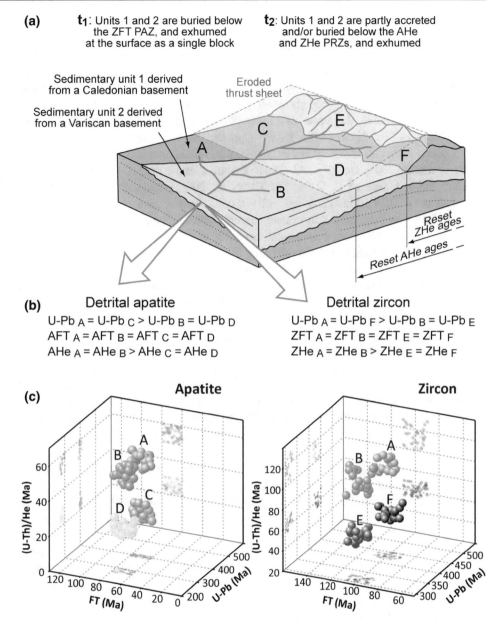

calibration sites or well-characterised samples with a known thermal history (e.g. House et al. 2002).

Finally, triple-dating can help to improve the understanding of FT and (U–Th)/He data and both methods in general. With exception of monazite, the temperature sensitivity of the (U–Th)/He method is generally slightly lower than the sensitivity of the FT method for similar minerals (Fig. 5.1). Therefore, (U–Th)/He ages should ideally be identical or younger than the FT ages obtained on the same mineral. With the exception of correlations between grain size and (U–Th)/He age that are sometimes observed (e.g. Reiners and Farley 2001), (U–Th)/He ages usually do not provide information on the style of cooling through the He partial retention zone. Therefore, in the absence of additional information, it is not evident whether the (U–Th)/He ages

are related to a distinct, geologically significant cooling event (and hence can be termed as "cooling" ages), or "apparent" ages without direct geological meaning (e.g. Stockli et al. 2000; Danišík et al. 2015; see Chap. 8 Malusà and Fitzgerald 2018a, b). In addition, single-grain (U–Th)/He ages often show high dispersion that may reflect a number of conditions—a protracted cooling through the He partial retention zone (Fitzgerald et al. 2006), inaccurate alpha ejection correction (Farley et al. 1996; Hourigan et al. 2005), radiation damage affecting He retentivity and closure temperature (Hurley 1952; Flowers et al. 2009; Guenthner et al. 2013; Danišík et al. 2017) or imperfection of dated crystals (e.g. undetected inclusions causing older than expected ages; Farley 2002; Ehlers and Farley 2003; Danišík et al. 2017). Some of these conditions reflect geologically meaningful processes while others result in data that should be evaluated carefully and perhaps disregarded. In this case, provision of U–Pb and FT ages on (U–Th)/He dated crystals provide further constraints on the cooling style and the origin of dispersion.

One of the challenges for modern low-temperature thermochronology remains the issue of "inverted" FT and (U–Th)/He ages (i.e. FT age < (U–Th)/He age) that have been reported from old, slowly cooled terrains (e.g. Hendriks 2003; Lorencak 2003; Belton et al. 2004; Hendriks and Redfield 2005; Green et al. 2006; Danišík et al. 2008; Ksienzyk et al. 2014). This "inverted" relationship, seemingly contradicting the closure temperature concept (Dodson 1973), called into question fundamental concepts of FT annealing and He diffusion, and the reliability of the methods, and prompted renewed interest in methodological research attempting to explain this discrepancy. Due to these apparently younger apatite FT ages versus (U–Th)/He ages, as well as the often large variation of (U–Th)/He single-grain ages from the same sample, enormous progress in the understanding of (U–Th)/He systematics has been achieved. Such work has revealed the importance of phenomena such as radiation damage (Reiners 2005; Shuster et al. 2006; Shuster and Farley 2009; Flowers 2009; Flowers et al. 2007, 2009; Guenthner et al. 2013; Danišík et al. 2017), zonation of parent nuclides (e.g. Meesters and Dunai 2002a, b; Hourigan et al. 2005; Fitzgerald et al. 2006; Danišík et al. 2017) or chemical composition (Djimbi et al. 2015). At the same time, there have also been considerable advances in FT thermochronology, both in methodology and modelling (e.g. Donelick et al. 2005; Enkelmann et al. 2005; Ketcham 2005; Ketcham et al. 2007a, b, 2009; Zattin et al. 2008; Jonckheere and Ratschbacher 2015, and Chap. 4; Gleadow et al. 2018). As a result, many of the (U–Th)/He data sets previously considered as discrepant could be explained. For example, it was shown that the accumulation of radiation damage can increase the closure temperature of the (U–Th)/He system to levels higher than the closure temperature of the FT system (e.g. Shuster et al. 2006; Guenthner et al. 2013; Gautheron and Tassan-Got 2010; Ketcham et al. 2013). Additionally, it was shown that, due to the faster response of the FT system to temperature change, the inverse FT—(U–Th)/He relationship could be a diagnostic indication of short duration heating events such as shear heating along faults (e.g. Tagami 2005) or wildfires (Reiners 2004) (see Chap. 8; Malusà and Fitzgerald 2018a, b). However, there are still situations where FT—(U–Th)/He age relationships are still a matter of ongoing discussion (see Chap 3, Ketcham 2018, Chap. 21, Kohn and Gleadow 2018, and also Hendriks and Redfield 2005, 2006; Söderlund et al. 2005; Green and Duddy 2006; Green et al. 2006; Shuster et al. 2006; Hansen and Reiners 2006; Flowers and Kelley 2011; Flowers and Farley 2012, 2013; Lee et al. 2013; Karlstrom et al. 2013;

Fox and Shuster 2014; Flowers et al. 2015, 2016; Gallagher 2016; Danišík et al. 2017). Several examples exist where the "inverted" FT and (U–Th)/He ages have not been satisfactorily explained, and it is not always straightforward whether the discrepancy arises from insufficient understanding of FT annealing, He diffusion or insufficient data quality (e.g. Hendriks and Redfield 2005, 2006; Green and Duddy 2006; Green et al. 2006; Kohn et al. 2009; Danišík et al. 2017). It should be noted that these studies reporting "inverted" FT and (U–Th)/He ages employed a single method or two methods applied to different grains, and this may introduce some bias into the results. In other cases, it was apparent that where (U–Th)/He ages were calculated as "mean ages" they were older than central apatite FT ages, rather than the totality of single-grain (U–Th)/He age data being evaluated in context of various factors such as grain size [eU] and a prolonged cooling history. Application of the in situ triple-dating approach has shown that apparent discrepancies between FT and (U–Th)/He ages do arise simply from such statistical misconceptions in conventional data treatment, where mean or single-grain (U–Th)/He ages were compared with the central apatite FT age, and not with the range of single-grain FT ages (e.g. Danišík et al. 2010a).

In addition to enhancing data interpretation, the routines used in FT dating can be applied to improve the quality of (U–Th)/He dating results. In some FT laboratories, it is common practice to evaluate the suitability of crystals for (U–Th)/He and U–Pb dating based on the sample quality as seen in the FT mounts at high-resolution ($\sim 1250 \times$). The high-resolution FT images provide information on crystal size, morphology, appearance and composition of inclusions, degree of radiation damage (e.g. Garver and Kamp 2002) and, perhaps most importantly, the distribution of U (Jolivet et al. 2003; Meesters and Dunai 2002a; Fitzgerald et al. 2006; Danišík et al. 2010a), which is critical for robust alpha ejection corrections (Farley et al. 1996; Hourigan et al. 2005). Although for apatite, U zonation is usually better revealed in induced tracks on mica external detectors, rather than spontaneous fission tracks in apatite grains. Instead of a random selection of grains for further geochronological analysis, such information can be utilised for a targeted grain selection strategy, aiming to represent all sub-populations present in the sample, which should lead to more realistic representation despite fewer grains being dated. Finally, long-term experience has shown that the quality of (U–Th)/He data obtained on samples previously dated by FT methods is better than the quality of data obtained on crystals handpicked and examined under a binocular or a low-magnification petrographic microscope.

Another practical application of FT imaging that simplifies a double-dating approach was reported by Evans et al. (2013). These authors applied SHRIMP U–Pb and conventional (U–Th)/He double-dating to zircons from a diamondiferous lamproite pipe and to detrital zircons from surrounding country rocks, in an attempt to test whether the lamproitic zircon crystals could be identified in detrital population based on their distinctive U–Pb/(U–Th)/He age pattern. The successful outcome of this experiment proved that the double-dating approach is a viable diamond exploration tool, the only practical limitation being the time and labour required to double-date a sufficient number of detrital grains. However, these authors showed that the analytical time of this approach could be dramatically reduced by employing chemical etching of fission tracks in zircon as a pre-screening procedure. Based on their different etching characteristics, this process distinguished lamproitic zircon from the majority of country rock zircons, making the procedure more time- and cost-efficient.

5.6   Concluding Remarks and Future Perspective

This chapter has described the principles, methodologies, applications and advantages of an in situ multi-dating approach in single crystals where the FT method is applied in combination with the U–Pb and/or (U–Th)/He methods. Notably for detrital samples, the first major advantage of multi-dating within a single crystal, as compared to applying a single method, is that data from multiple techniques provides more constraints on the thermal history. The high-temperature U–Pb geochronometer typically records processes at higher temperatures and at greater lithospheric depths, whereas lower-temperature FT and (U–Th)/He thermochronometers are sensitive to thermal changes in upper crust. The second major advantage is that the default provision of three ages from independent radiometric systems on single grains permits a direct internal consistency check of the results, which allows the researcher to identify analytical outliers and thus significantly improve the quality of data for geological interpretation.

Multi-dating has already proven to be a feasible and extremely powerful tool in detrital dating studies, as when detrital crystal geochronological data is obtained by single-dating methods, there may be some inherent ambiguities associated with a lack of geological context. Multi-dating may overcome this limitation by providing critical information about provenance, exhumation, deposition and post-depositional thermal history of source rocks. Finally, the triple-dating approach can be beneficial in solving apparent "inverted" FT and (U–Th)/He age relation issues and therefore offers a useful tool to address the occasional inconsistencies between FT and (U–Th)/He data, which can help to improve the understanding of FT and (U–Th)/He methods in general.

Several analytical procedures can used to obtain combined U–Pb, FT and (U–Th)/He ages on single grains. The most commonly applied approach employs EDM FT dating, U–Pb dating by LA-(MC)-ICPMS and conventional (U–Th)/He dating. This procedure is more efficient when employing FT dating using the LA-ICPMS instead of the EDM FT method where samples are irradiated in a nuclear reactor. An even more efficient approach with higher sample throughput combines FT dating by LA-ICPMS and in situ (U–Th)/He dating, whereby the U–Pb age is obtained as a by-product of LA-ICPMS analysis. Development of this promising concept is currently underway and initial results on the Durango apatite standard are encouraging, which suggests that this approach will be feasible in the future.

Future directions for the triple-dating approach should include the development and optimisation of methodologies and analytical instruments allowing rapid production of high-quality data. Thus far, the triple-dating approach has been successfully applied to zircon and apatite; however, multi-dating methodologies for other minerals (e.g. titanite, monazite, allanite) are yet to be developed and tested. The expected increase of triple-dating studies for detrital sample suites may call for development of new statistical approaches to data deconvolution and identification of principal components in multidimensional space. Last but not least, the reliability of triple-dating data sets and the capability to recover desired age information from detrital grains should be rigorously tested on synthetic samples or well-characterised natural test sites, before triple-dating is applied more broadly.

Acknowledgements This work was supported by the AuScope NCRIS2 program and Australian Scientific Instruments Pty Ltd. I would like to thank M. G. Malusà and P. G. Fitzgerald for editorial handling and help in conceiving Fig. 5.5, I. Dunkl for introducing me into FT world, N. Evans for training me in U–Th analysis, improvement of the manuscript, provision of apatite U–Pb data and constructive comments, D. Patterson for training me in Helium analysis and troubleshooting, C. Kirkland for stimulating discussions and processing of apatite U–Pb data, B. McDonald for development of RESOchron methods and help with LA-ICPMS analysis, T. Becker for help with AFM work, and C. May and C. Scadding from TSW Analytical for access to the solution ICPMS laboratory. I am grateful for the support of B. McInnes, M. Shelley, B. Godfrey, D. Gibbs, C. Gabay, A. Norris, P. Lanc and M. Hamel throughout the development of the RESOchron instrumentation. Constructive reviews by M. Zattin, B. Carrapa and the editors are acknowledged with thanks.

References

Barbarand J, Carter A, Wood I, Hurford T (2003) Compositional and structural control of fission-track annealing in apatite. Chem Geol 198(1):107–137

Belton D, Brown R, Kohn B, Fink D, Farley K (2004) Quantitative resolution of the debate over antiquity of the central Australian landscape: implications for the tectonic and geomorphic stability of cratonic interiors. Earth Planet Sci Lett 219(1):21–34

Bernet M (2018) Chapter 15: Exhumation studies of mountain belts based on detrital fission-track analysis on sand and sandstones. In: Malusà MG, Fitzgerald PG (eds) Fission-track thermochronology and its application to geology. Springer

Bernet M, Spiegel C (2004) Introduction: detrital thermochronology. Geol S Am S 378:1–6

Bernet M, Garver JI (2005) Fission-track analysis of detrital zircon. Rev Mineral Geochem 58(1):205–237

Bernet M, van der Beek P, Pik R, Huyghe P, Mugnier JL, Labrin E, Szulc A (2006) Miocene to recent exhumation of the central Himalaya determined from combined detrital zircon fission-track and U/Pb analysis of Siwalik sediments, western Nepal. Basin Res 18(4):393–412

Boyce J, Hodges K, Olszewski W, Jercinovic M (2005) He diffusion in monazite: Implications for (U–Th)/He thermochronometry. Geochem Geophys Geosys 6(12)

Boyce J, Hodges K, Olszewski W, Jercinovic M, Carpenter B, Reiners PW (2006) Laser microprobe (U–Th)/He geochronology. Geochim Cosmochim Ac 70(12):3031–3039

Boyce J, Hodges K, King D, Crowley JL, Jercinovic M, Chatterjee N, Bowring S, Searle M (2009) Improved confidence in (U–Th)/He thermochronology using the laser microprobe: an example from a Pleistocene leucogranite, Nanga Parbat, Pakistan. Geochem Geophys Geosys 10(9)

Brandon MT, Roden-Tice MK, Garver JI (1998) Late Cenozoic exhumation of the Cascadia accretionary wedge in the Olympic Mountains, northwest Washington State. Geol Soc Am Bull 110(8):985–1009

Burtner RL, Nigrini A, Donelick RA (1994) Thermochronology of Lower Cretaceous source rocks in the Idaho-Wyoming thrust belt. AAPG Bull 78(10):1613–1636

Campbell IH, Reiners PW, Allen CM, Nicolescu S, Upadhyay R (2005) He–Pb double dating of detrital zircons from the Ganges and Indus Rivers: implication for quantifying sediment recycling and provenance studies. Earth Planet Sci Lett 237(3):402–432

Carrapa B (2010) Resolving tectonic problems by dating detrital minerals. Geology 38(2):191–192

Carrapa B, DeCelles PG, Reiners PW, Gehrels GE, Sudo M (2009) Apatite triple dating and white mica ^{40}Ar/^{39}Ar thermochronology of syntectonic detritus in the Central Andes: a multiphase tectonothermal history. Geology 37(5):407–410

Carter A (2018) Chapter 14: Thermochronology on sand and sandstones for stratigraphic and provenance studies. In: Malusà MG, Fitzgerald PG (eds) Fission-track thermochronology and its application to geology. Springer

Carter A, Moss SJ (1999) Combined detrital-zircon fission-track and U–Pb dating: a new approach to understanding hinterland evolution. Geology 27(3):235–238

Carter A, Bristow C (2000) Detrital zircon geochronology: enhancing the quality of sedimentary source information through improved methodology and combined U–Pb and fission-track techniques. Basin Res 12(1):47–57

Carter A, Bristow C (2003) Linking hinterland evolution and continental basin sedimentation by using detrital zircon thermochronology: a study of the Khorat Plateau Basin, eastern Thailand. Basin Res 15(2):271–285

Chamberlain KR, Bowring SA (2001) Apatite–feldspar U–Pb thermochronometer: a reliable, mid-range (∼450 °C), diffusion-controlled system. Chem Geol 172(1):173–200

Cherniak D (1993) Lead diffusion in titanite and preliminary results on the effects of radiation damage on Pb transport. Chem Geol 110(1–3):177–194

Cherniak DJ (2010) Diffusion in accessory minerals: zircon, titanite, apatite, monazite and xenotime. Rev Mineral Geochem 72(1):827–869

Cherniak D, Watson E (2001) Pb diffusion in zircon. Chem Geol 172(1):5–24

Cherniak DJ, Watson EB (2003) Diffusion in zircon. Rev Mineral Geochem 53(1):113–143

Cherniak D, Lanford W, Ryerson F (1991) Lead diffusion in apatite and zircon using ion implantation and Rutherford backscattering techniques. Geochim Cosmochim Ac 55(6):1663–1673

Cherniak D, Watson EB, Grove M, Harrison TM (2004) Pb diffusion in monazite: a combined RBS/SIMS study. Geochim Cosmochim Ac 68(4):829–840

Chew DM, Donelick RA (2012) Combined apatite fission track and U–Pb dating by LA-ICP-MS and its application in apatite provenance analysis. Quant Miner Microanal Sediments Sed Rocks: Minerall Ass Can Short Course 42:219–247

Chew DM, Sylvester PJ, Tubrett MN (2011) U–Pb and Th-Pb dating of apatite by LA-ICPMS. Chem Geol 280(1):200–216

Cochrane R, Spikings RA, Chew D, Wotzlaw J-F, Chiaradia M, Tyrrell S, Schaltegger U, Van der Lelij R (2014) High temperature (>350 °C) thermochronology and mechanisms of Pb loss in apatite. Geochim Cosmochim Ac 127:39–56

Cox R, Košler J, Sylvester P, Hodych J Apatite fission-track (FT) dating by LAM-ICP-MS analysis. J Conf Abstr, 2000. p 322

Coyle D, Wagner G (1998) Positioning the titanite fission-track partial annealing zone. Chem Geol 149(1):117–125

Danišík M, Kuhlemann J, Dunkl I, Székely B, Frisch W (2007) Burial and exhumation of Corsica (France) in the light of fission track data. Tectonics 26(1)

Danišík M, Sachsenhofer RF, Privalov VA, Panova EA, Frisch W, Spiegel C (2008) Low-temperature thermal evolution of the Azov Massif (Ukrainian Shield—Ukraine)—Implications for interpreting (U–Th)/He and fission track ages from cratons. Tectonophysics 456(3):171–179

Danišík M, Pfaff K, Evans NJ, Manoloukos C, Staude S, McDonald BJ, Markl G (2010a) Tectonothermal history of the Schwarzwald ore district (Germany): an apatite triple dating approach. Chem Geol 278(1):58–69

Danišík M, Sachsenhofer R, Frisch W, Privalov V, Panova E, Spiegel C (2010b) Thermotectonic evolution of the Ukrainian Donbas Foldbelt revisited: new constraints from zircon and apatite fission track data. Basin Res 22(5):681–698

Danišík M, Kuhlemann J, Dunkl I, Evans NJ, Székely B, Frisch W (2012) Survival of ancient landforms in a collisional setting as revealed by combined fission track and (U–Th)/He thermochronometry: a case study from Corsica (France). J Geol 120(2):155–173

Danišík M, Fodor L, Dunkl I, Gerdes A, Csizmeg J, Hámor-Vidó M, Evans NJ (2015) A multi-system geochronology in the Ad-3 borehole, Pannonian Basin (Hungary) with implications for dating volcanic rocks by low-temperature thermochronology and for interpretation of (U–Th)/He data. Terra Nova 27(4):258–269

Danišík M, McInnes BI, Kirkland CL, McDonald BJ, Evans NJ, Becker T (2017) Seeing is believing: visualization of He distribution in zircon and implications for thermal history reconstruction on single crystals. Science Advances 3(2):e1601121

Djimbi DM, Gautheron C, Roques J, Tassan-Got L, Gerin C, Simoni E (2015) Impact of apatite chemical composition on (U–Th)/He thermochronometry: An atomistic point of view. Geochim Cosmochim Ac 167:162–176

Dodson MH (1973) Closure temperature in cooling geochronological and petrological systems. Contrib Mineral Petr 40(3):259–274

Donelick RA, Miller DS (1991) Enhanced TINT fission track densities in low spontaneous track density apatites using ^{252}Cf-derived fission fragment tracks: a model and experimental observations.

Int J Rad Appl Instr Part D Nuclear Tracks Rad Meas 18(3):301–307

Donelick RA, Ketcham RA, Carlson WD (1999) Variability of apatite fission-track annealing kinetics: II. Crystallographic orientation effects. Am Mineral 84(9):1224–1234

Donelick RA, O'Sullivan PB, Ketcham RA (2005) Apatite fission-track analysis. Rev Mineral Geochem 58(1):49–94

Donelick R, O'Sullivan P, Ketcham R, Hendriks B, Redfield T (2006) Relative U and Th concentrations from LA-ICP-MS for apatite fission-track grain-age dating. Geochim Cosmochim Ac 70(18): A143

Ehlers TA, Farley KA (2003) Apatite (U–Th)/He thermochronometry: methods and applications to problems in tectonic and surface processes. Earth Planet Sci Lett 206(1):1–14

Emmel B, Jacobs J, Crowhurst P, Daszinnies M (2007) Combined apatite fission-track and single grain apatite (U–Th)/He ages from basement rocks of central Dronning Maud Land (East Antarctica)—Possible identification of thermally overprinted crustal segments? Earth Planet Sci Lett 264(1):72–88

Enkelmann E, Jonckheere R, Wauschkuhn B (2005) Independent fission-track ages (φ-ages) of proposed and accepted apatite age standards and a comparison of φ-, Z-, ζ-and ζ 0-ages: implications for method calibration. Chem Geol 222(3):232–248

Evans NJ, McInnes BI, McDonald B, Danišík M, Jourdan F, Mayers C, Thern E, Corbett D (2013) Emplacement age and thermal footprint of the diamondiferous Ellendale E9 lamproite pipe, Western Australia. Mineralium Deposita 48(3):413–421

Evans N, McInnes B, McDonald B, Danišík M, Becker T, Vermeesch P, Shelley M, Marillo-Sialer E, Patterson D (2015) An in situ technique for (U–Th–Sm)/He and U–Pb double dating. J Analyt Atom Spect 30(7):1636–1645

Farley K (2000) Helium diffusion from apatite: general behavior as illustrated by Durango fluorapatite. J Geophys Res B 105 (B2):2903–2914

Farley KA (2002) (U–Th)/He dating: Techniques, calibrations, and applications. Rev Mineral Geochem 47(1):819–844

Farley K, Wolf R, Silver L (1996) The effects of long alpha-stopping distances on (U–Th)/He ages. Geochim Cosmochim Ac 60 (21):4223–4229

Fayon A (2011) Fission track dating of monazite: etching efficiencies as a function of U content. Paper presented at GSA Annual Meeting in Minneapolis, 9–12 October 2011

Fitzgerald P, Baldwin SL, Webb L, O'Sullivan PB (2006) Interpretation of (U–Th)/He single grain ages from slowly cooled crustal terranes: a case study from the Transantarctic Mountains of southern Victoria Land. Chem Geol 225(1):91–120

Fleischer RL, Price PB, Walker RM (1975) Nuclear tracks in solids: principles and applications. University of California Press

Flowerdew M, Millar IL, Curtis ML, Vaughan A, Horstwood M, Whitehouse MJ, Fanning CM (2007) Combined U–Pb geochronology and Hf isotope geochemistry of detrital zircons from early Paleozoic sedimentary rocks, Ellsworth-Whitmore Mountains block, Antarctica. Geol Soc Am Bull 119(3–4):275–288

Flowers RM (2009) Exploiting radiation damage control on apatite (U–Th)/He dates in cratonic regions. Earth Planet Sci Lett 277(1):148–155

Flowers R, Farley K (2012) Apatite ⁴He/³He and (U–Th)/He evidence for an ancient Grand Canyon. Science 338(6114):1616–1619

Flowers R, Farley K (2013) Response to Comments on "Apatite ⁴He/³He and (U–Th)/He evidence for an ancient Grand Canyon". Science 340(6129):143

Flowers RM, Kelley SA (2011) Interpreting data dispersion and "inverted" dates in apatite (U–Th)/He and fission-track datasets: an example from the US midcontinent. Geochim Cosmochim Ac 75 (18):5169–5186

Flowers R, Shuster D, Wernicke B, Farley K (2007) Radiation damage control on apatite (U–Th)/He dates from the Grand Canyon region, Colorado Plateau. Geology 35(5):447–450

Flowers RM, Ketcham RA, Shuster DL, Farley KA (2009) Apatite (U–Th)/He thermochronometry using a radiation damage accumulation and annealing model. Geochim Cosmochim Ac 73(8):2347–2365

Flowers RM, Farley KA, Ketcham RA (2015) A reporting protocol for thermochronologic modeling illustrated with data from the Grand Canyon. Earth Planet Sci Lett 432:425–435

Flowers RM, Farley KA, Ketcham RA (2016) Response to comment on "A reporting protocol for thermochronologic modeling illustrated with data from the Grand Canyon". Earth Planet Sci Lett 441:213

Fox M, Shuster DL (2014) The influence of burial heating on the (U–Th)/He system in apatite: Grand Canyon case study. Earth Planet Sci Lett 397:174–183

Galbraith R, Laslett G (1993) Statistical models for mixed fission track ages. Nuclear Tracks Rad Meas 21(4):459–470

Gallagher K (2012) Transdimensional inverse thermal history modeling for quantitative thermochronology. J Geophys Res: Sol Ea 117 (B2)

Gallagher K (2016) Comment on "A reporting protocol for thermochronologic modeling illustrated with data from the Grand Canyon" by Flowers, Farley and Ketcham. Earth Planet Sci Lett 441:211–212

Gallagher K, Brown R, Johnson C (1998) Fission track analysis and its applications to geological problems. Ann Rev Earth Planet Sci 26 (1):519–572

Gardés E, Jaoul O, Montel J-M, Seydoux-Guillaume A-M, Wirth R (2006) Pb diffusion in monazite: an experimental study of Pb²⁺ + Th⁴⁺ ⇔ 2Nd³⁺ interdiffusion. Geochim Cosmochim Ac 70 (9):2325–2336

Garver JI (2003) Etching zircon age standards for fission-track analysis. Rad Meas 37(1):47–53

Garver JI, Kamp PJ (2002) Integration of zircon color and zircon fission-track zonation patterns in orogenic belts: application to the Southern Alps. New Zealand. Tectonophysics 349(1):203–219

Gautheron C, Tassan-Got L (2010) A Monte Carlo approach to diffusion applied to noble gas/helium thermochronology. Chem Geol 273(3):212–224

Gleadow A (1981) Fission-track dating methods: what are the real alternatives? Nuclear Tracks 5(1–2):3–14

Gleadow A, Fitzgerald P (1987) Uplift history and structure of the Transantarctic Mountains: new evidence from fission track dating of basement apatites in the Dry Valleys area, southern Victoria Land. Earth Planet Sci Lett 82(1–2):1–14

Gleadow A, Lovering J (1974) The effect of weathering on fission track dating. Earth Planet Sci Lett 22(2):163–168

Gleadow A, Duddy I, Lovering J (1983) Fission track analysis: a new tool for the evaluation of thermal histories and hydrocarbon potential. Austr Petrol Expl Ass J 23:93–102

Gleadow A, Duddy I, Green PF, Lovering J (1986a) Confined fission track lengths in apatite: a diagnostic tool for thermal history analysis. Contrib Mineral Petr 94(4):405–415

Gleadow AJ, Duddy IR, Green PF, Hegarty KA (1986b) Fission track lengths in the apatite annealing zone and the interpretation of mixed ages. Earth Planet Sci Lett 78(2–3):245–254

Gleadow AJ, Gleadow SJ, Belton DX, Kohn BP, Krochmal MS, Brown RW (2009) Coincidence mapping-a key strategy for the automatic counting of fission tracks in natural minerals. Geol Soc Spec Publ 324(1):25–36

Gleadow AJ, Kohn BP, Lugo-Zazueta R, Alimanovic A (2012) The use of coupled image analysis and laser-ablation ICP-MS in fission track thermochronology. In: Goldschmidt Conference Abstracts 1765, Montreal, 24–29 June, 2012

Gleadow AJ, Kohn B, Seiler C (2018) Chapter 4: the future of fission-track thermochronology. In: Malusà MG, Fitzgerald PG (eds) Fission-track thermochronology and its application to geology. Springer

Green P, Duddy I (2006) Interpretation of apatite (U–Th)/He ages and fission track ages from cratons. Earth Planet Sci Lett 244(3):541–547

Green P, Duddy I, Laslett G, Hegarty K, Gleadow AW, Lovering J (1989a) Thermal annealing of fission tracks in apatite 4. Quantitative modelling techniques and extension to geological timescales. Chem Geol: Isotope Geosc Sect 79(2):155–182

Green PF, Duddy IR, Gleadow AJ, Lovering JF (1989b) Apatite fission-track analysis as a paleotemperature indicator for hydrocarbon exploration. In: Thermal history of sedimentary basins. Springer, pp 181–195

Green PF, Crowhurst PV, Duddy IR, Japsen P, Holford SP (2006) Conflicting (U–Th)/He and fission track ages in apatite: enhanced He retention, not anomalous annealing behaviour. Earth Planet Sci Lett 250(3):407–427

Guenthner WR, Reiners PW, Ketcham RA, Nasdala L, Giester G (2013) Helium diffusion in natural zircon: Radiation damage, anisotropy, and the interpretation of zircon (U–Th)/He thermochronology. Am J Sci 313(3):145–198

Hanchar JM, Hoskin PW (2003) Zircon–reviews in mineralogy and geochemistry, vol 53. Mineralogical society of America/geochemical society, p 500

Hansen K, Reiners PW (2006) Low temperature thermochronology of the southern East Greenland continental margin: evidence from apatite (U–Th)/He and fission track analysis and implications for intermethod calibration. Lithos 92(1):117–136

Hasebe N, Barbarand J, Jarvis K, Carter A, Hurford AJ (2004) Apatite fission-track chronometry using laser ablation ICP-MS. Chem Geol 207(3):135–145

Hawkins DP, Bowring SA (1999) U–Pb monazite, xenotime and titanite geochronological constraints on the prograde to post-peak metamorphic thermal history of Paleoproterozoic migmatites from the Grand Canyon, Arizona. Contrib Mineral Petr 134(2):150–169

Hendriks BWH (2003) Cooling and Denudation of the Norwegian and Barents Sea Margins, Northern Scandinavia: Constrained by Apatite Fission Track and (U–Th) He Thermochronology. Vrije universiteit

Hendriks B, Redfield T (2005) Apatite fission track and (U–Th)/He data from Fennoscandia: an example of underestimation of fission track annealing in apatite. Earth Planet Sci Lett 236(1):443–458

Hendriks BWH, Redfield TF (2006) Reply to: comment on "Apatite fission track and (U–Th)/He data from Fennoscandia: an example of underestimation of fission track annealing in apatite" by BWH hendriks and TF redfield. Earth Planet Sci Lett 248(1–2):569–577

Horne AM, van Soest MC, Hodges KV, Tripathy-Lang A, Hourigan JK (2016) Integrated single crystal laser ablation U/Pb and (U–Th)/He dating of detrital accessory minerals–Proof-of-concept studies of titanites and zircons from the Fish Canyon tuff. Geochim Cosmochim Ac 178:106–123

Hourigan JK, Reiners PW, Brandon MT (2005) U–Th zonation-dependent alpha-ejection in (U–Th)/He chronometry. Geochim Cosmochim Ac 69(13):3349–3365

House M, Kohn B, Farley K, Raza A (2002) Evaluating thermal history models for the Otway Basin, southeastern Australia, using (U–Th)/He and fission-track data from borehole apatites. Tectonophysics 349(1):277–295

Hurford AJ (1986) Cooling and uplift patterns in the Lepontine Alps South Central Switzerland and an age of vertical movement on the Insubric fault line. Contrib Mineral Petr 92(4):413–427

Hurford AJ (2018) Chapter 1: an historical perspective on fission-track thermochronology. In: Malusà MG, Fitzgerald PG (eds) Fission-track thermochronology and its application to geology. Springer

Hurley PM (1952) Alpha ionization damage as a cause of low helium ratios. Eos, Trans Am Geophys Union 33(2):174–183

Jennings E, Marschall H, Hawkesworth C, Storey C (2011) Characterization of magma from inclusions in zircon: apatite and biotite work well, feldspar less so. Geology 39(9):863–866

Jolivet M, Dempster T, Cox R (2003) Distribution of U and Th in apatites: implications for U–Th/He thermo chronology. CR Geosci, 899–906

Jonckheere R, Ratschbacher L (2015) Standardless fission-track dating of the Durango apatite age standard. Chem Geol 417:44–57

Karlstrom KE, Lee J, Kelley S, Crow R, Young RA, Lucchitta I, Beard LS, Dorsey R, Ricketts JW, Dickinson WR, Crossey L (2013) Comment on "Apatite ^{4}He/^{3}He and (U-Th)/He evidence for an ancient grand canyon". Science 340(6129):143. http://dx.doi.org/10.1126/science.1233982

Ketcham RA (2005) Forward and inverse modeling of low-temperature thermochronometry data. Rev Mineral Geochem 58(1):275–314

Ketcham R (2018) Chapter 3: fission track annealing: from geologic observations to thermal modeling. In: Malusà MG, Fitzgerald PG (eds) Fission-track thermochronology and its application to geology. Springer

Ketcham RA, Donelick RA, Carlson WD (1999) Variability of apatite fission-track annealing kinetics: III. Extrapolation to geological time scales. Am Mineral 84(9):1235–1255

Ketcham RA, Carter A, Donelick RA, Barbarand J, Hurford AJ (2007a) Improved measurement of fission-track annealing in apatite using c-axis projection. Am Mineral 92(5–6):789–798

Ketcham RA, Carter A, Donelick RA, Barbarand J, Hurford AJ (2007b) Improved modeling of fission-track annealing in apatite. Am Mineral 92(5–6):799–810

Ketcham RA, Donelick RA, Balestrieri ML, Zattin M (2009) Reproducibility of apatite fission-track length data and thermal history reconstruction. Earth Planet Sci Lett 284(3):504–515

Ketcham RA, Guenthner WR, Reiners PW (2013) Geometric analysis of radiation damage connectivity in zircon, and its implications for helium diffusion. Am Mineral 98(2–3):350–360

Kinny PD, Maas R (2003) Lu–Hf and Sm–Nd isotope systems in zircon. Rev Mineral Geochem 53(1):327–341

Kirkland C, Erickson T, Johnson T, Danišík M, Evans N, Bourdet J, McDonald B (2016a) Discriminating prolonged, episodic or disturbed monazite age spectra: An example from the Kalak Nappe Complex, Arctic Norway. Chem Geol 424:96–110

Kirkland C, Spaggiari C, Johnson T, Smithies R, Danišík M, Evans N, Wingate M, Clark C, Spencer C, Mikucki E (2016b) Grain size matters: implications for element and isotopic mobility in titanite. Precambrian Res 278:283–302

Kohn BP, Gleadow A (2018) Chapter 21: application of low-temperature thermochronology to craton evolution. In: Malusà MG, Fitzgerald PG (eds) Fission-track thermochronology and its application to geology. Springer

Kohn BP, Green PF (2002) Low temperature thermochronology: from tectonics to landscape evolution. Elsevier

Kohn BP, Lorencak M, Gleadow AJ, Kohlmann F, Raza A, Osadetz KG, Sorjonen-Ward P (2009) A reappraisal of low-temperature thermochronology of the eastern Fennoscandia Shield and radiation-enhanced apatite fission-track annealing. Geol Soc Spec Publ 324(1):193–216

Kohn BP, Chung L, Gleadow A (2018) Chapter 2: fission-track analysis: field collection, sample preparation and data acquisition. In: Malusà MG, Fitzgerald PG (eds) Fission-track thermochronology and its application to geology. Springer

Košler J, Sylvester PJ (2003) Present trends and the future of zircon in geochronology: laser ablation ICPMS. Rev Mineral Geochem 53(1):243–275

Krogh T (1982) Improved accuracy of U–Pb zircon ages by the creation of more concordant systems using an air abrasion technique. Geochim Cosmochim Ac 46(4):637–649

Ksienzyk AK, Dunkl I, Jacobs J, Fossen H, Kohlmann F (2014) From orogen to passive margin: constraints from fission track and (U–Th)/He analyses on Mesozoic uplift and fault reactivation in SW Norway. Geol Soc Spec Publ 390(SP390):327

Lee J, Stockli D, Kelley S, Pederson J, Karlstrom K, Ehlers T (2013) New thermochronometric constraints on the Tertiary landscape evolution of the central and eastern Grand Canyon, Arizona. Geosphere 9(2):216–228

Lisker F, Ventura B, Glasmacher U (2009) Apatite thermochronology in modern geology. Geol Soc Spec Publ 324(1):1–23

Liu W, Zhang J, Sun T, Wang J (2014) Application of apatite U–Pb and fission-track double dating to determine the preservation potential of magnetite–apatite deposits in the Luzong and Ningwu volcanic basins, eastern China. J Geochem Expl 138:22–32

Lorencak M (2003) Low temperature thermochronology of the Canadian and Fennoscandian Shields: Integration of apatite fission track and (U–Th)/He methods. University of Melbourne, School of Earth Sciences

Malusà MG, Fitzgerald PG (2018) Chapter 8: from cooling to exhumation: setting the reference frame for the interpretation of thermocronologic data. In: Malusà MG, Fitzgerald PG (eds) Fission-track thermochronology and its application to geology. Springer

Malusà MG, Fitzgerald PG (2018) Chapter 10: application of thermochronology to geologic problems: bedrock and detrital approaches. In: Malusà MG, Fitzgerald PG (eds) Fission-track thermochronology and its application to geology. Springer

Malusà MG, Garzanti E (2018) Chapter 7: the sedimentology of detrital thermochronology. In: Malusà MG, Fitzgerald PG (eds) Fission-track thermochronology and its application to geology. Springer

Malusà MG, Danišík M, Kuhlemann J (2016) Tracking the Adriatic-slab travel beneath the Tethyan margin of Corsica-Sardinia by low-temperature thermochronometry. Gondwana Res 31:135–149

Malusà MG, Wang J, Garzanti E, Liu ZC, Villa IM, Wittmann H (2017) Trace-element and Nd-isotope systematics in detrital apatite of the Po river catchment: implications for prove-nance discrimination and the lag-time approach to detrital thermochronology. Lithos 290–291:48–59

McDowell FW, McIntosh WC, Farley KA (2005) A precise ^{40}Ar–^{39}Ar reference age for the Durango apatite (U–Th)/He and fission-track dating standard. Chem Geol 214(3):249–263

McInnes BI, Evans NJ, Fu FQ, Garwin S (2005) Application of thermochronology to hydrothermal ore deposits. Rev Mineral Geochem 58(1):467–498

McInnes BI, Evans NJ, McDonald BJ, Kinny PD, Jakimowicz J (2009) Zircon U–Th-Pb-He double dating of the Merlin kimberlite field, Northern Territory, Australia. Lithos 112:592–599

Meesters A, Dunai T (2002a) Solving the production-diffusion equation for finite diffusion domains of the various shapes, part 1; implications for low temperature (U–Th)/He thermochronology

Meesters A, Dunai T (2002b) Solving the production-diffusion equation for finite diffusion domains of various shapes: Part II. Application to cases with α-ejection and nonhomogeneous distribution of the source. Chem Geol 186(3):347–363

Moore TE, O'Sullivan PB, Potter CJ, Donelick RA (2015) Provenance and detrital zircon geochronologic evolution of lower Brookian foreland basin deposits of the western Brooks Range, Alaska, and implications for early Brookian tectonism. Geosphere 11(1):93–122

Morton A, Yaxley G (2007) Detrital apatite geochemistry and its application in provenance studies. Geol S Am S 420:319–344

Parrish RR, Noble SR (2003) Zircon U–Th-Pb geochronology by isotope dilution—thermal ionization mass spectrometry (ID-TIMS). Rev Mineral Geochem 53(1):183–213

Price P, Walker R (1963) Fossil tracks of charged particles in mica and the age of minerals. J Geophys Res 68(16):4847–4862

Rahl JM, Reiners PW, Campbell IH, Nicolescu S, Allen CM (2003) Combined single-grain (U–Th)/He and U/Pb dating of detrital zircons from the Navajo Sandstone. Utah Geol 31(9):761–764

Rahn MK, Brandon MT, Batt GE, Garver JI (2004) A zero-damage model for fission-track annealing in zircon. Am Mineral 89(4):473–484

Reiners PW (2004) Thermochronology of wildfire and fault heating through single-grain (U–Th)/He and fission-track double-dating. In: Conference Abstracts of the GSA Annual Meeting, Denver, 7–10, 2004

Reiners PW (2005) Zircon (U–Th)/He thermochronometry. Rev Mineral Geochem 58(1):151–179

Reiners PW, Ehlers TA (2005) Low-temperature thermochronology: techniques, interpretations, and applications; Ed.: PW Reiners, TA Ehlers. Miner Soc Am. Washington

Reiners PW, Farley KA (1999) Helium diffusion and (U–Th)/He thermochronometry of titanite. Geochim Cosmochim Ac 63(22):3845–3859

Reiners PW, Farley KA (2001) Influence of crystal size on apatite (U–Th)/He thermochronology: an example from the Bighorn Mountains, Wyoming. Earth Planet Sci Lett 188(3):413–420

Reiners P, Campbell I, Nicolescu S, Allen C, Garver J, Hourigan J, Cowan D (2004) Double-and triple-dating of single detrital zircons with (U–Th)/He, fission-track, and U/Pb systems, and examples from modern and ancient sediments of the western US. In: AGU Fall Meeting Abstracts, 2004

Reiners PW, Spell TL, Nicolescu S, Zanetti KA (2004b) Zircon (U–Th)/He thermochronometry: He diffusion and comparisons with ^{40}Ar/^{39}Ar dating. Geochim Cosmochim Ac 68(8):1857–1887

Reiners PW, Campbell I, Nicolescu S, Allen CM, Hourigan J, Garver J, Mattinson J, Cowan D (2005) (U–Th)/(He-Pb) double dating of detrital zircons. Am J Sci 305(4):259–311

Schaltegger U, Schmitt A, Horstwood M (2015) U–Th–Pb zircon geochronology by ID-TIMS, SIMS, and laser ablation ICP-MS: recipes, interpretations, and opportunities. Chem Geol 402:89–110

Schoene B, Latkoczy C, Schaltegger U, Günther D (2010) A new method integrating high-precision U–Pb geochronology with zircon trace element analysis (U–Pb TIMS-TEA). Geochim Cosmochim Ac 74(24):7144–7159

Shen C-B, Donelick RA, O'Sullivan PB, Jonckheere R, Yang Z, She Z-B, Miu X-L, Ge X (2012) Provenance and hinterland exhumation from LA-ICP-MS zircon U–Pb and fission-track double dating of Cretaceous sediments in the Jianghan Basin, Yangtze block, central China. Sed Geol 281:194–207

Shuster DL, Farley KA (2009) The influence of artificial radiation damage and thermal annealing on helium diffusion kinetics in apatite. Geochim Cosmochim Ac 73(1):183–196

Shuster D, Flowers R, Farley K (2006) Radiation damage and helium diffusion kinetics in apatite. Geochim Cosmochim Ac 70(18):A590

Siebel W, Danišík M, Chen F (2009) From emplacement to unroofing: thermal history of the Jiazishan gabbro, Sulu UHP terrane, China. Mineral Petrol 96(3–4):163–175

Söderlund P, Juez-Larré J, Page LM, Dunai TJ (2005) Extending the time range of apatite (U–Th)/He thermochronometry in slowly cooled terranes: Palaeozoic to Cenozoic exhumation history of southeast Sweden. Earth Planet Sci Lett 239(3):266–275

Jt Stacey, Kramers J (1975) Approximation of terrestrial lead isotope evolution by a two-stage model. Earth Planet Sci Lett 26(2):207–221

Stockli DF (2005) Application of low-temperature thermochronometry to extensional tectonic settings. Rev Mineral Geochem 58(1):411–448

Stockli DF, Farley KA (2004) Empirical constraints on the titanite (U–Th)/He partial retention zone from the KTB drill hole. Chem Geol 207(3):223–236

Stockli DF, Farley KA, Dumitru TA (2000) Calibration of the apatite (U–Th)/He thermochronometer on an exhumed fault block, White Mountains. Calif Geol 28(11):983–986

Svojtka M, Košler J (2002) Fission-track dating of zircon by laser ablation ICPMS. In: Conference Abstracts of the Goldchmidt Conference, Geochim Cosmochim Ac, 2002. vol 15 A, pp A756–A756, Davos, 18–23 August, 2002

Tagami T (2005) Zircon fission-track thermochronology and applications to fault studies. Rev Mineral Geochem 58(1):95–122

Thomson SN, Gehrels GE, Ruiz J, Buchwaldt R (2012) Routine low-damage apatite U-Pb dating using laser ablation–multicollector–ICPMS. Geochem Geophys Geosys 13(2)

Tripathy-Lang A, Hodges KV, Monteleone BD, Soest MC (2013) Laser (U–Th)/He thermochronology of detrital zircons as a tool for studying surface processes in modern catchments. J Geophys Res: Earth 118(3):1333–1341

van Soest MC, Hodges KV, Wartho JA, Biren MB, Monteleone BD, Ramezani J, Spray JG, Thompson LM (2011) (U-Th)/He dating of terrestrial impact structures: the Manicouagan example. Geochem Geophys Geosys 12(5)

Vermeesch P (2004) How many grains are needed for a provenance study? Earth Planet Sci Lett 224(3):441–451

Vermeesch P (2018) Chapter 6: statistics for fission-track thermochronology. In: Malusà MG, Fitzgerald PG (eds) Fission-track thermochronology and its application to geology. Springer

Vermeesch P, Miller DD, Graham SA, De Grave J, McWilliams MO (2006) Multimethod detrital thermochronology of the Great Valley Group near New Idria, California. Geol Soc Am Bull 118(1–2):210–218

Vermeesch P, Sherlock SC, Roberts NM, Carter A (2012) A simple method for in-situ U–Th–He dating. Geochim Cosmochim Ac 79:140–147

Wagner G, Reimer G (1972) Fission track tectonics: the tectonic interpretation of fission track apatite ages. Earth Planet Sci Lett 14 (2):263–268

Wagner G, van den Haute P (1992) Fission-track dating. Enke, Stuttgart, p 285

Weise C, van den Boogaart KG, Jonckheere R, Ratschbacher L (2009) Annealing kinetics of Kr-tracks in monazite: implications for fission-track modelling. Chem Geol 260(1):129–137

Weisheit A, Bons P, Danišík M, Elburg M (2014) Crustal-scale folding: Palaeozoic deformation of the Mt Painter Inlier, South Australia. Geol Soc Spec Publ 394(1):53–77

Zattin M, Balestrieri ML, Hasebe N, Ketcham R, Seward D, Sobel E, Spiegel C (2008) Notes from the first workshop of the IGCP 543-low temperature thermochronology: applications and interlaboratory calibration. Episodes 31(3):356–357

Zattin M, Andreucci B, Thomson SN, Reiners PW, Talarico FM (2012) New constraints on the provenance of the ANDRILL AND-2A succession (western Ross Sea, Antarctica) from apatite triple dating. Geochem Geophys Geosys 13(10)

Zaun P, Wagner G (1985) Fission-track stability in zircons under geological conditions. Nucl Tracks Rad Meas 10(3):303–307

Zeitler P, Herczeg A, McDougall I, Honda M (1987) U–Th-He dating of apatite: a potential thermochronometer. Geochim Cosmochim Ac 51(10):2865–2868

Statistics for Fission-Track Thermochronology

6

Pieter Vermeesch

Abstract

This chapter introduces statistical tools to extract geologically meaningful information from fission-track (FT) data using both the external detector and LA-ICP-MS methods. The spontaneous fission of ^{238}U is a Poisson process resulting in large single-grain age uncertainties. To overcome this imprecision, it is nearly always necessary to analyse multiple grains per sample. The degree to which the analytical uncertainties can explain the observed scatter of the single-grain data can be visually assessed on a radial plot and objectively quantified by a chi-square test. For sufficiently low values of the chi-square statistic (or sufficiently high p values), the pooled age of all the grains gives a suitable description of the underlying 'true' age population. Samples may fail the chi-square test for several reasons. A first possibility is that the true age population does not consist of a single discrete age component, but is characterised by a continuous range of ages. In this case, a 'random effects' model can constrain the true age distribution using two parameters: the 'central age' and the '(over)dispersion'. A second reason why FT data sets might fail the chi-square test is if they are underlain by multimodal age distributions. Such distributions may consist of discrete age components, continuous age distributions, or a combination of the two. Formalised statistical tests such as chi-square can be useful in preventing overfitting of relatively small data sets. However, they should be used with caution when applied to large data sets (including length measurements) which generate sufficient statistical 'power' to reject any simple yet geologically plausible hypothesis.

6.1 Introduction

^{238}U is the heaviest naturally occurring nuclide in the solar system. Like all nuclides heavier than ^{208}Pb, it is physically unstable and undergoes radioactive decay to smaller, more stable nuclides. 99.9998% of the ^{238}U nuclei shed weight by disintegrating into eight He-nuclei (α-particles) and a ^{206}Pb atom, forming the basis of the U–Pb and (U–Th)/He clocks. The remaining 0.0002% of the ^{238}U undergoes spontaneous fission, forming the basis of FT geochronology (Price and Walker 1963; Fleischer et al. 1965). Because spontaneous fission of ^{238}U is such a rare event, the surface density of fission tracks (in counts per unit area) is 10–11 orders of magnitude lower than the atomic abundances of ^{238}U and ^{4}He, respectively. So whereas the U–Pb and (U–Th)/He methods are based on mass spectrometric analyses of billions of Pb and He atoms, FT ages are commonly based on manual counts of at most a few dozen features. Due to these low numbers, the FT method is a low precision technique. Whereas the analytical uncertainty of U–Pb and (U–Th)/He ages is expressed in % or ‰-units, it is not uncommon for single-grain FT age uncertainties to exceed 10% or even 100% (Sect. 6.2). Early attempts to quantify these uncertainties (McGee and Johnson 1979; Johnson et al. 1979) were criticised by Green (1981a, b), who subsequently engaged in a fruitful collaboration with two statisticians—Geoff Laslett and Rex Galbraith—to eventually solve the problem. Thanks to the combined efforts of the latter two people, it is fair to say that the statistics of the FT method are better developed than those of any other geochronological technique. Several statistical tools that were originally developed for the FT method have subsequently found applications in other dating methods. Examples of this are the radial plot (Sect. 6.3), which is routinely used in luminescence dating (Galbraith 2010b), random effects models

P. Vermeesch (✉)
London Geochronology Centre, University College London,
London, UK
e-mail: p.vermeesch@ucl.ac.uk

© Springer International Publishing AG, part of Springer Nature 2019
M. G. Malusà and P. G. Fitzgerald (eds.), *Fission-Track Thermochronology and its Application to Geology*,
Springer Textbooks in Earth Sciences, Geography and Environment, https://doi.org/10.1007/978-3-319-89421-8_6

(Sect. 6.4.2), which have been generalised to (U–Th)/He (Vermeesch 2010) and U–Pb dating (Rioux et al. 2012), and finite mixture models (Sect. 6.5), which were adapted for detrital U–Pb geochronology (Sambridge and Compston 1994).

The statistical analysis of fission tracks is a rich and diverse field of research, and this short chapter cannot possibly cover all its intricacies. Numerate readers are referred to the book by Galbraith (2005), which provides a comprehensive, detailed and self-contained review of the subject, from which the present chapter heavily borrows. The chapter comprises five sections, which address statistical issues of progressively higher order. Section 6.2 introduces the FT age equation using the external detector method (EDM), which offers the most straightforward and elegant way to estimate single-grain age uncertainties, even in grains without spontaneous fission tracks. Section 6.3 compares and contrasts different ways to visually represent multi-grain assemblages of FT data, including kernel density estimates, cumulative age distributions and radial plots. Section 6.4 reviews the various ways to estimate the 'average' age of such multi-grain assemblages, including the arithmetic mean age, the pooled age and the central age. This section will also introduce a chi-square test for age homogeneity, which is used to assess the extent to which the scatter of the single-grain ages exceeds the formal analytical uncertainties obtained from Sect. 6.2. This leads to the concept of 'overdispersion' (Sect. 6.4.2) and more complex distributions consisting of one or several continuous and/or discrete age components. Section 6.5 discusses three classes of mixed effects models to resolve discrete mixtures, continuous mixtures and minimum ages, respectively. It will show that these models obey the classic bias–variance trade-off, which will lead to a cautionary note regarding the use of formalised statistical hypothesis tests for FT interpretation. Finally, Sect. 6.6 will give the briefest of introductions to some statistical aspects of thermal history modelling, a more comprehensive discussion of which is provided in Chap. 3, (Ketcham 2018). In recent years, several fission-track laboratories around the world have abandoned the elegance and robustness of the EDM for the convenience of ICP-MS-based measurements. Unfortunately, the statistics of the latter is less straightforward and less well developed than that of the EDM. Section 6.7 presents an attempt to address this problem.

6.2 The Age Equation

The fundamental FT age equation is given by:

$$t = \frac{1}{\lambda_D} \ln\left(1 + \frac{\lambda_D}{\lambda_f} \frac{\rho_s}{[^{238}U]R}\right) \quad (6.1)$$

where λ_D is the total decay constant of ^{238}U (1.55125×10^{-10} year^{-1}; Jaffey et al. 1971), λ_f is the fission decay constant (7.9–8.7×10^{-17} year^{-1}; Holden and Hoffman 2000)[1], ρ_s is the density (tracks per unit area) of the spontaneous fission tracks on an internal crystal surface, $[^{238}U]$ is the current number of ^{238}U atoms per unit volume, and R is etchable range of the fission tracks, which is half the equivalent isotropic FT length. $[^{238}U]$ can be determined by irradiating the (etched) sample with thermal neutrons in a reactor. This irradiation induces synthetic fission of ^{235}U in the mineral, producing tracks that can be monitored by attaching a mica detector to the polished mineral surface and etching this monitor subsequent to irradiation. Using this external detector method (EDM), Eq. 6.1 can be rewritten as:

$$t = \frac{1}{\lambda_D} \ln\left(1 + \frac{1}{2}\lambda_D \zeta \rho_d \frac{\rho_s}{\rho_i}\right) \quad (6.2)$$

where ζ is a calibration factor (Hurford and Green 1983), ρ_i is the surface density of the induced fission tracks in the mica detector and ρ_d is the surface density of the induced fission tracks in a dosimeter glass of known (and constant) U-concentration. The latter value is needed to 'recycle' the calibration constant from one irradiation batch to the next, as neutron fluences might vary through time, or within a sample stack. ρ_s, ρ_i and ρ_d are unknown but can be estimated by counting the number of tracks N_* over a given area A_* (where '*' is either 's' for 'spontaneous', 'i' for 'induced' or 'd' for 'dosimeter'):

$$\hat{\rho}_s = \frac{N_s}{A_s}, \quad \hat{\rho}_i = \frac{N_i}{A_i} \text{ and } \hat{\rho}_d = \frac{N_d}{A_d} \quad (6.3)$$

It is customary for the spontaneous and induced fission tracks to be counted over the same area (i.e. $A_s = A_i$), either using an automated microscope stage (Smith and Leigh-Jones 1985; Dumitru 1993) or by simply repositioning the mica detector on the grain mount after etching (Jonckheere et al. 2003). Using these measurements, the estimated FT age ($\hat{t}$) is given by

$$\hat{t} = \frac{1}{\lambda_D} \ln\left(1 + \frac{1}{2}\lambda_D \hat{\zeta} \hat{\rho}_d \frac{N_s}{N_i}\right) \quad (6.4)$$

where $\hat{\zeta}$ is obtained by applying Eq. 6.4 to an age standard and rearranging. Equations 6.2 and 6.4 assume that the ratio of the etchable range (R) between the grain and the mica detector is the same for the sample and the standard.

---

[1]The uncertainty associated with the fission decay constant vanishes when λ_f is folded into the ζ-calibration constant. This is one of the main reasons why the ζ-method was developed (see Chap. 1, Hurford 2018).

Violation of this assumption leads to apparent FT ages of unclear geological significance. This is an important caveat as samples with shortened tracks are very common. See Sect. 6.6 and Chap. 3 (Ketcham 2018) for further details on how to deal with this situation. The standard error $s[\hat{t}]$ of the single-grain age estimate is given by standard first-order Taylor expansion:

$$s[\hat{t}]^2 \approx \left(\frac{\partial \hat{t}}{\partial \hat{\zeta}}\right)^2 s[\hat{\zeta}]^2 + \left(\frac{\partial \hat{t}}{\partial N_s}\right)^2 s[N_s]^2 + \left(\frac{\partial \hat{t}}{\partial N_i}\right)^2 s[N_i]^2 + \left(\frac{\partial \hat{t}}{\partial N_d}\right)^2 s[N_d]^2$$

(6.5)

where it is important to point out that all covariance terms are zero because $\hat{\zeta}$, N_s, N_i and N_d are independent variables.[2] To simplify the calculation of the partial derivatives, we note that $\ln(1 + x) \approx x$ if $x \ll 1$ so that, for reasonably low N_s/N_i values, Eq. 6.4 reduces to

$$\hat{t} \approx \frac{1}{2}\hat{\zeta}\hat{\rho}_d \frac{N_s}{N_i}$$

(6.6)

Using this linear approximation, it is easy to show that Eq. 6.5 becomes:

$$\left(\frac{s[\hat{t}]}{\hat{t}}\right)^2 \approx \left(\frac{s[\hat{\zeta}]}{\hat{\zeta}}\right)^2 + \left(\frac{s[N_s]}{N_s}\right)^2 + \left(\frac{s[N_i]}{N_i}\right)^2 + \left(\frac{s[N_d]}{N_d}\right)^2$$

(6.7)

The standard error of the calibration constant $\hat{\zeta}$ is obtained by repeated measurements of the age standard and will not be discussed further. The standard errors of N_s, N_i and N_d are governed by the Poisson distribution, whose mean equals its variance. This crucial property can be illustrated with a physical example in which a mica print attached to a dosimeter glass is subdivided into a number of equally sized squares (Fig. 6.1, left). Counting the number of induced fission tracks N_d in each square yields a skewed frequency distribution whose mean indeed equals its variance (Fig. 6.1, right). Applying this fact to Eq. 6.7, we can replace $s[N_s]^2$ with N_s, $s[N_i]^2$ with N_i and $s[N_d]^2$ with N_d to obtain the following expression for the standard error of the estimated FT age:

$$s[\hat{t}] \approx \hat{t}\sqrt{\left(\frac{s[\hat{\zeta}]}{\hat{\zeta}}\right)^2 + \frac{1}{N_s} + \frac{1}{N_i} + \frac{1}{N_d}}$$

(6.8)

Note that this equation breaks down if $N_s = 0$. There are two solutions to this problem. The easiest of these is to replace N_s and N_i with $N_s + 1/2$ and $N_i + 1/2$, respectively (Galbraith 2005, p. 80). A second (and preferred) approach is to calculate exact (and asymmetric) confidence intervals. See Galbraith (2005, p. 50) for further details about this procedure.

6.3 Fission-Track Plots

The single-grain uncertainties given by Eq. 6.8 tend to be very large. For example, a grain containing just four spontaneous fission tracks (i.e. $N_s = 4$) is associated with an analytical uncertainty of $\sqrt{4}/4 = 50\%$ even ignoring the analytical uncertainty associated with the ζ-calibration constant, the dosimeter glass, or the induced FT count. The single-grain age precision of the FT method, then, is orders of magnitude lower than that of other established geochronometers such as $^{40}Ar/^{39}Ar$ or $^{206}Pb/^{238}U$, which achieve percent or permil level uncertainties. To overcome this limitation and 'beat down the noise', it is important that multiple grains are analysed from a sample and averaged using methods described in Sect. 6.4. Multi-grain assemblages of FT data are also very useful for sedimentary provenance analysis and form the basis of a new field of research called 'detrital thermochronology' (Bernet 2018; Carter 2018). Irrespective of the application, it is useful for any multi-grain FT data set to first be assessed visually. This section will introduce three graphical devices to do this: cumulative age distributions, (kernel) density estimates and radial plots. To illustrate these graphical devices as well as the different summary statistics of Sect. 6.4, consider the four different geological scenarios shown in Fig. 6.2:

I. A rapidly cooled volcanic rock extruded at 15 Ma.
II. A slowly cooled intrusive rock exhibiting a range of Cl/F ratios resulting in a 150 Ma ± 20% range of apparent FT ages.
III. A detrital sample collected from a river draining two volcanic layers extruded at 15 and 75 Ma, respectively.
IV. A detrital sample collected from a river draining lithologies I and II.

6.3.1 The Cumulative Age Distribution (CAD)

The cumulative distribution function cdf(x) describes the fraction of the detrital age population whose age is less than or equal to x:

$$\text{cdf}(x) = P(t \le x)$$

(6.9)

---

[2]N_s, N_i are independent within a single grain, but of course not between different grains of the same sample, as the spontaneous and induced track counts both depend on the U-concentration, which tends to vary significantly from grain to grain (McGee and Johnson 1979; Johnson et al. 1979; Green 1981b; Galbraith 1981; Carter 1990).

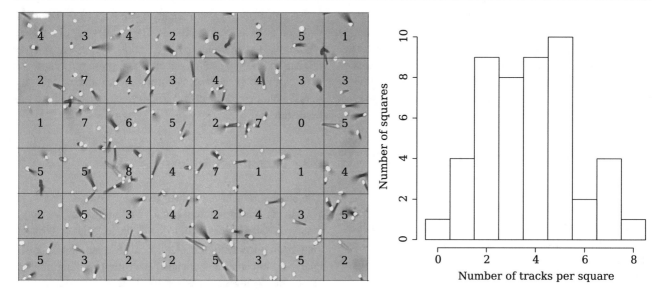

Fig. 6.1 Left: induced fission tracks recorded in a mica detector attached to a dosimeter glass. Numbers indicate the number of tracks counted in 48 150 × 150 µm-sized areas. Colours indicate single (yellow), double (blue) and triple (red) etch pits. Dosimeter glasses exhibit a uniform U-concentration so that the observed variation in the number of tracks is only due to Poisson statistics. Fission tracks were counted with FastTracks image recognition software (see Chap. 4, Gleadow 2018). Right: the frequency distribution of the FT counts, which has a mean of 3.7 and a variance of 3.5 counts per graticule, consistent with a Poisson distribution

Under Scenario I, the cdf consists of a simple step function, indicating that 0% of the grains are younger, and 100% are older than the extrusive age (Fig. 6.2, I-a). Under Scenario II, the cdf is spread out over a wider range, so that 90% of the ages are between 90 and 210 Ma (Fig. 6.2, II-a). Under Scenario III (Fig. 6.2, III-a), the cdf consists of two discrete steps at 15 and 75 Ma, the relative heights of which depend on the hypsometry of the river catchment and the spatial distribution of erosion (Vermeesch 2007). Finally, under Scenario IV, the cdf consists of a discrete step from 0 to 50% at 15 Ma, followed by a sigmoidal rise to 100% at 75 Ma (Fig. 6.2, IV-a).

In reality, the cdfs of Scenarios I–IV are, of course, unknown and must be estimated from sample data, by means of an empirical cumulative distribution function (ecdf), which may be referred to as a cumulative age distribution (CAD) in a geochronological context (Vermeesch 2007). A CAD is simply a step function in which the single-grain age estimates ($\hat{t}_j$, for $j = 1 \rightarrow n$) are plotted against their rank order:

$$CAD(x) = \sum_{j=1}^{n} 1(\hat{t}_j \leq x)/n \quad (6.10)$$

where 1(TRUE) = 1 and 1(FALSE) = 0. In contrast with the true cdfs, the measured CADs are invariably smoother, as the analytical uncertainties spread the dates out over a greater range. Because the uncertainties of FT ages are so large, the difference between the measured CADs and the true cdfs is very significant. Sections 6.4 and 6.5 of this chapter present several algorithms to extract the key parameters of the true age distribution (i.e. the cdfs) from the measurement distribution (CADs).

6.3.2 (Kernel) Density Estimates (KDEs)

A probability density function (pdf) is defined as the first derivative of the cdf:

$$pdf(x) = \frac{d[cdf(y)]}{dy}\bigg|_x \Leftrightarrow cdf(x) = \int_{-\infty}^{x} pdf(y)dy \quad (6.11)$$

Under Scenario I, the pdf is a discrete peak of zero width and infinite height, marking the timing of the volcanic eruption (Fig. 6.2, I-b). In contrast, under Scenario II, the pdf is a smooth (a)symmetric bell curve reflecting the spread in closing temperatures and, hence, ages, associated with the range of Cl/F-ratios present in apatites of this slowly cooled pluton (Fig. 6.2, II-b). Under Scenario III, the pdf consists of two discrete spikes corresponding to the two volcanic events (Fig. 6.2, III-b). Finally, under Scenario IV, the pdf effectively combines those of Scenarios I and II (Fig. 6.2, IV-b). The pdfs, like the cdfs discussed before, are unknown but can be estimated from sample data.

There are several ways to do this. Arguably the simplest of these is the histogram, in which the observations are grouped into a number of discrete bins. Kernel density estimates (KDEs) are a continuous alternative to the histogram, which are constructed by arranging the measurements from young

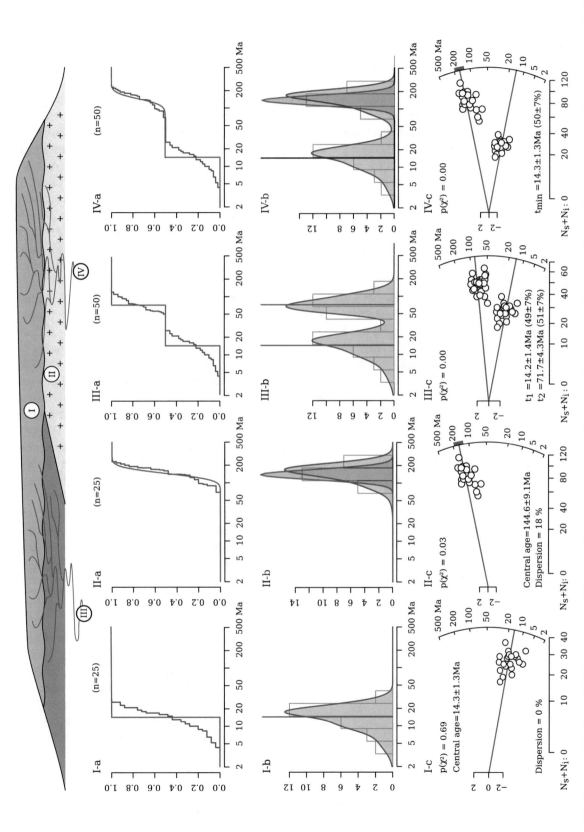

Fig. 6.2 Hypothetical landscape (hill) featuring four different sampling locations: I—a rapidly cooled volcanic rock extruded at 15 Ma; II—a slowly cooled plutonic rock containing 150 Ma ± 20% apatites; III—a modern river draining lithology I and an older (75 Ma) volcanic deposit; and IV—a second river draining lithologies I and II. For each of these scenarios, synthetic data were generated and plotted on three different graphical devices: a— theoretical cumulative distribution function (cdf, red) and synthetic cumulative age distribution (CAD, blue) of the samples at the four sampling locations; b—theoretical pdfs (red) and synthetic sample kernel density estimates (KDEs, blue) for the same synthetic data; c—radial plots of the synthetic data with model fits (Sects. 6.4 and 6.5)

to old along the time axis, adding a Gaussian 'bell curve' (or any other symmetric shape) on top of them and then summing those to create one continuous curve (Silverman 1986; Vermeesch 2012). The standard deviation of the Gaussian 'kernel' is called the 'bandwidth' of the estimator and may be chosen through a host of different approaches, a proper discussion of which falls outside the scope of this review (Abramson 1982; Silverman 1986; Botev et al. 2010; Vermeesch 2012). An important feature of all these algorithms is that the bandwidth monotonically decreases with increasing sample size. Please note that the so-called probability density plot (PDP, not to be confused with pdf!), in which the analytical uncertainty (or 0.6 times the analytical uncertainty, Brandon 1996) is used as a 'bandwidth' does not possess this feature. Therefore, PDPs are not proper density estimates, and consequently, their use is not recommended (Galbraith 1998; Vermeesch 2012). Like the CAD, which is a smooth version of the cdf, KDEs (and histograms) are smooth versions of the pdf. But whereas the CAD has only been smoothed once, histograms and KDEs are smoothed twice, once by the analytical uncertainties, and once by the width of the bins or kernels. Because the analytical uncertainties of FT data are so big, the components of FT age distributions are often spread out very widely, resulting in poorly resolved KDEs (blue curves in Figs. 6.2b).

6.3.3 Radial Plots

Single-grain fission-track age uncertainties are not only large, but generally also variable ('heteroscedastic'). Due to a combination of Poisson sampling statistics and variable U-concentrations, the analytical uncertainties propagated using Eq. 6.8 may vary over an order of magnitude within the same sample. Neither CADs nor KDEs (let alone PDPs) are able to capture this uncertainty. The radial plot is a graphical device that was specifically designed to address this issue (Galbraith 1988, 1990; Dunkl 2002; Vermeesch 2009). Given $j = 1 \ldots n$ numerical values z_j and their analytical uncertainties σ_j, the radial plot is a bivariate (x_j, y_j) scatterplot setting out a standardised estimate $(y_j = [z_j - z_0]/\sigma_j)$, where z_0 is some reference value) against the single-grain precision $(x_j = 1/\sigma_j)$. For FT data using the EDM,[3] it is convenient to use the following definitions for z_j and σ_j (Galbraith 1990):

$$z_j = \arcsin\sqrt{\frac{N_{sj} + 3/8}{N_{sj} + N_{ij} + 3/4}} \qquad (6.12)$$

and

$$\sigma_j = \frac{1}{2\sqrt{N_{sj} + N_{ij} + 1/2}} \qquad (6.13)$$

Precise measurements plot towards the right-hand side of the radial plot while imprecise measurements plot closer to the origin. A single-grain age may be read off by extrapolating a line from the origin (0,0) of the radial plot through the sample point (x_j, y_j) to a radial scale plotted at some convenient distance. Similarly, the analytical uncertainty can be obtained by extrapolating lines from the origin to the radial scale through the top and the bottom of an imaginary 2σ-error bar added to each sample point. Finally, drawing two parallel lines at 2σ distances from either side of the origin allow the analyst to visually assess whether all the single-grain ages within a sample agree within the analytical uncertainties.

Revisiting Scenario I of Fig. 6.2, the data points plot within a 2σ band on the radial plot, consistent with a single discrete age component (Fig. 6.2, I-c). Under Scenario II, the data are more dispersed and scatter beyond the 2σ band, reflecting the dispersion of the underlying geological ages (Fig. 6.2, II-c). Under Scenario III, the data are randomly scattered along two linear trajectories which represent the two volcanic events (Fig. 6.2, III-c). Finally, Scenario IV combines the radial patterns of Scenarios I and II, as expected (Fig. 6.2, IV-c). Of all the summary plots in Fig. 6.2, the radial plot contains the largest amount of quantitative information about the age measurements and about the underlying geological ages. Using the graphical design principles of Tufte (1983), the radial plot exhibits a far higher 'ink-to-information ratio' than the CAD, KDE or histogram. We will therefore use it as a basis from which to introduce the summary statistics discussed in the next section of this chapter.

6.4 Summary Statistics

The previous sections have shown that the presence of large and highly variable analytical uncertainties can easily obscure the underlying age distribution and all the geologically meaningful information encoded by it. The next two sections will introduce some useful summary statistics which can be used to disentangle that geologically meaningful information from the random noise produced by the Poisson counting uncertainties.

6.4.1 The Pooled Age

Let us begin with the single discrete age component in Scenario I of the previous section. Several approaches can be

---

[3]The remainder of this and the next three sections of this chapter will focus on the EDM. Alternative equations for ICP-MS-based fission-track data are provided in Sect. 6.7.

used to estimate this age from a set of noisy sample data. Panels I-a, I-b and I-c of Fig. 6.2 show that the single-grain age estimates follow an asymmetric probability distribution (symmetric when plotted on a logarithmic scale) which is skewed towards older ages. This is a consequence of the fact that, if N_{sj} and N_{ij} are sampled from two independent Poisson distributions with expected values ρ_s and ρ_i, respectively, then the conditional probability of N_{sj} on $N_{sj} + N_{ij}$ follows a binomial distribution:

$$P(N_{sj}|N_{sj} + N_{ij}) = \binom{N_{sj} + N_{ij}}{N_{sj}} \theta^{N_{sj}} (1-\theta)^{N_{ij}} \equiv f_j(\theta)$$

(6.14)

where $\theta \equiv \rho_s/(\rho_s + \rho_i)$ and $\binom{a}{b}$ is the binomial coefficient. Given a sample of n sets of FT counts, this leads to the following (log-)likelihood function for θ:

$$\mathcal{L}(\theta) = \sum_{j=1}^{n} \ln f_j(\theta)$$

(6.15)

where $f_j(\theta)$ is the probability mass function for the jth grain defined in Eq. 6.14. As a first approach to obtaining an 'average' age, one might be tempted to simply take the arithmetic mean of the single-grain age estimates. Unfortunately, the arithmetic mean does not cope well with outliers and asymmetric distributions and therefore yields poor estimates of the geological age. The geometric mean fares much better. It is closely related to the 'central age', which is discussed in Sect. 6.4.2. The 'pooled age' is obtained by maximising Eq. 6.15 to obtain a 'maximum likelihood' estimate ($\hat{\theta}$), and substituting $e^{\hat{\theta}}$ for N_s/N_i in Eq. 6.2, where $N_s = \sum_{j=1}^{n} N_{sj}$ and $N_i = \sum_{j=1}^{n} N_{ij}$. This is equivalent to taking the sum of all the spontaneous and induced tracks, respectively, and treating these as if they belonged to a single crystal. This procedure yields the correct age if the true ages are indeed derived from a single discrete age component (i.e. Scenario I). However, if there is any dispersion of the true FT ages, as is the case under Scenario II, then the pooled age will be biased towards values that are far too old. Whether this is the case or not can be verified using a formalised statistical hypothesis test. Galbraith (2005, p. 46) shows that in the absence of excess dispersion, the following statistic:

$$c^2 = \frac{1}{N_s N_i} \sum_{j=1}^{n} \frac{(N_{sj} N_i - N_{ij} N_s)^2}{N_{sj} + N_{ij}}$$

(6.16)

follows a chi-square distribution with $n-1$ degrees of freedom. The probability of observing a value greater than c^2

under this distribution is called the p value and can be used to formally test the assumption of zero dispersion. A 0.05 cut-off is often used as a criterion to abandon the single-grain age model of Scenario I and, hence, the pooled age.

6.4.2 Central Ages and 'Overdispersion'

A more meaningful age estimate for Scenario II is obtained using a two-parameter 'random effects' model, in which the true ρ_s/ρ_i-ratio is assumed to follow a log-normal distribution with location parameter μ and scale parameter σ (Galbraith and Laslett 1993):

$$\ln(\rho_s/\rho_i) \sim \mathcal{N}(\mu, \sigma^2)$$

(6.17)

This model gives rise to a two-parameter log-likelihood function:

$$\mathcal{L}(\mu, \sigma^2) = \sum_{j=1}^{n} \ln f_j(\mu, \sigma^2)$$

(6.18)

where the probability mass function $f_j(\mu, \sigma^2)$ is defined as:

$$f_j(\mu, \sigma^2) = \binom{N_{sj} + N_{ij}}{N_{sj}} \int_{-\infty}^{\infty} \frac{e^{\beta N_{sj}} (1 + e^{\beta})^{-N_{sj} - N_{ij}}}{\sigma \sqrt{2\pi} e^{(\beta - \mu)^2/(2\sigma^2)}} d\beta$$

(6.19)

in which the FT ratios are subject to two sources of variation: the Poisson uncertainty described by Eq. 6.15 and an '(over)dispersion' factor σ. Maximising Eq. 6.18 results in two estimates $\hat{\mu}$ and $\hat{\sigma}$ and their respective standard errors. Substituting $e^{\hat{\mu}}$ for N_s/N_i in Eq. 6.2 produces the so-called central age. $\hat{\sigma}$ quantifies the excess scatter of the single-grain ages that cannot be explained by the Poisson counting statistics alone. This dispersion can be just as informative as the central age itself, as it encodes geologically meaningful information about the compositional heterogeneity and cooling history of the sample. In the absence of excess dispersion (i.e. if $\hat{\sigma} = 0$) the central age equals the pooled age.

6.5 Mixture Models

A FT data set may fail the chi-square test introduced in the previous section for different reasons. The true ages may exhibit excess scatter according to Eq. 6.18. Or it may be so that there are more than one age component (Galbraith and Green 1990; Galbraith and Laslett 1993). These components could either be discrete age peaks (Scenario III) or they could be any combination of discrete and continuous age components (Scenario IV).

6.5.1 Finite Mixtures

Finite mixture models are a generalisation of the discrete age model of Scenario I in which the true ages are not derived from a single, but from multiple age populations (Galbraith and Green 1990). Scenario III is an example of this with two such components. In contrast with the common age model of Scenario I, which is completely described by a single parameter (θ, or the pooled age), and the random effects model, which comprises two parameters (μ and σ, or the central age and overdispersion), the finite mixture of Scenario III requires three parameters. These are the age of the first component, the age of the second component and the proportion of the grains belonging to the first component. The proportion belonging to the second component is simply the complement of the latter value. Generalising to N components, the log-likelihood function becomes:

$$\mathcal{L}(\pi_k, \theta_k, \; k = 1 \ldots N) = \sum_{j=1}^{n} \ln \left[\sum_{k=1}^{N} \pi_k f_j(\theta_k) \right]$$

$$\text{with } \pi_N = 1 - \sum_{k=1}^{N-1} \pi_k \tag{6.20}$$

where $f_j(\theta_k)$ is given by Eq. 6.14. Equation 6.20 can be solved numerically. Applying it to the single component data set of Scenario I again yields the pooled age as a special case. The detrital FT ages in Scenario III clearly fall into two groups so it is quite evident that there are two age components. Unfortunately, the situation is not always this clear. Due to the large single-grain age uncertainties discussed in Sect. 6.2, the boundaries between adjacent age components are often blurred, making it difficult to decide how many 'peaks' to fit. Several statistical approaches may be used to answer this question. One possibility is to use a log-likelihood ratio test. Suppose that we have solved Eq. 6.20 for the case of $N = 2$ age components and denote the corresponding maximum log-likelihood value as $\mathcal{L}_2$. We then consider an alternative model with $N = 3$ components. This results in two additional parameters (π_2 and θ_3) and a new maximum log-likelihood value, $\mathcal{L}_3$. We can assess whether the three-component model is a significant improvement over the two-component fit by comparing twice the difference between $\mathcal{L}_3$ and $\mathcal{L}_2$ to a chi-square distribution with two degrees of freedom (because we have added two additional parameters) and calculating the corresponding p value like before. An illustration of the log-likelihood ratio test is provided in Sect. 6.5.2. An alternative approach is to maximise the so-called Bayes Information Criterion (BIC), which is defined as

$$\text{BIC} = -2\mathcal{L}_{\max} + p \ln(n) \tag{6.21}$$

where $\mathcal{L}_{\max}$ is the maximum log-likelihood of a model comprising p parameters and n grains. A worked example of this method is omitted for brevity, and the reader is referred to Galbraith (2005, p. 91) for further details.

6.5.2 Continuous Mixtures

So far we have considered pdfs consisting of a single discrete age peak (Scenario I), a single continuous age distribution (Scenario II) and multiple discrete age peaks (Scenario III). The logical next step is to consider multiple continuous age distributions (Jasra et al. 2006). In principle such models can be obtained by maximising the following likelihood function

$$\mathcal{L}(\pi_k, \mu_k, \sigma_k^2, \; k = 1 \ldots N) = \sum_{j=1}^{n} \ln \left[\sum_{k=1}^{N} \pi_k f_j(\mu_k, \sigma_k^2) \right]$$

$$\text{with } \pi_N = 1 - \sum_{k=1}^{N-1} \pi_k$$

$$\tag{6.22}$$

where $f_j(\mu_k, \sigma_k^2)$ is given by Eq. 6.19. However, in reality this is often impractical due to the high number of parameters involved, which require exceedingly large data sets. In detrital geochronology, the analyst rarely knows that the data are underlain by a continuous mixture and so it is tempting to reduce the number of unknown parameters by simply assuming a discrete mixture. Unfortunately, this is fraught with problems as well since there is no upper bound on the number of discrete age components to fit to a continuous data set. To illustrate this point, let us reconsider the data set of Scenario II, this time applying a finite mixture model rather than the random effects model of Sect. 6.4.2.

For a small sample of $n = 10$ grains, the chi-square test for age homogeneity yields a p value of 0.47, which is above the 0.05 cut-off and thus provides insufficient evidence to reject the common age model (Table 6.1; Fig. 6.3a). Increasing the sample size to $n = 25$ results in a p value of 0.03, justifying the addition of extra model parameters (Fig. 6.2, I-c). Further increasing the sample size to $n = 100$ reduces the likelihood of the common age model (Eq. 6.15) and results in a p value of 0.0027, well below the 0.05 cut-off. Let us now replace the common age model with a two-component finite mixture model. For the same 100-grain sample, this increases the

Table 6.1 Application of the log-likelihood ratio test to a finite mixture fitting experiment shown in Fig. 6.3. Rows mark different sample sizes (with n marking the number of grains) drawn from Scenario II. Columns labelled as $\mathcal{L}_N$ show the log-likelihood of different model fits, where N marks the number of components. Columns labelled as $p(\chi_2^2)$ list the p values of a chi-square test with two degrees of freedom, which can be used to assess whether it is statistically justified to increment the number of fitting parameters (N) by one

	$\mathcal{L}_1$	$p(\chi_2^2)$	$\mathcal{L}_2$	$p(\chi_2^2)$	$\mathcal{L}_3$	$p(\chi_2^2)$	$\mathcal{L}_4$
$n = 10$	−422.4	0.67	−422.0	1.00	−422.0	1.00	−422.0
$n = 100$	−4598.3	0.001	−4591.4	0.45	−4590.6	1.00	−4590.6
$n = 1000$	−45,030.7	0.00	−44,966.5	0.00003	−44,956.1	0.67	−44,955.7

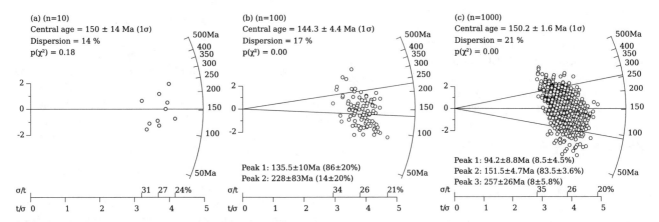

Fig. 6.3 Application of finite mixture modelling to the continuous mixture of Scenario II (Fig. 6.2). Increasing sample size from left (**a**) to right (**c**) provides statistical justification to fit more components using the log-likelihood ratio approach of Table 6.1. Note that the age of the youngest age component gets progressively younger with increasing sample size, from 150 Ma for sample (**a**) to 94 Ma for sample (**c**), and is therefore not a reliable estimator of the minimum age. $p(\chi^2)$ marks the p value of the chi-square test for age homogeneity and *not* the log-likelihood ratio tests of Table 6.1

log-likelihood from −4598.3 to −4591.4 (Table 6.1). Using the log-likelihood ratio test introduced in Sect. 6.5.1, that corresponds to a chi-square value of $2 \times (4598.3 − 4591.4) = 13.8$ and a p value of 0.001, lending support to the abandonment of the single age model in favour of the two-parameter model (Fig. 6.3b). However, doing the same calculation for a three-component model yields a log-likelihood of −4590.6 and a chi-square value of $2 \times (4591.4 − 4590.6) = 1.6$, resulting in an insignificant p value of 0.45 (Table 6.1). Thus, the 100-grain sample does *not* support the three-component model. It is only when the sample size is increased from 100 to 1000 grains that the chi-square test gains enough 'power' to justify the three-component finite mixture model (Fig. 6.3c). It is easy to see that this trend continues ad infinitum: with increasing sample size, it is possible to add ever larger numbers of components (Fig. 6.3).

One might object to this hypothetical example by noting that the finite mixture model is clearly inappropriate for a data set that is derived from a continuous mixture. But the key point is that *all* statistical models are inappropriate to some degree. Even the random effects model is a mathematical abstraction that does not exist in the real world. True age

distributions (pdfs) may be approximately (log)normal as in Scenario I, but they are never exactly so. Given a sufficiently large sample, formalised statistical hypothesis tests such as chi-square are always able to detect even the most minute deviation from any hypothetical age model and thereby provide statistical justification to add further parameters. This is important in the common situation where one is interested in the youngest age component of a fission-track age distribution, for example when one aims to calculate 'lag times' and estimate exhumation rates (Garver et al. 1999; Bernet 2018). It would be imprudent to estimate the lag-time by applying a general purpose multi-component mixture model and simply picking the youngest age component. This would provide a biased estimate of the minimum age, which would steadily drift towards younger values with increasing sample size (Fig. 6.3). Instead, it is better to use a simpler but more stable and robust model employing three[4] or four parameters to explicitly determine the minimum age component:

---

[4]Eq. 6.23 may be simplified by imposing the requirement that $\mu = e^\theta$, which significantly benefits numerical stability, while having only a minor effect on the accuracy of θ.

$$\mathcal{L}(\pi, \theta, \mu, \sigma^2) = \sum_{j=1}^{n} \ln\left[\pi f_j(\theta) + (1-\pi)f_j'(\mu, \sigma^2)\right] \quad (6.23)$$

where $f_j(\theta)$ is given by Eq. 6.14 and $f_j'(\mu, \sigma^2)$ is a truncated version of Eq. 6.19 (Galbraith and Laslett 1993). Applying this model to the synthetic example of Scenario IV correctly yields the age of the youngest volcanic unit regardless of sample size (Fig. 6.2, IV-c). In conclusion, statistical hypothesis tests such as chi-square can be used to prevent overinterpreting perceived 'clusters' of data that may arise from random statistical sampling fluctuations. But they must be used with caution, bearing in mind the simplifying assumptions which all mathematical models inevitably make and the dependence of test statistics and p values on sample size. Ignoring this dependence may lead to statistical models that might make sense in a mathematical sense, but have little or no geological relevance. This note of caution applies not only to mixture modelling but even more so to thermal history modelling, as will be discussed next.

6.6 Thermal History Modelling

So far in this chapter, we have made the implicit assumption (in Eq. 6.2) that all fission tracks have the same length. In reality, however, this is not the case and the length of (apatite) fission tracks varies anywhere between 0 and 16 μm as a function of the thermal history and chemical composition of a sample (Gleadow et al. 1986). If the compositional effects (notably the Cl/F ratio, Green et al. 1986) are well characterised, then the measured length distribution of horizontally confined fission tracks can be used to reconstruct the thermal history of a sample. Laboratory experiments show that the thermal annealing of fission tracks in apatite obeys a so-called fanning Arrhenius relationship, in which the degree of shortening logarithmically depends on both the amount and duration of heating (Green et al. 1985; Laslett et al. 1987; Laslett and Galbraith 1996; Ketcham et al. 1999, 2007):

$$\ln\left(1 - \sqrt[\Lambda]{L/L_0}\right) = c_0 - c_1 \frac{\ln(t) - \ln(t_c)}{1/T - 1/T_c} \quad (6.24)$$

where L_0 and L are the initial and measured track length, t and T are time and absolute temperature, respectively, and c_0, c_1, Λ, t_c and T_c are fitting parameters (Laslett and Galbraith 1996; Ketcham et al. 1999). Using these laboratory results, thermal history reconstructions are a two-step process. First, a large number of random thermal histories are generated, and for each of these, the fanning Arrhenius relationship is used to predict the corresponding FT length

distribution (Corrigan 1991; Lutz and Omar 1991; Gallagher 1995; Willett 1997; Ketcham et al. 2000; Ketcham 2005; Gallagher 2012). Then, these 'forward model' predictions are compared with the measured values and the 'best' matches are retained for geological interpretation. All the 'inverse modelling' software that has been developed over the years for the purpose of thermal history reconstructions essentially follows this same recipe. The most important difference between these algorithms is how they assess the goodness of fit and decide which candidate t–T paths to retain and which to reject.

One class of software, including the popular HeFTy program and its predecessor AFTSolve (Ketcham et al. 2000; Ketcham 2005), uses the p value of formalised hypothesis tests like the chi-square test described in the previous sections, to decide whether the measured FT length distribution is a 'good' ($p > 0.5$), 'acceptable' ($0.5 > p > 0.05$) or 'poor' ($p < 0.05$) fit. The problem with this approach is that, due to the dependence of p values on sample size, it inevitably breaks down for large data sets. This is because the statistical 'power' of statistical hypothesis test to resolve even the tiniest disagreement between the measured and the predicted length distribution, monotonically increases with increasing sample size (Vermeesch and Tian 2014).

A second class of inverse modelling algorithms (including QTQt, Gallagher 2012) does not employ formalised hypothesis tests or p values, but aims to extract the 'most likely' thermal history models among all possible t–T paths (Gallagher 1995; Willett 1997). These methods do not 'break down' when they are applied to large data sets. On the contrary, large data sets are 'rewarded' in the form of tighter 'credibility intervals' and higher resolution t–T paths. Furthermore, they are easily extended to multi-sample and multi-method data sets. However, with great power also comes great responsibility. Vermeesch and Tian (2014) show that QTQt always produces a 'best fitting' thermal history even for physically impossible data sets. To avoid this potential problem, it is of paramount importance that the model predictions are shown alongside the FT data (Gallagher 2012, 2016, Vermeesch and Tian 2014).

6.7 LA-ICP-MS-Based FT Dating

The EDM outlined in Sect. 6.2 continues to be the most widely used analytical protocol in FT dating. However, over the past decade, an increasing number of laboratories have abandoned it and switched to LA-ICP-MS as a means of determining the uranium concentration of datable minerals, thus reducing sample turnover time and removing the need to handle radioactive materials (Hasebe et al. 2004, 2009; Chew and Donelick 2012; Soares et al. 2014; Abdullin et al. 2016; Vermeesch 2017). The statistical analysis of ICP-MS-based FT data is less straightforward and less well developed than that of the EDM. As described in Sect. 6.2, the latter is based on simple ratios of Poisson variables, and forms the basis of a large edifice of statistical methods which cannot be directly applied to ICP-MS-based data. This section provides an attempt to address this issue.

6.7.1 Age Equation

The FT age equation for ICP-MS-based data is based on Eq. 6.1:

$$\hat{t} = \frac{1}{\lambda_D} \ln \left(1 + \frac{\lambda_D}{\lambda_f} \frac{N_s}{[\hat{U}] A_s R q} \right) \qquad (6.25)$$

where N_s is the number of spontaneous tracks counted over an area A_s, q is an 'efficiency factor' (~ 0.93 for apatite and ~ 1 for zircon, Iwano and Danhara 1998; Enkelmann and Jonckheere 2003; Jonckheere 2003; Soares et al. 2013) and $[\hat{U}]$ is the ^{238}U-concentration (in atoms per unit volume) measured by LA-ICP-MS. Equation 6.25 requires an explicit value for λ_f and assumes that the etchable range (R) is accurately known (Soares et al. 2014). Alternatively, these factors may be folded into a calibration factor akin to the EDM (Eq. 6.4):

$$\hat{t} = \frac{1}{\lambda_D} \ln \left(1 + \frac{1}{2} \lambda_D \hat{\zeta} \frac{N_s}{A_s [\hat{U}]} \right) \qquad (6.26)$$

in which $\hat{\zeta}$ is determined by analysing a standard of known FT age (Hasebe et al. 2004). Note that, in contrast with the 'absolute' dating method of Eq. 6.25, the ζ-calibration method of Eq. 6.26 allows $[\hat{U}]$ to be expressed in any concentration units (e.g. ppm or wt% of total U) or could even be replaced with the measured U/Ca-, U/Si- or U/Zr-ratios produced by the ICP-MS instrument. The standard error of the estimated age is given by

$$s[\hat{t}] \approx \hat{t} \sqrt{\left(\frac{s[\hat{\zeta}]}{\hat{\zeta}} \right)^2 + \left(\frac{s[\hat{U}]}{\hat{U}} \right)^2 + \frac{1}{N_s}} \qquad (6.27)$$

for the ζ-calibration approach (Eq. 6.26), where $s[\hat{U}]$ is the standard error of the uranium concentration measurement (or

the U/Ca-ratio measurement, say), which can be estimated using two alternative approaches as discussed in Sect. 6.7.2. Equation 6.27 can also be applied to the 'absolute' dating method (Eq. 6.25) by simply setting $s[\hat{\zeta}]/\hat{\zeta} = 0$.

6.7.2 Error Propagation of LA-ICP-MS-Based Uranium Concentrations

Uranium-bearing minerals such as apatite and zircon often exhibit compositional zoning, which must either be removed or quantified in order to ensure unbiased ages.

1. The effect of compositional zoning can be *removed* by covering the entire counting area with one large laser spot (Soares et al. 2014) or a raster (Hasebe et al. 2004). $s[\hat{U}]$ is then simply given by the analytical uncertainty of the LA-ICP-MS instrument, which typically is an order of magnitude lower than the standard errors of induced track counts in the EDM.
2. Alternatively, the uranium heterogeneity can be *quantified* by analysing multiple spots per analysed grain (Hasebe et al. 2009). In this case, it is commonly found that the variance of the different uranium measurements within each grain far exceeds the formal analytical uncertainty of each spot measurement. The following paragraphs will outline a method to measure that dispersion, even if some of the grains in a sample were only visited by the laser once.

The true statistical distribution of the U-concentrations within each grain is unknown but is likely to be log-normal:

$$\ln[\hat{U}_{jk}] \sim \mathcal{N}(\mu_j, \sigma_j^2) \qquad (6.28)$$

where $\hat{U}_{jk}$ is the kth out of n_j uranium concentration measurements, and μ_j and σ_j^2 are the (unknown) mean and variance of a normal distribution. Unfortunately, it is difficult to accurately estimate these two parameters from just a handful of spot measurements and it is downright impossible if $n_j = 1$. This problem requires a simplifying assumption such as $\sigma_j = \sigma \, \forall j$. In that case, we can estimate the parameters of Eq. 6.28 as follows:

$$\hat{\mu}_j = \sum_{k=1}^{n_j} \ln[\hat{U}_{jk}] / n_j \qquad (6.29)$$

and

$$\hat{\sigma}^2 = \sum_{j=1}^{n} \sum_{k=1}^{n_j} \left(\ln[\hat{U}_{jk}] - \hat{\mu}_j \right)^2 / \sum_{j=1}^{n} (n_j - 1). \qquad (6.30)$$

The (geometric) mean uranium concentration and standard error of the jth grain are then given by

$$\hat{U}_j = \exp[\hat{\mu}_j] \qquad (6.31)$$

and

$$s[\hat{U}_j] = \hat{U}_j \hat{\sigma} \qquad (6.32)$$

which may be directly plugged into Eqs. 6.25–6.27 to calculate FT ages and uncertainties. Note that this procedure ignores the analytical uncertainty of the individual U-measurements. A more sophisticated approach that combines the analytical uncertainties of the U-measurements with the dispersion of multiple spot measurements is provided by Vermeesch (2017).

6.7.3 Zero Track Counts

In contrast with the EDM, ICP-MS-based FT data do not offer an easy way to deal with zero track counts. One pragmatic solution to this problem is to approximate the ICP-MS-based uranium concentration measurement with an 'equivalent induced track density', using the following linear transformation:

$$\hat{N}_{ij} = \rho_j A_{sj}[\hat{U}_j] \qquad (6.33)$$

where A_{sj} is the area over which the spontaneous tracks of the jth grain have been counted and ρ_j plays a similar role as ρ_d in Eq. 6.2. From the requirement that the variance of a Poisson-distributed variable equals its mean (Sect. 6.2), it follows that:

$$\hat{N}_{ij} = \rho_j^2 A_{sj}^2 s[\hat{U}_j]^2 \qquad (6.34)$$

from which it is easy to determine ρ_j. The analytical uncertainty for the jth single-grain age is then given by:

$$s[\hat{t}_j] \approx \hat{t}_j \sqrt{\frac{1}{N_{sj}} + \frac{1}{\hat{N}_{ij}} + \left(\frac{s[\hat{\zeta}]}{\zeta} \right)^2} \qquad (6.35)$$

where N_{sj} indicates the number of spontaneous tracks measured in the jth grain, and $s[\hat{\zeta}]/\zeta = 0$ for the 'absolute' dating approach (Eq. 6.25). The zero track problem can then be solved using the methods mentioned in Sect. 6.2.

6.7.4 Plots and Models

To plot ICP-MS-based fission-track data on a radial plot, we can replace Eqs. 6.12 and 6.13 with

$$z_j = \ln(\hat{t}_j), \qquad (6.36)$$

$$\text{and } \sigma_j = \sqrt{ \left(\frac{s[\hat{\zeta}]}{\hat{\zeta}} \right)^2 + \left(\frac{s[\hat{U}]}{\hat{U}} \right)^2 + \frac{1}{N_s} } \qquad (6.37)$$

respectively (Galbraith 2010a). Alternatively, a square root transformation may be more appropriate for young and/or U-poor samples (Galbraith, *pers. commun.*):

$$z_j = \sqrt{\hat{t}_j}, \qquad (6.38)$$

$$\text{and } \sigma_j = s[\hat{t}_j] / \left(2\sqrt{\hat{t}_j} \right) \qquad (6.39)$$

The binomial likelihood function of Sect. 6.4.2 may be replaced with an alternative form assuming normal errors. Thus, Eq. 6.15 becomes:

$$\mathcal{L}(\theta) = \sum_{j=1}^{n} \ln\left[\mathcal{N}(z_j|\theta, \sigma_j'^2) \right] \qquad (6.40)$$

where $\mathcal{N}(A|B, C)$ stands for 'the probability density of observing a value A from a normal distribution with mean B and variance C', z_j is defined by Eq. 6.36 and σ_j' is given by

$$\sigma_j' = s[\hat{t}_j]/\hat{t}_j \qquad (6.41)$$

The chi-square statistic (Eq. 6.16) may be redefined as

$$c^2 = \sum_{j=1}^{n} \left(z_j/\sigma_j' \right)^2 - \left(\sum_{j=1}^{n} z_j/\sigma_j'^2 \right)^2 / \sum_{j=1}^{n} 1/\sigma_j'^2 \qquad (6.42)$$

(Galbraith 2010a). Finally, the random effects model of Eq. 6.18 can be replaced with:

$$\mathcal{L}(\mu, \sigma^2) = \sum_{j=1}^{n} \ln\left[\mathcal{N}(z_j|\mu, \sigma^2 + \sigma_j'^2) \right] \qquad (6.43)$$

Equations 6.40 and 6.43 can be readily plugged into Eqs. 6.20, 6.22 and 6.23 to constrain finite mixtures, continuous mixtures and minimum age models for ICP-MS-based data, respectively.

Acknowledgements The author would like to thank reviewers Rex Galbraith, Mauricio Bermúdez and editors Marco Malusà and Paul Fitzgerald for detailed feedback. The mica counts of Fig. 6.1 were performed by Yuntao Tian.

References

Abdullin F, Solé J, Meneses-Rocha, JdJ, Solari L, Shchepetilnikova V, Ortega-Obregón C (2016) LA-ICP-MS-based apatite fission track dating of the Todos Santos Formation sandstones from the Sierra de Chiapas (SE Mexico) and its tectonic significance. Int Geol Rev 58 (1):32–48

Abramson IS (1982) On bandwidth variation in kernel estimates—a square root law. Ann Stat 1217–1223

Bernet M (2018) Exhumation studies based on detrital fission track analysis on sand and sandstones. In: Malusá M, Fitzgerald P (eds) Fission track thermochronology and its application to geology, chapter 14. Springer, Berlin

Botev ZI, Grotowski JF, Kroese DP (2010) Kernel density estimation via diffusion. Ann Stat 38:2916–2957

Brandon M (1996) Probability density plot for fission-track grain-age samples. Radiat Meas 26(5):663–676

Carter A (1990) The thermal history and annealing effects in zircons from the Ordovician of North Wales. Int J Radiat Appl Instrum. Part D. Nucl Tracks Radiat Meas 17(3):309–313

Carter A (2018) Thermochronology on sand and sandstones for stratigraphic and provenance studies. In: Malusá M, Fitzgerald P (eds) Fission track thermochronology and its application to geology, chapter 13. Springer, Berlin

Chew DM, Donelick RA (2012) Combined apatite fission track and U-Pb dating by LA-ICP-MS and its application in apatite provenance analysis. Quant Min Microanal Sed Sediment Rocks: Mineral Assoc Canada, Short Course 42:219–247

Corrigan J (1991) Inversion of apatite fission track data for thermal history information. J Geophys Res 96(B6):10347–10360

Dumitru TA (1993) A new computer-automated microscope stage system for fission-track analysis. Nucl Tracks Radiat Meas 21 (4):575–580

Dunkl I (2002) TRACKKEY: a Windows program for calculation and graphical presentation of fission track data. Comput Geosci 28(1):3–12

Enkelmann E, Jonckheere R (2003) Correction factors for systematic errors related to the track counts in fission-track dating with the external detector method. Radiat Meas 36(1):351–356

Fleischer RL, Price PB, Walker RM (1965) Tracks of charged particles in solids. Science 149(3682):383–393

Galbraith R (1981) On statistical models for fission track counts. J Int Assoc Math Geol 13(6):471–478

Galbraith R (1988) Graphical display of estimates having differing standard errors. Technometrics 30(3):271–281

Galbraith RF (1990) The radial plot: graphical assessment of spread in ages. Nucl Tracks Radiat Meas 17:207–214

Galbraith R (1998) The trouble with "probability density" plots of fission track ages. Radiat Meas 29:125–131

Galbraith RF (2005) Statistics for fission track analysis. CRC Press

Galbraith R (2010a) Statistics for LA-ICPMS fission track dating. In: Thermo2010—12th international conference on thermochronology, Glasgow, p 175

Galbraith RF (2010b) On plotting OSL equivalent doses. Ancient TL 28:1–10

Galbraith RF, Green PF (1990) Estimating the component ages in a finite mixture. Nucl Tracks Radiat Meas 17:197–206

Galbraith R, Laslett G (1993) Statistical models for mixed fission track ages. Nucl Tracks Radiat Meas 21(4):459–470

Gallagher K (1995) Evolving temperature histories from apatite fission-track data. Earth Planet Sci Lett 136(3):421–435

Gallagher K (2012) Transdimensional inverse thermal history modeling for quantitative thermochronology. J Geophys Res: Solid Earth (1978–2012) 117(B2)

Gallagher K (2016) Comment on 'A reporting protocol for thermochronologic modeling illustrated with data from the Grand Canyon' by Flowers, Farley and Ketcham. Earth Planet Sci Lett 441:211–212

Garver JI, Brandon MT, Roden-Tice M, Kamp PJ (1999) Exhumation history of orogenic highlands determined by detrital fission-track thermochronology. Geol Soc, London, Spec Publ 154(1):283–304

Gleadow A (2018) Future developments in fission track thermochronology. In: Malusà M, Fitzgerald P (eds) Fission track thermochronology and its application to geology, chapter 4. Springer, Berlin

Gleadow A, Duddy I, Green PF, Lovering J (1986) Confined fission track lengths in apatite: a diagnostic tool for thermal history analysis. Contrib Miner Petrol 94(4):405–415

Green P (1981a) A criticism of the paper entitled "A practical method of estimating standard error of age in the fission track dating method" by Johnson, Mcgee and Naeser. Nucl Tracks 5(3):317–323

Green P (1981b) A new look at statistics in fission-track dating. Nucl Tracks 5(1–2):77–86

Green P, Duddy I, Gleadow A, Tingate P, Laslett G (1985) Fission-track annealing in apatite: track length measurements and the form of the Arrhenius plot. Nucl Tracks Radiat Meas (1982) 10 (3):323–328

Green P, Duddy I, Gleadow A, Tingate P, Laslett G (1986) Thermal annealing of fission tracks in apatite: 1. A qualitative description. Chem Geol: Isot Geosci Sect 59:237–253

Hasebe N, Barbarand J, Jarvis K, Carter A, Hurford AJ (2004) Apatite fission-track chronometry using laser ablation icp-ms. Chem Geol 207(3):135–145

Hasebe N, Carter A, Hurford AJ, Arai S (2009) The effect of chemical etching on LA-ICP-MS analysis in determining uranium concentration for fission-track chronometry. Geol Soc, London, Spec Publ 324(1):37–46

Holden NE, Hoffman DC (2000) Spontaneous fission half-lives for ground-state nuclide (technical report). Pure Appl Chem 72 (8):1525–1562

Hurford AJ (2018) A historical perspective on fission track thermochronology. In: Malusà M, Fitzgerald P (eds) Fission track thermochronology and its application to geology, chapter 1. Springer, Berlin

Hurford AJ, Green PF (1983) The zeta age calibration of fission-track dating. Chem Geol 41:285–317

Iwano H, Danhara T (1998) A re-investigation of the geometry factors for fission-track dating of apatite, sphene and zircon. In: Advances in fission-track geochronology. Springer, Berlin, pp 47–66

Jaffey A, Flynn K, Glendenin L, Bentley W, Essling A (1971) Precision measurement of half-lives and specific activities of ^{235}U and ^{238}U. Phys Rev C 4(5):1889

Jasra A, Stephens DA, Gallagher K, Holmes CC (2006) Bayesian mixture modelling in geochronology via markov chain monte carlo. Math Geol 38(3):269–300

Johnson NM, McGee VE, Naeser CW (1979) A practical method of estimating standard error of age in the fission track dating method. Nucl Tracks 3(3):93–99

Jonckheere R (2003) On the densities of etchable fission tracks in a mineral and co-irradiated external detector with reference to fission-track dating of minerals. Chem Geol 200(1):41–58

Jonckheere R, Ratschbacher L, Wagner GA (2003) A repositioning technique for counting induced fission tracks in muscovite external detectors in single-grain dating of minerals with low and inhomogeneous uranium concentrations. Radiat Meas 37(3):217–219

Ketcham RA (2005) Forward and inverse modeling of low-temperature thermochronometry data. Rev Mineral Geochem 58(1):275–314

Ketcham RA (2018) Fission track annealing: from geological observations to thermal modeling. In: Malusà M, Fitzgerald P (eds) Fission track thermochronology and its application to geology, chapter 3. Springer, Berlin

Ketcham RA, Donelick RA, Carlson WD (1999) Variability of apatite fission-track annealing kinetics: III. Extrapolation to geological time scales. Am Mineral 84(9):1235–1255

Ketcham RA, Donelick RA, Donelick MB et al (2000) AFTSolve: a program for multi-kinetic modeling of apatite fission-track data. Geol Mater Res 2(1):1–32

Ketcham RA, Carter A, Donelick RA, Barbarand J, Hurford AJ (2007) Improved modeling of fission-track annealing in apatite. Am Mineral 92(5–6):799–810

Laslett G, Galbraith R (1996) Statistical modelling of thermal annealing of fission tracks in apatite. Geochim Cosmochim Acta 60(24):5117–5131

Laslett G, Green PF, Duddy I, Gleadow A (1987) Thermal annealing of fission tracks in apatite 2. A quantitative analysis. Chem Geol: Isot Geosci Sect 65(1):1–13

Lutz TM, Omar G (1991) An inverse method of modeling thermal histories from apatite fission-track data. Earth Planet Sci Lett 104(2–4):181–195

McGee VE, Johnson NM (1979) Statistical treatment of experimental errors in the fission track dating method. J Int Assoc Math Geol 11(3):255–268

Price P, Walker R (1963) Fossil tracks of charged particles in mica and the age of minerals. J Geophys Res 68(16):4847–4862

Rioux M, Lissenberg CJ, McLean NM, Bowring SA, MacLeod CJ, Hellebrand E, Shimizu N (2012) Protracted timescales of lower crustal growth at the fast-spreading East Pacific Rise. Nat Geosci 5(4):275–278

Sambridge MS, Compston W (1994) Mixture modeling of multi-component data sets with application to ion-probe zircon ages. Earth Planet Sci Lett 128:373–390

Silverman B (1986) Density estimation for statistics and data analysis. Chapman and Hall, London

Smith M, Leigh-Jones P (1985) An automated microscope scanning stage for fission-track dating. Nucl Tracks Radiat Meas (1982) 10 (3):395–400

Soares CJ, Guedes S, Tello CA, Lixandrao Filho AL, Osorio AM, Alencar I, Dias AN, Hadler J (2013) Further investigation of the initial fission-track length and geometry factor in apatite fission-track thermochronology. Am Mineral 98:1381–1392

Soares C, Guedes S, Hadler J, Mertz-Kraus R, Zack T, Iunes P (2014) Novel calibration for LA-ICP-MS-based fission-track thermochronology. Phys Chem Mineral 41(1):65–73

Tufte ER (1983) The visual display of quantitative data. Graphics, Cheshire

Vermeesch P (2007) Quantitative geomorphology of the White Mountains (California) using detrital apatite fission track thermochronology. J Geophys Res (Earth Surf) 112(F11):3004

Vermeesch P (2009) RadialPlotter: a Java application for fission track, luminescence and other radial plots. Radiat Meas 44(4):409–410

Vermeesch P (2010) HelioPlot, and the treatment of overdispersed (U-Th-Sm)/He data. Chem Geol 271(3–4):108–111

Vermeesch P (2012) On the visualisation of detrital age distributions. Chem Geol 312–313:190–194

Vermeesch P (2017) Statistics for LA-ICP-MS based fission track dating. Chem Geol 456:19–27

Vermeesch P, Tian Y (2014) Thermal history modelling: HeFTy vs. QTQt. Earth-Sci Rev 139:279–290

Willett SD (1997) Inverse modeling of annealing of fission tracks in apatite; 1, A controlled random search method. Am J Sci 297 (10):939–969

The Sedimentology of Detrital Thermochronology

Marco G. Malusà and Eduardo Garzanti

Abstract

Detrital thermochronology is based on the radiometric dating of apatite, zircon and other minerals in sediments and sedimentary rocks. The objective of detrital thermochronology is to obtain quantitative information on sediment provenance and on the geologic evolution of the area from whence the sediment was generated. This chapter describes how the full potential of the detrital thermochronology approach can be exploited by applying simple sedimentology principles, in order to obtain provenance information that is largely independent from the physical and chemical modifications affecting sediment during transport, deposition and burial diagenesis. Simple strategies can be used to detect the effects of selective entrainment, which form placer and antiplacer deposits, and test the vulnerability of grain-age distributions to hydraulic sorting effects. The mineral fertility of eroded bedrock, which varies over orders of magnitude thus representing the largest source of potential bias in detrital thermochronology, can be readily measured by simple modifications to the standard procedures of mineral concentration. Multi-method studies are potentially biased by grain rounding and abrasion, as the removal of grain rims may lead to an incorrect interpretation of grain ages yielded by low-temperature thermochronometers. Bedload and suspended load have different transport time, and the instantaneous-transport-time assumption of exhumation studies based on the lag-time approach is not necessarily met.

7.1 Introduction

The detrital approach to geologic investigations utilises characteristics of sediments to obtain information on the area where the sediments were generated. Detrital studies include bulk sediment methods (e.g. the classic studies based on the modal composition of sand and sandstones—Zuffa 1985), multi-mineral methods (e.g. studies based on the interpretation of heavy mineral assemblages—Mange and Wright 2007) and single-mineral methods, which are based on the analysis of one mineral species displaying measurable variations in specific parameters (e.g. trace elements, isotopic ratios, etc.—Fedo et al. 2003; von Eynatten and Dunkl 2012). Single-mineral studies also include detrital thermochronology, which is based on the radiometric dating of apatite, zircon and other minerals.

This chapter illustrates the basic sedimentology principles that can be applied in detrital thermochronology to fully exploit the potentials of the single-mineral approach. It describes simple strategies to get the least biased information on the area from whence the sediment was generated, and to mitigate the risk of circular reasoning that vitiates detrital studies when based on untestable assumptions, leading to potentially incorrect geologic interpretations. Section 7.2 discusses the main variables controlling the detrital record in natural environments, and the potentials of the detrital thermochronology approach in providing provenance information that is largely unaffected by physical and chemical modifications during the sedimentary cycle (e.g. Komar 2007; Morton and Hallsworth 2007). Section 7.3 illustrates the basic principles of hydraulic sorting, how these principles can be used to improve mineral separation procedures in the laboratory, and describes strategies apt to evaluate the vulnerability of grain-age distributions to potential modifications induced by selective entrainment. Section 7.4 examines the potential implications of mechanical breakage and abrasion of mineral grains on multi-dating studies (e.g. Carter and Moss 1999), and the effects of metamictisation

M. G. Malusà (✉) · E. Garzanti
Department of Earth and Environmental Sciences, University of Milano-Bicocca, Piazza della Scienza 4, 20126 Milan, Italy
e-mail: marco.malusa@unimib.it

© Springer International Publishing AG, part of Springer Nature 2019
M. G. Malusà and P. G. Fitzgerald (eds.), *Fission-Track Thermochronology and its Application to Geology*,
Springer Textbooks in Earth Sciences, Geography and Environment, https://doi.org/10.1007/978-3-319-89421-8_7

for the preservation of detrital zircon. Section 7.5 summarises the information available on bedload and suspended load transfer time (e.g. Granet et al. 2010; Wittmann et al. 2011), and the implications for stratigraphic correlations and exhumation studies based on the analysis of the lag time between cooling and depositional ages (e.g. Garver et al. 1999). Section 7.6 evaluates the effects of mineral dissolution during weathering and diagenesis (e.g. Morton 2012), with particular emphasis on apatite and metamict zircon. Section 7.7 illustrates the mineral fertility bias, i.e. potential impact of the variable propensity of different parent bedrocks to yield detrital grains of specific minerals when exposed to erosion (Moecher and Samson 2006), and describes a simple approach for measuring the mineral fertility in bedrock starting from the analysis of modern sediment. Simple strategies to be applied from the early steps of sample processing to final data interpretation are summarised in Sect. 7.8 and in Table 7.1.

7.2 The Variables Controlling the Detrital Record

Clastic detritus generated by bedrock erosion, and eventually preserved in the stratigraphic record (Fig. 7.1), has generally experienced a long transport (commonly thousands of km), either as bedload or as suspended load, to finally reach a deep-sea sink (Graham et al. 1975; Ingersoll et al. 2003; Anfinson et al. 2016). During the sedimentary cycle of weathering and erosion, transport, deposition and burial, the detrital signature originally acquired in the source area (X in Fig. 7.1) may undergo relevant modifications by physical and chemical processes. Detrital grains are sorted during erosion, transport and deposition according to their size, density and shape (Y in Fig. 7.1), leading to major changes in the original compositional signature (Schuiling et al. 1985; Garzanti et al. 2009). Detrital grains also suffer abrasion and mechanical breakdown, and drastic modifications are induced by the different chemical stability of detrital minerals during weathering and burial diagenesis (Z in Fig. 7.1) (Johnsson 1993; Worden and Burley 2003). Physical modifications in sediments can be detected and modelled mathematically, but chemical modifications are much more difficult to deal with. Because solving for three variables (X, Y and Z in Fig. 7.1) requires an equal number of independent simultaneous equations, using the detrital record to obtain information on the source area (X) requires that both potential physical (Y) and chemical (Z) modifications to the original detrital signature are independently constrained.

Detrital thermochronology, when correctly applied, can be successfully employed to get unbiased information on the source area (see Table 7.1). Like other single-mineral

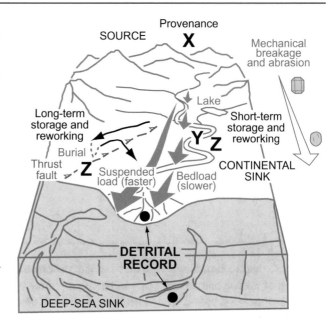

X PROVENANCE FINGERPRINT
+ fertility bias in single-mineral analysis

Y PHYSICAL MODIFICATIONS
◆ hydraulic sorting
◆ mechanical breakage and abrasion

Z CHEMICAL MODIFICATIONS
◆ weathering
◆ differential diagenetic dissolution

=   DETRITAL RECORD

Fig. 7.1 Main variables of the detrital record: the detrital signature acquired in the source area (X) is modified by physical (Y) and chemical (Z) processes during sediment transport towards the final sink and subsequent burial diagenesis. Because solving for three variables requires an equal number of independent equations, using the detrital record to obtain provenance information (X) requires that both physical and chemical modifications (Y and Z) are constrained independently

methods, detrital thermochronology minimises both the impact of differential mineral density during transport and deposition, and the impact of differential dissolution of minerals during weathering and diagenesis. If the thermochronologic fingerprint of mineral grains is independent from grain size, which can be easily tested, then the thermochronologic signal acquired in the source area is intrinsically robust with respect to hydrodynamic effects that sort detrital grains as a function of their size and density. If grain-age distributions are not independent from the grain size, then the effect of hydraulic sorting can be detected by using parameters (e.g. bulk grain density) that are fully independent from those used for provenance discrimination (i.e. the cooling age). Therefore, a correct detrital thermochronology approach provides information on the source

Table 7.1 Summary of the main natural processes and factors that may modify the original fingerprint of detritus

Contributing factor	Effect	Impact on detrital thermochronology	How to monitor or avoid	References	This work
Mineral fertility of parent bedrock	Controls the original mineral concentration in detritus	Source areas with high (or low) mineral fertility are overrepresented (or underrepresented) in the detrital thermochronology record	Measure the mineral fertility of eroded bedrock starting from the analysis of mineral concentration in modern sediment samples	Moecher and Samson (2006) Malusà et al. (2016a)	Sect. 7.7
Selective entrainment of mineral grains	Produces placer (and antiplacer) deposits with anomalous enrichment (and depletion) in denser mineral grains	May introduce a bias whenever a relationship exists between grain age and grain size	Record size and shape of dated mineral grains to test the vulnerability of age distributions to hydraulic sorting; detect placer deposits by measuring their bulk grain density	Komar (2007) Slingerland and Smith (1986)	Sect. 7.3
Mechanical grain breakdown during transport	Produces new mineral grains when inclusions are set free by the breakdown of their host mineral	Negligible	Measure, e.g. the downstream change in [zircon]/[Zr]	Dickinson (2008) Malusà et al. (2016a)	Sect. 7.4
Abrasion and rounding of grains during transport	May selectively remove grain rims in minerals with poor mechanical properties (e.g. in metamict zircon)	In multi-method studies may lead to an incorrect interpretation of ages yielded by low-temperature thermochronometers	Evaluate the downstream changes in grain shape parameters; compare double-dating results for euhedral and rounded grains	Kuenen (1959) Carter and Moss (1999)	Sect. 7.4
Temporary storage and reworking of sediment	Determines the incorporation of older sediment and a temporal delay between erosion and deposition especially in bedload	Limits the time resolution of stratigraphic correlations based on detrital minerals; may lead to underestimate exhumation rates based on lag-time analysis	Evaluate the sediment transport time and the impact of sediment recycling (e.g. by using the ratio of cosmogenic ^{26}Al and ^{10}Be)	Granet et al. (2010) Wittmann et al. (2011)	Sect. 7.5
Size-density sorting of grains during deposition	Causes an intrasample modal variability (i.e. different composition of different size classes of a sample)	Target minerals are potentially lost during mineral separation, if the size window for processing is not chosen correctly	Retrieve the grain-size parameters of the bulk sediment sample and model the sediment composition in different grain-size classes	Rubey (1933) Resentini et al. (2013)	Sect. 7.3
Chemical weathering	Determines the selective dissolution of unstable minerals in soils	Negligible. Apatite is unstable in acidic weathering environments, but fission-track ages are generally unaffected even in weathered samples	No specific strategy is required in detrital studies based on single-mineral methods	Gleadow and Lovering (1974) Morton (2012)	Sect. 7.6
Chemical dissolution during burial diagenesis and metamorphism	Determines the intrastratal dissolution of unstable mineral species and the growth of authigenic minerals	Precludes the measurement of mineral fertility, starting from the analysis of ancient sandstones; may determine the selective dissolution of metamict zircon grains	No specific strategy is required in detrital studies based on single-mineral methods	Morton and Hallsworth (2007) Malusà et al. (2013)	Sect. 7.6

area that is largely independent from physical and chemical modifications taking place during transport, deposition and diagenesis.

Single-mineral (and multi-mineral) studies are notably influenced by the variable mineral fertility of bedrock (Dickinson 2008). This issue does not affect the bulk sediment approach, although bulk sediment mineralogy may substantially differ even in sediments derived from the same bedrock lithology, when eroded under different climatic and geomorphic conditions (Johnsson 1993; Morton 2012). The mineral fertility in bedrock can be effectively measured by applying the basic principles of hydraulic sorting to the analysis of modern sediments (Malusà et al. 2016a). Parameters employed for mineral fertility measurements are independent from the cooling age of the provenance area. This prevents the common risk of circular reasoning (i.e. trying to solve a single equation for several variables) that typically characterises provenance studies whenever physical and chemical modifications to the original detrital signatures are not independently assessed.

7.3 Hydraulic Sorting and Sediment Composition

7.3.1 Settling Velocity and Size Shift

Dense minerals commonly used in detrital thermochronology (e.g. apatite, zircon and monazite) are usually associated in sorted sediments with coarser lower-density lithic grains and framework minerals (e.g. quartz and feldspars), according to the principle of *hydraulic equivalence* (Rubey 1933; Garzanti et al. 2008) (Fig. 7.2a). During settling in a fluid, grains with the same settling velocity are deposited together. The settling velocity reflects the balance between drag resistance and gravitational force, and thus depends on the density (δ_m) and size (D) of settling grains (Schuiling et al. 1985, Komar 2007) (Fig. 7.2b, c). The greater the difference in density between settling-equivalent grains and/or minerals, the greater is their difference in size, which is also referred to as the *size shift* (Fig. 7.2a) (Rittenhouse 1943). The size shift of apatite grains ($\delta_m = 3.20$ kg/dm^3) relative to quartz ($\delta_m = 2.65$ kg/dm^3) is much smaller than the size shift for zircon ($\delta_m = 4.65$ kg/dm^3) and monazite ($\delta_m = 5.15$ kg/dm^3) relative to quartz. Metamict zircon grains, due to their lower density (Ewing et al. 2003), show smaller size shifts relative to quartz than hydraulically equivalent non-metamict zircon grains (see Sect. 7.4.2 for further information on metamict zircon). Size shifts relative to quartz of feldspars and lithic grains, representing the main detrital components of siliciclastic sediments usually plotted on QFL ternary diagrams for provenance discrimination (e.g. Dickinson 1985), are minor and generally negligible (Fig. 7.2a).

Turbulence and viscosity have a variable impact on grain settling in water depending on the particle size (Fig. 7.2b). Drag resistance and settling velocity of pebbles are controlled by turbulence according to the Impact law. In clay and silt-sized grains, settling velocity is controlled by viscosity, according to the Stokes' law. In sand-sized grains, settling velocity is controlled by both viscosity and turbulence. Empirical formulas predicting size-shift values in water-laid sand (e.g. Cheng 1997, 2009) show that size shifts increase with grain size, from those predicted by the Stokes'

law to those predicted by the Impact law, according to a sigmoidal function (Fig. 7.2b).

Because grain-size distributions can be generally approximated by a log-normal function, size shifts can be effectively expressed using the ϕ scale derived from a logarithmic transformation of the Udden-Wentworth grain-size scale. The Udden-Wentworth grain-size scale divides sediments into different classes based on a constant ratio of 2 between successive class boundaries. When used with millimetres or microns, this scale is geometric, but after a logarithmic transformation ($\phi = -\log_2 d$, where d is the grain size in millimetres), the resulting ϕ scale is arithmetic (Fig. 7.2b). The negative sign is conventional and allows to avoid negative numbers in the sand range. Size shifts predicted by Cheng (1997)'s formula range from 0.2 to 0.35ϕ for apatite, to 0.55–1.0ϕ for zircon and 0.65–1.2ϕ for monazite, with wider ranges possibly expected for zircon grains depending on the degree of metamictisation (white belt in Fig. 7.2b). In wind-laid sand, the impact of viscosity on settling velocity becomes negligible, and size shifts are not dependent on grain size (Resentini et al. 2013). Beside density and size, grain shape also plays a major role in controlling settling of phyllosilicates, and micas settle slower than quartz in spite of their higher densities, just because of their platy shape (Doyle et al. 1983; Komar and Wang 1984; Le Roux 2005). The behaviour of micas, often employed in detrital thermochronology studies (e.g. Najman et al. 1997; Carrapa et al. 2004; Hodges et al. 2005) and largely transported as suspended load, is therefore more difficult to predict.

7.3.2 Grain-Size Parameters and Modelling of Compositional Variability

As an effect of size-density sorting during settling, different grain-size classes in a single bedload sample invariably display strongly different modal compositions. Such *intrasample compositional variability* in sediment can be modelled mathematically using specific numerical tools (e.g. MinSORTING; Resentini et al. 2013). MinSORTING is an Excel worksheet that calculates the size shift predicted for each mineral, by the Stokes' law for silt, Cheng (1997)'s formula for sand and the Impact law for gravel, and its distribution in different grain-size classes. The fundamental input parameters for this calculation are the fluid type (freshwater, sea water or air), which is supposedly known, and the mean grain size and sorting values of the bulk sediment. The procedure for measuring the *grain-size parameters* required for such a modelling is quite simple in loose sediment (Fig. 7.3a). After sieving the bulk sample, sediment trapped in each sieve is weighed (step 1), and the

Fig. 7.2 Settling equivalence and size shift: **a** Size relationships between quartz and settling-equivalent minerals having different densities (in kg/dm³, all of these spheres settle at 0.0267 m/s in freshwater); the size shift is the difference in size between settling-equivalent minerals (modified from Resentini et al. 2013, Malusà et al. 2016a); **b** Size shift relative to quartz predicted for apatite, zircon and monazite in bedload sediments of increasing grain size according to Cheng (1997)'s formula; **c** The settling velocity (*v*) of a detrital grain in a fluid reflects the balance between gravitational force (F_G) and drag resistance due to turbulence (F_T) and viscosity (F_V); D = grain diameter; g = gravity; δ_m = mineral density; δ_f = fluid density; η = fluid viscosity; C_D = drag coefficient (modified from Garzanti et al. 2008; Malusà et al. 2016a)

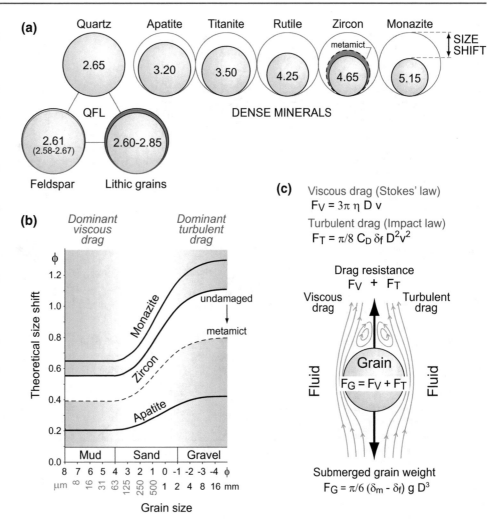

weight of different size classes is visualised as an histogram (step 2) or as a smoothed frequency curve, with the grain size decreasing along the *x*-axis away from the origin. Results are then converted into a smoothed cumulative frequency curve showing the percentage frequency of grains coarser than a particular value (step 3). The precision of this curve (calculated for sieving at 1ϕ intervals in Fig. 7.3a) can be improved by sieving the sample at 1/2 ϕ or 1/4 ϕ intervals. The smoothed cumulative frequency curve is the starting point to determine the relevant distribution percentiles (ϕ_n) (step 4) to be used as input parameters for the calculation of the *mean grain size* (D_m), *sorting* (σ) and *skewness* (S_k) of the bulk sediment, according to the formulas of Folk and Ward (1957) (step 5). Sorting is a measure of the spread of the grain-size distribution, i.e. of its standard deviation. Sorting is poor in the case of deposition by viscous flows, and best for deposits from tractive currents. Skewness is a measure of the symmetry of the distribution, which is best evaluated in smoothed frequency curves (Fig. 7.3b, c). Asymmetrical (skewed) distributions may show either a fine tail (positive skew) in the case of

excess fine material, or a coarse tail (negative skew) in the case of excess coarse material. In sorted sediments, denser minerals are concentrated in the finer tail of the size distribution, whereas the modal value of the distribution (i.e. the value of the mid-point of the most abundant size class) is largely controlled by low-density minerals such as quartz (Fig. 7.3b). Symmetrical distributions have indistinguishable mean grain size (D_m), median (ϕ_{50}) and modal values.

Starting from the mean grain size and sorting values of the bulk sediment measured by sieving, MinSORTING predicts the distribution of specific minerals in different grain-size classes of the sample. This information can be used to choose the most suitable size windows to maximise mineral recovery during separation in the laboratory, and to evaluate the fraction of the target mineral potentially falling outside of the chosen grain-size range. An example of this application is illustrated in Fig. 7.3d. In this figure, histograms show the distribution of apatite, zircon and monazite calculated in sediment samples having the same provenance (dissected continental block) but different mean grain size and sorting values. The grey belt indicates a fixed

Fig. 7.3 Modelling of intrasample compositional variability: **a** Procedure for measuring the grain-size parameters of a bulk sediment sample by sieving (steps 1 to 5, see text); **b** Conceptual composite smoothed frequency curve of a sorted sediment sample (in grey): the modal value is controlled by low-density minerals (in blue), whereas denser minerals (in red) are concentrated in the finer tail of the distribution; **c** Sorting, skewness and relationships between modal, median and mean grain size in symmetrical and asymmetrical distributions; **d** Example of distribution of apatite, zircon and monazite in different grain-size classes as modelled with MinSORTING (Resentini et al. 2013) for detritus derived from the erosion of a dissected continental block; upper row ($D_m = 2.5\ \phi$, $\sigma = 0.2$): all of the target mineral is concentrated in the size range (63–250 μm) chosen for mineral separation (in grey); intermediate row ($D_m = 2.5\ \phi$, $\sigma = 1.5$): nearly half of the mineral is lost in fractions finer or coarser than the selected size window; lower row ($D_m = 0.5\ \phi$, $\sigma = 0.2$): the target mineral is completely lost during separation, in spite of its original presence in the sediment sample

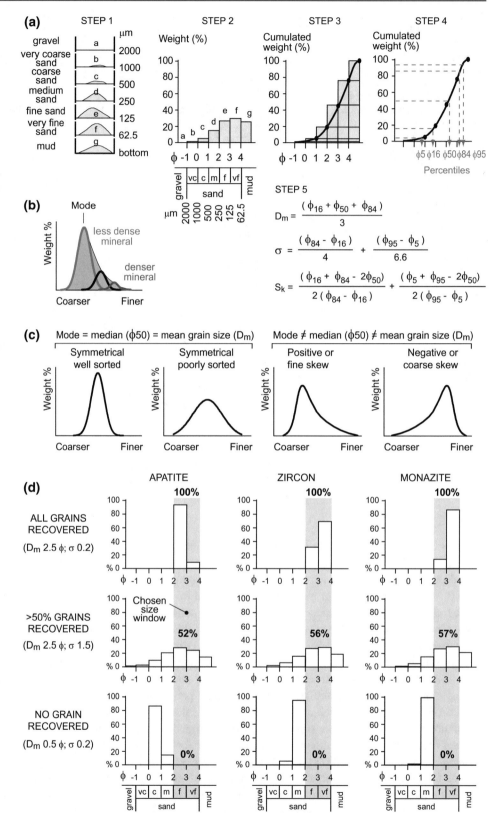

$$D_m = \frac{(\phi_{16} + \phi_{50} + \phi_{84})}{3}$$

$$\sigma = \frac{(\phi_{84} - \phi_{16})}{4} + \frac{(\phi_{95} - \phi_5)}{6.6}$$

$$S_k = \frac{(\phi_{16} + \phi_{84} - 2\phi_{50})}{2(\phi_{84} - \phi_{16})} + \frac{(\phi_5 + \phi_{95} - 2\phi_{50})}{2(\phi_{95} - \phi_5)}$$

size window (63–250 μm) that is often chosen for apatite and zircon separation in many laboratories. In the case of a fine-grained and very well-sorted sand ($D_m = 2.5\phi$; $\sigma = 0.2$), all apatite, zircon and monazite grains are expected to be concentrated in the 63–250 μm size window, and are potentially recovered during mineral separation from these size classes. However, these ideal conditions are not the rule. In poorly sorted samples with the same mean grain size ($D_m = 2.5\phi$; $\sigma = 1.5$), apatite, zircon and monazite grains are much more dispersed among size classes, if originally present in any size fraction of the eroded source rocks. By using the same fixed size window of 63–250 μm, half of the mineral amount is thus predicted to be lost in finer or coarser grain-size classes. The most relevant implications for mineral separation are shown in the third example, which considers a coarse-grained and very well-sorted sand ($D_m = 0.5\phi$; $\sigma = 0.2$). In this case, no target mineral grain, either apatite, zircon or monazite, will be recovered in the 63–250 μm size classes, despite its original presence in the sediment sample. This example underlines the importance of modelling intrasample compositional variability for a correct approach to mineral separation in sedimentary rocks, in order to avoid the risk associated with an incorrect choice of the analysed size window with respect to both sample grain size and mineral size shift. Modelling of compositional variability is often more problematic when dealing with cemented sandstones, because they may require either electric-pulse disaggregation or jaw crushing and disc milling before processing. In the latter case, grain-size relationships are lost, and modelling with MinSORTING is precluded.

7.3.3 Selective Entrainment and Generation of Placer and Antiplacer Deposits

Particle size (D) and density (δ_m) play a major role not only during grain settling, but also in controlling the process of selective entrainment during sediment transport (Komar and Li 1988; Komar 2007). In bedload sediment, coarser low-density mineral grains (e.g. quartz) are more easily entrained by tractive currents than smaller settling-equivalent grains of denser minerals (e.g. zircon in Fig. 7.4a). Coarser grains project higher above the bed, and thus experience greater flow velocities (v_D and v_T in Fig. 7.4a) and drag forces (F_D in Fig. 7.4a) than smaller denser grains. Coarser grains also have smaller pivoting angles (α in Fig. 7.4a), which favour the action of the drag force F_D compared to the counteracting vertical force, given by the difference between the submerged grain weight (F_G in Fig. 7.4a) and the lift force (F_L in Fig. 7.4a). Entrainment of detrital grains occurs when:

$$F_D > (F_G - F_L) \tan \alpha \qquad (7.1)$$

As a consequence of *selective entrainment*, the original neutral bedload is separated into a lag fraction strongly enriched in dense minerals, and entrained sediment that is instead depleted in dense minerals (Reid and Frostick 1985, Slingerland and Smith 1986). The former will be referred to as *placer* lag hereafter, whereas the latter will be referred to as *antiplacer* deposit (Fig. 7.4b). In placer lags, minerals such as garnet, zircon, rutile, monazite and magnetite can be enriched by orders of magnitude relative to the original neutral bedload, leading to a remarkable change in colour and density (>4.0 kg/dm^3 in magnetite placers).

7.3.4 Detection of Placer and Antiplacer Deposits

Different approaches can be used to detect placer and antiplacer deposits in detrital thermochronology studies. They are (i) the bulk grain density approach; (ii) the dense mineral concentration approach; (iii) the geochemical approach.

The Bulk Grain Density Approach The *bulk grain density* of a sediment sample can be measured to detect the anomalous concentrations of dense mineral grains due to selective entrainment processes. Bulk grain density can be measured in the laboratory with the aid of a hydrostatic balance. A quartered fraction of a bulk sediment sample is weighed first in air and next immersed in water, taking care to avoid floating due to surface tension. Grain density is calculated as:

$$\delta_{\text{sediment}} = W_{\text{in air}} / (W_{\text{in air}} - W_{\text{in water}}) \cdot \delta_{\text{water}} \qquad (7.2)$$

where δ_{water} is 0.9982 kg/dm^3 at 20 °C and W is the weight of the sample (Pratten 1981). The density of upper crustal rocks is generally in the range of 2.70 ± 0.05 kg/dm^3, and exceeds 2.80 kg/dm^3 only for mafic and ultramafic igneous and metamorphic rocks. Sediments derived from the erosion of these rocks are expected to have the same grain density of ~ 2.7 kg/dm^3 in the lack of hydraulic sorting effects. The upper diagram of Fig. 7.4c shows reference bulk grain density values expected in rocks found in different tectonic settings. Higher values in sediment point to hydraulic concentration of denser mineral grains by selective entrainment. The depletion of dense minerals in antiplacer deposits is more difficult to detect, because their grain density cannot be much lower than that of feldspar and quartz, i.e. ~ 2.65 kg/dm^3 (Fig. 7.4b, c), and thus not much different from the grain density of neutral bedload (Fig. 7.4c, upper diagram).

The Dense Mineral Concentration Approach Because the grain density of a sediment sample is controlled by the abundance of dense minerals, also the concentration of dense

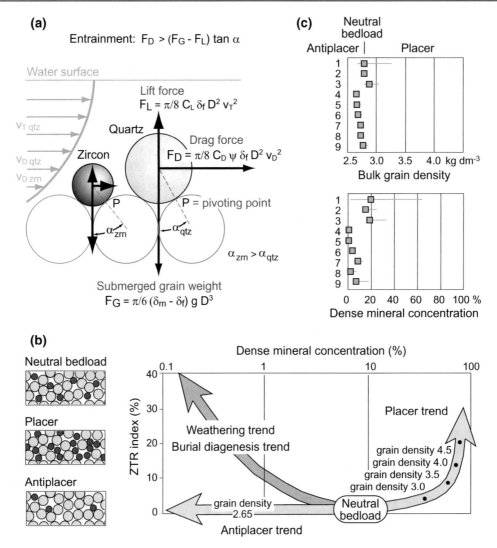

Fig. 7.4 Approaches to selective entrainment: **a** Coarser quartz grains are more easily entrained than smaller settling-equivalent zircon grains, because they have a smaller pivoting angle (α), project higher above the bed and experience greater flow velocities and drag forces (F_D) (based on Komar and Li 1988); v = current velocity (v_T, at level of grain's top; v_D, at level of effective drag force); ψ = sheltering factor (depending on the percentage of grain area exposed to fluid drag); C_L = lift coefficient; other keys as in Fig. 7.2c; **b** Neutral bedload, placers and antiplacers, and comparison between selective entrainment and weathering-diagenesis trends in sand and sandstones; the ZTR index (Hubert 1962) is the percentage of relatively durable zircon, tourmaline and rutile among transparent heavy minerals; selective entrainment in sand produces a progressive increase in dense mineral concentration in placers, and a progressive decrease in antiplacers; diagenetic dissolution in ancient sandstones produces a progressive decrease in dense mineral concentration and an increase in ZTR, which is not observed in antiplacer deposits (after Malusà et al. 2016a, and references therein); **c** Reference parameters for testing selective entrainment and anomalous concentrations of dense mineral grains in bulk sediment samples (grey squares = modal bulk grain density and dense mineral concentration values; grey bars = value spread; 1 to 9 = provenance: 1-2, undissected and dissected magmatic arc; 3, ophiolite; 4, recycled clastic; 5-6-7, undissected, transitional and dissected continental block; 8-9, subcreted and subducted axial belt; based on Malusà et al. 2016a, and references therein)

minerals retrieved during mineral separation can be used to detect selective entrainment effects, when compared to values typically observed in bedrock. In the lack of hydraulic effects, dense mineral concentration in modern sediments is generally <10%, but may reach values as high as 15–20% in sediments derived from magmatic arcs and obducted ophiolites (Fig. 7.4c, lower diagram) (Garzanti and Andò 2007).

Higher values may thus reveal the anomalous concentration of denser mineral grains by selective entrainment.

The Geochemical Approach Because chemical elements are preferentially hosted in specific minerals having different density and shape (McLennan et al. 1993), hydraulic sorting has an even stronger impact on the chemical composition of sediments than on its modal composition. In a sediment sample, zircon grains typically contribute most of Zr, Hf and a significant amount of Yb, Lu and U; monazite and allanite much of LREE and Th; apatite much of P; xenotime and garnet much of Y and HREE (Garzanti et al. 2010). The anomalous concentrations of these elements in the bulk sediment were employed to reveal anomalous concentrations of apatite, zircon or monazite by selective entrainment (e.g. Resentini and Malusà 2012). However, a large fraction (90% or more) of apatite or zircon originally present in bedrock may form inclusions in larger mineral grains, rather than individual mineral grains in detritus (Malusà et al. 2016a), and are thus selected by tractive currents according to the hydraulic properties of the host mineral. This implies that the concentrations of REE, Zr and Hf in bulk sediment are controlled not only by the hydraulic behaviour of denser minerals such as zircon and monazite, but also by the hydraulic behaviour of larger and less dense mineral grains (e.g. micas). Approaches based on grain density measurements are thus to be preferred over approaches based on geochemical analyses for the detection of placer deposits.

7.3.5 Selective Entrainment and Vulnerability Tests for Grain-Age Distributions

In detrital thermochronology studies, selective entrainment controlled by particle density and particle size may have a large impact on the final grain-age distributions observed in detritus. The single-mineral approach minimises the effect of variable particle density, as minor density variations are expected only for zircon grains with different levels of α-damage. However, selective entrainment may introduce a bias whenever a relationship exists between grain age and grain size. Therefore, detrital grains should be carefully tested for potential age-size relationships and, if any exists, particular care should be used to detect any potential effect of selective entrainment on the analysed samples. This can be done by using the bulk grain density approach described in Sect. 7.3.4.

The size and shape parameters of dated mineral grains can be easily measured, and should thus be recorded during thermochronologic analysis. Figure 7.5a shows a qualitative assessment of the age-size relationships in the fission-track (FT) datasets of two hypothetical sediment samples. For each grain, ages are plotted versus the *equivalent spherical diameter* (ESD), which is the cube root of the product of lengths of the three axes. The shorter axis measured in thin section is assumed to be equal to the intermediate axis for the sake of simplicity. If the grain age is independent from the ESD, i.e. from the grain size (upper diagram in Fig. 7.5a), the grain-age distribution is not vulnerable to hydraulic sorting effects.

Noteworthy, polymodal grain-age distributions cannot be compared by using their mean or median values (Fig. 7.5b), but require more specific statistic tests. An effective assessment of the age-size relationships of a detrital geochronology dataset, based on the Kolmogorov–Smirnov (K–S) method (Smirnov 1939; Young 1977), is shown in Fig. 7.5c. The K–S method considers the maximum distance $D_{n,n'}$ between the cumulative frequency curves ($F_{1,n}$ and $F_{2,n'}$) of two analysed samples

$$D_{n,n'} = \sup_x \left| F_{1,n}(x) - F_{2,n'}(x) \right| \qquad (7.3)$$

This distance is compared with the critical value K_α for a significance level $\alpha = 0.05$ (Eplett 1982; Hollander and Wolfe 1999), which is equal, for large datasets, to

$$K_{0.05} = 1.36 \cdot \sqrt{\frac{n + n'}{n \cdot n'}} \qquad (7.4)$$

where n and n' indicate the number of dated grains in the two samples. Results are shown in Fig. 7.5c as the difference between $K_{0.05}$ and $D_{n,n'}$, referred to as $V_{K\text{-}S}$. If $V_{K\text{-}S} > 0$, differences between distributions are statistically not significant (Malusà et al. 2013).

In the detrital zircon U–Pb dataset shown in Fig. 7.5c, the grain-age distributions in different size classes (i.e. fine sand, very fine sand and coarse silt) are not significantly different. Therefore, this dataset is not vulnerable to hydraulic sorting effects. The K–S test can be easily performed with the aid of a general purpose spreadsheet, without the need of more specific statistical software packages.

7.4 Mechanical Breakdown and Abrasion of Mineral Grains

During transport, detrital minerals may undergo mechanical breakage and progressive abrasion with the consequent potential removal of their external rims. Grains with euhedral shape may undergo progressive rounding, whereas small mineral inclusions liberated by the breakdown of their host grain may form new individual detrital grains, which will start to be sorted during subsequent tractive transport according to their size, density and shape. The potential impact of these processes in detrital thermochronology studies is discussed below.

7.4.1 Grain Rounding and Potential Impact on Zircon Double Dating

Detrital mineral grains may experience progressive abrasion and rounding during transport depending on their physical and mechanical properties. Studies of natural systems and

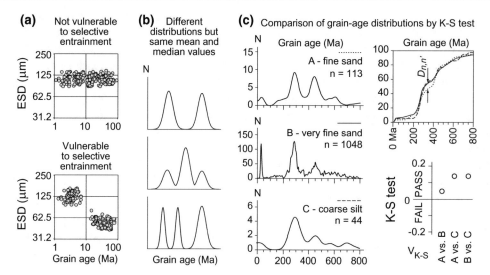

Fig. 7.5 Vulnerability tests for grain-age distributions: **a** Assessment of age-size relationships in detrital samples (ESD is the equivalent spherical diameter of the analysed grains); if the grain age is independent on grain size (upper diagram), the grain-age distribution is not vulnerable to hydraulic sorting effects (based on Resentini and Malusà 2012); **b** Polymodal grain-age distributions should not be compared by their mean or median values; **c** Quantitative assessment of the age-size relationships in a detrital sample, based on the Kolmogorov–Smirnov (K–S) method; $D_{n,n'}$ is the maximum distance between the cumulative frequency curves for grain ages in different grain-size classes; V_{K-S} is the difference between the critical value for a significance level $\alpha=0.05$, and $D_{n,n'}$; if $V_{K-S} > 0$, the grain-age distributions in different grain-size classes are not statistically different, and the dataset is not vulnerable to hydraulic sorting effects (modified after Malusà et al. 2013)

laboratory experiments show that mechanical wear is very effective in the aeolian environment, but much less effective during fluvial transport (Russell and Taylor 1937; Kuenen 1959, 1960; Garzanti et al. 2015). In modern river sands shed from the European Alps, the relative abundance of rounded zircon grains does not change from the Alpine mountain tributaries to the Po delta (Fig. 7.6a). Moreover, the percentage of rounded zircon grains systematically increases in older U–Pb age populations. This suggests that zircon rounding may reflect chemical abrasion by metamorphic fluids in the source rocks, possibly experienced during multiple cycles of diagenesis and metamorphism, rather than mechanical abrasion (Malusà et al. 2013). However, for older terranes such as the Yilgarn Craton, a weak positive correlation between zircon grain rounding and transport distance has been described (Markwitz et al. 2017).

The removal of external rims by grain abrasion may play a role in detrital thermochronology studies based on (U–Th)/He dating (Tripathy-Lang et al. 2013), because the correction factor for α-ejection that must be applied to raw He ages depends on grain size and U–Th heterogeneity (Reiners and Farley 2001; Hourigan et al. 2005). The potential impact of grain abrasion is particularly important in detrital thermochronology studies based on the double dating approach (e.g. U–Pb and FT dating on the same detrital grain) (Fig. 7.6b). Double dating is often employed to distinguish non-volcanic from volcanic zircon grains (e.g. Carter and

Moss 1999; Jourdan et al. 2013). Zircon grains derived from volcanic or subvolcanic rocks will display identical U–Pb and FT ages within error, because of rapid magma crystallisation in the upper crust where country rocks are at temperatures cooler than the partial annealing zone of the FT system (Malusà et al. 2011). Non-volcanic zircon grains will display instead FT ages that are younger than the corresponding U–Pb age. U–Pb analysis for double dating is best performed on zircon rims, because volcanic zircons commonly display inherited xenocrystic cores (e.g. Corfu et al. 2003), which crystallised well before the volcanic event recorded by the FT system. Whenever such rims are removed by abrasion, volcanic zircon grains will display only the U–Pb ages of the inherited core, which are older than the corresponding FT age. As a result, such FT ages could be incorrectly interpreted in terms of exhumation (see discussion in Chap. 8, Malusà and Fitzgerald 2018a), leading to incorrect geologic interpretations.

7.4.2 Preservation of Metamict Zircon During River Transport

Recent work suggested that metamict zircon grains could be selectively destroyed by abrasion during river transport (e.g. Fedo et al. 2003; Hay and Dempster 2009; Markwitz et al. 2017). However, this statement is not supported by

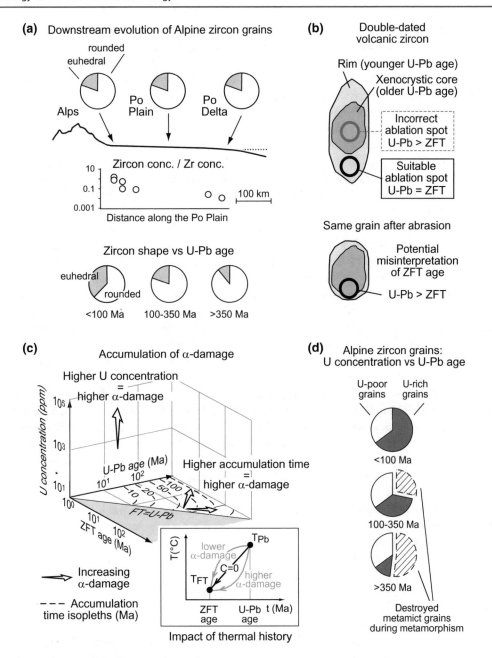

Fig. 7.6 Impact of mechanical breakdown and abrasion of detrital grains: **a** Downstream changes in the abundance of rounded zircon grains (upper pie charts) and in the ratio between individual zircon grain concentration and zirconium concentration in Po river sand derived from the European Alps (intermediate diagram), and percentage of rounded zircon grains in different age populations of the same dataset (lower pie charts) (modified after Malusà et al. 2013, 2016a); these data suggest that fluvial transport is not effective in promoting zircon rounding, and that the impact of mechanical breakdown of coarser mineral grains in the floodplain is not relevant for the production of additional individual zircon grains in sediment; however, the percentage of rounded zircon grains systematically increases in older U–Pb age populations, suggesting a close relationship between the roundness of zircon grains and their polycyclic origin; **b** Impact of grain rounding in double dating studies: grain rims in volcanic zircon display U–Pb ages indistinguishable from the zircon fission-track (ZFT) age, whereas the inherited xenocrystic cores may display much older U–Pb ages; whenever zircon rims are removed by abrasion, double dating may lead to an incorrect interpretation of ZFT ages; **c** α-damage accumulation in zircon is a function of U concentration and effective accumulation time t_α (the t_α isopleths in the diagram are modelled as a function of U–Pb and ZFT ages, assuming concavity C=0); α-damage may increase for any suitable combination of the above parameters in the diagram (white arrows) (simplified from Malusà et al. 2013); **d** Percentage of U-rich (>1000 ppm) zircon grains in different age populations found in Alpine detritus; the lack of old U-rich grains suggests selective dissolution of metamict zircon during diagenesis and metamorphism of source rocks

observational evidence, and experimental studies aimed at determining the relative mechanical stability of detrital minerals (e.g. Thiel 1940; Dietz 1973; Afanas'ev et al. 2008) are few and often led to contrasting results. Metamict zircon undergoes a change in structure with consequent decrease in density and hardness by a process of self-irradiation and accumulation of α-damage over time, an effect of high concentrations of radioactive elements such as U and Th in the crystal lattice (Ewing et al. 2003). Accumulated α-damage (Fig. 7.6c) is chiefly a function of U concentration (an equal mass of Th contributes only $\sim 5\%$) and of effective accumulation time (t_α), reduced by the effects of thermal annealing of α-damage at high temperature (Tagami et al. 1996; Nasdala et al. 2001). The t_α isopleths in the diagram of Fig. 7.6c are modelled as a function of the U–Pb and FT ages of each zircon grain, assuming monotonic cooling (concavity $C = 0$ in Fig. 7.6c) and a threshold temperature of effective α-damage retention of 400 °C, which is consistent with geologic evidence (Garver and Kamp 2002). As shown in Fig. 7.6c, accumulated α-damage increases: (i) along the z-axis of the diagram by increasing U concentration; (ii) in the basal plane of the diagram in any direction crosscutting the t_α isopleths towards higher t_α values; more in general (iii) for any suitable combination of these parameters in the three-dimensional space displayed in the diagram. Further variability is introduced by the diverse shape of the thermal path experienced by each zircon grain, which may imply a longer or shorter time interval for α-damage accumulation (lower frame in Fig. 7.6c).

The variability in α-damage accumulation expected in natural detrital zircon concentrates is therefore quite high. Focusing on the parameters controlling α-damage accumulation, Malusà et al. (2013) found that U-rich zircon grains (more metamict, on average, than U-poor grains having the same age) are present in all of the main U–Pb age populations detected in detritus of the Po river catchment (Fig. 7.6d). However, their abundance does not decrease downstream across the alluvial plain, which indicates that lower-density metamict zircon may survive long-distance river transport. In detrital zircon grains from the Murchison River, Western Australia, Markwitz et al. (2017) describe a trend of downstream increasing apparent mean density, but this trend may reflect the progressive addition of detritus from younger source terranes, rather than a progressive loss of metamict zircon grains during transport.

7.4.3 Liberation of Inclusions by Mechanical Breakdown

In natural sedimentary systems, the downstream grain-size reduction of detritus primarily occurs in the uppermost and higher relief parts of the catchment, and largely by selective deposition as a response to decreasing topographic gradient and competence of the fluvial current (e.g. Fedele and Paola 2007; Allen and Allen 2013). Gravel is generally deposited by mountain rivers in large fans at the transition between the mountainous hinterland and the floodplain lowlands, whereas sand and mud are transported mainly as bedload and suspended load, respectively, across the floodplain towards the river delta and the final sediment sink (Fig. 7.1). Mechanical breakdown of sand-sized grains during river transport is generally minor (Russel and Taylor 1937; Kuenen 1959), although it may liberate a conspicuous number of zircon and apatite inclusions that become individual grains.

The impact of this process may be quantified by comparing the concentration of individual zircon grains in bedload (measured according to the approach described in Sect. 7.7) with the zirconium concentration in the same bulk sediment sample as indicated by chemical analysis. The zirconium concentration of a sediment sample provides an upper threshold for its potential zircon concentration (Malusà et al. 2016a). Assuming that zirconium is exclusively hosted in zircon, and that the zirconium contributions by other rare minerals (baddeleyite, xenotime) and rock fragments (e.g. volcanic glass) are negligible, volume % zircon ≈ 1.15 wt% zirconium (Dickinson 2008). However, the potential zircon concentration provided by chemical analysis includes not only individual zircon grains, but also smaller zircon inclusions in larger grains. If the production of additional individual zircon grains by mechanical breakdown of the host mineral in the floodplain was relevant, then the ratio between individual zircon grain concentration and zirconium concentration should increase downstream along the floodplain. In the Po river catchment, where both datasets are available, such an increase is not observed (Fig. 7.6a).

7.5 Sediment Transfer Time and Implications for Detrital Thermochronology

Mineral grains used in detrital thermochronology may travel either mainly as *bedload* (e.g. apatite, zircon and monazite) or as *suspended load* (e.g. micas). Suspended load travels faster than bedload across the alluvial plain (e.g. Granet et al. 2010). The minimum time required by suspended load to reach the final sink is equal to the time needed by a flood wave to reach the closure section of a basin during a major flood event (i.e. hours to days depending on the size and physiography of the basin). The actual time is generally notably longer, because suspended load can be either temporarily deposited on levees, crevasse splays or even far from the active river channel, to be re-eroded and transported downstream during subsequent flood events. The transfer time of bedload is generally longer than for suspended load (Fig. 7.1), but for both bedload and suspended load even the order of magnitude of transfer time is difficult to quantify.

Granet et al. (2010) analysed the $^{238}U-^{234}U-^{230}Th$ radioactive disequilibria in river sediments derived from the Himalayas to infer that bedload transits slowly (i.e. >several 10^5 yr) in the plain, whereas the absence of significant variations in Th isotope composition indicates that suspended load travels much faster (<20–25 10^3 yr). There thus appears to be an order of magnitude of difference, or more, between the transfer time of suspended load and bedload. Depending on the geologic conditions, the bedload transfer time can be even much longer than observed in the Himalayan foreland basin, even in the lack of deep burial and subsequent reworking of sediment due to subsidence followed by basin inversion (Fig. 7.1). Wittmann et al. (2011) used the ratio of in situ-produced cosmogenic ^{26}Al and ^{10}Be in sediment of the Amazon cratonic basin to show that the modern Amazon River and its tributaries carry substantial amounts of sediment formerly stored for 1–3 Myr in the floodplain, thus being partly shielded from cosmic rays. In an active foreland basin, however, the incorporation of such old sediment during fluvial avulsion might be limited by rapid burial at depth (Wittmann et al. 2016). According to these observations, a substantially longer delay can be expected in the detrital thermochronology signals provided by apatite and zircon, transported as bedload, relative to mica entrained as suspended load. Moreover, the Amazon example shows that the bedload transfer time from source to sink can be relevant in detrital thermochronology studies based on the lag-time approach (e.g. Garver et al. 1999; Bernet et al. 2001), because the zero-transport-time assumption is not necessarily met.

The analysis of detrital minerals is often employed for the correlation of unfossiliferous clastic successions (Morton 2012). The grain-age distributions characterising different tectonic domains within an orogen can be quite variable, but they are progressively homogenised before reaching the final sink during transport along the floodplain (e.g. Malusà et al. 2013), as shown in Fig. 7.7a. The occurrence of highly variable grain-age fingerprints in modern river sediment (e.g. Bonich et al. 2017) is good news for provenance discrimination on a geochronologic basis, but bad news for long-distance correlation of continental clastic successions (Fig. 7.7a). By contrast, deep-sea turbidites are expected to display a homogeneous detrital signature over distances of several hundreds of kilometres, with apparent provenance changes only shown at different levels of the stratigraphic successions (e.g. Malusà et al. 2016b). Noteworthy, the transfer time of bedload sediment across the floodplain has implications not only for lag-time analysis, but also for the time resolution of correlations based on detrital minerals in

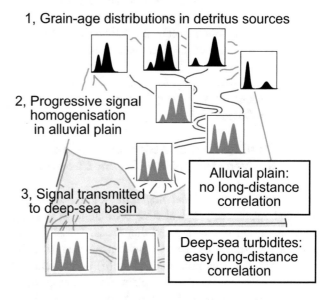

(a) Correlation of clastic successions

1, Grain-age distributions in detritus sources

2, Progressive signal homogenisation in alluvial plain

Alluvial plain: no long-distance correlation

3, Signal transmitted to deep-sea basin

Deep-sea turbidites: easy long-distance correlation

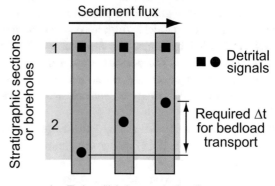

(b) Bedload velocity and time resolution in stratigraphic correlations

Sediment flux

Stratigraphic sections or boreholes

1

2

Detrital signals

Required Δt for bedload transport

1 - False "high resolution"
2 - True resolution limited by bedload velocity

Fig. 7.7 Signal homogenisation and impact of sediment transfer time from source to sink: **a** The highly variable detrital thermochronology fingerprints of source rocks are progressively homogenised during transport across the floodplain, and deep-sea turbidites may display homogeneous detrital signatures over thousands of kilometres; **b** Bedload transfer time across the floodplain limits the time resolution of stratigraphic correlations based on detrital minerals, which may be as low as >1 Myr

stratigraphic continental successions. The maximum time resolution of such correlations is in fact limited by the time required to transfer bedload sediment through all of the sampling sites under consideration (Fig. 7.7b). Based on available U-series isotope and cosmogenic data, the time resolution for stratigraphic correlations based on detrital minerals may thus be not much better than 1 Myr. This may be a problem for the analysis of Neogene or younger stratigraphic successions, but also implies that the presence of temporary or short-term sinks in a drainage basin (e.g. dams) has a negligible impact on the detrital record observed at the closure section of the river catchment.

7.6 Mineral Dissolution During Weathering and Diagenesis

7.6.1 Stability to Weathering of Minerals Relevant for Detrital Thermochronology

The original compositional fingerprint of a sediment is modified substantially by selective dissolution of unstable minerals during weathering and diagenesis. Dissolution of unstable minerals by the action of fluids may occur: (i) in the source rock and soils prior to erosion; (ii) during transient sediment exposure on the floodplain during transport; or (iii) after final deposition. Weathering prior to erosion has a direct impact on the mineral fertility of the source rock, i.e. on its propensity to yield a specific mineral when exposed to erosion (e.g. Moecher and Samson 2006, see Sect. 7.7). Depending on climatic and geomorphic conditions, minerals either may be removed quickly from the parent rock, or may instead be affected by prolonged chemical weathering in soils (Johnsson 1993). The potential impact of weathering prior to erosion is effectively taken into account when measuring the mineral fertility of bedrock as illustrated in Sect. 7.7.2.

In soil profiles developed at low latitudes (e.g. Thomas et al. 1999; Horbe et al. 2004), severe dissolution affects not only unstable minerals such as olivine or garnet, but also tourmaline and even zircon and quartz (Cleary and Conolly 1972; Nickel 1973; Velbel 1999; Van Loon and Mange 2007), traditionally considered as stable or even "ultrastable" (e.g. Pettijohn et al. 1972). In the strongly weathered quartzose bedload sand carried by equatorial rivers, zirconium concentration tends to be markedly lower than the upper continental crust standard (e.g. Dupré et al. 1996; Garzanti et al. 2013), which implies that metamict zircon may undergo dissolution in lateritic soils (Carroll 1953; Colin et al. 1993). Apatite is very soluble in acidic soils (Lång 2000; Morton and Hallsworth 2007), and a few decades may be sufficient for complete apatite dissolution in unfavourable conditions (e.g. in a peat bog, Le Roux et al. 2006). Apatite, when compared to much more stable species in the alluvial plain (e.g. tourmaline), is thus a sensitive indicator of weathering during alluvial storage (Morton and Hallsworth 1994). Gleadow and Lovering (1974) determined FT ages from apatite, titanite and zircon separated from fresh and highly weathered rock samples. They found no effect on thermochronologic ages even in extremely weathered

samples, apart from an apparent reduction in FT age in badly corroded apatites isolated from a residual clay. This suggests that the impact of chemical weathering on FT dating is probably negligible.

7.6.2 Stability of Detrital Minerals During Burial Diagenesis

The time and temperatures available for chemical reactions during a typical post-depositional path of a buried sandstone are incomparably longer and higher than during transit of sand from source to sink at the Earth's surface. Consequently, diagenetic effects are generally much more drastic than those of weathering (e.g. Andò et al. 2012; Morton 2012). More specifically, burial diagenesis is responsible for decreasing mineral diversity with increasing burial depth (Morton 1979; Cavazza and Gandolfi 1992; Milliken 2007). Burial depth and heat flow control the temperature and composition of the pore fluids, as well as the rate of pore fluid movement, which influence reactions such as smectite to illite transformation and the depth at which minerals become unstable (Morton 2012). Slow chemical reactions may proceed to completion if diagenesis is sufficiently long, but early cementation inhibits pore fluid movements, thus decreasing the rate of mineral dissolution (Bramlette 1941; Walderhaug and Porten 2007). During burial diagenesis, pyroxene, amphibole, epidote, titanite, staurolite and garnet are typically dissolved extensively or even completely in succession with increasing depth, and only zircon, tourmaline, rutile, apatite and Cr-spinel have good chances to survive in sandstones buried deeper than 3–4 km (Morton and Hallsworth 2007). In most ancient sandstones, heavy mineral assemblages only represent the strongly depleted durable residue of the much richer and more varied original detrital population (McBride 1985).

Dense minerals used routinely in detrital thermochronology, such as zircon, apatite, monazite and rutile, are among those relatively stable in burial conditions (Morton and Hallsworth 2007). Even though burial diagenesis generally leads to a sharp decrease in dense mineral concentration, the relative proportion of these minerals in ancient sandstones generally increases (dark grey arrow in Fig. 7.4b). The impact of selective mineral dissolution during burial diagenesis is generally overwhelming in provenance studies based on heavy mineral analysis (see, e.g., Garzanti and Malusà 2008, their Fig. 4), and also affects the relative proportion of quartz, feldspar and lithic grains in classic provenance studies based on the bulk sediment approach (McBride 1985), due to the lower stability of feldspar and most lithic grains compared to quartz. The impact of intrastratal dissolution is generally held to be negligible for detrital thermochronology studies (von

Eynatten and Dunkl 2012), although the higher susceptibility to dissolution of metamict zircon should be carefully considered.

7.6.3 Impact of Deep Burial on Metamict Zircon

Because of the change in structure induced by self-irradiation and accumulation of α-damage over time (Sect. 7.4.2), metamict zircon grains are much more reactive to chemical etching than non-metamict zircon grains, and thus potentially less stable during burial diagenesis and metamorphism. The detrital zircon record of the Po river drainage suggests a selective loss of up to ~40% metamict zircon (Malusà et al. 2013), also highlighted by a lack of old U-rich zircon grains as shown in the pie charts of Fig. 7.6d. Metamict zircon was not apparently destroyed by mechanical abrasion during river transport, as discussed in Sect. 7.4.2, and thus was most probably destroyed during burial diagenesis and/or regional metamorphism in the source rocks, in agreement with recent petrologic studies (e.g. Baldwin 2015; Kohn et al. 2015). The selective loss of metamict zircon during metamorphism implies that considerable amounts of Zr and U are made available for the recrystallisation of metamorphic overgrowths on other zircon crystals. The selective dissolution of old U-rich crystals also implies that polymetamorphic rocks are expected to show, on average, a lower percentage of old U-rich zircon crystals than their protoliths (Malusà et al. 2013). The ensuing quite variable U concentration in zircon grains shed from different source rocks has notable implications in terms of potential bias, as discussed in Chap. 16 (Malusà 2018).

7.7 Impact of Mineral Fertility

7.7.1 Fertility Variability in Bedrock

Single-mineral and multi-mineral approaches, when compared to bulk sediment approaches, are markedly influenced by the variable propensity of different parent bedrocks to yield detrital grains of specific minerals when exposed to erosion, which is also referred to as *mineral fertility* (Dickinson 2008; Hietpas et al. 2011). In detrital thermochronology studies, fertility controls the weight concentration of datable mineral grains in detritus, and represents by far the largest source of bias that is potentially introduced whenever single-mineral analyses are interpreted in terms of bulk sediment fluxes (Malusà et al. 2016a). Most detrital thermochronology studies assume negligible changes in mineral fertility (e.g. Malusà et al. 2009; Zhang et al. 2012; Glotzbach et al. 2013; Saylor et al. 2013; He et al. 2014), but

such constant-fertility assumption is generally untenable. Variations in mineral fertility commonly observed in natural rocks have a major impact on detrital grain-age distributions, and low fertility rocks are invariably underrepresented in the detrital thermochronology record (Fig. 7.8a) (Glotzbach et al. 2017; Malusà et al. 2017).

Mineral fertility measurements demonstrate that apatite and zircon fertility values can vary over three orders of magnitude, depending on the lithology and tectono-metamorphic evolution of eroded bedrock. In the European Alps, units chiefly consisting of metamorphic and plutonic rocks generally show higher apatite and zircon fertility than units largely consisting of sedimentary rocks (see Fig. 7.8b). The highest apatite fertility values (300–2600 mg/kg) are found in the medium-to-high-grade metamorphic rocks of the Lepontine dome (LD in Fig. 7.8b), within the area delimited by the isograds of diopside and sillimanite (Malusà et al. 2016a). High zircon fertility values characterise the Lepontine dome gneisses (10–70 mg/kg), migmatites and granitoid rocks of the Mont Blanc and Argentera Massifs (80–100 mg/kg), and Permo-Carboniferous metasediments of the Alpine accretionary wedge (50–90 mg/kg). The lowest apatite and zircon fertility values are found in sedimentary successions of the Southern Alps (apatite = 10–30 mg/kg; zircon = 0.5–6 mg/kg) and the Northern Apennines (apatite = 17–95 mg/kg; zircon = 0.5–7 mg/kg).

Apatite fertility in plutonic rocks is not exceptionally high, unlike commonly assumed. Values of 10–30 mg/kg characterise both the largest Alpine intrusive body (Adamello) and its country rocks. In the External Crystalline Massifs, apatite fertility is lower in the Mont Blanc Massif (MB in Fig. 7.8b), dominated by Upper Paleozoic intrusives, than in the Argentera Massif (AR in Fig. 7.8b), dominated by Lower Paleozoic migmatites. In the Menderes Massif of SW Turkey, apatite fertility is lower in Miocene granodiorites (~185 mg/kg) compared to their country rocks (~550 mg/kg) (Asti et al. 2018). Observations under an optical microscope (e.g. Rong and Wang 2016) suggest that apatite in plutonic rocks is generally very small (few microns), thus forming inclusions in larger minerals (e.g. feldspar) even in detritus, rather than individual mineral grains (Malusà et al. 2016a).

Apatite in metamorphic rocks is stable through a wide range of protoliths and metamorphic facies (Spear and Pyle 2002), and its fertility is largely controlled by the metamorphic texture. In low-grade metamorphic rocks, most of the apatite may be found as inclusions in poikiloblastic minerals (e.g. albite), whereas larger apatite crystals grow in medium-to-high-grade metamorphic rocks, thus providing larger amounts of datable apatite grains. Despite the evident relationships between bedrock geology and mineral fertility, major fertility variations can be observed even among tectonic units ascribed to the same palaeogeographic domain.

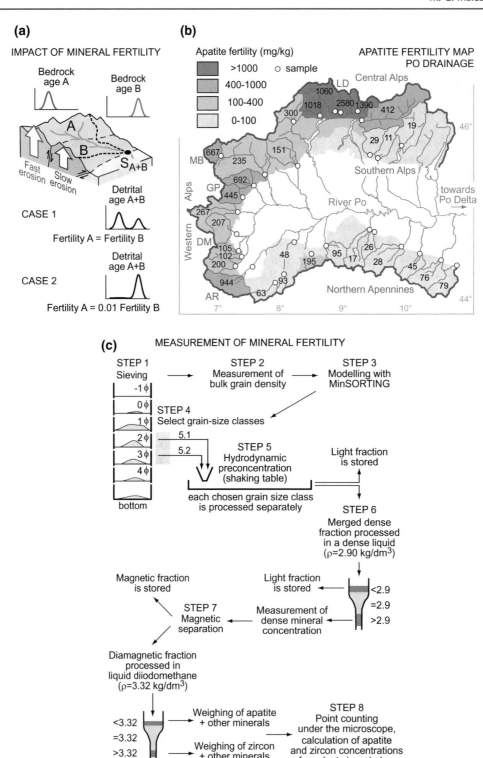

Fig. 7.8 Approach to mineral fertility bias: **a** Expected grain-age distributions in detritus derived from two eroding blocks (A and B) with distinct thermochronologic fingerprints (case 1: same mineral fertility in blocks A and B; case 2: lower mineral fertility in block A compared to block B); **b** Apatite fertility map of the Po river drainage (after Malusà et al. 2016a), showing that apatite fertility may vary over three orders of magnitude at least, depending on the lithology and the tectono-metamorphic evolution of eroded bedrock (red line = Po drainage boundary; dots = analysed samples); acronyms: AR, Argentera; DM, Dora-Maira; GP, Gran Paradiso; LD, Lepontine dome; MB, Mont Blanc; **c** Flow chart for the determination of mineral fertility in bedrock from the measurement of mineral concentration in a sediment sample (see text for details)

For example, the Dora-Maira and Gran Paradiso Internal Massifs (DM and GP in Fig. 7.8b) are both formed by subduction and eclogitization of European continental crust during the Paleogene, but apatite fertility is lower in the Dora-Maira (100–200 mg/kg) than in the Gran Paradiso (400–700 mg/kg). Similar variations are observed for zircon (DM = 0.4–1.2 mg/kg; GP = 5–8 mg/kg). The relationships between bedrock geology and mineral fertility are thus complex and hardly predictable. They depend not only on lithology, but more in general on the whole magmatic, sedimentary or metamorphic evolution of source rocks. Careful approaches to mineral fertility measurements are consequently encouraged.

7.7.2 Measurement of Mineral Fertility

In detrital studies, mineral fertility has been evaluated either by point counting under a microscope (e.g. Silver et al. 1981; Tranel et al. 2011), or by adopting a range of geochemical approaches (e.g. Cawood et al. 2003; Dickinson 2008). However, these approaches are either time-consuming or prone to introduce further unquantifiable bias (Malusà et al. 2016a). A more effective approach to mineral fertility determination is based on a simple consideration: if a rock is demolished by erosion and its fragments get finer downstream without significant modifications during transport and deposition, then sediment composition should faithfully reflect bedrock composition. The average mineral fertility of rocks exposed in the catchment can thus be determined by measuring the mineral concentration in the sediment derived from their erosion, provided that no significant modification was introduced by hydrodynamic processes in the sedimentary environment, and that no bias is introduced subsequently during mineral separation in the laboratory (Malusà et al. 2016a). These assumptions obviously need independent checking.

The recommended procedure for mineral fertility determination is synthesised in Fig. 7.8c. For this purpose, modern bedload sediment is collected along an active river channel within a mountain catchment, to avoid potential mixing of sediment reworked from the alluvial plain. All care should be taken to avoid sampling deposits affected by selective entrainment (i.e. placers and antiplacers) and even more to avoid concentrating dense minerals by panning in the field. Testing for modifications by hydraulic processes is discussed in Sect. 7.3.4. Notably, the mineral concentration measured in ancient sedimentary rocks should not be used to estimate the mineral fertility in their source rocks, because intrastratal dissolution during diagenesis almost invariably causes a significant depletion in unstable minerals, and a reduction in grain density (Sect. 7.6.2).

The chosen bedload sample is quartered in the laboratory and dry sieved at 1ϕ intervals (step 1 in Fig. 7.8c), in order to determine its mean grain size (D_m) and sorting values (σ) as described in Sect. 7.3.2 (Fig. 7.3a). The grain density of the bulk sample is measured on a small quartered fraction of the sample by using a hydrostatic balance (step 2 in Fig. 7.8c), as described in Sect. 7.3.4. MinSORTING (Resentini et al. 2013) is then used for modelling the distribution of the target mineral in different size classes (step 3), in order to choose the most suitable grain-size window for further mineral separation (see Sect. 7.3.2) and evaluate the fraction of target mineral potentially falling outside of the chosen grain-size range (step 4). A hydrodynamic preconcentration of the dense mineral fraction in each selected size class is then made with a shaking table (step 5). Because shaking tables exploit the basic principles of size-density sorting (Sect. 7.3), each different size class is processed separately to minimise the effects of grain size, and to effectively concentrate detrital grains according to their density. The dense fractions recovered after hydrodynamic preconcentration are then merged (step 6), and further purified in a dense liquid at 2.90 kg/dm^3 (e.g. sodium polytungstate). The dense fraction (>2.90 kg/dm^3) is recovered and weighed to obtain the dense mineral concentration of the sample for the analysed size window. The subsequent steps may vary according to the physical properties of the target mineral. In order to retrieve apatite and zircon, the dense fraction is refined with a Frantz magnetic separator under increasing field strength, following the recommendations of Sircombe and Stern (2002) (step 7). The diamagnetic fraction is further processed through liquid diiodomethane (3.32 kg/dm^3) to separate minerals denser than 3.32 kg/dm^3 (including zircon, $\sim$4.65 kg/dm^3) and minerals having densities between 2.9 and 3.32 kg/dm^3 (including apatite, $\sim$3.2 kg/dm^3). Quantities before and after each separation step are carefully weighed to detect any potential loss of material during processing. Point counting under the microscope allows one to determine the percentage of apatite and zircon grains in these concentrates (step 8), which also typically include other diamagnetic dense minerals such as andalusite, kyanite, sillimanite, pyrite and barite. The apatite and zircon concentrations for the selected size window are thus obtained. To get the apatite and zircon concentrations of the bulk sediment (step 9), we add to these values the amount of mineral expected to be found in size classes coarser and finer than the ones chosen for processing, as previously modelled by MinSORTING. Undatable apatite and zircon grains (e.g. grains smaller than 63 μm for FT analysis under an optical microscope) are not included in this calculation. Because the mineral concentration in the bulk sediment, calculated in step 9, is equal to the mineral fertility of the source rocks only if sediment composition was not modified by selective entrainment, the analysed samples need to be tested for anomalous concentrations of denser grains (step 10). This can be done by comparing the bulk grain density and/or the dense mineral concentration, measured during steps 2 and 6, with the reference values in eroded bedrock shown in Fig. 7.4c. Such a procedure for mineral fertility determination requires only minor modifications to the standard procedures adopted for the concentration of detrital apatite and zircon in most thermochronology laboratories (see Chap. 2, Kohn et al. 2018). The additional tasks required are largely justified by the information retrieved, which is essential for the correct interpretation of detrital thermochronology datasets as discussed in Chap. 10 (Malusà and Fitzgerald 2018b).

7.8 Summary and Recommendations

The effectiveness and resolution of detrital thermochronology studies are improved substantially by the knowledge and application of basic sedimentology principles, starting from sample processing to final data interpretation (Table 7.1). In order to exploit the full potential of the detrital thermochronology approach:

- minor modifications to standard concentration procedures should be adopted to measure the mineral fertility of eroded bedrock starting from the analysis of modern sediment samples;
- detrital samples should be fully characterised by sieving or equivalent to retrieve the relevant grain-size parameters of the bulk sediment sample;
- sediment composition in different size classes should be modelled to avoid mistakes during sample processing;
- the potential effects of selective entrainment should be tested, e.g. by measuring the bulk grain density of the sediment sample during mineral separation;
- the size and shape parameters of dated mineral grains should be systematically recorded to test the vulnerability of grain-age distributions to hydraulic sorting effects;
- the impact of rounding and abrasion in double-dated grains should be carefully evaluated to avoid incorrect interpretations of FT ages;
- the delay in detrital thermochronology signals in bedload and suspended load should be evaluated for its implications in stratigraphic correlations and lag-time analysis.

Acknowledgements We are grateful to researchers and graduate students in the laboratory of fission-track analysis at University of Milano-Bicocca for their contributions in establishing the approaches described in this work. Reviews by O. Anfinson and M. L. Balestrieri, and comments by P. G. Fitzgerald were of great help to improve the clarity of the manuscript.

References

Afanas' ev VP, Nikolenko EI, Tychkov NS et al (2008) Mechanical abrasion of kimberlite indicator minerals: experimental investigations. Russ Geol Geophys 49(2):91–97

Allen PA, Allen JR (2013) Basin analysis: principles and application to petroleum play assessment. Wiley, New York

Andò S, Garzanti E, Padoan M, Limonta M (2012) Corrosion of heavy minerals during weathering and diagenesis: a catalogue for optical analysis. Sed Geol 280:165–178

Anfinson OA, Malusà MG, Ottria G, Dafov LN, Stockli DF (2016) Tracking coarse-grained gravity flows by LASS-ICP-MS depth-profiling of detrital zircon (Aveto Formation, Adriatic Foredeep, Italy). Mar Petrol Geol 77:1163–1176

Asti R, Malusà MG, Faccenna C (2018) Supradetachment basin evolution unraveled by detrital apatite fission track analysis: the Gediz Graben (Menderes Massif, Western Turkey). Basin Res 30:502-521

Baldwin SL (2015) Highlights and breakthroughs. Zircon dissolution and growth during metamorphism. Am Mineral 100(5–6):1019–1020

Bernet M, Zattin M, Garver JI, Brandon MT, Vance JA (2001) Steady-state exhumation of the European Alps. Geology 29:35–38

Bonich MB, Samson SD, Fedo CM (2017) Incongruity of detrital zircon ages of granitic bedrock and its derived alluvium: an example from the Stepladder Mountains, Southeastern California. J Geol 125 (3)

Bramlette MN (1941) The stability of minerals in sandstone. J Sediment Res 11(1)

Carrapa B, Di Giulio A, Wijbrans J (2004) The early stages of the Alpine collision: an image derived from the upper Eocene–lower Oligocene record in the Alps-Apennines junction area. Sediment Geol 171:181–203

Carroll D (1953) Weatherability of zircon. J Sediment Res 23:106–116

Carter A, Moss SJ (1999) Combined detrital-zircon fission-track and U-Pb dating: a new approach to understanding hinterland evolution. Geology 27(3):235–238

Cavazza W, Gandolfi G (1992) Diagenetic processes along a basin-wide marker bed as a function of burial depth. J Sediment Res 62(2):261–272

Cawood PA, Nemchin AA, Freeman M, Sircombe K (2003) Linking source and sedimentary basin: detrital zircon record of sediment flux along a modern river system and implications for provenance studies. Earth Planet Sci Lett 210:259–268

Cheng NS (1997) Simplified settling velocity formula for sediment particle. J Hydraul Eng 123:149–152

Cheng NS (2009) Comparison of formulas for drag coefficient and settling velocity of spherical particles. Powder Tech 189:395–398

Cleary WJ, Conolly JR (1972) Embayed quartz grains in soils and their significance. J Sed Petr 42:899–904

Colin F, Alarcon C, Vieillard P (1993) Zircon: an immobile index in soils? Chem Geol 107:273–276

Corfu F, Hanchar JM, Hoskin PW, Kinny P (2003) Atlas of zircon textures. Rev Mineral Geochem 53(1):469–500

Dickinson WR (1985) Interpreting provenance relations from detrital modes of sandstones. Provenance of arenites. Springer, Netherlands, pp 333–361

Dickinson WR (2008) Impact of differential zircon fertility of granitoid basement rocks in North America on age populations of detrital zircons and implications for granite petrogenesis. Earth Planet Sci Lett 275:80–92

Dietz V (1973) Experiments on the influence of transport on shape and roundness of heavy minerals. Contrib Sediment 1:69–102

Doyle LJ, Carder KL, Steward RG (1983) The hydraulic equivalence of mica. J Sediment Res 53(2)

Dupré B, Gaillardet J, Rousseau D, Allègre CJ (1996) Major and trace elements of river-borne material: the Congo Basin. Geochim Cosmochim Ac 60:1301–1321

Eplett WJR (1982) The distributions of Smirnov type two-sample rank tests for discontinuous distributions functions. J Royal Stat Soc 44:361–369

Ewing RC, Meldrum A, Wang L, Weber WJ, Corrales LR (2003) Radiation effects in zircon. Rev Mineral Geochem 53:387–425

Fedele JJ, Paola C (2007) Similarity solutions for fluvial sediment fining by selective deposition. J Geophys Res-Earth 112:F02038

Fedo CM, Sircombe KN, Rainbird RH (2003) Detrital zircon analysis of the sedimentary record. Rev Mineral Geochem 53:277–303

Folk RL, Ward WC (1957) Brazos River bar: a study in the significance of grain size parameters. J Sediment Res 27:3–26

Garver JI, Kamp PJJ (2002) Integration of zircon color and zircon fission-track zonation patterns in orogenic belts: application to the Southern Alps, New Zealand. Tectonophysics 349:203–219

Garver JI, Brandon MT, Roden-Tice MK, Kamp PJJ (1999) Exhumation history of orogenic highlands determined by detrital fission track thermochronology. Geol Soc Spec Publ 154:283–304

Garzanti E, Andò S (2007) Heavy mineral concentration in modern sands: implications for provenance interpretation. Dev Sediment 58:517–545

Garzanti E, Malusà MG (2008) The Oligocene Alps: Domal unroofing and drainage development during early orogenic growth. Earth Planet Sci Lett 268:487–500

Garzanti E, Andò S, Vezzoli G (2008) Settling equivalence of detrital minerals and grain-size dependence of sediment composition. Earth Planet Sci Lett 273:138–151

Garzanti E, Andò S, Vezzoli G (2009) Grain-size dependence of sediment composition and environmental bias in provenance studies. Earth Planet Sci Lett 277:422–432

Garzanti E, Andò S, France-Lanord C, Censi P, Vignola P, Galy V, Lupker M (2010) Mineralogical and chemical variability of fluvial sediments: 1. Bedload sand (Ganga–Brahmaputra, Bangladesh). Earth Planet Sci Lett 299:368–381

Garzanti E, Padoan M, Andò S, Resentini A, Vezzoli G, Lustrino M (2013) Weathering and relative durability of detrital minerals in equatorial climate: sand petrology and geochemistry in the East African Rift. J Geol 121:547–580

Garzanti E, Resentini A, Andò S, Vezzoli G, Pereira A, Vermeesch P (2015) Physical controls on sand composition and relative durability of detrital minerals during long-distance littoral and eolian transport (coastal Namibia). Sedimentology 62:971–996

Gleadow AJW, Lovering JF (1974) The effect of weathering on fission track dating. Earth Planet Sci Lett 22(2):163–168

Glotzbach C, van der Beek P, Carcaillet J, Delunel R (2013) Deciphering the driving forces of erosion rates on millennial to million-year timescales in glacially impacted landscapes: an example from the Western Alps. J Geophys Res-Earth 118:1491–1515

Glotzbach C, Busschers FS, Winsemann J (2017) Detrital thermochronology of Rhine, Elbe and Meuse river sediment (Central Europe): implications for provenance, erosion and mineral fertility. Int J Earth Sci. https://doi.org/10.1007/s00531-017-1502-9

Graham SA, Dickinson WR, Ingersoll RV (1975) Himalayan-Bengal model for flysch dispersal in the Appalachian-Ouachita system. Geol Soc Am Bull 86:273–286

Granet M, Chabaux F, Stille P, Dosseto A, France-Lanord C, Blaes E (2010) U-series disequilibria in suspended river sediments and implication for sediment transfer time in alluvial plains: the case of the Himalayan rivers. Geochim Cosmochim Ac 74(10):2851–2865

Hay DC, Dempster TJ (2009) Zircon alteration, formation and preservation in sandstones. Sedimentology 56(7):2175–2191

He M, Zheng H, Bookhagen B, Clift P (2014) Controls on erosion intensity in the Yangtze River basin tracked by U-Pb detrital zircon dating. Earth-Sci Rev 136:121–140

Hietpas J, Samson S, Moecher D, Chakraborty S (2011) Enhancing tectonic and provenance information from detrital zircon studies: assessing terrane-scale sampling and grain-scale characterization. J Geol Soc London 168:309–318

Hodges KV, Ruhl KW, Wobus CW, Pringle MS (2005) $^{40}Ar/^{39}Ar$ thermochronology of detrital minerals. Rev Mineral Geochem 58:239–257

Hollander M, Wolfe D (1999) Nonparametric statistical methods. Wiley, New York

Horbe AMC, Horbe MA, Suguio K (2004) Tropical spodosols in northeastern Amazonas State, Brazil. Geoderma 119:55–68

Hourigan JK, Reiners PW, Brandon MT (2005) U-Th zonation-dependent alpha-ejection in (U-Th)/He chronometry. Geochim Cosmochim Ac 69:3349–3365

Hubert JF (1962) A zircon-tourmaline-rutile maturity index and the interdependence of the composition of heavy mineral assemblages with the gross composition and texture of sandstones. J Sediment Res 32:440–450

Ingersoll RV, Dickinson WR, Graham SA (2003) Remnant-ocean submarine fans: largest sedimentary systems on Earth. Geol S Am S 370:191–208

Johnsson MJ (1993) The system controlling the composition of clastic sediments. Geol S Am S 284:1–20

Jourdan S, Bernet M, Tricart P, Hardwick E, Paquette JL, Guillot S, Dumont T, Schwartz S (2013) Short-lived, fast erosional exhumation of the internal western Alps during the late early Oligocene: constraints from geothermochronology of pro-and retro-side foreland basin sediments. Lithosphere 5(2):211–225

Kohn MJ, Corrie SL, Markley C (2015) The fall and rise of metamorphic zircon. Am Mineral 100(4):897–908

Kohn B, Chung L, Gleadow A (2018) Chapter 2. Fission-track analysis: field collection, sample preparation and data acquisition. In: Malusà MG, Fitzgerald PG (eds) Fission-track thermochronology and its application to geology. Springer, Berlin

Komar PD (2007) The entrainment, transport and sorting of heavy minerals by waves and currents. Dev Sediment 58:3–48

Komar PD, Li Z (1988) Applications of grain-pivoting and sliding analyses to selective entrapment of gravel and to flow-competence evaluations. Sedimentology 35:681–695

Komar PD, Wang C (1984) Processes of selective grain transport and the formation of placers on beaches. J Geol 92:637–655

Kuenen PH (1959) Experimental abrasion; 3, fluviatile action on sand. Am J Sci 257:172–190

Kuenen PH (1960) Experimental abrasion 4: eolian action. J Geol 68:427–449

Lång LO (2000) Heavy mineral weathering under acidic soil conditions. Appl Geochem 15(4):415–423

Le Roux JP (2005) Grains in motion: a review. Sediment Geol 178:285–313

Le Roux G, Laverret E, Shotyk W (2006) Fate of calcite, apatite and feldspars in an ombrotrophic peat bog, Black Forest. Germany. J Geol Soc London 163(4):641–646

Malusà MG (2018) Chapter 16. A guide for interpreting complex detrital age patterns in stratigraphic sequences. In: Malusà MG, Fitzgerald PG (eds) Fission-track thermochronology and its application to geology. Springer, Berlin

Malusà MG, Fitzgerald PG (2018) Chapter 8. From cooling to exhumation: setting the reference frame for the interpretation of thermocronologic data. In: Malusà MG, Fitzgerald PG (eds) Fission-track thermochronology and its application to geology. Springer, Berlin

Malusà MG, Fitzgerald PG (2018) Chapter 10. Application of thermochronology to geologic problems: bedrock and detrital approaches. In: Malusà MG, Fitzgerald PG (eds) Fission-track thermochronology and its application to geology. Springer, Berlin

Malusà MG, Zattin M, Andò S, Garzanti E, Vezzoli G (2009) Focused erosion in the Alps constrained by fission-track ages on detrital apatites. Geol Soc Spec Publ 324:141–152

Malusà MG, Villa IM, Vezzoli G, Garzanti E (2011) Detrital geochronology of unroofing magmatic complexes and the slow erosion of Oligocene volcanoes in the Alps. Earth Planet Sci Lett 301(1):324–336

Malusà MG, Carter A, Limoncelli M, Villa IM, Garzanti E (2013) Bias in detrital zircon geochronology and thermochronometry. Chem Geol 359:90–107

Malusà MG, Resentini A, Garzanti E (2016a) Hydraulic sorting and mineral fertility bias in detrital geochronology. Gondwana Res 31:1–19

Malusà MG, Anfinson OA, Dafov LN, Stockli DF (2016b) Tracking Adria indentation beneath the Alps by detrital zircon U-Pb geochronology: Implications for the Oligocene-Miocene dynamics of the Adriatic microplate. Geology 44(2):155–158

Malusà MG, Wang J, Garzanti E, Liu ZC, Villa IM, Wittmann H (2017) Trace-element and Nd-isotope systematics in detrital apatite of the Po river catchment: Implications for provenance discrimination and the lag-time approach to detrital thermochronology. Lithos 290–291:48–59

Mange MA, Wright DT (eds) (2007) Heavy minerals in use. Elsevier, Amsterdam

Markwitz V, Kirkland CL, Mehnert A, Gessner K, Shaw J (2017) 3-D characterization of detrital zircon grains and its implications for fluvial transport, mixing, and preservation bias. Geochem Geophys Geosyst 18:4655–4673

McBride EF (1985) Diagenetic processes that affect provenance determinations in sandstones. In Zuffa GG (ed) Provenance of arenites. Dordrecht, Reidel, NATO ASI Series 148:95–113

McLennan SM, Hemming S, McDaniel DK, Hanson GN (1993) Geochemical approaches to sedimentation, provenance, and tectonics. Geol S Am S 284:21–40

Milliken KL (2007) Provenance and diagenesis of heavy minerals, Cenozoic units of the northwestern Gulf of Mexico sedimentary basin. Dev Sediment 58:247–261

Moecher DP, Samson SD (2006) Differential zircon fertility of source terranes and natural bias in the detrital zircon record: implications for sedimentary provenance analysis. Earth Planet Sci Lett 247:252–266

Morton AC (1979) Surface features of heavy mineral grains from Palaeocene sands of the central North Sea. Scot J Geol 15(4):293–300

Morton AC (2012) Value of heavy minerals in sediments and sedimentary rocks for provenance, transport history and stratigraphic correlation. Mineral Ass Canada Short Course Series 42:133–165

Morton AC, Hallsworth C (1994) Identifying provenance-specific features of detrital heavy mineral assemblages in sandstones. Sediment Geol 90(3):241–256

Morton AC, Hallsworth C (2007) Stability of detrital heavy minerals during burial diagenesis. Dev Sediment 58:215–245

Najman YMR, Pringle MS, Johnson MRW, Robertson AHF, Wijbrans JR (1997) Laser ^{40}Ar/^{39}Ar dating of single detrital muscovite grains from early foreland-basin sedimentary deposits in India: implications for early Himalayan evolution. Geology 25:535–538

Nasdala L, Wenzel M, Vavra G, Irmer G, Wenzel T, Kober B (2001) Metamictisation of natural zircon accumulation versus thermal annealing of radioactivity-induced damage. Contrib Mineral Petr 141:125–144

Nickel E (1973) Experimental dissolution of light and heavy minerals in comparison with weathering and intrastratal solution. Contrib Sediment 1:1–68

Pettijohn FJ, Potter PE, Siever R (1972) Sand and sandstone. Springer, New York

Pratten NA (1981) The precise measurement of the density of small samples. J Mater Sci 16:1737–1747

Reid I, Frostick LE (1985) Role of settling, entrainment and dispersive equivalence and of interstice trapping in placer formation. J Geol Soc London 142:739–746

Reiners PW, Farley KA (2001) Influence of crystal size on apatite (U–Th)/He thermochronology: an example from the Bighorn Mountains, Wyoming. Earth Planet Sci Lett 188(3):413–420

Resentini A, Malusà MG (2012) Sediment budgets by detrital apatite fission-track dating (Rivers Dora Baltea and Arc, Western Alps). Geol S Am S 487:125–140

Resentini A, Malusà MG, Garzanti E (2013) MinSORTING: An Excel® worksheet for modelling mineral grain-size distribution in sediments, with application to detrital geochronology and provenance studies. Comput Geosci 59:90–97

Rittenhouse G (1943) Transportation and deposition of heavy minerals. Geol Soc Am Bull 54:1725–1780

Rong J, Wang F (2016) Discussion about the origin of mineral textures in granite. In: Metasomatic textures in granites. Springer, Singapore

Rubey WW (1933) The size distribution of heavy minerals within a water laid sandstone. J Sediment Petr 3:3–29

Russell RD, Taylor RE (1937) Roundness and shape of Mississippi River sands. J Geol 45:225–267

Saylor JE, Knowles JN, Horton BK, Nie J, Mora A (2013) Mixing of source populations recorded in detrital zircon U-Pb age spectra of modern river sands. J Geol 121:17–33

Schuiling RD, DeMeijer RJ, Riezebos HJ, Scholten MJ (1985) Grain size distribution of different minerals in a sediment as a function of their specific density. Geol Mijnbouw 64:199–203

Silver LT, Williams IS, Woodhead JA (1981) Uranium in granites from the southwestern United States: actinide parent–daughter systems, sites and mobilization. U.S. Department of Energy Open–File Repository GJBX–45

Sircombe KN, Stern RA (2002) An investigation of artificial biasing in detrital zircon U-Pb geochronology due to magnetic separation in sample preparation. Geochim Cosmochim Ac 66:2379–2397

Slingerland R, Smith ND (1986) Occurrence and formation of water-laid placers. Annu Rev Earth Pl Sc 14:113–147

Smirnov NV (1939) On the estimation of the discrepancy between empirical curves of distribution for two independent samples. Bull Math Univ Moscow 2:3–14

Spear FS, Pyle JM (2002) Apatite, monazite, and xenotime in metamorphic rocks. Rev Mineral Geochem 48:293–335

Tagami T, Carter A, Hurford AJ (1996) Natural long termannealing of the zircon fission track system in Vienna Basin deep borehole samples: constraints upon the partial annealing zone and closure temperature. Chem Geol 130:147–157

Thiel GA (1940) The relative resistance to abrasion of mineral grains of sand size. J Sediment Res 10(3)

Thomas M, Thorp M, McAlister J (1999) Equatorial weathering, landform development and the formation of white sands in north western Kalimantan, Indonesia. CATENA 36:205–232

Tranel LM, Spotila JA, Kowalewski MJ, Waller CM (2011) Spatial variation of erosion in a small, glaciated basin in the Teton Range, Wyoming, based on detrital apatite (U-Th)/He thermochronology. Basin Res 23:571–590

Tripathy-Lang A, Hodges KV, Monteleone BD, Soest MC (2013) Laser (U-Th)/He thermochronology of detrital zircons as a tool for studying surface processes in modern catchments. J Geophys Res-Earth 118(3):1333–1341

Van Loon AJ, Mange AM (2007) "In situ" dissolution of heavy minerals through extreme weathering, and the application of the surviving assemblages and their dissolution characteristics to correlation of Dutch and German silver sands. Dev Sediment 58:189–213

Velbel MA (1999) Bond strength and the relative weathering rates of simple orthosilicates. Am J Sci 299(7–9):679–696

von Eynatten H, Dunkl I (2012) Assessing the sediment factory: the role of single grain analysis. Earth-Sci Rev 115:97–120

Walderhaug O, Porten KW (2007) Stability of detrital heavy minerals on the Norwegian continental shelf as a function of depth and temperature. J Sediment Res 77(12):992–1002

Wittmann H, Von Blanckenburg F, Maurice L, Guyot JL, Kubik PW (2011) Recycling of Amazon floodplain sediment quantified by cosmogenic ^{26}Al and ^{10}Be. Geology 39(5):467–470

Wittmann H, Malusà MG, Resentini A, Garzanti E, Niedermann S (2016) The cosmogenic record of mountain erosion transmitted across a foreland basin: source-to-sink analysis of in situ ^{10}Be, ^{26}Al and ^{21}Ne in sediment of the Po river catchment. Earth Planet Sci Lett 452:258–271

Worden RH, Burley SD (2003) Sandstone diagenesis: the evolution of sand to stone. Sandstone Diagenesis Recent Anc 4:3–44

Young IT (1977) Proof without prejudice: use of the Kolmogorov-Smirnov test for the analysis of histograms from flow systems and other sources. J Histochem Cytochem 25:935–941

Zhang JY, Yin A, Liu WC, Wu FY, Lin D, Grove M (2012) Coupled U-Pb dating and Hf isotopic analysis of detrital zircon of modern river sand from the Yalu River (Yarlung Tsangpo) drainage system in southern Tibet: Constraints on the transport processes and evolution of Himalayan rivers. Geol Soc Am Bull 124:1449–1473

Zuffa GG (ed) (1985) Provenance of arenites. Springer, Netherlands

Part II
The Geologic Interpretation of the Thermochronologic Record

From Cooling to Exhumation: Setting the Reference Frame for the Interpretation of Thermochronologic Data

8

Marco G. Malusà and Paul G. Fitzgerald

Abstract

The reference frame for the interpretation of fission-track (FT) data is a thermal reference frame. Using thermochronology to constrain exhumation largely depends on understanding the linkage between this reference frame and Earth's surface. The thermal frame of reference is dynamic, that is it is often neither stationary nor horizontal, as it is influenced by the shape of the topography, heat advection associated with rapid exhumation and mass redistribution across major faults. Here, we review the nomenclature and basic relationships related to cooling, uplift and exhumation and describe strategies to independently constrain the paleogeothermal gradient at the time of exhumation. In some cases, cooling may not be related to exhumation, but can be used instead to constrain the thermal evolution of the upper crust and the emplacement depth of magmatic rocks. In general terms, useful constraints on exhumation are often only directly provided by thermochronologic ages that are set during undisturbed exhumational cooling across the closure temperature isothermal surface. Thermochronologic ages from minerals crystallised at temperatures less than the closure temperature, e.g. in volcanic rocks and shallow intrusions, provide no direct constraint on exhumation.

M. G. Malusà (✉)
Department of Earth and Environmental Sciences, University of Milano-Bicocca, Piazza Delle Scienza 4, 20126 Milan, Italy
e-mail: marco.malusa@unimib.it

P. G. Fitzgerald
Department of Earth Sciences, Syracuse University, Syracuse, NY 13244, USA

8.1 Cooling, Exhumation, Rock and Surface Uplift

8.1.1 Nomenclature and Basic Relationships

Fission-track (FT) thermochronology constrains the cooling (the time–temperature path) of rocks in the upper crust within a temperature interval depending on the mineral and its composition (e.g. Fleischer et al. 1975; Hurford 1991; Wagner and van den Haute 1992). In a simple scenario, rocks will cool from a zone of total FT annealing (higher temperature) to a zone of total FT retention (lower temperature). Between these zones of total annealing and effectively zero annealing is the temperature interval where there is partial annealing, initially defined as the partial stability zone (e.g. Wagner et al. 1977, 1989; Gleadow and Duddy 1981; Gleadow et al. 1986). The term *partial annealing zone* (PAZ) was introduced in 1987 (Gleadow and Fitzgerald 1987) and is the term in common use today. The same concept of a temperature-controlled zone of partial loss generally applies to all geochronologic and thermochronologic methods. For noble gas thermochronology ($^{40}Ar/^{39}Ar$, (U–Th)/He), the term *partial retention zone* (PRZ) was introduced in 1998 for the $^{40}Ar/^{39}Ar$ method (Baldwin and Lister 1998) and subsequently adopted for (U–Th)/He dating (Wolf et al. 1998). In essence, the concept of PAZ and PRZ are the same, but the mechanism of loss (annealing of fission tracks vs diffusion of noble gas) is different in each case, so typically the FT community uses PAZ and the noble gas community uses PRZ. The concept of a *closure temperature* T_c and a *cooling age* applies in the case where rocks are cooling *monotonically* from high to low temperatures (Fig. 8.1) (Jäger 1967; Dodson 1973; Villa 1998). The closure temperature T_c can therefore be defined as the temperature of a rock at its thermochronologic cooling age (Dodson 1973). A cooling age depends on the rate of cooling, with T_c being higher for more rapid cooling (e.g. Reiners and Brandon 2006). It is different for different systems and

© Springer International Publishing AG, part of Springer Nature 2019
M. G. Malusà and P. G. Fitzgerald (eds.), *Fission-Track Thermochronology and its Application to Geology*,
Springer Textbooks in Earth Sciences, Geography and Environment, https://doi.org/10.1007/978-3-319-89421-8_8

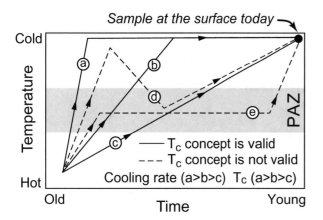

Fig. 8.1 "Closure temperature" and "cooling age" concepts apply to rocks that have cooled monotonically from higher to lower temperatures (a to c). In those cases, the closure temperature T_c is higher for more rapid cooling. The same concepts do not apply to rocks that have followed more complex cooling paths (d, e). PAZ denotes the partial annealing zone

minerals within those systems and is often variable depending on mineral composition and/or radiation damage (e.g. Rahn et al. 2004).

A thermochronologic age being representative of the time the rock cooled through a T_c isotherm does not apply if the rock does not cool monotonically through that isotherm. For example, if a sample cools into a PAZ or PRZ, is resident therein for a period of time before cooling again, or the sample cooled rapidly and is then partially thermally reset before later cooling. In these cases, the thermochronologic age cannot be recognised as a "cooling age"—as the "age" represents components from a composite cooling path (Fig. 8.1). It is a common mistake (or assumption) in some studies, often detrital studies employing thermochronologic data (see Chap. 10, Malusà and Fitzgerald 2018), that all ages represent "cooling ages" representative of the time since a sample cooled through T_c. That thermochronologic ages may be complex in their interpretation was recognised very early as thermochronology evolved. For example, early workers labelled graphs with "Apparent Age" (e.g. Gleadow 1990) or more recently substituting "Date" for "Age" for (U–Th)/He data (e.g. Flowers et al. 2009). In that case, "date" refers to a model prediction or analytical result, whereas "age" would refer to the geological interpretation of the data. (U–Th)/He data sets often show widely dispersed (U–Th)/He single grain ages due to a large number of factors, notably variable radiation damage due to different concentrations of U and Th between crystals, zonation and/or grain size variations, all magnified by slow cooling or residence within a PRZ (Reiners and Farley 2001; Meesters and Dunai 2002; Ehlers and Farley 2003; Fitzgerald et al. 2006; Flowers et al. 2009; Gautheron et al. 2009). However, we do not recommend the use of "date" and then "age" as defined by Flowers et al. (2009), as this introduces another term ("date") which is not

universally applied and is more likely to be more confusing that a simple "age" followed by the interpretation of the data, constraints on the time–temperature path, timing of events or episodes of exhumation, etc.

To distinguish simple from complex temperature–time (T–t) paths and determine whether a thermochronologic age can be simply interpreted as representing time since the sample crossed a T_c, some thermochronologic methods have kinetic parameters. They include confined track lengths in apatite FT thermochronology (AFT), kinetic proxies such as age spectra for some minerals in $^{40}Ar/^{39}Ar$ thermochronology, and multi-diffusion domain modelling of K-feldspar in $^{40}Ar/^{39}Ar$ thermochronology. These kinetic parameters allow evaluation of the rate of cooling, the T–t path and possible partial annealing/resetting, and thus whether an age represents a cooling age. Regardless of the complexity or simplicity of the T–t path, the reference frame for the interpretation of thermochronologic data is a thermal reference frame.

Our ability to use FT thermochronology to constrain *exhumation*, i.e. the motion of a column of rock towards Earth's surface (England and Molnar 1990; Stüwe and Barr 1998), depends on our ability to assess the linkage of this thermal reference frame with Earth's surface (Brown 1991; Fitzgerald et al. 1995; Braun 2002). In active geologic settings, the thermal reference frame is dynamic and is typically not horizontal (Parrish 1983; Stüwe et al. 1994; Mancktelow and Grasemann 1997; Huntington et al. 2007). In the simplest case, constraints on exhumation are potentially provided by the T_c approach. That is, the assumption that samples have cooled during their motion towards Earth's surface across a steady isothermal surface corresponding to the T_c of the chosen thermochronologic system (case 1 in Fig. 8.2a). If the depth of the T_c isothermal surface is known (often assuming, or using other methods to constrain a *paleogeothermal gradient*—see Sect. 8.3), an average exhumation rate can be calculated from the time of cooling across the T_c isothermal surface to the time of final exposure of the sample to Earth's surface (e.g. Wagner et al. 1977). In other words, in order to calculate an average exhumation rate using a single sample, the paleogeothermal gradient (i.e. the vertical distance between the T_c isotherm and Earth's surface) during exhumation has to be independently known. Reiners and Brandon (2006) indicated it was a reasonable assumption to apply the T_c approach to eroding mountain belts as the temperature range for a PAZ or PRZ is relatively narrow. However, in many mountain belts such as the Transantarctic Mountains (Gleadow and Fitzgerald 1987; Fitzgerald 1994, 2002; Miller et al. 2010) and the Alaska Range (Fitzgerald et al. 1993, 1995) to cite two well-known examples, it is obvious that higher elevation samples have resided for considerable periods of time within a PAZ before being exhumed, and for these higher elevation samples, the T_c approach is not valid.

Fig. 8.2 **a** Cooling (or heating)
recorded by low-temperature
thermochronometers may reflect
different scenarios either related
to exhumation or movement of
isotherms: Case 1—the motion
(exhumation) of rocks towards
Earth' surface. Case 2—in this
scenario, transient rising of
isothermal surfaces (i.e. an
increase in the geothermal
gradient) results in partial or
complete resetting of
low-temperature
thermochronometers, which may
be followed by thermal relaxation
(i.e. a decrease in the geothermal
gradient). Case 3—crystallisation
of minerals occurs at shallow
crustal depth above T_c isotherm;
b Rock uplift is equal to the
amount of exhumation plus the
amount of surface uplift (based on
England and Molnar 1990);
c Possible relationships between
exhumation and mountain growth
during uplift of buried rocks;
d Alternative mechanisms of
overburden removal during
exhumation; tectonic exhumation
is independent from topographic
relief and climate, and does not
necessarily require detritus
production

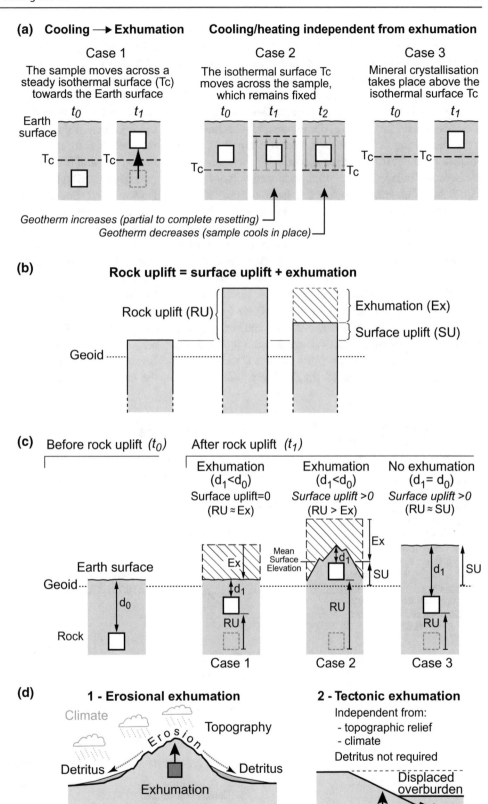

There are situations where cooling recorded by low-temperature thermochronometers is not necessarily linked to exhumation. For instance, the same cooling scenario as illustrated in case 1 of Fig. 8.2a may result from a T_c surface moving downwards across a rock sample that remains fixed relative to Earth's surface (case 2 in Fig. 8.2a). This may occur, for example, during thermal relaxation after a transient rising of isothermal surfaces (Braun 2002, 2016; Malusà et al. 2016). Another scenario is where magmatic rocks cool independently from exhumation (case 3 in Fig. 8.3a). If magma is emplaced at shallower crustal levels compared to the undisturbed T_c isothermal surface observed before intrusion (time t_0), magmatic rocks will cool, after intrusion, at temperatures below T_c even without moving towards the Earth's surface (time t_1), thus providing no direct constraint on exhumation (Malusà et al. 2011). Therefore, FT thermochronology can be used either to constrain exhumation (case 1 in Fig. 8.2a), or to constrain the thermal evolution of the upper crust and the emplacement depth of magmatic rocks (cases 2 and 3 in Fig. 8.2a).

FT thermochronology provides no direct constraint on *rock uplift* and *surface uplift* (England and Molnar 1990; Brown 1991; Fitzgerald et al. 1995; Corcoran and Dorè 2005). Rock uplift is defined as the motion of a column of rock (in the direction opposite to the gravity vector) relative to a fixed reference point such as the geoid (England and Molnar 1990) or the undeformed reference lithosphere (Sandiford and Powell 1990, 1991) (Fig. 8.2b). Surface uplift is instead defined as the displacement of Earth's surface relative to the geoid or the undeformed reference lithosphere and is a measure of the evolution of topography through time (England and Molnar 1990; Summerfield and Brown 1998). Positive surface uplift only occurs when *denudation* (chemical and/or mechanical removal of material) is less than rock uplift. Denudation and exhumation are often used as synonyms, but denudation refers more to the removal of material (erosion and/or tectonic processes), whereas exhumation can be thought of as the movement of the rock towards the surface or as the unroofing history of the rock (Walcott 1998; Ring et al. 1999). The geoid, i.e. the equipotential surface of gravity generally representing the reference frame for rock and surface uplift, has no direct linkage with the thermal reference frame that is relevant for the interpretation of FT data.

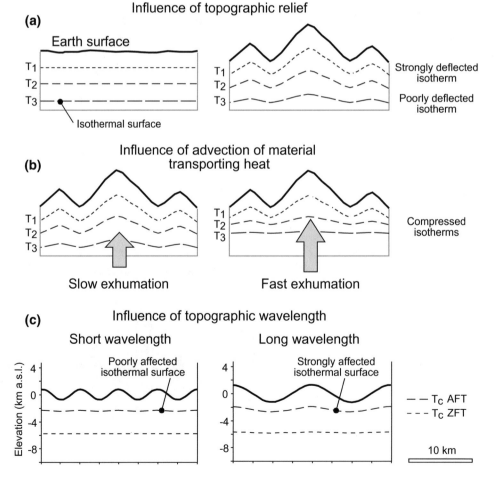

Fig. 8.3 a Isothermal surfaces mimic the shape of topography in a dampened fashion; **b** Heat advection during rapid exhumation determines the progressive elevation of isothermal surfaces and a transient increase in near-surface geothermal gradient; **c** Penetration of low-temperature isothermal surfaces beneath ridges is facilitated by longer topographic wavelengths (based on Stüwe et al. 1994; Mancktelow and Grasemann 1997; Braun 2002; Ehlers et al. 2005)

As shown in Fig. 8.2b, rock uplift is equal to the amount of exhumation plus the amount of surface uplift (England and Molnar 1990). This equation is quite simple, but is difficult to solve because surface and rock uplift are generally much more difficult to constrain than exhumation, even in those cases where the evidence of uplift is particularly clear, such as in collision orogens (e.g. Fitzgerald et al. 1995). In orogenic belts, rocks metamorphosed in the upper crust or even mid-crustal depths, as well as sedimentary rocks originally deposited in basins, may now be exposed at high elevations thus attesting long-term rock and surface uplift. But the temporal relationships between rock uplift, exhumation and topographic growth are generally undetermined (Fig. 8.2c). In principle, exhumation during rock uplift can either be associated or not with topographic growth (cases 1 and 2 in Fig. 8.2c); and, on the other hand, topographic growth during rock uplift can take place even without exhumation (case 3 in Fig. 8.2c).

The interplay of surface uplift, rock uplift and exhumation defines various stages in the denudational evolution of mountain belts, simply summarised as an immature stage of orogeny where surface uplift is greater than exhumation (during the rise of mountains), followed by a mature steady-state interval (surface uplift = 0) and then a long decaying stage where mountains are eroded away with isostatic response feeding the erosion (Spotila 2005). During immature to mature stages, tectonically driven rock uplift is the dominant mechanism promoting long-term erosion and exhumation, and the surface uplift during mountain rising is a consequence of rock uplift moderated by the erodibility of uplifting rocks (Burbank et al. 1996; Summerfield and Brown 1998; Malusà and Vezzoli 2006). In the decaying stage, eroded upper crustal rocks are progressively replaced by higher-density lower crustal and mantle rocks underneath, with consequent progressive reduction of both isostatic rebound and surface height (England and Molnar 1990; Stüwe and Barr 1998; Braun et al. 2006).

Rock uplift can be constrained by geodetic measurements using fixed benchmarks (Schlatter et al. 2005; Serpelloni et al. 2013), or by dating of marine and river terraces (Burbank 2002), but these measurements are limited to relatively recent periods of time, and extrapolation back through significant periods of geologic time (millions of years) is often problematic. Stable isotope paleoaltimetry has the potential to constrain how the topography of entire orogens has changed in the past (Poage and Chamberlain 2001; Roe 2005). But the interpretation of stable isotope data in terms of paleotopography is not straightforward, and requires a detailed understanding of the paleoclimatic conditions at the time of mountain growth (Blisniuk and Stern 2005). Stream profile analysis is often used to distinguish

steady state from transient landscapes and thus constrain rock uplift (Whipple and Tucker 1999, 2002; Miller et al. 2012), and this approach has been extrapolated back over 10s millions of years, depending on the situation. For example, in the case of rejuvenated topography in the Appalachian Mountains, Miller et al. (2013) extrapolated the timing of base-level change (rock uplift) back into the Miocene.

8.1.2 Erosional Versus Tectonic Exhumation

Exhumation, the movement of rocks towards Earth's surface requires the removal of rock overburden (England and Molnar 1990). The overlying rocks can be removed either tectonically or by denudation through erosion (Rahl et al. 2011) and landslides (Agliardi et al. 2013) (Fig. 8.2d). *Erosional exhumation* (1 in Fig. 8.2d) requires elevated topography and relief, which also determines the potential interaction between tectonics, surficial processes and climate (Burbank and Anderson 2001; Whipple 2009; Willett 2010). Topography and climate control the production and dispersal of detritus, and its accumulation in sedimentary basins (Schlunegger et al. 2001; Allen and Allen 2005). The exhumation history of eroded bedrock is thus potentially retrieved by the analysis of derived sedimentary rocks. In parent bedrock, erosional exhumation is typically associated with smoothly varying cooling age patterns in map view (e.g. Ring et al. 1999), even though age breaks may occur along faults accommodating differential erosional exhumation (e.g. Thomson 2002; Malusà et al. 2005).

By contrast, *tectonic exhumation* (2 in Fig. 8.2d) is independent from topographic relief and climate. During tectonic exhumation, rocks are exhumed because of overburden displacement, e.g. along low-angle normal faults (Foster et al. 1993; Foster and John 1999; Stockli et al. 2002; Ehlers et al. 2003). Production of detritus during tectonic exhumation is not required, although detritus may provide an intermittent record of exhumation (Asti et al. 2018). Unlike erosional exhumation, normal faulting generally shows an asymmetric distribution of cooling ages in map view, with a sharp discontinuity across the master detachment fault (Foster et al. 1993; Foster and John 1999; Armstrong et al. 2003; Fitzgerald et al. 2009). Although low-angle normal faults are not common in most collision orogens (Burbank 2002), the same concept of tectonic exhumation can be applied on a larger scale along convergent plate margins, where tectonic exhumation can be either promoted by divergence between upper plate and accretionary wedge, or by rollback of the lower plate (Brun and Faccenna 2008; Malusà et al. 2015; Liao et al. 2018).

8.2 The Evolving Shape and Spacing of Isothermal Surfaces

The morphology and spacing of isothermal surfaces in the upper crust may strongly influence FT data. Isotherms in the uppermost crust generally mimic the shape of the topography (Fig. 8.3), but in a dampened fashion dependent on the isotherm in question and the wavelength of the topography (Stüwe et al. 1994; Mancktelow and Grasemann 1997; Stüwe and Hintermüller 2000; Braun 2002). For short topographic wavelengths, low-temperature isothermal surfaces are generally unable to penetrate beneath ridges during exhumation, because of the lateral cooling effect of the steep valley flanks (Mancktelow and Grasemann 1997), but their penetration is progressively facilitated by increasing topographic wavelengths (Fig. 8.3c). The isothermal surfaces relevant for FT data are expected to be near horizontal in areas of low topographic relief (Fig. 8.3a) and variously deflected in active settings characterised by high relief. In this latter case, isothermal surfaces are elevated beneath ridges and depressed beneath valleys (Stüwe et al. 1994). This effect is more pronounced for the lower temperature isothermal surfaces (e.g. T_1 in Fig. 8.3a) than for the higher temperature isotherms (e.g. T_3 in Fig. 8.3a). In regions of high geothermal gradients, however, perturbations due to topography can also affect higher temperature isotherms (Braun 2002). Notably, the topography at the time of exhumation is much more relevant, for the interpretation of low-temperature thermochronologic data, than the present-day topography (Fitzgerald et al. 2006; Foeken et al. 2007; Wölfler et al. 2012). Topography may have remained steady state since the onset of exhumation, but this kind of assumption has to be supported by independent geologic evidence.

The *advection* of heat towards Earth's surface and the closer spacing (increased geothermal gradient) of isothermal surfaces is especially relevant during rapid exhumation (>0.3 km/Myr, Gleadow 1990; Gleadow and Brown 2000) in tectonically active regions (Mancktelow and Grasemann 1997; Braun 2002; Braun et al. 2006; Wölfler et al. 2012). Heat advection during exhumation leads to a progressive elevation of the isothermal surfaces and results in a transient increase in the near-surface geothermal gradient (Fig. 8.3b). Such an increase is followed by thermal relaxation at the end of the exhumation pulse. In contrast, in tectonically stable areas heat transfer from depth to the surface is dominated by *conduction* (Blackwell et al. 1989; Ehlers 2005). The conductive heat transfer rate is a function of the thermal conductivity of rock, which strongly varies according to the rock type (Braun et al. 2016). Transfer of heat by hydrothermal fluids, often localised along fault zones, may additionally favour localised transient disturbance of the background thermal field (Burtner and Negrini 1994; Ehlers and Chapman 1999; Whipp and Ehlers 2007).

Major faults may also have an impact on the shape of isothermal surfaces, possibly determined by the redistribution of mass through erosion and sedimentation from one side of the fault to the other (Ehlers 2005) (Fig. 8.4a, b). The erosional removal of material from Earth's surface results in the upward displacement of warmer rocks, whereas sedimentation has an opposite effect because the cool surface temperature is advected downward when sediment is deposited on Earth's surface. The potential downward advection induced by sedimentary burial is associated with oblique advection in the upthrown eroded block (Fig. 8.4a, b), which includes an horizontal component that is particularly important in the case of thrust faulting (ter Voorde et al. 2004; Huntington et al. 2007; Lock and Willett 2008). At the front of mountain ranges, erosion may thus lead to an increase of near-surface geothermal gradient in the hanging wall of the frontal thrusts, while foreland sedimentation may depress the geothermal gradient in the footwall (Ehlers 2005). The juxtaposition of enhanced and depressed thermal fields across major faults determines lateral heat flow and a curvature of isothermal surfaces that is a function of displacement rates and may be relevant for the interpretation of FT data (Fig. 8.4a, b). For very fast displacement rates (>2 km/Myr), isotherms may eventually become inverted across major thrust faults (Husson and Moretti 2002). The initial background thermal field observed before deformation is generally re-established when the fault is no longer active. Regardless, thermochronologic ages collected across thrust faults ideally over significant relief and with multiple systems can constrain when thrust faults are inactive (no age difference across the faults) and when thrusting began (age difference across the faults) (Lock and Willett 2008; Metcalf et al. 2009, Riccio et al. 2014). The impact of the thermal structure on FT data interpretation in high-relief areas is discussed in detail in Chap. 9 (Fitzgerald and Malusà 2018).

The minor deflections of high-temperature isotherms predicted beneath a rugged topography (Fig. 8.3a, b) might suggest that thermochronologic data relating to T_c higher than fission tracks might be interpreted within a rather simple thermal reference frame. However, numerical models that analyse the thermal evolution of accretionary wedges (e.g. Yamato et al. 2008; Jamieson and Beaumont 2013) suggest that this conclusion is likely incorrect. Figure 8.4c shows that the widely spaced isothermal surfaces characterising the accretionary wedge during subduction (time t_1) are strongly deformed, and compressed towards Earth's surface during synconvergent exhumation (time t_2), due to advection of heat transported by the exhuming deep rocks. Within this framework, the isothermal surfaces relevant for the interpretation of high-temperature thermochronologic

Fig. 8.4 **a, b** Impact of major faults on the shape of isothermal surfaces due to redistribution of mass through erosion and sedimentation (left); the background thermal field is re-established as soon as the fault is no longer active (right) (based on ter Voorde et al. 2004; Ehlers 2005; Huntington et al. 2007); **c** Progressive elevation of high-temperature isothermal surfaces within an accretionary wedge during synconvergent exhumation (based on Jamieson and Beaumont 2013)

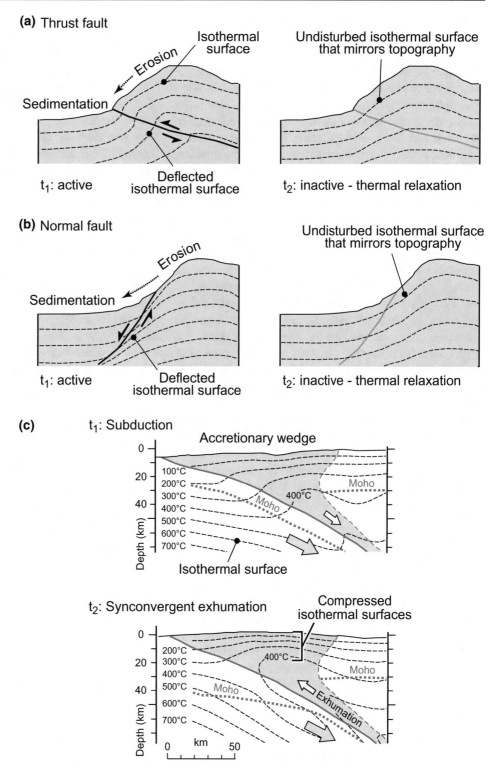

systems, such as $^{40}Ar/^{39}Ar$ on white mica (e.g. Carrapa et al. 2003; Hodges et al. 2005), are neither expected to be horizontal, nor fixed. Major large-scale changes in thermal structure are not specific to contractional settings, as they are also expected in extensional settings during rifting and subsequent break-up (Gallagher and Brown 1997; Whitmarsh et al. 2001), where geothermal gradients may increase during initial extension and rifting and then progressively decrease from values >80 to <30 °C/km (Morley et al. 1980; Malusà et al. 2016).

8.3 Changing Paleogeothermal Gradients in the Geologic Record

The paleogeothermal gradient is often a crucial parameter for the interpretation of low-temperature thermochronologic data, notably when using an "assumed" gradient to convert T_c to a depth and hence constrain an average exhumation rate from that depth to the surface. The paleogeothermal gradient is usually unknown, and depending on the situation, it is a common practice to assume some "normal and constant" geothermal gradient for this approach to calculate exhumation rates (e.g. 30 °C/km; Mancktelow and Grasemann 1997). The resultant time-averaged exhumation rate is directly dependent on this assumption, and hence, the reliability of thermochronologic interpretations is greatly improved whenever independent constraints provided by the geologic record can be incorporated into constraining the evolution of the paleogeothermal gradient. Notably, paleogeothermal gradients generally assumed in the literature for exhumation rate calculations in convergent settings are spread over the whole range of gradients, from ~ 15 to ~ 35 °C/km (Spear 1993), all of which can reasonably be expected to occur in continental orogenic belts (Chapman 1986).

In metamorphic rocks, the paleogeothermal gradients experienced during exhumation can be inferred from the analysis of pressure–temperature–time (P–T–t) paths, which may be independently constrained by petrologic and geochronologic data (Spear 1993; Miyashiro 1994; see Chap. 13, Baldwin et al. 2018). The average geothermal gradient between a rock sample at depth and Earth's surface is given by the ratio between the temperature and the depth of that sample at a given time (Fig. 8.5a). The temperature can be readily assessed on a P–T–t diagram, and the depth can be inferred from the same diagram with a simple pressure to depth conversion (e.g. Rubatto and Hermann 2001), assuming negligible deviations from lithostatic pressure due to tectonic overpressure (Mancktelow 2008; Reuber et al. 2016). Paleogeothermal gradients are thus visualised in the P–T–t diagram as the slope of the line connecting the P–T conditions of the analysed rock with those observed at Earth's surface. A multidisciplinary approach to P–T–t determination (e.g. the integration of petrologic and geochronologic data with fluid inclusion analysis) can be particularly useful in constraining the exhumation paths followed by rocks below the greenschist facies, which is the segment of the P–T–t path most relevant for the interpretation of low-temperature thermochronologic data. In the case of the Sesia-Lanzo unit of the Western Alps, such a multidisciplinary approach unravelled a progressive increase in paleogeothermal gradients from ~ 18 to ~ 30 °C/km after greenschist facies metamorphism (Malusà et al. 2006). If not

properly considered, such a variation in the reference thermal structure would have strongly affected the estimates of exhumation rates and fault offsets.

In the uppermost kilometres of the crust, modelled T–t paths derived from age and confined track-length data (Ketcham 2005) can also provide constraints on the evolution of the thermal gradient between samples collected in deep boreholes or steep vertical profiles (Gleadow and Brown 2000; Gallagher et al. 2005). In this case, the vertical distance (Δz) between samples is known and constant through time, whereas the evolving difference in temperature (ΔT) between different samples can be estimated by comparing their modelled T–t paths (Fig. 8.5b). This approach was applied, for example, to data from Denali in Alaska and the Pyrénées (Gallagher et al. 2005), and to the northern distal margin of the Alpine Tethys now exposed in Sardinia (Malusà et al. 2016). Improved constraints to paleogeothermal gradients in sedimentary basins are possibly provided by an integrated approach to the analysis of well data (Armstrong 2005) including, for example, multiple low-temperature thermochronometers (House et al. 2002), maximum paleotemperature indicators such as vitrinite

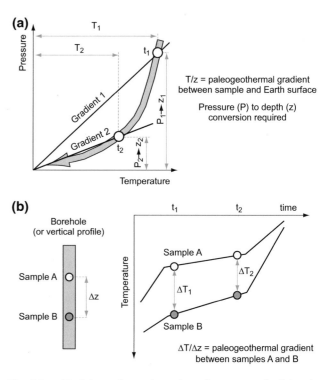

Fig. 8.5 **a** Evolving paleogeothermal gradients constrained by the analysis of pressure–temperature–time paths after pressure to depth conversion (e.g. Miyashiro 1994); **b** Evolution of thermal gradients between samples from deep boreholes or steep vertical profiles: the vertical distance (Δz) between samples is known, the difference in temperature (ΔT) is estimated from temperature–time paths (models) derived from age and confined track-length data (e.g. Gleadow and Brown 2000)

reflectance (Bray et al. 1992) and basin modelling (Osadetz et al. 2002).

8.4 When Cooling Is Not Related to Exhumation

During exhumation, rocks move towards Earth's surface ideally crossing the T_c isothermal surfaces of different thermochronologic systems. Cooling related to exhumation may thus produce, in a single sample, a range of progressively younger thermochronologic ages related to systems with progressively lower T_c (e.g. Wagner et al. 1977). When an array of samples from different elevations is analysed using the same dating method, exhumational cooling generally produces a normal age–elevation relationship, with older ages in samples located at higher elevation, and younger ages in samples located at lower elevation (Wagner and Reimer 1972; Gleadow and Fitzgerald 1987; Fitzgerald and Gleadow 1988). The faster the rate of exhumation, the smaller the difference in thermochronologic age expected in samples collected at different elevations from the same vertical profile. However, ages recorded by low-temperature thermochronometers are sometimes unrelated to exhumation. They may either reflect transient changes in the regional thermal structure of the upper crust (Braun 2016), episodes of crystal growth, post-magmatic cooling and mineral alteration (Malusà et al. 2011), or more localised transient thermal changes in rocks already exhumed above the PAZ (or the PRZ) due to hydrothermal fluid circulation, frictional heating along active faults (Tagami 2012) or wildfires (Reiners et al. 2007). These situations are discussed below.

8.4.1 Cooling Due to Thermal Relaxation

At the end of an orogenic event, isothermal surfaces previously uplifted towards Earth's surface normally undergo thermal relaxation (e.g. Braun 2016). Figure 8.6a shows the thermal structure under an eroding topography modelled for a 3 Myr fast erosional pulse, and the relationships between isothermal surfaces and select rock samples undergoing exhumation (black dots). The wavelength and amplitude of the topography are consistent with those observed in the Alpine region (Zanchetta et al. 2015) and are assumed to be steady state during the analysed time interval. Before fast exhumation, rock samples are inside the PAZ of the AFT system (time frame t_0). The main pulse of fast erosion (time frame t_1) determines the exhumation of rock samples and a compression of isothermal surfaces towards Earth's surface due to heat advection. When the main pulse of rapid

exhumation is completed, rock samples are still within the PAZ of the AFT system and have not recorded any significant temperature variation, yet. When erosion gets slower (time frame t_2), isothermal surfaces eventually move downward, re-establishing the thermal structure observed before the fast erosional pulse. Only at this stage rocks experience the fast cooling recorded by AFT data. Therefore, under specific conditions, a relevant time delay can be expected between exhumation and cooling recorded by low-temperature thermochronometers. Thus in the example of Fig. 8.6a, rock samples record the time of thermal relaxation, not the time of exhumation. Thermal relaxation after a major orogenic event has been recently invoked by Braun (2016) to explain two different thermochronologic data sets from the Himalaya (Bernet et al. 2006; Kellett et al. 2013). According to Braun (2016), these data sets may reflect the thermal re-equilibration of the Himalayan orogenic wedge after deactivation of the South Tibetan Detachment.

A regional rising of isothermal surfaces followed by thermal relaxation may be recorded not only in convergent settings, but also in extensional settings, for example in a passive continental margin that underwent rifting and subsequent continental break-up (Lemoine and de Graciansky 1988; Whitmarsh et al. 2001). Passive continental margins show normal-thickness continental crust in the proximal part and thinned continental crust in the distal part (Fig. 8.6b), which is generally buried beneath thick successions of post-rift sediments. Proximal passive margins are extensively studied worldwide (see Chap. 20, Wildman et al. 2018) and display a thermochronologic record typically dominated by exhumation due to erosion (Brown et al. 1990; Fitzgerald 1992; Gallagher et al. 1994; Gallagher and Brown 1997; Menzies et al. 1997; Gleadow et al. 2002), although some studies suggest the thermochronologic record was dominated by the thermal pulse accompanying rifting, such as in south-eastern Australia (Morley et al. 1980). Distal passive margins are far less studied, and the thermochronologic imprint they have acquired during rifting and continental break-up is generally obliterated (overprinted) by the overwhelming thermal effects of sedimentary burial during the post-rift evolution. However, in some situations, ancient distal margins have escaped post-rift thermal resetting and are now exposed above sea level. This is the case of Corsica–Sardinia in the Western Mediterranean, which represents a relict of the northern distal margin of the Mesozoic Alpine Tethys (Malusà et al. 2016). The complex thermochronologic and geologic record of Corsica–Sardinia can be effectively explained in terms of a regional rising of isothermal surfaces due to asthenospheric upwelling in the Jurassic, followed by thermal relaxation during continental break-up according to the conceptual model shown in Fig. 8.6b.

Fig. 8.6 **a** During fast erosion (t_1), isothermal surfaces are compressed towards Earth's surface due to heat advection; thermochronometers record cooling only when the original thermal structure is re-established (t_2) (modified from Zanchetta et al. 2015); **b** Pattern of thermochronologic ages in a distal passive margin due to rising of isothermal "bell-shaped" surfaces by asthenospheric upwelling followed by thermal relaxation after continental break-up. Diagnostic combinations of thermochronologic ages are expected according to crustal level and distance from the rift axis (modified from Malusà et al. 2016, schematic cross section based on Lemoine and de Graciansky 1988)

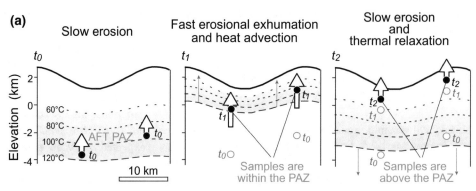

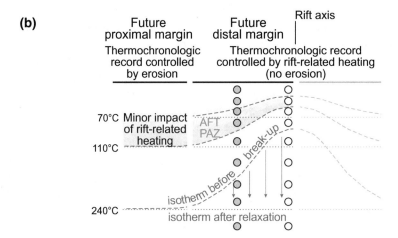

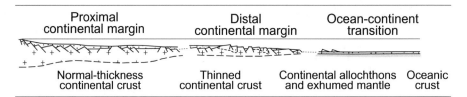

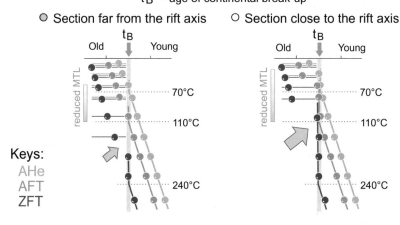

The thermal structure depicted in the upper panel of Fig. 8.6b shows isothermal surfaces roughly corresponding to the T_c of the (U–Th)/He system on apatite (AHe) and the FT system on apatite (AFT) and zircon (ZFT). It is obviously simplified compared to the real thermal structure of a passive margin, but reproduces well the age pattern observed in the geologic record of Corsica–Sardinia (Cavazza et al. 2001; Zarki-Jakni et al. 2004; Fellin et al. 2005, 2006; Danišík et al. 2007, 2012; Zattin et al. 2008; Malusà et al. 2016). As shown in the model, the rising isothermal surfaces attain a "bell shape" during rifting, lying at different depths according to their temperature. Inside each "bell", mineral ages are set during thermal relaxation after break-up, when isotherms move downwards and get flatter, in a similar fashion as in Fig. 8.6a. Below each bell, samples at the time of break-up still experience temperatures higher than the isotopic closure and thus mineral ages are set at a later stage, e.g. during later erosional exhumation. Above each bell, the temperature increase during asthenospheric upwelling is usually not sufficient to reset the thermochronologic system, and rocks may largely preserve the thermochronologic fingerprint acquired before the rifting. In this latter case, a reduction in mean FT length and/or a slight age rejuvenation is expected in samples that have cooled through the PAZ (or PRZ) during subsequent thermal relaxation.

Asthenospheric upwelling is expected to modify the normal age–elevation relationships produced during erosional exhumation, because of the systematic rejuvenation of thermochronologic ages in selected crustal levels. As a result, thermal relaxation after continental break-up can lead to near-invariant thermochronologic ages with elevation (lower panel in Fig. 8.6b), that might be misinterpreted as the evidence of fast exhumation. However, diagnostic combinations of thermochronologic ages from different systems are expected, according to the crustal level of each sample at the time of the rifting and according to the distance from the rift axis, thus reducing the risk of misinterpretation. For instance, near-identical AFT and AHe ages that are synchronous with the time of break-up, along with short mean FT lengths, are expected in rocks originally lying at shallow crustal levels and close to the rift axis. In contrast, near-identical AFT and ZFT ages are expected at deeper crustal levels, in association with younger AHe ages (Fig. 8.6b). Noteworthy, the near-identical AFT and ZFT ages synchronous with the time of break-up are set during sedimentation on top of this part of the crust, which rules out any contribution of cooling due to erosion. These age patterns thus provide no direct constraint on exhumation, but may provide a key for the reconstruction of ancient passive margins.

8.4.2 The Role of Magmatic Crystallisation

The role of magmatic crystallisation and post-magmatic cooling in the thermochronologic record is illustrated in the conceptual model of Fig. 8.7. This model shows the progressive setting of mineral ages (U–Pb on zircon, K–Ar on biotite, FT on zircon and apatite) after intrusion of magma at depth and growth of volcanoes at the surface, and subsequent erosional unroofing of volcanic and plutonic rocks along with their country rocks (Malusà et al. 2011).

Before magmatic intrusion (time t_0): Erosion is assumed to be negligible, and four crustal levels are distinguished in the country rock according to its temperature (levels 1–4 in Fig. 8.7). In this diagram, crustal levels are delimited by the PAZ of the AFT system (lower boundary of level 1), the PAZ of α-damaged ZFT system (lower boundary of level 2), and the T_c of the biotite K–Ar system (lower boundary of level 3). At time t_0, thermochronologic ages are set (recorded) at depths shallower than the corresponding isotopic closure, but they are not set at greater depths (i.e. ages are zero below T_c dependent on the system). For instance, AFT ages are set in level 1, but they are not set in levels 2-to-4, yet; ZFT ages are set in levels 1 and 2, but they are not set in levels 3 and 4. These FT ages may either record old crystallisation and/or exhumation events in the country rock, or histories of distant sediment sources, such as in the case of distinct FT age populations preserved in shallow-level sedimentary rocks (Bernet and Garver 2005).

Magma emplacement and crystallisation (time t_1): Plutons are intruded across levels 1 to 4, and volcanic rocks are emplaced at the surface. We define the *magmatic age* (t_i) as the crystallisation age of the magma. Because zircon U–Pb ages usually date crystal growth and thus the time of magma crystallisation (Dahl 1997; Mezger and Krogstad 1997), in a first-order approximation we can consider U–Pb zircon ages from magmatic rocks as magmatic ages. These ages are identical within error at any intrusion depth. And they are systematically younger than the U–Pb ages in adjacent country rocks (Fig. 8.7). AFT and ZFT ages typically record post-intrusion cooling, unless the magma is intruded into the upper crust (hypabyssal) where country rocks are resident at temperatures cooler than the PAZ. In this case, the time elapsed between crystallisation and the first retention of fission tracks is shorter than the resolution of the dating method (Jaeger 1968). Thus, FT ages from shallow intrusions and volcanic rocks should be considered magmatic ages as well. Such shallow intrusions, such as Mt Dromedary in Australia, are often used as age standards because all techniques give near-identical ages (Green 1985; see Chap. 1, Hurford 2018). These ages are the same at any

Fig. 8.7 Magmatic complex model showing the progressive setting of mineral ages after intrusion of magma at depth and growth of volcanoes at the surface, and subsequent erosional unroofing of volcanic and plutonic rocks along with their country rocks; different levels of the complex are characterised by different combinations of magmatic and exhumation ages (modified from Malusà et al. 2011). To the right, thermochronologic age pattern in country rocks within the contact aureole, where ages get progressively younger and approach the magmatic age in the vicinity of the intrusion (based on Calk and Naeser 1973)

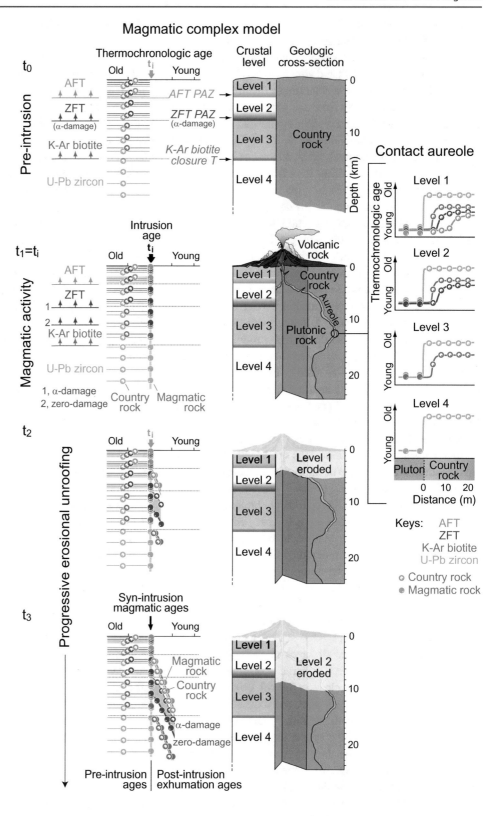

intrusion depth shallower than the PAZ and are younger than FT ages in the country rock, except for the immediate overprinted or partially overprinted thermal aureole around the pluton (Calk and Naeser 1973; Harrison and McDougall

1980; Schmidt et al. 2014). Within the thermal aureole, thermochronologic ages in country rocks get progressively younger and approach the magmatic age in the vicinity of the intrusion. The influence of magmatic heat in the country rock is expected to be greater in lower temperature systems than in higher temperature systems. The former are affected at a greater distance from the intrusion, but the latter are still affected at deeper crustal levels, where ages in lower temperature system are not set (right panel in Fig. 8.7).

Therefore, at time t_1, magmatic ages in the AFT system are set in level 1 shortly after intrusion, but they are not set in levels 2-to-4 where temperatures remain greater than T_c. Similarly for the ZFT system, ages are set in levels 1, 2 and perhaps uppermost level 3, but not in level 4. Note that T_c is higher for magmatic zircon with little α-damage than for zircon with greater amounts of α-damage in the country rock (Rahn et al. 2004). Biotite K–Ar (and Rb–Sr) ages are normally intermediate between zircon U–Pb and ZFT ages, because they have T_c intermediate between these systems. As for the FT systems, in the case of shallow intrusions and volcanic rocks, the time elapsed between biotite crystallisation and the closure of the K–Ar system is very short (Jaeger 1968). In the absence of biotite retrogression, biotite K–Ar ages in magmatic rocks are thus usually indistinguishable from the magmatic age in the whole depth range between level 1 and level 3. Noteworthy, all of these ages are set before the onset of erosional exhumation, which commences at time t_2 after the post-intrusion stabilisation of isothermal surfaces. Therefore, they provide no direct constraint on exhumation.

Erosional exhumation (from time t_2 onwards): Thermochronologic ages, in rocks located beneath the T_c isotherm depth at the time of intrusion, are set during subsequent erosional exhumation during time t_2 and t_3 (Fig. 8.7). These ages are always younger than the magmatic age, and because they are controlled by delayed closure of the thermochronologic system during undisturbed exhumational cooling, they will be termed *exhumation ages*. Exhumation ages constrain the upward motion of rocks towards Earth's surface, thus the ages get systematically younger with increasing depth and are generally indistinguishable in country rocks and encased magmatic rocks. The slope defined by exhumation ages in an array of samples from different depths is a function of the exhumation rate (see Chap. 9, Fitzgerald and Malusà 2018). In the case of the ZFT system, exhumation ages may span over a relatively large age range due to variations between grains in the amount of α-damage.

The different crustal levels identified within a magmatic complex are thus characterised by different combinations of magmatic and exhumation ages in country rocks and magmatic rocks. For example, crustal level 1 is characterised, for all thermochronologic systems, by magmatic ages in the intruded rocks and older pre-intrusion ages in country rocks, unless country rocks are proximal to the intrusion and reset or partially reset within the thermal aureole. Level 2 additionally shows AFT exhumation ages younger than the

magmatic age, and level 3 also includes ZFT exhumation ages older than the AFT ages, but younger than the magmatic ages still recorded by the biotite K–Ar and zircon U–Pb systems. In level 4, magmatic biotites are expected to yield K–Ar ages younger than the magmatic age. However, when biotites are rejuvenated by retrogression or late stage alteration (Roberts et al. 2001; Di Vincenzo et al. 2003; Villa 2010), an anomalous spread in K–Ar ages may mask the original age–depth relationship.

8.4.3 Localised Thermal Resetting: Impact of Wildfires, Frictional Heating and Hydrothermal Fluids

After exposure of rocks at Earth's surface, it is possible that the AHe and AFT thermochronologic systems, as well as the zircon (U–Th)/He system, may be completely or partially reset by wildfires (Wolf et al. 1998; Mitchell and Reiners 2003; Reiners et al. 2007). Wildfires generate short-duration, high-temperature thermal events that produce characteristic thermochronologic signatures in minerals such as apatite. These thermal events may affect AHe and AFT ages both in exposed bedrock and in detrital pebbles or soil. The contrasting activation energies of FT annealing and He diffusion lead to a kinetic crossover, whereby AFT ages are prone to be reset more rapidly than AHe ages (Mitchell and Reiners 2003). Wildfire heating may thus produce peculiar AFT-AHe relationships in single apatite grains, such as finite AHe ages and zero AFT ages, that can occur down to a penetration depth of ∼3 cm (Reiners et al. 2007). In bedrock thermochronology studies, potential effects of wildfires can be excluded by a proper sampling strategy, i.e. by avoiding samples from the outermost part of the analysed outcrop. However, this is not possible for detrital studies, unless they are based on the analysis of cobbles greater than ∼6 cm in diameter.

Short-duration, high-temperature thermal events can be also due to frictional heating during faulting and heat transfer into the surrounding rocks (Scholz 2002). This may lead to a temperature increase up to ∼1000 °C for several seconds and within several mm from the fault (Lachenbruch 1986; Murakami 2010). Spontaneous fission tracks in zircon are totally annealed when subject to a temperature of $850 ± 50$ °C for ∼4 s, suggesting that the ZFT system can be completely reset during brittle faulting (Otsuki et al. 2003; Tagami 2012). A larger rock volume around the fault, on the order of 1–100 m, can be thermally affected during and after faulting by hot fluid circulation, as suggested by the occurrence of hydrothermal veins around exhumed fault zones (Cox 2010). The effects of frictional and hydrothermal heating on low-temperature thermochronologic systems are discussed in more detail in Chap. 12 (Tagami 2018).

8.5 Conclusions

The reference frame for the interpretation of FT and other thermochronologic data is a thermal reference frame. Using FT thermochronology to constrain, under specific conditions, the exhumation of rocks largely depends on a proper understanding of the relationships between this reference frame and Earth's surface. This thermal reference frame is neither stationary, nor horizontal. It is strongly affected by the amplitude and wavelength of topography at the time of exhumation, by heat advection due to rapid exhumation and by mass redistribution across major faults. Major deflections may characterise not only low-temperature but also high-temperature isothermal surfaces. Paleo-geothermal gradients, i.e. the spacing of isothermal surfaces, are important for the interpretation of low-temperature thermochronologic data in terms of exhumation and can be constrained by the analysis of *P-T-t* paths and by modelled *T-t* paths derived from track-length distributions in samples collected along vertical profiles.

Noteworthy, ages recorded by low-temperature thermochronometers are sometimes not related to exhumation, and an a priori interpretation of thermochronologic ages exclusively in terms of exhumation may lead to incorrect geologic reconstructions. A significant time delay may occur between exhumation and cooling in case of fast erosional pulses, and low-temperature thermochronometers often record the time of thermal relaxation rather than the time of exhumation. Localised resetting of low-temperature thermochronologic systems can be due to transient wildfire heating, or to frictional and hydrothermal heating along fault zones. In general terms, simple constraints on exhumation are only provided by thermochronologic ages that are set during undisturbed exhumational cooling across the T_c isothermal surface, whereas thermochronologic ages that are set in minerals crystallised at shallower depth than T_c, e.g. in volcanic rocks or in shallow intrusions, provide no direct constraint on exhumation.

Acknowledgements This work benefited from insightful reviews by Phil Armstrong and Kurt Stüwe, and from comments by Suzanne Baldwin and students from Syracuse University.

References

Agliardi F, Crosta GB, Frattini P, Malusà MG (2013) Giant non-catastrophic landslides and the long-term exhumation of the European Alps. Earth Planet Sci Lett 365:263–274

Allen PA, Allen JR (2005) Basin analysis: principles and applications. Blackwell, London

Armstrong PA, Ehlers TA, Chapman DS et al (2003) Exhumation of the central Wasatch Mountains, Utah: 1. Patterns and timing of exhumation deduced from low-temperature thermochronology data. J Geophys Res-Sol Ea 108(B3)

Armstrong PA (2005) Thermochronometers in sedimentary basins. Rev Mineral Geochem 58(1):499–525

Asti R, Malusà MG, Faccenna C (2018) Supradetachment basin evolution unraveled by detrital apatite fission track analysis: the Gediz Graben (Menderes Massif, Western Turkey). Basin Res 30:502–521

Baldwin SL, Lister GS (1998) Thermochronology of the South Cyclades Shear Zone, Ios, Greece: Effects of ductile shear in the argon partial retention zone. J Geophys Res-Sol Ea 103(B4):7315–7336

Baldwin SL, Fitzgerald PG, Malusà MG (2018) Chapter 13. Crustal exhumation of plutonic and metamorphic rocks: constraints from fission-track thermochronology. In: Malusà MG, Fitzgerald PG (eds) Fission-track thermochronology and its application to geology. Springer, Berlin

Bernet M, Garver JI (2005) Fission-track analysis of detrital zircon. Rev Mineral Geochem 58(1):205–237

Bernet M, van der Beek P, Pik R, Huyghe P, Mugnier JL, Labrin E, Szulc A (2006) Miocene to recent exhumation of the central Himalaya determined from combined detrital zircon fission-track and U/Pb analysis of Siwalik sediments, western Nepal. Basin Res 18(4):393–412

Blackwell D, Steele J, Brott C (1989) Heat flow in the Pacific Northwest. In: Touloukian Y, Judd W, Roy R (eds) Physical properties of rocks and minerals. McGraw-Hill, New York, pp 495–502

Blisniuk PM, Stern LA (2005) Stable isotope paleoaltimetry: a critical review. Am J Sci 305(10):1033–1074

Braun J (2002) Quantifying the effect of recent relief changes on age–elevation relationships. Earth Planet Sci Lett 200(3):331–343

Braun J (2016) Strong imprint of past orogenic events on the thermochronological record. Tectonophysics 683:325–332

Braun J, van der Beek P, Batt G (2006) Quantitative thermochronology: numerical methods for the interpretation of thermochronological data. Cambridge University Press, Cambridge

Braun J, Stippich C, Glasmacher UA (2016) The effect of variability in rock thermal conductivity on exhumation rate estimates from thermochronological data. Tectonophysics 690:288–297

Bray RJ, Green PF, Duddy IR (1992) Thermal history reconstruction using apatite fission track analysis and vitrinite reflectance: a case study from the UK East Midlands and Southern North Sea. In: Hardman RFP (ed) Exploration Britain: geological insights for the next decade. Geol Soc, pp 3–25

Brown RW (1991) Backstacking apatite fission-track" stratigraphy": a method for resolving the erosional and isostatic rebound components of tectonic uplift histories. Geology 19(1):74–77

Brown RW, Rust DJ, Summerfield MA, Gleadow AJ, De Wit MC (1990) An Early Cretaceous phase of accelerated erosion on the south-western margin of Africa: evidence from apatite fission track analysis and the offshore sedimentary record. Int J Rad Appl Instr, Part D, Nucl Tracks Rad Meas 17(3):339–350

Brun JP, Faccenna C (2008) Exhumation of high-pressure rocks driven by slab rollback. Earth Planet Sci Lett 272:1–7

Burbank DW (2002) Rates of erosion and their implications for exhumation. Mineral Mag 66(1):25–52

Burbank DW, Anderson RS (2001) Tectonic geomorphology. Wiley, New York

Burbank DW, Leland J, Fielding E, Anderson RS, Brozovic N, Reid MR, Duncan C (1996) Bedrock incision, rock uplift and threshold hillslopes in the northwestern Himalayas. Nature 379 (6565):505–510

Burtner RL, Negrini A (1994) Thermochronology of the Idaho-Wyoming thrust belt during the Sevier Orogeny: a new, calibrated, multiprocess model. Am Assoc Petrol Geol Bull 78:1586–1612

Calk LC, Naeser CW (1973) The thermal effect of a basalt intrusion on fission tracks in quartz monzonite. J Geol 81(2):189–198

Carrapa B, Wijbrans J, Bertotti G (2003) Episodic exhumation in the Western Alps. Geology 31(7):601–604

Cavazza W, Zattin M, Ventura B, Zuffa GG (2001) Apatite fission-track analysis of Neogene exhumation in northern Corsica (France). Terra Nova 13(1):51–57

Chapman DS (1986) Thermal gradients in the continental crust. Geol Soc London Spec Publ 24(1):63–70

Corcoran DV, Doré AG (2005) A review of techniques for the estimation of magnitude and timing of exhumation in offshore basins. Earth-Sci Rev 72(3):129–168

Cox SF (2010) The application of failure mode diagrams for exploring the roles of fluid pressure and stress states in controlling styles of fracture-controlled permeability enhancement in faults and shear zones. Geofluids 10:217–233

Dahl PS (1997) A crystal-chemical basis for Pb retention and fission-track annealing systematics in U-bearing minerals, with implications for geochronology. Earth Planet Sci Lett 150(3):277–290

Danišík M, Kuhlemann J, Dunkl I, Székely B, Frisch W (2007) Burial and exhumation of Corsica (France) in the light of fission track data. Tectonics 26(1)

Danišík M, Kuhlemann J, Dunkl I, Evans NJ, Székely B, Frisch W (2012) Survival of ancient landforms in a collisional setting as revealed by combined fission track and (U-Th)/He thermochronometry: a case study from Corsica (France). J Geol 120(2):155–173

Di Vincenzo G, Viti C, Rocchi S (2003) The effect of chlorite interlayering on ^{40}Ar–^{39}Ar biotite dating: an ^{40}Ar–^{39}Ar laser-probe and TEM investigations of variably chloritised biotites. Contrib Mineral Petr 145:643–658

Dodson MH (1973) Closure temperature in cooling geochronological and petrological systems. Contrib Mineral Petr 40(3):259–274

Ehlers TA (2005) Crustal thermal processes and the interpretation of thermochronometer data. Rev Mineral Geochem 58(1):315–350

Ehlers TA, Chapman DS (1999) Normal fault thermal regimes: conductive and hydrothermal heat transfer surrounding the Wasatch fault, Utah. Tectonophysics 312(2):217–234

Ehlers TA, Farley KA (2003) Apatite (U-Th)/He thermochronometry: methods and applications to problems in tectonic and surface processes. Earth Planet Sci Lett 206:1–14

Ehlers TA, Willett SD, Armstrong PA, Chapman DS (2003) Exhumation of the central Wasatch Mountains, Utah: 2. Thermokinematic model of exhumation, erosion, and thermochronometer interpretation. J Geophys Res-Sol Ea 108(B3)

Ehlers TA, Chaudhri T, Kumar S, Fuller CW, Willett SD, Ketcham RA, Dunai TJ (2005) Computational tools for low-temperature thermochronometer interpretation. Rev Mineral Geochem 58(1):589–622

England P, Molnar P (1990) Surface uplift, uplift of rocks, and exhumation of rocks. Geology 18(12):1173–1177

Fellin MG, Picotti V, Zattin M (2005) Neogene to Quaternary rifting and inversion in Corsica: retreat and collision in the western Mediterranean. Tectonics 24(1)

Fellin MG, Vance JA, Garver JI, Zattin M (2006) The thermal evolution of Corsica as recorded by zircon fission-tracks. Tectonophysics 421(3):299–317

Fitzgerald PG (1992) The Transantarctic Mountains of southern Victoria Land: the application of apatite fission track analysis to a rift shoulder uplift. Tectonics 11(3):634–662

Fitzgerald PG (1994) Thermochronologic constraints on post-Paleozoic tectonic evolution of the central Transantarctic Mountains, Antarctica. Tectonics 13(4):818–836

Fitzgerald PG (2002) Tectonics and landscape evolution of the Antarctic plate since the breakup of Gondwana, with an emphasis on the West Antarctic Rift System and the Transantarctic Mountains. Roy Soc NZ Bull 35:453–469

Fitzgerald PG, Gleadow AJ (1988) Fission-track geochronology, tectonics and structure of the Transantarctic Mountains in northern Victoria Land, Antarctica. Chem Geol 73(2):169–198

Fitzgerald PG, Malusà MG (2018) Chapter 9. Concept of the exhumed partial annealing (retention) zone and age-elevation profiles in thermochronology. In: Malusà MG, Fitzgerald PG (eds) Fission-track thermochronology and its application to geology. Springer, Berlin

Fitzgerald PG, Stump E, Redfield TF (1993) Late Cenozoic uplift of Denali and its relation to relative plate motion and fault morphology. Science 259(5094):497–499

Fitzgerald PG, Sorkhabi RB, Redfield TF, Stump E (1995) Uplift and denudation of the central Alaska Range: a case study in the use of apatite fission track thermochronology to determine absolute uplift parameters. J Geophys Res-Sol Ea 100(B10):20175–20191

Fitzgerald PG, Baldwin SL, Webb LE, O'Sullivan PB (2006) Interpretation of (U-Th)/He single grain ages from slowly cooled crustal terranes: a case study from the Transantarctic Mountains of southern Victoria Land. Chem Geol 225(1):91–120

Fitzgerald PG, Duebendorfer EM, Faulds JE, O'Sullivan PB (2009) South Virgin–White Hills detachment fault system of SE Nevada and NW Arizona: applying apatite fission track thermochronology to constrain the tectonic evolution of a major continental detachment fault. Tectonics 28, https://doi.org/10.1029/2007tc002194

Fleischer RL, Price PB, Walker RM (1975) Nuclear tracks in solids: principles and applications. University of California Press

Flowers RM, Ketcham RA, Shuster DL, Farley KA (2009) Apatite (U-Th)/He thermochronometry using a radiation damage accumulation and annealing model. Geochim Cosmochim Ac 73(8):2347–2365

Foeken J, Persano C, Stuart FM, Ter Voorde M (2007) Role of topography in isotherm perturbation: Apatite (U-Th)/He and fission track results from the Malta tunnel, Tauern Window, Austria. Tectonics 26(3). https://doi.org/10.1029/2006tc002049

Foster DA, John BE (1999) Quantifying tectonic exhumation in an extensional orogen with thermochronology: examples from the southern Basin and Range Province. Geol Soc London Spec Publ 154(1):343–364

Foster DA, Gleadow AJ, Reynolds SJ, Fitzgerald PG (1993) Denudation of metamorphic core complexes and the reconstruction of the transition zone, west central Arizona: constraints from apatite fission track thermochronology. J Geophys Res-Sol Ea 98:2167–2185

Gallagher K, Brown R (1997) The onshore record of passive margin evolution. J Geol Soc London 154(3):451–457

Gallagher K, Hawkesworth CJ, Mantovani MSM (1994) The denudation history of the onshore continental margin of SE Brazil inferred from apatite fission track data. J Geophys Res-Sol Ea 99:18117–18145

Gallagher K, Stephenson J, Brown R, Holmes C, Fitzgerald PG (2005) Low temperature thermochronology and modeling strategies for multiple samples 1: vertical profiles. Earth Planet Sci Lett 237:193–208

Gautheron C, Tassan-Got L, Barbarand J, Pagel M (2009) Effect of alpha-damage annealing on apatite (U-Th)/He thermochronology. Chem Geol 266(3):157–170

Gleadow AJW (1990) Fission track thermochronology—reconstructing the thermal and tectonic evolution of the crust. Pacific Rim Congress III. Austr Inst Min Met, Gold Coast, Qld, pp 15–21

Gleadow, AJW, Brown RW (2000) Fission-track thermochronology and the long-term denudational response to tectonics. In:

Summerfield MA (ed) Geomorphology and global tectonics, John Wiley & Sons, p 57–75

Gleadow AJW, Duddy IR (1981) A natural long-term track annealing experiment for apatite. Nucl Tracks 5(1):169–174

Gleadow AJW, Fitzgerald PG (1987) Uplift history and structure of the Transantarctic Mountains: new evidence from fission track dating of basement apatites in the Dry Valleys area, southern Victoria Land. Earth Planet Sci Lett 82(1):1–14

Gleadow AJW, Duddy IR, Green PF, Lovering JF (1986) Confined fission track lengths in apatite: a diagnostic tool for thermal history analysis. Contrib Mineral Petr 94(4):405–415

Gleadow AJW, Kohn BP, Brown RW, O'Sullivan PB (2002) Fission track thermotectonic imaging of the Australian continent. Tectonophysics 349(1):5–21

Green PF (1985) Comparison of zeta calibration baselines for fission-track dating of apatite, zircon and sphene. Chem Geol 58 (1–2):1–22

Harrison TM, McDougall I (1980) Investigations of an intrusive contact, northwest Nelson, New Zealand - I. Thermal, chronological and isotopic constraints. Geochim Cosmochim Acta 44(12):1985–2003

Hodges KV, Ruhl KW, Wobus CW, Pringle MS (2005) ^{40}Ar/^{39}Ar thermochronology of detrital minerals. Rev Mineral Geochem 58 (1):239–257

House MA, Kohn BP, Farley KA, Raza A (2002) Evaluating thermal history models for the Otway Basin, southeastern Australia, using (U + Th)/He and fi ssion-track data from borehole apatites. Tectonophysics 349:277–295

Huntington KW, Ehlers TA, Hodges KV, Whipp DM (2007) Topography, exhumation pathway, age uncertainties, and the interpretation of thermochronometer data. Tectonics 26(4)

Hurford AJ (1991) Uplift and cooling pathways derived from fission track analysis and mica dating: a review. Geol Rundsch 80(2):349–368

Hurford AJ (2018) Chapter 1. An historical perspective on fission-track thermochronology. In: Malusà MG, Fitzgerald PG (eds) Fission-track thermochronology and its application to geology. Springer, Berlin

Husson L, Moretti I (2002) Thermal regime of fold and thrust belts—an application to the Bolivian sub Andean zone. Tectonophysics 345 (1):253–280

Jäger E (1967) Die Bedeutung der Biotit-Alterswerte. In: Jäger E et al (eds) Rb–Sr Altersbestimmungen an Glimmern der Zentralalpen: Bern, Kümmerly & Frey: Beitr. Geol. Karte der Schweiz, NF, 134, pp 28–31

Jaeger JC (1968) Cooling and solidification of igneous rocks. In: Hess HE, Poldervaart A (eds) Basalts: the poldervaart treatise on rocks of basaltic composition. John Wiley, New York, pp 503–536

Jamieson RA, Beaumont C (2013) On the origin of orogens. Geol Soc Am Bull 125(11–12):1671–1702

Kellett DA, Grujic D, Coutand I, Cottle J, Mukul M (2013) The South Tibetan detachment system facilitates ultra rapid cooling of granulite-facies rocks in Sikkim Himalaya. Tectonics 32(2):252–270

Ketcham RA (2005) Forward and inverse modeling of low-temperature thermochronometry data. Rev Mineral Geochem 58(1):275–314

Lachenbruch H (1986) Simple models for the estimation and measurement of frictional heating by an earthquake. USGS Open File Rep 86–508

Lemoine M, de Graciansky PC (1988) Histoire d'une marge continentale passive: les Alpes occidentales au Mésozoque—introduction. Bull Soc Geol Fr 8:597–600

Liao J, Malusà MG, Zhao L, Baldwin SL, Fitzgerald PG, Gerya T (2018) Divergent plate motion drives rapid exhumation of (ultra) high pressure rocks. Earth Planet Sci Lett 491:67–80

Lock J, Willett S (2008) Low-temperature thermochronometric ages in fold-and-thrust belts. Tectonophysics 456(3):147–162

Malusà MG, Vezzoli G (2006) Interplay between erosion and tectonics in the Western Alps. Terra Nova 18(2):104–108

Malusà MG, Fitzgerald PG (2018) Chapter 10. Application of thermochronology to geologic problems: bedrock and detrital approaches. In: Malusà MG, Fitzgerald PG (eds) Fission-track thermochronology and its application to geology. Springer, Berlin

Malusà MG, Polino R, Zattin M, Bigazzi G, Martin S, Piana F (2005) Miocene to present differential exhumation in the Western Alps: insights from fission track thermochronology. Tectonics 24(3). https://doi.org/10.1029/2008tc002370

Malusà MG, Philippot P, Zattin M, Martin S (2006) Late stages of exhumation constrained by structural, fluid inclusion and fission track analyses (Sesia–Lanzo unit, Western European Alps). Earth Planet Sci Lett 243(3):565–580

Malusà MG, Villa IM, Vezzoli G, Garzanti E (2011) Detrital geochronology of unroofing magmatic complexes and the slow erosion of Oligocene volcanoes in the Alps. Earth Planet Sci Lett 301(1):324–336

Malusà MG, Faccenna C, Baldwin SL, Fitzgerald PG, Rossetti F, Balestrieri ML, Ellero A, Ottria G, Piromallo C (2015) Contrasting styles of (U) HP rock exhumation along the Cenozoic Adria-Europe plate boundary (Western Alps, Calabria, Corsica). Geochem Geophys Geosyst 16(6):1786–1824

Malusà MG, Danišík M, Kuhlemann J (2016) Tracking the Adriatic-slab travel beneath the Tethyan margin of Corsica-Sardinia by low-temperature thermochronometry. Gondwana Res 31:135–149

Mancktelow NS (2008) Tectonic pressure: theoretical concepts and modelled examples. Lithos 103(1):149–177

Mancktelow NS, Grasemann B (1997) Time-dependent effects of heat advection and topography on cooling histories during erosion. Tectonophysics 270(3):167–195

Meesters A, Dunai TJ (2002) Solving the production–diffusion equation for finite diffusion domains of various shapes: Part II. Application to cases with α-ejection and nonhomogeneous distribution of the source. Chem Geol 186(3):347–363

Menzies M, Gallagher K, Yelland A, Hurford AJ (1997) Volcanic and nonvolcanic rifted margins of the Red Sea and Gulf of Aden: crustal cooling and margin evolution in Yemen. Geochim Cosmochim Ac 61(12):2511–2527

Metcalf JR, Fitzgerald PG, Baldwin SL, Muñoz JA (2009) Thermochronology of a convergent orogen: constraints on the timing of thrust faulting and subsequent exhumation of the Maladeta Pluton in the Central Pyrenean Axial Zone. Earth Planet Sci Lett 287(3):488–503

Mezger K, Krogstad EJ (1997) Interpretation of discordant U-Pb zircon ages: an evaluation. J Metamorph Geol 15(1):127–140

Miller SR, Fitzgerald PG, Baldwin SL (2010) Cenozoic range-front faulting and development of the Transantarctic Mountains near Cape Surprise, Antarctica: thermochronologic and geomorphologic constraints. Tectonics 29(1). https://doi.org/10.1029/2009tc002457

Miller SR, Baldwin SL, Fitzgerald PG (2012) Transient fluvial incision and active surface uplift in the Woodlark Rift of eastern Papua New Guinea. Lithosphere 4(2):131–149

Miller SR, Sak PB, Kirby E, Bierman PR (2013) Neogene rejuvenation of central Appalachian topography: evidence for differential rock uplift from stream profiles and erosion rates. Earth Planet Sci Lett 369:1–12

Mitchell SG, Reiners PW (2003) Influence of wildfires on apatite and zircon (U-Th)/He ages. Geology 31(12):1025–1028

Miyashiro A (1994) Metamorphic petrology. CRC Press, Boca Raton

Morley ME, Gleadow AJW, Lovering JF (1980) Evolution of the Tasman Rift: apatite fission track dating evidence from the southeastern Australian continental margin. Gondwana Five, Balkema, Rotterdam, pp 289–293

Murakami M (2010) Average shear work estimation of Nojima fault from fission-track analytical data. Earth Monthly 32:24–29 (in Japanese)

Osadetz KG, Kohn BP, Feinstein PB, O'Sullivan PB (2002) Thermal history of Canadian Williston basin from apatite fission-track thermochronology-implications for petroleum systems and geodynamic history. Tectonophysics 349:221–249

Otsuki K, Monzawa N, Nagase T (2003) Fluidization and melting of fault gouge during seismic slip: identification in the Nojima fault zone and implications for focal earthquake mechanisms. J Geophys Res-Solid Ea 108(B4)

Parrish RR (1983) Cenozoic thermal evolution and tectonics of the Coast Mountains of British Columbia: 1. Fission track dating, apparent uplift rates, and patterns of uplift. Tectonics 2(6):601–631

Poage MA, Chamberlain CP (2001) Empirical relationships between elevation and the stable isotope composition of precipitation and surface waters: considerations for studies of paleoelevation change. Am J Sci 301(1):1–15

Rahl JM, Haines SH, van der Pluijm BA (2011) Links between orogenic wedge deformation and erosional exhumation: evidence from illite age analysis of fault rock and detrital thermochronology of syn-tectonic conglomerates in the Spanish Pyrenees. Earth Planet Sci Lett 307:180–190

Rahn MK, Brandon MT, Batt GE, Garver JI (2004) A zero damage model for fission-track annealing in zircon. Am Mineral 89:473–484

Reiners PW, Brandon MT (2006) Using thermochronology to understand orogenic erosion. Annu Rev Earth Pl Sc 34:419–466

Reiners PW, Farley KA (2001) Influence of crystal size on apatite (U–Th)/He thermochronology: an example from the Bighorn Mountains. Wyoming. Earth Planet Sci Lett 188(3):413–420

Reiners PW, Thomson SN, McPhillips D, Donelick RA, Roering JJ (2007) Wildfire thermochronology and the fate and transport of apatite in hillslope and fluvial environments. J Geophys Res-Earth 112(F4)

Reuber G, Kaus BJ, Schmalholz SM, White RW (2016) Nonlithostatic pressure during subduction and collision and the formation of (ultra) high-pressure rocks. Geology 44(5):343–346

Riccio SJ, Fitzgerald PG, Benowitz JA, Roeske SM (2014) The role of thrust faulting in the formation of the eastern Alaska Range: thermochronological constraints from the Susitna Glacier Thrust Fault region of the intracontinental strike-slip Denali Fault system. Tectonics 33(11):2195–2217

Ring U, Brandon MT, Willett SD, Lister GS (1999) Exhumation processes. Geol Soc London Spec Publ 154(1):1–27

Roberts HJ, Kelley SP, Dahl PS (2001) Obtaining geologically meaningful ^{40}Ar–^{39}Ar ages from altered biotite. Chem Geol 172:277–290

Roe GH (2005) Orographic precipitation. Annu Rev Earth Pl Sc 33:645–671

Rubatto D, Hermann J (2001) Exhumation as fast as subduction? Geology 29(1):3–6

Sandiford M, Powell R (1990) Some isostatic and thermal consequences of the vertical strain geometry in convergent orogens. Earth Planet Sci Lett 98(2):154–165

Sandiford M, Powell R (1991) Some remarks on high-temperature—low-pressure metamorphism in convergent orogens. J Metamorphic Geol 9(3):333–340

Schlatter A, Schneider D, Geiger A, Kahle HG (2005) Recent vertical movements from precise levelling in the vicinity of the city of Basel, Switzerland. Int J Earth Sci 94(4):507–514

Schlunegger F, Melzer J, Tucker G (2001) Climate, exposed source-rock lithologies, crustal uplift and surface erosion: a theoretical analysis calibrated with data from the Alps/North Alpine Foreland Basin system. Int J Earth Sci 90(3):484–499

Schmidt JL et al (2014) Little Devil's postpile revisited: behavior of multiple thermochronometers in a contact aureole. In: Abstracts of the 14th international conference on thermochronology, Chamonix, France, 8–14 Sept 2014

Scholz CH (2002) The mechanics of earthquakes and faulting, 2nd edn. Cambridge University Press, Cambridge

Serpelloni E, Faccenna C, Spada G, Dong D, Williams SD (2013) Vertical GPS ground motion rates in the Euro-Mediterranean region: New evidence of velocity gradients at different spatial scales along the Nubia-Eurasia plate boundary. J Geophys Res-Sol Ea 118 (11):6003–6024

Spear FS (1993) Metamorphic phase equilibria and PTt paths. Min Soc Am Monograph 1, 518 p

Spotila JA (2005) Applications of low-temperature thermochronometry to quantification of recent exhumation in mountain belts. Rev Mineral Geochem 58(1):449–466

Stockli DF, Surpless BE, Dumitru TA, Farley KA (2002) Thermochronological constraints on the timing and magnitude of Miocene and Pliocene extension in the central Wassuk Range, western Nevada. Tectonics 21

Stüwe K, Barr TD (1998) On uplift and exhumation during convergence. Tectonics 17(1):80–88

Stüwe K, Hintermüller M (2000) Topography and isotherms revisited: the influence of laterally migrating drainage divides. Earth Planet Sci Lett 184(1):287–303

Stüwe K, White L, Brown R (1994) The influence of eroding topography on steady-state isotherms. Application to fission track analysis. Earth Planet Sci Lett 124(1–4):63–74

Summerfield MA, Brown RW (1998) Geomorphic factors in the interpretation of fission-track data. In: De Corte F (ed) van den Haute P. Advances in fission-track geochronology, Springer, pp 269–284

Tagami T (2012) Thermochronological investigation of fault zones. Tectonophysics 538–540:67–85

Tagami T. (2018) Chapter 12. Application of fission-track thermochronology to understand fault zones. In: Malusà MG, Fitzgerald PG (eds) Fission-track thermochronology and its application to geology. Springer, Berlin

Ter Voorde M, De Bruijne CH, Cloetingh S, Andriessen PAM (2004) Thermal consequences of thrust faulting: simultaneous versus successive fault activation and exhumation. Earth Planet Sci Lett 223(3):395–413

Thomson SN (2002) Late Cenozoic geomorphic and tectonic evolution of the Patagonian Andes between latitudes 42 S and 46 S: an appraisal based on fission-track results from the transpressional intra-arc Liquiñe-Ofqui fault zone. Geol Soc Am Bull 114:1159–1173

Villa IM (1998) Isotopic closure. Terra Nova 10(1):42–47

Villa IM (2010) Disequilibrium textures versus equilibrium modelling: geochronology at the crossroads. Geol Soc London Spec Publ 332:1–15

Wagner GA, Reimer GM (1972) Fission track tectonics: the tectonic interpretation of fission track apatite ages. Earth Planet Sci Lett 14 (2):263–268

Wagner GA, van den Haute P (1992) Fission-Track Dating. Kluwer Acad, Dordrecht

Wagner GA, Reimer GM, Jäger E (1977) Cooling ages derived by apatite fission track, mica Rb-Sr, and K-Ar dating: the uplift and cooling history of the central Alps. Mem Inst Geol Mineral Univ Padova 30:1–27

Wagner GA, Gleadow AJW, Fitzgerald PG (1989) The significance of the partial annealing zone in apatite fission-track analysis: projected track length measurements and uplift chronology of the Transantarctic Mountains. Chem Geol 79(4):295–305

Walcott RI (1998) Modes of oblique compression: late Cenozoic tectonics of the South Island of New Zealand. Rev Geophys 36 (1):1–26

Whipp DM, Ehlers TA (2007) Influence of groundwater flow on thermochronometer-derived exhumation rates in the central Nepalese Himalaya. Geology 35:851–854

Whipple KX (2009) The influence of climate on the tectonic evolution of mountain belts. Nat Geosci 2(2):97–104

Whipple KX, Tucker GE (1999) Dynamics of the stream-power river incision model: implications for height limits of mountain ranges, landscape response timescales, and research needs. J Geophys Res-Sol Ea 104(B8):17661–17674

Whipple KX, Tucker GE (2002) Implications of sediment-flux-dependent river incision models for landscape evolution. J Geophys Res-Sol Ea 107(B2)

Whitmarsh RB, Manatschal G, Minshull TA (2001) Evolution of magma-poor continental margins from rifting to seafloor spreading. Nature 413(6852):150–154

Wildman M, Beucher R, Cognè N (2018) Chapter 20. Fission track thermochronology applied to the geomorphological and geologic evolution of passive continental margins. In: Malusà MG, Fitzgerald PG (eds) Fission-track thermochronology and its application to geology. Springer, Berlin

Willett SD (2010) Late Neogene erosion of the Alps: a climate driver? Annu Rev Earth Pl Sc 38:411–437

Wolf RA, Farley KA, Kass DM (1998) Modeling of the temperature sensitivity of the apatite (U–Th)/He thermochronometer. Chem Geol 148(1):105–114

Wölfler A, Stüwe K, Danišík M, Evans NJ (2012) Low temperature thermochronology in the Eastern Alps: implications for structural and topographic evolution. Tectonophysics 541:1–18

Yamato P, Burov E, Agard P, Le Pourhiet L, Jolivet L (2008) HP-UHP exhumation during slow continental subduction: self-consistent thermodynamically and thermomechanically coupled model with application to the Western Alps. Earth Planet Sci Lett 271 (1):63–74

Zanchetta S, Malusà MG, Zanchi A (2015) Precollisional development and Cenozoic evolution of the Southalpine retrobelt (European Alps). Lithosphere 7(6):662–681

Zarki-Jakni B, van der Beek P, Poupeau G, Sosson M, Labrin E, Rossi P, Ferrandini J (2004) Cenozoic denudation of Corsica in response to Ligurian and Tyrrhenian extension: results from apatite fission track thermochronology. Tectonics 23(1)

Zattin M, Massari F, Dieni I (2008) Thermochronological evidence for Mesozoic-Tertiary tectonic evolution in the Eastern Sardinia. Terra Nova 20(6):469–474

Concept of the Exhumed Partial Annealing (Retention) Zone and Age-Elevation Profiles in Thermochronology

Paul G. Fitzgerald and Marco G. Malusà

Abstract

Low-temperature thermochronology is commonly applied to constrain upper crustal cooling histories as rocks are exhumed to Earth's surface via a variety of geological processes. Collecting samples over significant relief (i.e., vertical profiles), and then plotting age versus elevation, is a long-established approach to constrain the timing and rates of exhumation. An exhumed partial annealing zone (PAZ) or partial retention zone (PRZ) with a well-defined break in slope revealed in an age-elevation profile, ideally complemented by kinetic parameters such as confined track lengths, provides robust constraints on the timing of the transition from relative thermal and tectonic stability to rapid cooling and exhumation. The slope above the break, largely a relict of a paleo-PAZ usually with significant age variation with change in elevation, can be used to quantify fault offsets. The slope below the break is steeper and represents an apparent exhumation rate. We discuss attributes and caveats for the interpretation of each part of an age-elevation profile, and provide examples from Denali in the central Alaska Range, the rift-flank Transantarctic Mountains, and the Gold Butte block of southeastern Nevada, where multiple methods reveal exhumed PAZs and PRZs in the footwall of a major detachment fault. Many factors, including exhumation rates, advection of isotherms and topographic effects on near-surface isotherms, may affect the interpretation of data. Sampling steep profiles over short-wavelength topography and parallel to structures minimises misfits between age-elevation slopes and actual exhumation histories.

P. G. Fitzgerald (✉)
Department of Earth Sciences, Syracuse University, Syracuse, NY 13244, USA
e-mail: pgfitzge@syr.edu

M. G. Malusà
Department of Earth and Environmental Sciences, University of Milano-Bicocca, Piazza delle Scienza 4, 20126 Milan, Italy

9.1 Introduction

Thermochronology is the study of the thermal history of rocks and minerals. Thermochronologic ages are often interpreted as closure ages corresponding to a closure temperature T_c (Dodson 1973), which is defined as the temperature at which the system becomes closed, assuming monotonic cooling. Closure temperatures vary as a function of mineral kinetic parameters, cooling rate (T_c is higher for faster cooling), the composition of the mineral, and/or radiation damage of the crystal lattice (e.g., Gallagher et al. 1998; Reiners and Brandon 2006). In the case of low-temperature thermochronologic methods such as fission-track (FT) analysis, the simplest data interpretations are that thermochronologic ages record exhumation. This is because rocks cool as they move toward the surface of the Earth, i.e., as they are exhumed, either via tectonics and/or erosion.

Early FT studies (e.g., Naeser and Faul 1969) established that FT ages could be reset, or partially reset, due to temperature increases resulting from burial in a sedimentary basin or nearby igneous intrusions (e.g., Fleischer et al. 1965; Calk and Naeser 1973; Naeser 1976, 1981). Because temperature increases with depth, borehole studies (e.g., Naeser 1979; Gleadow and Duddy 1981) revealed that FT ages decreased with depth. In the case of apatite, borehole studies also revealed a zone of partial stability for fission-tracks corresponding to temperatures between ~ 120 and $60\ ^\circ C$ (Fig. 9.1). This zone was initially called the "partial stability field" (Wagner and Reimer 1972) or "field of partial stability" (Naeser 1981), but other terms such as "partial stability zone" (Gleadow and Duddy 1981) or "track annealing zone" (Gleadow et al. 1983) were also used. Gleadow and Fitzgerald (1987) introduced the term *partial annealing zone* or PAZ, a term that has remained in common usage. The PAZ for apatite FT (AFT) and its characteristics are well defined in some boreholes (Fig. 9.1). AFT ages decrease downhole while single grain age dispersion tends to increase. Confined track length distributions also show

© Springer International Publishing AG, part of Springer Nature 2019
M. G. Malusà and P. G. Fitzgerald (eds.), *Fission-Track Thermochronology and its Application to Geology*, Springer Textbooks in Earth Sciences, Geography and Environment, https://doi.org/10.1007/978-3-319-89421-8_9

Fig. 9.1 Classic form of the partial annealing zone (PAZ) is shown in this compilation diagram showing downhole temperature versus AFT age for samples from the Otway Basin in SE Australia (summarised from Gleadow and Duddy 1981; Gleadow et al. 1986; Green et al. 1989; Dumitru 2000). Decreasing AFT ages and changing track length distributions with increasing downhole temperature reflect the changing rate of annealing, from very slow above ~60 °C, to increasing within the PAZ and then "geologically instantaneous" in the zone of total annealing

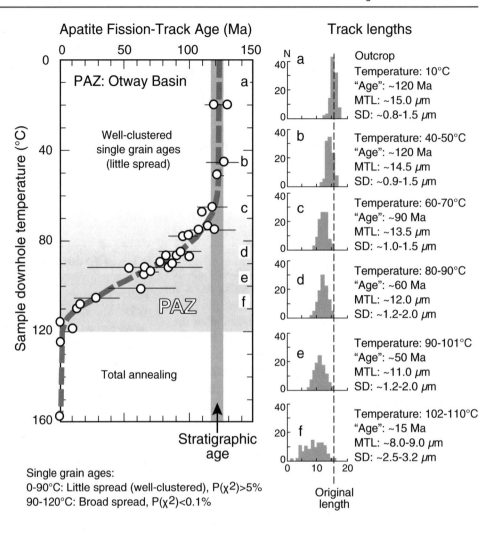

systematic trends—as distributions broaden downhole, mean lengths decrease and standard deviations increase (Gleadow and Duddy 1981; Gleadow et al. 1983; Green et al. 1986). These studies also documented the role of apatite composition in the rate of annealing, and hence the variable limits of the PAZ.

The PAZ can be defined as the temperature interval, corresponding to appropriate crustal depths, between where tracks are annealed geologically "instantaneously" and where they are retained without a significant loss in FT age on the geologic timescale. Annealing and track length reduction are more rapid at higher temperatures, but do continue, albeit much more slowly, even at ambient temperatures (e.g., Green et al. 1986). The same concept applies to other thermochronologic systems due to loss, or partial loss, via volume diffusion of daughter products, such as Ar in K–Ar systems (Baldwin and Lister 1998) and He in (U–Th)/He systems (Wolf et al. 1998). The term *partial retention zone* (PRZ) was introduced by these authors for these noble gas methods. We favour the use of PAZ for FT

thermochronology where annealing is the operative process for length reduction of individual tracks.

When samples are collected over significant relief, ages from different thermochronologic systems generally increase with increasing elevation. Age versus elevation plots, and the interpretation of age variations and slopes on these plots, marked a breakthrough for the earliest FT studies applied to tectonics (e.g., Wagner and Reimer 1972; Naeser 1976; Wagner et al. 1977). The slope of the age-elevation trend was initially interpreted as an "uplift rate," but it is now clear that this trend provides information on the rate of *exhumation*, where exhumation is the displacement of rocks with respect to Earth's surface (England and Molnar 1990). England and Molnar (1990) also defined the *surface uplift* as the displacement of the Earth's surface (typically the mean surface elevation over an area) with respect to the geoid, and the *rock uplift* as the displacement of a rock with respect to the geoid (see Chap. 8, Malusà and Fitzgerald 2018a). Exhumation, surface uplift, and rock uplift are linked by the following relationship:

$$\text{Surface uplift} = \text{rock uplift} - \text{exhumation} \qquad (9.1)$$

When the amount of exhumation and surface uplift can be independently constrained (i.e., because the paleo-mean land-surface elevation is known), rock uplift can also be constrained (Eq. 9.1). However, these situations are actually quite rare (e.g., Fitzgerald et al. 1995; Abbot et al. 1997). Moreover, it is important to emphasise that thermochronology does not constrain exhumation directly, because isotherms (the thermal reference frame) are controlled by a range of processes in the upper crust. The conversion from a thermal framework (which impacts the thermochronologic record) to a framework that constrains the timing and rates of exhumation requires assumptions about the dynamic thermal structure of the crust.

There are many factors that influence the slope of the age-elevation profile. Notably those related to the perturbation of the thermal structure of the crust associated with rapid exhumation, the shape of topography, or the way that samples are collected across the topography (e.g., Brown 1991; Brown et al. 1994; Stüwe et al. 1994; Mancktelow and Grasemann 1997). Many excellent papers and books discuss these topics in considerable detail (e.g., Braun 2002; Reiners et al. 2003; Braun et al. 2006; Reiners and Brandon 2006; Huntington et al. 2007; Valla et al. 2010; see also Chap. 8, Malusà and Fitzgerald 2018a; Chap. 10, Malusà and Fitzgerald 2018b; Chap. 17, Schildgen and van der Beek 2018). To plot thermochronologic ages versus elevation using an orthogonal coordinate system requires the fundamental assumption that data interpretation is one-dimensional (i.e., the particle path is vertical) and that samples are not collected laterally across the landscape. In essence, the vertical profile approach is the plotting of a three-dimensional data set in one dimension and coupled with the effects of advection of isotherms due to rapid exhumation often results in data misinterpretation.

In this chapter, we discuss the strategy of collecting samples over a significant elevation range to constrain the timing and rate of exhumation. We discuss the concept of the exhumed PAZ (or PRZ) and use a number of well-known examples to illustrate sampling strategies, common mistakes, factors, and assumptions that must be considered when interpreting thermochronologic data from age-elevation profiles.

9.2 Definition of a PAZ and PRZ

The classic form of a PAZ is shown clearly in AFT data plotted against depth (downhole temperature) from Otway Basin drill holes (Gleadow and Duddy 1981; Gleadow et al. 1983, 1986) (Fig. 9.1). The Otway Basin formed during the breakup between Australia and Antarctica and was filled with >3 km volcanogenic sediments of the Lower Cretaceous Otway Group. The upper part of the AFT profile is defined by AFT ages of ∼120 Ma, invariant down to a temperature of ∼60 °C. Ages then decrease progressively, within the PAZ itself as the rate of annealing increases, down to a zero age at depths where the present temperature is ∼120 °C. For the Otway Basin, the PAZ is therefore defined as between ∼120 and ∼60 °C. The chemical composition of the individual grains within the Otway Basin is variable, due to the volcanogenic provenance, with a mix of F-rich and Cl-rich grains (Green et al. 1985). Tracks within Cl-rich grains are more resistant to annealing (see Chap. 3, Ketcham 2018), so these grains reach a zero age at a higher temperature. The spread of single grain ages is the greatest at intermediate temperatures within the PAZ where the variable rates of annealing between grains of different chemical compositions are maximised (e.g., Green et al. 1986; Gallagher et al. 1998). Age dispersion within a PAZ is obvious when a downhole modeled-AFT age profile is plotted for Dpars of differing values (Fig. 9.2a). While not a direct proxy for chemical composition, Dpar—the diameter of track-etch figures measured parallel to the crystallographic c-axis—is commonly used as an indicator of the resistance of fission-tracks in apatite to thermal annealing (e.g., Burtner et al. 1994). In this example, the AFT age difference between a Dpar of 1 (fission-tracks are less resistant to annealing) and 3 (fission-tracks are more resistant to annealing) is ∼45 Ma (∼45% relative to the unannealed AFT age) at temperatures between 80 and 90 °C.

The variation of (U–Th)/He age with depth (temperature) is not discussed in detail in this chapter. However, formation of an apatite (U–Th)/He (AHe) age profile within a PRZ follows similar principles to those corresponding to an AFT PAZ. Note that there are slight differences in the shape of the modeled AHe PRZ versus the shape of the modeled-AFT PAZ at lower temperatures (Fig. 9.2), which reflect the different kinetics. In addition, the magnification of single grain age variations for AHe is even more pronounced within the PRZ, depending on the variation of effective uranium concentration [eU], grain size, zonation of the individual grains, and residence time within a PRZ (e.g., Reiners and Farley 2001; Farley 2002; Meesters and Dunai 2002; Fitzgerald et al. 2006; Flowers et al. 2009). Based on relative alpha particle production, [eU] is calculated as [U] + 0.235[Th] (e.g., Flowers et al. 2009). Figure 9.2b shows the age variation between a small grain with a low [eU] versus a large grain with a higher [eU] is ∼70 Ma (∼70% relative to unreset ages) at ∼65 °C (held constant for ∼100 Myr). This single grain age variation is a result of the grain size being the diffusion domain and the relative importance between loss via α-particle ejection and loss via volume diffusion, plus the effects of higher radiation

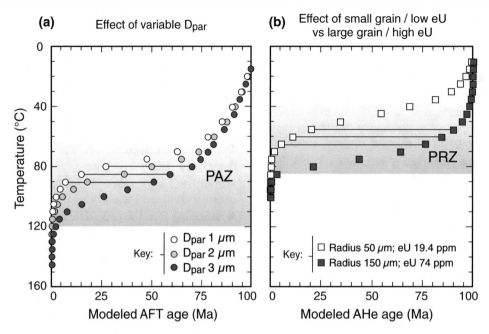

Fig. 9.2 a Effect of composition, using Dpar as a proxy, on the variation of AFT ages within a PAZ. In these modeled examples, the differences in ages are observed for samples with Dpar values of 1 and 3 μm. The greatest age dispersion lies in the interval 80–90 °C (marked by horizontal lines) where there is a ∼40 Myr (40%) age difference between the two extremes. Age dispersion due to chemical composition is magnified by residence within, or slow cooling through, the PAZ. Modeled ages were produced using HeFTy (Ketcham 2005) and the Ketcham et al. (2007) annealing algorithm. **b** In a similar fashion, AHe age dispersion within a PRZ is maximised when there is a mix of large grains/high [eU] (older grains) and small grains/low [eU] (younger grains). In this modeled scenario, the greatest difference in ages lies in the interval ∼55–65 ° C (marked by horizontal lines) where there is a ∼70 Myr (70%) age difference. Age dispersion due to variable [eU] and grain size (plus zoning and other factors—not modeled here) is magnified by residence within, or slow cooling through the PRZ. Modeled AHe ages were produced using HeFTy (Ketcham 2005), the Flowers et al. (2009) algorithm and the Ketcham et al. (2011) α-particle ejection parameters

damage, in effect storing He inside damage zones within the crystal lattice (Reiners and Farley 2001; Farley 2002; Flowers et al. 2009). Age dispersion is more pronounced within a PRZ if grain zonation is such that the rim is depleted relative to the core (e.g., Meesters and Dunai 2002; Fitzgerald et al. 2006).

Confined track lengths are an essential component to the application of the PAZ concept to AFT thermochronology as they provide a kinetic parameter used to constrain the thermal history of each sample (e.g., Gleadow et al. 1986). Interpretations can be both qualitative (e.g., long mean lengths ≥ 14 μm indicate simple rapid cooling vs. more complex distributions that may reflect residence within a PAZ, slow cooling, or reheating) but also provide a fundamental input parameter to inverse thermal modeling (e.g., Ketcham 2005; Gallagher 2012; see Chap. 3, Ketcham 2018). Confined track length distributions (CTLDs) from the Otway Basin samples reveal the now well-established classic pattern of a PAZ, i.e., decreasing mean lengths and increasing standard deviations with increasing temperatures downhole (Fig. 9.1). These distributions reflect a shortening of tracks that have resided at the same crustal level and hence same temperatures for the last ∼30 Myr (Gleadow

and Duddy 1981). Thus, in each case the maximum track length remains the same (reflecting the most recently formed confined track). But, as the rate of annealing increases with increasing temperature, the tracks anneal progressively faster and histograms broaden to reflect the annealing process. In this situation, track-length histograms can be imagined as conveyor belts; moving slowly from right to left when the temperature is low, and annealing slower near the upper part of the PAZ (i.e., leading to narrow distributions). Near the base of the PAZ (corresponding to higher temperatures), the conveyor belt is faster as the rate of annealing increases and CTLDs are broader. As mentioned above, compositional variation between apatite grains, particularly acute in Otway Basin sediments also leads to dispersion in downhole track lengths. In addition, anisotropic annealing of tracks in apatite (Green and Durrani 1977) where tracks perpendicular to the c-axis shorten faster than those parallel to the c-axis also leads to track length dispersion.

The shape and development of a PAZ (revealed in the AFT age-depth profile) are the result of a number of parameters including the thermal structure of the upper crust (i.e., the geothermal gradient) and the previous thermal and tectonic history (Figs. 9.1, 9.2 and 9.3). Typically, the characteristic AFT age-depth trend formed in a PAZ, for example, as shown in the Otway Basin, is interpreted as a time of "relative tectonic and thermal stability," with the upper part of the profile (<60 °C) reflecting the previous thermal history (i.e., formed in the geologic past). Within the PAZ, the slope of the AFT profile will vary as a function of the duration that samples have resided within the PAZ, the paleogeothermal gradient, and the relative thermal (and tectonic) stability at that time. The slope of the AFT-depth profile will be shallower if the PAZ forms over a long period of time (Fig. 9.3c). However, if samples are resident in the PAZ for only a short period of time, the variation in ages will not reflect the characteristic profile. If samples cool monotonically through the PAZ, then their AFT ages may be interpreted as closure ages representative of the time the sample cooled through the closure temperature T_c of the thermochronologic system under consideration (see Chap. 8, Malusà and Fitzgerald 2018a).

9.3 Vertical Profiles and the Exhumed PAZ/PRZ Concept

9.3.1 Recognition of an Exhumed PAZ

Thermochronologic data collected in age-elevation profiles from orogens may reveal simple linear relationships, and the T_c concept will apply (e.g., Reiners and Brandon 2006). However, depending on the level of erosion and the thermal/tectonic history, an exhumed PAZ may be revealed. This happens when a period of relative tectonic and thermal stability is followed by a period of rapid cooling and exhumation. Relative stability allows the development of the characteristic age-depth shape of a PAZ, which is then preserved in an age-elevation profile (Naeser 1979; Gleadow and Fitzgerald 1987) (Fig. 9.3a). The *break in slope* marks the base of a former PAZ, termed an *exhumed PAZ*, and approximates the onset of significant rapid cooling (Gleadow and Fitzgerald 1987; Fitzgerald and Gleadow 1988, 1990; Gleadow 1990). In essence, the break in slope represents a paleo ~110 °C isotherm, or whatever paleotemperature is appropriate given either the mineral composition or method under consideration. The significance of an inflection point in an age-elevation profile was first noted in the Rocky Mountains of Colorado (Naeser 1976). In an age-elevation profile from Mt Evans (4346 m), Chuck Naeser identified an inflection point at ~3300 m. This was interpreted as the former position of the ~105 °C isotherm, with the temperature based on Eielson (Alaska) drill hole age data (Naeser 1981; see Chap. 1, Hurford 2018). Naeser (1976) noted that "apatite below 3000 m reflects rapid uplift of the Rocky Mountains during the Laramide orogeny starting about 65 Myr ago," whereas "apatite above 3000 m was only partially annealed during the Cretaceous burial prior to the uplift."

The slope above the break in slope (i.e., the exhumed PAZ) is not steep and reflects the shape of the former PAZ formed during a time of relative tectonic and thermal stability (Fig. 9.3c), rather than an apparent exhumation rate. However, the steeper slope below the break in slope represents instead an apparent exhumation rate—taking into consideration all the caveats that can modify the slope, as discussed below. With respect to the formation of the classic age profile within a PAZ, it is worth noting that it is unlikely to have formed within a period of "complete and absolute" tectonic and thermal stability. The thermal frame of reference is dynamic, for example, because of cooling associated with exhumation and/or relaxation of isotherms and even heating associated with subsidence and burial (see Chap. 8, Malusà and Fitzgerald 2018a). A simple way to envision an exhumed PAZ is by comparing the conditions during the formation of a PAZ relative to the time following that. In essence, the recognition of an exhumed PAZ is dependent on our ability to recognise different components within a track length distribution; those tracks annealed (and shortened) while the sample is resident in a PAZ and then long tracks that result from rapid cooling. To a certain extent, the exhumed PAZ concept can be demonstrated by comparing the gentler slope of an exhumed PAZ with the steeper slope of the age-elevation profile beneath the break. For example, in the Denali profile in Alaska (see Sect. 9.4.1), the slope of the exhumed PAZ above the break is ~100 m/Myr compared to the slope (apparent exhumation rate) below the break (~1500 m/Myr) (Fitzgerald et al. 1995). In the Transantarctic Mountains, slopes of exhumed PAZs are ~15 m/Myr compared to slopes (apparent exhumation rate) of ~100 m/Myr below the break (e.g., Gleadow and Fitzgerald 1987; Fitzgerald 1992). The slopes are different, but the relative durations for each (slow cooling and formation of the PAZ vs. rapid cooling) are about the same (3:1).

Track length measurements are crucial for interpreting thermal histories in general, but are also very useful for identifying exhumed PAZs. CTLDs from below the break in slope generally reflects rapid cooling (mean track length >14 μm, standard deviation <1.5 μm). CTLDs from above the break in slope reflects longer residence within the PAZ, with distributions often being bimodal, reflecting the two components of tracks (Fig. 9.3a). These distributions typically have means of ~12–13 μm with standard deviations of >1.6 μm. Within an exhumed PAZ, there will be a greater spread of single grain ages, because slower cooling magnifies differences arising between grains of different retentivity (Fig. 9.2). This concept is of fundamental importance to detrital thermochronology, where interpretation of single grain age data usually assumes that all ages represent closure

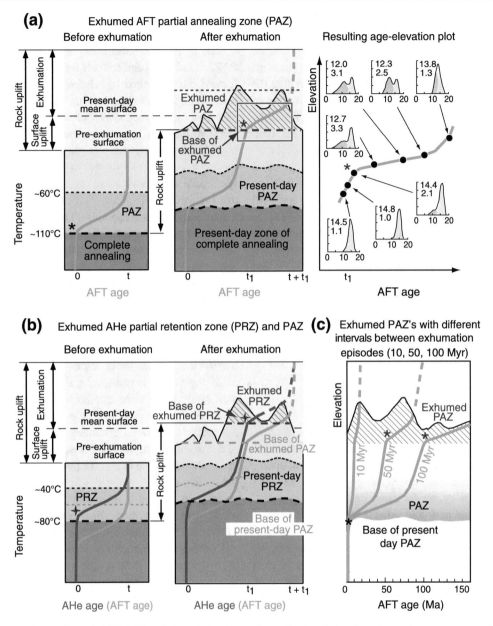

Fig. 9.3 **a** Concept of an exhumed AFT PAZ and the relationship between rock uplift, exhumation, and surface uplift. The left panel shows the characteristic shape of the AFT PAZ formed over the time period "t" during relative thermal and tectonic stability. Following a period of rock uplift, followed by exhumation and relaxation of the isotherms, an exhumed PAZ may be revealed in the age-elevation profile (middle panel). The asterisk marking the break in slope indicates the base of the exhumed PAZ and typically slightly underestimates the onset age of rock uplift as this point has to cool through the PAZ. Note that modeled isotherms mimic the surface topography. The right-hand panel shows the variation of track length distributions at various levels throughout the profile. Distributions below the break in slope contain only long tracks formed during rapid cooling with little time spent within the PAZ. In contrast, CTLDs

above the break in slope have shorter means and larger standard deviations as they contain two length components: pre-exhumation lengths (shortened due to annealing while resident within the PAZ) and long tracks that post-date the onset of rapid exhumation and have not been shortened due to annealing (modified from Fitzgerald et al. 1995). **b** The exhumed PAZ concept also applied to an exhumed AHe PRZ (marked by a cross) (modified from Fitzgerald et al. 2006). **c** An exhumed PAZ (or PRZ) will only be revealed in an age-elevation profile when there has been sufficient time between episodes of exhumation (e.g., 50 or 100 Myr in this example) to develop the classic form of a PAZ that represents a period of apparent thermal and tectonic stability. If the time period is too short, e.g., 10 Myr in this example, an exhumed PAZ will not be distinguishable (modified after Fitzgerald and Stump 1997)

ages, typically due to either rapid cooling via exhumation or cooling of volcanic rocks (e.g., Garver et al. 1999; Bernet and Garver 2005). In other words, the assumption in detrital studies is that all thermochronologic ages are geologically meaningful. However, age variations observed in an exhumed PAZ (or PRZ) shows that this is not always the situation. In many cases, only the age corresponding to the break in slope can be interpreted with respect to a geologic event. Lag-time calculations (thermochronologic age minus stratigraphic age) to constrain past exhumation rates may therefore be geologically meaningless if samples are eroded from within an exhumed PAZ (see below and Chap. 10, Malusà and Fitzgerald 2018b) because these ages do not represent closure ages and a large age dispersion exists within the exhumed PAZ.

A break in slope is drawn at the intersection of the lines representing the exhumed PAZ and the age-elevation profile below the break, although the base of a PAZ is actually a curve (Figs. 9.1, 9.2 and 9.3). This means that the time corresponding to the break in slope will slightly underestimate the timing of the onset of rapid cooling/exhumation. Also, samples close to the base of the PAZ will have to transit through the PAZ itself. Thus, if samples do not transit the PAZ rapidly, CTLDs will reflect more annealing (e.g., Stump and Fitzgerald 1992) and the age corresponding to the break in slope will slightly underestimate the true onset of rapid cooling/exhumation. Such a discrepancy may not be critical (i.e., not affect data interpretation) if an exhumed PAZ has a break in slope at ~50 Ma (for example). In this case, our ability to accurately locate the position of a break in slope is on the order of the precision of each age, so for a break in slope of ~50 Ma, it is on the order of ±5% for AFT. However, if a cooling event (and hence the time marked by the base of an exhumed PAZ) is Pliocene in age, distinguishing between the onset of cooling/exhumation at 6 or 4 Ma may be geologically significant.

In summary, the following factors affect the potential recognition of an exhumed PAZ (or PRZ) in an age-elevation profile:

- The period of relative thermal and tectonic stability prior to the initiation of more rapid cooling/exhumation. Sufficient time is in fact required to allow the shallow slope to develop (Fig. 9.3c).
- The magnitude of cooling/exhumation event that follows formation of the PAZ. If this is minor, then an exhumed PAZ will not be recognised. What is particularly important is the contrast in thermal and tectonic regimes between the formation of a PAZ and subsequent more rapid cooling/exhumation that preserves the classic PAZ form.
- When formation of a PAZ and subsequent cooling/exhumation occurred in the geologic record. If it occurred a long time ago, the precision of the dating method relative to the magnitude and duration of the geologic and/or tectonic event may not be adequate to reveal an exhumed PAZ, the exhumed PAZ may have been eroded away, or the proportion of shortened tracks (formed in the PAZ) to long tracks (subsequent to exhumation) may be insufficient to identify that period of relative stability,

even with modeling. In these cases, the interpretation of an age-elevation profile may be "slow or monotonic exhumation." Models may also suggest a "good-fit" T–t envelope that could be interpreted as "slow monotonous cooling," but data precision may be insufficient to reveal different events. An important point is that data and models must be interpreted within a geologic context. Geologic events are by their very nature episodic whether on the temporal scale of earthquakes or volcanoes, during times of accelerated mountain building or part of an orogenic cycle. The same analogy applies to different thermochronologic methods. Higher temperature methods tend to only reveal major events and often indicate "monotonous cooling," whereas the lower temperature methods may reveal individual events. Of course, lower temperature techniques are more susceptible to the effects of dynamic isotherms (e.g., Braun et al. 2006).

We have only discussed an exhumed PAZ with respect to the age of the break in slope as representing cooling due to exhumation, usually associated with rock uplift. However, in rapidly deforming orogens (e.g., Himalaya, Taiwan, and Southern Alps of New Zealand) a break in slope may also be revealed on a multi-method plot of age versus temperature, for analyses performed on the same sample (Kamp and Tippett 1993; Ching-Ying et al. 1990). Such a break is usually interpreted in terms of the onset of a major tectonic/exhumation/cooling event. However, because isotherms are perturbed during such rapid exhumation, the data may be equally well explained by a constant exhumation rate leading to an exponential decrease in temperature with time (Batt and Braun 1997; Braun et al. 2006).

The ultimate test of any thermochronologic interpretation involves comparison with the geologic record. For example, does an inflection point on either a single-method age versus elevation plot, or in a multi-method age versus temperature plot, represent relaxation of compressed isotherms or cooling due to erosion/exhumation due to uplift and creation of relief in an active (or recently inactive) orogen? If there is a nearby sedimentary basin with a large influx of detritus (e.g., conglomerates) that were the same age as that associated with the inflection point, this would provide evidence for an interpretation of cooling due to exhumation as opposed to relaxation of compressed isotherms.

9.3.2 Attributes and Information to Be Gained from an Exhumed PAZ

In previous sections, we discussed the characteristics of a PAZ and the recognition of an exhumed PAZ. In this section, we discuss the attributes of, and the information that can be obtained from an exhumed PAZ (Fig. 9.4), as further

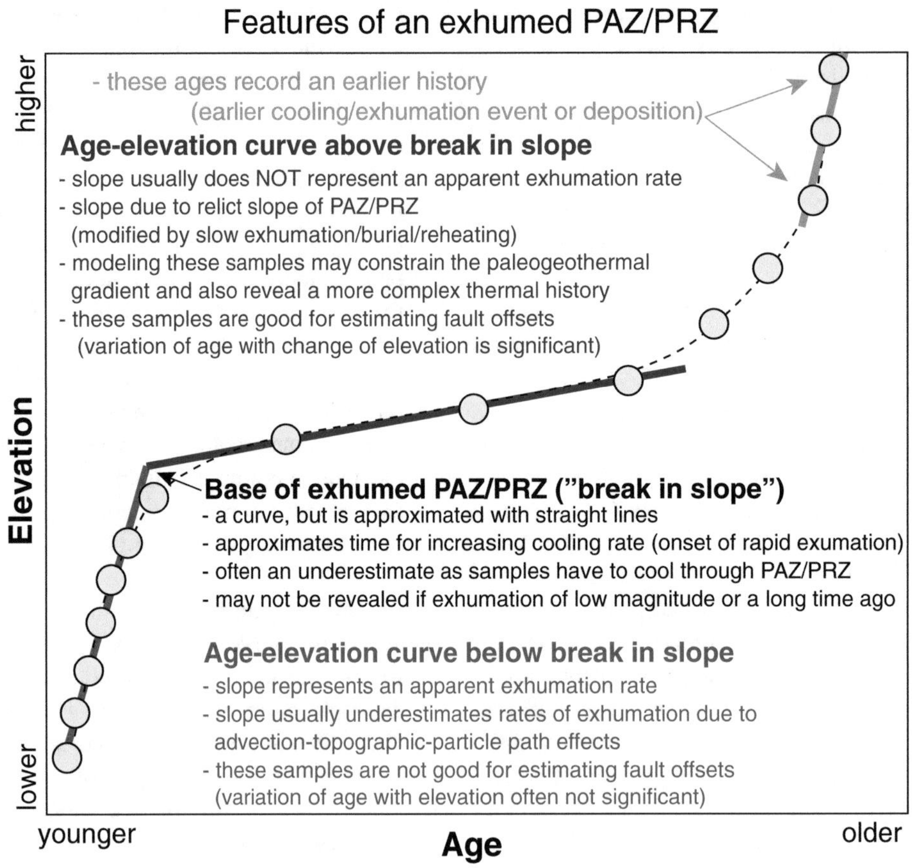

Features of an exhumed PAZ/PRZ

- these ages record an earlier history
  (earlier cooling/exhumation event or deposition)

Age-elevation curve above break in slope
- slope usually does NOT represent an apparent exhumation rate
- slope due to relict slope of PAZ/PRZ
  (modified by slow exhumation/burial/reheating)
- modeling these samples may constrain the paleogeothermal
  gradient and also reveal a more complex thermal history
- these samples are good for estimating fault offsets
  (variation of age with change of elevation is significant)

Base of exhumed PAZ/PRZ ("break in slope")
- a curve, but is approximated with straight lines
- approximates time for increasing cooling rate (onset of rapid exumation)
- often an underestimate as samples have to cool through PAZ/PRZ
- may not be revealed if exhumation of low magnitude or a long time ago

Age-elevation curve below break in slope
- slope represents an apparent exhumation rate
- slope usually underestimates rates of exhumation due to
  advection-topographic-particle path effects
- these samples are not good for estimating fault offsets
  (variation of age with elevation often not significant)

higher / lower — Elevation

younger — Age — older

Fig. 9.4 Summary of the attributes and information available from an exhumed PAZ/PRZ

illustrated in Sect. 9.4 with examples from Denali, the Transantarctic Mountains and the Gold Butte Block.

Timing of Exhumation As mentioned above, the time corresponding to the break in slope represents the transition from a relatively stable thermal and tectonic regime to a time of more rapid cooling, usually related to exhumation as a result of rock uplift. The break in slope will underestimate the "true" time of transition, because of the geometry of the curve and because rocks take time to transit the PAZ. Inverse thermal modeling (see Chap. 3, Ketcham 2018) for samples just above the break in slope will reveal a period of residence within the PAZ, followed by the onset of rapid cooling. Inverse thermal modeling for samples below the break in slope typically will reveal rapid cooling only, but not the onset of rapid cooling. During the transition from relative thermal and tectonic stability to rapid exhumation, the evolving thermal regime can be complex, before steady state may become established. Higher temperature thermochronometers are slower to respond and establish steady state than low-temperature systems (e.g., Reiners and Brandon 2006). In a modeling paper, Moore and England (2001) found that in the case of a sudden increase in the rate of cooling due to erosion, the advection of isotherms means that the recorded ages only show a gradual increase in the

rate of exhumation. Valla et al. (2010), in another modeling paper, discuss in detail the effect that changes in relief have on the resolution of thermochronologic data sets. They conclude that changes in relief can only be quantified and constrained if the rate of relief growth exceeds ~ 2–3 times the background exhumation rate. The bottom line is that if there are detectable variations in an age-elevation profile such as an exhumed PAZ, then these variations are significant and meaningful as regards the geologic and landscape (relief, topography) evolution.

Amount of Exhumation Because an exhumed PAZ corresponds to a time of relative stability, it is reasonable to assume that the geothermal gradient was also relatively stable. Thus, assuming a reasonable paleogeothermal gradient (typically between 20 and 30 °C/km representing a relatively stable continental geotherm) allows one to convert the paleotemperature at the base of the exhumed PAZ (e.g., ~ 110 °C) to the amount of rock removed. Depending on the elevation of the break in slope relative to the mean land-surface elevation, the amount of exhumation since the time of the break in slope can be estimated (see Brown 1991; Gleadow and Brown 2000). Thus, an average rate of exhumation since the timing of the break can be constrained and, if appropriate, this rate can be compared to the apparent exhumation rate based on the age-elevation slope below the break. If the stratigraphy (above the break) or rock removed can be reconstructed or independently constrained, then the paleogeothermal gradient can also be constrained, as for the Transantarctic Mountains in the Dry Valleys area (Gleadow and Fitzgerald 1987; Fitzgerald 1992). As the depth to the base of a PAZ is usually 3–5 km (being a function of paleogeothermal gradient), it is possible to design a sampling strategy, based on geological constraints or other information that is aimed at collecting an age-elevation profile that has an exhumed PAZ, purely because it does provide such a useful marker.

The Slope of the Exhumed PAZ The slope of this part of the profile (i.e., above the break in slope) does not typically represent an apparent exhumation rate. Viewed in one dimension, the slope depends on the factors discussed above (length of time over which the PAZ formed, paleogeothermal gradient) and the relative thermal stability both when the PAZ was formed and since rapid cooling began (approximated by the time corresponding to the break in slope). Inverse thermal modeling of samples over an elevation range can constrain the paleogeothermal gradient (see Chap. 8, Malusà and Fitzgerald 2018a). A factor in mathematically constraining the slope of the profile, both above and below the break in slope, is often the relatively large uncertainties (compared to the number of analyses) for AFT ages or the variation of single grain ages for AHe ages. This explains why age-elevation slopes are often determined as a line of best-fit "by eye," in addition to calculating least-square regression lines.

The Exhumed PAZ as a Tectonic Marker As mentioned above, low-temperature thermochronologic ages are very useful as tectonic makers because there may be systematic variation of age with elevation (e.g., Wagner and Reimer 1972). As such, this concept is often called "fission-track stratigraphy (Brown 1991). Notably, sample ages within an exhumed PAZ typically vary significantly with elevation changes and thus are more useful as tectonic markers in contrast to sample ages below a break in slope where ages may be concordant within error over considerable topographic relief (see Chap. 10, Malusà and Fitzgerald 2018b; Chap. 11, Foster 2018).

9.3.3 Attributes and Information to Be Gained from the Age-Elevation Profile Below the Break in Slope

The slope of the age-elevation profile below a break in slope is generally steep and CTLDs indicates rapid cooling. Samples spend little time within a PAZ as they cool rapidly

through it, and ages can be interpreted as closure temperature ages. As compared to the slope of the exhumed PAZ, this part of the profile is often more typical of what is often found in mountain belts, with the slope representing an *apparent exhumation rate*. Typically, the slope overestimates the real exhumation rate due to a combination of advection and isotherm compression, the effect of topography on near-surface isotherms, and where samples are collected across the topography or the topographic wavelength (Brown 1991; Stüwe et al. 1994; Brown and Summerfield 1997; Mancktelow and Grasemann 1997; Stüwe and Hintermüller 2000; Braun 2002, 2005; Ehlers and Farley 2003; Braun et al. 2006; Huntington et al. 2007; Valla et al. 2010).

The reference frame against which we plot the age of samples (i.e., the elevation, altitude or depth, E in Fig. 9.5a) is not necessarily representative of the thermal frame of reference where the samples acquire their thermochronologic signature (Z in Fig. 9.5b–e). When plotting age versus elevation, the assumption is that the reference frame, i.e., isotherms were horizontal, and that samples have been exhumed vertically. However, isotherms mimic topography, albeit in a dampened fashion, with the impact of relief on isotherms becoming less with greater depth (e.g., Stüwe et al. 1994; see Chap. 8, Malusà and Fitzgerald 2018a). The depth to relevant isotherms is greater under ridges and less under valleys. The slope (apparent exhumation rate) of an age versus elevation profile in this situation will overestimate the true exhumation rate because ΔZ is less than ΔE (Fig. 9.5b). In the situation where exhumation is rapid and isotherms are compressed, the slope (the apparent exhumation rate) of the age-elevation profile will also overestimate the true exhumation rate (Fig. 9.5c). The wavelength of the topography matters, although the relief does not (Reiners et al. 2003), with a greater wavelength (a wider valley) having a greater effect on the slope of the age-elevation profile than a shorter wavelength (a narrow valley). In the situation where advection has not compressed the isotherms, it is possible to correct the slope of the age-elevation profile for topographic effects. This correction can be made for different T_c isotherms and hence is relevant for different thermochronologic techniques. Using the admittance ratio (α: the ratio of isotherm depth to topographic relief; Braun 2002), Reiners et al. (2003) presented a method to correct the apparent slope. The wavelength of topography is measured, and with respect to an appropriate closure isotherm, the admittance ratio (which is independent of topographic relief) is graphically determined. The assumption in these situations is that the paleotopography at the time when the ages were recorded has remained the same as it is today. An example: if $T_c = 100$ °C and the wavelength of topography is 10 km, then $\alpha = 0.1$. The slope (apparent exhumation rate) of the age-elevation plot is multiplied by $(1 - \alpha)$ to yield a "true exhumation rate." If the measured slope is ~ 150 m/Myr, then the exhumation rate is $150 \times (1 - 0.1) = \sim 135$ m/Myr. The

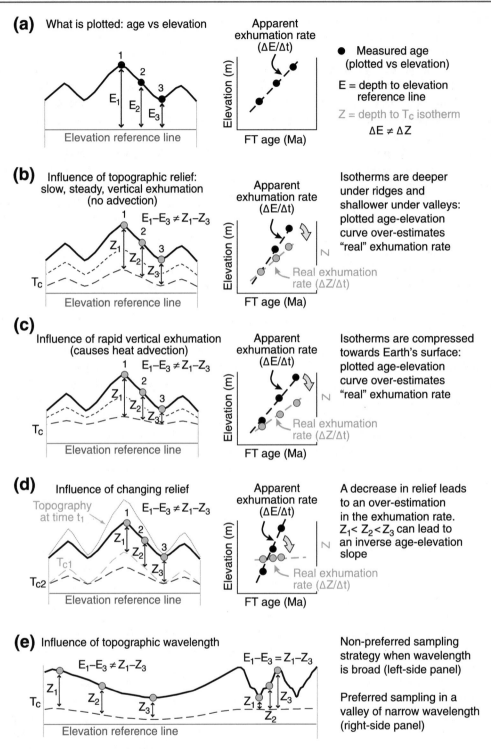

Fig. 9.5 Age-elevation scenarios, summarised from Stüwe et al. (1994), Mancktelow and Grasemann (1997), and Braun (2002). **a** The simple age-elevation model where ages are plotted against elevation. Implicit assumptions are that the thermal reference frame is horizontal, that ages reflect closure through a certain T_c, and that the particle path is vertical. The slope of the line is the "apparent exhumation rate." **b** The effects of topography on the subsurface isotherms. The depth to T_c isotherms is deeper under ridges than under valleys. As samples are not collected vertically, the plotted 1D age-elevation apparent exhumation rate will be an overestimate of the "real" exhumation rate. The topographic effect is more pronounced for low-temperature techniques than high temperature methods due to the greater deflection of the isotherms at shallow levels. **c** The effects of topography on the subsurface isotherms where exhumation is rapid, and isotherms are advected toward Earth's surface and compressed. The plotted 1D age-elevation apparent exhumation rate is an overestimate. **d** If surface topography changes dramatically over time, this will lead to an overestimation of the "true exhumation rate," plus in cases where $Z_1 < Z_2 < Z_3$, an inverse age-elevation relationship may also be recorded. **e** Vertical sampling profiles ideally should be collected from short-wavelength topography over the minimal horizontal distance

correction is greater for lower temperature techniques. Note however that the significance of being able to distinguish between a ∼150 m/Myr apparent exhumation rate and a ∼135 m/Myr exhumation rate, given the precision of individual ages and uncertainties on the age-elevation slope may be debatable. However, should there be age-elevation data from different techniques and with differing slopes; correcting these using this method may resolve the difference, with corrected slopes being similar.

The dimensionless Peclet number (*Pe*) can be used to quantitatively establish whether advection is the dominant form of heat transport (which will modify the age-elevation relationship) as compared to conduction (Batt and Braun 1997; Braun et al. 2006):

$$Pe = \dot{E}L/\kappa \tag{9.2}$$

where $\dot{E}$ = exhumation rate (km/Myr), L = thickness of the layer being exhumed (km), κ = thermal diffusivity of crustal rocks (km²/Myr). If *Pe* is much greater than 1, advection dominates, but if it is much less than 1, conduction will dominate.

There is a general rule to evaluating whether advection has played a role in the modification of the age-elevation slope. Not including the effects of topography, but if the slope of an age-elevation profile is less than ∼300 m/Myr (Parrish 1985; Brown and Summerfield 1997; Gleadow and Brown 2000), then advection is likely not a factor (also see Reiners and Brandon 2006, their Fig. 3).

In the situation where surface topography has evolved with time (Fig. 9.5d), where there has been a change of relief, the slope of the age-elevation profile will still overestimate the exhumation rate (Braun 2002). In extreme cases where $Z_1 < Z_2 < Z_3$, the slope of the age-elevation profile, collected over long-wavelength topography, may be inverted, and younger ages are found at higher elevations and older ages are found at lower elevations, even in the case where there is no observed fault offsets between samples. Note that in the same region, if a vertical sampling profile was collected from a cliff (representing short-wavelength topography), the slope of the age-elevation profile would be positive and the slope would represent an apparent mean exhumation rate (Braun 2002). Thus, collecting samples across topography with different wavelengths has the potential to yield different age-elevation relationships that may be interpreted quite differently, but yet have formed under similar conditions (Fig. 9.5e).

Development of sampling strategies and understanding various factors that affect the age-elevation relationship are critical for data interpretation. To augment geologic interpretations of thermochronologic data, quantitative modeling methods utilizing the heat transport equation can be used to constrain the rate of landscape evolution, that is, the shape of the surface topography and the rate at which it evolves (Braun 2002). Application of such methods usually requires sampling across different wavelengths on a regional scale. Quantitative modeling methods, such as the finite element code PeCube (Braun 2003) are powerful techniques, not only for interpreting existing data sets, but also for exploring the effects that exhumation (at different rates) and changing relief (constant, increasing, decreasing) have on the age-elevation relationship (e.g., Valla et al. 2010).

In the above discussion (and in Fig. 9.5), the assumption is that samples are exhumed vertically, and it is the rate of exhumation, the shape (and evolution) of topography, and the advection of isotherms that are the significant influences on the age-elevation relationship. However, samples are not always exhumed vertically, especially in areas of convergence where thrusting may play a role (e.g., ter Voorde et al. 2004; Lock and Willett 2008; Huntington et al. 2007; Metcalf et al. 2009). Notably, thrusting does not exhume rocks. It is erosion following the formation of topography during thrusting that exhumes and cools the rocks. Therefore, exhumation paths may have a significant lateral component. In this situation, the rock has a longer pathway to the surface, and this will affect the estimate of the exhumation rate. Huntington et al. (2007) explored this effect using a 3D finite element model, with topography modeled on the Himalaya. They also found that the shape of the isotherms as constrained by the shape of the orogen is important. Orogens are usually curvilinear in map view, with orogen perpendicular isotherms bent more than isotherms in a plane parallel to the orogen. With respect to vertical exhumation, the slope of the age-elevation profile from samples collected from gentle topographic slopes in the plane perpendicular to the orogen will overestimate considerably the "true" exhumation rate, dependent on rate of exhumation, topographic slope, and thermochronologic method used. For methods with lower T_c such as the AHe method, the apparent slope of the age-elevation profile can be much greater than the true exhumation rate. The discrepancy between the true modeled rate and the apparent slope is minimised when samples are collected from steeper topographic slopes and from age-elevation profiles oriented parallel to the trend of the mountain belt.

Different terminology may be used when samples are collected over considerable relief. "Vertical profile," "age-elevation profile," or "age-elevation relationships" (AER) are common. However, unless from a drill hole, samples collected over significant relief are never vertical, no matter how steep the side of a mountain may seem when sampled. Many factors, as discussed above, affect the slope and interpretation of an age-elevation profile, and it is important to design a sampling strategy appropriate to the questions being addressed. Some studies, for example, collect samples over a wide region, but may also plot ages versus elevation, to take advantage of the AER. In this case, care must be taken in data interpretation because samples do not likely fit the ideal age-elevation sampling criteria. The criteria can be summarised as: samples should be collected over significant relief, over a short horizontal distance, in short-wavelength topography and have, if possible, the profile oriented parallel to the trend of the mountains and not cross any known (or unknown) faults. In general, samples collected from young and active orogens (e.g., Himalaya—van der Beek et al. 2009) will yield young ages, heat advection must be taken into consideration as isotherms will be compressed, ages from thermochronologic methods with different T_c may be concordant, and an exhumed PAZ may only be preserved in the highest elevations dependent on the level of erosion. In this situation, the slope of the age-elevation profile is likely to greatly overestimate the true exhumation rate.

9.4 Examples of Exhumed PAZs and PRZs

In this section, we briefly introduce three examples where exhumed PAZs, and in some cases exhumed PRZs, are well exposed in age-elevation profiles. Each example offers data interpretation subtleties. The presentation of the data and the interpretations are based largely on the papers where these examples were first presented, but in some cases, there is more known about the geology or there is new thermochronologic information. Where appropriate, we briefly discuss the integration of new data while keeping within the objectives of the chapter.

Fig. 9.6 a Photograph of the southwestern flank of the Denali massif with sample locations (hollow dot means the sample is on the other side of the ridge) and AFT ages. Photograph by Paul Fitzgerald. b Tectonic sketch map of southern Alaska. YmP = Yakutat microplate. Modified from Haeussler (2008). c AFT age ($\pm 2\sigma$) versus elevation profile collected from the western flank of Denali. Representative CTLDs are shown with sample numbers, mean length (e.g., 13.8 µm) and standard deviation (e.g., 2.2 µm). The red asterisk marks the base of an exhumed PAZ and indicates the onset of rapid exhumation. CTLDs below the break in slope have long means (>14 µm) indicative of rapid cooling. In contrast, those above the break typically have shorter means, larger standard deviations, and more complex histograms often being bimodal, indicative of residence within the PAZ prior to rapid cooling (modified from Fitzgerald et al. 1993, 1995). d HeFTy inverse thermal models of select Denali samples. Good-fit paths lie within the magenta envelope and acceptable fit lie within the green envelope (see Ketcham 2005 for definitions). Modeling constraints are loose (a higher temperature T–t box and an ending low-temperature box) but c-axis projections for length measurements were not used on these original data from Fitzgerald et al. (1993,

9.4.1 Denali Profile ("Classic" Vertical Profile)

Denali, the highest mountain (6194 m) in the central Alaska Range and in North America, lies on the southern (concave) side of the McKinley restraining bend of the continental-scale right-lateral Denali fault system (Fig. 9.6a, b). Pacific–North American plate boundary forces in southern Alaska transfer stress inland causing slip along the Denali Fault (e.g., Plafker et al. 1992; Haeussler 2008; Jadamec et al. 2013). Stress is partitioned into fault-parallel and fault-normal components, with thrusting creating topography mainly at the restraining bends, to form the high mountains of the Alaska Range. Contrasting rheological properties of juxtaposed tectonostratigraphic terranes and suture zones formed during Mesozoic terrane accretion also play a role in the location of the highest topography and greatest exhumation (Fitzgerald et al. 2014). The Denali massif is largely defined by a granitic pluton, intruded at ∼60 Ma into country rock comprised largely of fine-grained Jurassic–Cretaceous metasediments (e.g., Reed and Nelson 1977; Dusel-Bacon 1994). The summit and upper ∼100 m of the mountain form a roof-pendant of these metasediments. The topography of the central Alaska Range is to a large extent defined by the erodibility of metasediments relative to the more resistant granitic plutons, as well as the relationship of the Denali Fault and the location of thrust faults (e.g., Fitzgerald et al. 1995, 2014; Haeussler 2008; Ward et al. 2012).

On the steep western flank of Denali, a vertical sampling profile covering ∼4 km of relief was collected from near the summit to the base of Mt Francis (Fig. 9.6a, c). As discussed above, sampling profiles are rarely vertical and even in such a steep and topographically impressive massif as Denali, the average slope along the sampling profile, between the summit and the base of Fission Ridge at 2500 m, is ∼24°. Approximately 45 samples were collected in this profile. An initial reconnaissance suite of 15 samples produced a

1995), and an average Dpar per sample (not per grain counted or each length measured) was applied. Yellow dots indicate the AFT age. Rapid cooling for higher elevation samples starts ∼1 Myr before the onset of rapid cooling/exhumation for those samples closer to the ∼6 Ma break in slope. e Diagram modified from Fitzgerald et al. (1995) showing the schematic late Miocene (prior to significant uplift) and present-day rock columns. By using geomorphic and sedimentological constraints, the late Miocene paleo-mean surface elevation (∼0.2 km) was estimated, which allows the amount of surface uplift to be calculated as ∼2.8 km (the present-day mean surface elevation is ∼3 km). The depth to the base of the paleo-PAZ was estimated as ∼4 km (using a pre-uplift paleogeothermal gradient of ∼25 °C/km) —and hence, as the base of the exhumed PAZ is now at ∼4.5, ∼8.5 km of rock uplift has occurred since the late Miocene. The amount of exhumation was estimated using Eq. 9.1 which yields ∼5.7 km (2.8 = 8.5–5.7 km). Exhumation using the method of Brown (1991) yields ∼5.75 km. Geological constraints and uncertainties on all of these figures are discussed in Fitzgerald et al. (1995). The summit of Denali was estimated to be 2.1–3 km below the surface in the late Miocene prior to the onset of significant uplift

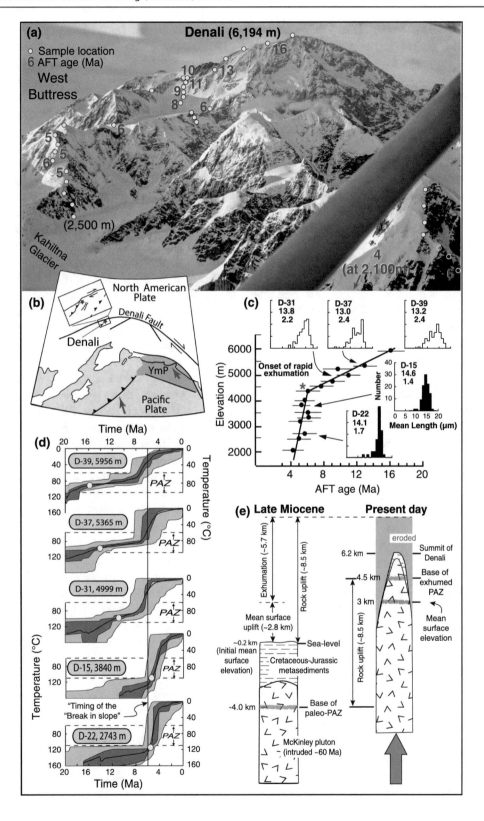

well-defined age-elevation profile (Fig. 9.6c), and the remaining samples were not processed. The uppermost sample, collected at the top of the Cassin Ridge, very close to the top of the pluton near the lithologic boundary with the metasediments, did not yield apatites, most likely because it was near the very edge of the pluton.

AFT ages decrease from ∼16 Ma near the summit of Denali (∼6 km elevation) to ∼4 Ma at ∼2 km elevation (Fig. 9.6c). A distinctive break in slope at ∼6 Ma occurs at an elevation of ∼4.5 km. Below this break in slope, for ages <6 Ma, CTLDs have means >14 μm and small standard deviations reflecting rapid cooling, although a few shorter tracks are present. Above this break in slope, samples with ages >6 Ma have CTLDs with means of ≤13.5 μm and larger standard deviations. These broader distributions, bimodal in cases, reflect shortening of older tracks while resident within a PAZ, in addition to the tracks formed after ∼6 Ma that are long and reflect rapid cooling. Fitzgerald et al. (1993, 1995) interpreted this break in slope as the base of an exhumed AFT PAZ, in effect representing the paleo ∼110 °C isotherm, and reflecting the transition from a time of relative thermal and tectonic stability to a time of rapid cooling, due to rapid denudation as a result of rapid rock uplift and formation of the central Alaska Range beginning ∼6 Ma (see discussion below).

Inverse thermal models undertaken using HeFTy (Ketcham 2005) are presented in Fig. 9.6d. These models confirm the qualitative interpretation above—long-term residence associated with slow cooling through the PAZ with rapid cooling beginning ∼6 Ma. The modeled rate of cooling for the older samples above the break in slope at ∼6 Ma is 2–3 °C/Myr, much slower than the modeled cooling rates for samples below the break, which are up to ∼50 °C/Myr. Samples below the break in slope will not necessarily indicate rapid cooling began at ∼6 Ma, for the very reason that these samples lie below the break in slope and hence started to cool after ∼6 Ma. The good-fit T–t envelopes for samples above the break suggest rapid cooling may have started closer to ∼7 Ma or even ∼8 Ma for the uppermost sample.

The elevation of the break in slope, in conjunction with a paleogeothermal gradient of ∼25 °C/km—estimated for the time of relative thermal and tectonic stability prior to the onset of rapid exhumation—was used to constrain the amount of exhumation at Denali since the late Miocene to ∼5.7 km. Gallagher et al. (2005) in their modeling study (see below) estimated the paleogeothermal gradient at Denali in the late Miocene as 24.7 °C/km. Fitzgerald et al. (1995) used geomorphic and sedimentological information to constrain the mean land-surface elevation in the late Miocene to ∼0.2 km. This allowed the late Miocene to recent amount and average rates of the following to be constrained at Denali (Fig. 9.6e): rock uplift (∼8.5 km with an average rate of ∼1.4 km/Myr), exhumation (∼5.7 km with an average rate of ∼1 km/Myr),

and surface uplift (∼2.8 km with an average rate of ∼0.5 km/Myr). The summit of Denali was approximately 2.1–3 km below the surface prior to late Miocene rapid exhumation, estimated by subtracting the difference between the elevation of the break in slope (4.5 km) and the summit elevation (6.2 km) from the depth below the surface of the base of the PAZ in the Miocene (3.8–4.7 km). A "horizontal" sampling transect collected near-perpendicular to the Denali Fault and across the central Alaska Range yielded progressively older AFT ages up to ∼37 Ma (Fitzgerald et al. 1995). With that trend, assuming that the slope of the age-elevation plot for samples >6 Ma maintained the same gentle slope for these older samples, the amount of rock uplift, exhumation, and surface uplift decreased to values of ∼3, ∼2, and ∼1 km on the southern side of the range.

The slope of the profile (∼160 m/Myr) above the break at ∼6 Ma represents an exhumed PAZ and as such does not represent an apparent exhumation rate. However, this slope was regarded by Fitzgerald et al. (1995) as being slightly too steep to represent a completely stable thermal and tectonic situation, and they interpreted this part of the profile as cooling at <3 °C/Myr. The apparent slope of the profile for samples younger than ∼6 Ma (<4.5 km elevation) is ∼1.5 km/Myr, steep enough that heat advection is likely significant. To confirm this, a Peclet number can be estimated (see Sect. 9.3.3). We use a layer thickness of ∼35 km, which represents crustal thickness south of the central Alaska Range, rather than the crustal thickness under the range which is ∼35–45 km (e.g., Veenstra et al. 2006; Brennan et al. 2011). This plus a thermal diffusivity of 25 km^2/Myr and an exhumation rate of ∼1 km/Myr yields a Peclet number of 1.4. Thus, the slope of the age-elevation profile below the break overestimates the exhumation rate, due to advection as well as topographic effects (see Fig. 9.5).

Gallagher et al. (2005) used the Denali data as a test case when modeling multiple samples in vertical profiles to constrain the maximum likelihood thermal history. They used a Markov chain Monte Carlo (MCMC) approach with a Bayesian test criteria to test for over-parameterisation and complexity to the thermal history. The results of that modeling confirmed the interpretation of age and track length data, i.e., slow cooling followed by an inferred increase in cooling rate between 7 and 5 Ma. When modeled individually, the higher elevation samples tended to imply the onset of rapid cooling were slightly earlier, as was the case for the HeFTy models presented in Fig. 9.6d.

The Denali AFT age-elevation profile has stood the test of time well. The simple interpretation of an exhumed PAZ representing a relatively stable thermal and tectonic setting prior to the onset of rapid cooling due to rock uplift and exhumation beginning in the late Miocene (∼6 Ma) remains essentially unchanged. However, more thermochronological data from other parts of the Alaska Range along the Denali

Fault reveal episodic cooling, with strong episodes beginning at ~ 25 and ~ 6 Ma, and slightly less predictable and often weaker episodes in the middle Miocene (~ 15 to ~ 10 Ma) (e.g., Haeussler et al. 2008; Benowitz et al. 2011, 2014; Perry 2013; Riccio et al. 2014; Fitzgerald et al. 2014). These episodic cooling events are inferred to result from the effects of plate boundary processes at the southern Alaska margin. These include collision of the Yakutat microplate, shallowing of the Yakutat slab that results in stronger coupling with the overriding North America plate, and changing relative plate motion between the Pacific and North American plates. In addition, in places Cretaceous cooling is revealed in low-temperature thermochronology data, possibly associated with early development of the Denali Fault. There is a strong ~ 50 and 40 Ma thermal signal in the regional data set that may be due to progressive east-to-west ridge subduction along the southern Alaskan margin (e.g., Trop and Ridgway 2007; Benowitz et al. 2012a; Riccio et al. 2014). At Denali itself, higher temperature thermochronological methods ($^{40}Ar/^{39}Ar$ multi-diffusion domain (MDD) modeling) reveal an episode of more rapid cooling beginning ~ 25 Ma and perhaps a lower magnitude event starting ~ 11 Ma (Benowitz et al. 2012b). As regards the Denali AFT profile, only the strong ~ 6 Ma event is revealed in the age-elevation profile, although the end of the ~ 25 Ma event may be revealed at ~ 20 Ma in the HeFTy model for sample D-39 from 5956 m (Fig. 9.6d). The possible ~ 11 Ma event is not significant enough to be revealed in the AFT age profile or within HeFTy models. However, this episode may contribute to the estimated pre-late Miocene cooling rate of 2–3 °C/Myr.

As new thermochronologic data are obtained from the Denali massif and the Alaska Range, more details about the temporal and spatial patterns of cooling and exhumation will be revealed. There are outstanding questions relating to the influence of various tectonic events, the stability of the McKinley restraining bend (Buckett et al. 2016) as well as the relative roles of terrane rheology and pre-existing structures versus fault geometry and partition of strain along the Denali Fault. The Denali AFT age-elevation profile provides a firm foundation for such studies.

9.4.2 Transantarctic Mountains: First Well-Defined Example of an Exhumed PAZ

The Transantarctic Mountains (TAM), stretching ~ 3000 km across Antarctica, are the world's longest non-contractional continental mountain range. The TAM formed along a fundamental lithospheric boundary between East and West Antarctica (e.g., Dalziel 1992). In the Ross Sea sector of Antarctica, the West Antarctic rift system lies on one side and cratonic East Antarctica on the other (Fig. 9.7a). The TAM reach elevations as high as ~ 4500 m and are typically 100–200 km wide. They can be envisaged as a number of asymmetric fault blocks dipping beneath the East Antarctic Ice Sheet, separated by transverse structures including transfer faults or accommodation zones. Outlet glaciers, typically large and draining the East Antarctic Ice Sheet, usually occupy these transverse structures.

The Dry Valleys region of southern Victoria (Fig. 9.7b) is permanently ice-free, exposing km-scale crustal sections in the valley walls. The TAM are faulted along their boundary with the West Antarctic rift system, where faults step-down ~ 2–5 km across the TAM Front (e.g., Barrett 1979; Fitzgerald 1992, 2002; Miller et al. 2010). Overall, the geology of the TAM appears relatively simple. This is because of the shallow inland-dipping nature of the range defined by Devonian-Triassic Beacon Supergroup strata and thick dolerite sills and basaltic volcanics of the Ferrar Dolerite and Kirkpatrick Basalt (the Ferrar large igneous province) intruded and extruded at ~ 180 Ma (e.g., Heimann et al. 1994). Unconformably, beneath these sedimentary and magmatic rocks are upper Proterozoic–Cambrian metamorphic rocks and Cambrian-Ordovician granites of the Granite Harbour Intrusives (e.g., Goodge 2007).

The TAM AFT data was instrumental in establishing the concept of an exhumed PAZ, notably the AFT age—elevation profile from Mt. Doorly in the Dry Valleys region (Fig. 9.7b, c) (Gleadow and Fitzgerald 1987). A ~ 800 m profile yielded AFT ages from ~ 83 to ~ 43 Ma. There is a

Fig. 9.7 a, b Map of Antarctica and part of the Dry Valleys area of southern Victoria Land indicating ice-free areas and faults (thin black lines), as well as location of the three vertical sampling profiles from this region. TAM = Transantarctic Mountains (black), WARS = West Antarctic rift system, SVL = southern Victoria Land, SHACK = Shackleton Glacier, WANT = West Antarctica, EANT = East Antarctica. **c–e** AFT age ($\pm 2\sigma$) versus elevation plots for Mt. Doorly, Mt. England and Mt. Barnes. The red asterisks mark the location of the break in slope indicative of the base of an exhumed PAZ and the onset of rapid cooling/exhumation. CTLDs are normalized to 100. **f** Cross section (x–x'—position shown in b) along the Mt. Doorly ridge (modified from Fitzgerald 1992) showing offset dolerite sills and idealized ~ 100 Ma AFT isochron delineating the structure across this part of the TAM Front. **g–h** AFT age ($\pm 2\sigma$) versus elevation plot for ▶ Mt. Munson in the Shackleton Glacier region, showing the onset of rapid cooling/exhumation at ~ 32 Ma and the offset of AFT ages due to faulting across the TAM Front. HeFTy inverse thermal models shown for selected samples from above the break in slope at Mt. Munson. Yellow dots = AFT ages. Models just above the break clearly show the onset of rapid cooling at ~ 32 Ma but this signal is lost only ~ 300 m above the break because the proportion of longer (rapidly cooled) to shorter (tracks residence for long periods of time in the PAZ) changes and hence for these higher elevation samples, the modeling simply indicates slow cooling since the AFT age of the sample. Figures are summarized from Gleadow and Fitzgerald (1987), Fitzgerald (1992, 2002) and Miller et al. (2010)

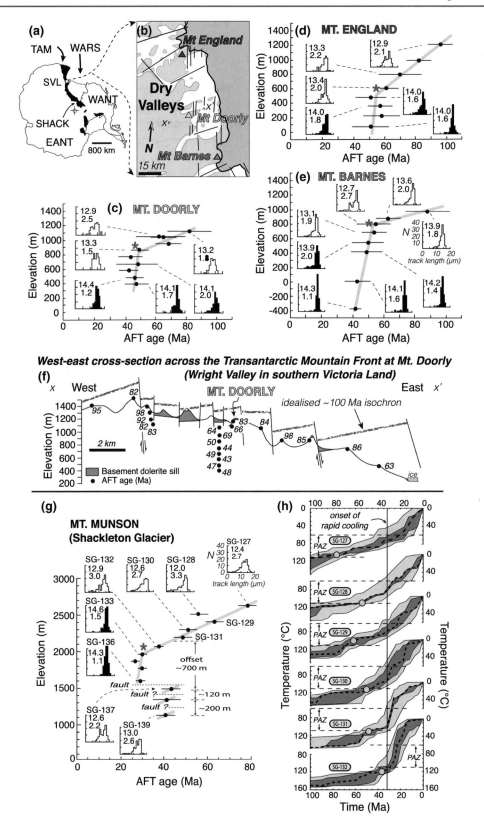

West-east cross-section across the Transantarctic Mountain Front at Mt. Doorly (Wright Valley in southern Victoria Land)

pronounced break in slope at an elevation of ∼800 m, corresponding to an AFT age of ∼50 Ma (Fig. 9.7c). CTLDs above the break in slope have the now-familiar shorter means (∼13 µm) with broader distributions reflected in larger standard deviations near ∼2 µm, with distributions often being bimodal. Below the break in slope, CTLDs have means >14 µm with narrow distributions reflected in their standard deviations, although there are once again still a few short tracks. Gleadow and Fitzgerald (1987) interpreted the break in slope as the base of a "fossil PAZ" that had been uplifted and preserved at higher elevations in the TAM, with uplift starting from ∼50 Ma. Samples above the break resided in the PAZ prior to ∼50 Ma, whereas samples below the break had an essentially zero age prior to the onset of uplift.

The Mt. Doorly age-elevation profile conclusively established the concept of an exhumed PAZ. This was due to a number of reasons:

- Gleadow and Fitzgerald (1987) renamed the "partial stability zone" as the PAZ, to denote what this pattern of ages and CTLDs represented, especially when compared to drill hole data where age and length patterns were similar (e.g., Naeser 1981; Gleadow and Duddy 1981; Gleadow et al. 1983).
- The rate of exhumation in the TAM is very slow, and the valley walls are glacially sculpted and therefore steep, so a vertical profile over only ∼800 m relief yielded a well-defined example of an exhumed PAZ.
- The Mt. Doorly profile had a greater concentration of samples (collected every ∼100 m of elevation), and this study was just underway when CTLDs were just starting to be routinely measured. So, while the variation of age with elevation is important and inflection points had been identified as paleoisotherms representing the base of a PAZ (Naeser 1976), age variation in conjunction with CTLDs made the interpretation of this age-elevation data more conclusive.
- The TAM study was in progress as the first thermal modeling programs—initially forward modeling, followed by inverse thermal modeling—were being developed. This allowed forward models to replicate the age-elevation trends and also the CTLDs. Interestingly, the interpretation of these data sets in the first papers was so obvious that these models were not presented, although various models started to be published in the early 1990s (e.g., Fitzgerald and Gleadow 1990; Fitzgerald 1992).

An important factor concerning interpretation and understanding of AFT data in the Dry Valleys region is the remarkable layer cake stratigraphy. Basement rocks are unconformably overlain by 2–3 km of sediments, both of which were then intruded in the Jurassic by thick (up to

300 m) sills and capped by basaltic lavas. Jurassic magmatism raised the geothermal gradient and completely reset fission-tracks during that time period for the Dry Valleys (Gleadow et al. 1984). However, some samples well inland in other parts of the TAM have been partially or not thermally reset by this Jurassic magmatism (e.g., Fitzgerald 1994; see also Chap. 13, Baldwin et al. 2018).

The level of erosion along the TAM Front is such that AFT has proven to be the best method to record the exhumation history of the TAM. The simple fault-block structure of the range usually means that the AFT results are predictable and reproducible. We demonstrate this by including two other vertical profiles from the Dry Valleys region. The Mt England (Fitzgerald 1992) and Mt. Barnes (Fitzgerald 2002) age-elevation profiles (Fig. 9.7d, e) are remarkably similar in form and age range with Mt. Doorly. All three profiles have similar CTLDs above and below the break in slope, with similar timing of the break in slope at 50–55 Ma, and with the same interpretation. The amount of exhumation and the average rates can be calculated using a variety of approaches: the depth to T_c from measured present-day geothermal gradients or assuming a typical "stable continental" paleogeothermal gradient, reconstruction of the stratigraphy above the break in slope, the AFT "stratigraphy" approach of Brown (1991) and the slope of the age-elevation profile below the break. The amount of exhumation since the early Cenozoic is on the order of ∼4.5 km near the edge of the TAM Front, at an average rate of ∼100 m/Myr.

The gentle slope of the age-elevation profile above the break, largely a relict of the shape of the PAZ formed prior to ∼50 Ma defining the AFT stratigraphy (see Sect. 9.3.2, also Chap. 10, Malusà and Fitzgerald 2018b; Chap. 11, Foster 2018; Chap. 13, Baldwin et al. 2018), can be used to constrain the location and offset of faults. This concept, well established now, but in its infancy then, was tested along the Mt. Doorly ridge (Fig. 9.7f) where offset dolerite sills are clearly visible and easily mapped, with vertical offsets well constrained (Gleadow and Fitzgerald 1987; Fitzgerald 1992). The slope of the upper part of the Doorly profile, the exhumed PAZ, is only ∼15 m/Myr so the variation of AFT age is significant for small elevation changes, well-suited to detect fault offsets. Reconstruction of the ∼100 Ma isochron, across mapped and unmapped faults, provides offset estimates on the displacements down to ∼200 m, taking into account uncertainties on AFT ages and variable dip of the faulted sill segments (Fig. 9.7f).

Other age-elevation profiles throughout the Dry Valleys also reveal exhumed PAZs. These profiles may have either only the gentle upper slope or the steeper lower part, dependent on their location across the TAM. Samples collected in isolation usually have ages and CTLDs that allows them to be placed in the context of the exhumed PAZ profile revealed in these three examples. However, the age pattern does become more complicated inland from the TAM Front because of earlier episodes of exhumation preserved in the AFT data set. Such patterns are revealed in profiles along the Ferrar Glacier west of Mt Barnes (Fitzgerald et al. 2006) and in other places along the TAM (see summary by Fitzgerald 2002 and also in Chap. 13, Baldwin et al. 2018). In some locations such as the Scott Glacier, there are multiple exhumed PAZs revealed in profiles across the mountains, which record episodic exhumation beginning in the Early Cretaceous, Late Cretaceous, and Early Cenozoic (Stump and Fitzgerald 1992; Fitzgerald and Stump 1997). Obviously, many factors including faulting, the orientation of the sampling profile with respect to structures, the steepness of the profile versus topographic effects and the possible variable effect of annealing during Jurassic magmatism may obfuscate the record.

We include in this section one more example of an exhumed PAZ from Mt. Munson (Fig. 9.7g), in the Shackleton Glacier region (Miller et al. 2010). This profile was collected starting just under the unconformity between the basement granites and the overlying Beacon Supergroup (see Chap. 13, Baldwin et al. 2018), and then down a ridge oriented across the grain of the orogen. Several small saddles along that ridge contain crush zones marking the locations of probable brittle faults. The age-elevation relationship reveals the classic form of an exhumed PAZ with CTLDs as discussed above, and a break in slope at $\sim$32 Ma. This break in slope is younger than those in the Dry Valleys, for two reasons (Fitzgerald 2002). The onset of rapid cooling marked by the break in slope becomes younger along the TAM, in a general southerly direction, from $\sim$55 to $\sim$40 Ma. Also, observed in several locations along the TAM, there is a coast-to-inland younging trend as a result of escarpment retreat (cf., Chap. 20, Wildman et al. 2018). Near the coast, the onset of rapid cooling is $\sim$40 Ma, decreasing to $\sim$32 Ma at Mt. Munson about 50 km inland. Faulting (downthrown toward the coast) is evident in the Mt. Munson age-elevation plot—as revealed by three samples with older AFT ages at lower elevations (<1500 m). These lower-elevation older samples have ages and CTLDs similar to those found at elevations above the break in slope (>2000 m), where ages are >32 Ma.

HeFTy inverse thermal models of the Mt. Munson profile (Fig. 9.7g) confirm the interpretation of an exhumed PAZ. But the main reason to present these is to show that in the models, the transition from long-term residence within the AFT PAZ (which could also be termed as slow cooling through the PAZ) prior to the onset of rapid cooling is only seen in samples lying just above the break in slope (e.g., SG-132). In samples only $\sim$300 m above the break (e.g., SG-130), the rapid cooling signal is not observed, but rather the models show monotonic cooling at an averaged rate. The lack of an observed rapid cooling signal beginning $\sim$32 Ma in these higher elevation/older samples results from the changing proportions of confined tracks in the distributions. In samples just above the break in slope, there is a much greater proportion of long tracks formed after $\sim$32 Ma, in comparison to tracks shortened while resident in the PAZ, so the model is able to constrain the onset of rapid cooling. In contrast, for samples $\sim$300 m higher in elevation above the break, AFT ages are already $\sim$60 Ma. In these samples, there is in essence a 50/50 split between tracks that underwent annealing for $\sim$30 Myr in the PAZ and those that record rapid cooling beginning $\sim$32 Ma, and modeling does not reveal the episodic cooling.

9.4.3 Gold Butte Block: Multiple Exhumed PAZs/PRZs

The Gold Butte Block (GBB) in southeastern Nevada lies within the eastern Lake Mead extensional domain, a zone of extension west of the Grand Canyon, Colorado Plateau, and east of the main part of the extended Basin and Range province (Fig. 9.8a–c). Bounding the west side of the GBB is the Lakeside Mine Fault, which forms the northern part of the $\sim$60 km long north–south striking South Virgin–White Hills detachment fault (e.g., Duebendorfer and Sharp 1998).

Fig. 9.8 a Location map for the Gold Butte Block (GBB) in southeastern Nevada. **b–c** Pre-extension ($\sim$20 Ma) restored cross section of the GBB and present-day cross section showing two models for the pre-extension geometry of the Lakeside Mine detachment fault. Model 1 is after Wernicke and Axen (1988) and Fryxell et al. (1992), and model 2 is from Karlstrom et al. (2010). Figure modified from Karlstrom et al. (2010), their Fig. 2. The $\sim$110 °C paleoisotherm is based on the restored location for the base of the restored AFT PAZ, and the $\sim$200 °C isotherm (in **b**, **c** and **e**) is based on K-feldspar MDD model results as outlined in Karlstrom et al. (2010). **d** AFT data plotted versus depth below the non-conformity between basement rocks and the Cambrian Tapeats Sandstone for models 1 and 2 of the pre-extension location of the Lakeside Mine detachment fault. This figure is modified from Fitzgerald et al. (2009), but with paleodepths (for model 1 which was used in their 2009 paper) slightly recalculated according to Karlstrom et al. (2010), their Fig. 2, and paleodepths for model 2 from that same figure. The ▶ age-paleodepth relationship clearly shows an exhumed PAZ, with characteristic CTLDs above and below the break in slope which marks the onset of rapid cooling due to tectonic exhumation as a result of extension beginning $\sim$17 Ma. **e** Age for different thermochronometers (AFT, ZFT, apatite, zircon and titanite (U–Th)/He—see text for references) plotted against paleodepths for model 1 and model 2. Uncertainties on ages are not included—to retain clarity for this figure. All methods, within error, constrain approximately the same age for the location of the base of the exhumed PAZs and PRZs, indicative of the onset of rapid cooling at $\sim$17 Ma. Paleodepths were slightly recalculated as described for (**d**). Paleogeothermal gradients are listed for each of the models, but only for the surface to the base of each PAZ/PRZ. Paleodepth estimation is fundamental to these paleogeothermal estimates, as discussed in the text

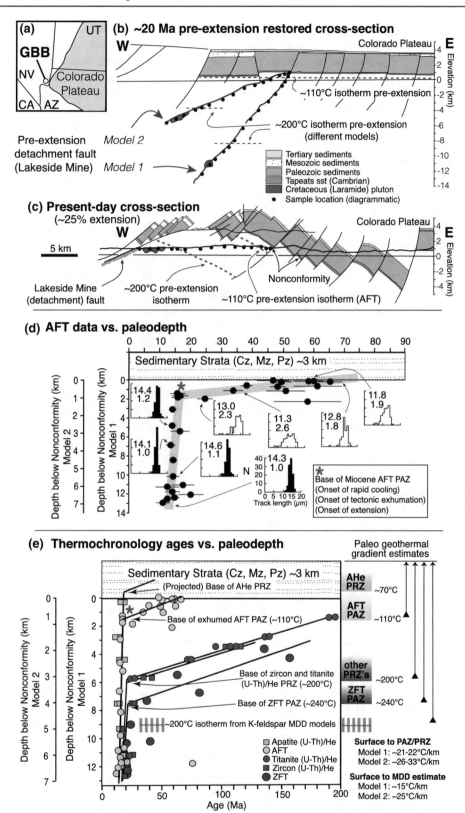

(a) GBB, UT, NV, Colorado Plateau, CA, AZ

Pre-extension detachment fault (Lakeside Mine)   *Model 2*
                                                   Model 1

(b) ~20 Ma pre-extension restored cross-section

W   Colorado Plateau E

~110°C isotherm pre-extension

~200°C isotherm pre-extension (different models)

Tertiary sediments
Mesozoic sediments
Paleozoic sediments
Tapeats sst (Cambrian)
Cretaceous (Laramide) pluton
• Sample location (diagrammatic)

Elevation (km): 4, 2, 0, -2, -4, -6, -8, -10, -12, -14

(c) Present-day cross-section
(~25% extension)

W   Colorado Plateau E

5 km

Lakeside Mine (detachment) fault   ~200°C pre-extension isotherm

Nonconformity

~110°C pre-extension isotherm (AFT)

Elevation (km): 4, 2, 0, -2, -4

(d) AFT data vs. paleodepth

0 10 20 30 40 50 60 70 80 90

Sedimentary Strata (Cz, Mz, Pz) ~3 km

Depth below Nonconformity (km) Model 2
Depth below Nonconformity (km) Model 1

14.4 / 1.2
14.1 / 1.0
13.0 / 2.3
14.6 / 1.1
11.3 / 2.6
12.8 / 1.8
11.8 / 1.9
14.3 / 1.0

N: 40, 30, 20, 10
0 5 10 15 20
Track length (µm)

★ Base of Miocene AFT PAZ
(Onset of rapid cooling)
(Onset of tectonic exhumation)
(Onset of extension)

(e) Thermochronology ages vs. paleodepth

Paleo geothermal gradient estimates

Sedimentary Strata (Cz, Mz, Pz) ~3 km

(Projected) Base of AHe PRZ

★ Base of exhumed AFT PAZ (~110°C)

Base of zircon and titanite (U-Th)/He PRZ (~200°C)

Base of ZFT PAZ (~240°C)

||||| ~200°C isotherm from K-feldspar MDD models

■ Apatite (U-Th)/He
○ AFT
● Titanite (U-Th)/He
■ Zircon (U-Th)/He
● ZFT

Depth below Nonconformity (km) Model 2
Depth below Nonconformity (km) Model 1

Age (Ma): 0, 50, 100, 150, 200

AHe PRZ   ~70°C
AFT PAZ   ~110°C
other PRZ's ~200°C
ZFT PAZ   ~240°C

Surface to PAZ/PRZ
Model 1: ~21–22°C/km
Model 2: ~26–33°C/km

Surface to MDD estimate
Model 1: ~15°C/km
Model 2: ~25°C/km

The GBB has been described as a tilted crustal section exposing ∼17 km of crust (Wernicke and Axen 1988; Fryxell et al. 1992; Brady et al. 2000) and as such has provided an ideal crustal section to which an almost complete suite of thermochronologic methods have been applied. The results from the various thermochronologic methods agree remarkably well, but there are some complications related to the determination of the paleodepth of samples. Estimates of the pre-extension paleogeothermal gradient are therefore also complicated, which we will also discuss briefly below.

One of the reasons the GBB is of such interest is because it was a key location for understanding the extensional structure of the crust (e.g., Wernicke and Axen 1988). During extension, the footwall of normal faults underwent isostatic uplift, the degree of which is important for evaluating contrasting models for the dip of normal faults during extension. The GBB was thought to represent an intact tilted crustal section with its western end exposing largely Proterozoic basement (paragneiss, orthogneiss, amphibolite, and various granitoids) from middle crustal levels. On its eastern side, the ∼50° easterly dipping Cambrian Tapeats Sandstone, the basal unit of the famous Grand Canyon stratigraphic section, unconformably overlies basement rocks. However, more recent structural reconstructions (Karlstrom et al. 2010) indicate that this titled crustal block model may be too simple. Their new model incorporates slip on initially subhorizontal detachment faults, such that the initial depths of the deepest footwall rocks under the Lakeside Mine Detachment are much shallower. Compared to the previous estimate of ∼17 km depth for the tilted crustal block model (shown as model 1 in Fig. 9.8), Karlstrom et al. (2010) proposed the pre-extension detachment extended to depths of ∼10 km (model 2). We plot in Fig. 9.8b the paleodepth of all samples, for model 1 (steeper initial dip of the detachment fault) and model 2 (shallower initial dip) according to the ∼20 Ma pre-extension structural reconstruction shown in Fig. 2 of Karlstrom et al. (2010).

AFT was the first thermochronologic method applied across the GBB. Samples were collected in basement rocks starting just below the unconformity and then west across the GBB and projected onto a line approximately parallel to the extension direction (Fitzgerald et al. 1991, 2009). The variation of AFT ages and CTLDs plotted versus paleodepth (Fig. 9.8d) reveals the classic form of an exhumed PAZ. The break in slope, middle Miocene in age (∼15–17 Ma) occurs ∼1.4 km below the unconformity indicating the onset of rapid cooling due to tectonic exhumation accompanying extension. Samples above the break were resident within or cooled slowly through the AFT PAZ. Thus, mean lengths are shorter (∼11–13 μm), distributions tend toward bimodality, and standard deviations are larger (∼2 μm or greater). In contrast, CTLDs below the break have longer means (>14 μm), unimodal distributions with smaller standard

deviations (∼1 μm) indicative of rapid cooling. Reiners et al. (2000) subsequently applied AHe dating across the GBB. No break in slope in the AHe ages was observed in samples collected from the basement (Fig. 9.8e), but a break would likely be expected within the overlying sedimentary strata. Reiners et al. (2000, 2002) did however observe inflection points in the zircon and titanite (U–Th)/He ages, interpreted as the base of exhumed PRZs. Reiners et al. (2000, 2002) used the reconstructed paleogeothermal gradient (∼20 °C/km) to constrain the T_c of these two systems, as did Bernet (2009) for zircon FT (ZFT) dating. The ZFT age-paleodepth pattern represented an exhumed zircon PAZ, with samples cooling slowly before rapid cooling began in the Miocene. Bernet (2009) also combined all the existing thermochronology data, noting a common breakpoint at ∼17 Ma, possibly slightly earlier (∼20 Ma) for the higher temperature methods, although this is probably statistically indistinguishable. Geologic studies in the region indicate that extension lasted from ∼17 to ∼14 Ma (Beard 1996; Brady et al. 2000; Lamb et al. 2010; Umhoefer et al. 2010). The thermochronologic constraints on the onset of extension agree well with the geologic studies, notably the sedimentary record in localised fault-bounded basins with interbedded ash layers, which offer a finer fidelity in terms of the timing of extension and tilting of strata.

Many of the thermochronologic studies described briefly for the GBB used the relationship of ages with paleodepth to constrain either: (i) the paleogeothermal gradient, using a Miocene mean-annual temperature of 10 °C, or (ii) the temperature of the base of a particular PAZ or PRZ, based on the estimate of paleogeothermal gradient and paleodepth of the break in slope. Thus, the location of a particular sample across the block and its paleodepth, depending on which structural model is used, is important. Outcrop exposures of the upper eastern end of the tilted section are quite complex, being offset in an en-echelon fashion. Mapped faults in the sedimentary section are also more easily identified than their possible extension into the underlying basement rocks. The determination of the map distance from the non-conformity to the sample and hence estimation of a pre-extension paleodepth slightly varies for each of the studies mentioned above, as explained by the authors in each case. Note that this middle Miocene paleogeothermal gradient was prior to the onset of extension and tectonic exhumation and erosion. During active tectonism, there would have been advection and a dynamic paleogeothermal gradient. In Fig. 9.8e, we plot the age (for the variety of techniques discussed) versus depth of each sample below the non-conformity. In essence, we have modified Fig. 3c from Fitzgerald et al. (2009), but recalculated the paleodepths according to Fig. 2 of Karlstrom et al. (2010). We use ∼3 km for the thickness of the pre-extension sedimentary section above the non-conformity. Model 1, with its steeper initial dip, yields paleogeothermal gradients of 21–23 °C/km, from the surface to the base of the various PAZs/PRZs, whereas model 2 with its shallower initial dip yields ∼26–33 °C/km. There is a greater disparity between the two models for the higher temperature methods, not only because of the varying model 1 versus model 2 paleodepths, but because of the greater uncertainty in sample location beneath the non-conformity. The estimates further diverge if the estimated ∼200 °C paleoisotherm constrained by MDD modeling of K-feldspar ^{40}Ar/^{39}Ar data is used (Karlstrom et al. 2010). The MDD data is complex but constrains a continuous temperature–time path of each sample between ∼250 and ∼175 °C. As discussed extensively in Karlstrom et al. (2010), there is a discrepancy between the location of the ∼200 °C isotherm constrained by these MDD modeling results and some of the other methods. Overall however the general age trends agree well. Some outlier older ages (i.e., those ages that are greater than ∼20 Ma and which lie structurally below the base of their respective PAZ/PRZs) across the GBB, such as a ∼77 Ma AFT age at the western (deeper) end (Fitzgerald et al. 1991) or some older ZFT ages to the west (Bernet 2009), plus 20–25 Ma zircon and titanite (U–Th)/He ages in the central part of the GBB, attest to structural complexity and possible incorporation of higher structural units within the fault zone while extension was ongoing.

The density of thermochronologic data across the GBB from a variety of different techniques, and we have not mentioned additional muscovite, biotite and hornblende ^{40}Ar/^{39}Ar data (Reiners et al. 2000; Karlstrom et al. 2010), or a more recent MDD K-feldspar study (Wong et al. 2014), has proven to be a great benchmark for the comparison and calibration of these different techniques. Notwithstanding the complications noted above, the concept of exhumed partial annealing and retention zones works well (Fig. 9.8e), notably as regards the timing for the onset of rapid cooling due to tectonic exhumation (∼17 Ma). Structural mapping plus outlier ages indicate the GBB is not a simple east-titled crustal block exposing mid-crustal level rocks on its western side. In detail, the GBB is more structurally complex, as reflected by varying estimates for the pre-extension paleogeothermal gradient which rely on structural models to constrain the paleodepth estimates of samples.

9.5 Summary and Conclusions

The collection of samples over significant relief and plotting age versus elevation to constrain the rates and timing of "uplift" (actually "exhumation") was among the earliest applications of the then-fledging low-temperature thermochronologic techniques, notably AFT dating (e.g., Wagner and Reimer 1972; Naeser 1976). This approach remains of fundamental importance in geologic and tectonic studies today, because rocks cool as they are exhumed. Thus,

samples collected over significant relief will reveal age variations, with ages generally increasing with increasing elevation. The thermal reference frame for data interpretation is dynamic and is modified by a number of factors:

- Surface topography modifies the shape of the isotherms under ridges and valleys, such that the isotherms are closer to the base of the valley than the top of the ridge. As a result, the slope of the age-elevation profile collected down the valley wall will overestimate the true exhumation rate.
- During rapid exhumation, isotherms are advected toward Earth's surface, and as a result, the slope of an age-elevation profile will overestimate the true exhumation rate.
- Changing relief will change the slope of an age-elevation profile, such that the measured slope will overestimate the true exhumation rate. When samples are collected across an evolving long-wavelength topography, an inverse age-elevation slope may sometimes be obtained.

The best sampling strategy to minimise these effects is to collect samples over significant relief, on steep terrain, parallel to the structural grain of an orogen (where the curvature of near-surface isotherms is less), and where faults are less likely to be crossed. Also, it is advisable to collect samples close enough together so that the form of a profile or variations in age versus elevation trends is clearly revealed, for example, in the case of an exhumed PAZ or PRZ. An important component of any study, both prior to sampling and during the interpretation stage, is to be aware of the factors that may change the slope of the profile and hence influence interpretation of the data. Defining age-elevation profiles as "vertical profiles" may signal an understanding of these factors, and hence, the application of a sampling strategy takes them into account. In some studies, describing and interpreting an age-elevation relationship where an appropriate sampling strategy was not employed have led to suspect interpretations that are potentially incorrect.

A PAZ/PRZ forms during a time of relative thermal and tectonic stability. The age-elevation slope within a PAZ/PRZ is a function of the paleogeothermal gradient and the time over which the PAZ/PRZ forms. It is not uncommon for the form of a PAZ/PRZ to develop during periods of slow cooling as samples move slowly through the PAZ/PRZ. Long-term residence within, or slow cooling through a PAZ/PRZ, magnifies the age variation between grains or samples, especially in case of varying compositions for AFT thermochronology, and/or zonation patterns or size differences between grains for apatite and zircon (U–Th)/He dating.

If, following a period of relative stability that allows development of a PAZ/PRZ in the rock record, there is an episode of rapid cooling/exhumation, an exhumed PAZ/PRZ may be revealed in an age-elevation profile. The break in slope (or inflection point) marks the base of the exhumed PAZ/PRZ, in effect a paleoisotherm. The timing of the break in slope marks the onset of the episode of rapid cooling/exhumation, although it is likely a slight underestimate. For AFT thermochronology, there is a distinctive pattern of CTLDs above and below the break in slope. Those above the break are broader and typically have shorter means (<13 μm), larger standard deviations (∼2 μm or greater) and are often bimodal with a systematic pattern of varying components of shorter versus longer tracks. CTLDs below the break are typically unimodal, have longer means (>14 μm) and smaller standard deviations (<1.5 μm).

We have illustrated three examples where age-elevation profiles have proven useful in developing and testing the concept of exhumed PAZ/PRZs and have shed light on the geologic and tectonic evolution of the study regions:

- In the classic Denali example, a vertical AFT age profile collected over ∼4 km relief reveals an exhumed PAZ, with a break in slope at ∼6 Ma and characteristic CTLDs above and below the break. The interpretation of data from this age-elevation profile is straightforward, and geomorphic and sedimentological constraints allow the pre-uplift mean land-surface elevation to be estimated, thus allowing rock uplift, surface uplift, and exhumation to be constrained at Denali and then extrapolated to the entire central Alaska Range.
- Age-elevation profiles for AFT data collected in the Dry Valleys region of the TAM were the first to clearly reveal an exhumed PAZ, notably with the assistance of CTLDs from above and below the break in slope. The steep glaciated terrain, the overall simple and well-exposed geology, the slow exhumation rate for samples below the break, density of samples plus multiple age-elevation profiles from a relatively small region, clearly established the concept and value of the exhumed PAZ approach. The significant age variation with elevation change for samples above the break in slope, tested against down-faulted dolerite sills along the Mt. Doorly ridge system, validated the concept of using AFT stratigraphy to constrain the relatively simple fault-block structure. Using one example from Mt. Munson in the Shackleton Glacier region, we show the applicability and reproducibility of the exhumed PAZ approach along the TAM, that in places reveals episodic exhumation with preservation of multiple exhumed PAZs separated by periods of relative thermal and tectonic stability.
- In the GBB of southeastern Nevada, ages from multiple techniques (AFT, ZFT, and (U–Th)/He on apatite, zircon and titanite) plotted versus pre-extension paleodepth in the footwall of a major detachment fault reveal exhumed PAZs and PRZs that mark the onset of rapid cooling due

to tectonic exhumation at ∼17 Ma. These data were either used to constrain pre-extension paleogeothermal gradients or alternatively constrain the temperature of the base of a particular PAZ/PRZ.

Acknowledgements PGF acknowledges research support from the Antarctic Research Centre of Victoria University of Wellington, the University of Melbourne, Syracuse University, and the National Science Foundation (Alaska, Antarctica and Gold Butte projects). PGF also thanks J. Pettinga and the Erskine Program at the University of Canterbury. Insightful and thorough reviews by Andrew Gleadow and Suzanne Baldwin and comments on various sections by Jeff Benowitz, Chilisa Shorten, and Thomas Warfel greatly improved this chapter.

References

Abbott LD, Silver EA, Anderson RS, Smith R, Ingle JC, Kling SA, Haig D, Small E, Galewsky J, Sliter W (1997) Measurement of tectonic surface uplift rate in a young collisional mountain belt. Nature 385:501–508

Baldwin SL, Lister GS (1998) Thermochronology of the South Cyclades shear zone, Ios, Greece; effects of ductile shear in the argon partial retention zone. J Geophys Res 103:7315–7336

Baldwin SL, Fitzgerald PG, Malusà MG (2018) Chapter 13. Crustal exhumation of plutonic and metamorphic rocks: constraints from fission-track thermochronology. In: Malusà MG, Fitzgerald PG (eds) Fission-track thermochronology and its application to geology. Springer, Berlin

Barrett PJ (1979) Proposed drilling in McMurdo Sound. Mem Nat Inst Polar Res, Spec Issue 13:231–239

Batt GE, Braun J (1997) On the thermomechanical evolution of compressional orogens. Geophys J Int 128:364–382

Beard LS (1996) Paleogeography of the Horse Spring Formation in relation to the Lake Mead fault system, Virgin Mountains, Nevada and Arizona. In: Bertatan KK (ed) Reconstructing the history of Basin and Range extension using sedimentology and stratigraphy, vol 303. Geological Society of America Special Paper, pp 27–60

Benowitz JA, Layer PW, Armstrong PA, Perry SE, Haeussler PJ, Fitzgerald PG, Vanlaningham S (2011) Spatial variations in focused exhumation along a continental-scale strike-slip fault: the Denali fault of the eastern Alaska Range. Geosphere 7:455

Benowitz JA, Haeussler PJ, Layer PW, O'Sullivan PB, Wallace WK, Gillis RJ (2012a) Cenozoic tectono-thermal history of the Tordrillo Mountains, Alaska: Paleocene-Eocene ridge subduction, decreasing relief, and late Neogene faulting. Geochem Geophys Geosys 13(4). https://doi.org/10.1029/2011gc003951

Benowitz JA, Bemis SP, O'Sullivan PB, Layer PW, Fitzgerald PG, Perry S (2012b) The Mount McKinley Restraining Bend: Denali Fault, Alaska. Geol Soc Am Abstr Programs 44(7):597

Benowitz JA, Layer PW, Vanlaningham S (2014) Persistent long-term (c. 24 Ma) exhumation in the Eastern Alaska Range constrained by stacked thermochronology. Geol Soc Lon Spec Publ 378:225–243

Bernet M (2009) A field-based estimate of the zircon fission-track closure temperature. Chem Geol 259:181–189

Bernet M, Garver JI (2005) Fission-track analysis of detrital zircon. Rev Mineral Geochem 58:205–238

Brady RJ, Wernicke B, Fryxell JE (2000) Kinematic evolution of a large-offset continental normal fault system, South Virgin Mountains, Nevada. Geol Soc Am Bull 112:1375–1397

Braun J (2002) Quantifying the effect of recent relief changes on age-elevation relationships. Earth Planet Sci Lett 200:331–343

Braun J (2003) Pecube: a new finite-element code to solve the 3D heat transport equation including the effects of a time-varying, finite amplitude surface topography. Comput Geosci 29:787–794

Braun J (2005) Quantitative constraints on the rate of landform evolution derived from low-temperature thermochronology. Rev Min Geochem 58:351–374

Braun J, van der Beek P, Batt G (2006) Quantitative thermochronology: numerical methods for the interpretation of thermochronological data. Cambridge University Press

Brennan P, Gilbert H, Ridgway KD (2011) Crustal structure across the central Alaska Range: Anatomy of a Mesozoic collisional zone. Geochem Geophys Geosyst 12:Q04010. https://doi.org/10.1029/2011GC003519

Brown R (1991) Backstacking apatite fission-track "stratigraphy": a method for resolving the erosional and isostatic rebound components of tectonic uplift histories. Geology 19:74–77

Brown RW, Summerfield MA (1997) Some uncertainties in the derivation of rates of denudation from thermochronologic data. Earth Surf Proc Land 22:239–248

Brown RW, Summerfield MA, Gleadow AJW (1994) Apatite fission track analysis: its potential for the estimation of denudation rates and implications for models of long-term landscape development. In: Kirby MJ (ed) Process models and theoretical geomorphology. Wiley, pp 23–53

Burkett CA, Bemis SP, Benowitz JA (2016) Along-fault migration of the Mount McKinley restraining bend of the Denali fault defined by late Quaternary fault patterns and seismicity, Denali National Park & Preserve, Alaska. Tectonophysics 693:489–506

Burtner RL, Nigrini A, Donelick RA (1994) Thermochronology of Lower Cretaceous source rocks in the Idaho-Wyoming thrust belt. AAPG Bull 78:1613–1636

Calk LC, Naeser CW (1973) The thermal effect of a basalt intrusion on fission tracks in quartz monzonite. J Geol 81:189–198

Ching-Ying L, Typhoon L, Lee CW (1990) The Rb-Sr isotopic record in Taiwan gneisses and its tectonic implications. Tectonophysics 183:129–143

Dalziel IWD (1992) Antarctica: a tale of two supercontinents. Annu Rev Earth Planet Sci 20:501–526

Dodson MH (1973) Closure temperatures in cooling geochronological and petrological systems. Contrib Mineral Petrol 40:259–274

Duebendorfer EM, Sharp WD (1998) Variation in extensional strain along-strike of the South Virgin-White Hills detachment fault: perspective from the northern White Hills, northwestern Arizona. Geol Soc Am Bull 110:1574–1589

Dumitru TA (2000) Fission-track geochronology. In: Noller JS, Sowers JM, Lettis WR (eds) Quaternary geochronology: methods and applications. Wiley, Hoboken, pp 131–155

Dusel-Bacon CE (1994) Metamorphic history of Alaska. In: Plafker G, Berg HC (eds) The geology of North America, v G-1 The Geology of Alaska. Geological Society of America, Boulder, CO, pp 495–533

Ehlers TA, Farley KA (2003) Apatite (U-Th)/He thermochronometry: methods and applications to problems in tectonic and surface processes. Earth Planet Sci Lett 206:1–14

England P, Molnar P (1990) Surface uplift, uplift of rocks, and exhumation of rocks. Geology 18:1173–1177

Farley KA (2002) (U-Th)/He dating: techniques, calibrations, and applications. In: Porcelli D, Ballentine CJ, Wieler R (eds) Noble gases in geochemistry and cosmochemistry, vol 47. Reviews Min Pet Soc Am, pp 819–844

Fitzgerald PG (1992) The Transantarctic Mountains of southern Victoria Land: the application of apatite fission track analysis to a rift shoulder uplift. Tectonics 11:634–662

Fitzgerald PG (1994) Thermochronologic constraints on post-Paleozoic tectonic evolution of the central Transantarctic Mountains, Antarctica. Tectonics 13:818–836

Fitzgerald PG (2002) Tectonics and landscape evolution of the Antarctic plate since Gondwana breakup, with an emphasis on the West Antarctic rift system and the Transantarctic Mountains. In: Gamble JA, Skinner DNB, Henrys S (eds) Antarctica at the close of a Millennium. Proceedings of the 8th international symposium on Antarctic Earth Science, vol 35. Royal Society of New Zealand Bulletin, pp 453–469

Fitzgerald PG, Gleadow AJW (1988) Fission-track geochronology, tectonics and structure of the Transantarctic Mountains in northern Victoria Land, Antarctica. Chem Geol 73:169–198

Fitzgerald PG, Gleadow AJW (1990) New approaches in fission track geochronology as a tectonic tool: examples from the Transantarctic Mountains. Nucl Tracks Radiat Meas 17:351–357

Fitzgerald PG, Stump E (1997) Cretaceous and Cenozoic episodic denudation of the Transantarctic Mountains, Antarctica: new constraints from apatite fission track thermochronology in the Scott Glacier region. J Geophys Res 102:7747–7765

Fitzgerald PG, Fryxell JE, Wernicke BP (1991) Miocene crustal extension and uplift in southeastern Nevada: constraints from apatite fission track analysis. Geology 19:1013–1016

Fitzgerald PG, Stump E, Redfield TF (1993) Late Cenozoic uplift of Denali and its relation to relative plate motion and fault morphology. Science 259:497–499

Fitzgerald PG, Sorkhabi RB, Redfield TF, Stump E (1995) Uplift and denudation of the central Alaska Range: a case study in the use of apatite fission-track thermochronology to determine absolute uplift parameters. J Geophys Res 100:20175–20191

Fitzgerald PG, Baldwin SL, O'Sullivan PB, Webb LE (2006) Interpretation of (U-Th)/He single grain ages from slowly cooled crustal terranes: a case study from the Transantarctic Mountains of southern Victoria Land. Chem Geol 225:91–120

Fitzgerald PG, Duebendorfer EM, Faulds JE, O'Sullivan PB (2009) South Virgin–White Hills detachment fault system of SE Nevada and NW Arizona: applying apatite fission track thermochronology to constrain the tectonic evolution of a major continental detachment fault. Tectonics 28. https://doi.org/10.1029/2007tc002194

Fitzgerald PG, Roeske SM, Benowitz JA, Riccio SJ, Perry SE, Armstrong PA (2014) Alternating asymmetric topography of the Alaska Range along the strike-slip Denali Fault: strain partitioning and lithospheric control across a terrane suture zone. Tectonics 33. https://doi.org/10.1002/2013tc003432

Fleischer RL, Price PB, Walker RM (1965) Effects of temperature, pressure, and ionization of the formation and stability of fission tracks in minerals and glasses. J Geophys Res 70:1497–1502

Flowers RM, Ketcham RA, Shuster DL, Farley KA (2009) Apatite (U-Th)/He thermochronometry using a radiation damage accumulation and annealing model. Geochim Cosmochim Acta 73:2347–2365

Foster DA (2018) Chapter 11. Fission-track thermochronology in structural geology and tectonic studies. In: Malusà MG, Fitzgerald PG (eds) Fission-track thermochronology and its application to geology. Springer, Berlin

Fryxell JE, Salton GG, Selverstone J, Wernicke B (1992) Gold Butte crustal section, South Virgin Mountains, Nevada. Tectonics 11:1099–1120

Gallagher K (2012) Transdimensional inverse thermal history modeling for quantitative thermochronology. J Geophys Res Solid Earth 117

Gallagher K, Brown RW, Johnson C (1998) Fission Track Analysis and its application to geological problems. Annu Rev Earth Planet Sci 26:519–572

Gallagher K, Stephenson J, Brown RW, Holmes C, Fitzgerald PG (2005) Low temperature thermochronology and modeling strategies for multiple samples 1: vertical profiles. Earth Planet Sci Lett 237:193–208

Garver JI, Brandon MT, Roden MMK, Kamp PJJ (1999) Exhumation history of orogenic highlands determined by detrital fission track thermochronology. Geol Soc London Spec Publ 154:283–304

Gleadow AJW (1990) Fission track thermochronology—reconstructing the thermal and tectonic evolution of the crust. In: Pacific Rim Congress, Gold Coast, Queensland, 1990. Australasian Institute of Mining Metallurgy, pp 15–21

Gleadow AJW, Brown RW (2000) Fission track thermochronology and the long term denudational response to tectonics. In: Summerfield MA (ed) Geomorphology and global tectonics. Wiley, NY, pp 57–75

Gleadow AJW, Duddy IR (1981) A natural long term annealing experiment for apatite. Nucl Tracks Radiat Meas 5:169–174

Gleadow AJW, Fitzgerald PG (1987) Uplift history and structure of the Transantarctic Mountains: new evidence from fission track dating of basement apatites in the Dry Valleys area, southern Victoria Land. Earth Planet Sci Lett 82:1–14

Gleadow AJW, Duddy IR, Lovering JF (1983) Fission track analysis: a new tool for the evaluation of thermal histories and hydrocarbon potential. APEA J 23:93–102

Gleadow AJW, McKelvey BC, Ferguson KU (1984) Uplift history of the Transantarctic Mountains in the Dry Valleys area, southern Victoria Land, Antarctica, from apatite fission track ages. NZ J Geol Geophys 27:457–464

Gleadow AJW, Duddy IR, Green PF, Hegarty KA (1986) Fission track lengths in the apatite annealing zone and the interpretation of mixed ages. Earth Planet Sci Lett 78:245–254

Goodge JW (2007) Metamorphism in the Ross orogen and its bearing on Gondwana margin tectonics. Geol Soc Am Spec Pap 419:185–203

Green PF, Durrani SA (1977) Annealing studies of tracks in crystals. Nucl Tracks Radiat Meas 1:33–39

Green PF, Duddy IR, Gleadow AJW, Tingate PR, Laslett GM (1985) Fission-track annealing in apatite: track length measurements and the form of the Arrhenius plot. Nucl Tracks Radiat Meas 10:323–328

Green P, Duddy I, Gleadow A, Tingate P, Laslett G (1986) Thermal annealing of fission tracks in apatite: 1. A qualitative description. Chem Geol Isotope Geosci 59:237–253

Green P, Duddy I, Laslett G, Hegarty K, Gleadow A, Lovering J (1989) Thermal annealing of fission tracks in apatite 4. Quantitative modelling techniques and extension to geological timescales. Chem Geol Isotope Geosci 79:155–182

Haeussler PJ (2008) An overview of the neotectonics of interior Alaska: far-field deformation from the Yakutat microplate collision. In: Freymueller JT, Haeussler PJ, Wesson RL, Ekström G (eds) Active tectonics and seismic potential of Alaska, vol 179. American Geophysical Union Monograph, pp 83–108. https://doi.org/10.1029/179gm05

Haeussler PJ, O'Sullivan PB, Berger AL, Spotila JA (2008) Neogene exhumation of the Tordrillo Mountains, Alaska, and correlations with Denali (Mount Mckinley). In: Freymueller JT, Haeussler PJ, Wesson RL, Ekström G (eds) Active tectonics and seismic potential of Alaska, vol 179. American Geophysical Union Monograph, pp 269–285. https://doi.org/10.1029/179gm15

Heimann A, Fleming TH, Elliot DH, Foland KA (1994) A short interval of Jurassic continental flood basalt volcanism in Antarctica as demonstrated by ^{40}Ar/^{39}Ar geochronology. Earth Planet Sci Lett 121:19–41

Huntington KW, Ehlers TA, Hodges KV, Whipp DM (2007) Topography, exhumation pathway, age uncertainties, and the interpretation of thermochronometer data. Tectonics 26

Hurford AJ (2018) Chapter 1. An historical perspective on fission-track thermochronology. In: Malusà MG, Fitzgerald PG (eds) Fission-track thermochronology and its application to geology. Springer, Berlin

Jadamec MA, Billen MI, Roeske SM (2013) Three-dimensional numerical models of flat slab subduction and the Denali fault driving deformation in south-central Alaska. Earth Planet Sci Lett 376:29–42

Kamp PJJ, Tippett JM (1993) Dynamics of Pacific plate crust in the South Island (New Zealand) zone of oblique continent-continent convergence. J Geophys Res: Solid Earth 98:16105–16118

Karlstrom KE, Heizler M, Quigley MC (2010) Structure and ^{40}Ar/^{39}Ar K-feldspar thermal history of the Gold Butte block: reevaluation of the tilted crustal section model. Geol Soc Am Spec Pap 463:331–352

Ketcham RA (2005) Forward and inverse modeling of low temperature thermochronometry data. Rev Mineral Geochem 58:275–314

Ketcham RA (2018) Chapter 3. Fission-track annealing: from geologic observations to thermal history modeling. In: Malusà MG, Fitzgerald PG (eds) Fission-track thermochronology and its application to geology. Springer, Berlin

Ketcham R, Carter A, Donelick R, Barbarand J, Hurford A (2007) Improved modeling of fission-track annealing in apatite. Am Mineral 92:799–810

Ketcham RA, Gautheron C, Tassan-Got L (2011) Accounting for long alpha-particle stopping distances in (U–Th–Sm)/He geochronology: refinement of the baseline case. Geochim Cosmochim Acta 75:7779–7791

Lamb MA, Martin KL, Hickson TA, Umhoefer PJ, Eaton L (2010) Stratigraphy and age of the Lower Horse Spring Formation in the Longwell Ridges area, southern Nevada: implications for tectonic interpretations. Geol Soc Am Spec Pap 463:171–201

Lock J, Willett S (2008) Low-temperature thermochronometric ages in fold-and-thrust belts. Tectonophysics 456:147–162

Malusà MG, Fitzgerald PG (2018a) Chapter 8. From cooling to exhumation: setting the reference frame for the interpretation of thermochronologic data. In: Malusà MG, Fitzgerald PG (eds) Fission-track thermochronology and its application to geology. Springer, Berlin

Malusà MG, Fitzgerald PG (2018b) Chapter 10. Application of thermochronology to geologic problems: bedrock and detrital approaches. In: Malusà MG, Fitzgerald PG (eds) Fission-track thermochronology and its application to geology. Springer, Berlin

Mancktelow NS, Grasemann B (1997) Time-dependent effects of heat advection and topography on cooling histories during erosion. Tectonophysics 270:167–195

Meesters AGCA, Dunai TJ (2002) Solving the production-diffusion equation for finite diffusion domains of various shapes (part II): application to cases with a-ejection and non-homogeneous distribution of the source. Chem Geol 186:347–363

Metcalf JR, Fitzgerald PG, Baldwin SL, Muñoz JA (2009) Thermochronology in a convergent orogen: constraints on thrust faulting and exhumation from the Maladeta Pluton in the Axial Zone of the Central Pyrenees. Earth Planet Sci Lett 287:488–503

Miller SR, Fitzgerald PG, Baldwin SL (2010) Cenozoic range-front faulting and development of the Transantarctic Mountains near Cape Surprise, Antarctica: thermochronologic and geomorphologic constraints. Tectonics 29. https://doi.org/10.1029/2009tc002457

Moore MA, England PC (2001) On the inference of denudation rates from cooling ages of minerals. Earth Planet Sci Lett 185:265–284

Naeser CW (1976) Fission track dating. USGS Open-File Report, pp 76–190

Naeser CW (1979) Thermal history of sedimentary basins: fission track dating of subsurface rocks. In: Scholle PA, Schluger PR (eds) Aspects of diagensis, vol 26, Spec Pub Soc Econ Geol Paleo Min, pp 109–112

Naeser CW (1981) The fading of fission-tracks in the geologic environment—data from deep drill holes. Nucl Tracks Radiat Meas 5:248–250

Naeser C, Faul H (1969) Fission track annealing in apatite and sphene. J Geophys Res 74:705–710

Parrish RR (1985) Some cautions which should be exercised when interpreting fission track and other dates with regard to uplift rate calculations. Nucl Tracks Radiat Meas 10:425

Perry S (2013) Thermotectonic evolution of the Alaska Range: low-temperature thermochronologic constraints. PhD thesis, Syracuse University, 204 p

Plafker G, Naeser CW, Zimmerman RA, Lull JS, Hudson T (1992) Cenozoic uplift history of the Mount McKinley area in the central Alaska Range based on fission track dating. USGS Bull 2041:202–212

Reed BL, Nelson SW (1977) Geologic map of the Talkeetna quadrangle, Alaska. USGS Misc. Field Studies Map 870-A

Reiners PW, Brandon MT (2006) Using thermochronology to understand orogenic erosion. Annu Rev Earth Planet Sci 34:419–466

Reiners PW, Farley KA (2001) Influence of crystal size on apatite (U–Th)/He thermochronology: an example from the Bighorn Mountains, Wyoming. Earth Planet Sci Lett 188:413–420

Reiners PW, Brady R, Farley KA, Fryxell JE, Wernicke B, Lux D (2000) Helium and argon thermochronometry of the Gold Butte Block, south Virgin Mountains, Nevada. Earth Planet Sci Lett 178:315–326

Reiners PW, Farley KA, Hickes HJ (2002) He diffusion and (U–Th)/He thermochronometry of zircon: initial results from Fish Canyon Tuff and Gold Butte. Tectonophysics 349:297–308

Reiners PW, Zhou Z, Ehlers TA, Changhai X, Brandon MT, Donelick RA, Nicolescu S (2003) Post-orogenic evolution of the Dabie Shan, eastern China, from (U-Th)/He and fission track thermochronology. Am J Sci 303:489–518

Riccio SJ, Fitzgerald PG, Benowitz JA, Roeske SM (2014) The role of thrust faulting in the formation of the eastern Alaska Range: thermochronological constraints from the Susitna Glacier Thrust Fault region of the intracontinental strike-slip Denali Fault system. Tectonics 33. https://doi.org/10.1002/2014tc003646

Schildgen T, van der Beek P (2018) Chapter 19. Application of low-temperature thermochronology to the geomorphology of orogenic systems. In: Malusà MG, Fitzgerald PG (eds) Fission-track thermochronology and its application to geology. Springer, Berlin

Stump E, Fitzgerald PG (1992) Episodic uplift of the Transantarctic Mountains. Geology 20:161–164

Stüwe K, Hintermüller M (2000) Topography and isotherms revisited: the influence of laterally migrating drainage divides. Earth Planet Sci Lett 184:287–303

Stüwe K, White L, Brown R (1994) The influence of eroding topography on steady-state isotherms: application to fission track analysis. Earth Planet Sci Lett 124:63–74

ter Voorde M, de Bruijne CH, Cloetingh SAPL, Andriessen PAM (2004) Thermal consequences of thrust faulting: simultaneous versus successive fault activation and exhumation. Earth Planet Sci Lett 223:395–413

Trop JM, Ridgway KD (2007) Mesozoic and Cenozoic tectonic growth of southern Alaska: a sedimentary basin perspective. Geol Soc Am Spec Pap 431:55–94

Umhoefer PJ, Beard LS, Martin KL, Blythe N (2010) From detachment to transtensional faulting: a model for the Lake Mead extensional domain based on new ages and correlation of subbasins. Geol Soc Am Spec Pap 463:371–394

Valla PG, Herman F, van der Beek PA, Braun J (2010) Inversion of thermochronological age-elevation profiles to extract independent estimates of denudation and relief history—I: theory and conceptual model. Earth Planet Sci Lett 295:511–522

van der Beek P, van Melle J, Guillot S, Pêcher A, Reiners PW, Nicolescu S, Latif M (2009) Eocene Tibetan plateau remnants preserved in the northwest Himalaya. Nat Geosci 2:364–368

Veenstra E, Christensen DH, Abers GA, Ferris A (2006) Crustal thickness variation in south-central Alaska. Geology 34: 781–784

Wagner GA, Reimer GM (1972) Fission track tectonics: the tectonic interpretation of fission track apatite ages. Earth Planet Sci Lett 14:263–268

Wagner GA, Reimer GM, Jäger E (1977) Cooling ages derived by apatite fission-track, mica Rb-Sr and K-Ar dating: the uplift and cooling history of the Central Alps. Mem Inst Geol Mineral Univ Padova 30:1–27

Ward DJ, Anderson RS, Haeussler PJ (2012) Scaling the Teflon Peaks: rock type and the generation of extreme relief in the glaciated western Alaska Range. J Geophys Res 117:1–20. https://doi.org/10. 1029/2011JF002068

Wildman M, Beucher R, Cogné N (2018) Chapter 20. Fission-track thermochronology applied to the evolution of passive continental margins. In: Malusà MG, Fitzgerald PG (eds) Fission-track thermochronology and its application to geology. Springer, Berlin

Wernicke B, Axen GJ (1988) On the role of isostasy in the evolution of normal fault systems. Geology 16:848–861

Wolf RA, Farley KA, Kass DM (1998) Modeling of the temperature sensitivity of the apatite (U-Th)/He thermochronometer. Chem Geol 148:105–114

Wong M, Roesler D, Gans PB, Zeitler PK, Idleman BD (2014) Field calibration studies of continuous thermal histories derived from multiple diffusion domain (MDD) modeling of $^{40}Ar/^{39}Ar$ K-feldspar analyses at the Grayback and Gold Butte Normal Fault Blocks, US Basin and Range. Am Geophys Union, Fall Meeting, abstract #EP21A-3521

Application of Thermochronology to Geologic Problems: Bedrock and Detrital Approaches

10

Marco G. Malusà and Paul G. Fitzgerald

Abstract

Low-temperature thermochronology can be applied to a wide range of geologic problems. In this chapter, we provide an overview of different approaches, underlying assumptions and suitable sampling strategies for bedrock and detrital thermochronologic analyses, with particular emphasis on the fission-track (FT) method. Approaches to bedrock thermochronology are dependent on the goals of the project and the regional geologic setting, and include application of: (i) multiple methods (e.g., FT, (U–Th)/He and U–Pb) on various mineral phases (e.g., apatite and zircon) from the same sample, (ii) single methods on multiple samples collected over significant relief or across a geographic region (regional approach) or (iii) multiple methods on multiple samples. The cooling history of rock samples can be used to constrain exhumation paths and provides thermochronologic markers to determine fault offset, timing of deformation and virtual tectonic configurations above the present-day topography. Detrital samples can be used to constrain erosion patterns of sediment source regions on both short-term (10^3–10^5 yr) and long-term (10^6–10^8 yr) timescales, and their evolution through time. The full potential of the detrital thermochronology approach is best exploited by the integrated analysis of samples collected from a stratigraphic succession, samples of modern sediment and independent mineral fertility determinations.

10.1 Bedrock Thermochronology Studies

Low-temperature thermochronology studies fall into two general categories: those where samples are collected from bedrock and those where samples are essentially detrital. Within each of these two groups there are also natural subdivisions such as: for bedrock studies, applying a key-locality/outcrop approach (e.g., collecting over significant relief (the vertical profile approach) to determine age-elevation relationships) and/or a more regional sampling approach, all of which are dependent on the objectives of the study, the available geologic information, if other methods are also to be applied and if subsequent thermal modelling might also be applied. The geologic problem being addressed ideally helps shape the sampling strategy. Typically, the part of a study that has the longest longevity is the data itself. For example, modelling methods change and evolve, but the data remains the same, all things being equal.

10.1.1 Multiple-Method Versus Age-Elevation Approach for the Analysis of Exhumation Rates

When bedrock samples are collected on a regional basis, an exhumation rate for single samples may be estimated by assuming that the ages represent closure temperature (T_c) ages, and by assuming (or knowing) the paleogeothermal gradient to calculate depth to T_c. Hence, the depth divided by the ages yields an apparent exhumation rate (e.g., Reiners and Brandon 2006). This approach is not applied quite so much to basement samples, especially if inverse thermal models are available (Chap. 3, Ketcham 2018). However, it still remains the fundamental approach for many detrital thermochronology studies (e.g., Garver et al. 1999) (see Sect. 10.2).

More common approaches in studies utilising low-temperature thermochronology on bedrock samples involve the combination of multiple methods with different T_c, and sampling along vertical profiles (Fig. 10.1). The

M. G. Malusà (✉)
Department of Earth and Environmental Sciences, University of Milano-Bicocca, Piazza della Scienza 4, 20126 Milan, Italy
e-mail: marco.malusa@unimib.it

P. G. Fitzgerald
Department of Earth Sciences, Syracuse University, Syracuse, NY 13244, USA

© Springer International Publishing AG, part of Springer Nature 2019
M. G. Malusà and P. G. Fitzgerald (eds.), *Fission-Track Thermochronology and its Application to Geology*,
Springer Textbooks in Earth Sciences, Geography and Environment, https://doi.org/10.1007/978-3-319-89421-8_10

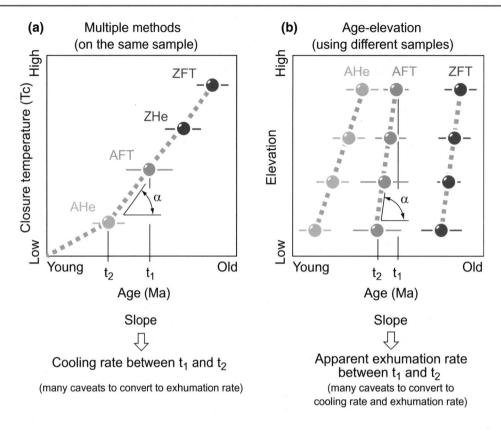

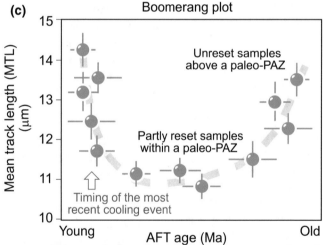

Fig. 10.1 Alternative approaches to constrain the rate of exhumation from bedrock samples. **a** Multiple-method approach. It is based on the analysis of different mineral/systems with progressively lower closure temperatures (T_c) on the same sample. The gradient of the slope in the diagram is a function of the average cooling rate between time t_1 and t_2. The paleogeothermal gradient during cooling has to be independently known in order to convert this average cooling rate to an average exhumation rate. **b** Age-elevation (vertical profile) approach. It is based on the analysis of an array of samples collected from top-to-bottom of significant relief and utilising the same dating method. When data are plotted on an age-elevation diagram, the gradient of the slope provides an apparent average exhumation rate for the whole time interval between the age yielded by the uppermost (t_1) and the lowermost (t_2) samples, or for individual discernable portions of the age-elevation profile that may have different slopes. Multiple methods can, and often are applied to the same array of samples to constrain the exhumation history over a longer time interval. **c** Boomerang plot that can be used to detect partly reset or unreset samples lying within or above a paleo-PAZ, and determine the onset of the most recent cooling event (based on Gallagher and Brown 1997). Acronyms: AFT, apatite fission track; AHe, apatite (U–Th)/He; ZFT, zircon fission track; ZHe, zircon (U–Th)/He

multiple-method approach (Fig. 10.1a) is based on the analysis of different thermochronologic systems in minerals retrieved from the same sample (e.g., Wagner et al. 1977; Hurford 1986; Moore and England 2001). Different mineral/systems with progressively lower T_c generally yield progressively younger thermochronologic ages, constraining a cooling history that is generally related to the exhumation of the sample towards the Earth's surface. In a diagram showing T_c versus the thermochronologic ages yielded by the analysed sample (Fig. 10.1a), the gradient of the slope is a function of the cooling rate in the selected time interval and is steeper for faster cooling. Converting this average cooling rate into an exhumation rate is not straightforward, as the paleogeothermal gradient during cooling has to be independently known (see Chap. 8, Malusà and Fitzgerald 2018). The multiple-method approach is generally more applicable for samples that have cooled across a wider temperature range (i.e., several hundreds of °C), which makes this approach suitable for the analysis of rocks exhumed from greater depths (see Chap. 13, Baldwin et al. 2018). Higher temperature thermochronologic methods are slower to respond to changes in exhumation rates (Moore and England 2001). Thus, if there is a step-wise increase in erosion rates at the surface, different methods on the same sample may give the appearance of a gradually increasing erosion rate with time as the T_c isotherms migrate at different rates to their new steady-state depths.

The *age-elevation (vertical profile) approach* is based on the analysis of an array of samples collected across a significant elevation range (and a short horizontal distance) employing the same dating method (e.g., Wagner and Reimer 1972; Naeser et al. 1983; Gleadow and Fitzgerald 1987; Fitzgerald and Gleadow 1988, 1990; Fitzgerald et al. 1995; Ruiz et al. 2009). During exhumation from depths greater than the T_c isothermal surface, samples collected from higher elevations cross the T_c isothermal surface before those from lower elevations, leading to a normal age-elevation relationship (i.e., samples collected at higher elevation have older ages). The gradient of the slope in an age-elevation diagram provides an apparent average exhumation rate for the time interval between the age yielded by the uppermost sample (t_1 in Fig. 10.1b) and the age yielded by the lowermost one (t_2 in Fig. 10.1b). The slope is generally steeper for faster exhumation rates. This estimate does not require any explicit assumption about the paleogeothermal gradient, but converting an apparent exhumation rate to a true exhumation rate using the age-elevation approach is not straightforward and requires a careful analysis of the relationships between the topography and the thermal reference frame (see Chap. 9, Fitzgerald and Malusà 2018).

An important assumption underlying the age-elevation approach is that the spatial relationships between samples have remained unchanged since the time of exhumation. Therefore, this approach should not be applied, for example, across faults that may have modified these original relationships: Samples from different fault blocks must be considered separately. Conversely, this approach can be used to detect faults between samples with contrasting or unexpected ages. Additionally, the difference in elevation between the lowest and the highest samples should be sufficiently large to allow a significant spread in thermochronologic ages, at least compared to the standard error associated with each age. This is particularly important for rapidly exhuming areas, where thermochronologic ages tend to be invariant with elevation. The most favourable conditions for a successful application of the age-elevation method are thus represented by a combination of high topographic relief and relatively slow exhumation rates, which allows for a larger spread of thermochronologic ages in the array of analysed samples. However, the time interval constrained by the age-elevation approach using one dating method only (t_1 to t_2 in Fig. 10.1b) may be quite short. In order to constrain the exhumation history over a longer time interval, multiple methods may be applied to the same array of samples (Fig. 10.1b). The time intervals constrained by different thermochronologic methods may either overlap, thus providing complementary constraints to specific segments of the exhumation path, or leave unconstrained time intervals between different methods, as often observed when the relief covered by samples is not sufficiently large. Different age-elevation slopes may be recorded for different thermochronologic methods even if the same exhumation rate is ongoing. This is because of the effects of topography and the bending of isothermal surfaces under ridges and valleys (e.g., Stüwe et al. 1994; Mancktelow and Grasemann 1997; Braun 2002), which can be investigated by suitable sampling strategies as discussed in Chaps. 9 (Fitzgerald and Malusà 2018) and 19 (Schildgen and van der Beek 2018).

The vertical spacing between samples is also important, and should be sufficiently small to highlight any change in slope, either ascribed to variations in exhumation rates or to the presence of a paleo-PAZ (Partial Annealing Zone) (see Chap. 9, Fitzgerald and Malusà 2018). For the calculation of an apparent exhumation rate, thermochronologic ages yielded by partly reset or unreset samples, i.e., by samples collected within or above a paleo-PAZ (as revealed, e.g., by grain age over dispersion or by confined track-length distributions) should not be included (see Chap. 6, Vermeesch 2018; Chap. 9, Fitzgerald and Malusà 2018).

Plots displaying the measured FT age against the mean track length (MTL) provide a useful approach to visualise the occurrence of partly reset or unreset samples lying within or above a paleo-PAZ (Fig. 10.1c). Samples collected from a range of initial paleodepths in a region that has undergone broadly coeval cooling will define a concave-up "boomerang" shape (Green 1986). This boomerang trend includes: (i) old ages with long MTLs corresponding to samples that have spent most time above the PAZ; (ii) intermediate ages with the shortest MTLs, corresponding to samples that have spent most time within the PAZ; and (iii) young ages with long MTLs that have experienced the highest initial paleotemperatures. The timing of the most recent cooling event is constrained by the transition from samples with "intermediate" FT ages and MTLs in the 12–13 μm range to samples with younger FT ages and MTLs in the range of 14–15 μm (e.g., Brown et al. 1994; Fitzgerald 1994; Gallagher and Brown 1997; Hendriks et al. 2007). These latter samples can be used to estimate an apparent exhumation rate using the age-elevation approach, provided these samples are collected from the same fault block and fulfil the criteria described in Chap. 9 (Fitzgerald and Malusà 2018).

10.1.2 Virtual Configuration of Rock Units by FT Thermochronology

Low-temperature thermochronology can also be used to envisage a paleo-thermal structure and infer the original thickness of overburden, which may include partly eroded stratigraphic successions or eroded thrust sheets (Fig. 10.2a–c). This is one application of *FT stratigraphy*, with "stratigraphy" represented by the variation of age with elevation or depth (in case of boreholes) (e.g., Brown 1991).

The radial plots in Fig. 10.2a illustrate the grain-age distributions normally expected in samples of sedimentary rocks collected at increasing depths down a borehole in a sedimentary basin. Samples that have experienced minor burial and have remained above the PAZ of the chosen thermochronologic system (black circles in Fig. 10.2a) will show either unimodal or polymodal grain-age distributions, with grain-age populations systematically older than the stratigraphic age of the sample. Samples that have experienced burial within the PAZ (grey circles in Fig. 10.2a) will display partly reset FT ages that with increasing temperature become progressively younger than the stratigraphic age of the sample. Samples located below the PAZ, i.e., at higher temperatures (white circles in Fig. 10.2a), will yield a zero FT age because the fission tracks have been totally annealed.

Subsequent to burial, and following rapid exhumation to the surface, let us now examine age/length trends on this exhumed stratigraphic succession (Fig. 10.2b). Sedimentary rocks originally located below the PAZ (white circles in Fig. 10.2b) yield unimodal grain-age distributions with a central age that is equal to, or slightly younger (although often not much younger) than this rapid exhumation.

Samples that have experienced burial within the PAZ (grey circles in Fig. 10.2b) will display FT ages ranging between the stratigraphic age of the sample and the time of exhumation. The timing of the transition between the completely reset and the partially reset samples typically marks the timing of the initiation of more rapid cooling and exhumation. Samples that have remained above the PAZ will provide information on older exhumation/crystallisation events in the sediment source areas, which can be investigated by using the detrital thermochronology approaches described in Sect. 10.2 (see also Chap. 14, Carter 2018; Chap. 15, Bernet 2018; Chap. 16, Malusà 2018 and Chap. 17, Fitzgerald et al. 2018 for more details). Clues for the interpretation of partially reset detrital samples can be found in Brandon et al. (1998) and in Chap. 18 (Schneider and Issler 2018).

The occurrence of partly or fully reset detrital samples in the uppermost levels of a stratigraphic sequence provides evidence of overburden removal by denudation (Fig. 10.2b). When the paleogeothermal gradient is reasonably assumed or independently known, the FT stratigraphy approach provides a useful way to estimate the depth of erosional denudation and the virtual configuration of eroded rock units above the present-day topography (Fig. 10.2c). Examples include studies in the central Alaska Range to estimate the amount of rock present above the summit of Denali prior to the onset of rapid exhumation in the Late Miocene (Fitzgerald et al. 1995, see also Chap. 9, Fitzgerald and Malusà 2018); in the Apennines to infer the virtual configuration of the Ligurian wedge on top of the Adriatic foredeep units (Zattin et al. 2002); in eastern Anti-Atlas to infer the past configuration of the Variscan belt on top of the Precambrian basement of the West African Craton (Malusà et al. 2007); in the Canadian Cordillera to infer the initial thickness of the Lewis thrust sheet (Feinstein et al. 2007); in SE France to infer the virtual configuration of the Alpine nappes on top of the Paleogene sedimentary succession of the European foreland (Labaume et al. 2008; Schwartz et al. 2017). The number of samples required for a reliable reconstruction of these virtual configurations is often quite high, and datasets will ideally include samples from different levels of the stratigraphic succession.

10.1.3 The Analysis of Fault Offsets Using Thermochronologic Markers

Low-temperature thermochronology often provides time constraints on fault activity and markers for the analysis of fault offsets (e.g., Gleadow and Fitzgerald 1987; Gessner et al. 2001; Raab et al. 2002; Fitzgerald et al. 2009; Niemi

Fig. 10.2 Analysis of virtual geologic configurations and fault offsets using thermochronologic markers. **a** Grain-age distributions expected for increasing depths down a borehole in a sedimentary basin: samples that have remained above the PAZ of the chosen thermochronologic system (black circles) show grain ages systematically older than the stratigraphic age of the sample; samples that have experienced burial within the PAZ (grey circles) display partly reset (annealed) FT ages that are younger than the stratigraphic age of the sample; samples located below the PAZ (white circles) yield zero FT ages because fission tracks are totally annealed. **b** The same stratigraphic succession after exhumation. Sedimentary rocks originally located below the PAZ (white circles) display unimodal grain-age distributions with a central age that records the time of last exhumation. Samples that have experienced burial within the PAZ (grey circles) display FT ages ranging between the stratigraphic age and the time of last exhumation. The occurrence of partly or fully reset samples in the uppermost levels of a stratigraphic succession provides evidence of overburden removal by erosion or tectonic processes. **c** Reconstructed configuration of eroded rock units based on the distribution of unreset, partly annealed and fully reset samples in a modern landscape. **d** Analysis of fault offset by FT stratigraphy (diagram is modified from Stockli et al. 2002). **e** Analysis of fault offsets by mineral-age and reverse mineral-age stratigraphy (based on Malusà et al. 2011); numbers refer to hypothetical crustal levels and letters refer to hypothetical stratigraphic units (see Chap. 16, Malusà 2018 for more details)

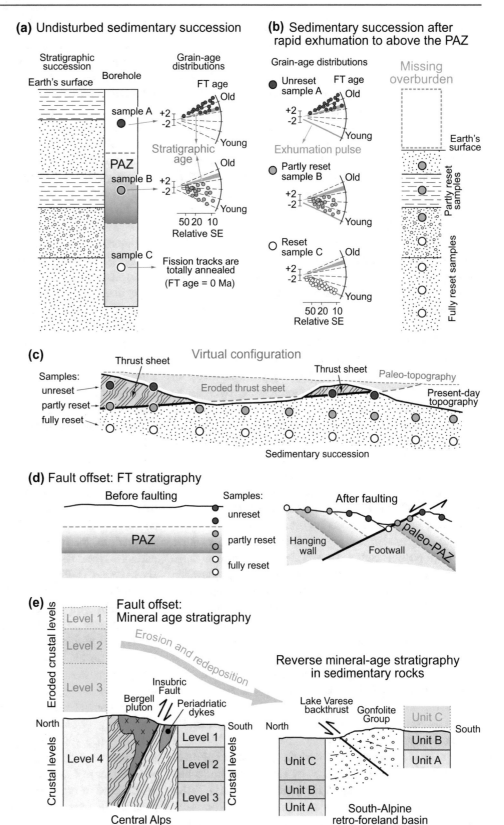

et al. 2013; Ksienzyk et al. 2014; Balestrieri et al. 2016, and Chap. 11, Foster 2018). Such markers are particularly useful in case of monotonous sequences of sedimentary, magmatic and metamorphic rocks, where geologic markers are absent, and in regions of poor outcrop where other markers are not available (e.g., forested or glaciated areas). Thermochronologic markers may include a paleo-PAZ and FT stratigraphy (Fig. 10.2d) (e.g., Brown 1991; Fitzgerald 1992; Foster et al. 1993; Bigot-Cormier et al. 2000; Stockli et al. 2002; Thiede et al. 2006; Richardson et al. 2008; Fitzgerald et al. 2009; Kounov et al. 2009; Miller et al. 2010), which potentially constrain fault activity in the uppermost few kilometres of the crust (Chap. 11, Foster 2018). Both the multiple-method and age-elevation approaches described in Sect. 10.1.1, when applied to distinct crustal blocks, are potentially able to highlight episodes of differential exhumation across major faults, thus constraining the age of fault activity (e.g., Gleadow and Fitzgerald 1987; Fitzgerald 1992, 1994; Foster et al. 1993; Foster and Gleadow 1996; Viola et al. 2001; West and Roden-Tice 2003; Malusà et al. 2005, 2009a; Niemi et al. 2013; Riccio et al. 2014). Given the precision of the ages on either side of the fault, which is typically ±5% (±1σ) for apatite FT (AFT) ages, faults with small offsets will not be revealed. Whether a fault offset is revealed often depends on the age-elevation relationship. If the slope is shallow (ages vary considerably with change in elevation) age-offsets due to faulting are often revealed, whereas when the slope is steep (ages do not vary much with elevation) faults are typically not revealed, as there is little age-offset across the fault unless the vertical fault offset is extremely large (i.e., several kilometres). Constraints on the fault offsets are usually more reliable when the local FT stratigraphy is well documented and multiple samples (if possible collected over a significant elevation range on either side of the fault to confirm the local age-elevation slope) are used. Observations of age trends such as these can be integrated by information provided by temperature-time paths, based on the modelling of confined track lengths in samples collected on the opposite sides of the fault (e.g., Thomson 2002; Balestrieri et al. 2016). If the paleogeothermal gradient is independently known, the multiple-method analysis of distinct fault blocks may provide reliable estimates of fault throws accommodated through time (e.g., Malusà et al. 2006).

In a well-known example, Fitzgerald (1992) tested the FT stratigraphy approach across the Transantarctic Mountain Front in the Dry Valleys area where the location and offset of normal faults are well-delineated by step-faulted horizontal dolerite sills (see also Chap. 9, Fitzgerald and Malusà 2018). There, reconstruction of the ~100 Ma isochron from samples collected systematically along the faulted ridge was used to reliably document fault offsets as small as ~200 m. This was, however, an ideal situation where the slope of the age-elevation relationship (an exhumed PAZ) was ~15 m/Myr; thus, an offset of 300–400 m represents a

discernable age difference of 20–25 Ma. Miller et al. (2010) undertook a similar study across the Transantarctic Mountain Front at Cape Surprise near the Shackleton Glacier, distinguishing between the presence of one master fault with offset of ~5 km or a series of smaller faults, as it proved to be.

Fault offsets can be also constrained by *mineral-age stratigraphy* (Malusà et al. 2011), which is based on the specific combinations of crystallisation and exhumation ages that are yielded by different thermochronometers (and minerals) at different depths in the crust (see Chap. 8, Malusà and Fitzgerald 2018, their Fig. 8.7). Such combinations allow one to define, on a thermochronologic basis, a stratigraphy of increasingly deeper crustal levels (Levels 1–4 in Fig. 10.2e) that may be able to constrain fault offsets in homogeneous rock sequences. Mineral-age stratigraphy can be particularly useful in cases where plutonic rocks dominate the geology (Fig. 10.2e). Use of mineral-age stratigraphy may also, when compared to FT stratigraphy, constrain greater offsets across faults because the data spans a greater temperature interval. In the case of the Insubric Fault in the European Alps, this approach provides evidence of >10–15 km fault offset of the uplifted crustal block north of the fault, relative to the downthrown southern side. On the north, the Periadriatic Bergell/Bregaglia pluton, emplaced in the Oligocene, is now unroofed down to a crustal level that was lying below the T_c of the K–Ar system on biotite at the time of intrusion (level 4 in Fig. 10.2e). This is attested by biotite K–Ar ages (26–21 Ma, Villa and von Blanckenburg 1991) that are much younger than the intrusion age provided by zircon U–Pb dating (30 ± 2 Ma, Oberli et al. 2004). In comparison, the block south of the Insubric Fault has Periadriatic dykes now exposed at the surface. These dykes show zircon U–Pb and AFT ages, yielding two clusters at 42–39 Ma and 35–34 Ma, that are generally indistinguishable within error (D'Adda et al. 2011; Malusà et al. 2011; Zanchetta et al. 2015), because they were emplaced at crustal levels lying within or just below the AFT PAZ at the time of intrusion (level 1 or 2 in Fig. 10.2e). These constraints integrate previous estimates of fault offsets based on the multiple-method approach (Hurford 1986).

The mineral-age stratigraphy observed in an orogen undergoing erosion may subsequently be observed as a "reverse" age stratigraphy in sedimentary basins fed from that eroding source. This reverse age trend can thus be used as a marker to detect tectonic repetitions and fault offsets in the absence of other stratigraphic markers (Fig. 10.2e, right panel). For example, application of this approach to the Oligocene–Miocene Gonfolite Group derived from the erosion of the Bergell pluton (Wagner et al. 1979; Bernoulli et al. 1989) provided evidence for a major and previously undetected thrust fault in the Gonfolite Group (Lake Varese backthrust in Fig. 10.2e) (Malusà et al. 2011).

Fig. 10.3 Time constraints to fault activity based on thermochronologic analysis. **a** Reverse fault (in red) accommodating differential uplift and exhumation before time t_3. The time of faulting is successfully constrained by AFT and AHe analyses in rock samples A and B collected on the opposite sides of the fault, and by modelling of confined track-length distributions. AHe ages in samples A and B are equal to t_3, and indistinguishable across the fault, which is in line with the cessation of fault activity not later than time t_3. **b** Reverse fault accommodating differential subsidence before time t_3, when the analysed samples C and D are at similar crustal level as samples A and B. AFT and AHe analyses of samples C and D provide no direct constraint to the age of faulting, whereas higher T_c thermochronometers would record events much older than t_1. The effects of advection on isothermal surfaces are not included for the sake of simplicity

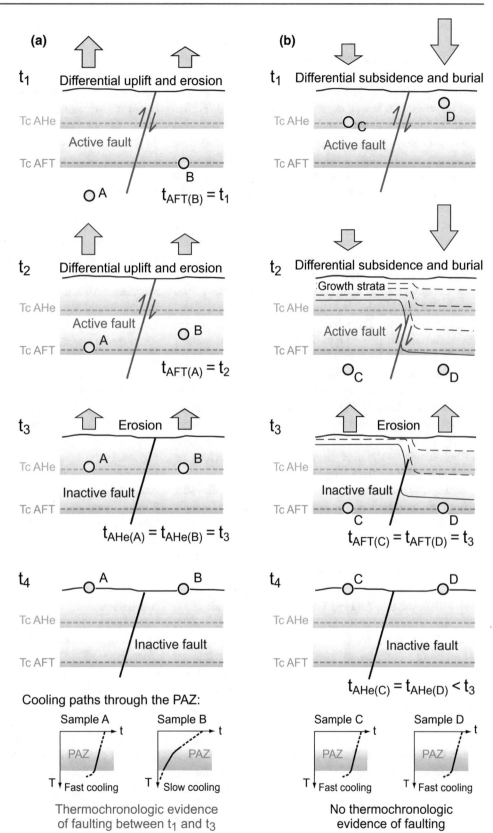

Thermochronologic analyses across major faults can ideally constrain the timing of deformation thanks to a detailed reconstruction of the differential exhumation paths experienced by different fault blocks through time. In the example of Fig. 10.3 (left panel), the activity of a reverse fault before time t_3 is well constrained by apatite (U–Th)/He (AHe) and AFT analyses in rock samples A and B collected on the opposite sides of the fault, and by inverse modelling using age and confined track-length data from the same samples. AHe ages in samples A and B are equal to t_3, and indistinguishable across the fault, which is in line with the cessation of fault activity not later than time t_3. Importantly, there is usually only clear thermochronologic evidence for fault offset when faults accommodate differential exhumation (Fig. 10.3a), but such evidence is not observed when faults accommodate differential subsidence (Fig. 10.3b). Such an observation, with respect to cooling rather than exhumation/subsidence, has long been established (e.g., Green et al. 1989). The situation of faults accommodating differential subsidence is typical, for instance, of rapidly subsiding foreland basins such as the Po Plain north of the Apennines (Pieri and Groppi 1981). In Fig. 10.3b, the reverse fault indicated in red is active at the same time interval as the fault in Fig. 10.3a, and when the analysed samples (C and D) at the time of faulting were at the same crustal level as samples A and B. However, the thermochronologic analysis of samples C and D provides no useful constraint on the timing of deformation.

In the case where fault motion is dominantly strike-slip, differential exhumation between fault blocks is usually minimal. As a consequence, thermochronologic evidence for faulting along strike-slip systems is generally observed particularly at restraining bends or step-overs, or where there is transpression. Well-known examples of these situations have been documented along the Denali Fault (e.g., Riccio et al. 2014), the San Andreas Fault (e.g., Niemi et al. 2013) and the Alpine Fault (e.g., Herman et al. 2007; Warren-Smith et al. 2016).

Sampling strategies to provide timing constraints on fault activity, and markers for the analysis of fault offsets (see Chap. 11, Foster 2018), must be designed to minimise the bias arising from the complexity of isotherms close to major faults (see Chap. 8, Malusà and Fitzgerald 2018). Large (crustally penetrative) faults are most likely to have a thermal perturbation (Ehlers 2005), and for these faults, it is advisable to collect samples tens to hundreds of metres from the main deformation zone. In contrast, some studies seek to exploit the thermal perturbation along the faults to constrain the fault activity (Tagami et al. 2001; Murakami and Tagami 2004), and samples are best collected from a closely spaced array across the fault plane (see Chap. 12, Tagami 2018).

10.1.4 FT Thermochronology as a Correlation Tool for Mesoscale Structural Data

In tectonically complex areas, the analysis of post-metamorphic deformation is particularly challenging, because deformation is often partitioned within different crustal blocks, fault orientation is not scale invariant, and the strain field inferred from mesoscale structures (i.e., outcrop scale) may differ from strain on a regional scale (Eyal and Reches 1983; Rebai et al. 1992; Martinez-Diaz 2002). Regional reconstructions thus require structural analysis at different scales, including field mapping of the network of major (km-scale) faults, and a dense distribution of time-constrained mesoscale strain axes that are derived from the kinematic analysis of fault-slip data in outcrops from high-strain and low-strain domains (Marrett and All-mendinger 1990). Within this framework, low-temperature thermochronology provides not only time constraints for the analysis of major faults, but also an effective tool to constrain the age of mesoscale deformation (e.g., Malusà et al. 2009a).

Fault-rock types and related deformation mechanisms show a generalised progression with depth (Fig. 10.4a); from gouge/breccias to cataclasites to mylonites, or to describe the mechanisms, from frictional-plastic to viscous flow (Sibson 1983; Snoke et al. 1998). The transition from frictional to viscous flow partly depends on strain rate, but it is dominantly controlled by lithology and temperature in the crust (Schmid and Handy 1991; Scholz 1998; Raterron et al. 2004). Therefore, the analysis of fault rocks in a specific lithology gives indications on the crustal level where mesoscale structures have formed. Because FT analysis provides chronological constraints on the passage of a rock through a given crustal level, the timing of both mesoscale structures and their associated strain axes can thus be constrained.

Crustal rocks are commonly dominated by quartz rheology; thus, the transition from frictional to viscous flow takes place at ~ 300 °C (Scholz 1998; Chen et al. 2013). The temperature of this change in deformational mechanism correlates well with the T_c of the FT system in low-damage zircon (Fig. 10.4a). Therefore, in general, mylonites will form at temperatures higher than the T_c of the zircon FT (ZFT) system. Cohesive cataclasites will form at temperatures that are higher than the T_c of the AFT system. Fault gouges and fault breccias are typical of the uppermost kilometres of the crust, i.e., of the temperature range best constrained by the AHe system (Fig. 10.4a). Because the frictional-to-viscous transition controls the thickness of the seismogenic zone (Priestley et al. 2008; Chen et al. 2013), the maximum depth extent for pseudotachylite formation in quartz-dominated crustal rocks also correlates with the T_c of

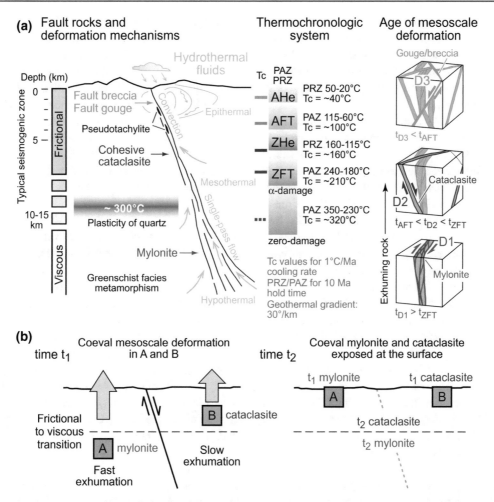

Fig. 10.4 **a** Fault-rock types and related deformation mechanisms show a generalised progression with depth. Mylonites in crustal rocks dominated by quartz rheology generally form at temperatures that are higher than the T_c of the ZFT system. Cohesive cataclasites generally form at temperatures that are higher than the T_c of the AFT system. Fault gouges and fault breccias are typical of the uppermost kilometres of the crust, i.e., of the temperature range best constrained by the AHe system. The panel on the right (3 cubes) conceptually illustrates the application of FT thermochronology to constrain the age of mesoscale deformation (and related strain axes) in a region where the exhumation history is independently constrained by AFT and ZFT data. During progressive exhumation, the rock mass is cut by a right-lateral shear zone marked by mylonitic rocks (phase D1), then by a normal fault marked by cohesive cataclasite (phase D2), and finally by a normal fault marked by gouge (phase D3). Phase D1 is probably older than the ZFT age, phase D2 may have taken place within a time range delimited by AFT and ZFT ages, whereas phase D3 is probably younger than the age provided by AFT (based on Malusà et al. 2009a). **b** Mylonites formed at time t_1 in rock mass A are now exposed at the surface (time t_2) adjacent to cataclasites (rock mass B). Both the mylonites and cataclasites formed at the same time but rock mass B formed at shallower levels, and was exhumed at slower rates. This example shows that fault-rock types alone cannot be used as a correlation tool for mesoscale structural data across different crustal blocks, because coeval deformation at the outcrop scale may be marked by different types of fault rocks. Integration with low-temperature thermochronology data is thus required for a correct interpretation of mesoscale structural data

the ZFT system (Fig. 10.4a). Fault rocks are often associated with veins and hydrothermal mineralisation that can provide additional constraints on the temperature of fault-rock formation (Wiltschko 1998).

Figure 10.4a conceptually illustrates the application of FT thermochronology to constrain the timing of mesoscale deformation and related strain axes, in a region where the exhumation history is independently constrained by AFT and ZFT ages. The rock mass (shown as a cube) in the right

panel, during progressive exhumation is cut by a right-lateral shear zone marked by mylonitic rocks (phase D1), then by a normal fault marked by cohesive cataclasite (phase D2), and finally by a normal fault marked by gouge (phase D3). When combined with FT data, we can conclude that phase D1 is likely older than the ZFT age, phase D2 may have taken place within a time range delimited by AFT and ZFT ages, whereas deformation phase D3 is likely younger than the AFT age. If pseudotachylites are found along these faults, they must be coeval or younger than the age provided by ZFT. The resolution of these constraints can be improved by using additional thermochronologic systems, and by considering not only rock types controlled by quartz rheology, but also other rock types controlled by the rheology of calcite, feldspar or olivine.

Interpreting mesoscale deformation on a regional scale solely based on deformation style, and without reliable time constraints provided by thermochronology, is potentially misleading. Deformation at the outcrop scale may be coeval in different crustal blocks, yet marked by different types of fault rock. Such a scenario may be possible if deformation occurred at the same time in different crustal blocks, but at different crustal levels (Malusà et al. 2009a). In the example of Fig. 10.4b, mylonites formed at time t_1 in one crustal block, are exposed at the surface together with cataclasites formed at the same time, but at shallower levels, in a nearby crustal block exhumed at slower rates. Therefore, deformation style should not be used for correlations of mesoscale structural data over large areas, unless integrated with low-temperature thermochronology data. The sampling strategy for this kind of study, that incorporates thermochronology and structural analysis, is similar to that required for the multiple-method approach (cf. Sect. 10.1.1). Samples should preferably be collected far enough away from major faults to avoid major perturbations of the background thermal field. Compared to the direct, typically more precise dating of synkinematic minerals grown along a fault plane (e.g., Freeman et al. 1997; Zwingmann and Mancktelow 2004), this approach generally requires a lower number of samples but does require multiple methods. However, it may provide constraints not only on single deformation steps, but on the whole deformation history recorded by sets of mesoscale faults.

10.2 Detrital Thermochronology Studies

Modern sediments and sedimentary rocks can provide invaluable information on the provenance and exhumation of the sediment sources (e.g., Baldwin et al. 1986; Cerveny et al. 1988; Brandon et al. 1998; Garver et al. 1999; Bernet et al. 2001; Willett et al. 2003; Ruiz et al. 2004; van der Beek et al. 2006). Sedimentary rocks for detrital thermochronology studies are typically collected through a stratigraphic succession. If sedimentary rocks have remained above the PAZ or partial retention zone (PRZ) of the chosen thermochronologic system since their deposition, the stratigraphic succession should reflect, in inverted order, the thermochronologic age structure observed in the sediment source area. Eroded detritus is generally distributed over a much wider area compared to that of the eroding source. Sedimentary successions thus record the erosional history of

a much thicker crustal section. Today, such crustal sections may be completely eroded away, and thus impossible to investigate by the bedrock approach.

Detrital grain-age distributions represent the source rocks. Depending on the geologic evolution of the source rocks, these distributions can be unimodal or polymodal. Polymodal distributions are usually deconvolved into different grain-age populations (e.g., Brandon 1996; Dunkl and Székely 2002; Vermeesch 2009), which are generally older than the depositional age of the analysed sample (Bernet et al. 2004a; Bernet and Garver 2005).

10.2.1 Approaches to Detrital Thermochronology Studies

Figure 10.5a provides a summary of potential approaches to detrital thermochronology studies by using either modern sediment samples (1 and 2 in Fig. 10.5a) and samples from a stratigraphic succession (3–5 in Fig. 10.5a). Detrital thermochronology can provide information on:

Average long-term erosion rates (1 in Fig. 10.5a) Analysing minerals from samples of modern river sediments allows one to construct grain-age distributions from each catchment sampled (e.g., Garver et al. 1999; Brewer et al. 2003). These grain-age distributions are representative, under certain conditions, of the FT stratigraphy within the drainage (e.g., Bernet et al. 2004a; Resentini and Malusà 2012). The age of the youngest detrital age component, if interpreted as an exhumation-related cooling age, can be used to provide preliminary, although sometimes imprecise constraints on the average long-term (10^6–10^8 yr) erosion rate of the source area, using the lag-time approach (Garver et al. 1999) and a zero stratigraphic age (see below). Notably, grain-age distributions are influenced by the drainage hypsometry. Ages will be generally older in grains eroded from summits, and younger in grains eroded from valleys, depending on the age-elevation relationship in bedrock. In the case of well-developed age-elevation relationships, it is hypothetically possible to constrain the elevation from where sediment grains were eroded (e.g., Stock et al. 2006; Vermeesch 2007). Grain-age distributions will also vary depending on the mineral fertility of different upstream eroding lithologies. The mineral fertility, which can be defined as the variable propensity of different parent rocks to yield detrital grains of specific minerals when exposed to erosion (Moecher and Samson 2006; Malusà et al. 2016), can be quite inhomogeneous. The main advantage of using the detrital approach to constrain an average long-term erosion rate is the possibility of getting a first-order picture of the erosion pattern over large areas by using a relatively low number of samples (e.g., Bernet et al. 2004a; Malusà and Balestrieri 2012; Asti et al. 2018). However, ages related

Fig. 10.5 Detrital thermochronologic analysis. **a** Cartoon showing the range of potential approaches and information retrieved from the thermochronologic analysis of modern sediment (1, 2) and sedimentary rocks (3, 4, 5); letters indicate sub-basins (A to D), hypothetical sampling sites (S) and relative peak size (*m*, *n*). Each approach is based on a range of assumptions that should be independently tested (see Table 10.1). Reliable constraints are best provided by integrating thermochronologic analysis of modern sediments, ancient sedimentary successions and independent mineral fertility determinations. **b**, **c** Alternative sampling strategies to perform sediment budgets based on the thermochronologic analysis of modern sediment samples (after Malusà et al. 2016); $D_{n,n'}$ is the maximum distance between cumulative frequency curves, K_α is the critical value for a significance level α (see Chap. 7, Malusà and Garzanti 2018)

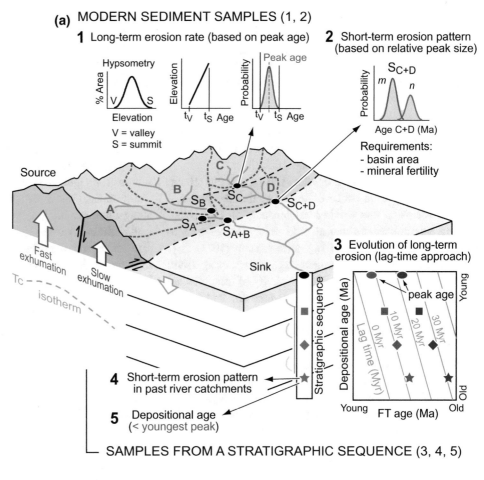

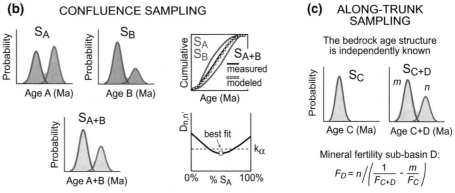

to exhumational cooling should be distinguished from cooling ages that are independent from exhumation, such as those provided by volcanic grains. These grains can be effectively detected by double-dating (see Chap. 5, Danišík 2018; Chap. 7, Malusà and Garzanti 2018), as shown in Chap. 15 (Bernet 2018).

Modern short-term erosion patterns (2 in Fig. 10.5a) Grain-age distributions in modern river sediments can be analysed to constrain sediment budgets and erosion patterns on timescales typical of bedload river transport (generally 10^3–10^5 yr). This approach exploits the size of grain-age

populations rather than the age of mineral grains, which is utilised instead as a provenance marker (Garver et al. 1999; Bernet et al. 2004b; Malusà et al. 2009b). Samples are either collected along the trunk of the river or at major confluences, and grain-age distributions are compared by statistical methods in order to determine the percentage of mineral grains derived from different sub-basins (see Sect. 10.2.2). This percentage, when integrated with the area of each sub-basin and its mineral fertility, which is independently derived (see Chap. 7, Malusà and Garzanti 2018), can be used to calculate the relative erosion rates of the sub-basins

(Malusà et al. 2016; 2017; Glotzbach et al. 2017, Braun et al. 2018). Relative erosion rates can be converted into absolute rates after integration with gauging data or cosmogenic-derived sediment fluxes for selected basins (e.g., Lupker et al. 2012; Wittmann et al. 2016). The same approach can be ideally applied to ancient alluvial sediments, using samples from lithified clastic sequences, whenever it can be reasonably assumed that the past mountain catchment was not markedly different from its present-day configuration (4 in Fig. 10.5a).

Evolution of long-term erosion rates (3 in Fig. 10.5a) The same approach utilised to constrain the average long-term erosion rate within a catchment (1 in Fig. 10.5a) can be extrapolated to ancient sedimentary successions to investigate the evolution of long-term erosion (10^6–10^8 yr timescales) in an adjacent orogen (e.g., Garver et al. 1999; Bernet et al. 2001; van der Beek et al. 2006). Samples are collected through a stratigraphic succession, and single minerals are analysed. The resulting grain-age distributions, when polymodal, are deconvolved into different grain-age populations (red and blue in Fig. 10.5), and these populations are then plotted on a lag-time diagram where the lag time is the difference between the cooling age and the depositional age (Garver et al. 1999), which is independently derived (e.g., by biostratigraphy or magnetostratigraphy). The lag time provides an estimate of the average exhumation rate of the analysed detrital grains from the depth of the T_c isothermal surface to the Earth's surface (Garver et al. 1999; Ruiz et al. 2004). The lag-time trend is often used to infer whether an eroding mountain belt is under a constructional, steady state or decay phase of evolution (Bernet et al. 2001; Carrapa et al. 2003; Spotila 2005; Ruiz and Seward 2006; Rahl et al. 2007; Zattin et al. 2010; Lang et al. 2016). More details and examples are provided in Chap. 15 (Bernet 2018) and Chap. 17 (Fitzgerald et al. 2018).

Depositional age (5 in Fig. 10.5a) Detrital thermochronology can be also used to constrain the depositional age of barren sedimentary successions (e.g., Carter et al. 1995; Rahn and Selbekk 2007). The depositional age of a sedimentary sample must be younger than the central age of the youngest grain-age population, provided that post-depositional annealing can be excluded.

10.2.2 Sediment Budgets Using Modern River Sediments

Constraining sediment budgets based on detrital thermochronology analysis of modern river sediments requires detrital sources with "thermochronologically distinct fingerprints". Such fingerprints are generally the result of different geologic or upper crustal exhumation histories (red and blue tectonic blocks in Fig. 10.5a). In a detrital sample

exclusively derived from one of these sources (e.g., the fast exhuming red block sampled in catchment C), the grain-age distribution depends on the catchment hypsometry and on the catchment age-elevation relationship. Detrital samples derived from the mixing of various sources (e.g., samples S_{A+B} and S_{C+D}, Fig. 10.5a) likely yield polymodal grain-age distributions including different grain-age populations. Two alternative sampling strategies are employed for an effective partitioning of mineral grains (i.e., to determine the relative proportion of grains from each source), and to perform a sediment budget starting from the analysis of grain-age distributions. These are the confluence sampling and the along-trunk sampling approaches (Malusà et al. 2016).

Confluence sampling approach This requires collecting samples from major tributaries (S_A and S_B in Fig. 10.5a) and from the trunk river downstream of their confluence (S_{A+B} in Fig. 10.5a). Partitioning of detrital grains is based on a linear combination of grain-age distributions upstream of the confluence (e.g., Bernet et al. 2004b). The best-fit solution is defined by the mixing proportion that minimises the misfit between the modelled distributions resulting from the linear combination, and the empirical grain-age distribution observed downstream of the confluence (squares and black line, respectively, in the cumulative probability diagram in Fig. 10.5b). The goodness of fit between modelled and empirical grain-age distributions can be evaluated using the parameter $D_{n,n'}$ of the Kolmogorov–Smirnov statistics (Dunkl and Székely 2002; Malusà et al. 2013; Vermeesch 2013), which shows a minimum in correspondence of the best-fit solution. In order to avoid meaningless linear combinations between indistinguishable grain-age distributions, a two-sample Kolmogorov–Smirnov test can be used to check whether grain-age distributions in samples S_A and S_B are statistically different or not. The confluence sampling approach requires at least three samples at each node (S_A, S_B, and S_{A+B}). It allows a direct measurement of mineral fertility in sub-basins A and B using the same samples collected for thermochronologic analysis (see Chap. 7, Malusà and Garzanti 2018) and does not require any independent information on the thermochronologic fingerprint of the parent bedrock.

Along-trunk approach This approach can be used when a river cuts across rock units that have distinct, independently known thermochronologic signatures that do not overlap. Thus, the age populations can be used to unequivocally discriminate the provenance of detrital grains (Resentini and Malusà 2012). The relative proportion of grains from each source, according to the along-trunk approach is based on the deconvolution of grain-age distributions in individual age components of specific size (m and n in Fig. 10.5a, c) (e.g., Brandon 1996; Dunkl and Székely 2002). Only two samples (S_C and S_{C+D}) are needed for unmixing the thermochronologic signal and characterising sub-basins C and D in terms of mineral fertility. Fertility of sub-basin C (F_C) is directly measured, fertility of sub-basin D (F_D) is calculated by the formula in Fig. 10.5c.

10.2.3 Assumptions in Detrital Thermochronology Studies

The approaches to detrital thermochronologic analysis illustrated in Fig. 10.5a rely on a range of assumptions that should be carefully considered and independently tested (see Table 10.1). When the age of a detrital population in a modern sediment sample is employed to constrain the long-term exhumation of bedrock exposed upstream of the sampling site (1 in Fig. 10.5a), the assumption is that the thermochronologic fingerprint of eroded bedrock reflects cooling during erosion. However, this assumption is not always valid. Thermochronologic ages may also reflect past transient changes in the thermal structure of the crust (Braun 2016), episodes of metamorphic or magmatic crystallisation within the eroded bedrock (Malusà et al. 2011; Kohn et al. 2015), or the cooling history of a different (distant) eroding source that provided detritus to sedimentary rocks now exposed within the drainage (Bernet and Garver 2005; von Eynatten and Dunkl 2012). Beside a careful evaluation of the geologic setting, the risk of misinterpretation can be minimised by multiple dating of detrital mineral grains (see Chap. 5, Danišík 2018), by the integration with thermochronologic analyses of cobbles (Chap. 17, Fitzgerald et al. 2018) and bedrock (Chap. 13, Baldwin et al. 2018), and by complementing detrital thermochronology analysis with thermal modelling (Ehlers et al. 2005; Braun et al. 2012).

The assumption of uniform mineral fertility in the catchment upstream of a sampling site, which is common to many detrital thermochronology studies (e.g., Glotzbach et al. 2013), should be carefully tested even in small basins, because grain ages sourced from elevation ranges characterised by low-fertility rocks would be under-represented in corresponding detrital grain-age distributions. This is particularly important when detrital and bedrock thermochronology data are compared to test if a mountain range is in steady state (e.g., Brewer et al. 2003; Ruhl and Hodges 2005). The analysis of the short-term erosion pattern based on the relative proportion of different grain-age populations in polymodal samples (2, Fig. 10.5a) relies on the additional assumption that the sediment is effectively mixed, and that the potential bias possibly introduced by hydraulic processes

during sediment transport and deposition is negligible. These assumptions can be tested using the strategies described in Chap. 7 (Malusà and Garzanti 2018).

The number of often untestable assumptions underlying the detrital thermochronologic approach largely increases in ancient sedimentary successions (see Table 10.1 and references therein). Compared to approaches 1 and 2 in Fig. 10.5a, lag-time analysis (3 in Fig. 10.5a) additionally implies that: (i) the time elapsed during sediment transport is negligible; (ii) the isothermal surfaces relevant for the chosen thermochronologic system have remained steady during exhumation; (iii) major changes in provenance can be excluded or can be detected and accounted for; (iv) the thermochronologic signal is not modified by post-depositional annealing.

Apart from transport by turbidity currents, the zero transport time assumption from erosion to deposition should be evaluated on a case-by-case basis. As an example, the transportation of apatite and zircon as river bedload in the Amazon basin may require a few millions of years, as attested by cosmogenic data (Wittmann et al. 2011). Thus, transportation time may become relevant for lag-time interpretation. The behaviour of isothermal surfaces during exhumation is generally difficult to predict (see Chap. 8, Malusà and Fitzgerald 2018) and may require an integrated approach including thermochronologic data from the source region and thermal modelling. A reliable assessment of detritus provenance generally requires a multiple-method approach to single-grain analysis (see Chap. 5, Danišík 2018; Chap. 14, Carter 2018) and a careful inspection of thermochronologic age trends along a stratigraphic succession (see Chap. 16, Malusà 2018). Lag-time analysis of detrital mineral grains is often utilised to infer the long-term evolution of an entire mountain range, but it is noteworthy that it may emphasise the cooling history of small parts of the mountain range that are characterised by highest mineral fertility (Malusà et al. 2017). The potential impact of post-depositional annealing may represent a problem not only for lag-time analysis, but also for the determination of the depositional age of a sedimentary rock by thermochronologic methods (5 in Fig. 10.5a). However, post-depositional annealing produces age trends that can be recognised when multiple samples are analysed along a stratigraphic succession (van der Beek et al. 2006, see Chap. 16, Malusà 2018 and Chap. 17, Fitzgerald et al. 2018). In this perspective, information on burial-related annealing derived from AFT length distributions gives this method a further advantage over other thermochronology methods that rely on age values only.

Table 10.1 Main assumptions for different approaches to detrital thermochronology and potential impact on data interpretation

Assumption and approach (identified by number code)	Alternative scenarios	References	Potential impact on data interpretation	How to test	This book
Thermochronologic ages in detritus reflect cooling of eroded bedrock during erosional exhumation (*1*, **3**)	Ages may reflect transient changes in the thermal structure of the crust, episodes of metamorphic or magmatic crystallisation, or cooling of distant sediment sources	Bernet and Garver (2005), Malusà et al. (2011), Braun (2016)	*Major*: misinterpretation of cooling ages in terms of exhumation of eroded bedrock	Multiple dating of mineral grains; integration with analysis of cobbles and bedrock, and thermal modelling	Chaps. 5, 8 and 17
Sediment is effectively mixed, grain-age distributions are not affected by hydraulic sorting (*1*, *2*, **3**, **4**)	A relationship between grain age and grain size may imply a potential vulnerability of grain-age distributions to hydraulic processes	Bernet et al. (2004a), Malusà et al. (2013)	*Moderate*: under-representation of specific grain-age populations in detritus	Assessment of age-size relationships in detritus, detection of placer lags	Chap. 7
Mineral fertility does not vary within the river catchment upstream of the sampling site (*1*)	Bedrock shows variable mineral fertility depending on the elevation (grain ages associated to low-fertility rocks are under-represented in detritus)	Brewer et al. (2003), Malusà et al. (2016)	*Moderate to major*: biased reconstruction of catchment FT stratigraphy; flawed evaluation of steady-state conditions	Inspection of bedrock geology; measurement of mineral fertility	Chap. 7
Mineral fertility does not vary regionally in a mountain belt, units are represented in detritus according to their size and erosion rate (*2*, **3**)	Mineral fertility varies over orders of magnitude across a mountain belt, tectonic units with lower mineral fertility are under-represented in the detrital record	Dickinson (2008), Malusà et al. (2017)	*Major*: lag-time analysis only constrains the evolution of high-fertility tectonic units; sediment budgets are incorrect	Inspection of bedrock geology; measurement of mineral fertility	Chap. 7
Sediment transport time from the source rocks to the final sink is negligible (**3**)	Minerals grains transported as bedload (e.g., apatite and zircon) may take a few millions of years to reach their final sediment sink	Garver et al. (1999), Wittmann et al. (2011)	*Moderate*: underestimation of exhumation rates based on lag-time analysis	Comparison between magmatic-age peaks and stratigraphic age of analysed samples	Chaps. 7 and 16
T_c isotherms have remained steady during exhumation (**3**)	Isotherms move upward due to heat advection triggered by fast erosion, and are then restored after thermal relaxation	Rahl et al. (2007), Braun (2016)	*Major*: unreliable erosion rate estimations; ages may record thermal relaxation, not erosion	Integration with thermal modelling	Chap. 8
No provenance change, lag-time pertains to a single eroding source (**3**)	Major provenance changes are recorded along the stratigraphic sequence	Ruiz et al. (2004), Glotzbach et al. (2011)	*Major*: incorrect recognition of lag-time trends if provenance changes remain undetected	Multi-dating approach and integration with other provenance constraints	Chaps. 5, 14 and 16
The drainage network has remained steady through time, same mineral fertility in bedrock eroded in the past and in bedrock eroded today (**4**)	Major drainage reorganisation, different bedrock geology in paleo- and modern river catchments	Garzanti and Malusà (2008)	*Major*: the past short-term erosion pattern based on sediment budgets is incorrect	Geomorphologic analysis; independent provenance constraints	Chap. 14
The detrital thermochronologic signal is preserved after deposition (**3**, **5**)	The detrital thermochronologic signal is modified by post-depositional annealing	van der Beek et al. (2006), Chirouze et al. (2012)	*Major*: incorrect constraints to the maximum depositional age; overestimation of exhumation rates based on lag-time analysis	Careful inspection of age trends along the stratigraphic succession	Chaps. 16 and 17

1 Average long-term erosion rates; *2* Modern short-term erosion patterns; **3** Evolution of long-term erosion rates by the lag-time approach; **4** Past short-term erosion patterns; **5** Determination of depositional age (in italics, approaches based on the analysis of modern sediments; in bold, approaches based on the analysis of sedimentary rocks)

10.3   Conclusions

Low-temperature thermochronology can be used to constrain a range of geologic processes using bedrock and/or detrital samples. In bedrock studies, the multiple-method and age-elevation approaches provide useful constraints on the exhumation history of rocks now exposed at the surface, provided that a range of assumptions are properly evaluated. The multiple-method approach provides direct constraints on the cooling history of a sample, but the paleogeothermal gradient must be independently known to convert an average cooling rate to an average exhumation rate. The age-elevation approach does not require explicit assumptions on the paleogeothermal gradient. A careful analysis of the relationships between the topography and the thermal reference frame is anyway required, as is an understanding of the various factors that may change the slope of an age-elevation profile. Multiple methods can be applied to the same array of vertical profile samples, which allows exhumation to be constrained over longer time intervals. FT stratigraphy provides a suitable reference frame to constrain virtual configurations of eroded rock units above the present-day topography. A range of thermochronologic markers can be used to highlight episodes of differential exhumation across a major fault and constrain the timing of fault activity. However, in some cases, major episodes of faulting are not revealed by low-temperature thermochronology. FT data combined with the analysis of fault rocks also provide useful constraints on the timing of mesoscale deformation, and a reliable tool for the correlation over large areas of mesoscale strain axes derived from fault-slip analysis.

Detrital thermochronologic analysis can be used to provide an averaged image of the thermochronologic age structure in bedrock and to constrain the erosion pattern of the sediment sources on short-term (10^3–10^5 yr) and long-term (10^6–10^8 yr) timescales. Detrital thermochronologic analysis can also be used to constrain the provenance of the dated grains. The analysis of modern sediment samples provides the baseline for the thermochronologic interpretation of older stratigraphic levels. The lag-time approach to detrital thermochronology provides information that can be used to investigate the evolutionary stage of an entire orogenic belt, and infer whether it is under a constructional, steady state or decay phase of evolution. Different approaches to detrital thermocronology are based on a range of assumptions that should be independently tested to avoid meaningless geologic interpretations. Depending on the objectives, the full potential of the detrital thermochronologic approach is best exploited by the integrated analysis of samples collected along a sedimentary succession, samples of modern river sediment, and independent mineral fertility determinations.

Acknowledgements This work benefited from constructive reviews by Maria Laura Balestrieri and Shari Kelley, comments on an early version of the manuscript by an anonymous reviewer, and comments by Suzanne Baldwin and students in her 2016 thermochronology class. PGF thanks the National Science Foundation for funding through the years as well as support from Jarg Pettinga and the Erksine Program at the University of Canterbury.

References

Asti R, Malusà MG, Faccenna C (2018) Supradetachment basin evolution unraveled by detrital apatite fission track analysis: the Gediz Graben (Menderes Massif, Western Turkey). Basin Res 30:502-521

Baldwin SL, Harrison TM, Burke K (1986) Fission track evidence for the source of Scotland District sediments, Barbados and implications for post-Eocene tectonics of the southern Caribbean. Tectonics 5:457–468

Baldwin SL, Fitzgerald PG, Malusà MG (2018) Chapter 13. Crustal exhumation of plutonic and metamorphic rocks: constraints from fission-track thermochronology. In: Malusà MG, Fitzgerald PG (eds) Fission-track thermochronology and its application to geology. Springer, Berlin

Balestrieri ML, Bonini M, Corti G, Sani F, Philippon M (2016) A refinement of the chronology of rift-related faulting in the broadly rifted zone, southern Ethiopia, through apatite fission-track analysis. Tectonophysics 671:42–55

Bernet M (2018) Chapter 15. Exhumation studies of mountain belts based on detrital fission-track analysis on sand and sandstones. In: Malusà MG, Fitzgerald PG (eds) Fission-track thermochronology and its application to geology. Springer, Berlin

Bernet M, Garver JI (2005) Fission-track analysis of detrital zircon. Rev Mineral Geochem 58(1):205–237

Bernet M, Zattin M, Garver JI, Brandon MT, Vance JA (2001) Steady-state exhumation of the European Alps. Geology 29:35–38

Bernet M, Brandon MT, Garver JI, Molitor B (2004a) Fundamentals of detrital zircon fission-track analysis for provenance and exhumation studies with examples from the European Alps. Geol Soc Am Spec Pap 378:25–36

Bernet M, Brandon MT, Garver JI, Molitor B (2004b) Downstream changes of Alpine zircon fission-track ages in the Rhône and Rhine Rivers. J Sediment Res 74:82–94

Bernoulli D, Bertotti G, Zingg A (1989) Northward thrusting of the Gonfolite Lombarda (South-Alpine Molasse) onto the Mesozoic sequence of the Lombardian Alps: implications for the deformation history of the Southern Alps. Eclogae Geol Helv 82:841–856

Bigot-Cormier F, Poupeau G, Sosson M (2000) Differential denudations of the Argentera Alpine external crystalline massif (SE France) revealed by fission track thermochronology (zircons, apatites). C R Acad Sci 5:363–370

Brandon MT (1996) Probability density plot for fission-track grain-age samples. Radiat Meas 26:663–676

Brandon MT, Roden-Tice MK, Garver JI (1998) Late Cenozoic exhumation of the Cascadia accretionary wedge in the Olympic Mountains, northwest Washington State. Geol Soc Am Bull 110:985–1009

Braun J (2002) Quantifying the effect of recent relief changes on age–elevation relationships. Earth Planet Sci Lett 200(3):331–343

Braun J (2016) Strong imprint of past orogenic events on the thermochronological record. Tectonophysics 683:325–332

Braun J, Gemignani L, van der Beek P (2018) Extracting information on the spatial variability in erosion rate stored in detrital cooling age distributions in river sands. Earth Surf Dynam 6:257–270

Braun J, van der Beek P, Valla P, Robert X, Herman F, Glotzbach C, Pedersen V, Perry C, Simon-Labric T (2012) Quantifying rates of landscape evolution and tectonic processes by thermochronology and numerical modeling of crustal heat transport using PECUBE. Tectonophysics 524–525:1–28

Brewer ID, Burbank DW, Hodges KV (2003) Modelling detrital cooling-age populations: insights from two Himalayan catchments. Basin Res 15:305–320

Brown RW (1991) Backstacking apatite fission-track "stratigraphy": a method for resolving the erosional and isostatic rebound components of tectonic uplift histories. Geology 19(1):74–77

Brown RW, Summerfield MA, Gleadow AJW (1994) Apatite fission track analysis: its potential for the estimation of denudation rates and implications of long-term landscape development. In: Kirkby MJ (ed) Process models and theoretical geomorphology. Wiley, Hoboken, pp 23–53

Carrapa B, Wijbrans J, Bertotti G (2003) Episodic exhumation in the Western Alps. Geology 31(7):601–604

Carter A (2018) Chapter 14. Thermochronology on sand and sandstones for stratigraphic and provenance studies. In: Malusà MG, Fitzgerald PG (eds) Fission-track thermochronology and its application to geology. Springer, Berlin

Carter A, Bristow CS, Hurford AJ (1995) The application of fission track analysis to the dating of barren sequences: examples from red beds in Scotland and Thailand. Geol Soc Spec Publ 89:57–68

Cerveny PF, Naeser ND, Zeitler PK, Naeser CW, Johnson NM (1988) History of uplift and relief of the Himalaya during the past 18 million years: evidence from fission-track ages of detrital zircons from sandstones of the Siwalik Group. New perspectives in basin analysis. Springer, New York, pp 43–61

Chen WP, Yu CQ, Tseng TL, Yang Z, Wang CY, Ning J, Leonard T (2013) Moho, seismogenesis, and rheology of the lithosphere. Tectonophysics 609:491–503

Chirouze F, Bernet M, Huyghe P, Erens V, Dupont-Nivet G, Senebier F (2012) Detrital thermochronology and sediment petrology of the middle Siwaliks along the Muksar Khola section in eastern Nepal. J Asian Earth Sci 44:94–106

D'Adda P, Zanchi A, Bergomi M, Berra F, Malusà MG, Tunesi A, Zanchetta S (2011) Polyphase thrusting and dyke emplacement in the central Southern Alps (northern Italy). Int J Earth Sci 100:1095–1113

Danišík M (2018) Chapter 5. Integration of fission-track thermochronology with other geochronologic methods on single crystals. In: Malusà MG, Fitzgerald PG (eds) Fission-track thermochronology and its application to geology. Springer, Berlin

Dickinson WR (2008) Impact of differential zircon fertility of granitoid basement rocks in North America on age populations of detrital zircons and implications for granite petrogenesis. Earth Planet Sci Lett 275:80–92

Dunkl I, Székely B (2002) Component analysis with visualization of fitting—PopShare, a windows program for data analysis. Goldschmidt conference abstracts 2002. Geochim Cosmochim Ac 66/A:201

Ehlers TA (2005) Crustal thermal processes and the interpretation of thermochronometer data. Rev Mineral Geochem 58:315–350

Ehlers TA, Chaudhri T, Kumar S, Fuller CW, Willett SD, Ketcham RA, Brandon MT, Belton DX, Kohn BP, Gleadow AJW, Dunai TJ, Fu FQ (2005) Computational tools for low-temperature thermochronometer interpretation. Rev Mineral Geochem 58(1):589–622

Eyal Y, Reches Z (1983) Tectonic analysis of the Dead Sea Rift region since the Late-Cretaceous based on mesostructures. Tectonics 2:167–185

Feinstein S, Kohn B, Osadetz K, Price RA (2007) Thermochronometric reconstruction of the prethrust paleogeothermal gradient and initial thickness of the Lewis thrust sheet, southeastern Canadian Cordillera foreland belt. Geol Soc Am Spec Pap 433:167–182

Fitzgerald PG (1992) The Transantarctic Mountains of southern Victoria Land: the application of apatite fission track analysis to a rift shoulder uplift. Tectonics 11(3):634–662

Fitzgerald PG (1994) Thermochronologic constraints on post-Paleozoic tectonic evolution of the central Transantarctic Mountains, Antarctica. Tectonics 13:818–836

Fitzgerald PG, Gleadow AJ (1988) Fission-track geochronology, tectonics and structure of the Transantarctic Mountains in northern Victoria Land, Antarctica. Chem Geol 73(2):169–198

Fitzgerald PG, Gleadow AJ (1990) New approaches in fission track geochronology as a tectonic tool: examples from the Transantarctic Mountains. Int J Rad Appl Instr Part D Nucl Tracks Rad Meas 17(3):351–357

Fitzgerald PG, Malusà MG (2018) Chapter 9. Concept of the exhumed partial annealing (retention) zone and age-elevation profiles in thermochronology. In: Malusà MG, Fitzgerald PG (eds) Fission-track thermochronology and its application to geology. Springer, Berlin

Fitzgerald PG, Sorkhabi RB, Redfield TF, Stump E (1995) Uplift and denudation of the central Alaska Range: a case study in the use of apatite fission track thermochronology to determine absolute uplift parameters. J Geophys Res Sol Earth 100(B10):20175–20191

Fitzgerald PG, Duebendorfer EM, Faulds JE, O'Sullivan PB (2009) South Virgin–White Hills detachment fault system of SE Nevada and NW Arizona: applying apatite fission track thermochronology to constrain the tectonic evolution of a major continental detachment fault. Tectonics 28

Fitzgerald PG, Malusà MG, Muñoz JA (2018) Chapter 17. Detrital thermochronology using conglomerates and cobbles. In: Malusà MG, Fitzgerald PG (eds) Fission-track thermochronology and its application to geology. Springer, Berlin

Foster DA (2018) Chapter 11. Fission-track thermochronology in structural geology and tectonic studies. In: Malusà MG, Fitzgerald PG (eds) Fission-track thermochronology and its application to geology. Springer, Berlin

Foster DA, Gleadow AJW (1996) Structural framework and denudation history of the flanks Kenya and Anza rifts, East Africa. Tectonics 15:258–271

Foster DA, Gleadow AJ, Reynolds SJ, Fitzgerald PG (1993) Denudation of metamorphic core complexes and the reconstruction of the transition zone, west central Arizona: constraints from apatite fission track thermochronology. J Geophys Res Sol Earth 98(B2):2167–2185

Freeman SR, Inger S, Butler RWH, Cliff RA (1997) Dating deformation using Rb–Sr in white mica: greenschist facies deformation ages from the Entrelor shear zone, Italian Alps. Tectonics 16:57–76

Gallagher K, Brown RW (1997) The onshore record of passive margin evolution. J Geol Soc Lond 154:451–457

Garver JI, Brandon MT, Roden-Tice MK, Kamp PJJ (1999) Exhumation history of orogenic highlands determined by detrital fission track thermochronology. Geol Soc Spec Publ 154:283–304

Garzanti E, Malusà MG (2008) The Oligocene Alps: domal unroofing and drainage development during early orogenic growth. Earth Planet Sci Lett 268:487–500

Gessner K, Ring U, Johnson C, Hetzel R, Passchier CW, Güngör T (2001) An active bivergent rolling-hinge detachment system: Central Menderes metamorphic core complex in western Turkey. Geology 29:611–614

Gleadow AJW, Fitzgerald PG (1987) Uplift history and structure of the Transantarctic Mountains: new evidence from fission track dating of basement apatites in the Dry Valleys area, southern Victoria Land. Earth Planet Sci Lett 82(1):1–14

Glotzbach C, Bernet M, van der Beek P (2011) Detrital thermochronology records changing source areas and steady exhumation in the Western European Alps. Geology 39(3):239–242

Glotzbach C, van der Beek P, Carcaillet J, Delunel R (2013) Deciphering the driving forces of erosion rates on millennial to million-year timescales in glacially impacted landscapes: an example from the Western Alps. J Geophys Res Earth 118:1491–1515

Glotzbach C, Busschers FS, Winsemann J (2017) Detrital thermochronology of Rhine, Elbe and Meuse river sediment (Central Europe): implications for provenance, erosion and mineral fertility. Int J Earth Sci. https://doi.org/10.1007/s00531-017-1502-9

Green PF (1986) On the thermo-tectonic evolution of Northern England: evidence from fission track analysis. Geol Mag 153:493–506

Green PF, Duddy IR, Laslett GM, Hegarty KA, Gleadow AJW, Lovering JF (1989) Thermal annealing of fission tracks in apatite 4. Quantitative modelling techniques and extension to geological timescales. Chem Geol Isot Geosci Sect 79(2):155–182

Hendriks B, Andriessen P, Huigen Y, Leighton C, Redfield T, Murrell G, Gallagher K, Nielsen SB (2007) A fission track data compilation for Fennoscandia. Norw J Geol 87:143–155

Herman F, Braun J, Dunlap WJ (2007) Tectonomorphic scenarios in the Southern alps of New Zealand. J Geophys Res Sol Earth 112 (B4)

Hurford AJ (1986) Cooling and uplift patterns in the Lepontine Alps South Central Switzerland and an age of vertical movement on the insubric fault line. Contrib Miner Petrol 92:413–427

Ketcham R (2018) Chapter 3. Fission track annealing: from geologic observations to thermal history modeling. In: Malusà MG, Fitzgerald PG (eds) Fission-track thermochronology and its application to geology. Springer, Berlin

Kohn MJ, Corrie SL, Markley C (2015) The fall and rise of metamorphic zircon. Am Miner 100(4):897–908

Kounov A, Viola G, De Wit M, Andreoli MAG (2009) Denudation along the Atlantic passive margin: new insights from apatite fission-track analysis on the western coast of South Africa. Geol Soc Spec Publ 324:287–306

Ksienzyk AK, Dunkl I, Jacobs J, Fossen H, Kohlmann F (2014) From orogen to passive margin: constraints from fission track and (U–Th)/He analyses on Mesozoic uplift and fault reactivation in SW Norway. Geol Soc Spec Publ 390

Labaume P, Jolivet M, Souquière F, Chauvet A (2008) Tectonic control on diagenesis in a foreland basin: combined petrologic and thermochronologic approaches in the Grès d'Annot basin (Late Eocene-Early Oligocene, French-Italian external Alps). Terra Nova 20:95–101

Lang KA, Huntington KW, Burmester R, Housen B (2016) Rapid exhumation of the eastern Himalayan syntaxis since the late Miocene. Geol Soc Am Bull 128:1403–1422

Lupker M, Blard PH, Lavé J, France-Lanord C, Leanni L, Puchol N, Charreau J, Bourlès D (2012) [10]Be-derived Himalayan denudation rates and sediment budgets in the Ganga basin. Earth Planet Sci Lett 333:146–156

Malusà MG (2018) Chapter 16. A guide for interpreting complex detrital age patterns in stratigraphic sequences. In: Malusà MG, Fitzgerald PG (eds) Fission-track thermochronology and its application to geology. Springer, Berlin

Malusà MG, Balestrieri ML (2012) Burial and exhumation across the Alps-Apennines junction zone constrained by fission-track analysis on modern river sands. Terra Nova 24:221–226

Malusà MG, Fitzgerald PG (2018) Chapter 8. From cooling to exhumation: setting the reference frame for the interpretation of thermocronologic data. In: Malusà MG, Fitzgerald PG (eds) Fission-track thermochronology and its application to geology. Springer, Berlin

Malusà MG, Garzanti E (2018) Chapter 7. The sedimentology of detrital thermochronology. In: Malusà MG, Fitzgerald PG (eds) Fission-track thermochronology and its application to geology. Springer, Berlin

Malusà MG, Polino R, Zattin M, Bigazzi G, Martin S, Piana F (2005) Miocene to present differential exhumation in the Western Alps: insights from fission track thermochronology. Tectonics 24(3): TC3004:1-23

Malusà MG, Philippot P, Zattin M, Martin S (2006) Late stages of exhumation constrained by structural, fluid inclusion and fission track analyses (Sesia–Lanzo unit, Western European Alps). Earth Planet Sci Lett 243(3):565–580

Malusà MG, Polino R, Cerrina Feroni A, Ellero A, Ottria G, Baidder L, Musumeci G (2007) Post-Variscan tectonics in eastern Anti-Atlas (Morocco). Terra Nova 19(6):481–489

Malusà MG, Polino R, Zattin M (2009a) Strain partitioning in the axial NW Alps since the Oligocene. Tectonics 28(TC002370):1–26

Malusà MG, Zattin M, Andò S, Garzanti E, Vezzoli G (2009b) Focused erosion in the Alps constrained by fission-track ages on detrital apatites. Geol Soc Spec Publ 324:141–152

Malusà MG, Villa IM, Vezzoli G, Garzanti E (2011) Detrital geochronology of unroofing magmatic complexes and the slow erosion of Oligocene volcanoes in the Alps. Earth Planet Sci Lett 301:324–336

Malusà MG, Carter A, Limoncelli M, Villa IM, Garzanti E (2013) Bias in detrital zircon geochronology and thermochronometry. Chem Geol 359:90–107

Malusà MG, Resentini A, Garzanti E (2016) Hydraulic sorting and mineral fertility bias in detrital geochronology. Gondwana Res 31:1–19

Malusà MG, Wang J, Garzanti E, Liu ZC, Villa IM, Wittmann H (2017) Trace-element and Nd-isotope systematics in detrital apatite of the Po river catchment: implications for provenance discrimination and the lag-time approach to detrital thermochronology. Lithos 290–291:48–59

Mancktelow NS, Grasemann B (1997) Time-dependent effects of heat advection and topography on cooling histories during erosion. Tectonophysics 270(3):167–195

Marrett R, Allmendinger RW (1990) Kinematic analysis of fault-slip data. J Struct Geol 12:973–986

Martinez-Diaz JJ (2002) Stress field variation related to fault interaction in a reverse oblique-slip fault: the Alhama de Murcia fault, Betic Cordillera, Spain. Tectonophysics 356:291–305

Miller SR, Fitzgerald PG, Baldwin SL (2010) Cenozoic range-front faulting and development of the Transantarctic Mountains near Cape Surprise, Antarctica: thermochronologic and geomorphologic constraints. Tectonics 29(1)

Moecher DP, Samson SD (2006) Differential zircon fertility of source terranes and natural bias in the detrital zircon record: implications for sedimentary provenance analysis. Earth Planet Sci Lett 247:252–266

Moore MA, England PC (2001) On the inference of denudation rates from cooling ages of minerals. Earth Planet Sci Lett 185:265–284

Murakami M, Tagami T (2004) Dating pseudotachylyte of the Nojima fault using the zircon fission-track method. Geophys Res Lett 31(12)

Naeser CW, Bryant B, Crittenden MD, Sorensen ML (1983) Fission-track ages of apatite in the Wasatch Mountains, Utah: an uplift study. Geol Soc Am Mem 157:29–36

Niemi NA, Buscher JT, Spotila JA, House MA, Kelley SA (2013) Insights from low-temperature thermochronometry into transpressional deformation and crustal exhumation along the San Andreas fault in the western Transverse Ranges, California. Tectonics 32:1602–1622

Oberli F, Meier M, Berger A, Rosenberg CL, Gieré R (2004) U–Th–Pb and 230Th/238U disequilibrium isotope systematics: precise accessory mineral chronology and melt evolution tracing in the Alpine Bergell intrusion. Geochim Cosmochim Acta 68:2543–2560

Pieri M, Groppi G (1981) Subsurface geological structure of the Po Plain, Italy. Progetto Finalizzato Geodinamica 414:1–13

Priestley K, Jackson J, McKenzie D (2008) Lithospheric structure and deep earthquakes beneath India, the Himalaya and southern Tibet. Geophys J Int 172:345–362

Raab MJ, Brown RW, Gallagher K, Carter A, Weber K (2002) Late Cretaceous reactivation of major crustal shear zones in northern Namibia: constraints from apatite fission track analysis. Tectonophysics 349:75–92

Rahl JM, Ehlers TA, van der Pluijm BA (2007) Quantifying transient erosion of orogens with detrital thermochronology from syntectonic basin deposits. Earth Planet Sci Lett 256:147–161

Rahn MK, Selbekk R (2007) Absolute dating of the youngest sediments of the Swiss Molasse basin by apatite fission track analysis. Swiss J Geosci 100(3):371–381

Raterron P, Wu Y, Weidner DJ, Chen J (2004) Low-temperature olivine rheology at high pressure. Phys Earth Planet Inter 145:149–159

Rebai S, Philip H, Taboada A (1992) Modern tectonic stress field in the Mediterranean region: evidence for variations in stress directions at different scales. Geophys J Int 110:106–140

Reiners PW, Brandon MT (2006) Using thermochronology to understand orogenic erosion. Annu Rev Earth Planet Sci 34:419–466

Resentini A, Malusà MG (2012) Sediment budgets by detrital apatite fission-track dating (Rivers Dora Baltea and Arc, Western Alps). Geol Soc Am Spec Pap 487:125–140

Riccio SJ, Fitzgerald PG, Benowitz JA, Roeske SM (2014) The role of thrust faulting in the formation of the eastern Alaska Range: thermochronological constraints from the Susitna glacier thrust fault region of the intracontinental strike-slip Denali fault system. Tectonics 33(11):2195–2217

Richardson NJ, Densmore AL, Seward D et al (2008) Extraordinary denudation in the Sichuan basin: insights from low-temperature thermochronology adjacent to the eastern margin of the Tibetan Plateau. J Geophys Res Solid Earth 113(B4)

Ruhl KW, Hodges KV (2005) The use of detrital mineral cooling ages to evaluate steady state assumptions in active orogens: an example from the central Nepalese Himalaya. Tectonics 24

Ruiz G, Seward D (2006) The Punjab foreland basin of Pakistan: a reinterpretation of zircon fission-track data in the light of Miocene hinterland dynamics. Terra Nova 18:248–256

Ruiz G, Seward D, Winkler W (2004) Detrital thermochronology—a new perspective on hinterland tectonics, an example from the Andean Amazon Basin, Ecuador. Basin Res 16:413–430

Ruiz G, Carlotto V, Van Heiningen PV, Andriessen PAM (2009) Steady-state exhumation pattern in the Central Andes–SE Peru. Geol Soc Spec Publ 324:307–316

Schildgen TF, van der Beek PA (2018) Chapter 19. Application of low-temperature thermochronology to the geomorphology of orogenic systems. In: Malusà MG, Fitzgerald PG (eds) Fission-track thermochronology and its application to geology. Springer, Berlin

Schmid SM, Handy MR (1991) Towards a genetic classification of fault rocks: geological usage and tectonophysical implications. In: Muller DW et al (eds) Controversies in modern geology. Academic Press, Cambridge, pp 339–361

Schneider DA, Issler DR (2018) Chapter 18. Application of low-temperature thermochronology to hydrocarbon exploration. In: Malusà MG, Fitzgerald PG (eds) Fission-track thermochronology and its application to geology. Springer, Berlin

Scholz CH (1998) Earthquakes and friction laws. Nature 391:37–42

Schwartz S, Gautheron C, Audin L, Dumont T, Nomade J, Barbarand J, Pinna-Jamme R, van der Beek P (2017) Foreland exhumation controlled by crustal thickening in the Western Alps. Geology G38561-1

Sibson RH (1983) Continental fault structure and the shallow earthquake source. J Geol Soc Lond 140:741–767

Snoke AW, Tullis J, Todd VR (1998) Fault-related rocks: a photographic atlas. Princeton University Press, Princeton

Spotila JA (2005) Applications of low-temperature thermochronometry to quantification of recent exhumation in mountain belts. Rev Miner Geochem 58(1):449–466

Stock GM, Ehlers TA, Farley KA (2006) Where does sediment come from? Quantifying catchment erosion with detrital apatite (U–Th)/ He thermochronometry. Geology 34:725–728

Stockli DF, Surpless BE, Dumitru TA, Farley KA (2002) Thermochronological constraints on the timing and magnitude of Miocene and Pliocene extension in the central Wassuk Range, western Nevada. Tectonics 21

Stüwe K, White L, Brown R (1994) The influence of eroding topography on steady-state isotherms. Application to fission-track analysis. Earth Planet Sci Lett 124(1–4):63–74

Tagami T. (2018) Chapter 12. Application of fission-track thermochronology to understand fault zones. In: Malusà MG, Fitzgerald PG (eds) Fission-track thermochronology and its application to geology. Springer, Berlin

Tagami T, Hasebe N, Kamohara H, Takemura K (2001) Thermal anomaly around the Nojima Fault as detected by fission-track analysis of Ogura 500 m borehole samples. Isl Arc 10(3–4):457–464

Thiede RC, Arrowsmith JR, Bookhagen B, McWilliams M, Sobel ER, Strecker MR (2006) Dome formation and extension in the Tethyan Himalaya, Leo Pargil, northwest India. Geol Soc Am Bull 118:635–650

Thomson SN (2002) Late cenozoic geomorphic and tectonic evolution of the Patagonian Andes between latitudes 42 S and 46 S: an appraisal based on fission-track results from the transpressional intra-arc Liquiñe-Ofqui fault zone. Geol Soc Am Bull 114:1159–1173

van der Beek P, Robert X, Mugnier JL, Bernet M, Huyghe P, Labrin E (2006) Late Miocene–recent exhumation of the central Himalaya and recycling in the foreland basin assessed by apatite fission-track thermochronology of Siwalik sediments, Nepal. Basin Res 18:413–434

Vermeesch P (2007) Quantitative geomorphology of the White Mountains (California) using detrital apatite fission track thermochronology. J Geophys Res Earth 112(F3)

Vermeesch P (2009) RadialPlotter: a Java application for fission track, luminescence and other radial plots. Rad Meas 44:409–410

Vermeesch P (2013) Multi-sample comparison of detrital age distributions. Chem Geol 341:140–146

Vermeesch P (2018) Chapter 6. Statistics for fission-track thermochronology. In: Malusà MG, Fitzgerald PG (eds) Fission-track thermochronology and its application to geology. Springer, Berlin

Villa IM, von Blanckenburg F (1991) A hornblende ^{39}Ar–^{40}Ar age traverse of the Bregaglia tonalite (southeast Central Alps). Schweiz Mineral Petrogr Mitt 71:73–87

Viola G, Mancktelow NS, Seward D (2001) Late Oligocene-Neogene evolution of Europe-Adria collision: new structural and geochronological evidence from the Giudicarie fault system (Italian Eastern Alps). Tectonics 20:999–1020

von Eynatten H, Dunkl I (2012) Assessing the sediment factory: the role of single grain analysis. Earth Sci Rev 115:97–120

Wagner GA, Reimer GM (1972) Fission track tectonics: the tectonic interpretation of fission track apatite ages. Earth Planet Sci Lett 14(2):263–268

Wagner GA, Reimer GM, Jäger E (1977) Cooling ages derived by apatite fission track, mica Rb–Sr, and K–Ar dating: the uplift and cooling history of the central Alps. Mem Ist Geol Miner Univ Padova 30:1–27

Wagner GA, Miller DS, Jäger E (1979) Fission track ages on apatite of Bergell rocks from central Alps and Bergell boulders in Oligocene sediments. Earth Planet Sci Lett 45(2):355–360

Warren-Smith E, Lamb S, Seward D, Smith E, Herman F, Stern T (2016) Thermochronological evidence of a low-angle, mid-crustal detachment plane beneath the central South Island, New Zealand. Geochem Geophys Geosyst 17:4212–4235

West DP, Roden-Tice MK (2003) Late Cretaceous reactivation of the Norumbega fault zone, Maine: evidence from apatite fission-track ages. Geology 31(7):649–652

Willett SD, Fisher D, Fuller C, En-Chao Y, Chia-Yu L (2003) Erosion rates and orogenic-wedge kinematics in Taiwan inferred from fission-track thermochronometry. Geology 31:945–948

Wiltschko DV (1998) Analysis of veins in low temperature environments—introduction for structural geologists. Geol Soc Am. Short Course, 96 pp

Wittmann H, Von Blanckenburg F, Maurice L, Guyot JL, Kubik PW (2011) Recycling of Amazon floodplain sediment quantified by cosmogenic ^{26}Al and ^{10}Be. Geology 39(5):467–470

Wittmann H, Malusà MG, Resentini A, Garzanti E, Niedermann S (2016) The cosmogenic record of mountain erosion transmitted across a foreland basin: source-to-sink analysis of in situ ^{10}Be, ^{26}Al and ^{21}Ne in sediment of the Po river catchment. Earth Planet Sci Lett 452:258–271

Zanchetta S, Malusà MG, Zanchi A (2015) Precollisional development and Cenozoic evolution of the Southalpine retrobelt (European Alps). Lithosphere 7(6):662–681

Zattin M, Picotti V, Zuffa GG (2002) Fission-track reconstruction of the front of the Northern Apennine thrust wedge and overlying Ligurian unit. Am J Sci 302(4):346–379

Zattin M, Talarico FM, Sandroni S (2010) Integrated provenance and detrital thermochronology studies on the ANDRILL AND-2A drill core: late Oligocene-early miocene exhumation of the Transantarctic Mountains (southern Victoria Land, Antarctica). Terra Nova 22:361–368

Zwingmann H, Mancktelow N (2004) Timing of Alpine fault gouges. Earth Planet Sci Lett 223:415–425

Fission-Track Thermochronology in Structural Geology and Tectonic Studies

David A. Foster

Abstract

Apatite fission-track (AFT) and zircon fission-track (ZFT) data along with other low-temperature thermochronologic data are widely used in the fields of structural geology and tectonics to determine the timing/duration of events, the amount of exhumation in mountain belts, rates of slip on faults, and the geometries of fault networks. In this chapter, I review applications of AFT and ZFT data in extensional tectonic settings. Examples of data sets and interpretations are summarized from the Cenozoic-Recent North American Basin and Range Province. These data constrain displacements of normal faults, rates of slip on faults, paleogeothermal gradients, and the original dip of low-angle normal faults.

11.1 Introduction

Fission-track (FT) data on apatite and zircon (AFT and ZFT, respectively) are important tools for structural geology and tectonics, as was recognized from the earliest studies in mountain belts (Wagner and Reimer 1972). These low-temperature thermochronometers have been widely used to constrain the magnitude and timing of faulting in the brittle upper crust through displaced age–elevation profiles, and the rates of displacement of crust being exhumed through closure temperature (T_c) isotherms by faulting and erosion. The direct application of FT thermochronology in structural geology is more common for extensional or transtensional tectonic settings where cooling is a result of "tectonic" exhumation along with erosion (e.g., Fitzgerald et al. 1991; Foster et al. 1991). There are, however, many excellent examples of FT data being applied to thrust and fold belt systems (e.g., O'Sullivan et al. 1993; McQuarrie et al. 2008; Espurt et al. 2011; Mora et al. 2014) to determine when particular thrusts or thrust systems were active. When thrusting forms topographic relief and induces rapid erosion and cooling, the FT data constrain the timing of thrusting (e.g., Metcalf et al. 2009). The delay in time between thrusting and erosion in some orogenic belts, however, limits direct measurement of fault parameters from cooling ages in many shortening systems. This chapter is, therefore, focused on examples from rifts, fault blocks, and metamorphic core complexes, particularly in environments where tectonic exhumation usually dominates over erosion.

11.2 Faulted Partial Annealing Zone (PAZ) Profiles to Constrain Fault Geometry

In regions lacking traditional sedimentary or volcanic stratigraphic markers, geologic structure and fault block geometry may be constrained using relative displacements of a reference AFT or ZFT age/track length relief profile, if regional consistency in the profile is established (e.g., Gleadow and Fitzgerald 1987; Brown 1991; Fitzgerald 1992; Foster and Gleadow 1992; Foster et al. 1993; Stockli et al. 2003). Reference profiles are constructed by measuring FT ages and track lengths from samples collected over a wide range of elevations in steep mountainous terrain or from drill holes. Paleo-PAZ inflection(s) in the FT age–elevation profile and steep gradients in the reference profile reveal the upper crustal thermal history of an area (e.g., Foster and Gleadow 1993) and form the basis for a pseudo-stratigraphy (Brown 1991) (see also Chap. 10, Malusà and Fitzgerald 2018a, b). Vertical offsets in the reference profile are due to normal faulting and block tilting that has occurred after the PAZ profile was formed in an originally stable region. An AFT age and mean track length stratigraphy can be viewed as being composed of layers of crust with similar (to within ±10–20 °C) thermal histories

D. A. Foster (✉)
Department of Geological Sciences, University of Florida, Gainesville, FL 32611, USA
e-mail: dafoster@ufl.edu

© Springer International Publishing AG, part of Springer Nature 2019
M. G. Malusà and P. G. Fitzgerald (eds.), *Fission-Track Thermochronology and its Application to Geology*,
Springer Textbooks in Earth Sciences, Geography and Environment, https://doi.org/10.1007/978-3-319-89421-8_11

for temperatures below ~100–120 °C prior to faulting (Brown 1991). Vertical offsets between reference age–elevation profiles that are beyond two-sigma errors are used to constrain displacements on faults; sample traverses perpendicular to structural trends may elucidate block tilting and horizontal displacement. This approach is generally able to detect faults with displacements of the order of hundreds of meters or more and is most applicable to regions that had been geologically stable and slowly eroded for timescales of $>10^7$ year prior to extension (Brown et al. 1994). A comprehensive review of AFT PAZ profiles is given in Chap. 9, (Fitzgerald and Malusà 2018).

11.3 Kenya and Anza Rifts Example of Fault Geometry

The plot in Fig. 11.1a shows a reference AFT age–elevation profile for basement rocks exposed east and west of the Kenya Rift in central Kenya (Foster and Gleadow 1993, 1996). The reference profile is composed of data from samples collected from steep mountain fronts at intervals of <100 m. The filled boxes are data from the Cherangani Hills, a range west of the Kenya Rift that reaches elevations of >3000 m, and the remaining data are from the Mathews Range and Karisia Hills, which are east of the Kenya Rift (Foster and Gleadow 1996). The samples from the Karisia Hills are elevated 1100 m to reconstruct displacement due to Cenozoic faulting (Foster and Gleadow 1992). The reference profile exhibits three segments where the age–elevation relationship is linear with a rather steep slope (and long track lengths for the lowest segment). These are separated by two preserved PAZ intervals, which were established before periods of more rapid erosion at about 120 Ma and about 65 Ma. A similar age–elevation relationship is found throughout Kenya and Tanzania, establishing the regional nature of the FT stratigraphy (e.g., Foster and Gleadow 1996; Noble et al. 1997; Spiegel et al. 2007; Toores Acosta et al. 2015).

Figure 11.1b shows a plot of the topography along an east–west section at 1°10′N latitude. The section crosses the Miocene-Recent Kenya Rift and the Cretaceous–Paleogene Anza Rift. The variation in AFT age with elevation in the mountain ranges, and with distance along the traverses between the ranges, is superimposed on the topographic section from the reference stratigraphy in the inset box. The three segments with steep slopes (and consistent mean track lengths), and the two paleo-PAZs in Fig. 11.1a, define five FT stratigraphic markers. Offsets in segments of the reference profile indicate the presence of faults with displacements estimated by restoring the formerly continuous layers. The FT stratigraphy also reveals the tilt direction for the footwall blocks bounded by the normal faults, which in this case is consistent across the section. These data suggest that the normal faults were related to extension that formed the Anza Rift, and that the modern East Africa Rift, in Kenya, is superimposed on structures related to previous rifting. In this example, the AFT stratigraphy reveals the broad structural framework of the basement areas east and west of the Kenya Rift where traditional stratigraphic markers are lacking. It is important to note that the small-scale structure may be significantly more complex than shown in Fig. 11.1b, because the resolution of the FT data reveals only structures with displacements of the order of hundreds of meters.

11.4 Normal Fault Slip Rates

Low-temperature thermochronometers are an effective way to determine the rate of slip on normal faults, which is a key parameter for understanding the development of fault systems, particularly low-angle normal faults bounding metamorphic core complexes (e.g., Foster et al. 1993; John and Foster 1993; Foster and John 1999; Wells et al. 2000; Campbell-Stone et al. 2000; Carter et al. 2004, 2006; Stockli 2005; Brichau et al. 2006; Fitzgerald et al. 2009). Lateral gradients in apparent age along strike in the displacement direction of normal faults are related to the progressive quenching of footwall rocks as they moved through the PAZ (or the partial retention zone—PRZ) for the thermochronometer (Fig. 11.2), as long as the footwall was below (at higher temperature) the base of the PAZ before slip occurred (Foster et al. 2010). The inverse of the slope on a plot of apparent age versus distance in the slip direction reveals the slip rate. For accurate slip rate estimates, the T_c isotherm for the thermochronometer must have remained approximately horizontal or fixed during the interval of slip revealed by the data (Ketcham 1996; Ehlers et al. 2003). This assumption was investigated for low-angle normal faults using 2-D conductive cooling models by Ketcham (1996) and found to be reasonable with a few million years after the onset of extension, because the isotherms reach a steady-state position by that time. Before the first few million years, the isotherms advance along the detachment surface causing uncertainties in the slip rate (Ketcham 1996). Advection of isotherms will result in an underestimate of the slip rate (Ehlers et al. 2001; Fitzgerald et al. 2009). The thermochronometric data alone give time-averaged slip rates on timescales of about a million years and do not rule out significantly faster or slower rates of detachment slip over shorter timescales. Ehlers et al. (2003) showed that combining low-temperature thermochronologic data, along with thermal–kinematic modeling of the evolving isotherms, revealed changes in slip rate on the steeply dipping (45°–60°) Wasatch normal fault.

Fig. 11.1 a Composite AFT age versus elevation regional profile for central Kenya (after Foster and Gleadow 1996). This age–elevation profile establishes an AFT stratigraphy, which is used to reconstruct offsets on normal faults and constrain tilting of fault blocks. **b** Offsets of the regional AFT age–elevation profile show the gross framework of normal faults that bound the basement mountain ranges in central Kenya (after Foster and Gleadow 1996). Without the FT data, it would not be possible to determine the regional-scale normal faults or block tilting related to the Anza Rift, because no regional sedimentary or volcanic stratigraphy exists within the areas where mainly Precambrian basement rocks are exposed

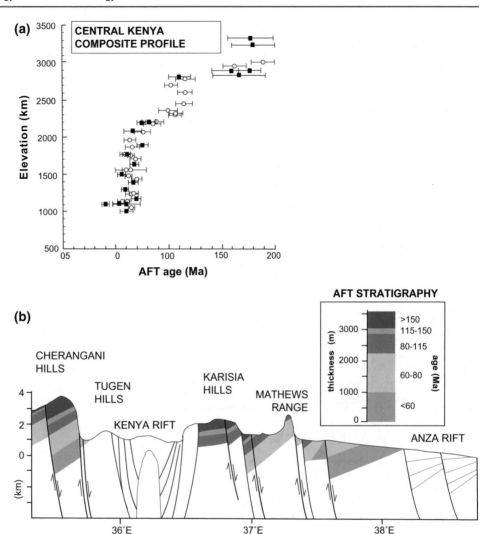

11.5 Bullard Detachment Example to Constrain Slip Rates

The Bullard detachment is a large-scale, low-angle normal fault that separates footwall rocks of the Miocene Buckskin-Rawhide and Harcuvar metamorphic core complexes in Arizona, USA (Spencer and Reynolds 1991; Scott et al. 1998). Spencer and Reynolds (1991) estimated about 90 km of displacement for this detachment system based on reconstructing distinctive sedimentary and plutonic units in the hanging wall and footwall. Figure 11.3 is a plot of the variation of AFT age with distance in the slip direction for Buckskin-Rawhide and Harcuvar metamorphic core complex footwall rocks, immediately beneath the projection of the detachment (Foster et al. 1993). The inverse slope of the Buckskin-Rawhide core complex data suggests a detachment fault slip rate of 7.7 ± 3.6 km/Myr (±2 sigma). Regressions of the slip rate and two-sigma errors, for these and other slip rates, were calculated using methods of York (1969) for

non-correlated errors, considering two-sigma errors in age and sample location/projection. For the regressions of the AFT data, in those cases where the number of data points is relatively small the two-sigma errors are considered to approximate the uncertainties.

All of the samples from the Buckskin-Rawhide footwall give cooling ages at least 5 or 6 million years younger than the onset of extension and have long mean track lengths, indicating rapid cooling. A value of 6.5 ± 3.0 km/Myr for slip rate along that detachment is given by results from the adjacent Harcuvar Mountains core complex. The trend in the Harcuvar Mountains is influenced by one relatively precise AFT age of about 21 Ma from the structurally shallowest sample. Removing this one sample, because of the possibility it may have cooled before the depth of the 110 °C isotherm was stationary (e.g., Ketcham 1996), gives a slip rate of 7.7 ± 3.1 km/Myr. As expected, this rate is similar to that from the Buckskin Mountains because both footwalls were unroofed from beneath the same detachment system (Spencer and Reynolds 1991).

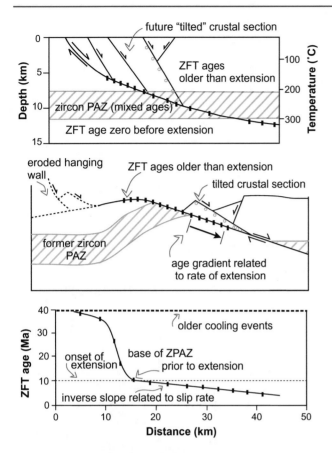

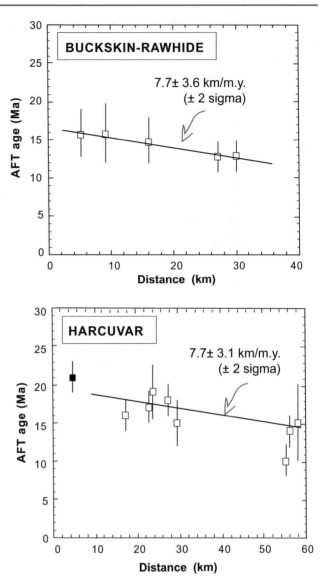

Fig. 11.2 Example showing the exhumation of a zircon PAZ by slip on a low-angle normal fault system (after Foster and John 1999; Wells et al. 2000; Stockli 2005; Foster et al. 2010). The upper diagram shows the zircon PAZ at depth before extension, with the future locations of the normal fault system. The filled boxes represent sample locations above, within, and below the PAZ. The gray circles represent potential sample locations in a future tilted crustal section bounded by normal faults. The middle plot shows the exhumation of the PRZ and crust beneath the zircon PAZ by displacement on a low-angle normal fault. The lower plot shows the distribution of ZFT ages along the trace of the detachment in the direction of slip 10 million years after extension. The samples that were above the zircon PAZ (ZPAZ) reveal cooling ages older than the time of extension, those that were within the ZPAZ give mixed ages that progressively get younger with depth, the samples that were beneath the ZPAZ at deeper depths show a gradual decrease in age with distance, the inverse of which gives the slip rate on the fault. A similar concept also applies to other thermochronologic systems with different PAZ or PRZ temperature intervals

Fig. 11.3 Plots of AFT age against distance in the displacement direction of the Bullard detachment in the Buckskin-Rawhide and Harcuvar metamorphic core complex footwalls

Spencer and Reynolds (1991) independently estimated an extension rate of 8–9 km/Myr along the Buckskin-Rawhide detachment, based on the amount of slip ($\sim$90 km) that had occurred between 23 and 25 Ma (when syn-extensional basins started to form) and about 15 Ma (when the lower plate rocks cooled >100 °C). The timing of initial exposure of footwall rocks at the southwestern and northeastern ends of the Buckskin-Rawhide core complex, based on distinctive clasts in conglomerates, also indicates a slip rate of $\sim$7 km/Myr for this fault (Scott et al. 1998).

The relatively large errors for the Miocene AFT ages result in significant two-sigma errors for slip rate examples in Fig. 11.3. Carter et al. (2004) and Singleton et al. (2014) showed that a more precise measurement of slip rate of this fault could be constrained using (U-Th)/He data. Although the results of these two studies are not in agreement on the slip rate (for reasons listed below), they are within error of the values in Fig. 11.3. Displacement rates for other normal faults in the Basin and Range using FT data referenced in this section range from <1 km/Myr to >10 km/Myr.

The example in Fig. 11.3 appears relatively clear with the exception of the relatively large errors. There are notable examples in the literature where AFT and/or other low-temperature thermochronologic data from normal fault systems are much more complicated and reasonable estimates of slip rate are not possible to constrain. Many detachment faults are composite structures that form from the merger of discrete segments active at different times (e.g., Lister and Davis 1989). This occurs when fault segments are transferred from the footwall to the hanging wall, or from the hanging wall to the footwall, and where secondary breakaway zones develop (e.g., Lister and Davis 1989). Data sets from composite detachments may show overlapping segments or repeated age/distance relationships parallel to the slip direction (e.g., Pease et al. 1999; Campbell-Stone et al. 2000; Stockli et al. 2006; Fitzgerald et al. 2009; Singleton et al. 2014) that need to be assessed separately. Hydrothermal flow within active normal fault systems may result in rapid quenching of the footwall or heating along the detachment (e.g., Morrison and Anderson 1998). The data quality from some detachments is poor due to low uranium concentration, numerous fluid inclusions, and other factors which may mean the slip rate cannot be constrained (e.g., Fitzgerald et al. 1993). Finally, the footwalls of some normal fault systems are too deeply eroded and/or folded by isostatic rebound to project sample locations to the level of the former fault surface (e.g., Foster and Raza 2002).

11.6 Paleogeothermal Gradient

An elusive, but important parameter, for understanding extensional tectonics is the value of the geothermal gradient before extension started. Knowing the geothermal gradient prior to the onset of extension is important for elucidating processes that drive extension in a particular region. Active or passive extension processes, and those driven by magmatism or thermal weakening of the lithosphere, are partly related to different geothermal gradients. Elevated geothermal gradients commonly accompany extension due to mantle upwelling and decompression melting. Gradients prior to extension, however, are often much different and relate to a previous tectonic regime, but are invaluable for calculating the paleodepths of T_c isotherms before exhumation.

There are several examples where FT studies from tilted fault blocks have yielded information about the geothermal gradient prior to extension (Foster et al. 1991; Fitzgerald et al. 1991, 2009; Howard and Foster 1996; Stockli et al. 2003; see also Chap 9, Fitzgerald and Malusà 2018). Paleogeothermal gradient data may then be used to determine, for example, the amount of exhumation that has taken place in metamorphic core complexes and the original dip of faults. Samples collected along traverses from tilted crustal sections (thick fault blocks) parallel to the movement direction of normal faults reveal thermal histories from increasingly deeper paleodepths. Paleoisotherms are identified at the depth where isotopic systems recording pre-extension cooling ages, at shallow depths, give way, at deeper levels, to cooling ages that mark rapid exhumation during extension. The transition where the age–depth curve forms a break-in slope and intersects the age versus paleodepth curve at the time that extension started represents the base of an exhumed PAZ (Fitzgerald et al. 1991; Howard and Foster 1996). A geothermal gradient may be calculated when two paleoisotherms are revealed by thermochronologic data (see Chap. 8, Malusà and Fitzgerald 2018a, b), or one paleoisotherm combined with the location of an unconformity with known depth beneath the surface.

In the Basin and Range Province, examples of this method include studies of the Gold Butte fault block in Utah (Fitzgerald et al. 1991, 2009; Reiners et al. 2000; Bernet 2009; Karlstrom et al. 2010), the Grayback fault block in Arizona (Howard and Foster 1996), and the White Mountains fault block in California/Nevada (Stockli et al. 2003). In each of these cases, the depth below an unconformity (of known depth) for the PAZ for AFT and ZFT (or PRZ for (U-Th)/He ages) is relatively well constrained. The depth of the base of an exhumed PAZ/PRZ relating to the depth of a particular geotherm (e.g., 110° for base of the apatite PAZ) within a relatively intact fault block then allows the calculation of the paleogeothermal gradient.

In the three southern Basin and Range examples listed above, the pre-extension gradients were found to have been relatively low (≤ 20 °C/km, e.g., Grayback) or normal (≤ 25 °C/km, Gold Butte). This has important implications for the tectonic setting of the region in Paleogene time. Relatively, normal paleogeothermal gradients indicate that magmatism was unlikely to have weakened the crust prior to extension and even lower geothermal gradients (<20 °C/km) are more typical of subduction zone settings rather than orogenic highlands. In the case of the US Cordillera, the relatively low Paleogene geothermal gradients may be related to cooling of the lithosphere over a flat slab segment of the subducting Farallon Plate (e.g., Dumitru et al. 1991).

11.7 Grayback Fault Block Example to Constrain Paleogeothermal Gradients

The Grayback fault block in the Tortilla Mountains, Arizona (Fig. 11.4), exposes a Proterozoic through Paleocene granitic crustal section about 12-km thick (Howard and Foster 1996). The crustal section was tilted eastward during Oligocene to Miocene extension, which led to the exhumation of core complexes in the Santa Catalina, Rincon, Tortolita, and Picacho Mountains (Dickinson 1991). Stratigraphy of the overlying Tertiary rocks indicates that

Fig. 11.4 **a** Cross-section of the Grayback fault block in southern Arizona with AFT and ZFT (*italic*) ages (after Howard and Foster 1996). **b** Plot of AFT and ZFT ages against paleodepth for the Grayback block, Arizona (after Howard and Foster 1996)

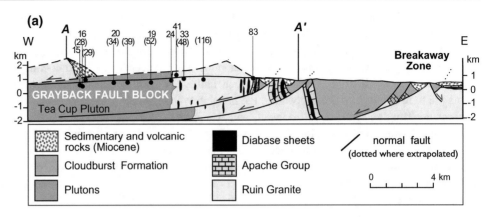

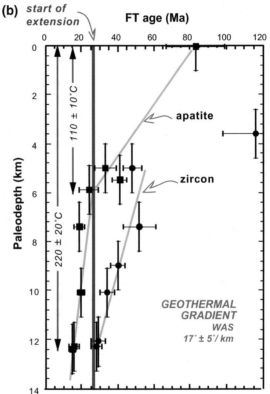

tilting of the Grayback block took place between 25 and 16 Ma. The block was titled to near-vertical or a slightly overturned orientation based on the dip of Proterozoic Apache Group and diabase dikes that were horizontal prior to tilting (Howard 1991).

AFT and ZFT ages from the Grayback fault block are shown in a plot of age against paleodepth (Fig. 11.4b). The AFT ages decrease westward (deeper paleodepths) from ~83 Ma at the unconformity to a break-in slope at ~24 Ma and ~5–6 km depth. Mean track lengths for samples between 0 and 6 km depth are <13 μm, indicating relatively slow cooling through the apatite PAZ. Below ~5–6 km depth, the AFT ages decrease from ~24–15 Ma and have long mean track lengths (>14 μm),

indicating more rapid cooling. The ZFT ages also decrease to the west and become concordant with the initiation of extension at paleodepths of 12.1–12.3 km.

The break-in slope in the AFT age transect represents the position of the base of the apatite PAZ (~110 °C) (Gleadow and Fitzgerald 1987) prior to Tertiary tilting. All of the apatite samples below the break-in slope were cooled rapidly from temperatures where tracks were totally annealed, based on the long mean track lengths. The form of the ZFT age profile with no Oligocene–Miocene break-in slope suggests that all of the samples were at or colder than and structurally above the temperature of total annealing (~220–250 °C for 10^7 year timescales, e.g., Brandon et al. 1998; Bernet 2009). However, the fact that the ages of the two deepest samples

Fig. 11.5 Geological map of the Chemehuevi Mountains, California (modified from John and Foster 1993; Foster and John 1999). The section A–B is the projection through the southern part of the footwall in Fig. 11.6. The thick dashed lines are paleoisotherms for the southern and central parts of the footwall to the Chemehuevi detachment fault when extension started at ∼22 Ma. Paleotemperatures for sample points constraining the isotherms were calculated from thermal histories of the footwall rocks obtained from $^{40}Ar/^{39}Ar$ and FT data, where three to five minerals with different T_c were analyzed from each sample. The isotherms show a gradual increase in temperature to the northeast in the known direction of tectonic transport. Isotherms are not shown for the northern part of the footwall, because of syn-extensional plutons in that area

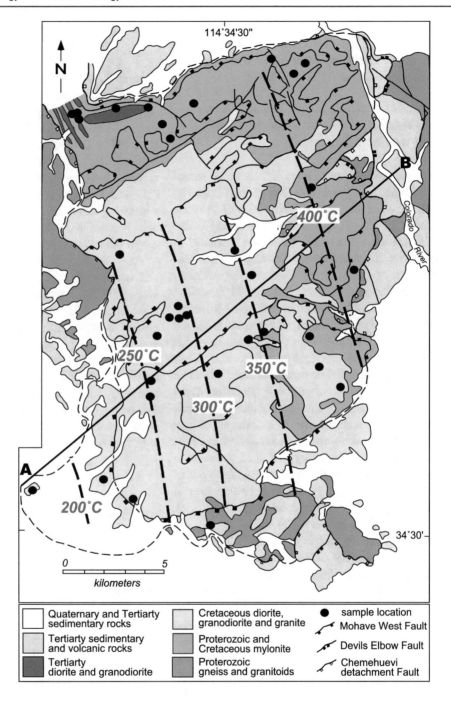

are concordant with, but not younger than, start of tilting suggests the samples at 12.3 km were at 220 ± 30 °C at ∼25 Ma.

Howard and Foster (1996) calculated the paleogeothermal gradient for the Grayback block from the difference in depth of the paleoisotherms at 5.7 ± 0.4 km (110 ± 10 °C) and 12.15 ± 0.7 km (220 ± 30 °C). This gives a gradient of 17.1 ± 5.3 °C/km. The errors include values of known and estimated errors in annealing temperatures and the projections. A gradient calculated between the surface and the 110 ± 10 °C isotherm, assuming a surface temperature of 15 ± 10 °C for the late Oligocene, gives a gradient of 16.7 ± 4.9 °C/km. The mean of these two estimates gives a paleogeothermal gradient of 17 ± 5 °C/km.

11.8 Dip Angles of Faults Prior to Tilting and Isostatic Rebound

One of the fundamental parameters in structural reconstructions of normal and detachment fault systems is the dip of normal faults, when they were initiated and while they are active. This information is essential for calculating extensional strain on local and regional scales and is needed to assess models for the formation of low-angle detachment faults in the light of apparent contradictions with Andersonian fault mechanics (e.g., John and Foster 1993; Wernicke 1995). Tilting of fault blocks and isostatic rebound commonly reduces the dip of faults during and after displacement (e.g., Wernicke and Axen 1988; Spenser and Reynolds 1991). Reconstructing the original dips of faults that do not intersect or offset stratigraphic or other geometric indicators in crystalline rocks is particularly difficult. The controversy regarding the original dips of low-angle detachment faults through the seismogenic crust serves as an example of how difficult it is at times to reconstruct extensional fault systems. Thermochronology data, including FT data, have contributed greatly to the understanding of normal fault systems and initial dips of faults (e.g., Foster et al. 1990; John and Foster 1993; Lee 1995; Foster and John 1999; Pease et al. 1999; Stockli 2005; Fitzgerald et al. 2009).

There are several approaches to constrain fault dips with thermochronologic data. The calculation requires knowing: (1) the direction of slip on the fault, (2) the location of at least two temperatures at a particular time (before or during faulting) along the fault trace as defined by FT or other thermochronometers, and (3) either an assumption about the geothermal gradient or a measure of the paleogeothermal gradient from a titled hanging-wall crustal section. The geothermal gradient should be constrained to within about ± 10 °C/km for a reasonable estimate of dip (John and Foster 1993; Foster and John 1999; Stockli 2005; Fitzgerald et al. 2009). The variation in paleotemperature along the slip direction of a detachment fault at a particular time is related to the dip of the fault provided that: (1) lateral variations in the thermal gradient (due to voluminous intrusions or hydrothermal flow in the footwall) can be ruled out, and (2) that the samples are from below a single fault system. The second may introduce uncertainty for detachment systems with secondary breakaway faults that result in greater extension in down-dip section.

11.9 Chemehuevi Detachment Example to Constrain Fault Dip

FT and $^{40}Ar/^{39}Ar$ data from rocks in the footwall of the Chemehuevi detachment in Southeastern California constrain the initiation angle of this regional detachment fault system (Foster et al. 1990; John and Foster 1993; Foster and John 1999). Contoured values of mineral cooling age from biotite ($^{40}Ar/^{39}Ar$), K-feldspar ($^{40}Ar/^{39}Ar$), ZFT, AFT, and titanite FT decrease north eastward in the slip direction and define the locations of paleoisotherms in the footwall before and during extension (Fig. 11.5). These thermochronologic data indicate a moderate paleotemperature field gradient across the footwall prior to faulting.

At ~ 22 Ma, granitic rocks exposed in the southwestern and northeastern portions of the footwall were at ≤ 200 and ≥ 400 °C, respectively, and were separated by a distance of some 23 km along the known slip direction. This gradual increase in temperature with depth is attributed to the gentle warping of originally subhorizontal isothermal surfaces and constrains the exposed part of the Chemehuevi detachment fault to have had an initial dip of 15° to 30° using a range of geothermal gradients (Fig. 11.6). Syn-extensional plutons are not present in the southern part of the footwall, so the smooth gradient in paleotemperature is not likely to be due to local variations in the geotherm.

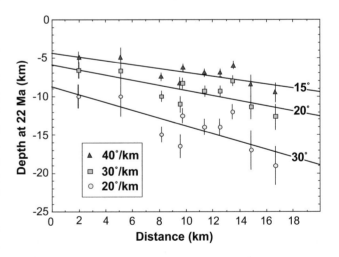

Fig. 11.6 Plot of calculated paleodepth for samples from the southern Chemehuevi Mountains projected to a southwest–northeast cross-section in the direction of slip on the Chemehuevi detachment fault. Paleodepth for each point is calculated from the temperature of each sample at 22 Ma, based on the thermochronologic data assuming three different geothermal gradients. Lines indicate regressions of the data for each gradient and give regional average initial dips of the Chemehuevi detachment

11.10 Conclusions

Many other excellent examples of applications of FT data to structural geology and tectonics exist in the literature. The examples summarized in this chapter provide a general outline for using FT data in normal fault studies. In all cases, a relatively large data set is needed to reduce uncertainty and some of the more powerful applications combine FT, (U-Th)/He, and/or ^{40}Ar/^{39}Ar data from the same samples or from the same structure.

Acknowledgements I would like to thank Stephanie Brichau and Paul Fitzgerald for helpful reviews of the original manuscript and the many collaborators that contributed to the studies summarized in this chapter.

References

Bernet M (2009) A field-based estimate of the zircon fission-track closure temperature. Chem Geol 259:181–189

Brichau S, Ring U, Ketcham RA, Carter A, Stockli D, Brunel M (2006) Constraining the long-term evolution of the slip rate for a major extensional fault system in the central Aegean, Greece, using thermochronology. Earth Planet Sci Lett 241:293–306

Brandon MT, Roden-Tice MK, Garver JI (1998) Late Cenozoic exhumation of the Cascadia accretionary wedge in the Olympic Mountains, NW Washington State. Geol Soc Amer Bull 110:985–1009

Brown RW (1991) Backstacking apatite fission-track "stratigraphy": a method for resolving the erosional and isostatic rebound components of tectonic uplift histories. Geology 19:74–77

Brown RW, Summerfield MA, Gleadow AJW (1994) Apatite fission track analysis: its potential for the estimation of denudation rates and implications for models of long-term landscape development. In: Kirkby MJ (ed) Process models and theoretical geomorphology. Wiley, New York, pp 23–53

Campbell-Stone E, John BE, Foster DA, Geissman JW, Livaccari RF (2000) Mechanisms for accommodation of Miocene extension, low-angle normal faulting, magmatism, and secondary breakaway faulting in the southern Sacramento Mountains, southeastern California. Tectonics 19:566–587

Carter TJ, Kohn BP, Foster DA, Gleadow AJW (2004) How the Harcuvar Mountains metamorphic core complex became cool: evidence from apatite (U-Th)/He thermochronometry. Geology 32:985–988

Carter TJ, Kohn BP, Foster DA, Gleadow AJW, Woodhead JD (2006) Late-stage evolution of the Chemehuevi and Sacramento detachment faults from apatite (U-Th)/He thermochronology—evidence for mid-Miocene accelerated slip. Geol Soc Am Bull 118:689–709

Dickinson WR (1991) Tectonic setting of faulted tertiary strata associated with the Catalina core complex in southern Arizona. Geol Soc Am Sp Paper 264

Dumitru TA, Gans PB, Foster DA, Miller EL (1991) Refrigeration of the western Cordilleran lithosphere during Laramide shallow-angle subduction. Geology 19:1145–1148

Ehlers TA, Armstrong PA, Chapman DS (2001) Normal fault thermal regimes and the interpretation of low-temperature thermochronometers. Phys Earth Planet Int 126:179–194

Ehlers TA, Willett SD, Armstrong PA, Chapman DS (2003) Exhumation of the central Wasatch Mountains, Utah: 2. thermokinematic model of exhumation, erosion, and thermochronometer interpretation. J Geophys Res 108(B3):2173

Espurt N, Barbarand J, Roddaz M, Brusset S, Baby P, Saillard M, Hermoza W (2011) A scenario for late Neogene Andean shortening transfer in the Camisea Subandean zone (Peru, 12°S): implications for growth of the northern Andean Plateau. Geol Soc Am Bull 123:2050–2068

Fitzgerald PG (1992) The Transantarctic Mountains of southern Victoria Land: the application of apatite fission track analysis to a rift shoulder uplift. Tectonics 11:634–662

Fitzgerald PG, Malusà MG (2018) Chapter 9. Concept of the exhumed partial annealing (retention) zone and age-elevation profiles in thermochronology. In: Malusà MG, Fitzgerald PG (eds) Fission-track thermochronology and its application to geology. Springer

Fitzgerald PG, Fryxel JE, Wernicke BP (1991) Miocene crustal extension and uplift in southeastern Nevada: constraints from apatite fission track analysis. Geology 19:1013–1016

Fitzgerald PG, Reynolds SJ, Stump E, Foster DA, Gleadow AJW (1993) Thermochronologic evidence for timing of denudation and rate of crustal extension of the South Mountain metamorphic core complex and Sierra Estrella, Arizona. Nucl Tracks 21:555–563

Fitzgerald PG, Duebendorfer EM, Faulds JE, O'Sullivan P (2009) South virgin—white hills detachment fault system of SE Nevada and NW Arizona: applying apatite fission track thermochronology to constrain the tectonic evolution of a major continental detachment system. Tectonics 28 TC2001

Foster DA, Gleadow AJW (1992) The morphotectonic evolution of rift-margin mountains in central Kenya: constraints from apatite fission-track thermochronology. Earth Planet Sci Lett 113:157–171

Foster DA, Gleadow AJW (1993) Episodic denudation in East Africa: a legacy of intracontinental tectonism. Geophys Res Lett 20:2395–2398

Foster DA, Gleadow AJW (1996) Structural framework and denudation history of the flanks of the Kenya and Anza rifts, East Africa. Tectonics 15:258–271

Foster DA, John BE (1999) Quantifying tectonic exhumation in an extensional orogen with thermochronology: examples from the southern Basin and Range Province. Geol Soc (London) Sp Pub 154:356–378

Foster DA, Raza A (2002) Low-temperature thermochronological record of exhumation of the Bitterroot metamorphic core complex, northern Cordilleran Orogen. Tectonophysics 349:23–36

Foster DA, Harrison TM, Miller CF, Howard KA (1990) The ^{40}Ar/^{39}Ar thermochronology of the eastern Mojave Desert, California and adjacent western Arizona with implications for the evolution of metamorphic core complexes. J Geophys Res 95:20, 005–20, 024

Foster DA, Miller DS, Miller CF (1991) Tertiary extension in the Old Woman Mountains area, California: evidence from apatite fission track analysis. Tectonics 10:875–886

Foster DA, Gleadow AJW, Reynolds SJ, Fitzgerald PG (1993) The denudation of metamorphic core complexes and the reconstruction of the Transition Zone, west-central Arizona: constraints from apatite fission-track thermochronology. J Geophys Res 98:2167–2185

Foster DA, Grice WC, Kalakay TJ (2010) Extension of the Anaconda metamorphic core complex: ^{40}Ar/^{39}Ar thermochronology with implications for Eocene tectonics of the northern Rocky Mountains and the Boulder batholith. Lithosphere 2:232–246

Gleadow AJW, Fitzgerald PG (1987) Uplift history and structure of the Transantarctic Mountains: new evidence from fission track dating of basement apatites in the Dry Valleys area, southern Victoria Land. Earth Plan Sci Lett 82:1–14

Howard KA (1991) Intrusion of horizontal dikes: tectonic significance of Middle Proterozoic diabase sheets widespread in the upper crust throughout the southwestern US. J Geophys Res 96:12461–12478

Howard KA, Foster DA (1996) Thermal and unroofing history of a thick, tilted Basin and Range crustal section, Tortilla Mountains, Arizona. J Geophys Res 101:511–522

John BE, Foster DA (1993) Structural and thermal constraints on the initiation angle of detachment faulting in the southern Basin and Range: the Chemehuevi Mountains case study. Geol Soc Am Bull 105:1091–1108

Karlstrom KE, Heizler M, Quigley MC (2010) Structure and ^{40}Ar/^{39}Ar K-feldspar thermal history of the Gold Butte block: reevaluation of the tilted crustal section model. Geol Soc Am Special Paper 463:331–352

Ketcham RA (1996) Thermal models of core-complex evolution in Arizona and New Guinea: implications for ancient cooling paths and present-day heat flow. Tectonics 15:933–951

Lee J (1995) Rapid uplift and rotation of mylonitic rocks from beneath a detachment fault: insights from potassium feldspar ^{40}Ar/^{39}Ar thermochronology, northern Snake Range, Nevada. Tectonics 14:54–77

Lister GS, Davis GA (1989) The origin of metamorphic core complexes and detachment faults formed during Tertiary continental extension in the Colorado River region, U.S.A. J Struct Geol 11:65–93

Malusà MG, Fitzgerald PG (2018) Chapter 8. From cooling to exhumation: setting the reference frame for the interpretation of thermocronologic data. In: Malusà MG, Fitzgerald PG (eds) Fission-track thermochronology and its application to geology. Springer

Malusà MG, Fitzgerald PG (2018) Chapter 10. Application of thermochronology to geologic problems: bedrock and detrital approaches. In: Malusà MG, Fitzgerald PG (eds) Fission-track thermochronology and its application to geology. Springer

McQuarrie N, Barns JB, Ehlers TA (2008) Geometric, kinematic, and erosional history of the central Andean Plateau, Bolivia (15–17˚S). Tectonics 27 TC3007

Metcalf JR, Fitzgerald PG, Baldwin SL, Muñoz J-A (2009) Thermochronology of a convergent orogen: constraints on the timing of thrust faulting and subsequent exhumation of the Maladeta Pluton in the Central Pyrenean Axial Zone. Earth Plan Sci Lett 287:488–503

Mora A, Ketcham RA, Higuera-Diaz IC, Bookhagen B, Jimenez L, Rubiano J (2014) Formation of passive-roof duplexes in the Colombian Subandes and Peru. Lithosphere 6:456–472

Morrison J, Anderson JL (1998) Footwall refrigeration along a detachment fault: implications for thermal evolution of core complexes. Science 279:63–66

Noble W, Foster D, Gleadow A (1997) The post-Pan-African thermal and extensional history of crystalline basement rocks in eastern Tanzania. Tectonophysics 275:331–350

O'Sullivan PB, Green PF, Bergman SC, Decker J, Duddy IR, Gleadow AJW, Turner DL (1993) Multiple phases of Tertiary uplift and erosion in the National Wildlife Refuge, Alaska, revealed by apatite fission track analysis. AAPG Bull 77:359–385

Pease V, Foster D, O'Sullivan P, Wooden J, Argent J, Fanning C (1999) The Northern Sacramento Mountains, Part II: Exhumation history and detachment faulting. In: Mac Niocaill C, Ryan PD (eds) Continental Tectonics. Geol Soc (London) Sp Pub 164:199–237

Reiners PW, Brady R, Farley KA, Fryxell JE, Wernicke B, Lux D (2000) Helium and argon thermochronometry of the gold butte block, south Virgin Mountains, Nevada. Earth Planet Sci Lett 178:315–326

Spiegel C, Kohn BP, Belton DX, Gleadow AJW (2007) Morphotectonic evolution of the central Kenya rift flanks: implications for late Cenozoic environmental change in East Africa. Geology 35:427–430

Scott RJ, Foster DA, Lister GS (1998) Tectonic implications of rapid cooling of denuded lower plate rocks from the Buckskin-Rawhide metamorphic core complex, west-central Arizona. Geol Soc Am Bull 110:588–614

Singleton JS, Stockli DF, Gans PB, Prior MG (2014) Timing, rate, and magnitude of slip on the Buckskin-Rawhide detachment fault, west central Arizona. Tectonics 33:1596–1615

Spencer JE, Reynolds SJ (1991) Tectonics of mid-Tertiary extension along a transect through west-central Arizona. Tectonics 10:1204–1221

Stockli DF, Dumitru TA, McWilliams MO, Farley KA (2003) Cenozoic tectonic evolution of the White Mountains, California and Nevada. Geol Soc Am Bull 115:788–816

Stockli DF (2005) Application of low-temperature thermochronometry to extensional tectonic settings. In: Reiners PW, Ehlers TA (eds) Low-temperature thermochronology: techniques, interpretations, and applications. Rev Min Geochem 58:411–448. Mineralogical Society of America, Chantilly, Virginia

Stockli DF, Brichau S, Dewane TJ, Hager C, Schroeder J (2006) Dynamics of large-magnitude extension in the Whipple Mountains metamorphic core complex. Geochim Cosmochim Acta 70:A616

Torres Acosta V, Bande A, Sobel ER, Parra M, Schildgen TF, Stuart F, Strecker MR (2015) Cenozoic extension in the Kenya Rift from low-temperature thermochronology: links to diachronous spatiotemporal evolution of rifting in East Africa. Tectonics 34:2367–2386

York D (1969) Least-squares fitting of a straight line with correlated errors. Earth Planet Sci Lett 5:320–324

Wagner GA, Reimer GM (1972) Fission track tectonics: the tectonic interpretation of fission track apatite ages. Earth Planet Sci Lett 14:263–268

Wells ML, Snee LW, Blythe AE (2000) Dating of major normal fault systems using thermochronology: an example from the Raft River detachment, Basin and Range, western United States. J Geophys Res 105:16, 303–16, 327

Wernicke B (1995) Low-angle normal faults and seismicity: a review. J Geophys Res 100:20159–20174

Wernicke B, Axen G (1988) On the role of isotasy in the evolution of normal fault systems. Geology 16:848–851

Application of Fission-Track Thermochronology to Understand Fault Zones

12

Takahiro Tagami

Abstract

The timing and thermal effects of fault motions can be constrained by fission-track (FT) thermochronology and other thermochronological analyses of fault zone rocks. Materials suitable for such analyses are produced by fault zone processes, such as: (1) mechanical fragmentation of host rocks, grain-size reduction of fragments and recrystallisation of grains to form mica and clay minerals, (2) secondary heating/melting of host rocks as the result of friction and (3) mineral vein formation as a consequence of fluid flow associated with fault motion. The geothermal structure of fault zones is primarily controlled by three factors: (a) the regional geothermal structure around the fault zone that reflects the background thermotectonic history of the study area, (b) frictional heating of wall rocks by fault motion and consequent heat transfer into surrounding rocks and (c) thermal effects of hot fluid flow in and around the fault zone. The thermal sensitivity of FTs is briefly reviewed, with a particular focus on the fault zone thermal processes, i.e., flash and hydrothermal heating. Based on these factors, representative examples as well as key issues, including sampling strategy, are highlighted for using thermochronology to analyse fault zone materials, such as fault gouges, pseudotachylytes and mylonites. The thermochronologic analyses of the Nojima fault in Japan are summarised, as an example of multidisciplinary investigations of an active seismogenic fault system. Geological, geomorphological and seismological implications of these studies are also discussed.

12.1 Introduction

Seismic activity and faulting are manifestations of geodynamic processes of the dynamic Earth. They involve a series of fault zone processes and products, such as gouge and pseudotachylyte formation, as a result of repeated fault motions on a geologic timescale at different crustal depths. Because those processes are accompanied by temperature changes along the fault zone, geo- and thermochronological techniques are potentially useful to detect and date the thermal episodes under active and ancient faulting regimes. In particular, fission-track (FT), K–Ar (Ar/Ar) and U–Th methods have successfully been applied to fault zone rocks at a variety of tectonic settings (e.g. Flotte et al. 2001; van der Pluijm et al. 2001; Mulch et al. 2002; Boles et al. 2004; Murakami and Tagami 2004; Sherlock et al. 2004; Zwingmann and Mancktelow 2004; Murakami et al. 2006a; Rolland et al. 2007; Haines and van der Pluijm 2008; see Tagami 2012 for a review). In addition, (U–Th)/He analyses have recently been conducted on zircon separates from fault zones (Yamada et al. 2012; Maino et al. 2015; see also Ault et al. 2015 for haematite analysis).

Compared to other techniques, the advantages of the FT (and also the (U–Th)/He) method where applied to fault zones are: (1) greater thermal sensitivity for secondary heating episodes associated with faulting; (2) higher mechanical/chemical survivability of mineral grains used for FT analysis (e.g. zircon) in the core of fault zones and (3) widespread occurrence of mineral grains used for FT analysis (i.e. apatite and zircon) within continental crust rocks. As a result, FT analysis revealed thermal histories in or near fault zones in various regions, for example, the Median Tectonic Line in Japan (Tagami et al. 1988), the Alpine fault in New Zealand (Kamp et al. 1989), the Pejo fault system in the Italian Eastern Alps (Viola et al. 2003), the San Gabriel Fault in California (d'Alessio et al. 2003) and the Nojima Fault in Japan (Murakami and Tagami 2004).

T. Tagami (✉)
Division of Earth and Planetary Sciences, Kyoto University, 606-8502 Kyoto, Japan
e-mail: tagami@kueps.kyoto-u.ac.jp

© Springer International Publishing AG, part of Springer Nature 2019
M. G. Malusà and P. G. Fitzgerald (eds.), *Fission-Track Thermochronology and its Application to Geology*, Springer Textbooks in Earth Sciences, Geography and Environment, https://doi.org/10.1007/978-3-319-89421-8_12

This chapter reviews the state-of-the-art of FT thermochronology of seismogenic fault zones and illustrates the relevant scientific backgrounds of fault zone processes and materials, geothermal regime of fault zones, thermal stability of FTs in fault settings and key studies of application to different fault lithologies.

12.2 Fault Zone Processes and Materials

12.2.1 Faults and Fault Zone Rocks

The schematic cross-section of a typical fault zone[1] is shown in Fig. 12.1a, where the geologic features of fault rocks are presented along with thermomechanical characteristics (see also Chap. 10, Malusà and Fitzgerald 2018). Fault zone rocks are classified as gouges, cataclasites, pseudotachylytes, mylonites, etc., based primarily on their textures (Sibson 1977). The formation process of fault zone materials includes three primary categories: (1) mechanical fragmentation of host rocks, coupled with grain-size reduction and recrystallisation to form secondary minerals, such as clay; (2) frictional heating by fault motions that occasionally leads to melting of wall rocks and (3) fluid flow and chemical precipitation to form mineral veins. These processes are described below in more detail.

12.2.2 Grain-Size Reduction and Recrystallisation

Frictional fault motion, which is controlled by brittle fracture at shallow crustal levels, causes damage and abrasion of host-rock surfaces. As a result, loose particles with angular shapes, called gouge, are formed. As the faulting progresses, the gouge shears as a granular material, resulting in a systematic reduction in the average grain size.

Fault gouges generally contain a variety of neoformed clay minerals, such as illite and smectite. In igneous and metamorphic environments, in situ clay mineralisation plays a major role in forming authigenic clays in gouge. Fluid flow is likely a key factor to promote mineralogical reactions, and the authigenic illites formed are useful for K–Ar (Ar/Ar) dating.

At deeper crustal levels where rock-forming minerals are semi-brittle to plastic, ductile faulting (or shearing) is dominant and likely forms mylonites. Recrystallisation of biotite takes place at ∼ 350–400 °C (Simpson 1985), whereas white micas are formed by synkinematic crystallisation at

<hr>

[1]"Fault zone" is a term to describe the spatial extent along a fault where rocks are significantly deformed by the faulting.

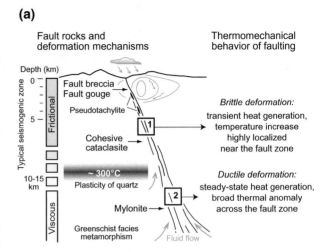

(a)

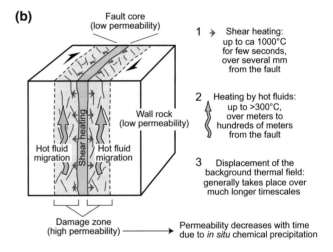

(b)

Fig. 12.1 Synoptic model of a fault zone. **a** Fault rocks and deformation mechanisms against depth (and temperature). **b** Permeability structure and thermal regime (after Scholz 1988; Tagami 2012, as modified from Fig. 10.4a—Malusà and Fitzgerald 2018)

the expense of feldspars at ∼ 400 °C (Rolland et al. 2007). Those newly formed micas are widely used for Ar/Ar thermochronology (see Tagami 2012 for more details).

12.2.3 Frictional Heating and Pseudotachylyte Formation

The mechanical work of faulting primarily comprises three factors: (a) frictional heating, (b) surface energy of gouge formation and (c) elastic radiation (Scholz 2002). Factors (b) and (c) are in general much less important than frictional heating, and thus the mechanical work is primarily expended by the generation of frictional heat. The generation of heat on a fault plane can accordingly be approximated as:

$$\tau v = q \qquad (12.1)$$

where τ is the mean shear stress acting on a fault sliding at velocity v, and q is the heat flow generated by the fault motion (Scholz 2002).

The thermomechanical behaviour of faulting is classified into two general regimes: (a) under brittle conditions, a transient heat pulse is generated by rapid coseismic fault slip, with $v = \sim 10$–100 cm/s; (b) under ductile conditions, heat generation can be regarded as steady-state, due to a long-term fault motion averaged over geological time, with $v = \sim 1$–10 cm/year (Scholz 2002).

In the brittle regime, heat generation is highly localised near the fault zone owing to the greater v coupled with low thermal conductivity of rocks. In this case, fault zone rocks can occasionally be molten to form glassy vein-shaped rocks, called pseudotachylytes (Sibson 1975). In contrast, the latter type of heating is expected to form a broader, regional thermal anomaly across the fault zone due to the smaller v and constant heat generation. Some regional metamorphic aureoles in convergent plate boundaries (or transcurrent shear zones) may be attributable to such long-term heating at depths (Scholz 1980). In both regimes, the time and magnitude of heating can be assessed quantitatively by thermochronology using FT and other techniques, although the rock sampling strategy is substantially different between the two.

The energy budget of earthquakes can also be estimated by thermochronological methods. The mechanical work of fault motion, W_f, is a function of mean shear stress, which is difficult to measure directly on the fault plane. Thus, an available approach to constrain W_f is to measure the generated heat flow q during an earthquake. This can be achieved either by: (a) detection of the temperature anomaly across the fault zone by drilling into the fault soon after the earthquake (Brodsky et al. 2010, and references therein), or (b) geothermometric analyses of fault zone rocks that experienced frictional heating, such as vitrinite reflectance or FT analysis (e.g. O'Hara 2004).

12.2.4 Fluid Flow and Mineral Vein Formation

A variety of mineral veins in fault zones, such as calcite, quartz and ore deposits, are typically formed as fracture fillings under an extensional regime, as a consequence of fluid flow and in situ chemical precipitation. At crustal temperatures exceeding 200–300 °C, the healing and sealing of fractures effectively proceeds to produce veins, probably within the conventional lifetime of hydrothermal systems (Cox 2005). Hence, the formation of extensional vein arrays in seismogenic crustal depths can closely represent the time of brittle failure and permeability enhancement that induce fluid flow. During the seismic cycle of a fault system, a significant slip event likely produces large ruptures in the fault zone, reduces the fault strength, enhances fluid flow and eventually leads to veins formation (e.g., Sibson 1992). This seismic episode is followed by an interseismic period when the fault strength is progressively recovered as the healing and sealing of ruptures progresses. This evolution is called fault-valve behaviour (Sibson 1992).

The time of brittle failure and permeability enhancement of a fault zone can be constrained by dating mineral veins. U–Th disequilibrium analysis of carbonate veins has successfully been applied to date neo-tectonic fault systems (Flotte et al. 2001; Boles et al. 2004; Verhaert et al. 2004; Watanabe et al. 2008; Nuriel et al. 2012). In addition, hot fluid flow caused by brittle faulting should be recorded in adjacent wall rocks as a thermal anomaly, which may be detected by low-temperature thermochronology of host minerals, e.g. FT analysis on apatite and zircon separated from fault zone rocks.

12.3 Thermal Regime of Fault Zones

This section provides a brief overview of temperature variations in a fault zone in space and time. The thermal regime is a key parameter governing the response of applied thermochronometers in terms of thermal stability (or retentivity/diffusivity) of accumulated daughter nuclides (or lattice damages) in the target mineral.

12.3.1 Regional Geothermal Structure and Background Thermal History

The geothermal structure of solid Earth can be approximated by a one-dimensional temperature profile against depth from the Earth's surface. The temperature increases toward the centre of the Earth, with a higher geothermal gradient within the lithosphere. Tectonics perturbs the first-order geothermal structure of the upper continental crust with significant departures from its average geothermal gradients, generally assumed to be on the order of 30 °C/km (see discussion in Chap. 8, Malusà and Fitzgerald 2018). The thermal regime of a fault zone is controlled by three factors (Fig. 12.1b): (a) the regional geothermal structure and background thermal history of the study area, (b) frictional heating of the wall rocks during faulting in the brittle regime and (c) heating of the wall rocks by hot fluid flow in and around the fault zone.

Where the fault motion has some vertical component, the two blocks separated by the fault show differential uplift/subsidence movements with respect to each other. If uplift is accompanied by exhumation, the rocks within the uplifted block effectively cool down, due to the resultant long-term downward motion of geotherms to adjust for the new state of geothermal equilibrium. Conversely, if subsidence is accompanied by sediment deposition, the rocks within the subsided block are effectively heated, as a result of burial and long-term upward motion of geotherms (note that the thermal effects of thrust faulting can be grossly different (e.g. Metcalf et al. 2009)). As the fault motion continues, the amount of fault slip accumulates and thus the difference in the thermal signature progressively increases between rocks that were once juxtaposed each other across the fault boundary. The difference eventually may become large enough to be resolved by an appropriate thermochronometric technique, which then places constraints on the timing and magnitude of vertical components of the accumulated fault motions. Low-temperature thermochronology using (U–Th)/He, FT and/or K–Ar (^{40}Ar/^{39}Ar) techniques is particularly useful for reconstructing such regional thermal histories. Note that such modifications to the background thermal structure occur over timescales that are much longer than those characterising thermal processes within a fault zone, i.e. frictional heating in the brittle regime and hot fluid flow. Also note that, on a regional scale, the background thermal history may vary depending on a variety of factors, such as surface topography, spatial variation of geothermal structure, tectonic tilting and ductile deformation at depth (see Chap. 8, Malusà and Fitzgerald 2018).

12.3.2 Frictional Heating of Wall Rocks by Fault Motion

Frictional heating in the brittle deformation regime is characterised by episodic temperature increase up to ~1000 °C, within a typical time period of several seconds and a spatial range of several mm from the fault (Fig. 11.1b). Lachenbruch (1986) quantified the production of frictional heat and its conductive transfer into wall rocks, using the heat conduction models of Carslaw and Jaeger (1959). Suppose a fault slips along a fault zone of width $2a$ during a time interval $0 < t < t^*$, where t^* is a duration of the slip. Within the fault zone ($0 < x < a$, where x is a distance to the fault plane), the temperature rise ΔT during faulting ($t < t^*$) is given by

$$\Delta T = \frac{\tau}{\rho c}\frac{v}{a}\left\{t\left[1 - 2\iint erfc\frac{a-x}{\sqrt{4\alpha t}} - 2\iint erfc\frac{a+x}{\sqrt{4\alpha t}}\right]\right\} \tag{12.2}$$

where ρ is the density, c is the specific heat and α is the thermal diffusivity (Fig. 12.2), whereas the temperature rise after faulting ($t^* < t$) is given by

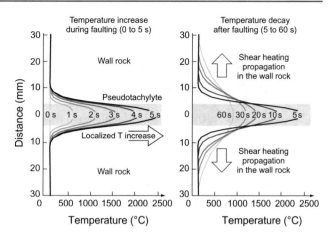

Fig. 12.2 Profiles of temperature increase during faulting (left) and temperature decay after a faulting event (right), for a shear work of 50 MPa m. The temperature increase ΔT was calculated using the equations of Carslaw and Jaeger (1959) and Lachenbruch (1986), assuming a duration of the local seismic slip of 5 s. The temperature exceeds 1000 °C in a faulting zone of 8 mm width (grey box) for ~30 s, following the onset of faulting (after Murakami 2010)

$$\Delta T = \frac{\tau}{\rho c}\frac{v}{a}\left\{t\left[1 - 2\iint erfc\frac{a-x}{\sqrt{4\alpha t}} - 2\iint erfc\frac{a+x}{\sqrt{4\alpha t}}\right]\right.$$
$$- (t - t^*)\left[1 - 2\iint erfc\frac{a-x}{\sqrt{4\alpha(t-t^*)}}\right.$$
$$\left.\left. - 2\iint erfc\frac{a+x}{\sqrt{4\alpha(t-t^*)}}\right]\right\} \tag{12.3}$$

Instead, outside of the fault zone ($x > a$), the temperature rise ΔT during faulting ($t < t^*$) is given by

$$\Delta T = \frac{2\tau}{\rho c}\frac{v}{a}\left\{t\left[\iint erfc\frac{x-a}{\sqrt{4\alpha(t)}} - \iint erfc\frac{x+a}{\sqrt{4\alpha(t)}}\right]\right\} \tag{12.4}$$

and the temperature elevation after faulting ($t^* < t$) is given by

$$\Delta T = \frac{2\tau}{\rho c}\frac{v}{a}\left\{t\left[\iint erfc\frac{x-a}{\sqrt{4\alpha t}} - \iint erfc\frac{x+a}{\sqrt{4\alpha t}}\right]\right.$$
$$- (t - t^*)\left[\iint erfc\frac{x-a}{\sqrt{4\alpha(t-t^*)}}\right.$$
$$\left.\left. - \iint erfc\frac{x+a}{\sqrt{4\alpha(t-t^*)}}\right]\right\} \tag{12.5}$$

The mean shear stress τ can then be estimated using these equations, where geothermometric analysis of known kinetics is carried out (O'Hara 2004; Tagami 2012).

According to Lachenbruch (1986), if the slip duration t^* is negligibly small relative to post-seismic observation time $(t - t^*)$, and also if our observation time t is sufficiently large compared to the time constant λ of the shear zone $(\lambda = a^2/4\alpha)$, Eqs. (12.2)–(12.5) can be simplified (for any x, $t \gg t^*$, and $t \gg \lambda$) to

$$\Delta T = (\tau u/\rho c)\,(\pi \alpha t)^{-1/2} \exp\left(-x^2/4\alpha t\right) \quad (12.6)$$

where u is the slip distance ($u = vt^*$). The mean shear stress τ can accordingly be calculated for certain x and t conditions by substituting individual appropriate values to ρ, c and α, and by measuring ΔT and u. This approach was applied to the case of temperature anomaly measurement across a fault zone, conducted by drilling into the fault soon after the earthquake (Brodsky et al. 2010).

12.3.3 Hot Fluid Flow in and Around the Fault Zone

Fluid flow within a fault zone can be inferred from the occurrence of mineral veins formed by in situ chemical precipitation (see Sect. 12.2.4). The spatial range of the effective flow is primarily on the order of 1–100 m normal to the fault, based on natural occurrences of mineral veins (e.g. Boles et al. 2004; Watanabe et al. 2008; Cox 2010). The permeability structure of fault zones, which is a key to control the flow, in general consists of three regions: (a) the fault core, which comprises fault gouge and breccia, both characterised by low permeability; (b) the damage zone, which consists of fractured rocks and has high permeability that offers effective pathways for seepage flows and (c) the bedrock protolith with low permeability (Evans et al. 1997; Seront et al. 1998). The mean permeability of a fault zone is inferred to decrease with time, as a result of narrowing/closure of pathways due to the continued fluid flow with chemical precipitation to form veins. The permeability likely recovers if the fault zone experiences new seismic activity and resultant reopening of the pathways. This temporal model was first tested by the Nojima Fault Zone Probe Project (see review in Tagami 2012).

If the seepage flow in a fault zone is dominated by upward components, the fault zone is heated by flows from deeper crustal levels and is hotter than the environmental temperature (Fig. 12.1b). Hot springs are often found near active fault systems, and some of them likely have deep origins as indicated by their geochemical signatures (e.g., Fujimoto et al. 2007). Some of the large faults continue from the surface to >10 km depths (Scholz 2002), and the temperature of upcoming fluids is estimated to exceed 300 °C under the assumption of a normal geothermal gradient of $\sim$30 °C/km, and neglecting heat loss during seepage flow.

12.4 FT Stability During Flash and Hydrothermal Heating

This section gives a brief overview of the kinetic formulation of FT annealing and its application to fault zone settings. If a host rock undergoes a temperature increase, fission tracks that have been accumulated are shortened progressively and eventually erased by thermal annealing. The reduction of FT lengths is a function of heating time and temperature, and the temperature interval of partial annealing is substantially variable among different minerals. FT annealing is more precisely quantified by using the reduction of etched track length than the etched track density (see Chap. 3, Ketcham 2018), and the shape of the track-length distribution is diagnostic of the thermal history of the rock. Accordingly, horizontal confined track lengths are routinely analysed to determine the annealing kinetic functions, an example of which is:

$$\mu = 11.35\left[1 - \exp\left\{-6.502 + 0.1431\,\frac{(\ln t + 23.515)}{(1000/T - 0.4459)}\right\}\right] \quad (12.7)$$

where μ is the mean FT length in zircon, after annealing at T Kelvin for t hours (Tagami et al. 1998). For more details, see other comprehensive reviews (e.g. Donelick et al. 2005; Tagami 2005; Tagami and O'Sullivan 2005).

The thermal regime of fault zones is characterised by frictional heating and hot fluid flow, in addition to the regional background geothermal structure. Hence, for reliable FT thermochronologic analysis of fault zones, additional consideration is needed to assess the influence of both high-temperature short-term heating and hydrothermal heating on the FT annealing kinetics (which is usually determined in a dry atmospheric environment in a conventional laboratory and extrapolated to geological timescales).

Laboratory heating experiments to simulate thermal perturbations at hydrothermally pressurised conditions due to hot fluid flow were performed on zircon (Brix et al. 2002; Yamada et al. 2003) using a hydrothermal synthetic apparatus. It was found using the same zircon sample and analytical procedure that the observed FT length reduction in the atmosphere is indistinguishable from that under hydrothermal conditions (Yamada et al. 2003). The results of Brix et al. (2002) using Fish Canyon Tuff zircon also exhibit consistency of annealing kinetics between dry and hydrothermal conditions. It thus suggests that the conventional zircon annealing kinetics can also be applied to hydrothermal heating conditions in nature, such as fault zones and sedimentary basins settings.

Frictional heating along a fault in the brittle regime is a short-term geological phenomenon with an effective heating duration on the order of seconds, which is significantly shorter than the conventional laboratory heating of $\sim 10^{-1}$ to 10^4 h. Thus, high-temperature and short-term heating (i.e. flash heating) experiments were specially designed and conducted using a graphite furnace coupled with infrared radiation thermometers (Murakami et al. 2006b). The observed track-length reduction in zircon by 3.6–10 s of heating at 599–912 °C is, overall, slightly greater than that predicted by the zircon FT annealing kinetics, based upon the heating for $\sim 10^{-1}$ to 10^4 h at ~ 350–750 °C (Yamada et al. 1995; Tagami et al. 1998). Yamada et al. (2007) proposed revised kinetic models that integrate results of both the flash heating and conventional laboratory to geological heating. It should be noted here that spontaneous tracks in zircon are totally annealed at 850 ± 50 °C for ~ 4 s, suggesting that the zircon FT system can be completely reset during the ordinary pseudotachylyte formation in nature (Otsuki et al. 2003).

A note should be added concerning FT annealing kinetics of apatite. The kinetics has been established in dry atmospheric environments in conventional laboratories and also tested and calibrated on geological timescales by analysing drilled cores of sedimentary basins. However, the influences of both flash and hydrothermal heating on the kinetics are not well known yet, and thus specially designed experiments are strongly needed for apatite. These experiments will provide a more robust basis for interpreting apatite FT data in terms of frictional heating and hot fluid flow in fault zones.

12.5 Rock Sampling Strategy

The rock sampling strategy for fault zone studies may be different from other applications primarily due to the characteristic heat source associated with faulting in the brittle regime (cf. Chap. 10, Malusà and Fitzgerald 2018, and Chap. 11, Foster 2018). As argued above, three thermal factors relevant to the fault zone generally yield characteristic thermal histories (i.e. temperature–time paths) with different spatial ranges (Fig. 12.1b): (a) background thermal history that reflects long-term tectonics of crustal basement, with an ordinary spatial range of ~ 1–100 km from the fault; (b) episodic frictional heating up to an order of 1000 °C (i.e., occasionally above the melting temperature of wall rocks), with a typical time period of several seconds and spatial range of several mm from the fault (in case of brittle deformation) and (c) heating by hot fluid flow, with temperatures below the melting point of host rock, within ~ 100 m of the fault zone. In theory, we can place constraints on each of these three factors by analysing a series of rocks in and around the fault zone using thermochronological techniques. In planning the research strategy toward this

goal, we need to take into account the following geological and analytical aspects:

- First of all, we need to choose appropriate thermochronometers with thermal sensitivities suited to the expected temperature and time ranges of the thermal history of interest.
- If we want to analyse heating events localised to the fault zone, i.e. factors (b) and (c), the background thermal history of the protolith needs to be simple, so that we can readily extract the thermal history signals of interest from the background noise. In addition, the background thermal history needs to be the same along the fault; otherwise, each locality chosen for analysis will have different boundary conditions.
- To conduct thermochronologic analysis with good resolution, it is also desirable to have a large age contrast between the background thermal history and the heating event(s) of the fault zone.
- In order to reveal localised heating events, it is essential to choose the optimal traverse(s) across the fault zone that will satisfy the above conditions. Hence, adequate knowledge is required also on the spatial geometry of the fault plane.

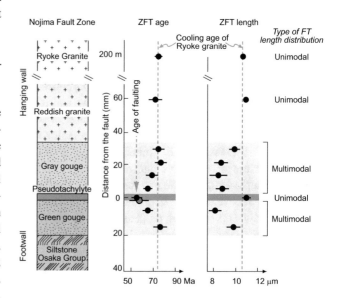

Fig. 12.3 Lithology of the sampled fault rock section in the Hirabayashi trench of the Nojima Fault, Japan, along with a plot of mean age, mean length and type of length distributions of fission tracks in zircon (ZFT). Another sample from the pseudotachylyte layer is also plotted (grey circle). The grey-dashed lines represent the mean age and length of a sample from the Ryoke host rock collected at ~ 200 m away from the fault. The sample from the pseudotachylyte layer has an age significantly younger than that of initial cooling of the samples from the Ryoke host rock. Error bars are ± 1 SE (after Murakami and Tagami 2004)

To constrain the three factors (a)–(c) with good confidence, we need to collect rock samples at different spatial intervals toward the fault plane. At localities >100 m distant from the fault, the interval is not necessarily short and thus rocks can be sampled as for ordinary thermochronologic studies. As approaching the fault, however, we need to make the sample interval shorter in order to determine the possible spatial change of thermochronologic data. In particular, where we intend to detect the thermal effects of the factor (b) (i.e. frictional heating), the interval should be kept as short as several mm within ∼10 cm of the fault plane (d'Alessio et al. 2003; Murakami and Tagami 2004) (Figs. 12.3 and 12.4). Such rock sampling requires a special caution for handling brittle fault rocks, such as cataclasites and mylonites. If possible, the section across the fault plane could be cut into blocks by a portable rock saw, so that the sample blocks can be brought back to the laboratory for precise and contamination-free sampling (e.g. manual cutting by metal blades, handpicking pieces by tweezers). Otherwise, fault rocks need to be precisely divided and sampled in situ, with special care to avoid possible contamination. In these regards, the ideal condition may be to sample rocks from a continuous drill core section across the

fault. Alternatively, trenching the fault will offer an opportunity to sample fault zone rocks in a three-dimensional geometry. In contrast, the sampling strategy for fault zones in the ductile regime is substantially different from that of the brittle regime mentioned above. This is because the expected heat generation in the ductile regime is steady state over geological time, with a faulting velocity of ∼1–10 cm/year (see Sect. 12.2.3), and this likely results in a broader, regional thermal anomaly across the fault zone. As will be documented in Sect. 12.6.2, the spatial extent of the thermal anomaly may reach 10 km away from the fault., Hence, the spatial interval between individual rock sampling localities need not be short, as for sampling within the brittle regime, and thus rocks can be sampled as the ordinary thermochronologic studies. It is noted, however, that the thermal anomaly formed by ductile faulting may be difficult to distinguish from the background thermal history using thermochronology, because the two thermal processes may lead to similar spatial distributions of thermochronologic data.

12.6 Key Studies

12.6.1 The Nojima Fault

The Nojima Fault runs along the northwestern coast of the Awaji Island, Hyogo Prefecture, Japan and is a high-angle reverse fault dipping 83° SE with a right-lateral slip component. A >10 km long surface rupture was formed along the active Nojima Fault, as a result of the 1995 Kobe earthquake (Hyogoken-Nanbu earthquake; M7.2). Shortly after the earthquake, the Nojima Fault Zone Probe Project was initiated as a multidisciplinary geoscience program (Oshiman et al. 2001; Shimamoto et al. 2001; Tanaka et al. 2007), involving drilling a series of boreholes that penetrated the fault at depth. Two boreholes penetrated the Nojima Fault at different depths with nearly complete core recoveries: at 625.27 m by the Geological Survey of Japan 750 m (GSJ-750) borehole at the Hirabayashi (northern) site; and at 389.4 m by the University Group 500 m (UG-500) borehole at the Toshima (southern) site. In addition, the Nojima Fault was also trenched at Hirabayashi, where the fault rocks exposed include, from the hanging wall to the footwall, a granitic cataclasite, a 2–10 mm wide pseudotachylyte layer and the siltstone of the Osaka Group. At Hirabayashi, the fault rupture formed in 1995 is located about 10 cm below the pseudotachylyte.

Using the Hirabayashi trench, Murakami and Tagami (2004) carried out FT analysis on zircon separates from a 50-cm-wide grey fault rock that, from the footwall toward the hanging wall, consists of (1) greenish-grey gouge of the footwall (NT-LG; ∼20 mm wide), (2) pseudotachylyte

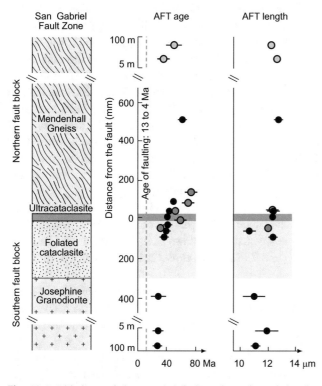

Fig. 12.4 Lithology of the sampled fault rock section of the San Gabriel fault zone, southern California, along with a plot of mean age and length of fission tracks in apatite (AFT). Data from two transects across the fault are shown in different symbols, neither of which show any significant reductions by the fault activity, even in samples within just 2 cm of the ultracataclasite. Error bars are ±1 SE uncertainty (after d'Alessio et al. 2003)

(NT-Pta; ~2–10 mm wide), (3) grey gouge of the hanging wall (NT-UG; ~30 mm wide) and (4) reddish granite (NF-HB1; ~20 mm wide) (Fig. 12.3). Otsuki et al. (2003) estimated the temperature of the pseudotachylyte formation as ~750–1280 °C, based primarily on the observation of melting of K-feldspar and plagioclase. Both FT ages and track lengths systematically vary with distance to the pseudotachylyte, with the youngest age of 56 ± 4 (1SE) Ma from the pseudotachylyte layer. While the age of background regional cooling is 74 ± 3 Ma, four gouge samples from the hanging wall (i.e. NT-UG 1–4) and two gouge samples from the footwall yielded ages of 65–76 Ma, with progressive younging toward the pseudotachylyte. The track-length distribution changes when approaching the pseudotachylyte from unimodal (long tracks) to widespread multimodal (long and short tracks), and eventually to unimodal again (long tracks). These data suggest that the zircon FT system in the pseudotachylyte layer was totally reset and subsequently cooled at ~56 Ma, with a thermal perturbation occurring in the surrounding fault zone rocks. This interpretation is supported by a combination of (a) the temperature estimate for the Nojima pseudotachylyte formation of ~750–1280 °C, on the basis of feldspar melting textures (Otsuki et al. 2003) and (b) laboratory flash heating experiments of zircon FT system (Murakami et al. 2006b).

The borehole rocks from GSJ-750 and UG-500 were also analysed by zircon FT thermochronology (Tagami et al. 2001; Murakami et al. 2002; Tagami and Murakami 2007). Zircon age and length data suggest: (a) an ancient heating event causing a temperature increase into the zircon partial annealing zone (PAZ) within ~25 m in both the footwall and hanging wall at the Hirabayashi borehole (GSJ-750) and (b) an ancient heating into the zircon PAZ within 3 m of the fault in the hanging wall only at Toshima (UG-500). The age of the last cooling after the secondary heating was estimated on partially annealed samples near the fault using the Monte Trax inverse modelling (Gallagher 1995) as 35.0 ± 1.1 (1SE), 38.1 ± 1.7 and 31.3 ± 1.4 Ma at the Hirabayashi borehole (GSJ-750), and 4.4 ± 0.3 Ma at Toshima (UG-500). The maximum temperature experienced during the secondary heating is not uniquely determined since the degree of FT annealing also depends on the heating duration. The source of the secondary heating is likely heat transfer and dispersion via fluids within the fault zone because the spatial range of the annealing zone (i.e. ~25 m and ~3 m from the fault) is too large to be attributed to simple heat conduction. This interpretation is also favoured by the fact that the degree of FT annealing is positively correlated with deformation/alteration of the borehole rocks. The result of the in situ heat dispersion calculation indicates that in situ frictional heat is not sufficient to explain the degree of FT annealing, and hence some additional heat is required, for example, upward flow of hot fluid along the fault zone from deeper crustal levels.

Zwingmann et al. (2010a) reported authigenic illite K–Ar ages from six granitic samples outcropping at Hirabayashi, including three fault gouge samples in close proximity to the pseudotachylyte layer. The six ages of the <2 μm fractions fall in the range of 56.8 ± 1.5 (1SE) to 42.2 ± 1.0 Ma, whereas the <0.1 and <0.4 μm fractions of the three gouge samples are anomalously younger at 30.3 ± 0.9 to 9.1 ± 1.6 Ma. The former ages (57–42 Ma) are younger than the time of regional cooling of the granitic protolith (74 ± 3 Ma) and interpreted as the time of brittle faulting. The latter ages (30–9 Ma) on the finer fractions, which have lower effective closure temperatures, probably reflect a secondary loss of radiogenic ^{40}Ar, likely as a result of thermal overprints near the fault caused by hot fluid flow. Five UG-500 core samples were also analysed at Toshima, with three ages from 50.7 ± 1.2 to 45.0 ± 0.9 Ma for the <0.4 and <2 μm fractions (less K-feldspar contamination), suggesting a time of brittle faulting similar to that of the Hirabayashi outcrop.

In addition, Watanabe et al. (2008) measured U–Th radioactive disequilibrium on calcite veins from 1484 m depth from the UG-1800 m borehole at Toshima (~1 km southwest of the UG-500 drilling locality). The presence of radioactive disequilibrium in $^{234}U/^{238}U$ suggests that the age of calcite precipitation was younger than 1 Ma, which constrains the time of fluid infiltration into the fault zone. Furthermore, electron spin resonance (ESR) analyses conducted on borehole rocks from the UG-500, documented that the ESR intensity (Al and E′ centres) is significantly reduced within ~3 mm from the fault plane (Fukuchi and Imai 2001; Matsumoto et al. 2001). This is likely caused by frictional heating of the Nojima Fault, suggesting that the ESR method can potentially be applied to dating recent movements of an active fault system.

Based on the thermochronologic and other constraints mentioned above, a plausible evolution for the Nojima Fault is reconstructed as follows (see Tagami and Murakami (2007) and Zwingmann et al. (2010a) for more details):

- By ~56 Ma, the fault had already initiated as an in-plane fault offset at crustal depths. The depth may have been >15 km based on fluid inclusion data (Boullier et al. 2001).
- From ~56 to 42 Ma, brittle faulting took place (or had continued) in some of the segments of the ancient Nojima fault.
- At ~35 and 4 Ma at Hirabayashi and Toshima, respectively, fault activity was accompanied by heat transfer and dispersion via fluids circulating within the fault zone.

- At ~1.2 Ma, the present Nojima fault system was formed by the reactivation of the ancient Nojima Fault. Note that such reactivation phenomena have been widely recognised elsewhere (e.g. Holdsworth et al. 1997).

Another important tectonic constraint derived from zircon FT data of the Nojima pseudotachylyte is that the mean shear stress τ can be estimated by combining Eqs. (12.2) to (12.5) of frictional heat production and conductive transfer and Eq. (12.7) of zircon FT annealing (Murakami 2010). The logic of this approach is:

- Temperature profiles are given through time t for a specific τ by assigning or assuming parameters of Eqs. (12.2) to (12.5), i.e. the density ρ, specific heat c, thermal diffusivity α, fault zone width a, sliding velocity v and slip duration t^*.
- The thermal history at each locality can be derived from the series of temperature profiles above, by specifying a certain distance from the fault centre, x.
- For a certain combination of τ and x values, a mean FT length μ is predicted using Eq. (12.7), by integrating the annealing effects for each of the divided time intervals of the thermal history.
- Accordingly, for a certain τ value, a spatial profile of predicted μ is constructed against x.
- For a range of τ values, the spatial profiles of μ are computed and compared with measured FT length data, in order to search for the best fitting τ value.

Consequently, the net shear work (i.e. $\tau v t^*$) was estimated as ~50 MPa m for the faulting event at ~56 Ma (Murakami 2010), which likely favours the "strong" fault model that supports higher shear stresses (Scholz 2002).

12.6.2 Other Examples

Fault Gouges Although fault gouges are widespread along fault cores in most active fault systems, the application of the FT techniques to these rocks is quite limited and has not been very successful in constraining the areal extent of thermal perturbation or the age of faulting. D'Alessio et al. (2003) conducted apatite FT analysis on the samples adjacent to and within the San Gabriel fault zone, southern California, which was likely active from 13 to 4 Ma and has since been exhumed from depths of 2 to 5 km. At the studied locality, the San Gabriel fault consists of an ultracataclasite zone of 1–8 cm width that juxtaposes the Medenhall gneiss to the north with the Josephine granodiorite to the south (Fig. 12.4). Apatite FT ages and lengths show no significant

reductions by the fault activity, even in samples within just 2 cm from the ultracataclasite. The absence of any measurable FT annealing implies that either each slip was never larger than 4 m, or the average apparent coefficient of friction was <0.4, on the basis of the forward modelling of heat generation, heat transport and apatite FT annealing.

Wolfler et al. (2010) applied the apatite FT and (U–Th)/He techniques to the samples from drill cores transecting the Lavanttal fault system, Eastern Alps. Apatite FT ages and lengths exhibit slight reductions toward the fault cores, whereas apatite (U–Th)/He ages also show younger ages in the fault cores. These results suggest that the samples were reheated either by frictional heating and/or by hot fluid flow within the fault zone.

Instead, illite K–Ar and ^{40}Ar/^{39}Ar analyses have widely been applied to date fault gouges under a variety of tectonic settings. The age of authigenic illite, formed in situ within the fault zone, should directly date the time of (some stages of) fault zone activity. After some pioneering studies, van der Pluijm and collaborators succeeded in quantifying the ratio of authigenic and detrital micas for individual clay size fractions by using quantitative X-ray analysis of clay grain-size populations (e.g. van der Pluijm et al. 2001, 2006; Solum et al. 2005; Haines and van der Pluijm 2008). In addition, Zwingmann and collaborators further demonstrated the applicability of illite K–Ar dating by analysing Alpine fault gouges (e.g., Zwingmann and Mancktelow 2004; Zwingmann et al. 2010b).

Pseudotachylytes Dating the glassy matrix of pseudotachylytes has been attempted by several studies, including ^{40}Ar/^{39}Ar thermochronology (Vredefort dome, South Africa, by Reimold et al. 1990; North Cascade Mountains, western USA, by Magloughlin et al. 2001; Alpine Fault, New Zealand, by Warr et al. 2003; More-Trondelag Fault, Central Norway, by Sherlock et al. 2004), glass FT thermochronology (Alpine Fault, New Zealand, by Seward and Sibson 1985) and Rb–Sr geochronology (Quetico and Rainy Lake-Seine River fault, western Superior Canadian Shield, by Peterman and Day 1989). As shown in those studies, constraining the ages of the glassy matrix of pseudotachylyte likely encountered three potential pitfalls: i.e. we do not know whether or not:

- Complete age resetting occurs as the result of frictional heating during pseudotachylyte formation. This is primarily an issue of diffusion kinetics of the radiogenic isotope during flash heating. In the case of ^{40}Ar/^{39}Ar thermochronology, however, the presence of inherited argon at crustal depths may further violate the key assumptions of dating pseudotachylyte, i.e. the host-rock argon is lost to an infinite reservoir during near-instantaneous frictional melting (Sherlock et al. 2004).

- The isotopic system has been affected by later thermal events. Possible breakdown of the radiometric closed system is probably an issue of diffusion kinetics of the radiogenic isotope in hydrothermal heating environments, but could also be caused by devitrification of the glassy matrix of pseudotachylytes, as often observed in nature.

- The glassy matrix of the pseudotachylyte is completely free of pieces of country rock. If not, the complete resetting of the age will be extremely difficult for some of the thermochronologic techniques, such as $^{40}Ar/^{39}Ar$ method.

An alternative thermochronologic approach was adopted by Murakami and Tagami (2004), who carried out zircon FT analyses on a pseudotachylyte and its host rocks of the Nojima fault zone, southwest Japan (Fig. 12.3) (see Sect. 12.6.1). This new approach was subsequently applied to a variety of pseudotachylytes formed in different tectonic settings (e.g. Takagi et al. 2007, Tsergo Ri Landslide, Nepal; Takagi et al. 2010, Median Tectonic Line, southwest Japan). Both FT and U–Pb analyses were conducted on zircons separated from a pseudotachylyte layer and surrounding granitic fault rocks of the ancient Asuke shear zone, central Japan (Murakami et al. 2006a). The FT age of the pseudo-tachylyte is 53 ± 4 (1SE) Ma, which is significantly younger than the 73 ± 4 Ma age of the host rock away from the fault that gives the time of regional cooling. According to track-length information, the zircon FT data from the pseu-dotachylyte were interpreted to have been totally reset and subsequently cooled at ~ 53 Ma. Furthermore, U–Pb anal-ysis shows a range of ages from ~ 67 to 76 Ma, which confirms that all of the host rocks formed approximately at the same time throughout the section.

Mylonites To constrain the time of ductile deformation in mylonites (Fig. 12.1a), application of the K–Ar (and $^{40}Ar/^{39}Ar$) systems to micas has predominantly been used in a variety of tectonic settings (e.g. Mulch et al. 2002; Sher-lock et al. 2004; Rolland et al. 2007 and references therein). FT thermochronology has been applied to mylonites in ductile fault zones, as will be mentioned below, which may constrain steady-state heat generation by long-term faulting averaged over geological time ($v = \sim 1$–10 cm/year; see Sect. 12.2.3). In contrast to frictional heating at shallower depths within a brittle regime that generates short-term heat pulses, ductile deformation at deeper levels is characterised by long-term heating accompanied by a broader, regional thermal anomaly across the fault zone. This may result in regional metamorphic aureoles across convergent plate boundaries (or transcurrent shear zones) (Scholz 1980). In this regard, the Alpine fault, New Zealand, where the obli-que convergence of two tectonic plates is ongoing (see Chap. 13, Baldwin et al. 2018), has been intensively studied by thermochronologic techniques. A systematic decrease in K–Ar ages over ~ 10 km was found toward the fault and was interpreted to represent an argon depletion aureole formed by long-term frictional heating at depth (Scholz et al. 1979, and references therein). This interpretation was based on a tectonic model suggesting that there was constant exhumation over the past 5 Myr throughout the study area. A later FT thermochronologic study, however, suggested the total amount of late Cenozoic exhumation exhibits an exponential increase toward the Alpine fault (Kamp et al. 1989). These authors estimated a differential, asymmetric uplift pattern across the fault, on the basis of apatite and zircon FT ages that show systematic trends to become younger toward the Alpine fault. They suggested that, as a result of the asymmetric exhumation, a 13–25 km wide regional metamorphic belt was exposed immediately to the east of the Alpine fault. More recent studies in the Southern Alps of New Zealand also interpreted the thermochronologic data in terms of exhumation (e.g., Little et al. 2005; Ring and Bernet 2010; Warren-Smith et al. 2016).

The Median Tectonic Line, southwest Japan, represents another example of long-term faulting that involves mylonite formation within a ductile regime. An apatite and zircon FT thermochronologic study was conducted on the granitic rocks of the Ryoke Belt, one of the regional metamorphic belts along the Median Tectonic Line (Tagami et al. 1988). A systematic age decrease of apatite FT ages over ~ 3–10 km toward the Median Tectonic Line was interpreted to be a consequence of long-term shear heating in the ductile fault zone. Recent FT and (U–Th)/He studies, however, have revealed regional differential exhumation histories of tec-tonic blocks that are bounded by neo-tectonic fault systems (Sueoka et al. 2012, 2016). Therefore, the previously observed age decrease toward the Median Tectonic Line may also be attributed to such differential uplift, rather than to a long-term shear heating process.

12.7 Summary and Future Perspectives

Fault zone rocks are the consequence of long-term, repeated motions of faults that reflect both spatial and temporal varia-tions in tectonic stress regimes. Age determination of fault movements, therefore, plays a key role in understanding the geotectonics of faults and, particularly, in assessing the past seismic activity of an active fault system. The technical and methodological advancements of FT and other thermochronological methods over the last few decades enable us to date a variety of fault zone materials formed during the development of a fault: e.g. (i) FT analysis of zircons from pseudotachylyte layers, fault gouge and associated deformed rocks, using the annealing kinetics based on the conventional laboratory heating experiments and also verified by flash and hydrothermal heating experiments; and (ii) K–Ar (^{40}Ar/^{39}Ar) dating on authigenic illite within a fault gouge, coupled with the evaluation of the influence of detrital contamination.

Furthermore, a series of drill cores into seismogenic fault zones, such as the Nojima Fault Zone Probe Project, offered the opportunity to systematically sample fresh fault zone rocks. These factors have provided a notable promotion of "fault zone thermochronology".

Further perspectives with respect to both the methodology and application of these techniques will be gained by:

- Additional laboratory flash and hydrothermal heatings as well as mechanical shearing experiments for a series of low-temperature thermochronological systems, such as apatite and zircon FT and (U–Th)/He, illite K–Ar (^{40}Ar/^{39}Ar) and quartz and feldspar ESR, TL and OSL systems. These will provide a more robust basis to interpret the fault zone thermochronologic data from a variety of fault zone rock types that represent different faulting depths and temperatures.
- A comparison of analysed data between thermochronologic methods having significantly different activation energies, such as zircon FT and (U–Th)/He. This may help to constrain the effective duration and temperature of fault zone thermal events, which will help to identify their heat source(s).
- A more systematically combined usage of apatite and zircon FT and (U–Th)/He, illite K–Ar, carbonate U–Th, and quartz and feldspar ESR, TL and OSL analyses on well-documented active fault systems, which will shed further light on our understanding of the thermomechanical processes in (paleo-) seismogenic-zone faults.

Acknowledgements The author thankfully acknowledges Ann Blythe and Meinert Rahn for their constructive and critical reviews of the manuscript, and Marco G. Malusà and Paul G. Fitzgerald for helpful editing of the chapter, particularly about the illustrations. The author thanks Masaki Murakami, Horst Zwingmann, Yumy Watanabe and Akito Tsutsumi for their helpful comments and arguments during the writing of the present chapter.

References

Ault AK, Reiners PW, Evans JP, Thomson SN (2015) Linking hematite (U–Th)/He dating with the microtextural record of seismicity in the Wasatch fault damage zone, Utah, USA. Geology 43:771–774

Baldwin SL, Fitzgerald PG, Malusà MG (2018) Chapter 13. Crustal exhumation of plutonic and metamorphic rocks: constraints from fission-track thermochronology. In: Malusà MG, Fitzgerald PG (eds) Fission-track thermochronology and its application to geology. Springer, Berlin

Boles JR, Eichhubl P, Garven G, Chen J (2004) Evolution of a hydrocarbon migration pathway along basin-bounding fault: evidence from fault cement. AAPG Bull 88:947–970

Boullier AM, Ohotani T, Fujimoto K, Ito H, Dubois M (2001) Fluid inclusions in pseudotachylytes from the Nojima fault, Japan. J Geophys Res 106:21965–21977

Brix MR, Stockhert B, Seidel E, Theye T, Thomson SN, Kuster M (2002) Thermobarometric data from a fossil zircon partial annealing zone in high pressure-low temperature rocks of eastern and central Crete, Greece. Tectonophysics 349:309–326

Brodsky EE, Mori J, Fulton PM (2010) Drilling into faults quickly after earthquakes. EOS Am Geophys Un 91:237–238

Carslaw HS, Jaeger JC (1959) Conduction of heat in solids. Oxford University Press, Oxford, 510 p

Cox SF (2005) Coupling between deformation, fluid pressures and fluid flow in ore-producing hydrothermal environments. In: Economic geology 100th anniversary volume, pp 39–75

Cox SF (2010) The application of failure mode diagrams for exploring the roles of fluid pressure and stress states in controlling styles of fracture-controlled permeability enhancement in faults and shear zones. Geofluids 10:217–233

d'Alessio MA, Blythe AE, Burgmann R (2003) No frictional heat along the San Gabriel Fault, California; evidence from fission-track thermochronology. Geology 31:541–544

Donelick RA, O'Sullivan PB, Ketcham RA (2005) Apatite fission-track analysis. Rev Mineral Geochem 58:49–94

Evans JP, Forster CB, Goddard JV (1997) Permeability of fault-related rocks, and implications for hydraulic structure of fault zone. J Struct Geol 19:1393–1404

Flotte N, Plagnes V, Sorel D, Benedicto A (2001) Attempt to date Pliocene normal faults of the Corinth-Patras Rift (Greece) by U/Th method, and tectonic implications. Geophys Res Lett 28:3769–3772

Foster DA (2018) Chapter 11. Fission-track thermochronology in structural geology and tectonic studies. In: Malusà MG, Fitzgerald PG (eds) Fission-track thermochronology and its application to geology. Springer, Berlin

Fujimoto K, Ueda A, Ohtani T, Takahashi M, Ito H, Tanaka H, Boullier AM (2007) Borehole water and hydrologic model around the Nojima fault, SW Japan. Tectonophysics 443:174–182

Fukuchi T, Imai N (2001) ESR and ICP analyses of the DPRI 500 m drill core samples penetrating through the Nojima Fault, Japan. Isl Arc 10:465–478

Gallagher K (1995) Evolving temperature histories from apatite fission-track data. Earth Planet Sci Lett 136:421–435

Haines SH, van der Pluijm BA (2008) Clay quantification and Ar–Ar dating of synthetic and natural gouge: application to the Miocene Sierra Mazatan detachment fault, Sonora, Mexico. J Struct Geol 30:525–538

Holdsworth RE, Butler CA, Roberts AM (1997) The recognition of reactivation during continental deformation. J Geol Soc London 154:73–78

Kamp PJJ, Green PF, White SH (1989) Fission track analysis reveals character of collisional tectonics in New Zealand. Tectonics 8:169–195

Ketcham R (2018) Chapter 3. Fission track annealing: from geologic observations to thermal modeling. In: Malusà MG, Fitzgerald PG (eds) Fission-track thermochronology and its application to geology. Springer, Berlin

Lachenbruch H (1986) Simple models for the estimation and measurement of frictional heating by an earthquake. USGS Open File Rep 86-508

Little TA, Cox S, Vry JK, Batt G (2005) Variations in exhumation level and uplift rate along the oblique-slip Alpine fault, central Southern Alps, New Zealand. GSA Bull 117:707–723

Magloughlin JF, Hall CM, van der Pluijm BA (2001) ^{40}Ar–^{39}Ar geochronometry of pseudotachylytes by vacuum encapsulation: North Cascade Mountains, Washington, USA. Geology 29:51–54

Maino M, Casini L, Ceriani A, Decarlis A, Di Giulio A, Seno S, Setti M, Stuart FM (2015) Dating shallow thrusts with zircon (U–Th)/He thermochronometry—the shear heating connection. Geology 43:495–498

Malusà MG, Fitzgerald PG (2018) Chapter 8. From cooling to exhumation: setting the reference frame for the interpretation of thermocronologic data. In: Malusà MG, Fitzgerald PG (eds) Fission-track thermochronology and its application to geology. Springer, Berlin

Malusà MG, Fitzgerald PG (2018) Chapter 10. Application of thermochronology to geologic problems: bedrock and detrital approaches. In: Malusà MG, Fitzgerald PG (eds) Fission-track thermochronology and its application to geology. Springer, Berlin

Matsumoto H, Yamanaka C, Ikeya M (2001) ESR analysis of the Nojima fault gouge, Japan, from the DPRI 500 m borehole. Isl Arc 10:479–485

Metcalf JR, Fitzgerald PG, Baldwin SL, Munoz JA (2009) Thermochronology of a convergent orogeny: constraints on the timing of thrust faulting and subsequent exhumation of the Maladeta Pluton in the central Pyrenean axial zone. Earth Planet Sci Lett 287:488–503

Mulch A, Cosca MA, Handy MR (2002) In-situ UV-laser ^{40}Ar/^{39}Ar geochronology of a micaceous mylonite: an example of defect-enhanced argon loss. Contrib Mineral Petrol 142:738–752

Murakami M (2010) Average shear work estimation of Nojima fault from fission-track analytical data. Earth Monthly (in Japanese) 32:24–29

Murakami M, Tagami T (2004) Dating pseudotachylyte of the Nojima fault using the zircon fission-track method. Geophys Res Lett 31

Murakami M, Tagami T, Hasebe N (2002) Ancient thermal anomaly of an active fault system: zircon fission-track evidence from Nojima GSJ 750 m borehole samples. Geophys Res Lett 29

Murakami M, Kosler J, Takagi H, Tagami T (2006a) Dating pseudotachylyte of the Asuke Shear Zone using zircon fission-track and U–Pb methods. Tectonophysics 424:99–107

Murakami M, Yamada R, Tagami T (2006b) Short-term annealing characteristics of spontaneous fission tracks in zircon: a qualitative description. Chem Geol 227:214–222

Nuriel P, Rosenbaum G, Zhao JX, Feng Y, Golding SD, Villement B, Weinberger R (2012) U–Th dating of striated fault planes. Geology 40:647–650

O'Hara K (2004) Paleostress estimated on ancient seismogenic faults based on frictional heating of coal. Geophys Res Lett 31

Oshiman N, Ando M, Ito H, Ikeda R (eds) (2001) Geophysical probing of the Nojima fault zone. Isl Arc 10:197–198

Otsuki K, Monzawa N, Nagase T (2003) Fluidization and melting of fault gouge during seismic slip: identification in the Nojima fault zone and implications for focal earthquake mechanisms. J Geophys Res 108

Peterman ZE, Day W (1989) Early Proterozoic activity on Archean faults in the western Superior province—evidence from pseudotachylite. Geology 17:1089–1092

Reimold WU, Jessberger EK, Stephan T (1990) ^{40}Ar–^{39}Ar dating of pseudotachylite from the Vredefort dome, South Africa: a progress report. Tectonophysics 171:139–152

Ring U, Bernet M (2010) Fission-track analysis unravels the denudation history of the Bonar Range in the footwall of the Alpine fault, South Island, New Zealand. Geol Mag 147:801–813

Rolland Y, Corsini M, Rossi M, Cox SF, Pennacchioni G, Mancktelow N, Boullier AM (2007) Comment on "Alpine thermal and structural evolution of the highest external crystalline massif: The Mont Blanc" by P. H. Leloup, N. Arnaud, E. R. Sobel, and R. Lacassin. Tectonics 26

Scholz CH (1980) Shear heating and the state of stress on faults. J Geophys Res 85:6174–6184

Scholz CH (1988) The brittle-plastic transition and the depth of seismic faulting. Geol Runds 77:319–328

Scholz CH (2002) The mechanics of earthquakes and faulting, 2nd edn. Cambridge University Press, Cambridge, UK, 471 p

Scholz CH, Beavan J, Hanks TC (1979) Frictional metamorphism, argon depletion, and tectonic stress on the Alpine fault, New Zealand. J Geophys Res 84:6770–6782

Seront S, Wong TF, Caine JS, Forster CB, Bruhn RL, Fredrich JT (1998) Laboratory characterization of hydromechanical properties of a seismogenic normal fault system. J Struct Geol 20:865–881

Seward D, Sibson RH (1985) Fission-track age for a pseudotachylite from the Alpine fault zone, New Zealand. New Z J Geol Geophys 28:553–557

Sherlock SC, Watts LM, Holdsworth R, Roberts D (2004) Dating fault reactivation by Ar/Ar laserprobe: an alternative view of apparently cogenetic mylonite-pseudotachylite assemblages. J Geol Soc London 161:335–338

Shimamoto T, Takemura K, Fujimoto K, Tanaka H, Wibberley CAJ (2001) Nojima fault zone probing by core analyses. Isl Arc 10:357–359

Sibson RH (1975) Generation of pseudotachylite by ancient seismic faulting. Geophys J Royal Astron Soc 43:775–794

Sibson RH (1977) Fault rocks and fault mechanisms. J Geol Soc London 133:191–213

Sibson RH (1992) Implications of fault-valve behavior for rupture nucleation recurrence. Tectonophysics 211:283–293

Simpson C (1985) Deformation of granitic rocks across the brittle-ductile transition. J Struct Geol 5:503–512

Solum JG, van der Pluijm BA, Peacor DR (2005) Neocrystallization, fabrics, and age of clay minerals from an exposure of the Moab fault, Utah. J Struct Geol 27

Sueoka S, Kohn BP, Tagami T, Tsutsumi H, Hasebe N, Tamura A, Arai S (2012) Denudational history of the Kiso Range, central Japan, and its tectonic implications: constraints from low-temperature thermochronology. Isl Arc 21:32–52

Sueoka S, Tsutsumi H, Tagami T (2016) New approach to resolve the amount of Quaternary uplift and associated denudation of the mountain ranges in the Japanese Islands. Geosci Front 7:197–210

Tagami T (2005) Zircon fission-track thermochronology and applications to fault studies. Rev Mineral Geochem 58:95–122

Tagami T (2012) Thermochronological investigation of fault zones. Tectonophysics 538–540:67–85

Tagami T, Galbraith RF, Yamada R, Laslett GM (1998) Revised annealing kinetics of fission tracks in zircon and geological implications. In: van den Haute P, De Corte F (eds) Advances in fission-track geochronology. Kluwer, Dordrecht, The Netherlands, pp 99–112

Tagami T, Hasebe N, Kamohara H, Takemura K (2001) Thermal anomaly around Nojima fault as detected by the fission-track analysis of Ogura 500 m borehole samples. Isl Arc 10:457–464

Tagami T, Lal N, Sorkhabi RB, Nishimura S (1988) Fission track thermochronologic analysis of the Ryoke Belt and the Median Tectonic Line Southwest Japan. J Geophys Res 93:13705–13715

Tagami T, Murakami M (2007) Probing fault zone heterogeneity on the Nojima fault: constraints from zircon fission-track analysis of borehole samples. Tectonophysics 443:139–152

Tagami T, O'Sullivan PB (2005) Fundamentals of fission-track thermochronology. Rev Mineral Geochem 58:19–47

Takagi H, Arita K, Danhara T, Iwano H (2007) Timing of the Tsergo Ri Landslide, Langtang Himal, determined by fission track age for pseudotachylite. J Asian Earth Sci 29:466–472

Takagi H, Shimada K, Iwano H, Danhara T (2010) Oldest record of brittle deformation along the Median Tectonic Line: fission-track age for pseudotachylite in the Taki area, Mie Prefecture. J Geol Soc Japan 116:45–50

Tanaka H, Chester FM, Mori JJ, Wang CY (2007) Drilling into fault zones. Tectonophysics 443:123–125

van der Pluijm BA, Hall CM, Vrolijk PJ, Pevear DR, Covey MC (2001) The dating of shallow faults in the Earth's crust. Nature 412:172–175

van der Pluijm BA, Vrolijk PJ, Pevear DR, Hall CM, Solum J (2006) Fault dating in the Canadian Rocky Mountains: evidence for late Cretaceous and early Eocene orogenic pulses. Geology 34:837–840

Verhaert G, Muchez P, Sintubin M, Similox-Tohon D, Vandycke S, Keppens E, Hodge EJ, Richards DA (2004) Origin of paleofluids in a normal fault setting in the Aegean region. Geofluids 4:300–314

Viola G, Mancktelow NS, Seward D, Meier A, Martin S (2003) The Pejo fault system; an example of multiple tectonic activity in the Italian Eastern Alps. GSA Bull 115:515–532

Warr LN, van der Pluijm BA, Peacor DR, Hall CM (2003) Frictional melt pulses during a ∼1.1 Ma earthquake along the Alpine fault, New Zealand. Earth Planet Sci Lett 209:39–52

Warren-Smith E, Lamb S, Seward D, Smith E, Herman F, Stern T (2016) Thermochronological evidence of a low-angle, mid-crustal detachment plane beneath the central South Island, New Zealand. Geochem Geophys Geosyst 17:4212–4235

Watanabe Y, Nakai S, Lin A (2008) Attempt to determine U–Th ages of calcite veins in the Nojima fault zone, Japan. Geochem J 42:507–513

Wolfler A, Kurz W, Danisik M, Rabitsch R (2010) Dating of fault zone activity by apatite fission track and apatite (U–Th)/He thermochronometry: a case study from the Lavanttal fault system (Eastern Alps). Terra Nova 22:274–282

Yamada K, Hanamuro T, Tagami T, Shimada K, Takagi H, Yamada R, Umeda K (2012) The first (U–Th)/He dating of a pseudotachlylyte collected from the Median Tectonic Line, southwest Japan. J Asian Earth Sci 45:17–23

Yamada K, Tagami T, Shimobayashi N (2003) Experimental study on hydrothermal annealing of fission tracks in zircon. Chem Geol 201:351–357

Yamada R, Murakami M, Tagami T (2007) Statistical modelling of annealing kinetics of fission tracks in zircon; reassessment of laboratory experiments. Chem Geol 236:75–91

Yamada R, Tagami T, Nishimura S, Ito H (1995) Annealing kinetics of fission tracks in zircon: an experimental study. Chem Geol 122:249–258

Zwingmann H, Mancktelow N (2004) Timing of Alpine fault gouges. Earth Planet Sci Lett 223:415–425

Zwingmann H, Yamada K, Tagami T (2010a) Timing of brittle deformation within the Nojima fault zone, Japan. Chem Geol 275:176–185

Zwingmann H, Mancktelow N, Antognini M, Lucchini R (2010b) Dating of shallow faults: new constraints from the AlpTransit tunnel site (Switzerland). Geology 38:487–490

Crustal Exhumation of Plutonic and Metamorphic Rocks: Constraints from Fission-Track Thermochronology

Suzanne L. Baldwin, Paul G. Fitzgerald and Marco G. Malusà

Abstract

The thermal evolution of plutonic and metamorphic rocks in the upper crust may be revealed using fission-track (FT) analyses and other low-temperature thermochronologic methods. The segment of pressure–temperature–time–deformation (*P-T-t-D*) rock paths potentially constrained by FT data corresponds to the lower greenschist facies, prehnite–pumpellyite, and zeolite facies of metamorphic rocks and also includes regions where diagenetic alteration occurs. When plutonic and metamorphic rocks are exhumed, thermal perturbations caused by fluid alteration, and crystallisation below relevant closure/annealing temperatures at relatively shallow crustal depths, may preclude a simplistic interpretation of thermochronologic ages in terms of monotonic cooling. However, FT ages and track-length measurements provide kinetic data that allow interpretation of *T-t* paths, even in cases where assumptions based on bulk closure temperatures are violated. We show that geologically well-constrained sampling strategies, and application of multiple thermochronologic methods on cogenetic minerals from plutonic and metamorphic rocks, may provide the most promising means to document the timing, rates, and mechanisms of crustal processes. Case studies are presented for: (1) (ultra)high-pressure (U)HP metamorphic terranes (e.g., Papua New Guinea, Western Alps, Western Gneiss Region, Dabie–Sulu), (2) an extensional orogen (Transantarctic Mountains), (3) a compressional orogen (Pyrenees), and (4) a transpressional plate boundary zone (Alpine fault zone, New Zealand).

13.1 Introduction

Plutonic and metamorphic rocks form at depth beneath the Earth's surface. Plutonic rocks crystallise at depth from magmas (i.e., silicate melts). Prior to crystallising, magmas transport heat and mass by flow at temperatures and pressures that depend upon the magma's bulk composition. Crystallisation involves both nucleation and crystal growth processes, with rates dependent upon the temperature–time (*T-t*) history. In contrast, metamorphic rocks form by solid-state crystallisation of protoliths (igneous, metamorphic or sedimentary rocks) that have been subjected to changes in temperatures and pressures. Metamorphic minerals and their textures change primarily in response to temperature that together with available fluids drive metamorphic reactions. The result is that original mineral assemblages may be transformed to more stable assemblages at new pressure and temperature conditions.

Major perturbations of crustal geothermal gradients are required to form igneous and metamorphic rocks, so it cannot be assumed a priori that these rocks achieved equilibrium as a result of steady-state conditions (e.g., Spear 1993). In active plate boundary zones, where most igneous and metamorphic rocks form, geothermal gradients are spatially complex and change as plate boundaries evolve. Transient geothermal gradients result from heat sources (e.g., intruding magmas, exothermic reactions) and heat sinks (subducting slabs, endothermic reactions). For example, at divergent plate boundaries rising asthenosphere causes decompression melting, which results in steepening of the geothermal gradient and high-temperature metamorphism of the country rock. At convergent plate boundaries, subducting cold lithosphere leads to high-*P*/low-*T* metamorphism and results in low geothermal gradients relative to steady-state geothermal gradients (Fig. 13.1a). If active deformation is associated with rapid exhumation, geothermal gradients are likely to change due to heat advection as rocks move rapidly from depth towards the surface. Our ability to

S. L. Baldwin (✉) · P. G. Fitzgerald
Department of Earth Sciences Syracuse University, Syracuse, NY 13244, USA
e-mail: sbaldwin@syr.edu

M. G. Malusà
Department of Earth and Environmental Sciences, University of Milano-Bicocca, Piazza Delle Scienza 4, 20126 Milan, Italy

© Springer International Publishing AG, part of Springer Nature 2019
M. G. Malusà and P. G. Fitzgerald (eds.), *Fission-Track Thermochronology and its Application to Geology*,
Springer Textbooks in Earth Sciences, Geography and Environment, https://doi.org/10.1007/978-3-319-89421-8_13

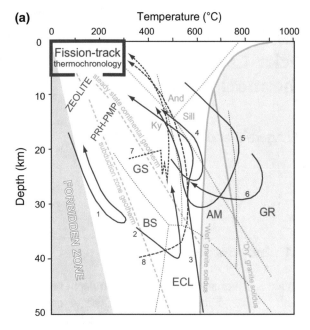

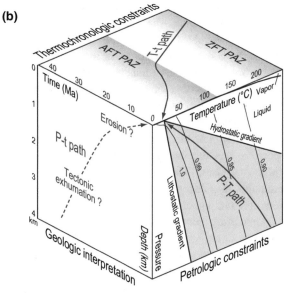

Fig. 13.1 a Depth–temperature diagram showing examples of *P-T* paths for metamorphic rocks (in blue) (after Philpotts and Ague 2009): 1, Franciscan Complex (Ernst 1988); 2, Western Alps (Ernst 1988); 3, Dora-Maira (Rubatto and Hermann 2001); 4 and 5, central Massachusetts (Tracy and Robinson 1980); 6, Adirondacks, NY (Bohlen et al. 1985); 7 and 8, upper and lower units of the Tauern window, Eastern Alps (Selverstone et al. 1984; Selverstone and Spear 1985). Note that the lowest temperature parts of all *P-T* paths are not constrained. Blue box indicates *P-T* space relevant for constraining histories using FT thermochronology. Metamorphic facies (in red): AM, amphibolite; BS, blueschist; ECL, eclogite; GR, granulite; GS, greenschist; PRH-PMP, prehnite–pumpellyite. Reaction curves for Al_2SiO_5, wet and dry solidi indicated by light blue dotted lines. **b** *T-t-depth* space for rock *P-T-t* paths corresponding to very low grade and diagenetic conditions. *P(depth)-T* conditions are determined from fluid inclusions (lines of constant density for the H_2O system, in g/cm^3, after Goldstein and Reynolds 1994). The hypothetical *T-t* path includes ZFT and AFT with partial annealing zones (PAZ) indicated. The dashed line on the left-hand panel shows an example of a *P(depth)-t* path associated with shallow crustal exhumation mechanisms

constrain crustal exhumation histories of plutonic and metamorphic rocks largely depends on our understanding of the dynamic thermal reference frame used to interpret thermochronologic data (see Chap. 8, Malusà and Fitzgerald 2018a) and an understanding of the range of chemical and physical processes that can potentially affect plutonic and metamorphic rocks during exhumation.

This chapter discusses the final exhumation paths of plutonic and metamorphic rocks, as they make their way to the surface, and the importance of using FT thermochronology to constrain and quantify the timescales, rates, and mechanisms of crustal motion on geologic timescales. It is written from a "rock exhumation trajectory" perspective, following plutonic and metamorphic rocks from deep crustal levels where constraints on exhumation are generally obtained using high-temperature thermochronologic techniques and petrologic data, towards shallow crustal levels where low-temperature thermochronologic techniques are applicable. There are many common assumptions associated with techniques used to constrain exhumation from deep crustal levels as compared to those used to constrain exhumation from shallow crustal levels. However, important differences exist, such as the role of mineral (re)crystallisation in the deep crust versus the influence of topography on isotherms at shallow crustal levels. We present case studies from different tectonic settings to illustrate how FT thermochronology on minerals from metamorphic and plutonic rocks can be interpreted within a geologic framework. Our synthesis takes into account potential complications due to processes (e.g., heat advection, hydrothermal alteration) that may affect rocks during crustal exhumation.

13.2 Thermochronologic Data Interpretation of Plutonic and Metamorphic Rocks

13.2.1 An Integrated Approach to P-T-t-D Path Determination

Mineral assemblages and textures preserved in plutonic and metamorphic rocks provide a record of changing pressure (*P*), temperature (*T*), and deformation (D) during transit from depth to the surface. Mineral assemblages and textures are a function of bulk rock compositions, rheology, volatile contents, and *P-T* conditions. Principles of physical chemistry and phase equilibria applied to natural rocks and synthetic materials by experimentalists and thermodynamic modellers allow petrologists to assess *P-T* conditions (e.g., Spear 1993; Powell and Holland 2010; Sawyer et al. 2011). A rock's *P-T* path can be constructed by connecting regions in *P-T* space where the stability of mineral assemblages, compositions, or changes in compositions (e.g., in the case of zoned minerals), and their textures, are known (e.g., Spear

1993). Reactions used to quantify metamorphic pressures and temperatures typically occur diachronously, and radiometric dating techniques can be applied to minerals to determine the ages associated with segments of the *P-T* paths (Fig. 13.1a). Thermobarometric data provided by petrologic analysis and *T-t* information provided by thermochronology can be integrated to define *P-T-t* paths that shed light on geologic processes controlling crustal rock exhumation (e.g., Baldwin and Harrison 1992; Duchêne et al. 1997; Malusà et al. 2011; Baldwin 1996).

Accessory phases have proven especially useful for linking isotopic ages to petrologic and textural information (e.g., Kohn 2016). Most radiometric data (Rb–Sr, ^{40}Ar/^{39}Ar, U–Pb, Sm–Nd, Lu–Hf) can be interpreted with respect to mineral (re)crystallisation to infer the timing and rates of crustal processes such as metamorphism and ductile deformation. Field, macro-, micro-, and nano-structural analysis provide the structural context required for correlating mineral assemblages from different outcrops and to add rheologic constraints in the construction of *P-T-t-D* paths. In the low-temperature range—corresponding to the lower greenschist, prehnite–pumpellyite, and zeolite facies of metamorphic rocks and including diagenesis—time constraints provided by FT thermochronology are particularly useful to define the final portion of the exhumation path (Fig. 13.1b) (e.g., Malusà et al. 2006). However, many published exhumation paths do not incorporate data that allow paths to be extended to the lowest temperature ranges (Fig. 13.1a). In such cases, information, potentially provided by full integration of petrologic and thermochronologic data sets, remains unexploited.

13.2.2 Processes, Timescales, and Rates

If it can be demonstrated that rocks cooled monotonically from high to low temperatures, and minerals represent equilibrium assemblages, application of geothermometers and thermochronometers with equilibration temperatures equal to isotopic closure temperatures (T_c) can be simply applied (e.g., Hodges 1991). However, petrologic evidence, such as mineral inclusion suites, mineral zoning patterns, and microstructures often reveals that equilibrium has not been achieved during exhumation, rendering thermochronologic interpretations based on simple T_c models invalid. During transit to the surface, most minerals in plutonic and metamorphic rocks only partially retain their radiogenic daughter nuclides, either due to metamorphic (re) crystallisation which is often accompanied by deformation or due to diffusive loss of radiogenic daughter products. Therefore, knowledge of the minerals' petrogenesis provides constraints on rate-limiting daughter product loss mechanisms (e.g., volume diffusion, dissolution/precipitation,

syn-kinematic recrystallisation) and aids in thermochronologic data interpretation. Because apatite FT (AFT) thermochronology is usually interpreted with respect to temperatures less than ~120 °C, and zircon FT (ZFT) thermochronology less than ~300 °C, taking into account (re)crystallisation of minerals within these temperature ranges is often neglected in AFT and ZFT thermochronologic data interpretation. However, metamorphic rims can form on pre-existing zircons at temperatures as low as ~250 °C (e.g., Rasmussen 2005; Hay and Dempster 2009) complicating isotopic data interpretation on zircons with demonstrable growth zones (Zirakparvar et al. 2014). Especially in cases where FT data are integrated with U–Pb ages, zircon petrogenesis must be known to ensure accurate geologic interpretations are made.

Distinguishing between the timing of mineral and rock formation, and cooling related to exhumation, is particularly important for the analysis of plutonic and metamorphic rocks. Timescales for magmatic cooling may range over orders of magnitude, from millions of years (e.g., in the case of slowly cooled batholiths) to <100,000 years (e.g., Petford et al. 2000). Modelled timescales of regional metamorphism during continent–continent collision (e.g., England and Thompson 1984) are orders of magnitude greater than timescales derived from garnet growth zones based on diffusion modelling (e.g., Dachs and Proyer 2002; Ague and Baxter 2007; Spear 2014) and from numerical modelling of thermochronologic data (e.g., Camacho et al. 2005; Viete et al. 2011). Short-lived orogenic events (<1 Myr; Dewey 2005) may result in rapid rock exhumation at rates comparable to plate tectonic rates (i.e., cm/year; e.g., Zeitler et al. 1993; Rubatto and Hermann 2001; Baldwin et al. 2004).

13.2.3 Approaches Used to Determine Rock Exhumation Rates

Two approaches have commonly been used to determine exhumation rates from thermochronologic data (e.g., Purdy and Jager 1976; Blythe 1998; McDougall and Harrison 1999 and references therein). These are generally known as the multiple method and age–elevation approaches (see Chap. 10, Malusà and Fitzgerald 2018b). The first approach utilises multiple thermochronologic methods applied to minerals from the same sample. Cooling rates are calculated using differences in bulk T_c divided by the difference in apparent ages corresponding to the minerals analysed. Cooling rates are then converted to exhumation rates assuming a geothermal gradient. This bulk closure temperature approach—interpolation of *T-t* points obtained from analyses and assuming a nominal T_c (Dodson 1973)—has many built-in assumptions which are usually violated when considering the exhumation of metamorphic and plutonic rocks (e.g., Harrison and Zeitler 2005). Assumptions made when using this approach include: (a) diffusion is the loss mechanism operative over geologic time, (b) kinetic parameters are known, and (c) geothermal gradients remained constant and/or are known during the time period investigated.

The second common approach involves age determination on a suite of samples collected over a large elevation range (i.e., "vertical profiles"; see Chap. 9; Fitzgerald and Malusà 2018). The simple interpretation of the slope on an age–elevation profile is that it represents an apparent exhumation rate. However, due to advection of isotherms and topographic effects, the slope on an age–elevation profile typically provides an overestimate of the exhumation rate (e.g., Gleadow and Brown 2000; Braun 2002; Huntington et al. 2007). In some cases, the age–elevation profile may reveal an exhumed partial annealing zone (PAZ) or partial retention zone (PRZ). In these cases, a distinctive break in slope is interpreted to mark the base of a former PAZ/PRZ, and the slope below the break in slope marks an increase in cooling rate, usually associated with an increase in exhumation rate (see Chap. 9; Fitzgerald and Malusà 2018). In AFT thermochronology, ages and track-length distributions are used to determine thermal histories and cooling rates (see Chap. 3; Ketcham 2018). Modelled AFT thermal histories can be extended to higher temperatures through integration of modelled $^{40}Ar/^{39}Ar$ step heat data on cogenetic K-feldspar (e.g., Lovera et al. 2002; see examples below).

13.3 Application of FT Thermochronology to the Exhumation of (U)HP Terranes

Blueschist and eclogite-facies metamorphic rocks form when lithosphere is subducted faster than it can thermally equilibrate, and isotherms are depressed leading to characteristic high-*P/T* geothermal gradients (Fig. 13.1a). The discovery of coesite (the high-pressure SiO_2 polymorph) in eclogite-facies metamorphic rocks (Chopin 1984; Smith 1984) led to development of the field of UHP metamorphism (e.g., Coleman and Wang 1995; Hacker 2006; Gilotti 2013). Evidence of UHP metamorphism has been documented in more than twenty terranes, in regions of present or former plate convergence (e.g., Guillot et al. 2009; Liou et al. 2009). It is now accepted that UHP rocks form when oceanic and continental lithosphere is subducted to mantle depths, as confirmed by geophysical evidence (Zhao et al. 2015; Kufner et al. 2016). However, there is no consensus concerning how UHP rocks are exhumed from mantle depths to the surface (e.g., Malusà et al. 2015; Ducea 2016 and references therein).

Low-temperature thermochronology usually constrains rock exhumation from shallow crustal levels. Since the final stage of (U)HP exhumation may occur tens or hundreds of millions of years after the main exhumation phase (i.e., from mantle depths), low-temperature thermochronologic ages may not necessarily be interpreted relative to the timing of (U)HP exhumation, especially in the case of pre-Cenozoic UHP terranes (Fig. 13.2b). We emphasise that the timing of final exhumation within the subduction channel, as constrained by FT data, is essential for an accurate tectonic interpretation of petrologic and thermochronologic data from subduction complexes. Depending upon the paleogeothermal gradients, AFT ages may correspond to the timing of cooling and exhumation from depths ranging from ∼15 km (e.g., in the case of syn-subduction exhumation with gradients of 10 °C/km) to ∼4 km (e.g., in the case of post-subduction exhumation with gradients of ∼30 °C/km). Independent constraints on paleogeothermal gradients (see Chap. 8, Malusà and Fitzgerald 2018a, b) are thus crucial for a reliable analysis of (U)HP rock exhumation. FT thermochronology may also be used to determine when different lithologic units (e.g., comprising a tectonic mélange) are amalgamated to form a composite terrane. Below, we summarise low-temperature constraints on (U)HP terranes and explain why these data are essential to assess timing, rates, and mechanisms of final (U)HP rock exhumation.

13.3.1 Cenozoic (U)HP Terranes

Eastern Papua New Guinea (PNG) and the Western Alps are among the best-studied examples of Cenozoic (U)HP terranes. The PNG (U)HP terrane is exhuming in a region of active rifting within the obliquely convergent Australian–Woodlark plate boundary zone (Baldwin et al. 2004, 2008). Domes of high-grade migmatitic gneisses (e.g., Davies and Warren 1988; Gordon et al. 2012), comprised of protoliths derived largely from Australian continental crust (Zirakparvar et al. 2012), are separated from oceanic lithospheric fragments by mylonitic shear-zone carapace (e.g., Hill et al. 1992; Little et al. 2007). Seismically active normal faults flank the domes (e.g., Abers et al. 2016) and are interpreted to have formed within an accretionary wedge along the former subduction thrust now marked by serpentinite (e.g., Baldwin et al. 2012). The location of intermediate depth earthquakes in proximity to exhumed coesite eclogite (Abers et al. 2016) suggests that rock exhumation from UHP depths may be ongoing. The timing of UHP metamorphism in eastern PNG (∼7–8 Ma) is based on concordant ages obtained on cogenetic minerals from coesite eclogite using three methods: in situ zircon ion probe U–Pb (Monteleone et al. 2007), garnet Lu–Hf (Zirakparvar et al. 2011), and phengite $^{40}Ar/^{39}Ar$ (Baldwin and Das 2015). Most metamorphic zircon growth occurred during exhumation (Monteleone et al. 2007; Gordon et al. 2012; Zirakparvar et al. 2014) as confirmed by zircon petrologic models (Kohn et al.

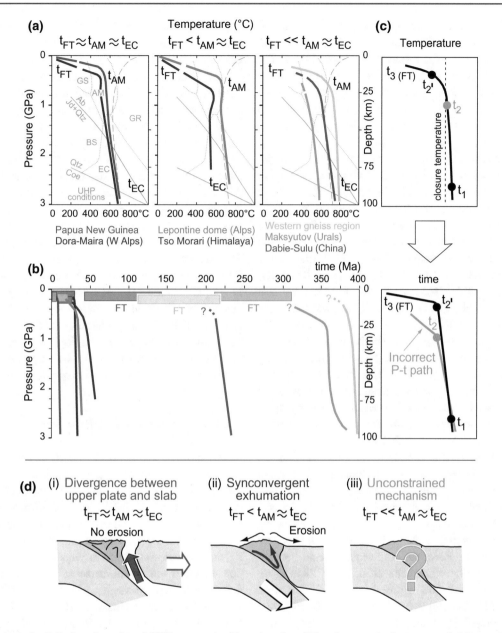

Fig. 13.2 **a** Schematic *P-T* plots for selected UHP terranes, with colours corresponding to terranes indicated: Papua New Guinea (Baldwin and Das 2015), Dora-Maira (Chopin et al. 1991; Gebauer et al. 1997; Rubatto and Hermann 2001), Lepontine (Becker 1993; Gebauer 1996; Brouwer et al. 2004; Nagel 2008), Tso Morari (de Sigoyer et al. 2000; Schlup et al. 2003), Western Gneiss Region (Rohrman et al. 1995; Carswell et al. 2003; Kzienzyk et al. 2014), Maksyutov (Lennykh et al. 1995; Leech and Stockli 2000), Dabie–Sulu (Reiners et al. 2003; Hu et al. 2006; Liou et al. 2009); t_{FT} indicates age constraints based on FT analysis. Timing of amphibolite facies metamorphism (t_{AM}), timing of eclogite-facies metamorphism (t_{EC}) based on U–Pb, $^{40}Ar/^{39}Ar$, and Lu–Hf isotopic data. **b** Schematic *P-t* paths of UHP terranes to illustrate differences in the length of time associated with final exhumation to the surface relative to the timing of UHP metamorphism. **c** Upper panel: schematic *P-T* paths (shown in black) illustrate the importance of having geobarometric constraints associated with exhumation paths. The timing of peak UHP conditions (t_1), retrograde overprint (t_2 and $t_{2'}$), and final exhumation (t_3(FT))

based on FT analyses. $t_{2'}$ is the age recorded by a mineral at low P conditions as a result of a late syn-kinematic recrystallisation event (e.g., a late greenschist facies foliation marked by micas) or of a localised thermal event (e.g., due to hydrothermal fluids). Lower panel: this shows how it is possible to obtain an incorrect *P-t* path if the timing of a retrograde overprint $t_{2'}$ (e.g., late zircon growth or mica recrystallisation) is incorrectly identified. **d** Schematic cross sections illustrating possible mechanisms for UHP exhumation related to: (i) divergence between the upper plate and the subducting slab leading to rapid rock exhumation within the forearc; erosional exhumation plays a minor role during exhumation. FT ages close to the timing of amphibolite facies retrogression and peak eclogite-facies conditions are predicted. (ii) Syn-convergent exhumation where erosional processes play a significant role in the exhumation of rocks within the forearc. FT ages are less than the timing of amphibolite facies retrogression and peak eclogite-facies conditions. (iii) Exhumation mechanisms are undetermined for cases in which FT ages are significantly younger than isotopic ages associated with (U)HP metamorphism

2015). An AFT age of 0.6 ± 0.2 Ma (2σ) was obtained from the coesite locality (Baldwin et al. 1993) and provides constraints on the lowest temperature portions of the *P-T-t-D* path. In general, AFT ages are challenging to obtain in these rocks, due to low apatite abundances in some rock types, low [U], and a few tracks. In eastern PNG, confined tracks are very rare, but have been imaged using heavy ion implantation to provide etchant pathways (see Chap. 2; Kohn et al. 2018). AFT ages are often close to zero, with high errors and a few track-length distributions to model, but the data are geologically meaningful and interpretable (Fitzgerald et al. 2015). Depth estimates based on preservation of coesite, together with the timing of UHP metamorphism and AFT data, indicate that average minimum exhumation rates are >1 cm/year (Baldwin et al. 1993, 2004, 2008; Hill and Baldwin 1993; Monteleone et al. 2007). (U)HP exhumation models for eastern PNG remain a topic of debate (e.g., Ellis et al. 2011; Petersen and Buck 2015), but final exposure of (U)HP rocks at the surface was likely facilitated by microplate rotation (Webb et al. 2008) and consequent divergence between the oceanic upper plate and the subducting slab (Fig. 13.2d). This kinematic scenario would have favoured the rise, from >90 km depths, of buoyant, low density, migmatitic gneisses containing mafic eclogite, via ductile flow within the subduction channel (Malusà et al. 2015, Liao et al. 2018).

The role of FT thermochronology in understanding the mechanisms of (U)HP rock exhumation is even more important in the case of the Western Alps, where (U)HP rocks have resided at shallow crustal levels during the past 30 Myr. The Western Alps formed as a result of Cretaceous to Paleogene subduction of the Tethyan oceanic lithosphere and of the adjoining European continental margin beneath the Adriatic microplate (Lardeaux et al. 2006; Zhao et al. 2015). UHP rocks are now exposed in a 20–25 km wide metamorphic belt that includes eclogitised continental crust (e.g., the Dora-Maira unit; Chopin et al. 1991) and metaophiolites (e.g., Frezzotti et al. 2011). The exhumation paths of these units are well constrained by petrologic and thermochronologic data (see Malusà et al. 2011 for a synthesis). Peak metamorphism at $P = 2.8$–3.5 GPa and $T = 700$–750 °C (e.g., Schertl et al. 1991; Compagnoni et al. 1995) is dated to 40–35 Ma using U–Pb ion probe analyses on zircon rims and titanite, and Sm–Nd isochron analyses (e.g., Gebauer et al. 1997; Rubatto et al. 1998; Amato et al. 1999; Rubatto and Hermann 2001). Subsequent exhumation took place at rates faster than subduction rates (Malusà et al. 2015) (Fig. 13.2b). Apatite FT and (U–Th)/He (AHe) data (e.g., Malusà et al. 2005; Beucher et al. 2012) provide constraints on the final part of the (U)HP exhumation path, attesting to rapid exhumation close to the surface by the

early Oligocene, as confirmed by the biostratigraphic age of sedimentary rocks locally overlying the Western Alps eclogites (Vannucci et al. 1997).

The Western Alps example, like eastern PNG, thus illustrates the short duration between the timing of peak (U)HP metamorphism and subsequent exhumation to the Earth's surface. In this case, exhumation also occurred during the same subduction cycle that produced the (U)HP rocks, likely a result of divergent motion between the Adriatic upper plate and the European slab (Malusà et al. 2011; Solarino et al. 2018; Liao et al. 2018, Fig. 13.2d). In contrast, the Lepontine dome of the Central Alps records slower crustal exhumation (Brouwer et al. 2004; Nagel 2008), similar to the exhumational record provided by the Tso Morari eclogites in the Himalaya (de Sigoyer et al. 2000; Schlup et al. 2003). The exhumation path of the Lepontine dome is consistent with predictions of syn-convergent exhumation numerical models (e.g., Yamato et al. 2008; Jamieson and Beaumont 2013). The integration of thermochronologic and petrologic data sets thus reveals along-strike differences in exhumation patterns and mechanisms preserved in the Alpine orogenic rock record.

13.3.2 Pre-Cenozoic (U)HP Terranes

In the case of pre-Cenozoic (U)HP terranes such as the Dabie–Sulu of eastern China (e.g., Liou et al. 2009), the Maksyutov Massif of Russia (e.g., Lennykh et al. 1995), and the Western Gneiss (U)HP terrane of Norway (e.g., Carswell et al. 2003), FT data are even more essential to distinguish the timing and mechanisms of exhumation. This is because FT data permit assessment of whether or not final exhumation occurred during the same subduction cycle that produced the (U)HP rocks (e.g., Rohrman et al. 1995; Leech and Stockli 2000; Reiners et al. 2003; Hu et al. 2006; Kzienzyk et al. 2014). In the Western Gneiss (U)HP terrane, geochronologic data (Lu–Hf, Sm–Nd, Rb–Sr, U–Pb) have been interpreted to date the timing of (U)HP metamorphism ~430–400 Ma (Carswell et al. 2003; DesOrmeau et al. 2015). Together with thermobarometric constraints, a two-stage exhumation history for the Norwegian (U)HP terrane has been proposed. Initial exhumation, from mantle depths to lower crustal depths, was followed by stalling of the terrane at depths where mineral assemblages were overprinted during high-temperature amphibolite facies metamorphism (Walsh and Hacker 2004). Extensional processes are inferred to have led to the final exhumation to the surface. Presently, the Western Gneiss terrane is an elevated passive margin (see Chap. 20, Wildman et al. 2018). By quantifying contributions from crustal isostasy and dynamic topography to the present-day topography, Pedersen et al. (2016) propose that high topography existed since the Caledonian orogeny (i.e., ~490–390 Ma). However, there are regional variations in Jurassic to Cretaceous AFT ages that vary as a function of elevation (Rohrman et al. 1995). Such long durations, between the timing of UHP metamorphism and ages recorded by AFT, indicate that final exhumation is not related to the same subduction cycle that formed the Western Gneiss terrane (Fig. 13.2b). Without better certainty regarding linkages between the higher and lower pressure segments of rock exhumation paths, the mechanism responsible for UHP exhumation during, or shortly after, the Caledonian subduction cycle still remains largely unconstrained.

In the Dabie–Sulu (U)HP terrane, petrologic and thermochronologic studies reveal that Triassic–Jurassic UHP metamorphism was followed by Cretaceous plutonism (Hacker et al. 1998, 2000; Ratschbacher et al. 2000). Low-temperature thermochronologic data (i.e., ^{40}Ar/^{39}Ar K-feldspar, AFT, (U–Th)/He on zircon (ZHe), and apatite) yielded a range of ages spanning more than 115 Myr. These data were interpreted to result from slow cooling and used to infer steady-state exhumation rates (0.05–0.07 km/Myr) (Reiners et al. 2003). Liu et al. (2017) further detail the complex thermal histories of the Sulu (U)HP terrane and report AFT and AHe ages as young as 65–40 Ma. As in the Western Gneiss Region, the long duration between UHP metamorphism and final cooling of these terranes indicates that final exhumation was not related to the subduction event that formed the Dabie–Sulu UHP terrane.

A comparison of *P-T-t* paths for selected (U)HP terranes (Fig. 13.2b) suggests that similar exhumation rates from mantle depths can be inferred, based on slopes of depth–time plots to crustal levels. The low-temperature histories of (U)HP rocks revealed by FT thermochronology can be used to distinguish between tectonic and erosional exhumation mechanisms in the upper crust (Fig. 13.2d). We caution, however, that if thermochronologic data linking segments of *P-T-t* paths from mid-crustal to shallow crustal depths are misinterpreted (e.g., based on incorrect assumptions about the pressure/inferred depth of mineral crystallisation), exhumation rates may be overestimated (segment $t_2 - t_3$ in Fig. 13.2c). For example, if ^{40}Ar/^{39}Ar white mica ages are interpreted as "cooling ages" (i.e., t_2 in Fig. 13.2c), when in fact white mica crystallised below its T_c for argon, (i.e., $t_{2'}$ in Fig. 13.2c), estimated exhumation rates following mica (re) crystallisation will be incorrect. Such complications are more likely in older (U)HP terranes that have experienced a protracted evolution, with the potential for hydrothermal alteration in the upper crust.

13.4 Application of FT Thermochronology to Extensional Orogens: The Transantarctic Mountains

Plutonic and metamorphic rocks may preserve a record of deep orogenic processes hundreds of millions of years prior to their final exhumation to the surface. Therefore, information provided by classic petrologic and geochronologic approaches while pertinent to an earlier orogenic event may not be relevant to understanding late-stage mountain-building events and landscape evolution. The Transantarctic Mountains (TAM) case study provides an example of a protracted crustal evolution characterised by slow cooling, followed by episodic exhumation associated with rift flank formation during extensional orogenesis. The ~3500-km-long TAM mark the physiographic and lithospheric divide between East and West Antarctica (Dalziel 1992; Fig. 13.3a). The mountain belt bisects the continent and is ~100–200 km wide, with elevations locally exceeding 4500 m. The TAM define the western edge of the Mesozoic–Cenozoic intracontinental West Antarctic Rift System and the eastern margin of the East Antarctic craton, thereby providing a geomorphic barrier for the East Antarctic Ice Sheet. The TAM are related to formation of the West Antarctic rift and are inferred to represent an erosional remnant of a collapsed plateau (Bialas et al. 2007), with the rift flank associated with flexure of strong East Antarctic lithosphere (e.g., Stern and ten Brink 1989).

The overall geology of the TAM is relatively simple (e.g., Elliot 1975). Basement rocks are composed primarily of Late Proterozoic–Cambrian metamorphic rocks and Cambrian–Ordovician granitoids of the Granite Harbour Intrusive Suite (Fig. 13.3a). Basement rocks were deformed during the Cambrian–Ordovician Ross Orogeny that preceded and accompanied intrusion of granitoids (e.g., Goodge 2007). Following the Ross Orogeny, 16–20 km of rock exhumation resulted in formation of the low-relief Kukri Erosion Surface (Gunn and Warren 1962; Capponi et al. 1990). Basement rocks were subsequently unconformably overlain by Devonian–Triassic glacial, alluvial, and shallow marine sediments of the Beacon Supergroup (e.g., Barrett 1991). During the Jurassic, extensive basaltic magmatism (Ferrar large igneous province) occurred along the TAM, as well as in adjoining parts of Gondwana, South Africa, South America, and southern Australia (e.g., Elliot 1992; Elliot and Fleming 2004). Dolerite sills (up to 300 m thick) intruded both basement and sedimentary cover. Step heat experiments on feldspars from the sills yielded ^{40}Ar/^{39}Ar ages of 177 Ma (Heimann et al. 1994). Mafic volcanism (i.e., the Kirkpatrick Basalt; Elliot 1992) was contemporaneous with dolerite sill

Fig. 13.3 a Map of Antarctica and schematic cross section (A-A′) of the TAM in the Shackleton–Beardmore–Byrd glacier region showing simplified geology. Shallowly dipping rocks of the TAM extend beneath the East Antarctic Ice Sheet. Normal faults in the TAM front expose more deeply exhumed plutonic rocks of the Cambrian–Ordovician Granite Harbour Intrusives (modified after Barrett and Elliot 1973; Lindsay et al. 1973; Fitzgerald 1994). Black regions are TAM with approximate locations indicated: BG = Beardmore Glacier, NVL and SVL = northern and southern Victoria Land, SC = Scott Glacier, SH = Shackleton Glacier, TH = Thiel Mountains.
b Schematic composite temperature–time plot for samples below the Kukri Erosion Surface (purple) and from the TAM front (i.e., at deeper crustal levels; red).
c Composite AFT age—crustal depth profiles for the central TAM, Beardmore glacier region illustrating differential cooling, and exhumation patterns revealed by AFT ages. After Fitzgerald (1994), Fitzgerald and Stump (1997), and Blythe et al. (2011) for the Byrd Glacier

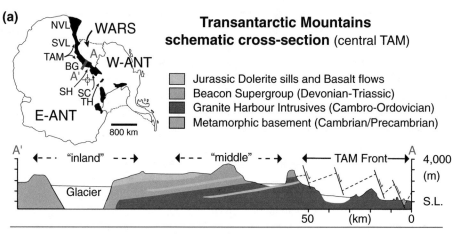

(a)

Transantarctic Mountains schematic cross-section (central TAM)

☐ Jurassic Dolerite sills and Basalt flows
☐ Beacon Supergroup (Devonian-Triassic)
☐ Granite Harbour Intrusives (Cambro-Ordovician)
☐ Metamorphic basement (Cambrian/Precambrian)

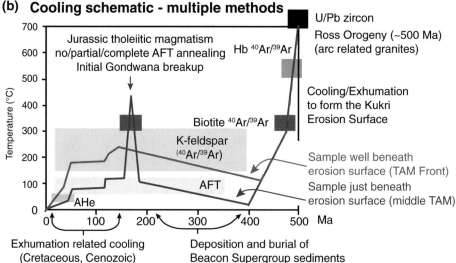

(b) Cooling schematic - multiple methods

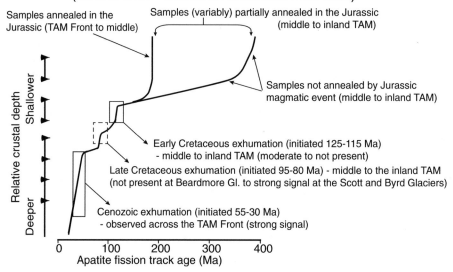

(c)   Composite AFT age – elevation profile
(central TAM – based on Beardmore Gl. data)

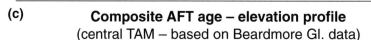

emplacement. The present-day outcrop pattern of the TAM generally reflects its simple tilt block structure dipping inland (Fig. 13.3a). Outcrops of Kirkpatrick Basalt are limited to the inland parts of the range, whereas basement representing deeper crustal levels is exposed primarily along the coastal sector, extending inland along major outlet glaciers. In a few coastal locations such as Cape Surprise in the central TAM (Barrett 1965; Miller et al. 2010), Beacon Supergroup rocks are down-faulted by 3–5 km. Beacon Supergroup rocks have also been recovered offshore southern Victoria Land at a depth of 825 m below seafloor in the Cape Roberts drillhole#3 (Cape Roberts Science Team 2000). In most cases, AFT ages on basement rocks were (i) completely reset as a result of the thermal effects of Jurassic magmatism (Fig. 13.3b) or (ii) were resident at depths below the base of the PAZ prior to Cretaceous and younger exhumation (e.g., Gleadow and Fitzgerald 1987). However, along the inland flank of the TAM, un-reset or partially reset AFT ages (Fig. 13.3c) have been documented (Fitzgerald and Gleadow 1988; Fitzgerald 1994).

Following Jurassic tholeiitic magmatism, and prior to Late Cenozoic alkaline volcanism of the McMurdo Volcanic Group (LeMasurier and Thomson 1990), a ~ 160 Myr gap in the onshore geologic record of the TAM exists. Coring of sedimentary basins in the Ross Sea recovered sediment as old as Upper Eocene (Barrett 1996; Cape Roberts Science Team 2000). However, because no core older than Upper Eocene has been recovered from adjacent sedimentary basins, and the onshore geologic record is missing, studies of the uplift and exhumation history of the TAM have relied primarily on the application of thermochronology, largely AFT thermochronology on basement granitoids (e.g., Gleadow and Fitzgerald 1987). More recently, detrital thermochronology on drill core from the West Antarctic rift provides additional contributions to our understanding of the TAM exhumation history (e.g., Zattin et al. 2012).

13.4.1 Sampling Strategy, Data, and Interpretation

The TAM front (Barrett 1979) is marked by a major normal fault zone, extending ~ 20–30 km inland from the coast and resulting in 2–5 km of displacement down to the coast (Fitzgerald 2002). The amount of exhumation decreases inland as inferred from the geological outcrop pattern and overall architecture of the TAM (Fig. 13.3a). The level of exhumation, combined with spectacular outcrops of Ross Orogen granites, often rich in accessory minerals, means that AFT has proven to be the best method to constrain the exhumation history of the TAM (Fig. 13.3b). The sampling strategy involved collecting granitic samples over significant relief across the range. AFT data revealed multiple exhumed

PAZs, defined by breaks in slope (see Chap. 9, Fitzgerald and Malusà 2018) in age–elevation profiles across the mountains. These data were interpreted to indicate periods of exhumation separated by periods of relative thermal and tectonic stability, i.e., episodic exhumation (Gleadow and Fitzgerald 1987; Fitzgerald and Gleadow 1990; Stump and Fitzgerald 1992). Samples above the break in slope contain shorter confined mean track lengths with larger standard deviations, a result of prolonged durations spent in the PAZ where track lengths are partially annealed (i.e., shortened). As the amount of exhumation decreases inland across the TAM (and the elevation of the range increases), AFT ages become older. The timing of the breaks in slope, representing the base of exhumed PAZs, also becomes older inland as the amount of exhumation decreases. These data reveal the timing, amount, and rate of rock exhumation in the TAM (e.g., Gleadow and Fitzgerald 1987; Fitzgerald and Gleadow 1990; Fitzgerald, 1992, 1994, 2002; Stump and Fitzgerald 1992; Balestrieri et al. 1994, 1997; Gleadow et al. 1984; Fitzgerald and Stump 1997; Lisker 2002; Miller et al. 2010). Exhumation rates, determined from the slope of age–elevation profiles below the break in slope, indicate rates typically <200 m/Myr. Because exhumation is so slow, heat is transported primarily via conduction, and advection has not modified the slope of the profile (e.g., Brown and Summerfield 1997). While there are many caveats to take into account when using the slope of an age–elevation profile to constrain the exhumation rate (e.g., Braun 2002, see also Chap. 9, Fitzgerald and Malusà 2018), corrections for topographic effects in the TAM are likely to be minimal (e.g., Fitzgerald et al. 2006).

The age trends and exhumation history are dependent on the location of a sample (or age profile) along the TAM, as well as its location across the range (Fig. 13.3c). Late Jurassic exhumation revealed in the Thiel Mountains, and well inland of the present-day rift flank (Fitzgerald and Baldwin 2007) is in general followed by periods of Early and Late Cretaceous exhumation. The major period of exhumation accompanying rock uplift that formed the TAM began in the Early Cenozoic (Gleadow and Fitzgerald 1987; Fitzgerald and Gleadow 1988; Fitzgerald 1992, 2002), but periods of more rapid exhumation in the Oligocene and Early Miocene have also been documented. The onset of early Cenozoic exhumation is variable along the TAM, younging from north to south: ~ 55 Ma in northern Victoria Land and southern Victoria Land, ~ 50 Ma in the Beardmore Glacier area and the Shackleton Glacier, and ~ 45 Ma in the Scott Glacier region. In places, an inland-younging trend of AFT ages is also apparent (e.g., in the Shackleton Glacier; Miller et al. 2010; in southern Victoria Land; Fitzgerald 2002). This inland-younging trend is interpreted to result from escarpment retreat at a rate of ~ 2 km/Myr, with the retreat rate apparently slowing dramatically ~10 Myr following onset of early Cenozoic exhumation (Miller et al. 2010). Exhumation rates also vary across the TAM, decreasing inland as the overall amount of rock uplift decreases.

13.4.2 Comparison with Other Thermochronologic Data Sets, and Tectonic Implications

Application of multiple thermochronologic methods on cogenetic minerals has confirmed that AFT data and inverse thermal models, on samples collected over varying elevations, provide the most information on the formation of the TAM. For example, $^{40}Ar/^{39}Ar$ data on K-feldspars from the Thiel Mountains (Fitzgerald and Baldwin 2007) yield Paleozoic ages which are significantly younger than granitoid crystallisation ages (Fig. 13.3b). The $^{40}Ar/^{39}Ar$ K-feldspar data are interpreted to date the timing of cooling associated with erosional exhumation that led to the formation of the Kukri Erosion Surface. In the Ferrar Glacier region of southern Victoria Land, AHe single grain ages on an age–elevation profile collected in granitic rocks yielded considerable intrasample variation that could be correlated with cooling rate, but in combination with AFT data indicated episodes of exhumation in the Cretaceous and Eocene (Fitzgerald et al. 2006). Detrital geochronology from glacial deposits yields Paleozoic and Mesozoic ages, with variable ZHe (480–70 Ma) and AHe (200–70 Ma) ages (Welke et al. 2016). Detrital data from drillholes offshore southern Victoria Land (Zattin et al. 2012; Olivetti et al. 2013) support the onshore AFT interpretations but also add information about provenance and younger exhumation events to the south along the TAM.

To summarise, AFT thermochronology successfully reveals the timing and patterns of Late Jurassic, Early Cretaceous, Late Cretaceous, and Cenozoic exhumation events in the TAM. These studies confirmed that erosional exhumation that formed the Kukri peneplain was not the mechanism responsible for the formation and landscape evolution of the TAM. Instead, episodic exhumation can be related to regional tectonic events including:

- Jurassic rifting and accompanying widespread basaltic magmatism (Ferrar large igneous province) that variably reset AFT ages;
- Plateau collapse and the initial break-up between Australia and Antarctica in the Early Cretaceous;
- Extension between East and West Antarctica in the Late Cretaceous accommodated on low-angle extensional faults (in the Ross Embayment and Marie Byrd Land);
- Southwards propagation of a seafloor spreading rift tip, from the Adare Trough into continental crust underlying

the western Ross Sea in the Early Cenozoic (e.g., Fitzgerald and Baldwin 1997; Fitzgerald 2002; Bialas et al. 2007).

13.5 Application of FT Thermochronology to Compressional Orogens: The Pyrenees

Thermal histories of plutonic and metamorphic rocks inferred from compressional orogens are often complicated (e.g., Dunlap et al. 1995; ter Voorde et al. 2004; Lock and Willett al. 2008; Metcalf et al. 2009). This is because thrusting does not exhume rocks, thrust burial may reset or partially reset thermochronologic systems, and rocks may undergo multiple periods of cooling and exhumation. Thrusting may also be in-sequence or out-of-sequence. Thus, a full understanding of the geologic and structural evolution is usually required before optimal sampling strategies can be developed. In this case study of the central Pyrenees, we illustrate how integration and modelling of thermochronologic data on cogenetic minerals from plutonic rocks collected in vertical profiles reveal a geologic evolution spanning 300 Myr. The results are interpreted with respect to magma crystallisation and cooling, exhumation, burial, heating during thrusting, burial and final exhumation (re-excavation) to the surface.

The Pyrenees mountains began to form in the Late Cretaceous as a result of convergence between the European and Iberian plates (Fig. 13.4a) (e.g., Munoz 2002). The core of the range (i.e., the Axial Zone) consists of an antiformal south-vergent duplex structure, composed of imbricate thrust sheets of Hercynian basement (Fig. 13.4b). The Axial Zone is flanked to the north and south by fold-and-thrust belts. Prior to the onset of convergence in the Late Cretaceous, the region now occupied by the Pyrenean mountain range was the site of Triassic and Early Cretaceous rift basins (e.g., Puigdefabregas and Souquet 1986). During the Late Cretaceous, some of the rift basins and much of the Axial Zone were below sea level, as indicated by Upper Cenomanian shallow-water carbonates that grade into deeper marine sediments and turbidites north of the Axial Zone (Seguret 1972; Berastegui et al. 1990). In the Maastrichtian, the foreland basins shallowed to tidal conditions and received continental fluvial sediments sourced by basement rocks. Initial convergence and crustal thickening were accommodated prior to the development of significant topography above sea level (McClay et al. 2004). Deformation within the orogen proceeded from north to south such that thrust sheets or portions of a thrust sheet (footwall, hanging wall, proximal to the fault, distal to the fault) preserve different aspects of the Pyrenean orogenesis. Exhumation in the Pyrenees is dominantly erosional (e.g., Morris et al. 1998);

thus, age patterns determined from low-temperature thermochronology (e.g., AFT and AHe) are usually interpreted with respect to the emergence and erosion of topography, and/or changes in base level following thrusting. Late Paleozoic biotite $^{40}Ar/^{39}Ar$ ages (Fig. 13.4c) document the timing of crystallisation of Hercynian intrusives, with variable degrees of partial resetting interpreted to result from Pyrenean orogenesis (e.g., Jolivet et al. 2007). $^{40}Ar/^{39}Ar$ K-feldspar age spectra were interpreted to result from argon loss via volume diffusion due to thrust burial and heating. Therefore, it is the low-temperature thermochronologic methods that document the timing and duration of thrusting, burial, and exhumation during intracontinental convergence.

13.5.1 Multi-method Thermochronology on Cogenetic Minerals from Vertical Profiles

In developing a sampling strategy, it is important to first recognise that the thermal evolution of footwall and hanging wall rocks within imbricate thrust sheets (e.g., the antiformal south-vergent duplex structure in the Pyrenees) varies as a function of position within the thrust system (ter Voorde et al. 2004; Metcalf et al. 2009). As intracontinental convergence proceeds, rocks at different structural positions will preserve a record of different maximum and minimum temperatures during burial due to thrust loading. The thermal history revealed by thermochronologic analysis of minerals will therefore vary with structural position (Fig. 13.4b). As long as displacement rates are sufficiently slow to allow for conductive thermal equilibration (e.g., Husson and Moretti 2002), the timing and relative magnitude of thermal events should agree. However, the maximum and minimum temperatures recorded by low-temperature thermochronologic methods will vary systematically, dependent upon the sample's structural position.

Here, we use a thermochronologic study of cogenetic minerals from granitoid samples, collected over ~ 1450 m relief within the Maladeta Pluton of the Pyrenean Axial Zone, to illustrate how application of AFT, AHe, and $^{40}Ar/^{39}Ar$ methods reveals the burial and exhumation history during thrusting and nappe emplacement (Metcalf et al. 2009). The Maladeta Massif lies within the Orri thrust sheet, presently occupying the immediate footwall of the Gavarnie Thrust, a major Alpine-age thrust fault (Fig. 13.4b). Biotite and K-feldspar from the highest elevations of the Maladeta Pluton (2850 m) in the central Axial Zone yielded maximum $^{40}Ar/^{39}Ar$ ages of ~ 280 Ma, close to the age of intrusion and interpreted to date the timing of rapid cooling during the Hercynian orogeny (Fig. 13.4c). All $^{40}Ar/^{39}Ar$ step heat experiments on K-feldspars yielded disturbed age spectra (i.e., age gradients), with the degree of partial $^{40}Ar*$ loss

varying as a function of sample elevation, and consistent with each sample's structural position in the footwall of the Gavarnie Thrust (Metcalf et al. 2009). Thus, the highest elevation sample experienced the least amount of $^{40}Ar*$ partial loss, while the lowest elevation sample experienced the greatest amount of $^{40}Ar*$ loss. Minimum $^{40}Ar/^{39}Ar$ K-feldspar ages associated with each age spectrum were interpreted to result from argon loss via volume diffusion due to thrust burial and heating.

AFT thermochronology on samples from the Maladeta profile (Fig. 13.4d) yielded ages and track-length distributions that varied as a function of elevation (Fitzgerald et al. 1999). The upper part of the profile (i.e., samples at highest elevations; 1945–2850 m) gave concordant AFT ages, with mean track lengths ≥ 14 μm for confined track-length distributions. Data from this part of the Maladeta profile were interpreted to result from rapid cooling due to exhumation between ~ 35 and ~ 32 Ma at rates of 1–3 km/Myr. The lower part of the profile (i.e., samples at 1125–1780 m elevations) yielded younger AFT ages that decrease with decreasing elevation. These samples were interpreted as reflecting slower exhumation and partial annealing due to burial of the southern flank of the Pyrenees by syn-tectonic conglomerates shed off the eroding Axial Zone thrust sheets (Coney et al. 1996). The form of the lower part of the age–elevation profile when interpreted within the geologic framework implies that there must have been Late Miocene re-excavation of the syn-tectonic conglomerates that filled the foreland basin and that were overlying the fold-and-thrust belt. Fillon and van der Beek (2012) undertook thermo-kinematic modelling to evaluate various tectonic and geomorphic scenarios using this AFT data as well as AHe ages from this region (Gibson et al. 2007; Metcalf et al. 2009). Their best-fit models, started at 40 Ma, indicated there was rapid exhumation between ~ 37 and 30 Ma at rates of >2.5 km/Myr followed by infilling of topography by syn-tectonic conglomerates with re-excavation and incision of the southern Pyrenean wedge beginning ~ 9 Ma.

While AFT and AHe thermochronology are discussed above constrain thermal histories from ~ 120 to ~ 40 °C, K-feldspar $^{40}Ar/^{39}Ar$ data and multi-diffusion domain (MDD) models extend the thermal histories into the higher temperature range of 350–150 °C (Lovera et al. 1989, 1997, 2002). Assuming that argon retention in nature and argon loss in the laboratory are controlled by thermally activated volume diffusion, argon data from step heat experiments can be inverted to yield continuous cooling histories (Lovera et al. 2002). Although K-feldspars from the Maladeta Pluton have experienced a complex geologic history, MDD models of $^{40}Ar/^{39}Ar$ K-feldspar data yielded continuous T-t histories between the higher and lower temperature thermochronologic constraints. The combined K-feldspar MDD, AFT, and AHe best-fit thermal models for each sample form

Fig. 13.4 **a** Simplified geologic map of the Pyrenean orogen with **b** ECORS cross section (A-A') indicated (modified from Fitzgerald et al. 1999; Munoz 2002; Verges et al. 2002; Metcalf et al. 2009). **c** Compilation of the thermal constraints for the Maladeta Pluton (modified from Metcalf et al. 2009, with additional information from Fillon and van der Beek 2012). **d** Simplified AFT age—elevation profile from the Maladeta Massif (modified from Fitzgerald et al. 1999, with additional information from Fillon and van der Beek 2012)

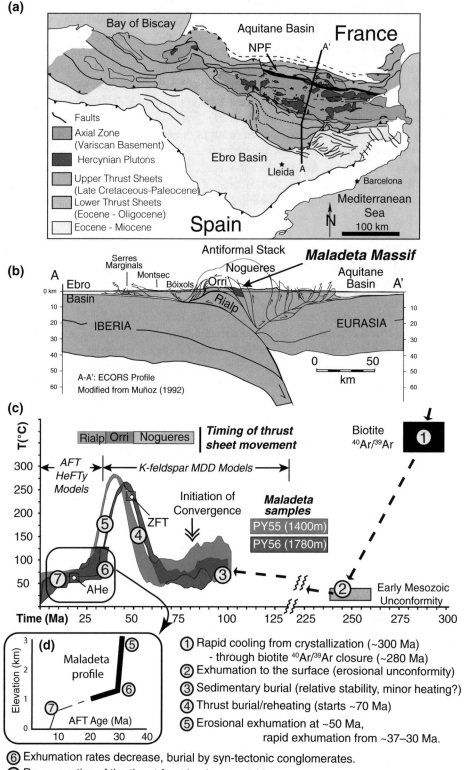

overlapping thermal history "dovetails" (e.g., PY55 and PY56; Metcalf et al. 2009; Fig. 13.4c) that are interpreted to date the timing of imbricate thrusting to form the Axial Zone antiformal stack. Ages and models obtained using different techniques are both internally consistent and most importantly agree with all available geologic observations. For example, the onset of heating and maximum temperatures, as indicated by thermal models, correlate with structural position and lateral distance from the Gavarnie Thrust and are also consistent with the geologic history of progressive burial of the Maladeta Pluton under a south-vergent thrust sheet (Munoz 2002).

13.5.2 Tectonic Interpretation and Methodologic Implications

We can summarise the thermochronologic data from the Maladeta Pluton and integrate it with geologic constraints to determine evolution of the pluton spanning 300 Myr. The thermal and geologic history includes magma crystallisation and cooling during the Hercynian orogeny, followed by cooling and exhumation to the surface. Mesozoic sediment deposition led to burial of plutonic rocks. Convergence of Iberia with Europe during the Alpine orogeny led to thrusting, heating due to overthrusting, exhumation to the surface in a number of phases, reburial by syn-tectonic conglomerates, and then final re-excavation in the Late Miocene (Fig. 13.4c, d). Following magma crystallisation at ∼300 Ma, initial cooling to below ∼325–400 °C is recorded by ∼280 Ma biotite $^{40}Ar/^{39}Ar$ ages (Metcalf et al. 2009). Subsequent cooling, as plutonic rocks were exhumed to the surface, is constrained in part by the Late Paleozoic–Early Mesozoic erosional unconformity preserved in the northern Maladeta Pluton (Zwart 1979). During the Mesozoic, plutonic rocks remained largely below sea level as shallow marine sediments were deposited. Burial and heating of the Maladeta Pluton in the footwall of the Gavarnie Thrust are recorded in both K-feldspar $^{40}Ar/^{39}Ar$ data and MDD thermal models, as well as reset Cenozoic AFT ages in a region that was at the surface in the Late Paleozoic–Early Mesozoic (Munoz 1992). The onset of erosional exhumation in the Maladeta at ∼50 Ma is recorded by K-feldspar $^{40}Ar/^{39}Ar$ MDD thermal models with accelerated exhumation from 37 to 30 Ma confirmed by AFT age–elevation relationships and modelling (Fitzgerald et al. 1999; Metcalf et al. 2009; Fillon and van der Beek 2012). From ∼30 Ma to the present, a decrease in exhumation rate is recorded by AFT thermal models and age–elevation relationships for both AFT and AHe data, with subsequent re-excavation of the southern flank of the Pyrenees beginning at ∼9 Ma. No single mineral/method reveals the complete thermal history that can be interpreted with respect to the timing and duration of thrusting, burial, and exhumation during intra-continental convergence. In this case, AFT and AHe data from both the hanging wall and footwall of the Gavarnie Thrust only provide minimum age constraints on thrust fault activity and underestimate the onset of thrust fault activity by as much as 30 Myr. The complex thermal histories revealed by multi-method thermochronology on cogenetic minerals from vertical (age–elevation) profiles also illustrate that mineral ages from these plutonic samples cannot be simply interpreted with respect to bulk closure temperatures. This Pyrenean example illustrates the necessity of combining multiple techniques as well as thermal modelling to fully reveal and interpret the geodynamic evolution of intracontinental convergent orogens.

13.6 Application of FT Thermochronology to Transpressional Plate Boundary Zones: The Alpine Fault of New Zealand

Continental transform plate boundary zones are characterised by dominantly highly localised strike-slip shear zones. Their orientation changes as they evolve, and in cases where plate motion has a significant oblique component, spectacular mountain ranges may form. In this case study, we highlight how FT thermochronology has been used to document the geodynamic evolution of the plate boundary zone in the South Island of New Zealand. The interpretation of thermochronologic data in this rapidly evolving dynamic plate boundary is complicated due to heat advection and potential (re)crystallisation associated with fluid–rock interaction. As new data (i.e., temperature, fluid pressure) from active plate-bounding faults are obtained (Sutherland et al. 2017), FT data interpretations may require re-evaluation, particularly in cases where there is evidence for late-stage fluids that transport heat and may have caused (partial) annealing of fission tracks.

13.6.1 Tectonic and Geologic Setting

The South Island of New Zealand straddles the Australian-Pacific plate boundary zone and is actively undergoing oblique continent–continent convergence (e.g., Walcott 1998). In the North Island and north-eastern part of the South Island, oceanic crust of the Pacific (PAC) plate subducts westwards beneath the Australian (AUS) plate. In the south western most part of the South Island, subduction polarity reverses, and the AUS plate subducts eastwards beneath the PAC plate. Both subduction systems are linked by a wide, dextrally transpressional fault zone in the South Island that has evolved since the latest Oligocene to Early Miocene (e.g., Cox and Sutherland 2007) with the Alpine

Fig. 13.5 Digital elevation model of the AUS-PAC transpressional plate boundary zone in the South Island of New Zealand made using GeoMap app (http://www.geomapapp.org; Ryan et al. 2009). Cross sections of the central portion of the Southern Alps (A-A') and southern segment of the Southern Alps (B-B') after Warren-Smith et al. (2016). Nested regions of reset [40]Ar/[39]Ar hornblende ages (*i* (in yellow)), and biotite ages (*ii* (dashed), and *iii* (dashed)); after Little et al. 2005), and plots of AFT and ZFT ages versus distance from the Alpine Fault after Warren-Smith et al. 2016 and Tippett and Kamp 1993). Time–temperature envelopes derived from MDD models of K-feldspar [40]Ar/[39]Ar data for samples WCG-3 and WCG-1 from West Coast granites (AUS plate affinity) from Batt et al. (2004)

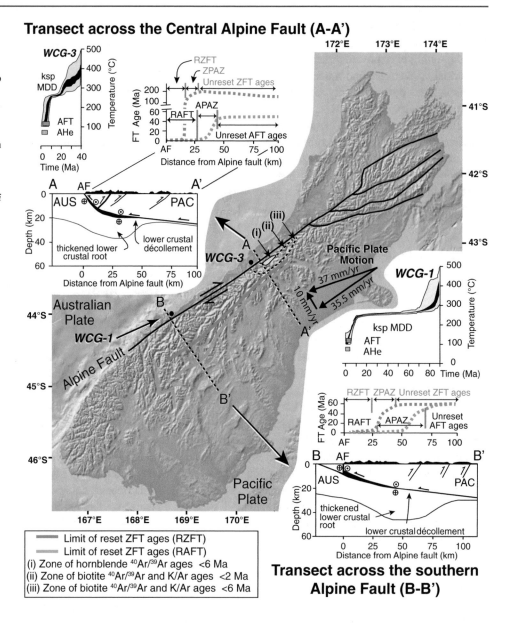

Transect across the Central Alpine Fault (A-A')

Transect across the southern Alpine Fault (B-B')

Limit of reset ZFT ages (RZFT)
Limit of reset ZFT ages (RAFT)
(i) Zone of hornblende [40]Ar/[39]Ar ages <6 Ma
(ii) Zone of biotite [40]Ar/[39]Ar and K/Ar ages <2 Ma
(iii) Zone of biotite [40]Ar/[39]Ar and K/Ar ages <6 Ma

fault zone marking the continental transform (Fig. 13.5). While the majority of the plate motion is accommodated on the Alpine Fault, slip is distributed and accommodated on faults across the entire South Island, as indicated by active seismicity and geodetic studies (e.g., Beavan et al. 2007; Wallace et al. 2006). Both geology and geodesy constrain the horizontal components of the displacement field, including velocities, strain and strain rates. Present-day AUS-PAC relative plate motion indicates that deformation is broadly partitioned into a strike-slip component of 33–40 mm/year and a fault-normal compressive component of 8–10 mm/year (Beavan et al. 2007). The Southern Alps, one of the fastest rising and eroding mountain ranges in the world, consists of (meta)greywacke that was progressively thickened to form a crustal monocline within the dextrally transpressive Alpine fault zone. Geodetic data for the central

portion of the Southern Alps region, corresponding to the Alpine fault zone and straddling the area of highest topographic relief (i.e., the Mt. Cook region), indicate surface vertical uplift rate estimates ranging from 5 to 8 mm/year (Beavan et al. 2002, 2010; Houlie and Stern 2012), comparable to rock uplift rates and exhumation rates derived from thermochronology, as discussed below.

Basement rocks of the South Island are divided broadly into a Western Province consisting mainly of granite and gneiss of AUS plate affinity, and an Eastern Province of PAC affinity consisting primarily of metamorphosed Permian to Lower Cretaceous Torlesse greywacke and the Haast Schist Belt comprising the Otago and Alpine schists (e.g., Cox and Sutherland 2007). The transpressive AUS-PAC plate boundary zone is a relatively broad anastomosing network of high strain zones (e.g., Toy et al. 2008, 2010) in which slivers of both hanging wall Alpine Schist (PAC affinity) and footwall Western Province rocks (AUS affinity) have been incorporated and heterogeneously deformed. Details of the early evolution of the modern orogen (i.e., the Southern Alps of PAC provenance) have yet to be fully revealed (e.g., Cox and Sutherland 2007). However, application of multiple thermochronologic methods on cogenetic K-feldspar and apatite from rocks of the Western Province (i.e., of AUS provenance located west of the Alpine Fault) has demonstrated that the early evolution of the Alpine fault zone is preserved in the footwall of the Alpine Fault (e.g., Batt et al. 2004) (Fig. 13.5, samples WCG-1 and WCG-3).

A steeply dipping metamorphic belt is exposed in the hanging wall (PAC) of the Alpine Fault where the metamorphic grade of Alpine Schist generally increases westwards towards the fault, reaching the oligoclase zone of the amphibolite facies (e.g., Cooper 1972, 1974). Temperatures and pressures reached by hanging wall greywackes were inferred assuming metamorphic assemblages achieved equilibrium (Grapes and Wattanabe 1992), corresponding to metamorphic mineral isograds (e.g., garnet, biotite; Little et al. 2005). However, metamorphic mineral(s) crystallised over a range of P-T conditions, where the availability of aqueous fluids enhanced reaction rates, triggering new mineral growth and recrystallisation of protoliths. As (re) crystallisation continued, complete to partial resetting of isotopic systematics within the minerals occurred. For example, zoned Late Cretaceous garnets have rims that overgrew the Alpine Fault mylonitic foliation (Vry et al. 2004). Fabrics preserve polyphase deformational histories in Alpine Fault mylonites (Toy et al. 2008), as indicated by porphyroclastic biotite (inherited from the Alpine Schist) and neocrystallised biotite within the mylonite zone in the hanging wall of the Alpine Fault (Toy et al. 2010).

13.6.2 Thermochronologic Data and Geologic Interpretation

For more than 35 years, thermochronologic studies have contributed to understanding the AUS-PAC plate boundary evolution and the landscape evolution of the Southern Alps (Fig. 13.5). Early studies documented that radiometric ages vary across the structural trend of the mountains (Sheppard et al. 1975; Adams 1980; Adams and Gabites 1985; Kamp et al. 1989; Tippett and Kamp 1993). Thermochronologic data have commonly been interpreted as ages corresponding to bulk T_c (e.g., Batt et al. 2000; Little et al. 2005). In the case of more retentive thermochronologic systems (e.g., $^{40}Ar/^{39}Ar$ mineral ages), age variations have also been suggested to be a result of variable post-metamorphic cooling involving partial Ar loss during Neogene exhumation (Adams and Gabites 1985; Chamberlain et al. 1995)

and/or "excess Ar" (Batt et al. 2000). FT ages from the Alpine Schist are generally interpreted to indicate the timing of Neogene cooling and exhumation (e.g., Kamp et al. 1989; Batt et al. 1999). Map compilations have been made that indicate the amount of exhumation in the Southern Alps (Tippett and Kamp 1993; Batt et al. 2000). These studies have interpreted isotopic ages as the timing of exhumation *from below the related closure depth* (i.e., the depth at which the ambient crustal temperature exceeds the respective T_c), assuming a "pre-uplift geothermal gradient".

Transects across the central and southern Alpine Fault (A-A′ and B-B′ in Fig. 13.5) reveal reset AFT and ZFT ages east of the Alpine Fault with the youngest ages (Middle Miocene and younger) adjacent to the Alpine Fault (e.g., Kamp et al. 1989; Tippett and Kamp 1993; Batt et al. 2000; Herman et al. 2009; Warren-Smith et al. 2016). With progressive increase in distance from the Alpine Fault (25–100 km), AFT and ZFT ages gradually increase from reset to partially annealed samples and then older (i.e., un-reset) samples, reaching Early Cenozoic and Mesozoic ages, respectively. These data have been interpreted to reflect a higher rock uplift rate and deeper exhumation closer to the Alpine Fault. The greatest amount of exhumation occurs within a narrow ∼50-km-long segment centred on the Franz Josef Glacier region where the highest peaks occur. In the central portion of the Southern Alps (A-A′ in Fig. 13.5), a narrow zone of reset FT ages has been identified that coincides with where the fault is steeper, where back-thrusting has built up topography, and where erosional exhumation is enhanced. In the central portion, the lower crustal root is thinner as compared to the southern portion of the Southern Alps. In the southern segment (B-B′ in Fig. 13.5), a wider zone of reset FT ages occurs, where the fault dip is shallower, the deformation zone is wider, and strain is partitioned over a larger region.

On the AUS (western) side of the plate boundary zone, temperature–time plots compiled using MDD models based on $^{40}Ar/^{39}Ar$ K-feldspar data together with AFT and AHe data (Batt et al. 2004) are shown for central (WG-3) and southern sections (WG-1) of the Alpine fault zone (Fig. 13.5). Also indicated (close to WG-3) are regions east of the Alpine Fault where $^{40}Ar/^{39}Ar$ hornblende and biotite ages are <6 Ma (Chamberlain et al. 1995; Little et al. 2005). Despite complexity in the data, and differences in presentation of thermochronologic data sets, some comparisons can be made for these locations. For example, gneisses and granites from the AUS side of the central portion of the fault zone contain K-feldspar that resided for shorter duration within the argon PRZ as compared to K-feldspars from the AUS side of the southern segment of the fault zone (Fig. 13.5). K-feldspar from the southern segment of the AUS plate preserves more of the pre-20 Ma history.

However, considerable scatter in isotopic ages from adjacent samples using the same mineral/method has rendered interpretation challenging (e.g., Warren-Smith et al. 2016) and also calls into question simple T_c interpretations and exhumation rate calculations based on assumed temperature to depth conversions. For example, Ring et al. (2017) used total fusion illite ^{40}Ar/^{39}Ar ages (1.36 ± 0.27 Ma, 1.18 ± 0.47 Ma), along with ZFT (0.79 ± 0.11 and 0.81 ± 0.17 Ma) and ZHe ages (0.35 ± 0.03 and 0.4 ± 0.06 Ma) from fault gouge to construct a cooling history assuming bulk T_c for each mineral/method pair. However, illite from fault gouge directly above the current trace of the Alpine Fault yielded complex ^{40}Ar/^{39}Ar laser spectra with apparent ages, corresponding to a significant percentage of ^{39}Ar released, within error of zero. Alternative interpretations, invoking partial (re)crystallisation and partial loss of radiogenic daughter products, are possible.

Toy et al. (2010) argue, based on Alpine fault zone materials now exposed at the surface, that geothermal gradients in the crust above the structural brittle–viscous transition are ∼40 °C/km and decrease to ∼10 °C/km below the structural brittle–viscous transition. Geothermal gradients evolved over time and were locally modified due to heat advection resulting from focused fluid flow, as documented by temperature and fluid pressure data from the Alpine Fault (e.g., Sutherland et al. 2017, and references therein). The Sutherland et al. study measured an average geothermal gradient of 125 ± 55 °C/km in a borehole drilled in the hanging wall of the Alpine Fault. Such high temperatures are sufficient to reset AFTs at relatively shallow depths and indicate that the present-day AFT PAZ is at a depth of only 400–800 m at this location. Exhumation-related fluid flow has been used to explain the pairing of seismic and electrical conductivity anomalies observed in the Southern Alps in New Zealand (e.g., Jiracek et al. 2007; Stern et al. 2007), low-frequency earthquake activity (Chamberlain et al. 2014), as well as the formation of abundant vein-infilled back shears in the Alpine Schist (e.g., Wightman and Little 2007). These results provide further evidence for extensive hydration in the brittle part of the Alpine Fault, with sufficiently large fluid fluxes capable of advecting heat and elevating thermal gradients on a local scale. Advective heat flow may also trigger recrystallisation (via dissolution–reprecipitation), of thermochronologically relevant mineral phases. Apatite is susceptible to metasomatic (fluid-induced) alteration over a wide range of pressures and temperatures, and even surface conditions (Harlov et al. 2005; Harlov 2015). Zircon is also prone to diagenetic and low-temperature metamorphic growth driven by fluids, especially in radiation-damaged zones of zircon crystals (Rubatto 2017). Given sufficient fluid and time, metasomatism is a viable mechanism to reset thermochronometers

(Hay and Dempster 2009). Such petrologic considerations may help to explain the poor correlations between thermochronologic data and topography, and/or local faults, correlations that were hampered by imprecise data with poor reproducibility (e.g., Herman et al. 2009).

Surface uplift rates in the central Southern Alps have been estimated to range from 5 to 10 mm/yr (Wellman 1979; Bull and Cooper 1986; Norris and Cooper 2001). Early estimates of the amount of exhumation using FT data (Tippett and Kamp 1993; Kamp and Tippett 1993) were overestimated as compared to mass balance calculations based on plate convergence (Walcott 1998). It was subsequently realised that overestimates of the amount of exhumation had assumed that rock *P-T-t-D* paths during orogenesis were vertical, when in fact rock trajectories had significant horizontal components (Willett et al. 1993; Koons 1995; Walcott 1998). The style of orogenesis (see cross sections in Fig. 13.5) in the Southern Alps meant that rocks follow paths for long distances (and hence long durations) parallel or near-parallel to relevant isotherms, as compared to the distance and durations followed by rock paths perpendicular to relevant isotherms. In addition, isotherms are not everywhere parallel to the surface, and geothermal gradients evolve with time as heat is advected upwards towards the Alpine Fault. Rapid exhumation of hot, tectonically advected rocks along the Alpine Fault has resulted in transient, localised geothermal gradients of >125 °C/km in the upper 3–4 km of the crust (Sutherland et al. 2017).

Additional factors complicate determination of exhumation rates in the Southern Alps. Firstly, mineral equilibria modelling indicates that erosional exhumation of greywacke produces a continual supply of new fluid at temperatures as low as 400 °C and pressures <2 kbar, corresponding to <7 km depths (Vry et al. 2010). This means that there may be abundant fluids within the upper crust available to transport heat (e.g., Toy et al. 2010). Secondly, the presence of fluids may facilitate (re)crystallisation of micas at temperatures below their T_c for argon. Micas may therefore recrystallise at much shallower depths than inferred "closure" depths calculated from assumed T_c and assumed steady-state geothermal gradients. If crystallisation occurred at shallower depths than those assumed for T_c and steady-state geothermal gradients, exhumation rates will be overestimated (Fig. 13.2c). Thirdly, microstructures and fluid inclusion data from the central Alpine fault zone indicate that quartz veins formed at relatively shallow crustal depths, with little variation in depths to relevant isotherms inferred for both hanging wall and fault rocks (Toy et al. 2010). In zones where thermochronologic data yield Alpine-related exhumation ages ($\leq$6 Ma in the Southern Alps), geobarometry is required to constrain the depth of crystallisation before exhumation rates can be calculated (Fig. 13.2).

Regional thermochronologic studies may mask effects due to localised recrystallisation, for example, due to late-stage hydrothermal alteration. What is generally lacking in studies on the Southern Alps is an understanding of rock particle paths obtained from *P-T-t-D* analyses on key samples. It is clear, however, that FT data are crucial to determine the timing of exhumation and brittle deformation as the Alpine fault zone evolved within the AUS-PAC plate boundary zone. To summarise, in the active AUS-PAC plate boundary in the South Island of New Zealand, partitioning of strain, erosion, mass wasting as well as the orographic effect of the Southern Alps continues to impact the landscape evolution of the range. Exhumation-related fluid flow may enhance syn-kinematic recrystallisation of minerals. Independent geobarometric data, to constrain the depth of mineral crystallisation, may be required before mineral ages can be interpreted with respect to the geodynamic evolution. While a wealth of thermochronologic data exists in the literature, sample sites are often scattered, and simple interpretations based on assumed T_c may not be valid, especially given abundant evidence for fluid flow, and documented high geothermal gradients within the active plate-bounding fault zone.

13.7 Conclusions

Thermochronologic studies of plutonic and metamorphic rocks contribute quantitative data that provide insight into deep Earth processes. Successful application of thermochronologic methods to tectonics and geodynamics has been demonstrated through use of geologically and petrologically well-constrained sampling strategies, multiple methods applied to cogenetic minerals, and modelling using kinetic parameters to obtain continuous temperature–time histories. Case studies highlight the importance of FT thermochronology to determine the final exhumation of plutonic and metamorphic rocks within different tectonic and geodynamic settings:

- In (U)HP metamorphic terranes, the integration of petrologic data and multiple thermochronologic methods document prograde, peak, and retrograde *P-T-t-D* rock paths. FT thermochronology constrains the timing of final exhumation, thereby allowing assessment of whether (U)HP rocks were exhumed to the surface within the same subduction cycle that produced eclogite-facies rocks, and the mechanism(s) by which rocks were exhumed to near-surface P-T conditions.
- In extensional orogens, such as the TAM, AFT thermochronologic studies of samples collected in vertical profiles, across and along the range, offer the best approach to constrain the timing and rate of episodic

cooling during rift flank development and landscape evolution.

- In intraplate collisional orogens, such as the Pyrenees mountains, best results are provided using a sampling strategy employing application of multiple low-temperature thermochronologic methods on cogenetic samples collected over a large range in elevation. This approach can constrain the timing of thrusting during orogenesis and the timing of subsequent exhumation. Data from age–elevation profiles, forward and inverse thermal modelling, and thermo-kinematic modelling are complementary, consistently revealing the sequence of orogenic events.
- In active transpressive plate boundary zones, such as the AUS-PAC plate boundary zone, FT thermochronology provides key constraints on timescales of orogenesis, geodynamic, and landscape evolution in the Southern Alps of New Zealand. However, the potential impact of hydrothermal fluid advection, on the (partial) resetting and annealing of fission tracks, may require re-evaluation of some geodynamic interpretations.

Acknowledgements SLB and PGF acknowledge support from the U. S. National Science Foundation. SLB and PGF thank J. Pettinga and the Erskine Program at the University of Canterbury. SLB thanks the Thonis family endowment. Thorough reviews by A. Blythe, M. Danišík, J. Gonzalez, T. Warfel, M. Jimenez, J.M. Brigham, N. Perez Consuegra, and R. Glas are greatly appreciated.

References

Abers GA, Eilon Z, Gaherty JB, Jin G, Kim YH, Obrebski M, Dieck C (2016) Southeast Papuan crustal tectonics: imaging extension and buoyancy of an active rift. J Geophys Res Solid Earth 121:951–971

Adams CJ (1980) Uplift rates and thermal structure in the Alpine fault zone and Alpine schists, Southern Alps, New Zealand. Geol Soc London Spec Publ 9:211–222

Adams CJ, Gabites JE (1985) Age of metamorphism and uplift in the Haast schist group at Haast pass, Lake Wanaka and Lake Hawea, South Island, New Zealand. New Z J Geol Geophys 28:85–96

Ague JJ, Baxter EF (2007) Brief thermal pulses during mountain building recorded by Sr dif-fusion in apatite and multicomponent diffusion in garnet. Earth Planet Sci Lett 261:500–516

Amato JM, Johnson CM, Baumgartner LP, Beard BL (1999) Rapid exhumation of the Zermatt-Saas ophiolite deduced from high-precision Sm, Nd and Rb–Sr geochronology. Earth Planet Sci Lett 171:425–438

Baldwin SL (1996) Contrasting P-T-t histories for blueschists from the western Baja terrane and the Aegean: effects of synsubduction exhumation and backarc extension. In: Bebout GE, Scholl DW, Kirby SH, Platt JP (eds) Subduction top to bottom, American Geophysical Union, Washington, DC. https://doi.org/10.1029/GM096p0135

Baldwin SL, Das JP (2015) Atmospheric Ar and Ne returned from mantle depths to the Earth's surface by forearc recycling. Proc Nat Acad Sci 112:14174–14179

Baldwin SL, Harrison TM (1992) The P-T-t history of serpentinite matrix mélange from west-central Baja California. Geol Soc Am Bull 104:18–31

Baldwin SL, Lister GS, Hill EJ, Foster DA, McDougall I (1993) Thermochronologic con-straints on the tectonic evolution of active metamorphic core complexes, D'Entrecasteaux Islands, Papua New Guinea. Tectonics 12:611–628

Baldwin SL, Monteleone BD, Webb LE, Fitzgerald PG, Grove M, Hill EJ (2004) Pliocene eclogite exhumation at plate tectonic rates in eastern Papua New Guinea. Nature 431:263–267

Baldwin SL, Webb LE, Monteleone BD (2008) Late Miocene coesite-eclogite exhumed in the Woodlark Rift. Geology 36:735–738

Baldwin SL, Fitzgerald PG, Webb LE (2012) Tectonics of the New Guinea region. Annu Rev Earth Planet Sci 40:495–520

Balestrieri ML, Bigazzi G, Ghezzo C, Lombardo B (1994) Fission track dating of apatites from the Granite Harbour Instrusive suite and uplift-denduation history of the Transantarctic Mountains in the area between the Mariner and David Glaciers (Northern Victoria Land, Antarctica). Terra Antartica 1:82–87

Balestrieri ML, Bigazzi G, Ghezzo C (1997) Uplift—denudation of the Transantarctic Mountains between the david and the mariner glaciers, Northern Victoria Land (Antarctica): Constraints by apatite fission-track analysis. In: Ricci CA (ed) The Antarctic region: geological evolution and processes. Terra Antartica Publication, Siena, pp 547–554

Barrett PJ (1965) Geology of the area between the Axel Heiberg and Shackleton Glaciers, Queen Maud Mountains, Antarctica. New Z J Geol Geophys 8:344–370

Barrett PJ (1979) Proposed drilling in McMurdo Sound—1979 Memoir of the National Institute of Polar Research. Special Issue 13:231–239

Barrett PJ (1991) The Devonian to Triassic Beacon Supergroup of the Transantarctic Mountains and correlatives in other parts of Antarctica. In: Tingey RJ (ed) The geology of Antarctica, vol 17. Oxford Monographs on Geology and Geophysics. Clarendon Press, Oxford, pp 120–152

Barrett PJ (1996) Antarctic paleoenvironment through Cenozoic times —a review. Terra Antarct 3:103–119

Barrett PJ, Elliot DH (1973) Reconnaissance geologic map of the Buckley Island Quadrangle, Transantarctic Mountains. Antarctica, United States Geological Survey, Reston, Va

Batt GE, Kohn BP, Braun J, McDougall I, Ireland TR (1999) New insight into the dynamic development of the Southern Alps, New Zealand, from detailed thermochronological investigation of the Mataketake Range pegmatites. Geol Soc London Spec Publ 154:261–282

Batt GE, Braun J, Kohn BP, McDougall I (2000) Thermochronological analysis of the dynamics of the Southern Alps, New Zealand. Geol Soc Am Bull 112:250–266

Batt GE, Baldwin SL, Cottam M, Fitzgerald PG, Brandon M (2004) Cenozoic plate boundary evolution in the South Island of New Zealand: New thermochronological constraints. Tectonics 23: TC4001

Beavan J, Tregoning P, Bevis M, Kato T, Meertens C (2002) Motion and rigidity of the Pacific Plate and implications for plate boundary deformation. J Geophys Res Solid Earth 107

Beavan J, Ellis S, Wallace LM, Denys P (2007) Kinematic constraints from GPS on oblique convergence of the Pacific and Australian plates, central South Island, New Zealand. In: Okaya D, Stern TA, Davey FJ (eds) A Continental Plate Boundary: Tectonics at South Island, New Zealand, vol 175. American Geophysical Union. Washington, DC, pp 75–94

Beavan J, Denys P, Denham M, Hager B, Herring T, Molnar P (2010) Distribution of present-day vertical deformation across the Southern Alps, New Zealand, from 10 years of GPS data. Geophys Res Lett 37

Becker H (1993) Garnet peridotite and eclogite Sm–Nd mineral ages from the Lepontine dome (Swiss Alps): New evidence for Eocene high-pressure metamorphism in the central Alps. Geology 21:599–602

Berástegui X, García JM, Losantos M (1990) Structure and sedimentary evolution of the Organyà basin (Central South Pyrenean Unit, Spain) during the Lower Cretaceous. Bull Soc Géol Fr 8:251–264

Beucher R, Beek P, Braun J, Batt GE (2012) Exhumation and relief development in the Pelvoux and Dora-Maira massifs (western Alps) assessed by spectral analysis and inversion of thermochronological age transects J Geophys Res Earth Surface 117

Bialas RW, Buck WR, Studinger M, Fitzgerald PG (2007) Plateau collapse model for the Transantarctic Mountains-West Antarctic Rift system: insights from numerical experiments. Geology 35:687

Blythe AE (1998) Active tectonics and ultrahigh-pressure rocks. In Hacker BR, Liou JG (eds) When continents collide: geodynamics and geochemistry of ultrahigh-pressure rocks. Springer Netherlands, pp 141–160

Blythe AE, Huerta AD, Utevsky E (2011) Evaluating the Mesozoic West Antarctic Plateau col-lapse hypothesis: results from apatite fission-track and (U–Th)/He analyses from Byrd Glacier Outlet. In: AGU Fall Meeting Abstracts, 2011

Bohlen SR, Valley JW, Essene EJ (1985) Metamorphism in the Adirondacks. I. petrology, pressure and temperature. J Petrol 26:971–992

Braun J (2002) Quantifying the effect of recent relief changes on age-elevation relationships. Earth Planet Sci Lett 200:331–343

Brouwer FM, Van De Zedde DMA, Wortel MJR, Vissers RLM (2004) Late-orogenic heating during exhumation: Alpine PTt trajectories and thermomechanical models. Earth Planet Sci Lett 220:185–199

Brown RW, Summerfield MA (1997) Some uncertainties in the derivation of rates of denudation from thermochronologic data. Earth Surf Proc Land 22:239–248

Bull WB, Cooper AF (1986) Uplifted marine terraces along the Alpine fault, New Zealand. Sci-ence 234:1225–1228

Camacho A, Lee JKW, Hensen BJ, Braun J (2005) Short-lived orogenic cycles and the eclogitization of cold crust by spasmodic hot fluids. Nature 435:1191

Cape Roberts Science Team (2000) Studies from the Cape Roberts Project, Ross Sea Antarctica. Initial report on CRP-3 vol 7. Terra Antartica, vol 1/2. Terra Antartica Publication, Siena, Italy

Capponi G, Messiga B, Piccardo GB, Scambelluri M, Traverso G, Vannucci R (1990) Meta-morphic assemblages in layered amphibolites and micaschists from the Dessent Formation (Mountaineer Range, Antarctica). Mem Soc Geol Ital 43:87–95

Carswell DA, Brueckner HK, Cuthbert SJ, Mehta K, O'Brien PJ (2003) The timing of stabilisa-tion and the exhumation rate for ultra-high pressure rocks in the Western Gneiss Region of Norway. J Metam Geol 21:601–612

Chamberlain CP, Zeitler PK, Cooper AF (1995) Geochronologic constraints of the uplift and metamorphism along the Alpine Fault, South Island, New Zealand. New Z J Geol Geophys 38:515–523

Chamberlain CP, Shelly DR, Townend J, Stern TA (2014) Low-frequency earthquakes reveal punctuated slow slip on the deep extent of the Alpine fault, New Zealand. Geochem Ge-ophys Geosyst 15:2984–2999

Chopin C (1984) Coesite and pure pyrope in high-grade blueschists of the Western Alps: a first record and some consequences. Contrib Mineral Petr 86:107–118. https://doi.org/10.1007/BF00381838

Chopin C, Henry C, Michard A (1991) Geology and petrology of the coesite-bearing terrain, Dora Maira massif, Western Alps. Eu J Miner 3:263–291

Coleman RG, Wang X (1995) Overview of the geology and tectonics of UHPM. Ultrahigh pressure metamorphism, pp 1–32

Compagnoni R, Hirajima T, Chopin C (1995) Ultra-high-pressure metamorphic rocks in the Western Alps. Ultrahigh pressure metamorphism, pp 206–243

Coney PJ, Muñoz JA, McClay K, Evenchick CA (1996) Syn-tectonic burial and post-tectonic exhumation of an active foreland thrust belt, southern Pyrenees, Spain. J Geol Soc 153:9–16

Cooper AF (1972) Progressive metamorphism of metabasic rocks from the Haast Schist Group of southern New Zealand. J Petrol 13:457–492

Cooper AF (1974) Multiphase deformation and its relationship to metamorphic crystallisation at Haast River, South Westland, New Zealand. New Z J Geol Geophys 17:855–880

Cox SC, Sutherland R (2007) Regional geological framework of South Island, New Zealand, and is significance for understanding the active plate boundary. In: Okaya D, Stern TA, Davey FJ (eds) A Continental Plate Boundary: Tectonics at South Island, New Zealand, vol 175. American Geophysical Union, Washington, DC, pp 19–46. https://doi.org/10.1029/175gm03

Dachs E, Proyer A (2002) Constraints on the duration of high-pressure metamorphism in the Tauern Window from diffusion modelling of discontinuous growth zones in eclogite garnet. J Metam Geol 20:769–780

Dalziel IWD (1992) Antarctica: a tale of two supercontinents. Annu Rev Earth Planet Sci 20:501–526

Davies HL, Warren RG (1988) Origin of eclogite-bearing, domed, layered metamorphic com-plexes ("core complexes") in the D'Entrecasteaux Islands, Papua New Guinea. Tectonics 7:1–21

de Sigoyer J, Chavagnac V, Blichert-Toft J, Villa IM, Luais B, Guillot S, Cosca M, Mascle G (2000) Dating the Indian continental subduction and collisional thickening in the northwest Himalaya: Multichronology of the Tso Morari eclogites. Geology 28:487–490

DesOrmeau JW, Gordon SM, Kylander-Clark ARC, Hacker BR, Bowring SA, Schoene B, Sam-perton KM (2015) Insights into (U) HP metamorphism of the Western Gneiss Region, Norway: a high-spatial resolution and high-precision zircon study. Chem Geol 414:138–155

Dewey JF (2005) Orogeny can be very short. Proc Nat Acad Sci 102:15286–15293

Dodson MH (1973) Closure temperatures in cooling geochronological and petrological systems. Contrib Mineral Petr 40:259–274

Ducea MN (2016) RESEARCH FOCUS: understanding continental subduction: a work in progress. Geology 44:239–240

Duchene S, Lardeaux J-M, Albarède F (1997) Exhumation of eclogites: insights from depth-time path analysis. Tectonophysics 280:125–140

Dunlap WJ, Teyssier C, McDougall I, Baldwin S (1995) Thermal and structural evolution of the intracratonic Arltunga Nappe Complex, central Australia. Tectonics 14:1182–1204

Elliot DH (1975) Tectonics of Antarctica: a review. Am J Sci 275:45–106

Elliot DH (1992) Jurassic magmatism and tectonism associated with Gondwanaland break-up; an Antarctic perspective. Geol Soc London Spec Publ 68:165–184

Elliot DH, Fleming TH (2004) Occurrence and dispersal of magmas in the Jurassic Ferrar large igneous province, Antarctica. Gondwana Res 7:223–237

Ellis SM, Little TA, Wallace LM, Hacker BR, Buiter SJH (2011) Feedback between rifting and diapirism can exhume ultrahigh-pressure rocks. Earth Planet Sci Lett 311:427–438

England PC, Thompson AB (1984) Pressure-temperature-time paths of regional metamor-phism I. Heat transfer during the evolution of regions of thickened continental crust. J Petrol 25:894–928

Ernst WG (1988) Tectonic history of subduction zones inferred from retrograde blueschist PT paths. Geology 16:1081–1084

Fillon C, van der Beek P (2012) Post-orogenic evolution of the southern Pyrenees: constraints from inverse thermo-kinematic modelling of low-temperature thermochronology data. Basin Res 24:418–436

Fitzgerald PG (1992) The Transantarctic Mountains of southern Victoria Land: the application of apatite fission track analysis to a rift shoulder uplift. Tectonics 11:634–662

Fitzgerald PG (1994) Thermochronologic constraints on post-Paleozoic tectonic evolution of the central Transantarctic Mountains, Antarctica. Tectonics 13:818–836

Fitzgerald PG (2002) Tectonics and landscape evolution of the Antarctic plate since Gondwana breakup, with an emphasis on the West Antarctic rift system and the Transantarctic Mountains. In: Gamble JA, Skinner DNB, Henrys S (eds) Antarctica at the close of a Millennium. In: Proceedings of the 8th international symposium on Antarctic earth science. The Royal Society of New Zealand Bulletin, 35 edn. Royal Society of New Zealand, pp 453–469

Fitzgerald PG, Baldwin SL (2007) Thermochronologic constraints on Jurassic rift flank denudation in the Thiel Mountains, Antarctica. In: Cooper AK, Raymond CR et al. (eds) Antarctica: a keystone in a changing world. USGS open-file report 2007

Fitzgerald PG, Baldwin SL (1997) Detachment fault model for the evolution of the Ross Embayment: geologic and fission track constraints from DSDP site 270. In: Ricci CA (ed) The Antarctic region: geological evolution and processes. Terra Antarctica Publication, Siena, pp. 555–564

Fitzgerald PG, Gleadow AJW (1988) Fission-track geochronology, tectonics and structure of the Transantarctic Mountains in northern Victoria Land, Antarctica. Chem Geol Isotope Geosci Sect 73:169–198

Fitzgerald PG, Gleadow AJW (1990) New approaches in fission track geochronology as a tectonic tool: examples from the Transantarctic Mountains. Nucl Tracks 17:351–357

Fitzgerald PG, Malusà MG (2018) Chapter 9: concept of the exhumed partial annealing (retention) zone and age-elevation profiles in thermochronology. In: Malusà MG, Fitzgerald PG (eds) Fission-track thermochronology and its application to geology. Springer

Fitzgerald PG, Stump E (1997) Cretaceous and Cenozoic episodic denudation of the Transantarctic Mountains, Antarctica: new constraints from apatite fission track thermochronology in the Scott Glacier region. J Geophys Res 102:7747–7765

Fitzgerald PG, Muñoz JA, Coney PJ, Baldwin SL (1999) Asymmetric exhumation across the Pyrenean orogen: implications for the tectonic evolution of collisional orogens. Earth Planet Sci Lett 173:157–170

Fitzgerald PG, Baldwin SL, O'Sullivan PB, Webb LE (2006) Interpretation of (U–Th)/He single grain ages from slowly cooled crustal terranes: a case study from the Transantarctic Mountains of southern Victoria Land. Chem Geol 225:91–120

Fitzgerald PG, Baldwin SL, Bermúdez MB, Webb LE, Little TA, Miller SR, Malusa MG, Seward D (2015) Exhumation of the Papuan New Guinea (U)HP terrane: constraints from low temperature thermochronology. XI International Eclogite Conference, Dominican Republic. http://www.ruhr-uni-bochum.de/eclogite/iec11/IEC-2015-abstract-volume.pdf

Frezzotti ML, Selverstone J, Sharp ZD, Compagnoni R (2011) Carbonate dissolution during subduction revealed by diamond-bearing rocks from the Alps. Nature Geosci 4:703

Gebauer D (1996) A P-T-t Path for an (ultra?) High-Pressure ultramafic/mafic rock association and its felsic country-rocks based on SHRIMP dating of magmatic and metamorphic zircon domains. example: Alpe Arami (Central Swiss Alps). Earth Proc: Read Isotopic Code 307–329

Gebauer D, Schertl HP, Brix M, Schreyer W (1997) 35 Ma old ultrahigh-pressure metamorphism and evidence for very rapid exhumation in the Dora Maira Massif, Western Alps. Lithos 41:5–24

Gibson M, Sinclair HD, Lynn GJ, Stuart FM (2007) Late- to post-orogenic exhumation of the central Pyrenees revealed through combined thermochronological data and thermal modeling. Basin Res 19:323–334

Gilotti JA (2013) The realm of ultrahigh-pressure metamorphism. Elements 9:255–260

Gleadow AJW, Brown RW (2000) Fission track thermochronology and the long term denuda-tional response to tectonics. In: Summerfield MA (ed) Geomorphology and global tectonics. John Willey and Sons, New York, pp 57–75

Gleadow AJW, Fitzgerald PG (1987) Uplift history and structure of the Transantarctic Mountains: new evidence from fission track dating of basement apatites in the Dry Valleys area, southern Victoria Land. Earth Planet Sci Lett 82:1–14

Gleadow AJW, McKelvey BC, Ferguson KU (1984) Uplift history of the Transantarctic Mountains in the Dry Valleys area, southern Victoria Land, Antarctica, from apatite fission track ages. New Z J Geol Geophys 27:457–464

Goldstein RH, Reynolds TJ (1994) Systematics of fluid inclusions in diagenetic minerals. SEPM Short Course 31, Tulsa, 199 pp

Goodge JW (2007) Metamorphism in the Ross orogen and its bearing on Gondwana margin tectonics. Geol S Am S 419:185–203

Gordon SM, Little TA, Hacker BR, Bowring SA, Korchinski M, Baldwin SL, Kylander-Clark ARC (2012) Multi-stage exhumation of young UHP–HP rocks: timescales of melt crystallization in the D'Entrecasteaux Islands, southeastern Papua New Guinea. Earth Planet Sci Lett 351–352:237–246

Grapes R, Watanabe T (1992) Metamorphism and uplift of Alpine schist in the Franz Josef-Fox Glacier area of the Southern Alps, New Zealand. J Metam Geol 10:171–180

Guillot S, Hattori K, Agard P, Schwartz S, Vidal O (2009) Exhumation processes in oceanic and continental subduction contexts: a review. In: Subduction Zone Geodynamics. Springer, pp 175–205

Gunn BM, Warren G (1962) Geology of Victoria Land between the Mawson and Mulock Glaciers, Antarctica vol 70–71. New Z Geol Survey Bull, Lower Hutt

Hacker BR (2006) Pressures and temperatures of ultrahigh-pressure metamorphism: implications for UHP tectonics and H_2O in subducting slabs. Int Geol Rev 48:1053–1066

Hacker BR, Ratschbacher L, Webb LE, Ireland T, Walker D, Shuwen D (1998) U/Pb zircon ages constrain the architecture of the ultrahigh-pressure Qinling-Dabie Orogen, China. Earth Planet Sci Lett 161:215–230

Hacker BR, Ratschbacher L, Webb LE, McWilliams MO, Ireland T, Calvert A, Dong S, Wenk HR, Chateigner D (2000) Exhumation of ultrahigh-pressure continental crust in east central China: late Triassic-Early Jurassic tectonic unroofing. J Geophys Res Solid Earth 105:13339–13364

Harlov DE (2015) Apatite: a fingerprint for metasomatic processes. Elements 11:171–176

Harlov DE, Wirth R, Förster H-J (2005) An experimental study of dissolution–reprecipitation in fluorapatite: fluid infiltration and the formation of monazite. Contrib Mineral Petr 150:268–286

Harrison TM, Zeitler PK (2005) Fundamentals of noble gas thermochronometry. Rev Mineral Geochem 58:123–149

Hay D, Dempster T (2009) Zircon behaviour during low-temperature metamorphism. J Petrol 50:571–589

Heimann A, Fleming TH, Elliot DH, Foland KA (1994) A short interval of Jurassic continental flood basalt volcanism in Antarctica as demonstrated by $^{40}Ar/^{39}Ar$ geochronology. Earth Planet Sci Lett 121:19–41

Herman F, Cox S, Kamp P (2009) Low-temperature thermochronology and thermokinematic modeling of deformation. Tectonics

Hill EJ, Baldwin SL (1993) Exhumation of high-pressure metamorphic rocks during crustal extension in the D'Entrecasteaux region, Papua New Guinea. J Metam Geol 11:261–277

Hill EJ, Baldwin SL, Lister GS (1992) Unroofing of active metamorphic core complexes in the D'Entrecasteaux Islands, Papua New Guinea. Geology 20:907–910

Hodges KV (1991) Pressure-temperature-time paths. Annu Rev Earth Planet Sci 19:207–236

Houlié N, Stern TA (2012) A comparison of GPS solutions for strain and SKS fast directions: Implications for modes of shear in the mantle of a plate boundary zone. Earth Planet Sci Lett 345:117–125

Hu S, Kohn BP, Raza A, Wang J, Gleadow AJW (2006) Cretaceous and Cenozoic cooling history across the ultrahigh pressure Tongbai-Dabie belt, central China, from apatite fission-track thermochronology. Tectonophysics 420:409–429

Huntington KW, Ehlers TA, Hodges KV, Whipp DM (2007) Topography, exhumation pathway, age uncertainties, and the interpretation of thermochronometer data. Tectonics 26

Husson L, Moretti I (2002) Thermal regime of fold and thrust belts—an application to the Bolivian subAndean zone. Tectonophysics 345:253–280

Jamieson RA, Beaumont C (2013) On the origin of orogens. Geol Soc Am Bull 125:1671–1702

Jiracek GR, Gonzalez VM, Grant Caldwell T, Wannamaker PE, Kilb D (2007) Seismogenic, electrically conductive, and fluid zones at continental plate boundaries in New Zealand, Himalaya, and California, USA. In: O'kaya D, Stern TA, Davey F (eds) A Continental Plate Boundary: Tectonics at South Island, New Zealand, vol 175. American Geophysical Union, Washington DC, pp 347–369

Jolivet M, Labaume P, Brunel M, Arnaud N, Campani M (2007) Thermochronology constraints for the propagation sequence of the south Pyrenean basement thrust system (France-Spain). Tectonics 26: TC5007

Kamp PJJ, Tippett JM (1993) Dynamics of Pacific plate crust in the South Island (New Zealand) zone of oblique continent-continent convergence. J Geophys Res Solid Earth 98:16105–16118

Kamp PJJ, Green PF, White SH (1989) Fission track analysis reveals character of collisional tectonics in New Zealand. Tectonics 8:169–195

Ketcham R (2018) Chapter 3. Fission track annealing: from geologic observations to thermal history modeling. In: Malusà MG, Fitzgerald PG (eds) Fission-track thermochronology and its application to geology. Springer

Kohn B, Chung L, Gleadow A (2018) Chapter 2. Fission-track analysis: field collection, sample preparation and data acquisition. In: Malusà MG, Fitzgerald PG (eds) Fission-track thermochronology and its application to geology. Springer

Kohn MJ (2016) Metamorphic chronology-a tool for all ages. Am Mineral 101:25–42

Kohn MJ, Corrie SL, Markley C (2015) The fall and rise of metamorphic zircon. Am Mineral 100:897–908

Koons PO (1995) Modeling the topographic evolution of collisional belts. Annu Rev Earth Planet Sci 23:375–408

Ksienzyk AK, Dunkl I, Jacobs J, Fossen H, Kohlmann F (2014) From orogen to passive margin: constraints from fission track and (U–

Th)/He analyses on Mesozoic uplift and fault reactivation in SW Norway. Geol Soc London Spec Publ 390(SP390):327

Kufner S-K, Schurr B, Sippl C, Yuan X, Ratschbacher L, Ischuk A, Murodkulov S, Schneider F, Mechie J, Tilmann F (2016) Deep India meets deep Asia: Lithospheric indentation, delamination and break-off under Pamir and Hindu Kush (Central Asia). Earth Planet Sci Lett 435:171–184

Lardeaux J-M, Schwartz S, Tricart P, Paul A, Guillot S, Béthoux N, Masson F (2006) A crustal-scale cross-section of the south-western Alps combining geophysical and geological imagery. Terra Nova 18:412–422

Leech ML, Stockli DF (2000) The late exhumation history of the ultrahigh-pressure Maksyutov Complex, south Ural Mountains, from new apatite fission track data. Tectonics 19:153–167

LeMasurier WE, Thomson JW (eds) (1990) Volcanoes of the Antarctic Plate and Southern Oceans. Antarctic research series, vol 48. American Geophysical Union, Washington, DC

Lennykh VI, Valizer PM, Beane R, Leech M, Ernst WG (1995) Petrotectonic evolution of the Maksyutov Complex, Southern Urals, Russia: implications for ultrahigh-pressure meta-morphism. Int Geol Rev 37:584–600

Liao J, Malusà MG, Liang Z, Baldwin SL, Fitzgerald PG, Gerya T (2018) Divergent plate motion drives rapid exhumation of (ultra) high pressure rocks. Earth Planet Sci Lett 491:67–80. https://doi.org/10.1016/j.epsl.2018.03.024

Lindsay JF, Gunner J, Barrett PJ (1973) Reconnaissance geologic map of the Mount Elizabeth and Mount Kathleen quadrangles, Transantarctic Mountains, Antarctica. US Geological Survey Washington, DC, 1:250,000

Liou JG, Ernst WG, Zhang RY, Tsujimori T, Jahn BM (2009) Ultrahigh-pressure minerals and metamorphic terranes—the view from China. J Asian Earth Sci 35:199–231

Lisker F (2002) Review of fission track studies in northern Victoria Land, Antarctica; passive margin evolution versus uplift of the Transantarctic Mountains. Tectonophysics 349:57–73

Little TA, Cox SE, Vry JK, Batt G (2005) Variations in exhumation level and uplift rate along the obliqu-slip Alpine fault, central Southern Alps. New Zealand. Geol Soc Am Bull 117:707

Little TA, Baldwin SL, Fitzgerald PG, Monteleone BM (2007) Continental rifting and meta-morphic core complex formation ahead of the Woodlark Spreading Ridge, D'Entrecasteaux Islands, Papua New Guinea. Tectonics 26: TC1002. doi:1010.1029/2005TC001911

Liu LP, Li Z-X, Danišík M, Li S, Evans N, Jourdan F, Tao N (2017) Thermochronology of the Sulu ultrahigh-pressure metamorphic terrane: implications for continental collision and lithospheric thinning. Tectonophysics 712:10–29

Lock J, Willett S (2008) Low-temperature thermochronometric ages in fold-and-thrust belts. Tectonophysics 456:147–162

Lovera OM, Richter FM, Harrison TM (1989) The 40Ar/39Ar thermochrometry for slowly cooled samples having a distribution of diffusion domain sizes. J Geophys Res 94:17917–17935

Lovera OM, Grove M, Harrison TM, Mahon KI (1997) Systematic analysis of K-feldspar 40Ar 39Ar step heating results: I. Significance of activation energy determinations. Geochim Cosmochim Ac 61:3171–3192

Lovera OM, Grove M, Harrison TM (2002) Systematic analysis of K-feldspar ^{40}Ar/^{39}Ar step heating results II: Relevance of laboratory argon diffusion properties to nature. Geochim Cosmochim Ac 66:1237–1255

Malusà MG, Fitzgerald PG (2018a) Chapter 8. From cooling to exhumation: setting the reference frame for the interpretation of thermochronologic data. In: Malusà MG, Fitzgerald PG (eds) Fission-track thermochronology and its application to geology. Springer

Malusà MG, Fitzgerald PG (2018b) Chapter 10. Application of thermochronology to geologic problems: bedrock and detrital approaches. In: Malusà MG, Fitzgerald PG (eds) Fission-track thermochronology and its application to geology. Springer

Malusà MG, Polino R, Zattin M, Bigazzi G, Martin S, Piana F (2005) Miocene to present differential exhumation in the Western Alps: insights from fission track thermochronology. Tectonics 24:1–23 TC3004

Malusà MG, Philippot P, Zattin M, Martin S (2006) Late stages of exhumation constrained by structural, fluid inclusion and fission track analyses (Sesia–Lanzo unit, Western European Alps). Earth Planet Sci Lett 243:565–580

Malusà MG, Faccenna C, Garzanti E, Polino R (2011) Divergence in subduction zones and exhumation of high pressure rocks (Eocene Western Alps). Earth Planet Sci Lett 310:21–32

Malusà MG, Faccenna C, Baldwin SL, Fitzgerald PG, Rossetti F, Balestrieri ML, Danišík M, El-lero A, Ottria G, Piromallo C (2015) Contrasting styles of (U) HP rock exhumation along the Cenozoic Adria-Europe plate boundary (Western Alps, Calabria, Corsica). Geochem Geophys Geosyst 16:1786–1824

McClay K, Muñoz J-A, García-Senz J (2004) Extensional salt tectonics in a contractional orogen: a newly identified tectonic event in the Spanish Pyrenees. Geology 32:737–740

McDougall I, Harrison TM (1999) Geochronology and thermochronology by the ^{40}Ar/^{39}Ar method vol 9. Oxford Monographs on Geology and Geophysics, 2nd edn. Oxford University Press, New York

Metcalf JR, Fitzgerald PG, Baldwin SL, Muñoz JA (2009) Thermochronology in a convergent orogen: constraints on thrust faulting and exhumation from the Maladeta Pluton in the Axial Zone of the Central Pyrenees. Earth Planet Sci Lett 287:488–503

Miller SR, Fitzgerald PG, Baldwin SL (2010) Cenozoic range-front faulting and development of the Transantarctic Mountains near Cape Surprise, Antarctica: Thermochronologic and geo-morphologic constraints. Tectonics 29. https://doi.org/10.1029/2009tc002457

Monteleone BD, Baldwin SL, Webb LE, Fitzgerald PG, Grove M, Schmitt A (2007) Late Miocene-Pliocene eclogite-facies metamorphism, D'Entrecastreaux Islands, SE Papua New Guinea. J Metam Geol 25:245–265

Morris RG, Sinclair HD, Yelland AJ (1998) Exhumation of the Pyrenean orogen: implications for sediment discharge. Basin Res 10:69–85

Muñoz JA (1992) Evolution of a continental collision belt: ECORS Pyrenees crustal balanced cross-section. In: McClay K (ed) Thrust Tectonics. Chapman and Hall, London, pp 235–246

Muñoz JA (2002) The Pyrenees Alpine tectonics; I, The Alpine system north of the Betic Cor-dillera. In: Gibbons W, Moreno T (eds) The geology of Spain. The Geological Society, London, p 649

Nagel TJ (2008) Tertiary subduction, collision and exhumation recorded in the Adula nappe, central Alps. Geol Soc London Spec Publ 298:365–392

Norris RJ, Cooper AF (2001) Late Quaternary slip rates and slip partitioning on the Alpine Fault, New Zealand. J Struct Geol 23:507–520

Olivetti V, Balestrieri ML, Rossetti F, Talarico FM (2013) Tectonic and climatic signals from apatite detrital fission track analysis of the Cape Roberts Project core records, South Victoria Land, Antarctica. Tectonophysics 594:80–90

Pedersen VK, Huismans RS, Moucha R (2016) Isostatic and dynamic support of high topography on a North Atlantic passive margin. Earth Planet Sci Lett 446:1–9

Petersen KD, Buck WR (2015) Eduction, extension, and exhumation of ultrahigh-pressure rocks in metamorphic core complexes due to subduction initiation. Geochem Geophys Geosyst 16:2564–2581

Petford N, Cruden AR, McCaffrey KJW, Vigneresse JL (2000) Granite magma formation, transport and emplacement in the Earth's crust. Nature 408:669

Philpotts A, Ague J (2009) Principles of igneous and metamorphic petrology. Cambridge University Press

Powell R, Holland T (2010) Using equilibrium thermodynamics to understand metamorphism and metamorphic rocks. Elements 6:309–314

Puigdefàbregas C, Souquet P (1986) Tectonostratigraphic cycles and depositional sequences of the Mesozoic and Tertiary from the Pyrenees. Tectonophysics 129:173–203

Purdy JW, Jager E (1976) K–Ar ages on rock forming minerals from Central Alps. Mem Univ Padova 30

Rasmussen B (2005) Zircon growth in very low grade metasedimentary rocks: evidence for zirconium mobility at ∼250 °C. Contrib Mineral Petr 150:146–155

Ratschbacher L, Hacker BR, Webb LE, McWilliams M, Ireland T, Dong S, Calvert A, Chateigner D, Wenk HR (2000) Exhumation of the ultrahigh-pressure continental crust in east central China: Cretaceous and Cenozoic unroofing and the Tan-Lu fault. J Geophys Res Solid Earth 105:13303–13338

Reiners PW, Zhou Z, Ehlers TA, Changhai X, Brandon MT, Donelick RA, Nicolescu S (2003) Post-orogenic evolution of the Dabie Shan, eastern China, from (U–Th)/He and fission track thermochronology. Am J Sci 303:489–518

Ring U, Uysal IT, Glodny J, Cox SC, Little T, Thomson SN, Stubner K, Bozkaya O (2017) Faultgouge dating in the Southern Alps, New Zealand. Tectonophysics 717:321–338

Rohrman M, Beek P, Andriessen P, Cloetingh S (1995) Meso-Cenozoic morphotectonic evolution of southern Norway: Neogene domal uplift inferred from apatite fission track thermo-chronology. Tectonics 14:704–718

Rubatto D (2017) Zircon: the metamorphic mineral. Rev Min Geochem 83:261–295

Rubatto D, Hermann J (2001) Exhumation as fast as subduction? Geology 29:3–6

Rubatto D, Gebauer D, Fanning M (1998) Jurassic formation and Eocene subduction of the Zermatt-Saas-Fee ophiolites: implications for the geodynamic evolution of the Central and Western Alps. Contrib Mineral Petr 132:269–287

Ryan WBF, Carbotte SM, Coplan JO, O'Hara S, Melkonian A, Arko R, Weissel RA, Ferrini V, Goodwillie A, Nitsche F, Bonczkowski J, Zemsky R (2009) Global multi-resolution topography synthesis. Geochem Geophys Geosyst 10. https://doi.org/10.1029/2008gc002332

Sawyer EW, Cesare B, Brown M (2011) When the continental crust melts. Elements 7:229–234

Schertl H-P, Schreyer W, Chopin C (1991) The pyrope-coesite rocks and their country rocks at Parigi, Dora Maira Massif, Western Alps: detailed petrography, mineral chemistry and PT-path. Contrib Mineral Petr 108:1–21

Schlup M, Carter A, Cosca M, Steck A (2003) Exhumation history of eastern Ladakh revealed by 40Ar/39Ar and fission-track ages: the Indus River-Tso Morari transect. NW Himalaya J Geol Soc 160:385–399

Seguret M (1972) Etude tectonique des nappes et séries décollées de la partie centrale du ver-sant sud des Pyrénées Pub Ustela, Géol Struct, pp 155

Selverstone J, Sprear F (1985) Metamorphic P-T Paths from pelitic schists and greenstones from the south-west Tauern Window, Eastern Alps. J Metam Geol 3:439–465

Selverstone J, Spear FS, Franz G, Morteani G (1984) High-pressure metamorphism in the SW Tauern Window, Austria: PT paths from hornblende-kyanite-staurolite schists. J Petrol 25:501–531

Sheppard DS, Adams CJ, Bird GW (1975) Age of metamorphism and uplift in the Alpine Schist Belt, New Zealand. Geol Soc Am Bull 86:1147–1153

Smith DC (1984) Coesite in clinopyroxene in the Caledonides and its implications for geodynamics. Nature 310:641–644

Solarino S, Malusà MG, Eva E, Guillot S, Paul A, Schwartz S, Zhao L, Aubert C, Dumont T, Pondrelli S, Salimbeni S, Wang Q, Xu X, Zheng T, Zhu R (2018) Mantle wedge exhumation beneath the Dora-Maira (U)HP dome unravelled by local earthquake tomography (Western Alps). Lithos 296–299:623–636

Spear FS (1993) Metamorphic phase equilibria and pressure-temperature-time paths. Min Soc Am, Washington, DC

Spear FS (2014) The duration of near-peak metamorphism from diffusion modelling of garnet zoning. J Metam Geol 32:903–914

Stern T, ten Brink US (1989) Flexural uplift of the Transantarctic Mountains. J Geophys Res 94:10315–10330

Stern TA, Okaya D, Kleffmann S, Scherwath M, Henrys S, Davey FJ (2007) Geophysical exploration and dynamics of the Alpine Fault zone. In: Okaya D, Stern TA, Davey FJ (eds) A Continental Plate Boundary: Tectonics at South Island, New Zealand, vol 175. American Geophysical Union, Washington, DC, pp 207–233. https://doi.org/10.1029/175gm11

Stump E, Fitzgerald PG (1992) Episodic uplift of the Transantarctic Mountains. Geology 20:161

Sutherland R, Townend J, Toy V, Upton P, Coussens J, Allen M, Baratin L-M, Barth N, Becroft L, Boese C (2017) Extreme hydrothermal conditions at an active plate-bounding fault. Nature 546:137–140

ter Voorde M, de Bruijne CH, Cloetingh SAPL, Andriessen PAM (2004) Thermal consequences of thrust faulting: simultaneous versus successive fault activation and exhumation. Earth Planet Sci Lett 223:395–413

Tippett JM, Kamp PJJ (1993) Fission track analysis of the late Cenozoic vertical kinematics of continental Pacific crust, South Island, New Zealand. J Geophys Res Solid Earth 98:16119–16148

Toy VG, Prior DJ, Norris RJ (2008) Quartz fabrics in the Alpine Fault mylonites: Influence of pre-existing preferred orientations on fabric development during progressive uplift. J Struct Geol 30:602–621

Toy VG, Craw D, Cooper AF, Norris RJ (2010) Thermal regime in the central Alpine Fault zone, New Zealand: constraints from microstructures, biotite chemistry and fluid inclusion data. Tectonophysics 485:178–192

Tracy RJ, Robinson P (1980) Evolution of metamorphic belts: information from detailed petrologic studies. The Caledonides in the USA 2:189–196

Vannucci G, Piazza M, Pastorino P, Fravega P (1997) Le facies a coralli coloniali e rodoficee calcaree di alcune sezioni basali della Formazione di Molare (Oligocene del Bacino Terziario del Piemonte. Italia nord-occidentale Mem Atti Soc Toscana Sci Nat, Ser A 104:1–27

Vergés J, Fernàndez M, Martínez A (2002) The Pyrenean orogen: pre-, syn-, and post-collisional evolution. J Virtual Expl 8:55–84

Viete DR, Hermann J, Lister GS, Stenhouse IR (2011) The nature and origin of the Barrovian metamorphism, Scotland: diffusion length scales in garnet and inferred thermal time scales. J Geol Soc 168:115–132

Vry JK, Baker J, Maas R, Little T, Grapes R, Dixon M (2004) Zoned (Cretaceous and Cenozoic) garnet and the timing of high grade metamorphism, Southern Alps, New Zealand. J Metam Geol 22:137–157

Vry J, Powell R, Golden KM, Petersen K (2010) The role of exhumation in metamorphic dehydration and fluid production. Nature Geosci 3:31

Walcott RI (1998) Modes of oblique compression: late Cenozoic tectonics of the South Island of New Zealand. Rev Geophys 36:1–26

Wallace LM, Beavan J, McCaffrey R, Berryman K, Denys P (2006) Balancing the plate motion budget in the South Island, New Zealand using GPS, geological and seismological data. Geophys J Int. https://doi.org/10.1111/j.1365-246X.2006.03183.x

Walsh EO, Hacker BR (2004) The fate of subducted continental margins: two-stage exhumation of the high-pressure to ultrahigh-pressure Western Gneiss Region, Norway. J Metam Geol 22:671–687

Warren-Smith E, Lamb S, Seward D, Smith E, Herman F, Stern T (2016) Thermochronological evidence of a low-angle, mid-crustal detachment plane beneath the central South Island, New Zealand. Geochem Geophys Geosyst 17:4212–4235

Webb LE, Baldwin SL, Little TA, Fitzgerald PG (2008) Can microplate rotation drive subduction inversion? Geology 36:823–826

Welke B, Licht K, Hennessy A, Hemming S, Pierce Davis E, Kassab C (2016) Applications of detrital geochronology and thermochronology from glacial deposits to the Paleozoic and Mesozoic thermal history of the Ross Embayment, Antarctica. Geochem Geophys Geosyst 17:2762–2780

Wellman H (1979) An uplift map for the South Island of New Zealand, and a model for uplift of the Southern Alps. Royal Soc New Z Bull 18:13–20

Wightman RH, Little TA (2007) Deformation of the Pacific Plate above the Alpine Fault ramp and its relationship to expulsion of metamorphic fluids: an array of backshears. In: Okaya D, Stern T, Davey F (eds) A Continental Plate Boundary: Tectonics at South Island, New Zealand, vol 175. American Geophysical Union, Washington DC, pp 177–205

Wildman M, Cogné N, Beucher R (2018) Fission-track thermochronology applied to the evolution of passive continental margins. In:

Malusà MG, Fitzgerald PG (eds) Fission-track thermochronology and its application to geology. Springer

Willett S, Beaumont C, Fullsack P (1993) Mechanical model for the tectonics of doubly vergent compressional orogens. Geology 21:371–374

Yamato P, Burov E, Agard P, Le Pourhiet L, Jolivet L (2008) HP-UHP exhumation during slow continental subduction: self-consistent thermodynamically and thermomechanically coupled model with application to the Western Alps. Earth Planet Sci Lett 271:63–74

Zattin M, Andreucci B, Thomson SN, Reiners PW, Talarico FM (2012) New constraints on the provenance of the ANDRILL AND 2A succession (western Ross Sea, Antarctica) from apatite triple dating. Geochem Geophys Geosyst 13

Zeitler PK, Chamberlain CP, Smith HA (1993) Synchronous anatexis, metamorphism, and rapid denudation at Nanga Parbat (Pakistan Himalaya). Geology 21:347

Zhao L, Paul A, Guillot S, Solarino S, Malusà MG, Zheng T, Aubert C, Salimbeni S, Dumont T, Schwartz S (2015) First seismic evidence for continental subduction beneath the Western Alps. Geology 43:815–818

Zirakparvar NA, Baldwin SL, Vervoort JD (2011) Lu-Hf garnet geochronology applied to plate boundary zones: insights from the (U)HP terrane exhumed within the Woodlark Rift. Earth Planet Sci Lett 309:56–66

Zirakparvar NA, Baldwin SL, Schmitt AK (2014) Zircon growth in (U) HP quartzo-feldspathic host gneisses exhumed in the Woodlark Rift of Papua New Guinea. Geochem Geophys Geosyst 15:1258–1282

Zirakparvar NA, Baldwin SL, Vervoort JD (2012) The origin and evolution of the Woodlark Rift of Papua New Guinea. Gondwana Res. https://doi.org/10.1016/j.gr.2012.06.013

Zwart HJ (1979) The geology of the Central Pyrenees. Leidse Geol Mededelingen 50:1–74

Thermochronology on Sand and Sandstones for Stratigraphic and Provenance Studies

14

Andrew Carter

Abstract

Clastic detritus preserved within a sedimentary basin represents a natural reservoir of geological information that can be used to constrain sediment deposition age and develop a picture of the sediment routing system and source terrain(s) in terms of location, age, composition and tectonic and climate stability. This chapter charts the development and applications of fission-track (FT) analysis to solve stratigraphic and provenance problems. Many of the interpretative tools and strategies developed for FT data are also applicable to detrital (U–Th)/He and other geochronological data. Provenance interpretations based on double and triple-dating strategies may be further improved by combining with mineral trace element data.

14.1 Historical Background

The goal of early provenance studies was much the same as today. In addition to constraining sediment depositional age, studies employed a set of samples through a stratigraphic sequence in a basin to reconstruct changes in the sediment routing systems, make stratigraphic correlations, define inputs from different source areas and reconstruct the uplift and erosion histories of the source rocks. The first fission-track (FT) papers appeared in the late 1970s, whereas (U–Th)/He provenance studies did not arrive until after 2000 (e.g. Rahl et al. 2003), and for this reason, most of the interpretative developments stem from FT thermochronology.

The first FT dating papers appeared in 1964 (Fleischer et al. 1964, 1965) but the standard FT methodology, as practiced today, only came into being following technical advances and a more harmonised approach made through the late 1970s into the early 1980s. The international fission-track workshop,

convened in Pisa, Italy in September 1980 and attended by over forty scientists, marked a turning point in method development and calibration as it confronted a number of persisting issues that needed to be resolved to enable the technique to become more widely accepted within the geochronological community (see Chap. 1, Hurford 2018). These included refinement of etching techniques to optimise track revelation and registration (Green and Durrani 1978), understanding the natural causes of track annealing (Burchart et al. 1979), neutron irradiation dosimetry and the value of the spontaneous-fission decay constant of ^{238}U (Hurford and Green 1981). The first inter-laboratory comparison exercise was also conducted at this time to assess inter-laboratory accuracy and precision (Naeser et al. 1981). As a consequence, most of the FT papers published between 1970 and 1985 were focused on methodological issues rather than application to geological problems. The few FT provenance studies published throughout the early to mid-1970s were mainly concerned with providing a maximum age for the timing of deposition of unfossiliferous sandstones (Gleadow and Lovering 1974; McGoldrick and Gleadow 1977) and stratigraphic ages of volcanic ashes (Naeser et al. 1974; Izett et al. 1974; Gleadow 1980; Johnson et al. 1982).

Many of the early volcanic ash studies were based on fission tracks in glass (e.g. Boellstorff and Steineck 1975) but when glass and zircon FT ages were compared from the same sample, it was realised that tracks were less stable in glass (Seward 1979) and similar to apatite it was not deemed suitable for provenance studies because in most cases source information would have been lost to thermal resetting. Thus when Hurford and Carter (1991) reviewed the few provenance-related papers published prior to 1990, zircon was the most widely used mineral due to its greater thermal stability. To a large extent, this has not changed and apatite remained less widely used than zircon until the advent of double and triple-dating approaches that combined analyses of FT, U–Pb and (U–Th)/He on the same grains. Prior to this development few studies had attempted to use detrital apatites to constrain provenance. Notable exceptions were ocean drilling cores from locations where sedimentation rates were

A. Carter (✉)
Department of Earth and Planetary Sciences, Birkbeck, University of London, Malet Street, London, WC1E 7HX, UK
e-mail: a.carter@ucl.ac.uk

© Springer International Publishing AG, part of Springer Nature 2019
M. G. Malusà and P. G. Fitzgerald (eds.), *Fission-Track Thermochronology and its Application to Geology*,
Springer Textbooks in Earth Sciences, Geography and Environment, https://doi.org/10.1007/978-3-319-89421-8_14

low and depths of burial shallow (Duddy et al. 1984; Corrigan and Crowley 1990, 1992; George and Hegarty 1995; Clift et al. 1996, 1997, 1998).

Working with zircon, the early provenance studies soon recognised the potential for data bias due to a range of factors that included small grains and counting areas, complex uranium zonation or uncountable grains due to extremely high track densities, defects or inclusions. A further complication stems from the need to make multiple grain mounts and apply different durations of etching (Naeser 1987; Garver 2003) to sample the range of ages present in any given detrital sample. Bias remains an issue as discussed more recently by Malusà et al. (2013) and in Chap. 16, Malusà (2018).

Early provenance papers used simple histogram plots to identify mixed ages, but it became apparent that the interpretation of detrital, mixed age data sets required more robust treatment in the form of statistical tools to display the observed FT ages and to identify and extract the constituent component ages. The next section reviews the development of these statistical tools and interpretative strategies.

14.2 Development of Interpretative Strategies

Fission track datasets used to determine the provenance or stratigraphic age of sand and sandstones typically contain a mixture of grains ages (overdispersed). The external detector method (EDM) is ideally suited to this because a fission track age can be obtained from individual grains. The EDM is not prone to experimental bias from variations in uranium as is the case for the population method (see Chap. 1, Hurford 2018); hence, the variation within an EDM data set is more likely to be natural rather than experimental. It is also necessary to be able to statistically estimate the most likely ages, and define the proportions and number of age components.

During the early FT studies, it was not uncommon to count a relatively low (6–12) number of grains per sample. However, it was recognised that this failed to provide sufficient sampling to capture any natural variation and determine whether the grain ages belonged to a homogenous population, consistent with a Poisson distribution. A χ^2 test was used (Galbraith 1981) as a standard test for homogeneity, i.e. test if the individual data are consistent with a common ratio of spontaneous and induced counts. In many provenance studies, a common outcome is evidence of over-dispersion with respect to Poisson variation in the spontaneous and induced track counts, which is consistent with a heterogeneous mix of grain ages. However, the χ^2 test

can only indicate evidence for or against the null hypothesis. A low (<0.05) ρ-value indicates heterogeneity within the data set but it does not quantify the level of variation or show the nature of the distribution of the data. Realisation of the limitations of the χ^2 test led to the development of methods for graphical display of the grain data and probability models for quantifying the extent of variation and identification of component populations. These are described below.

Early provenance studies plotted single-grain ages as histograms, but as this approach ignores the associated estimation errors, it was realised that such plots cannot represent true age variation. To try and overcome this Hurford et al. (1984) and Kowallis et al. (1986) applied a Gaussian density function to each estimate to generate a continuous frequency curve or total probability density function. Whilst this appeared to account for grain ages with differing precisions by representing each estimate as a narrow density function, the summary plot is an average density function that mixes both good and poor information (see Galbraith 1998). Hurford et al. (1984) even experimented with adjusting a smoothing factor to obtain greater resolution between overlapping peaks, a procedure that was further developed by Brandon (1996). However, Galbraith (1998) pointed out that since the probability density plots are not a kernel density estimate in the true sense such treatments are not valid, a point recently emphasised in the treatment of all types of detrital geochronological data by Vermeesch (2012).

To avoid problems associated with probability density plots, Rex Galbraith devised the radial plot (Galbraith 1988, 1990) where the measurement error of each counted grain has a standard deviation on the Y-scale. Radial plots are now widely used, including within the luminescence community who use a hybrid type of display known as the Abanico plot (Fig. 14.1) that combines both kernel density estimate and radial plot to display the distribution of ages of differing precision along with a picture of the age frequency distribution (Dietze et al. 2016). A number of software packages that uses these and other methods are freely available to researchers including radial/density plotter (Vermeesch 2009) and the Abanico R package (Dietze et al. 2016). Density plotter uses the statistical models for mixed FT ages and minimum age (Galbraith and Green 1990; Galbraith and Laslett 1993), expanded upon in Galbraith (2005). These methods also underpin Binomfit, a Windows[TM] program by Mark Brandon (1992) for the estimation of ages and uncertainties for concordant and mixed grain age distributions. The plotting program, PopShare (Dunkl 2002), employs a different mixture modelling algorithm called the Simplex method (Cserepes 1989) although it is unclear which function PopShare minimises. These methods all

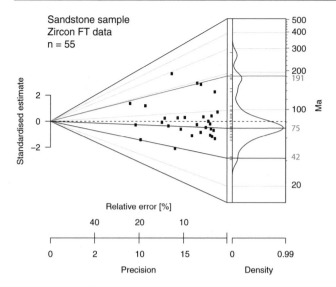

Fig. 14.1 Example of an Abanico plot (Dietze et al. 2016) that combines the radial plot and kernel density plots in a single diagram

assume Gaussian distributions, which led to Sambridge and Compston (1994) devising a mixture modelling method that included non-Gaussian statistics to reduce the influence of outliers. Later, Jasra et al. (2006) developed Bayesian mixture models and used Markov chain Monte Carlo (MCMC) methods to fit the models, which is available as a standalone program written by Kerry Gallagher (BayesMix) or as an R package. Such tools enable provenance studies to identify and extract components ages from populations of mixed ages.

14.3 Methodologies to Enhance the Interpretation of FT Provenance Data

From the earliest days of FT stratigraphic and provenance studies, it was recognised that interpretations of detrital data sets were strengthened by the inclusion of other types of geological data including zircon crystal morphology (Pupin 1980) and colour linked to α-radiation damage (Fielding 1970; Garver and Kamp 2002). Throughout the late 1980s and 1990s, a succession of provenance papers were published exploring applications of detrital FT data (Carter 1999) tackling problems such as stratigraphic ages of biostratigraphically barren clastic sediments (Carter et al. 1995), dating weathered tuffs (Winkler et al. 1990) and identifying sources of basin sediments (and indirectly paleosediment routing systems, Cerveny 1986; Garver and Brandon 1994; Carter et al. 1995), but interpretations often suffered from a lack of resolution to pinpoint sediment sources.

14.3.1 Zircon Double and Triple Dating

A major stumbling block for studies of basin sediments sourced from large areas such as the foreland basins of the Andes or Himalaya is that regional exhumation from metamorphic temperatures often gave a non-unique zircon FT age. This hindered identification of specific source locations within a mountain belt, and hence, it was impossible to distinguish between volcanic and exhumational cooling ages. To overcome this, Carter and Moss (1999) performed the first zircon double dating on Mesozoic fluvial sediments deposited in the Khorat Basin of eastern Thailand. Double dating is now an integral part of detrital studies in orogenic belts including the Himalayas (Carter et al. 2010; Najman et al. 2010), Pyrenees (Whitchurch et al. 2011), Andes (Thomson and Hervé 2002), Alaska (Perry et al. 2009) and the Chinese Loess Plateau (Stevens et al. 2013). Rahl et al. (2003) adapted the approach by combining (U–Th)/He and U–Pb dating on zircons in a study of the Lower Jurassic Navajo Sandstone in Utah to constrain sediment routing, recycling and identify links to sources within the Appalachian orogeny. Campbell et al. (2005) applied the same approach to study the sources and level of recycling of sediment within the bedloads of the Ganges and Indus rivers. Triple dating of single grains of zircon by combining FT, (U–Th)/He and U–Pb was first reported by Reiners et al. (2004) (see Chap. 5, Danišík 2018 for further details).

One key issue when dealing with double data sets that include FT data is that individual FT ages are relatively imprecise compared to single-grain U–Pb ages. Due to the large single-grain uncertainties, the FT age of a sample is based on a population (typically 20–30) of counts that should fall within a normal Poisson distribution. Within a Poisson distribution, individual counts (grains) scatter around the true age and the amount of scatter can look significant when plotted against the comparatively precise U–Pb grain ages giving the impression that double data are bad. This is illustrated in Fig. 14.2 that compares zircon double-dating results for the Tardree Rhyolite zircon (Ganerød et al. 2011), which is sometimes used as an internal FT age standard. The FT data have a central age that matches (within error) the true age and the population of counts all fall within a Poisson distribution, i.e. the data are not over-dispersed. However, not all of the individual ages lie within error of the 1:1 line; hence, double-dating interpretations should avoid low numbers of FT grain ages and not make interpretations based on a single-grain FT age.

14.3.2 Apatite Double and Triple Dating

Apatite can also be dated by the uranium–lead method (Thomson et al. 2012) and has been developed as a

Fig. 14.2 Zircon double dating of the Tardree Rhyolite. This volcanic rock has the same U–Pb and FT age that record the time of formation. The individual FT data appear significantly scattered compared to the U–Pb ages. This scatter is consistent with a normal Poisson distribution that gives a FT central age consistent with the true age

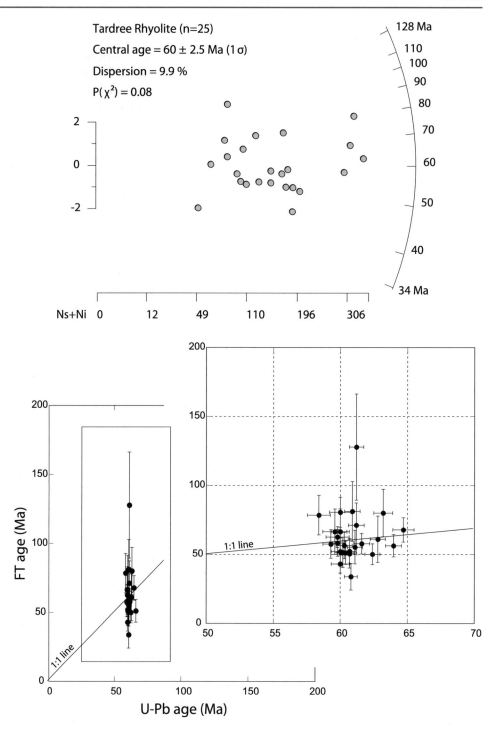

thermochronometer that can distinguish cooling paths over the interval between ~375 and 570 °C (Cochrane et al. 2014). Combination with FT (double dating) provides a powerful tool for discerning provenance, and a useful introduction to laser ablation ICP-MS analyses for both FT and U–Pb analysis can be found in Chew et al. (2011, 2012). For greater resolution of thermal history, some studies have resorted to obtaining triple dates (FT, U–Pb and (U–Th)/He) from the same apatite grain. Carrapa et al. (2009) were the

first to report triple dating in a study directed at understanding Andean Mountain building in the Paleozoic and Cenozoic. In a study on detrital grains extracted from sedimentary rocks obtained from drillcore in the Victoria Land Basin of Antarctica, Zattin et al. (2012) applied triple dating on apatite to constrain a period of Oligocene exhumation in the Transantarctic Mountains and place constraints on basin evolution and provenance. U–Pb dating was performed on selected grains within the FT mount, and these were then removed for helium analyses (see Chap. 5, Danišík 2018 for further details).

14.4 Development of Novel Methods for Detrital Zircon Provenance

Where next? The content of Ti-in-zircon is a function of crystallisation conditions (Watson et al. 2006). Average, zircons from mafic igneous rocks have higher Ti concentrations than those from felsic rocks, and therefore, some have considered this as a potential tool for discrimination between sources of detrital zircon. However, it has been found that mafic rocks can have values similar to felsic rocks and vice versa (Fu et al. 2008), and therefore, the method cannot be used to blindly link a detrital zircon to a felsic or basic source. Further, the Ti-in-zircon thermometer is based on a calibration that assumes rutile as the Ti-buffering phase, which cannot be assumed in detrital systems. It can, however, be used to discriminate sources when the source area is well described, and in this regard, it may prove a useful provenance tool as a complement to zircon FT data along with (U–Th)/He and U–Pb ages.

Hafnium isotopes are increasingly measured on samples/grains with U–Pb ages to determine if the zircon provenance was juvenile, evolved or mixed (Patchett et al. 1981). The Lu–Hf isotopic system can be combined with U–Pb dating on single grains to provide more robust constraints on crustal residence age as demonstrated for rocks in the Himalayas (Richards et al. 2005). Carter (2007) presented an example of hafnium isotope data measured on zircon FT and U–Pb dated grains from Himalayan sands in Bangladesh that demonstrated the potential of this combined approach for discerning sources within distinct tectonostratigraphic units. Although this example demonstrated the viability of measuring hafnium on FT-dated zircons, no study has yet adopted this approach.

14.5 Development of Novel Methods for Detrital Apatite Provenance

Heavy mineral associations and geochemical variations have proven essential aids to provenance interpretations (Yim et al. 1985; Baldwin et al. 1986; Kowallis 1986; Garver and Brandon 1994; Lonergan and Johnson 1998; Ruiz et al. 2004). An interesting stratigraphic example used cathodoluminescence related to chemical concentrations and FT densities in apatite as a tracer to correlate hydrothermal tin–tungsten veins in the Panasqueira mines of Portugal (Knutson et al. 1985). Because apatites contain wide variations in trace elements, there is significant potential to exploit elemental variations to provide additional source information.

Much of our understanding of trace elements in apatites stems from studies of the compositional controls on FT annealing and helium diffusion. Gleadow and Duddy (1981) first made a connection between over-dispersed apatite grain ages and a possible compositional control, later confirmed by the studies of Green et al. (1985, 1986) on samples from the same drill holes in the Otway Basin that revealed a correlation between apparent apatite FT ages and grain chlorine content. Although apatite chlorine content is important, it rarely accounts for all of the observed variation in grain ages within an over-dispersed data set. The likely reason for this is that a wide range of elements (over half of the periodic table) may be substituted into the Ca, P or anion sites of apatite $Ca_{10}(PO_4)_6$ (F,Cl,OH) (Elliott 1994); thus other trace elements, if suitably abundant, may also influence the annealing sensitivity of fission tracks, e.g. Mn, Sr and Fe (Ravenhurst et al. 1993; Burtner et al. 1994; Carlson et al. 1999), rare-earth elements and SiO_2 (Carpéna 1998). Likewise, composition also has a role in controlling helium diffusion (Mbongo Djimbi et al. 2015). From a provenance, perspective variation in trace elements is a positive as it may provide a means with which to define the provenance and type of host rock.

Dill (1994) was one of the first to consider the utility of using apatite trace element signatures to determine the origin of detrital apatites in clastic rocks by investigating the origin of apatites in Permian and Triassic red beds from southeast Germany. His study was inconclusive, in part due to the uncertain influences of diagenetic alteration. A more detailed study by Belousova et al. (2002) examined the composition of over 700 apatites from a range of rock types. Results showed how different rock types have distinctive absolute and relative abundances of many trace elements (including rare-earth elements, Sr, Y, Mn, Th), and that chondrite-normalised trace element patterns could be used to develop a discriminant tree to recognise the original host rock of an apatite. More recently, Bruand et al. (2016) showed how trace element analysis of apatite inclusions within zircon and titanite provide useful constraints on magmatic history and source whole rock chemistry of their host magmas.

14.5.1 Combined Apatite FT and Trace Element Analyses

While conventional provenance studies have started to use apatite chemistry, comparatively little work has been done to combine apatite compositional information with thermochronometric data to define the original host rock, possibly

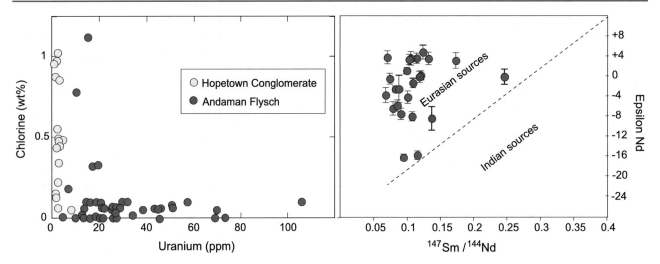

Fig. 14.3 Compositions of apatites from the Paleogene Andaman Flysch exposed on South Andaman Island are compared against apatite from a stratigraphically older unit (Hopetown conglomerate). Clasts in the Hopetown conglomerate show local arc sources. On the left, the plot compares apatite chlorine and uranium concentrations and shows two distinct sources. Low uranium chlorine-rich grains are typical of volcanic sources. It has been suggested that the Andaman Flysch is Bengal Fan material eroded from the Himalayas. Apatite Nd compositions, plotted on the right as εNd units against $^{147}Sm/^{144}Nd$, belong to the Eurasian plate field and rule out Indian plate sources

because as yet there is no comprehensive database of apatite compositions (Morton and Yaxley 2007). Allen et al. (2007) in a study of the provenance of Paleogene rocks on the Andaman Islands plotted apatite chlorine content against uranium content (Fig. 14.3, left panel) to show that apatites from an older rock unit (Hopetown Conglomerate, Mithakhari Group) were not the same as in a younger rock unit (Andaman Flysch), i.e. the two rocks did not share the same sources of apatite. To further discriminate source regions, this study also compared apatite single-grain Nd measured on samples from the Andaman Flysch formation with sands from the Himalayas. The right panel of Fig. 14.3 shows the Andaman Flysch data plotted as εNd units (Y-axis) against $^{147}Sm/^{144}Nd$ (X-axis). A compilation of published bulk rock data (Henderson et al. 2010) showed that plotting the data in this way enables discrimination between sources within the Indian and Eurasian plates. It is clear from Fig. 14.3 that the Andaman Flysch apatites came from sources within Eurasia and not India.

The Sm/Nd ratios of apatites typically range from 0.2 to 0.5 (Belousova et al. 2002) and are similar to those of average continental crust (~0.2, Taylor and McLennan 1985). Foster and Carter (2007) developed a novel provenance technique that combined FT and in situ Sm–Nd isotopic measurement of detrital apatites to constrain the source rocks. First applied on modern river sands in the Himalayas, it was possible to tie apatite FT age populations to specific Himalayan tectonostratigraphic units that were being eroded at different rates (Carter and Foster 2009). The method works well for apatites in which a relatively short period has elapsed since they were last equilibrated with the whole rock.

There is clearly significant potential for increasing the breadth of compositional data from apatite grains used in

detrital thermochronometric analyses to enhance provenance interpretations. Figure 14.4 demonstrates how existing methods that combine thermochronometry and composition data can be used to distinguish between a volcanic arc source, plutonic rocks and older metamorphic crust. The next step is to be able to determine the igneous rock type. In this regard, apatite has significant potential as it is a relatively early crystallising phase and has a high partition coefficient for REE. Abundances of F, Mn, Sr and REE are known to vary with rock type (e.g. between peraluminous I-type and S-type rocks) and have been judged to be suitable provenance indicators (Belousova et al. 2002; Chu et al. 2009; Malusà et al. 2017). Use of REE and trace element patterns based on in situ measurement by LA-ICP-MS and electron microprobe analysis to develop a discriminant tree of source rock types would seem the most obvious place to start and combined dating and trace element studies are likely to become more commonplace in the future.

14.6 Concluding Remarks

This short review has outlined the methodological developments and interpretative strategies, built largely on the development of FT analysis, that underpin detrital thermochronology. Although fission track has been the principal thermochronometer used in provenance studies, with improved understanding of helium diffusion and the advent of in situ technologies wider use of detrital (U–Th)/He analyses should increase the fidelity of provenance interpretations. While thermochronology data are undeniably important, combining age data with geochemical

Fig. 14.4 Example of the application of single-grain techniques coupled with fission tracks in detrital studies. FT analysis of a detrital sample collected from a river draining a young volcanic source (A) and granitic pluton (C) exhumed together with its gneissic country rock (B) only allows the discrimination of two age peaks. The differentiation of the relative contribution of (B) and (C) can be achieved by looking at the different apatite Sm/Nd and zircon Lu/Hf signatures in the two sources. Note that the relative contributions from (A), (B) and (C), as obtained by analysing different mineral species, differ due to non-homogeneous fertility of apatite and zircon in the source rocks (see Chap. 7, Malusà and Garzanti 2018)

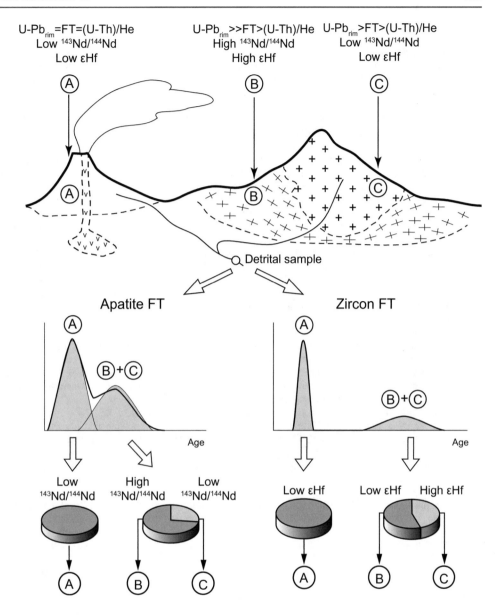

information obtained from the same grains may further strengthen provenance interpretations. The latter remains relatively unexplored, and there is considerable scope for the development of new methodologies.

Acknowledgements The author thanks Dave Chew, Alberto Resentini (Fig. 14.4) and Marco G. Malusà for their timely and constructive reviews.

References

Allen R, Carter A, Najman Y, Bandopadhyay PC et al (2007) New constraints on the sedimentation and uplift history of the Andaman-Nicobar accretionary prism, South Andaman Island. Geol Soc Am Spec Pap 436:223–256

Baldwin SL, Harrison TM, Burke K (1986) Fission track evidence for the source of accreted sandstones, Barbados. Tectonics 5:457–468

Belousova EA, Griffin WL, O'Reilly SY, Fisher NI (2002) Apatite as an indicator mineral for mineral exploration: trace-element compositions and their relationship to host rock type. J Geochem Explor 76:45–69

Boellstorff JD, Steineck PL (1975) The stratigraphic significance of fission track ages from volcanic ashes in the marine late Cenozoic of Southern California. Earth Plan Sci Lett 27:143–154

Brandon MT (1992) Decomposition of fission-track grain-age distributions. Amer J Sci 292:535–564

Brandon M (1996) Probability density plot for fission-track grain-age samples. Radiat Meas 26:663–676

Bruand E, Storey C, Fowler M (2016) An apatite for progress: inclusions in zircon and titanite constrain petrogenesis and provenance. Geology 44:91–94

Burchart J, Butkiewicz T, Dakowski M, Gałazka-Friedman J (1979) Fission track retention in minerals as a function of heating time during isothermal experiments: a discussion. Nucl Tracks 3:109–117

Burtner RL, Nigrini A, Donelick RA (1994) Thermochronology of lower Cretaceous source rocks in the Idaho-Wyoming thrust belt. AAPG Bull 78:1613–1636

Campbell IH, Reiners PW, Allen CM, Nicolescu S, Upadhyay R (2005) He-Pb double dating of detrital zircons from the Ganges and Indus rivers: implications for sediment recycling and provenance studies. Earth Plan Sci Lett 237:402–432

Carlson WD, Donelick RA, Ketcham RA (1999) Variability of apatite fission-track annealing kinetics: I. Experimental results. Am Mineral 84:1213–1223

Carpéna J (1998) Uranium-235 fission track annealing in minerals of the apatite group: an experimental study. In: van den Haute P, De Corte F (eds) Advances in fission-track geochronology. Kluwer Academic Publishers, Dordrecht, pp 81–92

Carter A, Moss SJ (1999) Combined detrital-zircon fission-track and U-Pb dating: a new approach to understanding hinterland evolution. Geology 27:235–238

Carter A, Bristow C, Hurford AJ (1995) The application of FT analysis to the dating of barren sequences: examples from red beds in Scotland and Thailand. Geol Soc Lond Spec Pub 89:57–68

Carter A (1999) Present status and future avenues of source region discrimination and characterisation using fission-track analysis. Sed Geol 124:31–45

Carter A, Foster G (2009) Improving constraints on apatite provenance: Nd measurement on FT dated grains. Geol Soc Spec Publ 324:1–16

Carter A (2007) Heavy minerals and detrital fission-track thermochronology. Dev Sedimentol 58:851–868

Carter A, Najman N, Bahroudi A, Bown P, Garzanti E, Lawrence RD (2010) Locating earliest records of orogenesis in western Himalaya: evidence from Paleogene sediments in the Iranian Makran and Pakistan Katawaz basin. Geology 38:807–810

Carrapa B, DeCelles PG, Reiners PW, Gehrels GE, Sudo M (2009) Apatite triple dating and white mica $^{40}Ar/^{39}Ar$ thermochronology of syntectonic detritus in the central Andes: a multiphase tectonothermal history. Geology 37:407–410

Cerveny PF (1986) Uplift and erosion of the Himalaya over the past 18 million years: evidence from fission track dating of detrital zircons and heavy mineral analysis. M.Sc. thesis, Dartmouth College, Hanover, New Hampshire

Chew DM, Sylvester PJ, Tubrett MN (2011) U-Pb and Th-Pb dating of apatite by LA-ICPMS. Chem Geol 280:200–216

Chew DM, Donelick RA (2012) Combined apatite fission track and U-Pb dating by LA-ICPMS and its application in apatite provenance analysis. Mineral Ass Canada Short Course 42:219–247

Chu M-F, Wang K-L, Griffin WL, Chung S-L, O'Reilly SY, Pearson NJ, Iizuka Y (2009) Apatite composition: tracing petrogenetic processes in transhimalayan granitoids. J Petrol 50:1829–1855

Clift PD, Carter A, Hurford AJ (1996) Constraints on the evolution of the East Greenland Margin: evidence from detrital apatite in offshore sediments. Geology 24:1013–1016

Clift PD, Carter A, Hurford AJ (1998) The erosional history of north-east Atlantic passive margins and constraints on the influence of a passing plume. J Geol Soc 155:787–800

Clift PD, Lorenzo J, Carter A, Hurford AJ (1997) Transform tectonics and thermal rejuvenation on the Côte d'Ivoire-Ghana margin, west Africa. J Geol Soc 154:483–489

Cochrane R, Spikings RA, Chew D, Wotzlaw J-F, Chiaradia M, Tyrrell S, Schaltegger U, Van der Lelij R (2014) High temperature (>350 °C) thermochronology and mechanisms of Pb loss in apatite. Geochim Cosmochim Acta 127:39–56

Corrigan JD, Crowley KD (1990) Fission-track analysis of detrital apatites from sites 717 and 718, leg 116, central Indian Ocean. Proc Ocean Drill Prog Sci Results 116:75–92

Corrigan JD, Crowley KD (1992) Unroofing of the Himalayas: A view from apatite fission-track analysis of Bengal fan sediments. Geophys Res Lett 19:2345–2348

Cserepes L (1989) Numerical mathematics—for geophysicist students. Tankönyvkiadó, Budapest, p 358

Danišík M (2018) Integration of fission-track thermochronology with other geochronologic methods on single crystals (Chapter 5). In: Malusà MG, Fitzgerald PG (eds) Fission-track thermochronology and its application to geology. Springer

Dietze M, Kreutzer S, Burow C, Fuchs MC, Fischer M, Schmidt C (2016) The abanico plot: visualising chronometric data with individual standard errors. Quart Geochron 31:12–18

Dill HG (1994) Can REE patterns and U-Th variations be used as a tool to determine the origin of apatite in clastic rocks. Sed Geol 92:175–196

Duddy IR, Gleadow AJW, Keene JB (1984) FT dating of apatite and sphene from palaeogene sediments of deep sea drilling project leg 81, site 555. Init Rep Deep Sea Drill Proj 81:725–729

Dunkl I, Székely B (2002) Component analysis with visualization of fitting—PopShare, a Windows program for data analysis. Goldschmidt Conf Abs, Geochim Cosmochim Acta 66A:201

Elliott JC (1994) Structure and chemistry of the apatites and other calcium orthophosphates. Studies in inorganic chemistry, vol 18. Elsevier, Amsterdam. 389 pp

Fielding PE (1970) The distribution of uranium, rare earths and colour centres in a crystal of natural zircon. Am Mineral 55:428–440

Fleischer RL, Price PB, Symes EM, Miller DS (1964) Fission track ages and track-annealing behavior of some micas. Science 143:349–351

Fleischer RL, Price PB, Walker RM (1965) Effects of temperature, pressure and ionisation on the formation and stability of fission tracks in minerals and glasses. J Geophys Res 70:1497–1502

Foster GL, Carter A (2007) Insights into the patterns and locations of erosion in the Himalaya—a combined fission-track and in situ Sm–Nd isotopic study of detrital apatite. Earth Planet Sci Lett 257:407–418

Fu B, Page FZ, Cavoise AJ, Fournelle J, Kita NT, Lackey JS, Wilde S, Valley JW (2008) Ti-in-zircon thermometry: applications and limitations. Contrib Mineral Petrol 156:197–215

Galbraith RF (1981) On statistical model for fission track counts. Math Geol 13:471–488

Galbraith RF (1988) Graphical display of estimates having differing standard errors. Technometrics 30:271–281

Galbraith RF (1998) The trouble with "probability density" plots of fission track ages. Rad Measur 29:125–131

Galbraith RF (1990) The radial plot: graphical assessment of spread in ages. Nucl Tracks Radiat Meas 17:207–214

Galbraith RF, Green PF (1990) Estimating the component ages in a finite mixture. Nucl Tracks Radiat Meas 17:197–206

Galbraith RF, Laslett GM (1993) Statistical models for mixed fission track ages. Nucl Tracks 21:459–470

Galbraith RF (2005) Statistics for fission track analysis. Interdisciplinary Statistics Series. Chapman and Hall/CRC, 224 pp

Ganerød M, Chew DM, Smethurst M, Troll VR, Corfu F, Meade F. Prestvik T (2011) Geochronology of the tardree rhyolite complex, Northern Ireland: implications for North Atlantic magmatism and zircon fission track and (U-Th)/He studies. Chem Geol 286:222–228

Garver JI, Kamp PJJ (2002) Integration of zircon color and zircon fission track zonation patterns in orogenic belts: application of the Southern Alps, New Zealand. Tectonophysics 349:203–219

Garver JI (2003) Etching age standards for fission track analysis. Radiat Meas 37:47–54

Garver JI, Brandon MT (1994) Erosional denudation of the British Columbia Coast Ranges as determined from fission-track ages of detrital zircon from the Tofino basin, Olympic Peninsula, Washington. Geol Soc Amer Bull 106:1398–1412

George AD, Hegarty KA (1995) FT analysis of detrital apatites from sites 859, 860 and 862, Chile triple junction. Proc Ocean Drill Program Sci Results 141:181–186

Green PF, Durrani SA (1978) A quantitative assessment of geometry factors for use in fission track studies. Nucl Track Detect 2:207–213

Gleadow AJW, Duddy IR (1981) A natural long-term track annealing experiment for apatite. Nucl Tracks 5:169–174

Green PF, Duddy IR, Gleadow AJW, Tingate PR, Laslett GM (1985) Fission-track annealing in apatite: track length measurements and the form of the Arrhenius plot. Nucl Tracks 10:323–328

Green PF, Duddy IR, Gleadow AJW, Tingate PR, Laslett GM (1986) Thermal annealing of fission tracks in apatite: 1. A qualitative description. Chem Geol Isot Geosci Sect 59:237–253

Gleadow AJW (1980) Fission track age of the KBS Tuff and associated hominid remains in northern Kenya. Nature 284:225–230

Gleadow AJW, Lovering JF (1974) The effect of weathering on fission track dating. Earth Planet Sci Lett 22:163–168

Henderson AL, Foster GL, Najman Y (2010) Testing the application of in situ Sm–Nd isotopic analysis on detrital apatites: a provenance tool for constraining the timing of India-Eurasia collision. Earth Planet Sci Lett 297:42–49

Hurford AJ (2018) An historical perspective on fission-track thermochronology (Chapter 1) In: Malusà MG, Fitzgerald PG (eds) Fission-track thermochronology and its application to geology. Springer

Hurford AJ, Green PF (1981) Standards, dosimetry and the uranium-238 λ_f decay constant: a discussion. Nucl Tracks 5:73–75

Hurford AJ, Fitch FJ, Clarke A (1984) Resolution of the age structure of the detrital zircon populations of two lower Cretaceous sandstones from the Weald of England by fission track dating. Geol Mag 121:269–277

Hurford AJ, Carter A (1991) The role of fission track dating in discrimination of provenance. Geol Soc Spec Publ 57:67–78

Izett GA, Naeser CW, Obradovich J (1974) Fission-track age of zircons from an ash bed in the pico formation (Pliocene and Pleistocene) near Ventura, California. Geol Soc Am Abstr Programs 6:197

Jasra A, Stephens DA, Gallagher K, Holmes CC (2006) Analysis of geochronological data with measurement error using Bayesian mixtures. Math Geol 38:269–300

Johnson GD, Zeitler P, Naeser CW, Johnson NM, Summers DM, Frost CD, Opdyke NP, Tahirkheli RAK (1982) The occurrence and fission-track ages of late Neogene and Quaternary volcanic sediments, Siwalik group, northern Pakistan. Palaeogeog Palaeoclim Palaeoecol 37:63–93

Knutson C, Peacor DR, Kelly WC (1985) Luminescence, colour and fission track zoning in apatite crystals of the Panasqueira tin-tungsten deposit, Beira-Baixa, Portugal. Am Mineral 79:829–837

Kowallis BJ, Heaton JS, Bringhurst K (1986) Fission-track dating of volcanically derived sedimentary rocks. Geology 14:19–22

Lonergan L, Johnson C (1998) A novel approach for reconstructing the denudation histories of mountain belts with an example from the Betic Cordillera (S. Spain). Basin Res 10:353–364

Malusà MG (2018) A guide for interpreting complex detrital age patterns in stratigraphic sequences (Chapter 16). In: Malusà MG, Fitzgerald PG (eds) Fission-track thermochronology and its application to geology. Springer

Malusà MG, Garzanti E (2018) The sedimentology of detrital thermochronology (Chapter 7). In: Malusà MG, Fitzgerald PG (eds) Fission-track thermochronology and its application to geology. Springer

Malusà MG, Carter A, Limoncelli M, Garzanti E, Villa IM (2013) Bias in detrital zircon geochronology and thermochronometry. Chem Geol 359:90–107

Malusà MG, Wang J, Garzanti E, Liu ZC, Villa IM, Wittmann H (2017) Trace-element and Nd-isotope systematics in detrital apatite of the Po river catchment: implications for provenance discrimination and the lag-time approach to detrital thermochronology. Lithos 290–291:48–59

Mbongo Djimbi D, Gautheron C, Roques J, Tassan-Got L, Gerin C, Simoni E (2015) Impact of apatite chemical composition on (U-Th)/He thermochronometry: an atomistic point of view. Geochim Cosmochim Acta 167:162–176

McGoldrick PJ, Gleadow AJW (1977) Fission-track dating of lower Palaeozoic sandstones at Tatong, North Central Victoria. J Geol Soc Australia 24:461–464

Morton A, Yaxley G (2007) Detrital apatite geochemistry and its application in provenance studies. In: Arribas J, Critelli S, Johnson MJ (eds) Sedimentary provenance and petrogenesis perspectives from geochemistry. Geol Soc Am Spec Pap 420:319–344

Naeser CW, Izett GA, Wilcox RE (1974) Zircon fission-track ages of pearlette family ash beds in Meade County, Kansas. Geology 1:187–189

Naeser CW, Zimmermann RA, Cebula GT (1981) Fission-track dating of apatite and zircon: an interlaboratory comparison. Nucl Tracks 5:65–72

Naeser ND, Zeitler PK, Naeser CW, Cerveny PF (1987) Provenance studies by fission-track dating of zircon-etching and counting procedures. Nucl Tracks Rad Meas 13:121–126

Najman Y, Appel E, Boudagher-Fadel M, Bown P, Carter A, Garzanti E, Godin L, Han J, Liebke U, Oliver G, Parrish R, Vezzoli G (2010) The timing of India-Asia collision: sedimentological, biostratigraphic and palaeomagnetic constraints. J Geophys Res 115:B12416

Patchett P, Kouvo O, Hedge C, Tatsumoto M (1981) Evolution of the continental crust and mantle heterogeneity: evidence from Hf isotopes. Contrib Mineral Petr 75:263–267

Perry SE, Garver JI, Ridgway KD (2009) Transport of the Yakutat terrane, Southern Alaska: evidence from sediment petrology and detrital zircon fission-track and U/Pb double dating. J Geol 117:156–173

Pupin JP (1980) Zircon and granite petrology. Contrib Mineral Petrol 73:207–220

Rahl JM, Reiners PW, Campbell IH, Nicolescu S, Allen CM (2003) Combined single-grain (U-Th)/He and U/Pb dating of detrital zircons from the Navajo Sandstone, Utah. Geology 31(9):761

Ravenhurst CE, Roden MK, Willet SD, Miller DS (1993) Dependence of fission track annealing on apatite crystal chemistry. Nucl Tracks Rad Meas 21:622

Reiners PW, Campbell IS, Nicolescu S, Allen CA, Garver JI, Hourigan JK, Cowan DS (2004) Double- and triple-dating of single detrital zircons with (U-Th)/He, fission-track, and U/Pb systems, and examples from modern and ancient sediments of the western U.S. American Geophysical Union, Fall Meeting 2004, abstract T51D-01

Richards A, Argles T, Harris N, Parrish R, Ahmad T, Darbyshire F, Dragantis E (2005) Himalayan architecture constrained by isotopic tracers from clastic sediments. Earth Planet Sci Lett 236:773–796

Ruiz GMH, Seward D, Winkler W (2004) Detrital thermochronology—a new perspective on hinterland tectonics, an example from the Andean Amazon basin, Ecuador. Basin Res 16:413–430

Sambridge MS, Compston W (1994) Mixture modeling of multi-component data sets with application to ion-probe zircon ages. Earth Planet Sci Lett 128:373–390

Seward D (1979) Comparison of zircon and glass fission-track ages from tephra horizons. Geology 7:479–482

Stevens T, Carter A, Watson TP, Vermeesch P, Andò S, Bird AF, Lu H, Garzanti E, Cottam MA, Sevastjanova I (2013) Genetic linkage between the Yellow River, the Mu Us desert and the Chinese Loess Plateau. Quat Sci Rev 78:355–368

Taylor SR, McLennan SM (1985) The continental crust: its composition and evolution. Blackwell Scientific Oxford, 312 pp

Thomson SN, Hervé F (2002) New time constraints for metamorphism at the ancestral Pacific Gondwana margin of southern Chile (42°S–52°S). Rev Geol Chile 29:255–271

Thomson SN, Gehrels GE, Ruiz J, Buchwaldt R (2012) Routine low-damage apatite U-Pb dating using laser ablation–multicollector–ICPMS. Geochem Geophys Geosyst 13:Q0AA21

Vermeesch P (2009) Radial plotter: a Java application for fission track, luminescence and other radial plots. Radiat Meas 44:409–410

Vermeesch P (2012) On the visualisation of detrital age distributions. Chem Geol 312:190–194

Watson EB, Wark DA, Thomas JB (2006) Crystallization thermometers for zircon and rutile. Contrib Mineral Petrol 151:413–433

Whitchurch AL, Carter A, Sinclair HD, Duller RA, Whittaker AC, Allen PA (2011) Sediment routing system evolution within a diachronously uplifting orogen: insights from detrital zircon fission track and U-Pb thermochronological analyses from the south-central Pyrenees. Am J. Science 311:1–43

Winkler W, Hurford AJ, Perch-Nielsen K, Odin GS (1990) Fission track and nannofossil ages from a Palaeocene bentonite in the Schlieren Flysch (Central Alps, Switzerland). Schweiz Mineral Petrogr Mitt 70:389–396

Yim W-S, Gleadow AJW, van Moort JC (1985) Fission track dating of alluvial zircons and heavy mineral provenance in Northeast Tasmania. J Geol Soc 142(351):356

Zattin M, Andreucci B, Thomson SN, Reiners PW, Talarico F (2012) New constraints on the provenance of the ANDRILL AND-2A succession (western Ross Sea, Antarctica) from apatite triple dating. Geochem Geophys Geosyst 13:Q10016

Exhumation Studies of Mountain Belts Based on Detrital Fission-Track Analysis on Sand and Sandstones

15

Matthias Bernet

Abstract

Fission-track (FT) analysis of detrital apatite and zircon from modern sediments and ancient sandstone is a commonly used approach for studying and quantifying the long-term exhumation history of convergent mountain belts. Being aware of potential bias in the age spectra because of sampling, sample preparation and statistical data treatment such as peak-fitting, FT ages from sediments and sedimentary rocks of known depositional age can be readily transferred into long-term average exhumation or erosion rates using the lag-time concept. Double dating of single grains with the FT and U–Pb methods provides additional valuable provenance information, for example, for identifying volcanically derived grains, which may obscure the exhumation signal. Applying both apatite and zircon FT dating on the same samples allows combining the study of source area exhumation and the thermal evolution of sedimentary basins.

15.1 Introduction

The record of erosional exhumation of a mountain belt is preserved in the sediments and sedimentary rocks deposited in the adjacent basin. Because sediments are primarily transported by rivers from the source areas to the basins, collecting samples from modern river sediments and ancient fluvial and marine sandstone holds great potential for determining the exhumation history of an orogenic mountain belt. Apatite and zircon fission-track (FT) dating, possibly combined with U–Pb analyses on the same grains, are now commonly used techniques for determining sediment provenance and rates of exhumation. One of the first studies that used a detrital thermochronology approach to study

exhumation of the north-western Himalayas was by Zeitler et al. (1982). They used FT analysis of detrital zircon from foreland basin deposits to investigate the long-term exhumation record of the Himalayas and presented the precursor of what developed later into the lag-time concept for detrital thermochronology (Brandon and Vance 1992; Garver et al. 1999).

Many detrital FT studies followed in the Himalaya (e.g. Cerveny et al. 1988; Bernet et al. 2006; van der Beek et al. 2006; Steward et al. 2008; Chirouze et al. 2012, 2013; Lang et al. 2016), but also in the European Alps (e.g. Spiegel et al. 2000, 2004; Bernet et al. 2001, 2004a, b, 2009; Trautwein et al. 2002; Cederbom et al. 2004; Malusà et al. 2009; Glotzbach et al. 2011; Jourdan et al. 2013; Bernet 2013), and other mountain belts around the world (e.g. Brandon and Vance 1992; Garver and Brandon 1994a, b; Carter and Moss 1999; Soloviev et al. 2001; Garver and Kamp 2002; Carter and Bristow 2003; Stewart and Brandon 2004; Garver et al. 2005; Enkelmann et al. 2008, 2009; Parra et al. 2009; Bermúdez et al. 2013, 2017). This chapter provides a summary of the pertinent aspects of detrital FT thermochronology, the lag-time concept, estimation of exhumation rates and the combination with U–Pb dating on single grains, including some precautions for sampling and data interpretation.

15.2 The Exhumation Signal in the Detrital Record

In convergent mountain belts, exhumation of deep-seated rocks is driven by a combination of normal faulting and erosion to remove the overburden (Platt 1993; Ring et al. 1999). Because exhumation means bringing rocks closer to the surface (England and Molnar 1990), thrust faults and parallel strike-slip faulting do not exhume rocks. In tectonically active regions, exhumation rates can reach values of more than 10 km/Myr but are for most orogens somewhere

M. Bernet (✉)
Institut des Sciences de la Terre, CNRS, Université Grenoble Alpes, Grenoble, France
e-mail: matthias.bernet@univ-grenoble-alpes.fr

© Springer International Publishing AG, part of Springer Nature 2019
M. G. Malusà and P. G. Fitzgerald (eds.), *Fission-Track Thermochronology and its Application to Geology*,
Springer Textbooks in Earth Sciences, Geography and Environment, https://doi.org/10.1007/978-3-319-89421-8_15

between 0.1 and 1 km/Myr (Burbank 2002; Montgomery and Brandon 2002). The exhumation signal is transferred from source areas to sedimentary basins in the form of detrital apatite and zircon FT cooling ages. The exhumation record is more or less preserved during burial, depending on thermal conditions and the partial annealing zones of the apatite and zircon FT systems. Samples from proximal locations to the source tend to provide local information on exhumation rates, whereas distally collected samples reflect a strongly mixed exhumation signal from the larger source area on an orogen scale (Fig. 15.1; Bernet et al. 2004b; Bernet and Garver 2005).

15.2.1 Modern Sediments

For any exhumation study, the analysis of modern river, beach or glacial sediments is important for obtaining the present-day exhumation signal. This signal is easier to interpret than the exhumation signal from ancient sedimentary rocks, as the size of the present-day drainage area and the source rock lithologies are known, and the detrital age signal can in many cases be compared to bedrock FT ages (e.g. Bernet et al. 2004a, b; see Chap. 10, Malusà and Fitzgerald 2018a, b). Furthermore, parameters such as short-term erosion rates (from sediment yield or cosmogenic radionuclide analysis), topography, seismicity, precipitation patterns, vegetation cover as well as numerical modelling can be included for completing exhumation studies (e.g. Bermúdez et al. 2013). The present-day exhumation signal provides a baseline for interpreting the exhumation signal from ancient sandstone samples.

Modern sediments are commonly sampled from medium- to coarse-grained sand (250 µm–1 mm) deposits in the river channel or on the beach, or from glacial deposits. Resentini et al. (2013) show that the most suitable grain size depends on sediment sorting and the size difference between the apatite and zircon grains relative to quartz (also see Chap. 7, Malusà and Garzanti 2018). However, this may be difficult to determine in the field, and not all projects allow for the collection of pilot samples before sampling for thermo- and geochronological analyses. If the objective is to do only thermo- and geochronology, samples are often collected from placer deposits, and 2–4 kg of sediment is sometimes panned in the field to increase the apatite and zircon yield, even if this may potentially introduce a small sampling bias. If the full heavy mineral spectrum (e.g. Mange and Wright 2007) or mineral fertility (Malusà et al. 2016) is to be determined from the same sample, then only bulk sediment samples should be taken. River sediments tend to be efficiently mixed, and detrital FT age spectra tend to well reflect the bedrock age distribution in the source area (Zeitler et al. 1982; Bernet et al. 2004a, b; Bermúdez et al. 2013). It is useful to note the petrology of pebbles at the sampling location, as they provide a quick overview on the variety of source rock lithology.

15.2.2 Ancient Sandstones

Medium- to coarse-grained clastic sandstones can as well be analysed with detrital FT analysis, and depending on the petrologic composition (arkose, lithic arenite, quartz arenite, etc.) and mineral fertility, 2–7 kg should be collected in the field. Apatites are not very stable under (tropical) weathering conditions (Morton 2012). Therefore, heavily weathered sandstone should be avoided. Equally, shales and mudstone should be avoided as the apatite and zircon grains of such samples are too small for FT analysis, because small grains are difficult to polish during sample preparation and the exposed surface areas are too small for track counting. Petrographic thin section and heavy mineral analyses of the sampled sandstone can provide independent provenance information for the exhumation study.

15.3 The Lag-Time Concept

The lag-time concept is used for converting FT ages into exhumation rates. Following Garver et al. (1999), lag time is here defined as the difference between the apparent cooling age and the time of deposition (Fig. 15.1). The transport time in the fluvial system can be considered negligible (Heller et al. 1992), if it is shorter than the error of the cooling age. The lag time, therefore, integrates the time needed for exhuming rock from the closure depth of the used dating technique to the surface, erosion of the rock, fluvial and/or glacial transport and deposition in the basin. For applying the lag-time concept, it is important that the time of deposition of ancient sandstone is reasonably well known (uncertainty not more than ±1 Myr), in order to obtain first-order exhumation rate estimates. Biostratigraphic information can help, if concise biozones have been defined and are constrained by stratigraphic correlation to absolute ages based on magnetostratigraphic constraints or absolute age dating of, for example, volcanic ash layers. If a strict stratigraphic control is not available, the lag-time concept cannot be applied.

In response to changes in exhumation rates, lag times will vary (Rahl et al. 2007). Changes from slow (~ 0.1 km/Myr) to relatively fast ($> \sim 1.0$ km/Myr) exhumation or vice versa may cause an important perturbation of the thermal structure, and advection or relaxation of isotherms (Rahl et al. 2007; Braun 2016; see Chap 8, Malusà and Fitzgerald 2018a). This will result in changes in the geothermal gradient and consequently influence lag times.

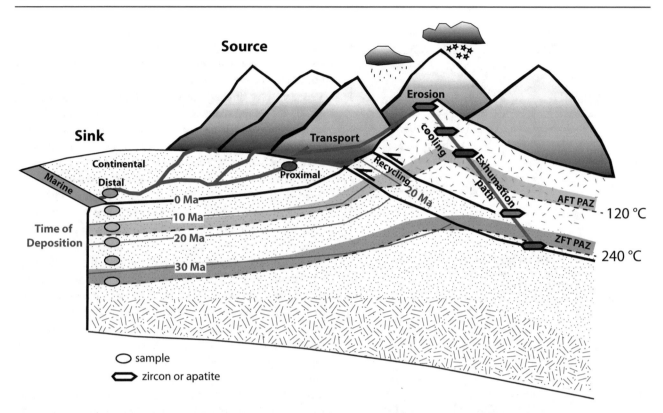

Fig. 15.1 Source-to-sink sediment routing and the lag-time concept. Lag time is the difference between the FT cooling age and the time of deposition. Proximal and distal refer to sampling location. Green points indicate samples from the stratigraphic record

15.3.1 Single Grain-Ages, Peak Ages, Central Ages and Minimum Ages

Once the time of deposition is known through a well-established stratigraphic framework, the question is which FT age should be used and plotted against the depositional age on a lag-time plot: single grain-ages, peak ages, central ages or minimum ages (Fig. 15.2)? For many detrital exhumation studies, up to 100 grains are analysed per sample. Non-reset samples will in most cases have an over-dispersed grain-age spectrum, with ages ranging from a few millions to hundreds of millions of years. The problem with plotting such data is that the individual errors of single grain-ages tend to be large and the plots are difficult to read because of the overlap. Furthermore, a single FT grain-age has a limited meaning in a detrital age distribution, because it presents just one data point within a Poisson distribution (assuming no wider dispersion due to grain chemistry or accumulated radiation damage) (see Chap. 14, Carter 2018). Hence, to define source exhumation rates, age components needed to be extracted.

For better structuring the detrital FT data, the observed grain-age distributions can be decomposed by statistical means into major grain-age components or peaks (e.g. Galbraith and Green 1990; Brandon, 1992, 1996). For this, different software packages exist, such as Binomfit (see

Ehlers et al. 2005), RadialPlotter (Vermeesch 2009) or Bayes MixQT (Gallagher et al. 2009). A detrital sample may contain between one and four statistically significant age peaks, which can be plotted on a lag-time plot (Fig. 15.2). Caution is needed when using these programs. Age peaks need to be inspected in relation to the central age, age dispersion, chi-square test results and the other age peaks of the sample. For example, one has to check if there is significant overlap between two age peaks or are peaks well separated within the same sample. In case of age peak overlap, one has to decide if the two separate peaks are geologically justified or if they may be an artefact of statistical analysis, possibly based on data with large uncertainties due to low track counts and/or low uranium concentrations. The proper identification of age peaks is important, particularly when they are used for determining source area exhumation rates. Naylor et al. (2015) used synthetic data and a Monte Carlo bootstrap method to show that it is possible to map the systematic bias in the peak age modelling as a function of the true ages (Fig. 15.3). The bootstrap modelling of synthetic detrital FT data based on real FT data may help to quantify systematic bias and improve data interpretation, for example, avoiding the interpretation of changes in peak age lag times as geologically meaningful when in reality they may be a statistical artefact (Naylor et al. 2015).

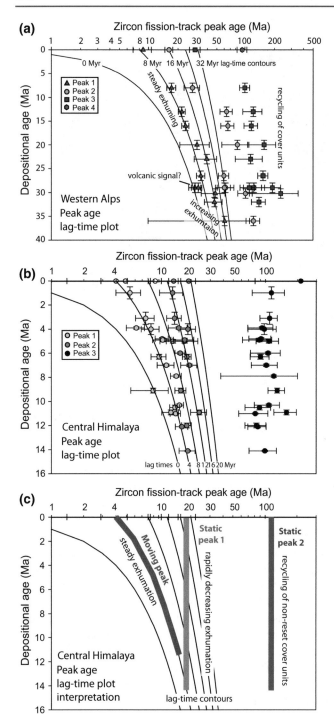

Fig. 15.2 Lag-time plots of detrital zircon FT data from the foreland basins of Western Alps and Himalaya, also including the lag-time trends with static and moving peaks (Bernet et al. 2006, 2009; Jourdan et al. 2013)

Furthermore, the question arises—What is the geological significance of an age peak in a detrital grain-age distribution? In general, a peak age does not necessarily correspond

to a specific tectonic or thermal event in the past, because the apatite and zircon grains were derived from an eroding landscape which will provide a continuum of cooling ages. For example, Fig. 15.4 shows zircon FT data of the small Vénéon River drainage in the Western Alps, which drains among others the southern flank of the La Meije massif. The bedrock zircon FT ages of the La Meije massif show a continuous range of cooling ages from about 14 Ma in the Romanche valley at 1400 m elevation to about 27 Ma at 3000 m elevation, near the peak of the mountain (van der Beek et al. 2010), as a result of exhumation cooling. The detrital zircon FT data show three age peaks at 14, 21.5 and 55.2 Ma (Bernet 2013). It would be wrong to interpret the three detrital age peaks as representing three distinct tectonic events. In fact, zircon grains belonging to the 14 Ma peak correspond to the bedrock zircon FT age at the elevation where the river is currently incising. Zircon grains belonging to the 21 Ma peak were likely derived from about 1800–3000 m elevation. This shows that erosion of relatively slowly cooled rocks exposed along moderately high relief topography can provide a detrital cooling age spectrum which may be resolved as two distinct age peaks by peak-fitting, even if nothing else happened than moderately slow erosional exhumation for tens of millions of years. The zircon grains belonging to the 55.2 Ma peak were derived from cover rocks that experienced only partial annealing during Alpine metamorphism. Note that the 55.2 Ma age has no geological significance in terms of Alpine evolution. It only indicates that exhumation has been slow (<0.1 km/Myr) and the cover rocks have not yet been fully removed.

Looking at the zircon FT age peaks from ancient sandstone in the Western Alps foreland basin (Fig. 15.2), it is obvious that partially annealed or non-annealed zircon grains are common but contain no direct information about the Alpine orogeny. Furthermore, in stratigraphically younger samples, and distal modern river or beach samples, the effect of sediment recycling from inverted foreland basin deposits that were integrated into the orogenic mountain belt needs to be taken into account. This is shown, for example, in the Rhône River drainage and Rhône delta detrital zircon FT data of Bernet et al. (2004a, b). Similarly, Fig. 15.2 also shows an example of the exhumation signal from the Himalayan foreland basin (Bernet et al. 2006). Consequently, for detecting tectonic and/or thermal events in a mountain belt from detrital FT data requires a long-term record to see the lag-time evolution over tens of millions of years, as discussed further below.

The central age can be used for estimating an average age of an over-dispersed age distribution in a detrital sample instead of the more commonly used pooled age for bedrock

Input of grain data and closure age model

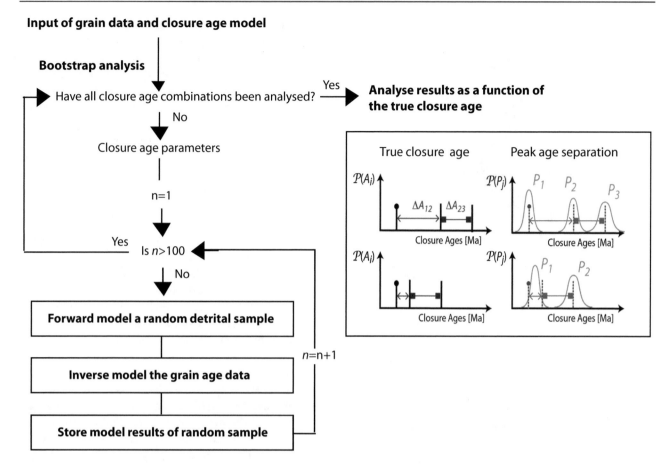

Fig. 15.3 Quantifying uncertainty with the Monte Carlo bootstrap method, simplified after Naylor et al. (2015). Well-separated true closure age populations can be detected by peak-fitting, whereas near-true closure ages maybe statistically better represented as one peak (Naylor personal comm.)

samples (Galbraith and Laslett 1993). However, it is important to check the grain-age distribution in detail to see if the over-dispersion is related to a single outlier or to a mixture of different age populations. Detrital apatite FT data may contain many zero-track grains with possibly high single grain-age uncertainties. The grain-age distributions need to be examined carefully, and the use of radial plots is advisable to visualise the age spread (see Chap. 6, Vermeesch 2018). The central age may serve for estimating drainage basin average exhumation rates, even if this signal may be very variable.

The minimum age is an estimation of the first coherent age component that can be determined in a grain-age distribution (Galbraith and Laslett 1993). In many cases, the minimum age is identical to the first peak age determined with peak-fitting and can provide a proxy for the fast exhumation rates at the time of deposition. In the case of a volcanic contribution, the minimum age can be used as a proxy for the depositional age (Soloviev et al. 2001).

15.3.2 Distinguishing the Exhumation Signal from a Volcanic Signal

In orogens with volcanic activity, the exhumation signal may be obscured by the contribution of volcanically derived apatite and zircon grains, with cooling ages close to or identical to the depositional age. The crystal shape can be an indication, but it is better to perform FT and U–Pb double dating on single grains (Carter and Moss 1999) to clearly identify volcanic grains, as they have the same crystallisation and cooling ages. In contrast, grains cooled during exhumation have in general different crystallisation and cooling ages. Reiners et al. (2005) have described the theoretical background for (U–Th)/He and U–Pb double dating, and the same is applicable for FT and U–Pb double dating. Jourdan et al. (2013) have applied this approach to the Western Alps (Fig. 15.4). In the Western Alps, the last volcanic activity occurred at about 30 Ma with local andesite deposits (von Blanckenburg et al. 1998; Malusà et al. 2011).

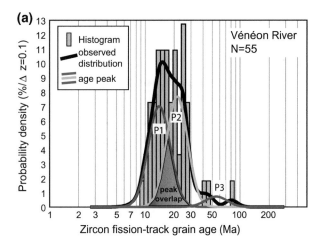

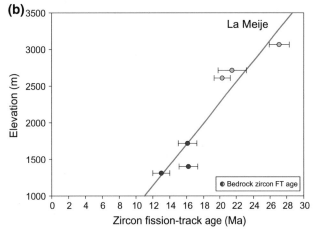

Fig. 15.4 Probability density plot of detrital zircon FT data from the Vénéon River (Bernet 2013) as compared to bedrock FT ages from different elevations within La Meije massif (van der Beek et al. 2010)

The lag-time plot with the zircon FT peak ages from the Western Alps foreland basin, reported in Fig. 15.2, shows a peak age with a very short lag-time between 30 and 28 Ma. Does this mean very fast exhumation or volcanic activity, or both at that time? Zircon double dating confirmed a contribution of volcanically derived Oligocene zircons. These zircon grains were subsequently removed from the data set to obtain a pure exhumation signal, which suggests that rapid exhumation prevailed for a short time during the early Oligocene in the Western Alps, contemporaneous to the volcanic activity (Jourdan et al. 2013). Double dating of zircon and apatite has become a standard tool and is now applied to a growing number of studies (see Chap. 5, Danišík 2018).

15.3.3 Lag-Time Trends: Moving and Static Peaks

The lag-time concept is useful for tracing the long-term evolution of exhumation. In the case of initially increasing

exhumation rates, a shortening of lag time may be observed up-section. This may represent the removal of non- or partially annealed sedimentary cover rocks during the constructional phase of a mountain belt and adjustment of the upper crustal thermal structure to the new steady-state (see Chap. 16, Malusà 2018). During this transient situation, it is difficult to determine exhumation rates from lag times until the system has equilibrated to the new conditions (Rahl et al. 2007). When peak ages become continuously younger up-section over millions of years of deposition, this is referred to as a moving peak (Fig. 15.2). When the peak ages change at the same rate as the depositional ages, the lag times remain constant. This is an indication of continuous exhumation of the mountain belt (e.g. Bernet et al. 2006, 2009; Glotzbach et al. 2011). In the case that the peak ages remain constant up-section, this is referred to as a static peak, as illustrated, for example, by the Himalayan detrital zircon FT record (Fig. 15.2). Static peaks may indicate: a large-scale tectonic or thermal event or the end of fast exhumation in the past, during which a thick section of the crust was rapidly cooled and now sheds grains with always the same cooling ages into the sedimentary system (Braun 2016); or recycling of foreland basin deposits and re-sedimentation in stratigraphically younger sections; or a combination of the two (Garver et al. 1999; Chirouze et al. 2013). These alternative scenarios may be difficult to distinguish on a thermochronologic ground alone, and additional geological and sediment petrographic information may be needed together with thermal modelling.

15.3.4 Estimating Exhumation Rates from Lag Time

FT cooling ages can be transferred into exhumation rates in more or less sophisticated ways (e.g. Garver et al. 1999), and so can lag times of moving peaks be used for estimating exhumation rates (see discussion in Chap. 10, Malusà and Fitzgerald 2018b). Lag-times of static peaks are not useful for determining exhumation rates as they only indicate a slowing of exhumation, in case the cooling ages are synorogenic, or simply removal of non-rest cover units, if the cooling ages are pre-orogenic (Braun 2016). Exhumation rates determined from lag times should be regarded as long-time average exhumation rate estimates, because they integrate exhumation since the time from initial cooling below the closure temperature in the upper crust to the time of deposition. Surface erosion rates may have been variable on shorter timescales during exhumation, but this signal is smoothed out in the detrital FT data, given the millions of years of lag-time until the grain is deposited in the basin. A simple approach for estimating exhumation rates from peak age lag times is using the 1D steady-state thermal

Fig. 15.5 Zircon FT and U–Pb double dating results from the Western Alps pro-side foreland basin (Clumanc and Saint Lions conglomerates, Barrême basin; Jourdan et al. 2013)

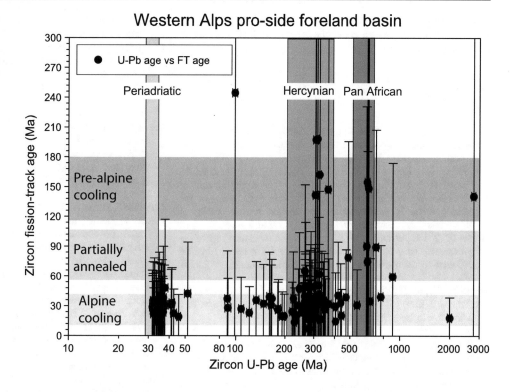

advection model of the Age2Edot code of M. Brandon (Reiners and Brandon 2006). For most exhumation studies, this level of first-order estimates is sufficient given the uncertainties of the peak ages and the depositional ages. Nevertheless, Naylor et al. (2015) warned that exhumation rate estimates based on peak ages can be systematically biased and may have large uncertainties, resulting in non-unique interpretations. In such a case, the bias introduced by over-fitting the data with too many peaks would propagate into the lag-time and exhumation rate estimation.

Glotzbach et al. (2011) used thermo-kinematic 3D PeCUBE modelling (Braun 2003) to show that the detrital apatite FT data from Miocene to present-day Western Alps foreland basin sediments have the resolution to detect a potential doubling of erosion rates between 5 and 2.5 Ma, but these data would not detect if the rate change occurred at 1 Ma. PeCUBE modelling has the advantage that it can take changes in the crustal thermal structure into account, to provide more robust interpretations of exhumation rates (Braun et al. 2012).

15.3.5 Negative Lag Time

Negative lag times occur when the FT peak age is younger than the age of deposition. This is more commonly the case in detrital apatite with increasing burial depth, due to partial annealing because of burial heating. Such data need to be interpreted in terms of basin evolution instead of source area exhumation (see Chap. 16, Malusà 2018). A static peak with negative lag times and increasing paleo-burial depth can be used to constrain the timing of basin inversion (e.g. Cederbom et al. 2004; van der Beek et al. 2006; Chirouze et al. 2013).

15.4 Conclusions

Detrital FT analysis of sand and sandstone is a powerful tool for studying the exhumation history of orogenic mountain belts, because the FT data can be used with the lag-time concept for determining exhumation rates in the past. One has to be aware of potential sampling, preparation and statistical bias when interpreting grain-age data and modelling exhumation rates. Nonetheless, the observed detrital grain-age spectra reflect drainage basin FT cooling ages and, if these cooling ages reflect exhumation, detrital thermochronologic analysis of sand and sandstone provides useful information on the exhumation history of the source rocks. For determining geologically sound long-term exhumation scenarios, detrital FT analysis can be combined with other geological methods, such as single grain-double dating in combination with U–Pb dating, sediment petrographic provenance information, stratigraphic, structural, seismic or climatic data (Fig. 15.5).

Acknowledgements This chapter benefited from reviews by Andy Carter and Eva Enkelmann and editorial handling by Marco G. Malusà and Paul G. Fitzgerald.

References

Bermúdez MA, van der Beek P, Bernet M (2013) Strong tectonic and weak climatic control on exhumation rates in the Venezuelan Andes. Lithosphere 25:3–16

Bermúdez MA, Hoorn C, Bernet M, Carrillo E, van der Beek P, Garver JI, Mora JL, Mehrkian K (2017) The detrital record of Late-Miocene to Pliocene surface uplift and exhumation of the Mérida Andes in the Maracaibo and Barinas foreland basins of Venezuela. Basin Res. https://doi.org/10.1111/bre.12154

Bernet M (2013) Detrital zircon fission-track thermochronology of the present-day Isere River drainage system in the Western Alps: no evidence for increasing erosion rates at 5 Ma. Geosciences 3:528–542

Bernet M, Garver JI (2005) Fission-track analysis of detrital zircon. In: Reiners P, Ehlers T (eds) Low-temperature thermochronology. Rev Mineral Geochem 58:205–238

Bernet M, Zattin M, Garver JI, Brandon MT, Vance JA (2001) Steady-state exhumation of the European Alps. Geology 29:35–38

Bernet M, Brandon MT, Garver JI, Molitor B (2004a) Fundamentals of detrital zircon fission-track analysis for provenance and exhumation studies. Geol S Am S 378:25–36

Bernet M, Brandon MT, Garver JI, Molitor B (2004b) Downstream changes of Alpine zircon fission-track ages in the Rhône and Rhine rivers. J Sediment Res 74:82–94

Bernet M, van der Beek P, Pik R, Huyghe P, Mugnier JL, Labrin E, Szulc A (2006) Miocene to Recent exhumation of the central Himalaya determined from combined detrital zircon fission-track and U/Pb analysis of Siwalik sediments, western Nepal. Basin Res 18:393–412

Bernet M, Brandon MT, Garver JI, Balestrieri ML, Ventura B, Zattin M (2009) Exhuming the Alps through time: clues from detrital zircon fission-track ages. Basin Res 21:781–798

Brandon MT (1992) Decomposition of fission-track grain age distributions. Am J Sci 26:535–564

Brandon MT (1996) Probability density plot for fission-track grain-age samples. Rad Meas 26:663–676

Brandon MT, Vance JA (1992) Tectonic evolution of the Cenozoic Olympic subduction complex, Washington State, as deduced from fission-track ages for detrital zircons. Am J Sci 292:565–636

Braun J (2003) Pecube: a new finite element code to solve the heat transport equation in three dimensions in the earth's crust including the effects of a time-varying, finite amplitude surface topography. Comput Geosci 29:787–794

Braun J (2016) Strong imprint of past orogenic events on the thermochronological record. Tectonophysics 683:325–332

Braun J, van der Beek P, Valla P, Robert X, Herman F, Glotzbach C, Pedersen V, Perry C, Simon-Labric T, Prigent C (2012) Quantifying rates of landscape evolution and tectonic processes by thermochronology and numerical modeling of crustal heat transport using PECUBE. Tectonophysics 524–525:1–28

Burbank DW (2002) Rates of erosion and their implications for exhumation. Min Mag 66:25–52

Carter A (2018) Thermochronology on sand and sandstones for stratigraphic and provenance studies (Chapter 14). In: Malusà MG, Fitzgerald PG (eds) Fission-track thermochronology and its application to geology. Springer, Berlin

Carter A, Bristow CS (2003) Linking hinterland evolution and continental basin sedimentation by using detrital zircon thermochronology: a study of the Khorat Plateau basin, eastern Thailand. Basin Res 15:271–285

Carter A, Moss SJ (1999) Combined detrital zircon fission-track and U-Pb dating: a new approach to understanding hinterland evolution. Geology 27:235–238

Cederbom CE, Sinclair HD, Schlunegger F, Rahn MK (2004) Climate-induced rebound and exhumation of the European Alps. Geology 32:709–712

Cerveny PF, Naeser ND, Zeitler PK, Naeser CW, Johnson NM (1988) History of uplift and relief of the Himalaya during the past 18 million years: evidence from fission-track ages of detrital zircons from sandstones of the Siwalik Group. In: Kleinspehn K, Paola C (eds) New perspectives in basin analysis. Springer, New York, pp 43–61

Chirouze F, Bernet M, Huyghe P, Erens V, Dupont-Nivet G, Senebier F (2012) Detrital thermochronology and sediment petrology of the middle Siwaliks along the Muksar Khola section in eastern Nepal. J Asian Earth Sci 44:117–135

Chirouze F, Huyghe P, van der Beek P, Chauvel C, Chakraborty T, Dupont-Nivet G, Bernet M (2013) Tectonics, exhumation and drainage evolution of the Eastern Himalaya since 13 Ma from detrital geochemistry and thermochronology, Kameng River Section, Arunachal Pradesh. Geol Soc Am Bull 125

Danišík M (2018) Integration of fission-track thermochronology with other geochronologic methods on single crystals (Chapter 5). In: Malusà MG, Fitzgerald PG (eds) Fission-track thermochronology and its application to geology. Springer, Berlin

Ehlers TA, Chaudhri T, Kumar S, Fuller C, Willett SD, Ketcham R, Brandon MT (2005). Computational tools for low-temperature thermochronometer interpretation. In: Reiners PW and Ehlers TA (eds), Low-Temperature Thermochronology. Techniques, Interpretations and Applications. Rev Mineral Geochem, Mineralogical Society of America 58:589–622. https://doi.org/10.2138/rmg.2005.58.22

England P, Molnar P (1990) Surface uplift, uplift of rocks, and exhumation of rocks. Geology 18:1173–1177

Enklemann E, Garver JI, Pavlis TL (2008) Rapid exhumation of ice-covered rocks of the Chugach–St. Elias orogen, Southeast Alaska. Geology 36:915–918

Enklemann E, Zeitler PK, Pavlis TL, Garver JI, Ridgway KD (2009) Intense localized rock uplift and erosion in the St Elias orogen of Alaska. Nat Geosci. https://doi.org/10.1038/NGEO502

Galbraith RF, Green PF (1990) Estimating the component ages in a finite mixture. Nucl Tracks Rad Meas 17(197):206

Galbraith RF, Laslett GM (1993) Statistical models for mixed fission track ages. Nucl Tracks Rad Meas 21:459–470

Gallagher K, Charvin K, Nielsen S, Sambridge M, Stephenson J (2009) Markov chain Monte Carlo (MCMC) sampling methods to determine optimal models, model resolution and model choice for Earth Science problems. Mar Petrol Geol 26:525–535

Garver JI, Brandon MT (1994a) Fission-track ages of detrital zircon fro, Cretaceous strata, southern British Colombia: implications for the Baja BC hypothesis. Tectonics 13:401–420

Garver JI, Brandon MT (1994b) Erosional denudation of the British Colmbia coast ranges as determined fro, fission-track ages of detrital zircon from the Tofino basin, Olympic Peninsula, Washington. Geol Soc Am Bull 106:1398–1412

Garver JI, Kamp PJJ (2002) Integration of zircon color and zircon fission-track zonation patterns in orogenic belts: application to the Southern Alps, New Zealand. Tectonophysics 349:203–219

Garver JI, Brandon MT, Roden-Tice MK, Kamp PJJ (1999) Exhumation history of orogenic highlands determined by detrital fission-track thermochronology. Geol Soc Spec Publ 154:283–304

Garver JI, Reiners PW, Walker LJ, Ramage JM, Perry SE (2005) Implications for timing of Andean uplift from thermal resetting of radiation damaged zircon in the Cordillera Huayhuash, Northern Peru. J Geol 113:117–138

Glotzbach C, Bernet M, van der Beek P (2011) Detrital thermochronology records changing source areas and steady exhumation in the Western and Central European Alps. Geology 39:239–242

Heller PL, Tabor RW, O'Neil JR, Pevear DR, Shafiquillah M, Winslow NS (1992) Isotopic provenance of Paleogene sandstones from the accretionary core of the Olympic Mountains, Washington. Geol Soc Am Bull 104:140–153

Jourdan S, Bernet M, Tricart P, Hardwick E, Paquette JL, Guillot S, Dumont T, Schwartz S (2013) Short-lived fast erosional exhumation of the internal Western Alps during the late Early Oligocene: constraints from geo-thermochronology of pro- and retro-side foreland basin sediments. Lithosphere 5:211–225

Lang KA, Huntington KW, Burmester R, Housen B (2016) Rapid exhumation of the eastern Himalayan syntaxis since the late Miocene. Geol Soc Am Bull. https://doi.org/10.1130/B31419.1

Malusà MG (2018) A guide for interpreting complex detrital age patterns in stratigraphic sequences (Chapter 16). In: Malusà MG, Fitzgerald PG (eds) Fission-track thermochronology and its application to geology. Springer, Berlin

Malusà MG, Fitzgerald PG (2018) From cooling to exhumation: setting the reference frame for the interpretation of thermocronologic data (Chapter 8). In: Malusà MG, Fitzgerald PG (eds) Fission-track thermochronology and its application to geology. Springer, Berlin

Malusà MG, Fitzgerald PG (2018) Application of thermochronology to geologic problems: bedrock and detrital approaches (Chapter 10). In: Malusà MG, Fitzgerald PG (eds) Fission-track thermochronology and its application to geology. Springer, Berlin

Malusà MG, Garzanti E (2018) The sedimentology of detrital thermochronology (Chapter 7). In: Malusà MG, Fitzgerald PG (eds) Fission-track thermochronology and its application to geology. Springer, Berlin

Malusà MG, Zattin M, Andò S, Garzanti E, Vezzoli G (2009) Focused erosion in the Alps constrained by fission-track ages on detrital apatites. Geol Soc Spec Publ 324:141–152

Malusà MG, Villa IM, Vezzoli G, Garzanti E (2011) Detrital geochronology of unroofing magmatic complexes and the slow erosion of Oligocene volcanoes in the Alps. Earth Planet Sci Lett 301:324–336

Malusà MG, Resentini A, Garzanti E (2016) Hydraulic sorting and mineral fertility bias in detrital geochronology. Gondwana Res 31:1–19

Mange MA, Wright DT (eds) (2007) Heavy minerals in use. Elsevier

Montgomery DR, Brandon MT (2002) Topographic controls on erosion rates in tectonically active mountain ranges. Earth Planet Sci Lett 201:481–489

Morton AC (2012) Value of heavy minerals in sediments and sedimentary rocks for provenance, transport history and stratigraphic correlation. Min Ass Canada Short Course 42:133–165

Naylor M, Sinclair HD, Bernet M, van der Beek P, Kirstein LA (2015) Bias in detrital fission track grain-age populations: Implications for reconstructing changing erosion rates. Earth Planet Sci Lett 422:94–104

Parra M, Mora A, Sobel ER, Strecker MR, González R (2009) Episodic orogenic front migration in the northern Andes: Constraints from low-temperature thermochronology in the Eastern Cordillera, Colombia. Tectonics 28:TC4004

Platt JP (1993) Exhumation of high-pressure rocks: a review of concepts and processes. Terra Nova 5:119–133

Rahl JM, Ehlers TA, van der Pluijm BA (2007) Quantifying transient erosion of orogens with detrital thermochronology from syntectonic basin deposits. Earth Planet Sci Lett 256:147–161

Reiners PW, Brandon MT (2006) Using thermo-chronology to understand orogenic erosion. Annu Rev Earth Pl Sc 34:419–466

Reiners PW, Campbell IH, Nicolescu S, Allen CM, Hourigan JK (2005) (U-Th)/(He-Pb) double dating of detrital zircons. Am J Sci 305:259–311

Resentini A, Malusà MG, Garzanti E (2013) MinSORTING: An Excel® worksheet for modelling mineral grain-size distribution in sediments, with application to detrital geochronology and provenance studies. Comput Geosci 59:90–97

Ring U, Brandon MT, Willett SD, Lister GS (1999) Exhumation processes. Geol Soc London Spec Publ 157:1–27

Soloviev AV, Garver JI, Shapiro MN (2001) Fission-track dating of detrital zircon from sandstone of the Lesnaya Group, northern Kamchatka. Strat Geol Correl 9:293–303

Spiegel C, Kuhlemann J, Dunkl I, Frisch W, von Eynatten H, Balogh K (2000) The erosion history of the Central Alps: evidence from zircon fission-track data of the foreland basin sediments. Terra Nova 12:163–170

Spiegel C, Siebel W, Kuhlemann J, Frisch W (2004) Toward a comprehensive provenance analysis: a multi-method approach and its implications for the evolution of the Central Alps. Geol Soc Am S 378:37–50

Stewart RJ, Brandon MT (2004) Detrital-zircon fission-track ages for the "Hoh Formation": implications for late Cenozoic evolution of the Cascadia subduction wedge. Geol Soc Am Bull 116:60–75

Stewart RJ, Hallet B, Zeitler PK, Malloy MA, Allen CM, Trippett D (2008) Brahmaputra sediment flux dominated by highly localized rapid erosion from the easternmost Himalaya. Geology 36:711–714

Trautwein B, Dunkl I, Kuhlemann J, Frisch W (2002) Cretaceous Tertiary Rhenodanubian flysch wedge (Eastern Alps): clues to sediment supply and basin configuration from zircon fission-track data. Terra Nova 13:382–393

van der Beek P, Robert X, Mugnier JL, Bernet M, Huyghe P, Labrin E (2006) Late Miocene—Recent denudation of the central Himalaya and recycling in the foreland basin assessed by detrital apatite fission-track thermochronology of Siwalik sediments. Nepal. Basin Res 18:413–434

van der Beek, P, Valla P, Herman F, Braun J, Persano C, Dobson, KJ, Labrin E (2010) Inversion of thermochronological age-elevation profiles to extract independent estimates of denudation and relief history-II: application to the French Alps. Earth Planet Sci Lett 296:9–22

Vermeesch P (2009) RadialPlotter: a Java application for fission track, luminescence and other radial plots. Radiat Meas 44:409–410

Vermeesch P (2018) Statistics for fission-track thermochronology (Chapter 6). In: Malusà MG, Fitzgerald PG (eds) Fission-track thermochronology and its application to geology. Springer, Berlin

von Blanckenburg F, Kagami H, Deutsch A, Oberli F, Meier M, Wiedenbeck M, Bart S, Fischer H (1998) The origin of Alpine plutons along the Periadriatic Lineament. Schweiz Mineral Petr Mitt 78:55–66

Zeitler PK, Johnson NM, Briggs ND, Naeser CW (1982) Uplift history of the NW Himalaya as recorded by fission-track ages on detrital zircon. In: Proceedings of the Symposium on Mesozoic and Cenozoic Geology, China, pp 481–494

A Guide for Interpreting Complex Detrital Age Patterns in Stratigraphic Sequences

Marco G. Malusà

Abstract

Thermochronologic age trends in sedimentary rocks collected through a stratigraphic sequence provide invaluable insights into the provenance and exhumation of the sediment sources. However, a correct recognition of these age trends may be hindered by the complexity of many detrital thermochronology datasets. Such a complexity is largely determined by the complexity of the thermochronology of eroded bedrock that may record, depending on the thermochronologic system under consideration, cooling during exhumation, episodes of magmatic crystallisation, metamorphic mineral growth and/or late-stage mineral alteration in single or multiple source areas. This chapter illustrates how different geologic processes produce different patterns of thermochronologic ages in detritus. These basic age patterns are variously combined in the stratigraphic record and provide a key for the geologic interpretation of complex detrital thermochronology datasets. Grain-age distributions in sedimentary rocks may include stationary age peaks and moving age peaks. Stationary age peaks provide no direct constraint on exhumation, as they relate to episodes of magmatic crystallisation, metamorphic growth or thermal relaxation in the source rocks. Moving age peaks are generally set during exhumation and can be used to investigate the long-term erosional evolution of mountain belts using the lag-time approach. Post-depositional annealing due to burial produces age peaks that become progressively younger down section. The appearance of additional older age peaks moving up section may provide evidence for a major provenance change. When interpreting detrital thermochronologic age trends, the

potential bias introduced by natural processes in the source-to-sink environment and inappropriate procedures of sampling and laboratory processing should be taken into account.

16.1 Introduction

Sedimentary successions reflect, in inverted order, the thermochronologic age structure observed in the sediment source areas (Garver et al. 1999; Ruiz et al. 2004; van der Beek et al. 2006). The simplest scenario of rock cooling through the closure temperature (T_c) of the chosen thermochronologic system during erosional exhumation (e.g. Braun et al. 2006; Herman et al. 2013; Willett and Brandon 2013) results in progressively younger thermochronologic ages up section through the sedimentary succession (e.g. Reiners and Brandon 2006). This simple concept is the basis for the geologic interpretation of most detrital thermochronology datasets, notably those where single grain ages are determined from sedimentary units within a stratigraphic sequence, and the resulting grain-age distributions are deconvolved into grain-age populations that are used to infer the long-term erosional evolution of a mountain belt (e.g. Bernet and Spiegel 2004; Ruhl and Hodges 2005; Enkelmann et al. 2008; Carrapa 2009; see discussion in Chap. 15, Bernet 2018).

However, as is well known from thermochronology applied to basement rocks, interpretation of all thermochronologic ages as representing cooling through the T_c isothermal surface during exhumation is sometimes too simplistic (e.g. Gleadow 1990; Villa 1998; Williams et al. 2007; see also Chap. 13, Baldwin et al. 2018 and Chap. 17, Fitzgerald et al. 2018). Moreover, detrital thermochronology datasets are sometimes very complex, and this complexity may hinder a correct identification of thermochronologic age trends. Complexity in detrital thermochronology datasets may arise from contributions of any of the following:

M. G. Malusà (✉)
Department of Earth and Environmental Sciences, University of Milano-Bicocca, Piazza Delle Scienza 4, 20126 Milan, Italy
e-mail: marco.malusa@unimib.it

© Springer International Publishing AG, part of Springer Nature 2019
M. G. Malusà and P. G. Fitzgerald (eds.), *Fission-Track Thermochronology and its Application to Geology*,
Springer Textbooks in Earth Sciences, Geography and Environment, https://doi.org/10.1007/978-3-319-89421-8_16

- Original complexity of thermochronologic data from the provenance region. Such complexities may arise from a range of processes, depending on the thermochronologic system under consideration, that may include not only annealing and diffusion during exhumation (Reiners and Brandon 2006) or during transient changes of the thermal state of the crust (Braun 2016), but also episodes of magmatic crystallisation, growth of metamorphic minerals and late-stage mineral alteration (Carter and Moss 1999; Malusà et al. 2011, 2012; Jourdan et al. 2013). The thermochronologic fingerprint of the provenance region may also be complicated by the presence of unreset sedimentary successions, reflecting the cooling history of older eroding sources (Bernet and Garver 2005; Rahl et al. 2007).

- Mixing of detritus from multiple source areas that are characterised by different geologic or upper crustal exhumation histories.

- Modifications of the original thermochronologic signal due to hydraulic processes during source-to-sink sediment transport and deposition (Schuiling et al. 1985; Komar 2007; see discussion in Chap. 7, Malusà and Garzanti 2018).

- Potential post-depositional annealing due to thick sedimentary burial (e.g. Ruiz et al. 2004; van der Beek et al. 2006).

When detrital thermochronologic datasets are particularly complex, geologists may be tempted to interpret only part of the data, typically those considered more significant. However, geologic interpretations that only explain part of a dataset are prone to be incorrect. A paradigmatic example of this situation, for a sedimentary succession derived from the progressive erosion of the Central Alps (Malusà et al. 2011), is illustrated in Chap. 17 (Fitzgerald et al. 2018).

This chapter illustrates how different processes contributing to the original thermochronologic complexity of a provenance region, when considered separately, may lead to different and largely predictable thermochronologic age patterns in detritus. These basic age patterns, described in detail in Sect. 16.2, are variously combined in the stratigraphic record and provide a key for a reliable geologic interpretation of complex detrital thermochronology datasets from samples collected through a stratigraphic sequence. In large part, this chapter is based on lessons learnt from studies in the European Alps and adjacent sedimentary basins (e.g. Malusà et al. 2011, 2012, 2013, 2016a, 2017). Results from the European Alps have a general validity and are presented as conceptual schemes in order to facilitate the application of these concepts to other study areas.

Usually, in detrital thermochronology, grain-age distributions are held to represent a faithful mirror of thermochronologic ages in eroded bedrock (Bernet et al. 2004; Resentini and Malusà 2012). However, potential modifications to the original thermochronologic fingerprint of detritus may occur due to a range of sources of bias, including natural processes in the source-to-sink environment and inappropriate procedures of sampling, laboratory treatment and analysis (e.g. Sláma and Košler 2012; Malusà et al. 2013, 2016a). Any underestimations (or overestimations) of specific grain-age populations with respect to the original thermochronologic fingerprint of bedrock may have an impact on geologic interpretations. Therefore, they should be carefully considered and accounted for. Simple strategies for bias minimisation are described in Sect. 16.3.

16.2 Basic Detrital Age Patterns in Stratigraphic Successions

16.2.1 Moving Age Peaks from Exhumational Cooling

When thermochronologic ages in bedrock are set during erosional exhumation and cooling through the T_c of a thermochronologic system, the sedimentary rocks produced by bedrock erosion will show unimodal grain-age distributions with thermochronologic ages that get progressively younger up section (e.g. Garver et al. 1999; Ruiz et al. 2004; Reiners and Brandon 2006). Ages will be generally older in grains eroded from summits, and younger in grains eroded from valleys, depending on the age-elevation relationship in the bedrock exposed within the catchment (see Chap. 10, Malusà and Fitzgerald 2018b). The thermochronologic age trends observed in sedimentary rocks will be shifted towards older ages for thermochronologic systems with progressively higher T_c (see Fig. 16.1a). Examples of geologic interpretations of detrital thermochronology datasets based on the application of these simple principles are illustrated in Chap. 15 (Bernet 2018).

16.2.2 Stationary and Moving Age Peaks from the Unroofing of Magmatic Complexes

The progressive erosion of a volcanic–plutonic complex (see Chap. 8, Malusà and Fitzgerald 2018a and references therein) provides a suitable starting point to illustrate the basic age patterns expected in the detrital record when both erosional exhumation and crystallisation of new minerals are taken into account (Malusà et al. 2011). Let us consider the intrusion of magma at depth within metamorphic country rocks and coeval formation of volcanoes at the surface,

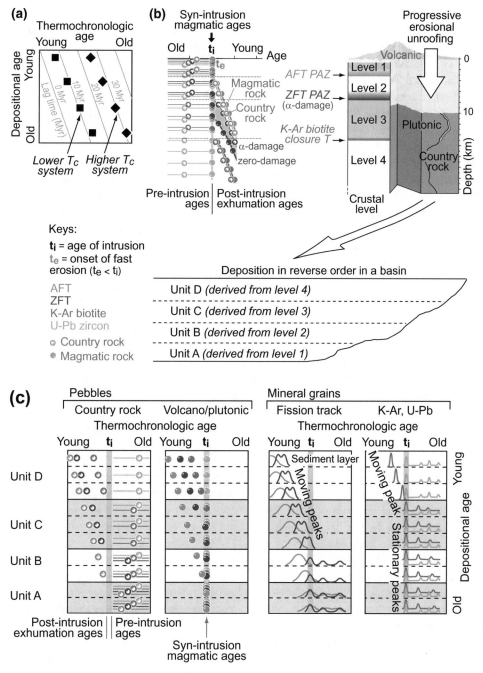

Fig. 16.1 Basic detrital age patterns and their interpretation (I). **a** Detrital thermochronologic ages recording the progressive cooling of eroded bedrock through the T_c isothermal surface define moving age peaks, with progressively younger thermochronologic ages up section. Thermochronologic systems with higher T_c have older ages but similar trends. **b** Magmatic complex model (Malusà et al. 2011): the sedimentary succession derived from the progressive erosion of a magmatic complex associated with a single intrusive event is expected to reflect, in reverse order, the mineral age stratigraphy observed in the source rocks. Crustal level 1 is eroded first and deposited as unit A, followed in succession by erosion of level 2 (unit B), then level 3 (unit C) and finally level 4 (unit D). Mineral age stratigraphy in the source rocks includes: (i) ages that are set before magmatic crystallisation and are exclusively found in country rocks (pre-intrusion ages); (ii) ages that are set during magmatic intrusion/volcanism and are exclusively found in volcanic and plutonic rocks (syn-intrusion magmatic ages); and (iii) ages that are set during progressive erosion (due to cooling/exhumation) of the magmatic complex and are found both in plutonic and country rocks (see Chap. 8, Malusà and Fitzgerald 2018a for more details on the formation of the age structure shown in (**b**)). **c** The erosion of the magmatic complex in (**b**) produces pebbles of both volcanic/plutonic rocks and country rocks, and sand-sized detritus that includes single grains of apatite, zircon and biotite originally belonging either to eroded magmatic rocks or to country rocks. In such detritus, syn-intrusion magmatic ages define grain-age populations that are constant up section (stationary age peaks) and provide no direct constraint on exhumation. Post-intrusion exhumation ages define grain-age populations that become increasingly younger up section (moving age peaks) and because these thermochronologic ages were set during exhumation, they provide direct constraints on the rock motion towards the Earth's surface. Thus, they can be used to investigate the long-term erosional evolution of mountain belts using the lag-time approach

during a single magmatic pulse at time t_i (see Fig. 16.1b). This is followed by the progressive erosional unroofing of these volcanic and plutonic rocks along with their country rocks at time $t_e < t_i$. The resulting age structure, described in detail in Chap. 8 (Malusà and Fitzgerald 2018a), allows the identification of different crustal levels on the basis of the depth of relevant T_c isotherms before the onset of erosion. The hypothetical four crustal levels 1–4 of Fig. 16.1b are delimited by the high-temperature boundaries of the partial annealing zones (PAZ) of the apatite fission-track (AFT) and α-damaged zircon fission-track (ZFT) systems, and by the T_c of the K–Ar system in biotite. The age pattern characterising levels 1–4 includes: (i) ages that are set before magmatic intrusion and are exclusively yielded by country rocks; (ii) ages that are set during crystallisation of a magmatic intrusion and volcanism at the surface and are exclusively yielded by magmatic rocks; and (iii) ages that are set during progressive erosion of the magmatic complex and are yielded both by magmatic and country rocks. Note that ages within a contact aureole may be completely and/or partially reset (Calk and Naeser 1973; Harrison and McDougall 1980), but grains derived from the contact aureole are typically volumetrically insignificant and are not included in this conceptual analysis for the sake of simplicity.

Erosion of this volcanic–plutonic complex produces detritus of various grain sizes and various lithologies. Clasts may be of volcanic, plutonic or country rock, and sand may include grains of apatite, zircon, biotite or other minerals of mixed magmatic and country-rock provenance. Such detritus accumulates in reverse order in the sedimentary basin, forming a sedimentary succession consisting of four hypothetical stratigraphic units (A–D in Fig. 16.1b), each one exclusively derived from the progressive erosion of the four levels identified in the source. The detrital age pattern resulting from this simple scenario is shown in Fig. 16.1c.

The oldest stratigraphic unit A, which is entirely eroded from level 1, contains clasts of country rock yielding ages older than t_i, as well as clasts of volcanic and shallow intrusive rocks. The clasts of these volcanic and shallow intrusive rocks yield magmatic crystallisation ages $(=t_i)$ that are identical within error in all thermochronologic systems, and are invariant up section in the whole unit A. Cooling in the magmatic rocks is in fact too fast, and the relative error of ZFT and AFT ages too large to highlight a trend of up section decreasing thermochronologic ages from these clasts. Sandstones derived from the erosion of country rocks, volcanic rocks and shallow intrusive rocks of level 1 consistently show an age peak around the magmatic age t_i that is constant up section, and is referred to as *stationary age peak* hereafter. This stationary age peak (also referred to as the static peak in Chap. 15) coexists with scattered older ages, also observed within these sandstones, that are set during the pre-intrusion history of the country rock or even inherited from its protolith. All of these ages are set before the onset of erosion; therefore, they provide no direct constraint on exhumation.

Detritus in unit B is entirely eroded from level 2. The level of exhumation is such that AFT ages (sourced from below the apatite paleo-PAZ) are now identical in clasts sourced from plutonic and country rocks within each sediment layer and are invariably younger than t_i. AFT ages become increasingly younger up section because they reflect cooling during exhumation. Associated U–Pb, K–Ar and ZFT ages from magmatic rock clasts define a stationary age peak indistinguishable from the magmatic crystallisation age t_i, but within country-rock clasts, these ages are instead scattered and older than t_i. In sandstones, zircon and biotite grains define composite age distributions in the ZFT, K–Ar and U–Pb systems, with a stationary age peak corresponding to the magmatic crystallisation age t_i. Apatite grains, however, show unimodal AFT age distributions with a single peak that is younger than t_i, and which becomes increasingly younger up section. This peak is hereafter referred to as *moving age peak*. Thermochronologic ages belonging to this moving age peak are set during exhumation. Therefore, they provide direct constraints on rock motion towards Earth's surface.

Stratigraphic unit C, which is entirely eroded from level 3, yields exhumation FT ages not only on apatite, but also on α-damaged zircon within country-rock clasts. In sandstones of unit C, AFT and ZFT age patterns define moving peaks younger than t_i. Bimodal zircon age peaks are possibly observed whenever zircon grains with low and high levels of α-damage are found in the same sample. Single biotite and zircon grains in sandstones define composite K–Ar and U–Pb age distributions with a stationary peak corresponding to the magmatic crystallisation age t_i. Finally, in stratigraphic unit D that is entirely derived from level 4, exhumation ages are the rule with the exception of U–Pb zircon ages. FT and K–Ar ages in sandstones of unit D define moving age peaks younger than the magmatic age (t_i), but zircon U–Pb ages younger than the magmatic age could be occasionally observed as a result of post-magmatic zircon recrystallisation (Baldwin 2015; Kohn et al. 2015).

Therefore, the detrital thermochronology record shown in Fig. 16.1c includes stationary age peaks and moving age peaks. Stationary age peaks are invariant up section and are formed by thermochronologic ages that are set during magmatic crystallisation, and before the onset of erosional exhumation. They therefore provide no direct constraint on exhumation. Moving age peaks get progressively younger up section and are formed by thermochronologic ages that are set during progressive rock cooling during exhumation. These moving age peaks may be used to investigate the long-term erosional evolution of mountain belts using the lag-time approach, taking into account all the usual caveats as discussed in Chap. 10 (Malusà and Fitzgerald 2018b) and Chap. 15 (Bernet 2018).

Noteworthy, magmatic crystallisation ages are found not only in volcanic grains (see Chap. 15, Bernet 2018), but also in mineral grains derived from plutonic rocks that have crystallised at depth shallower than the T_c isothermal surface of the thermochronologic system under consideration (i.e. ~ 4 and ~ 7 km depth, respectively, for the AFT and ZFT systems assuming a paleogeothermal gradient of 30 °C/km). For a correct interpretation of detrital thermochronology data, these mineral grains should be carefully detected by double dating (see Chap. 5, Danišík 2018; Chap. 15, Bernet 2018). Because magmatic zircon grains may preserve older inherited cores, U–Pb analysis of double-dated grains is best performed on grain rims (see Chap. 7, Malusà and Garzanti 2018).

16.2.3 Delayed Response to Erosion

The first appearance of a moving age peak (of a particular thermochronologic method/mineral) after the onset of erosion takes place when the whole rock pile with a thermochronologic fingerprint (relative to each thermochronologic method/mineral) acquired before the onset of erosion is completely removed (Fig. 16.2a). The delay in such response depends on the T_c of the thermochronologic system under consideration and on the erosion rate (Rahl et al. 2007). Low-T_c systems such as (U–Th)/He on apatite (AHe) and AFT more readily respond to abrupt changes in erosion rate, whereas higher-T_c systems are less sensitive to rapid changes in erosion rate. In Fig. 16.2a, moving age peaks defined by detrital AFT ages first appear in unit B, those defined by detrital ZFT ages only appear in unit C, whereas those defined by detrital K–Ar ages only appear in unit D. Delays between the onset of rapid cooling/exhumation and the deposition of mineral grains yielding these higher-temperature thermochronologic ages may be greater than 10 Myr even for relatively fast erosion rates (on the order of 1 mm/y) (Malusà et al. 2011). In other words, this delay means that the detrital signal of rapid cooling/exhumation, dependent on the method/mineral employed, may not be part of the stratigraphic sequence derived from this rapid cooling/exhumation episode, but may be eroded and deposited in a subsequent episode. For example, let us consider the ZFT thermochronometer and a hypothetical mountain belt that has experienced rapid unroofing in the Paleocene, with erosional removal of a ~ 5-km-thick crustal section and consequent sediment deposition in a foreland basin. This episode of rapid exhumation is followed by tectonic quiescence and by renewed erosion and deposition in the foreland basin since the Miocene. In the foreland basin, zircon grains from Paleocene strata will not yield ZFT ages

supporting Paleocene exhumation in the source area. Instead, ZFT ages attesting rapid Paleocene exhumation will be found in Miocene or younger strata, with a time delay of several Myr.

16.2.4 The Lag-Time Interpretation of Age Peaks from Unroofing Magmatic Complexes

In Fig. 16.2b, the biotite K–Ar grain-age distribution derived from the model of Fig. 16.1b is interpreted in a classic lag-time diagram. Lag time is always >0, unless there are modifications of the thermochronologic signal by alteration or post-depositional burial-related partial or complete resetting (e.g. Garver et al. 1999; Ruiz et al. 2004). When interpreted in terms of lag time, both the stationary age peaks and the moving age peaks derived from the erosion of a magmatic complex appear to define a lag-time trend. The lag-time trend 1, defined by stationary age peaks, might be erroneously used to infer a decay phase of evolution of the eroding source. However, ages defining these stationary age peaks are set before the onset of exhumation. Therefore, the increasing lag time observed in this case provides no direct constraint on the erosional evolution of the source area. By contrast, the constant lag-time trend 2 that is defined by a moving age peak is supportive of steady-state erosion, in line with the erosional evolution imposed by the model.

The conceptual model of Fig. 16.1b is based on the assumption of one single episode of intrusive/volcanic activity, but this is not typical of most large plutonic–volcanic complexes that will experience multiple intrusive phases. The model of Fig. 16.2c includes multiple magmatic pulses: the first episode occurs before the onset of erosional exhumation (t_{i1}), the second pulse occurs when the erosion of crustal level 1 is completed (t_{i2}), and the third one occurs after the erosion of crustal level 2 (t_{i3}). The resulting detrital ZFT age pattern in units A–D includes (i) peaks exclusively formed by magmatic crystallisation ages (marked by crosses), (ii) peaks exclusively formed by exhumation ages (marked by full dots) and (iii) peaks formed by both exhumation and magmatic ages (empty dots). Peaks formed by magmatic crystallisation ages provide no direct constraint on exhumation and should be excluded from lag-time analysis (see Chap. 15, Bernet 2018). Importantly, magmatic ages invariably form the youngest ZFT age peaks of the distribution in units A–C, i.e. not only in sedimentary rocks coeval with magmatic activity, but also in much younger sedimentary rocks deposited in the final sink. This underlines the importance of a multi-method dating approach to the analysis of the lag time, in order to avoid the potential misinterpretation of magmatic crystallisation ages in terms

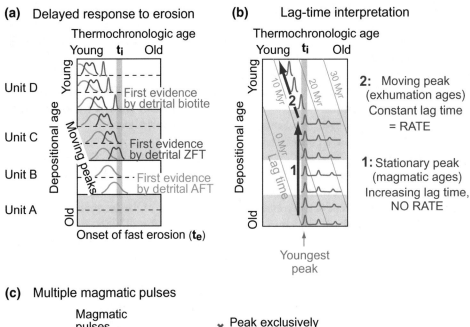

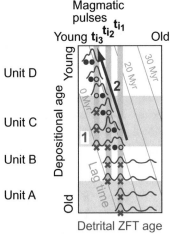

Fig. 16.2 Basic detrital age patterns and their interpretation (II). **a** Delayed response to erosion. Moving age peaks recording progressive erosion in the hinterland do not appear in detritus until the source rocks with an older thermochronologic fingerprint are removed by erosion. Time delay depends on the T_c of the thermochronologic system under consideration and on the erosion rate. Low-T_c systems such as AFT are more sensitive to changes in erosion rate than higher-T_c systems. **b** Lag-time interpretation of the biotite K–Ar grain-age distribution shown in Fig. 16.1 (green lines in the diagram are lines of equal lag time). The youngest grain-age population of the distribution defines two different lag-time trends (1 and 2) in older and younger strata. However, ages

defining the stationary age peak for trend 1 are set before the onset of exhumation. Only ages set during exhumation and defining the moving age peak 2 can be used to investigate the long-term erosional evolution of the source areas using the lag-time approach. **c** Magmatic complex model for multiple magmatic pulses and lag-time interpretation of associated detrital ZFT data. Magmatic pulses in the model occur before the onset of erosional exhumation (t_{i1}) and after erosion of crustal levels 1 (t_{i2}) and 2 (t_{i3}). The resulting detrital ZFT age pattern includes a number of peaks exclusively formed by different magmatic crystallisation ages. These peaks, indicated in red and representing the youngest peaks of the distribution in units A–C, should be excluded from lag-time analysis

of exhumation. Ideally, the provenance source region is well known, and information from the detrital record can be compared to the geologic/thermochronologic record of the source region to better constrain the erosional exhumation.

The occurrence of a stationary age peak that is older than the moving age peak within the same sediment layer (and for the same thermochronologic system) is supportive of mixing of detritus from different source areas.

16.2.5 Extrapolation to Unroofing Metamorphic Belts

Stationary age peaks may also be found in detritus shed from a metamorphic dominated source (e.g. Carrapa et al. 2003) due to the presence of young micas that have grown, during low-grade metamorphism, at temperatures below the isotopic closure of the $^{40}Ar/^{39}Ar$ system. The magmatic complex model demonstrates that simple constraints (e.g. using the lag-time approach) on exhumation are only provided by those thermochronologic ages that are set during monotonic cooling associated with exhumation across the T_c isothermal surface. By contrast, thermochronologic ages that are set in minerals (re)crystallised at shallower depths and at temperatures less than T_c provide no direct constraint on exhumation. The same "magmatic complex model" concept can be extrapolated to metamorphic minerals such as micas (Malusà et al. 2012). In metamorphic rocks, micas

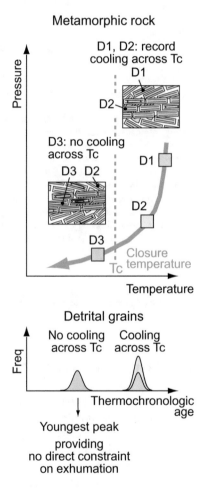

Fig. 16.3 Basic detrital age patterns and their interpretation (III). Conceptual pressure-temperature exhumation path of a metamorphic rock, with micas grown along different foliation planes during deformation stages D1–D3. D1 and D2 indicate micas grown at higher temperature than the isotopic closure of the $^{40}Ar/^{39}Ar$ system. These micas will cross the T_c isothermal surface of the $^{40}Ar/^{39}Ar$ system during exhumation. D3 indicates micas that have instead grown at lower temperature than the diffusion-only isotopic closure of the $^{40}Ar/^{39}Ar$ system. These micas will not cross the T_c isothermal surface during exhumation. When eroded, micas D1 and D2 will potentially define a moving age peak in detritus and therefore can be used to investigate the long-term exhumation history of the source rocks using the lag-time approach. Micas D3 will define instead a stationary age peak that provides no direct constraint on exhumation (Malusà et al. 2012)

commonly show microtextures typical of petrological disequilibrium (e.g. Challandes et al. 2008; Glodny et al. 2008); thus, different generations of micas generally define foliations formed under different P–T conditions during exhumation towards Earth's surface (Fig. 16.3). When mica is grown at temperatures greater than the isotopic closure (D1 and D2 in Fig. 16.3), the mineral will cross the T_c isothermal surface during exhumation. In detritus, the resulting $^{40}Ar/^{39}Ar$ ages should decrease regularly with decreasing depth in the stratigraphic column and could be used to infer exhumation rates using the lag-time approach, provided that retrogression and fluid-assisted recrystallisation are negligible (e.g. Villa 1998). Conversely, micas grown at lower temperature than the diffusion-only isotopic closure temperature (D3 in Fig. 16.3) will constrain the age of mineral growth only. Micas D3 therefore yield $^{40}Ar/^{39}Ar$ ages that provide no direct constraint on exhumation unless coupled with pressure estimates (e.g. Massonne and Schreyer 1987). However, pressure estimates may require that mineral assemblages are in petrologic equilibrium, which cannot be ascertained in the case of detrital mineral grains. Distinguishing micas of different generations in detritus is obviously an arduous task. However, when using micas in detrital studies, the potential occurrence of micas grown at temperatures lower than T_c should be carefully considered. In some detrital studies, the youngest mica populations are used to constrain the exhumation history of eroded bedrock, yet this may not always be appropriate as the youngest mica individual crystal ages in a rock may reflect late mineral growth rather than $^{40}Ar/^{39}Ar$ ages related to exhumation.

The occurrence, in the same detrital sample, of multiple moving age peaks recording exhumational cooling is supportive of mixing of detritus from source areas that are characterised by different upper crustal exhumation histories.

16.2.6   Stationary Age Peaks Due to Thermal Relaxation

Stationary age peaks may be also found in a stratigraphic section as a result of a major event of thermal relaxation in eroded bedrock. An example of this may be the case of detritus eroded from the Himalaya, where stationary age peaks may mirror the thermal relaxation expected after a major orogenic event in the hinterland (Braun 2016). Major relaxation of isothermal surfaces is also expected in distal passive margins after continental break-up (Malusà et al. 2016b). Also in these cases, stationary age peaks in detritus may provide no direct constraint on the exhumation of the source rocks. Unlike the pattern of Fig. 16.1c (magmatic complex model), stationary age peaks due to thermal relaxation are generally not found in the deepest levels of a stratigraphic succession, because the relevant "elevated" isotherms (unlike magma) do not reach the Earth's surface (see Chap. 8, Malusà and Fitzgerald 2018a).

16.2.7 Impact of Mineral Alteration

Moving age peaks are not always linked to exhumation, but may also be the result of mineral alteration during exhumation of the source region or burial diagenesis of derived sedimentary rocks. For example, in magmatic clasts derived from the Bregaglia/Bergell pluton (Giger 1990), K–Ar ages in biotite define apparent moving age peaks, whereas lower T_c thermochronometers define a stationary age peak indistinguishable from the age of intrusion (Malusà et al. 2011, their Fig. 8). This moving K–Ar age peak is not related to exhumation, but is an artefact caused by anomalous K–Ar ages associated with substoichiometric K concentrations, which provide evidence for chloritisation of biotite. Alteration of minerals may lead to thermochronologic ages younger than the depositional age and may explain, for example, the enigmatic occurrence of detrital mica with negative lag times where there has been no deep burial to cause thermally reset or partially reset ages (e.g. White et al. 2002).

16.2.8 Impact of Sediment Recycling

Minor modifications to the detrital age patterns illustrated in Fig. 16.1c are expected in the case of sediment recycling, which introduces in detritus a number of mineral grains having thermochronologic ages older than expected. As a result, moving age peaks get increasingly skewed, with an older tail representing the recycled grains (Fig. 16.4a). However, sediment recycling is not expected to affect the shape and age of stationary age peaks.

16.2.9 Impact of Post-depositional Annealing or Resetting Due to Burial

Detrital age patterns may be also affected by post-depositional annealing or diffusion (depending on the thermochronologic system under consideration) towards the base of thick sedimentary successions. Post-depositional annealing is usually easily detected, because both original magmatic and exhumation ages, as well as FT peaks in single detrital minerals, become increasingly younger down section (e.g. Ruiz et al. 2004; van der Beek et al. 2006), showing the opposite trend than that observed in un-reset sedimentary successions (Fig. 16.4b). Such a reversal is not

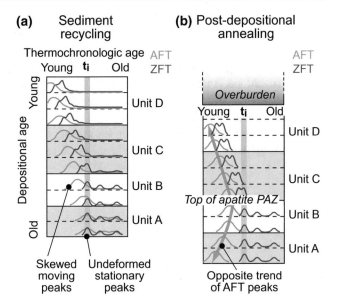

Fig. 16.4 Basic detrital age patterns and their interpretation (IV). **a** Sediment recycling introduces in detritus a number of mineral grains with thermochronologic ages older than expected. As a result, moving age peaks get increasingly skewed with an older tail representing the recycled grains, whereas stationary age peaks are not affected. **b** Post-depositional annealing towards the base of thick sedimentary successions produces grain-age peaks that become increasingly younger with depth (i.e. down section) showing the opposite trend than that observed in un-reset sedimentary successions. Such a reversal is not observed in higher-T_c systems within the same stratigraphic level (after Malusà et al. 2011)

observed in higher-T_c systems within the same stratigraphic level (Malusà et al. 2011).

16.2.10 Detection of Provenance Changes

The provenance of a grain-age population can be preliminarily constrained by thermochronologic data from bedrock and detritus following a simple principle: thermochronologic ages in detritus within a stratigraphic succession, unless reset by burial, must be equal to or older than the thermochronologic ages now observed in bedrock within the potential source areas. Bedrock units showing older ages can be safely excluded as a potential source (e.g. Garzanti and Malusà 2008). However, provenance constraints exclusively based on cooling ages are often poor for older stratigraphic units and must be integrated by complementary analyses (see Chap. 14, Carter 2018). Major changes in provenance can be easily unravelled by inspecting the trends of detrital thermochronologic ages observed along a stratigraphic succession. If provenance does not change, the detrital thermochronology record should follow one of the trends illustrated in Figs. 16.1, 16.2, 16.3 and 16.4. The sudden appearance of older grain-age populations moving up section along a stratigraphic succession is strong evidence for a major provenance change, e.g. due to drainage reorganisation (Ruiz et al. 2004; Glotzbach et al. 2011; Asti et al. 2018).

16.3 Managing Bias in Detrital Thermochronology

The thermochronologic fingerprint of detritus should ideally reflect the thermochronologic imprint of eroded bedrock. Such an imprint can be quite complicated, due to the combined effects of a range of processes, including annealing and diffusion (depending on the thermochronologic system under consideration) either during exhumation or during transient changes in the thermal state of the crust, episodes of magmatic crystallisation, metamorphic mineral growth, late-stage alteration of the source rock prior to erosion, and the potential occurrence of unreset sedimentary successions reflecting the cooling history of older eroding sources. However, the thermochronologic signal inherited from the eroded bedrock is prone to be affected by a number of potential sources of bias during sediment transport from source to sink and during laboratory treatment and data processing. In order to avoid potential misinterpretations of detrital thermochronology datasets, these sources of potential bias should be carefully considered and minimised, as discussed below.

16.3.1 Bias Due to Natural Processes and Laboratory Treatment

During sediment transport prior to deposition, detrital grains are sorted by tractive currents according to their size, shape and density (Schuiling et al. 1985). If the age distribution in dated minerals shows a relationship with grain size, hydraulic sorting may have had an impact on grain-age distributions in detritus, which could be considerably different from the original distribution in eroded bedrock (Malusà et al. 2016a). In the case of a relationship between grain age and grain size, a common occurrence in (U–Th)/He dating (e.g. Reiners and Farley 2001), bias may be also introduced during sample processing. Examples include: gold panning during sampling (Bernet and Garver 2005); an incorrect use of the shaking table (Wilfley or Gemeni table) during hydrodynamic concentration of dense minerals (Sláma and Košler 2012); non-random grain selection during handpicking and analysis (Cawood et al. 2003); electrostatic charging in plastic ware (e.g. sample vials) that may lead to a selective loss of smaller grains; sieving during sample

preparation so that only a limited size range is analysed (Malusà et al. 2013).

An inappropriate procedure of magnetic separation may also introduce a bias, if mineral age has a relationship with magnetic properties (Sircombe and Stern 2002). Bias may also occur during the preparation of grain mounts for FT dating, if a relationship exists between grain age and grain shape or size. When mounted in epoxy or Teflon for polishing and etching, elongated grains tend to align themselves parallel to the c-axis, and most of them are thus suitably oriented for FT counting (see Chap. 2, Kohn et al. 2018). On the other hand, equidimensional grains tend instead to be randomly oriented, which implies that grain mounts involving a mixture of elongated and equidimensional detrital grains, will generally have a larger proportion of elongated grains suitable for counting than equidimensional grains. This problem is generally negligible for ZFT datasets, because most zircon grains typically have an elongated shape, but may be relevant for AFT detrital datasets.

Potential relationships between grain age and grain size (or grain shape) should be carefully evaluated in detrital thermochronology studies, as illustrated for example in Chap. 7 (Malusà and Garzanti 2018). Datasets showing no apparent relationship between these parameters are inherently robust with respect to the above sources of bias. In the case where age varies with grain size, bias should be instead minimised, e.g. by avoiding sampling of placer deposits and gold panning, by handpicking to remove impurities rather than to select datable grains, and by processing of different grain-size classes separately in shaking tables, in order to prevent any selective loss of mineral grains from specific grain-age populations.

16.3.2 Intrinsic Bias Specific to Zircon

ZFT datasets are also affected by other, more specific sources of intrinsic bias that propagates to other datasets whenever zircon grains are subject to double or triple dating. Potential bias is introduced, for instance, by uncountable overlapping fission tracks in U-rich grains (e.g. Ohishi and Hasebe 2012; Gombosi et al. 2014). A comparison between detrital zircon U–Pb and ZFT datasets from the same region (Fig. 16.5a) shows that U-rich grains ([U] >1000 ppm), although detected in samples analysed by LA-ICP-MS, are missing in the corresponding ZFT dataset, because these grains are undatable by the FT method. For example, undatable U-rich zircon grains in the case illustrated in Fig. 16.5a exceed 40% of the dataset. As a result, a bias is likely introduced in detrital ZFT datasets whenever different source areas (A and B in Fig. 16.5b) shed zircon grains with different U concentrations, for example if they have different

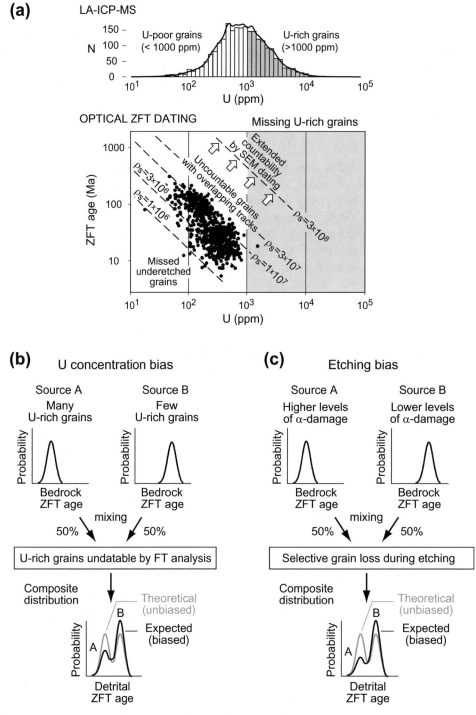

Fig. 16.5 **a** Comparison between detrital zircon U–Pb (LA-ICP-MS) and ZFT datasets from the European Alps. U-rich grains ([U] > 1000 ppm) that exceed 40% of the dataset are detected in samples analysed by LA-ICP-MS, but they are missing in the corresponding ZFT dataset. This is because the very high density of fission tracks means they are undatable by standard counting (modified from Malusà et al. 2013; ZFT data from Bernet et al. 2004). Countability can be improved by applying weaker/shorter etches and counting fission tracks at higher magnifications, for example under a scanning electron microscope (SEM). **b** U concentration bias. Two source areas (A and B) shed an equal amount of zircon grains to a detrital sample, but the percentage of U-rich grains (>1000 ppm) is higher in the source A than in the source

B, for example due to the different metamorphic history of these source rocks. Because U-rich grains are systematically missed during analysis, the detrital source A is expected to be under-represented, or even missed, in the detrital FT record. **c** Etching bias. Two source areas (A and B) shed an equal amount of zircon grains to a detrital sample, but the percentage of older zircon grains with higher levels of α-damage is higher in source A than in source B, for example due to the different age of the protoliths of these source rocks. Because zircon grains with high levels of α-damage are expected to be selectively overetched and eventually lost during routine etching, the detrital source A is expected to be under-represented, or even missed, in the detrital FT record (modified from Malusà et al. 2013)

metamorphic histories (Malusà et al. 2013). Because U-rich grains ([U] > 1000 ppm) are systematically not counted, detrital sources supplying such U-rich grains are therefore systematically under-represented, or even missed, in the detrital ZFT age record (*U concentration bias*, Malusà et al. 2013). Such potential U concentration bias can be assessed by measuring the distribution of U concentration in zircon grains from different source areas, at least in the case of double dating that involves LA-ICP-MS analysis. Even though bias due to overlapping tracks is largely independent of etching, countability can be improved by applying weaker/shorter etches and counting fission tracks either under a scanning electron microscope (Montario and Garver 2009; Gombosi et al. 2014) or an atomic-force microscope (Ohishi and Hasebe 2012). This type of bias does not affect FT dating of minerals with much lower U concentration, such as apatite.

Further bias in detrital ZFT datasets may also occur due to the differential etching response of zircon grains with different amount of α-damage (e.g. Gleadow et al. 1976; Kasuya and Naeser 1988; Tagami et al. 1990, 1996). Accumulated α-damage is a function of U concentration and effective accumulation time, also considering the effects of thermal annealing of α-damage at high temperature (Tagami et al. 1996; Garver and Kamp 2002). Zircons that are old and have more accumulated α-damage require a shorter etching time to reveal fission tracks for counting, with respect to young zircons where α-damage is much less and the crystal lattice less damaged. Thus, older zircon grains with higher levels of α-damage may be selectively overetched and eventually lost during routine etching, even in case of multiple etching times to reveal the full spectra of ZFT ages (e.g. Bernet and Garver 2005). Whenever different source areas (A and B in Fig. 16.5c) shed zircon grains with different levels of α-damage, for example because of the different age of their protoliths, detrital sources shedding zircon grains with old U–Pb ages and high levels of α-damage are expected to be systematically under-represented in the detrital ZFT record (*etching bias*, Malusà et al. 2013), or even missed when forming small age populations close to the detection limit (see Sect. 16.3.3). In this extreme case, both U concentration bias and etching bias may have major implications for lag-time analysis.

Notably, zircon grains with higher levels of α-damage are more easily entrained by tractive currents than non-metamict zircon grains, because they are less dense (see Chap. 7, Malusà and Garzanti 2018). Therefore, if zircon grains with different provenance and different levels of α-damage are mixed together before reaching the final sink, hydraulic sorting during sediment transport may introduce further bias in the detrital ZFT record.

16.3.3   Bias Introduced During Data Processing

Bias in detrital thermochronology datasets is also potentially introduced by inappropriate methods of data visualisation, or by decomposition of grain-age distributions into individual age components, because different deconvolution methods have a different sensitivity to outliers and assume different shapes for parent populations, for example Gaussian, log-normal or Laplacian (Sambridge and Compston 1994; Brandon 1996; Jasra et al. 2006). Detrital thermochronology data can be presented in a number of ways: (i) in a simple age histogram with the frequency or proportion of ages plotted against binned ages; (ii) in a radial plot (Galbraith 1990); (iii) as a probability density plot that takes into account analytical error but lacks any theoretical basis as a probability density estimator; (iv) as a kernel density plot that plots ages in an age frequency plot and then imposes a kernel of some bandwidth to constrain the shape of the distribution, although this shape and hence identification of age peaks may be dependent on the bandwidth selected (Vermeesch 2012). These issues may be significant and are discussed in more detail in Chap. 6 (Vermeesch 2018).

As datasets get larger and detrital data are compared from sample to sample, or region to region in order to constrain similarities or likeness between populations, both the visualisation of data is important as are methods of comparing sample/data analytics (Malusà et al. 2013; Vermeesch 2013; Andersen et al. 2017). In detrital datasets, minor age components are possibly underestimated or even missed whenever the number of analysed grains is not large enough (see, e.g. Dodson et al. 1988; Vermeesch 2004; Andersen 2005). The detection limit of an age component can be defined (Vermeesch 2004) as the maximum size of a single population likely to remain undetected after n analyses (Fig. 16.6). In the approach proposed by Andersen (2005), no assumptions are made about the number, nature or distribution of the populations to be detected. In contrast, in a more conservative approach proposed by Vermeesch (2004), the detection limit is defined as the size of m equally distributed, equally abundant populations, one of which is likely to escape detection in n analyses. In both cases, the probability level associated with the detection limit (e.g. 50 or 95%) has to be specified. The optimal number of grains that should be dated in a detrital sample according to the above approaches is reported in Fig. 16.6. If the grain-age populations are uniformly distributed in the sample, 117 grains should be dated to be 95% certain that no population $\geq 5\%$ was missed (Vermeesch 2004). As a general rule, it is important to indicate what is the size of the smallest population fraction that we can be 95% certain that was not missed. According to Andersen (2005), the 95% limit can be regarded as a relatively safe indicator that a

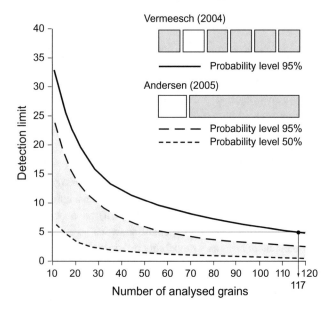

Fig. 16.6 Detection limits of an age component in a detrital grain-age distribution (i.e. the maximum size of a single population likely to remain undetected after *n* analyses) for a probability level of 50 and 95% (modified from Vermeesch 2004; Andersen 2005). According to the approach proposed by Andersen (2005), no assumptions are made about the number, nature or distribution of the populations to be detected. According to a more conservative approach proposed by Vermeesch (2004), the detection limit is defined as the size of *m* equally distributed, equally abundant populations (6 squares), one of which is likely to escape detection (the white square) in *n* analyses (top right). The 95% limit can be regarded as a safe indicator that a population of the corresponding true abundance will be detected in *n* analyses; the 50% limit represents the threshold for populations that are more probably overlooked than observed in *n* analyses (Andersen 2005). If the grain-age populations are uniformly distributed in the sample, 117 grains should be dated to be 95% certain that no population $\geq 5\%$ is missed (Vermeesch 2004); if the grain-age populations are not uniformly distributed, the required number of grains to be dated could be lower (~ 60)

population of the corresponding true abundance will be detected in *n* analyses, whereas the 50% limit represents the threshold for populations that are more probably overlooked than observed in *n* analyses. Counting statistics is particularly important when provenance studies discuss the absence of certain age fractions. If the purpose of the study is to prove the presence of one or more specific age fractions in a detrital population, once these fractions have been found, there may be no reason to date more grains (Vermeesch 2004).

16.4 Summary and Recommendations

As discussed in this chapter, sedimentary successions usually reflect, in inverted order, the thermochronologic age structure observed in the sediment source areas. The simplest

scenario is rock cooling across a T_c isothermal surface during erosional exhumation, resulting in progressively younger thermochronologic ages moving up section through a sedimentary succession. However, this approach is sometimes too simplistic, as it does not consider the full range of geologic processes that determine the final thermochronologic fingerprint of sediments. These processes may include (re)crystallisation, diffusion and alteration in the source rocks, potential modifications during transport and deposition, and post-depositional annealing. Different processes produce different age patterns in sediment and sedimentary rocks. A correct recognition of these patterns allows a reliable geologic interpretation of the detrital thermochronology record. Clues for geologic interpretation of thermochronologic age patterns in sediments and sedimentary rocks are summarised below:

- The detrital thermochronology record may include stationary age peaks and moving age peaks. Stationary age peaks provide no direct constraint on exhumation, as they relate to episodes of magmatic crystallisation, late-metamorphic mineral growth or thermal relaxation in the source rocks. Moving age peaks are generally set during exhumation and can be used to investigate the long-term erosional evolution of mountain belts using the lag-time approach.
- The first appearance of a moving age peak occurs in detritus well after the episode of exhumation recorded by low-temperature thermochronometers. The time delay depends on the T_c of the thermochronologic system under consideration, and on the erosion rate.
- Ages reflecting magmatic crystallisation or late-metamorphic mineral growth may form the youngest peak of a detrital grain-age distribution. These ages and the corresponding peaks are particularly prone to be misinterpreted in terms of cooling during exhumation.
- The occurrence of a stationary age peak that is older than the moving age peak within the same detrital sample, or the occurrence of two (or more) moving age peaks within the same sample, are supportive of mixing of detritus from source areas characterised by different upper crustal exhumation histories and thermochronologic fingerprints.
- Post-depositional annealing or resetting due to burial produces age peaks that become progressively younger down section, showing the opposite trend than that observed in unreset sedimentary successions.
- The appearance of older grain-age populations moving up section through a stratigraphic succession may provide evidence of a major provenance change.
- If grain age shows dependence with grain size, relevant bias may be introduced by natural processes in the source-to-sink environment, or by inappropriate

procedures of sampling, laboratory treatment and analysis. This situation requires appropriate strategies for bias minimisation.

- Detrital sources supplying a majority of U-rich grains, or older zircon grains with higher levels of α-damage, are prone to be under-represented in detrital ZFT datasets because of the impact of U concentration bias and etching bias. This bias may propagate to other datasets in case of double or triple dating of detrital zircon grains.
- Detection limits of age components in detrital grain-age distributions should be indicated and fully considered during data interpretation, especially when detrital thermochronology studies discuss the absence of specific grain-age populations.

Acknowledgements This work benefited from insightful discussions with I. M. Villa and from constructive reviews by M. L. Balestrieri, S. Kelley and P. G. Fitzgerald.

References

Andersen T (2005) Detrital zircons as tracers of sedimentary provenance: limiting conditions from statistics and numerical simulation. Chem Geol 216:249–270

Andersen T, Kristoffersen M, Elburg MA (2017) Visualizing, interpreting and comparing detrital zircon age and Hf isotope data in basin analysis—a graphical approach. Basin Res. https://doi.org/10.1111/bre.12245

Asti R, Malusà MG, Faccenna C (2018) Supradetachment basin evolution unraveled by detrital apatite fission track analysis: the Gediz Graben (Menderes Massif, Western Turkey). Basin Res 30:502–521

Baldwin SL (2015) Highlights and breakthroughs. Zircon dissolution and growth during metamorphism. Am Mineral 100(5–6):1019–1020

Baldwin SL, Fitzgerald PG, Malusà MG (2018) Chapter 13. Crustal exhumation of plutonic and metamorphic rocks: constraints from fission-track thermochronology. In: Malusà MG, Fitzgerald PG (eds) Fission-track thermochronology and its application to geology. Springer, Berlin

Bernet M (2018) Chapter 15. Exhumation studies of mountain belts based on detrital fission-track analysis on sand and sandstones. In: Malusà MG, Fitzgerald PG (eds) Fission-track thermochronology and its application to geology. Springer, Berlin

Bernet M, Garver JI (2005) Fission-track analysis of detrital zircon. Rev Mineral Geochem 58(1):205–237

Bernet M, Spiegel C (eds) (2004) Detrital thermochronology. Geol S Am S 378

Bernet M, Brandon MT, Garver JI, Molitor B (2004) Fundamentals of detrital zircon fission-track analysis for provenance and exhumation studies with examples from the European Alps. Geol S Am S 378:25–36

Brandon MT (1996) Probability density plot for fission-track grain-age samples. Radiat Meas 26:663–676

Braun J (2016) Strong imprint of past orogenic events on the thermochronological record. Tectonophysics 683:325–332

Braun J, van der Beek P, Batt G (2006) Quantitative thermochronology: numerical methods for the interpretation of thermochronological data. Cambridge University Press, Cambridge

Calk LC, Naeser CW (1973) The thermal effect of a basalt intrusion on fission tracks in quartz monzonite. J Geol 81(2):189–198

Carrapa B (2009) Tracing exhumation and orogenic wedge dynamics in the European Alps with detrital thermochronology. Geology 37:1127–1130

Carrapa B, Wijbrans J, Bertotti G (2003) Episodic exhumation in the Western Alps. Geology 31(7):601–604

Carter A (2018) Chapter 14. Thermochronology on sand and sandstones for stratigraphic and provenance studies. In: Malusà MG, Fitzgerald PG (eds) Fission-track thermochronology and its application to geology. Springer, Berlin

Carter A, Moss SJ (1999) Combined detrital zircon fission-track and U–Pb dating: a new approach to understanding hinterland evolution. Geology 27:235–238

Cawood PA, Nemchin AA, Freeman M, Sircombe K (2003) Linking source and sedimentary basin: detrital zircon record of sediment flux along a modern river system and implications for provenance studies. Earth Planet Sci Lett 210:259–268

Challandes N, Marquer D, Villa IM (2008) P-T-t modelling, fluid circulation, and ^{39}Ar-^{40}Ar and Rb-Sr mica ages in the Aar Massif shear zones (Swiss Alps). Swiss J Geosci 101:269–288

Danišík M (2018) Chapter 5. Integration of fission-track thermochronology with other geochronologic methods on single crystals. In: Malusà MG, Fitzgerald PG (eds) Fission-track thermochronology and its application to geology. Springer, Berlin

Dodson MH, Compston W, Williams IS, Wilson JF (1988) A search for ancient detrital zircons in Zimbabwean sediments. J Geol Soc London 145:977–983

Enkelmann E, Garver JI, Pavlis TL (2008) Rapid exhumation of ice-covered rocks of the Chugach–St. Elias orogen, Southeast Alaska. Geology 36:915–918

Fitzgerald PG, Malusà MG, Muñoz JA (2018) Chapter 17. Detrital thermochronology using conglomerates and cobbles. In: Malusà MG, Fitzgerald PG (eds) Fission-track thermochronology and its application to geology. Springer, Berlin

Galbraith RF (1990) The radial plot: graphical assessment of spread in ages. Nucl Tracks Radiat Meas 17:207–214

Garver JI, Kamp PJJ (2002) Integration of zircon color and zircon fission-track zonation patterns in orogenic belts: application to the Southern Alps, New Zealand. Tectonophysics 349:203–219

Garver JI, Brandon MT, Roden-Tice MK, Kamp PJJ (1999) Exhumation history of orogenic highlands determined by detrital fission track thermochronology. Geol Soc Spec Publ 154:283–304

Garzanti E, Malusà MG (2008) The Oligocene Alps: domal unroofing and drainage development during early orogenic growth. Earth Planet Sci Lett 268:487–500

Giger M (1990) Geologische und Petrographische Studien an Geröllen und Sedimenten der Gonfolite Lombarda Gruppe (Südschweiz und Norditalien) und ihr Vergleich mit dem Alpinen Hinterland. PhD Thesis, University of Bern

Gleadow AJW (1990) Fission track thermochronology—reconstructing the thermal and tectonic evolution of the crust. In: Pacific Rim Congress III, Austr Inst Min Met, Gold Coast, Queensland, pp 15–21

Gleadow AJW, Hurford AJ, Quaife RD (1976) Fission track dating of zircon: improved etching techniques. Earth Planet Sci Lett 33:273–276

Glodny J, Kühn A, Austrheim H (2008) Diffusion versus recrystallization processes in Rb–Sr geochronology: isotopic relics in eclogite facies rocks, Western Gneiss Region, Norway. Geochim Cosmochim Acta 72:506–525

Glotzbach C, Bernet M, van der Beek P (2011) Detrital thermochronology records changing source areas and steady exhumation in the Western European Alps. Geology 39:239–242

Gombosi DJ, Garver JI, Baldwin SL (2014) On the development of electron microprobe zircon fission-track geochronology. Chem Geol 363:312–321

Harrison TM, McDougall I (1980) Investigations of an intrusive contact, northwest Nelson, New Zealand—I. Thermal, chronological and isotopic constraints. Geochim Cosmochim Acta 44 (12):1985–2003

Herman F, Seward D, Valla PG, Carter A, Kohn B, Willett SD, Ehlers TA (2013) Worldwide acceleration of mountain erosion under a cooling climate. Nature 504:423–426

Jasra A, Stephens DA, Gallagher K, Holmes CC (2006) Analysis of geochronological data with measurement error using Bayesian mixtures. Math Geol 38:269–300

Jourdan S, Bernet M, Tricart P, Hardwick E, Paquette JL, Guillot S, Dumont T, Schwartz S (2013) Short-lived fast erosional exhumation of the internal Western Alps during the late Early Oligocene: constraints from geo-thermochronology of pro- and retro-side foreland basin sediments. Lithosphere 5:211–225

Kasuya M, Naeser CW (1988) The effect of α-damage on fission-track annealing in zircon. Nucl Tracks Radiat Meas 14:477–480

Kohn MJ, Corrie SL, Markley C (2015) The fall and rise of metamorphic zircon. Am Mineral 100(4):897–908

Kohn B, Chung L, Gleadow A (2018) Chapter 2. Fission-track analysis: field collection, sample preparation and data acquisition. In: Malusà MG, Fitzgerald PG (eds) Fission-track thermochronology and its application to geology. Springer, Berlin

Komar PD (2007) The entrainment, transport and sorting of heavy minerals by waves and currents. Dev Sedimentol 58:3–48

Malusà MG, Fitzgerald PG (2018a) Chapter 8. From cooling to exhumation: setting the reference frame for the interpretation of thermocronologic data. In: Malusà MG, Fitzgerald PG (eds) Fission-track thermochronology and its application to geology. Springer, Berlin

Malusà MG, Fitzgerald PG (2018b) Chapter 10. Application of thermochronology to geologic problems: bedrock and detrital approaches. In: Malusà MG, Fitzgerald PG (eds) Fission-track thermochronology and its application to geology. Springer, Berlin

Malusà MG, Garzanti E (2018) Chapter 7. The sedimentology of detrital thermochronology. In: Malusà MG, Fitzgerald PG (eds) Fission-track thermochronology and its application to geology. Springer, Berlin

Malusà MG, Villa IM, Vezzoli G, Garzanti E (2011) Detrital geochronology of unroofing magmatic complexes and the slow erosion of Oligocene volcanoes in the Alps. Earth Planet Sci Lett 301:324–336

Malusà MG et al. (2012) Geochronology of detrital minerals: single-grain petrology, stratigraphy and the slow erosion of Oligocene Alps. In: Abstracts of the 13th international conference on thermochronology, Guilin, China, 24–28 Aug 2012

Malusà MG, Carter A, Limoncelli M, Villa IM, Garzanti E (2013) Bias in detrital zircon geochronology and thermochronometry. Chem Geol 359:90–107

Malusà MG, Resentini A, Garzanti E (2016a) Hydraulic sorting and mineral fertility bias in detrital geochronology. Gondwana Res 31:1–19

Malusà MG, Danišík M, Kuhlemann J (2016b) Tracking the Adriatic-slab travel beneath the Tethyan margin of Corsica–Sardinia by low-temperature thermochronometry. Gondwana Res 31:135–149

Malusà MG, Wang J, Garzanti E, Liu ZC, Villa IM, Wittmann H (2017) Trace-element and Nd-isotope systematics in detrital apatite of the Po river catchment: implications for provenance discrimination and the lag-time approach to detrital thermochronology. Lithos 290–291:48–59

Massonne HJ, Schreyer W (1987) Phengite geobarometry based on the limiting assemblage with K-feldspar, phlogopite, and quartz. Contrib Mineral Petrol 96(2):212–224

Montario MJ, Garver JI (2009) The thermal evolution of the Grenville Terrane revealed through U-Pb and fission-track analysis of detrital zircon from Cambro-Ordovician quartz arenites of the Potsdam and Galway Formations. J Geol 117(6):595–614

Ohishi S, Hasebe N (2012) Observation of fission-tracks in zircons by atomic force microscope. Radiat Meas 47(7):548–556

Rahl JM, Ehlers TA, van der Pluijm BA (2007) Quantifying transient erosion of orogens with detrital thermochronology from syntectonic basin deposits. Earth Planet Sci Lett 256:147–161

Reiners PW, Brandon MT (2006) Using thermochronology to understand orogenic erosion. Annu Rev Earth Planet Sci 34:419–466

Reiners PW, Farley KA (2001) Influence of crystal size on apatite (U–Th)/He thermochronology: an example from the Bighorn Mountains, Wyoming. Earth Planet Sci Lett 188:413–420

Resentini A, Malusà MG (2012) Sediment budgets by detrital apatite fission-track dating (Rivers Dora Baltea and Arc, Western Alps). Geol S Am S 487:125–140

Ruhl KW, Hodges KV (2005) The use of detrital mineral cooling ages to evaluate steady state assumptions in active orogens: an example from the central Nepalese Himalaya. Tectonics 24(4)

Ruiz G, Seward D, Winkler W (2004) Detrital thermochronology–a new perspective on hinterland tectonics, an example from the Andean Amazon Basin, Ecuador. Basin Res 16:413–430

Sambridge MS, Compston W (1994) Mixture modeling of multi-component data sets with application to ion-probe zircon ages. Earth Planet Sci Lett 128:373–390

Schuiling RD, DeMeijer RJ, Riezebos HJ, Scholten MJ (1985) Grain size distribution of different minerals in a sediment as a function of their specific density. Geol Mijnbouw 64:199–203

Sircombe KN, Stern RA (2002) An investigation of artificial biasing in detrital zircon U–Pb geochronology due to magnetic separation in sample preparation. Geochim Cosmochim Acta 66:2379–2397

Sláma J, Košler J (2012) Effects of sampling and mineral separation on accuracy of detrital zircon studies. Geochem Geophys Geosyst 13 (Q05007):1–17

Tagami T, Ito H, Nishimura S (1990) Thermal annealing characteristics of spontaneous fission tracks in zircon. Chem Geol 80:159–169

Tagami T, Carter A, Hurford AJ (1996) Natural long term annealing of the zircon fission track system in Vienna Basin deep borehole samples: constraints upon the partial annealing zone and closure temperature. Chem Geol 130:147–157

van der Beek P, Robert X, Mugnier JL, Bernet M, Huyghe P, Labrin E (2006) Late Miocene–recent exhumation of the central Himalaya and recycling in the foreland basin assessed by apatite fission-track thermochronology of Siwalik sediments, Nepal. Basin Res 18:413–434

Vermeesch P (2004) How many grains are needed for a provenance study? Earth Planet Sci Lett 224:441–451

Vermeesch P (2012) On the visualisation of detrital age distributions. Chem Geol 312:190–194

Vermeesch P (2013) Multi-sample comparison of detrital age distributions. Chem Geol 341:140–146

Vermeesch P (2018) Chapter 6. Statistics for fission-track thermochronology. In: Malusà MG, Fitzgerald PG (eds) Fission-track thermochronology and its application to geology. Springer, Berlin

Villa IM (1998) Isotopic closure. Terra Nova 10(1):42–47

White NM, Pringle M, Garzanti E, Bickle M, Najman Y, Chapman H, Friend P (2002) Constraints on the exhumation and erosion of the High Himalayan Slab, NW India, from foreland basin deposits. Earth Planet Sci Lett 195(1):29–44

Willett SD, Brandon MT (2013) Some analytical methods for converting thermochronometric age to erosion rate. Geochem Geophys Geosyst 14:209–222

Williams ML, Jercinovic MJ, Hetherington CJ (2007) Microprobe monazite geochronology: understanding geologic processes by integrating composition and chronology. Annu Rev Earth Planet Sci 35:137–175

Detrital Thermochronology Using Conglomerates and Cobbles

17

Paul G. Fitzgerald, Marco G. Malusà and Joseph A. Muñoz

Abstract

Detrital thermochronology data from cobbles, either within modern sediments or basin stratigraphy, can provide excellent constraints on the exhumation history of the adjacent orogen or hinterland. This approach is especially powerful if multiple techniques are applied to each cobble. With cobbles, all grains have a common thermal history; thus, for apatite fission-track (AFT) thermochronology, ages are better defined than those for single-grains, and track-length measurements permit meaningful thermal models. Well-constrained basin stratigraphy is required to generate lag-time plots that combined with thermal modelling may constrain the rate and timing of cooling/exhumation events in the orogen, as well as later basin inversion. Limitations in this approach include the number of cobbles that need to be analysed to provide a representative sampling of the source region and to capture the pre-depositional exhumation history. Caveats also apply regarding closure temperature assumptions, integrated exhumation rates, time from erosion to basin deposition, variable provenance of the cobbles, paleo-relief in the source area, as well as burial-related partial annealing. Two examples illustrate this detrital cobble approach applied to basin stratigraphy. On the southern flank of the Pyrenean orogen, cobble AFT thermochronology from Eocene-to-Oligocene syntectonic conglomerates records three episodes of cooling/exhumation in the hinterland due to progressive movement of thrust sheets, followed by burial and Late

Miocene re-excavation. The cobble thermochronologic record complements in-situ thermochronologic data from the orogen. South of the European Alps, thermochronology on conglomeratic clasts in the Gonfolite basin record exhumation in the Bergell region of the Central Alps. There, a stationary peak at ∼30 Ma records the emplacement of plutonic and volcanic rocks, whereas a moving peak starting at ∼25 Ma that decreases in age up-section records exhumation-related cooling.

17.1 Introduction

Detrital geochronology and detrital thermochronology typically involve determining the age of single-grains in orogenic sediments collected from sedimentary basins (ideally from a well-dated stratigraphic section) or from recent sediments to both identify provenance and quantify the thermochronological evolution (cooling and exhumation) of the source region (e.g. Garver et al. 1999). Detrital geo-thermochronology utilises dating techniques such as U–Pb (zircon, apatite, monazite), $^{40}Ar/^{39}Ar$ (mica) and fission track (zircon, apatite). Detrital grains have usually been eroded from orogenic highlands, transported and then deposited in a proximal or distal basin. The three main applications of detrital thermochronology, all closely related, are provenance analysis, landscape evolution and exhumation studies (e.g. Bernet and Spiegel 2004; and Chap. 15, Bernet 2018). Detrital thermochronology provenance studies link distinctive age patterns to sediment source areas (e.g. Baldwin et al. 1986; Hurford and Carter 1991; Carrapa et al. 2004; Resentini and Malusà 2012; Gehrels 2014, and Chap. 14, Carter 2018) and can also be used to constrain orogen development by tracing sediment transport pathways (e.g. Cawood and Nemchin 2001; Malusà et al. 2016a) as well locating positions of drainage divides (e.g. Spiegel et al. 2001). One of the advantages of using detrital thermochronology is that one is not limited to sampling exposed bedrock. Basin sediments

P. G. Fitzgerald (✉)
Department of Earth Sciences, Syracuse University, Syracuse, NY 13244, USA
e-mail: pgfitzge@syr.edu

M. G. Malusà
Department of Earth and Environmental Sciences University of Milano-Bicocca, 20126 Milan, Italy

J. A. Muñoz
Departament de Dinàmica de la Terra i de l'Oceà, Universitat de Barcelona, 08028 Barcelona, Spain

© Springer International Publishing AG, part of Springer Nature 2019
M. G. Malusà and P. G. Fitzgerald (eds.), *Fission-Track Thermochronology and its Application to Geology*, Springer Textbooks in Earth Sciences, Geography and Environment, https://doi.org/10.1007/978-3-319-89421-8_17

may therefore record an earlier part of the exhumation history, from higher structural levels that were eroded off the orogen in its early stages of development. Conversely, the detrital record may miss younger exhumation events, as these deeper structural levels may not have yet eroded off the orogen (see Chap. 16, Malusà 2018). In special cases, such as when an ice-sheet covers the landscape except for the highest peaks, detrital sediments may capture the record of more recent exhumation, whereas bedrock thermochronology from high ridges may only record much older ages. Such a situation occurs in the St. Elias Mountains of southeastern Alaska where active erosion is occurring under the ice and the youngest thermochronology ages are found within modern glacial detritus (Enkelmann et al. 2015). With respect to detrital thermochronology and exhumation levels of an orogen that may or may not be sampled, a similar concept applies with in-situ bedrock thermochronology. There, low-temperature thermochronologic methods have younger ages and constrain younger cooling/exhumation events, whereas higher-temperature techniques have older ages and constrain older cooling/exhumation events. A more complete record of the thermal, exhumation, deposition, geological and tectonic record of an orogen will likely involve data from both in-situ (bedrock) thermochronology and detrital thermochronology from associated basins. On that theme, there are many assumptions in the detrital › (e.g. Garver et al. 1999; Rahl et al. 2007; and Chap. 8 and Chap. 10, Malusà and Fitzgerald 2018a, b; and Chap. 15, Bernet 2018), many of which are relevant to applying thermochronology to cobbles.

In this chapter, we summarise the approach of using apatite fission-track (AFT) thermochronology on cobbles collected in a stratigraphic framework. We compare this approach briefly with the more conventional detrital thermochronology approach on sand or sandstone (Sect. 17.3) and illustrate cobble thermochronology using an example from syntectonic conglomerates deposited in the South Pyrenean Foreland Basin on the southern flank of the central Pyrenees (Beamud et al. 2011; Rahl et al. 2011). Crucial to this approach and important in the Pyrenees study is the use of both the lag-time approach (Garver et al. 1999) to constrain rates of exhumation in the hinterland, and inverse modelling (e.g. Ketcham 2005) of AFT age and track-length data collected from each cobble to constrain the timing of events. The results of the Pyrenees cobble data are compared to in-situ low-temperature thermochronology constraints from granitic massifs within the central Pyrenean orogen. We also present another example from the Bergell region of the Alps (e.g. Wagner et al. 1977, 1979; Hurford 1986; Rahn 2005; Malusà et al. 2011, 2016a; Mahéo et al. 2013; Fox et al. 2014).

The composition of sedimentary deposits, including conglomerate clasts has long been used by sedimentologists and stratigraphers, linking clasts back to their original lithotectonic units or tectonostratigraphic terranes (e.g. DeCelles 1988). However, using thermochronology on individual conglomerate clasts in foreland basin deposits in a similar fashion is more recent, evolving from early thermochronology studies in the Alps. Wagner et al. (1979) obtained AFT ages from three boulders in upper Oligocene molasse deposits in the central Alps and compared these to data, including age-elevation profiles, collected from the nearby Bergell Massif north of the Insubric Fault. Using the AFT ages to estimate the paleo-elevation position of the boulders, Wagner et al. (1979) were able to estimate the amount and rate of "uplift" (exhumation) since they were eroded from the massif. Despite the prediction of Wagner et al. (1979) that this would become an important approach to study the vertical extent of mountain belts in the geological past, it has not been widely applied. Dunkl et al. (2009) developed the pebble population dating method, using two examples from the eastern Alps, but this method was mainly aimed at the provenance of these deposits rather than the exhumation history of the hinterland. Also in the eastern Alps, Brügel et al. (2004) used geochemistry and multiple thermochronometers on gneiss pebbles in Miocene conglomerates to constrain the provenance as well as the cooling and exhumation history of the hinterland from where they originated. More recently, Falkowski et al. (2016) applied multiple techniques to glacial cobbles sourced from ice-covered terrain in southeastern Alaska to constrain both the provenance and exhumation history of this mountainous ice-covered terrain.

17.2 Detrital Thermochronology: The Stratigraphic Approach

The stratigraphic approach to detrital thermochronology has the advantage of being able to constrain exhumation rates through time (e.g. Bernet et al. 2004). Multiple or single thermochronometers can be applied (e.g. Chap. 14, Carter 2018), with different methods providing different information and having different caveats as regards the interpretation of the data. U–Pb dating on zircons (e.g. Amidon et al. 2005a, b) provides information more on the provenance than the exhumation history, whereas zircon fission-track dating (ZFT) may provide information on both (e.g. Cerveny et al. 1988; Garver et al. 1999). Hypothetically, the spread in single-grain ages can also be used to constrain paleo-relief. This requires a high-precision technique, a positive paleo-"age-elevation relationship" that is not too steep (Stock and Montgomery 1996; Reiners 2007) and assuming that mineral fertility variations in the drainage area are negligible (Malusà et al. 2016b; and Chap. 7, Malusà and Garzanti 2018). Thermochronologic methods that incorporate a kinetic parameter to constrain cooling rates also allow better constraints on the exhumation history of the source region as well as the post-depositional thermal history of basin strata. Examples of such methods with kinetic parameters are $^{40}Ar/^{39}Ar$ thermochronology using K-feldspar temperature cycling experiments and multi-diffusion domain modelling (Lovera et al. 2002) and AFT thermochronology with track length and apatite compositional measurements or proxies such as Dpar (e.g. Ketcham et al. 2007; Gallagher 2012). The kinetic parameters facilitate generation of inverse thermal models that constrain best-fit temperature–time $(T - t)$ pathways. Note that lower-temperature thermochronologic methods are more sensitive to shorter-term variations in exhumation rates, as the higher-temperature methods take longer to respond to changes in erosion rate at the surface (e.g. Bernet and Spiegel 2004).

When analysing detrital samples, ~ 100 grains are usually dated in order to minimise the risk of missing relevant grain age populations (e.g. Stewart and Brandon 2004; Vermeesch 2004; Andersen 2005; see Chap. 16, Malusà 2018). In contrast to conventional in-situ thermochronology, dated detrital single grains do not necessarily have a common source or thermal (exhumation) history, so grains are typically plotted in a probability distribution with significant age populations or peaks determined using either a Gaussian or binomial peak fitting method (e.g. Hurford et al. 1984; Galbraith and Green 1990; Brandon 1996; Bernet et al. 2001, 2004; Dunkl and Székely 2002; Vermeesch 2012). In addition, because dated AFT single-grain ages do not have a common source or thermal (exhumation) history, a combined confined track-length distribution may be meaningless, unless the sample is completely reset in the sedimentary basin. A typical issue with AFT thermochronology is that the relatively low [U] means relatively few numbers of fission tracks and hence large errors (often exceeding 50% on single-grain ages). Thus, individual peak populations may be poorly defined (e.g. Garver et al. 1999). ZFT dating does not usually have this problem; the higher [U] and also higher closure temperature means individual grains have more tracks and hence lower single-grain age errors. An additional problem with fission tracks in zircons is that crystals with different amounts of radiation damage have different etching times (to reveal the fission tracks) and if this not accounted for, bias may result (e.g. Bernet and Garver 2005; Malusà et al. 2013).

There are four stages (Fig. 17.1a) involved in conceptualising processes related to detrital thermochronology analyses and the factors that influence the data and the interpretation of the data (as synthesised from Garver et al. 1999; Bernet and Spiegel 2004; Bernet and Garver 2005; van der Beek et al. 2006; and Chap. 7, Malusà and Garzanti 2018).

(i, ii) Exhumation and erosion: As samples are exhumed toward the surface in the orogenic hinterland, they cool through the partial annealing zone (PAZ, Gleadow and Fitzgerald 1987) or partial retention zone (PRZ, Baldwin and Lister 1998; Wolf et al. 1998) of whatever thermochronologic system is being applied. A PAZ (fission-track methods) or PRZ (other methods) is a temperature interval between higher temperatures where there is total annealing or total loss via diffusion of the daughter product, and cooler temperatures where ages typically become uniform and often reflect earlier cooling and exhumation of a crustal section. As temperature increases with depth, and thermochronological ages are often plotted with respect to depth, PAZs and PRZs are often envisaged as crustal entities, with their width, thickness and shape dependent on the dynamic versus static thermal regime. The closure temperature concept (Dodson 1973) assumes that samples have cooled monotonically through a closure temperature and the "closure age t_c" reflects the time when the thermochronologic system closed and the daughter product began accumulating. Exhumation is typically accomplished via erosion, which is associated with tectonic uplift and usually faulting (e.g. Brown 1991; Ring et al. 1999). Using detrital thermochronology to quantify exhumation of the eroded orogenic highland using the lag-time approach (see below) relies on the closure temperature concept, although the cooling may not be monotonic and partial annealing of fission tracks or partial loss of the daughter product may be significant as well. The closure temperature, depending on the particular thermochronologic system, may vary according to composition of the mineral, [eU], radiation damage and rate of cooling. For AFT thermochronology, composition may be established grain by grain using Dpar or microprobe analysis to calculate Cl and F concentrations. Additionally, the r_{mr0} parameter (Ketcham et al. 1999, 2007), which is a measure of the relative resistance to annealing for an apatite, may be applied to constrain different kinetic grain populations (see Chap. 3, Ketcham 2018; Chap. 18, Schneider and Issler 2018). Applying a closure temperature to an age (e.g. ~ 120 °C for fission tracks in apatite cooling at ~ 10 °C/Myr: Reiners and Brandon 2006) and assuming or constraining the paleo-geothermal gradient and a paleo-mean annual temperature allows one to estimate the depth to closure, and hence an average or integrated exhumation rate (e.g. Garver et al. 1999). This simple calculation does not take into account the rate of cooling. If cooling is rapid, then the closure temperature is higher and thus the estimation of the closure depth is an underestimate, and hence, estimation of the exhumation rate will also be an underestimate. Rapid exhumation (on the order of >300 m/Myr) also results in advection and compression of the isotherms (e.g. Mancktelow and Grasemann 1997; Gleadow and Brown 2000; Braun et al. 2006). This is extremely important because when advection compresses isotherms in the upper crust, it will affect a reliable

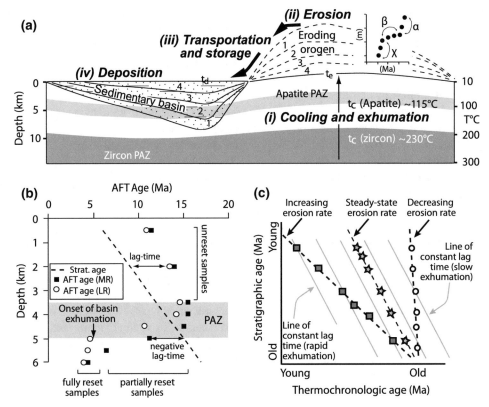

Fig. 17.1 Concept diagram explaining detrital thermochronology (modified from Bernet and Garver 2005; van der Beek et al. 2006; Rahl et al. 2007). **a** Samples collected from basin stratigraphy (strata 4 to 1) can be used to constrain integrated exhumation rates using the lag-time principle. Samples are exhumed from a hinterland orogen (eroded layers 1 through 4), assumed to have cooled through a closure temperature (t_c), in this case AFT and ZFT, arriving at the surface to be eroded at t_e. Lag-time = t_c (cooling age) − t_d (time of deposition or stratigraphic age). The time from erosion (t_e) to deposition (t_d) is assumed to be geologically instantaneous ($t_e - t_d = 0$ Ma). van der Beek et al. (2006) assumed a hinterland and sedimentary basin geotherm of 20 °C/km when creating the original of this figure. We incorporate a stylised age-elevation plot into the eroding orogen. This simple plot reflects a likely age-elevation profile, including an exhumed PAZ that may be observed if samples were collected over significant relief in the orogen itself. In this plot samples labelled "α" reflect an earlier history. These samples cooled into the upper crust during an earlier phase of cooling/exhumation but were not eroded at that time. Samples "β" were resident in a PAZ (or PRZ) prior to exhumation, so these samples represent an exhumed PAZ or PRZ. Samples labelled "χ" cooled as a result of exhumation related to the formation of this orogen. In the case of samples "β", the closure temperature concept does not apply and these samples therefore do not yield a meaningful lag-time, whereas for samples "χ", the age does

reflect the time the sample cooled through a closure temperature and the lag-time can be converted to an exhumation rate using assumptions discussed in the text. For samples "α" that also cooled relatively quickly, the closure temperature concept does apply, but because these samples were cooled during an earlier period of exhumation, the lag-time conversion is meaningless (it will underestimate the real exhumation rate). **b** Modelled AFT ages (from van der Beek et al. 2006) plotted against depth to demonstrate how sample AFT age may increase with depth (to older stratigraphic units) before samples within a sedimentary basin are partially or completely annealed. For partially or completely annealed samples, the lag-time principle to constrain exhumation rate does not apply. MR = more retentive apatite grains (Dpar = 2) and LR = less retentive grains (Dpar = 1). **c** Lag-time plots (stratigraphic age vs. thermochronologic age) for three scenarios (modified from Rahl et al. 2007). (i) Steady-state erosion (exhumation) in which samples each take the same time to reach the surface from an assumed closure depth (determined assuming a steady-state geotherm) and each successive sample is successively younger (thermochronologic age) with decreasing stratigraphic age (depth) but the "lag-time" remains constant. (ii) Increasing erosion rate where thermochronologic ages "young" up-section faster than the stratigraphic ages and lag-time decreases. (iii) Decreasing erosion rate, thermochronologic ages "young" up-section slower than the stratigraphic ages and the "lag-time" increases

estimation of the paleo-geothermal gradient and hence depth of closure and average exhumation rate (Braun et al. 2006; Rahl et al. 2007; see also Chap. 8, Malusà and Fitzgerald 2018a, b). Rahl et al. (2007) stress that in the case of transient erosion (exhumation), the calculation of erosion rates using the assumption of thermal steady state will generally result in error and that assumptions of steady erosion in

mountain belts should be used with caution. While sometimes difficult to observe depending on the tectonic situation as well as the thermochronologic technique and sampling strategy applied, exhumation often varies both spatially and temporally within an orogen, such as in the example of the Pyrenean orogen discussed below. Using a simple lag-time plot to determine the average or integrated exhumation rate also assumes a simple exhumation pathway. However, an average rate may conceal episodes of rapid exhumation. In addition, if samples have been resident in the orogenic hinterland at levels shallower in the crust than the PAZ/PRZ (samples within zone α in Fig. 17.1a), or are resident within a PAZ/PRZ before exhumation starts (samples within zone β in Fig. 17.1a), the resultant ages do not reflect simple exhumation cooling. For example, samples may be cooled (exhumed) to shallow crustal levels during one exhumation episode but not exposed at the surface until the next exhumation episode when they are eroded, transported and deposited. The lag-time conversion to an exhumation rate (discussed below) for these samples thus approximates an average rate of the two exhumation episodes including the period in between. In this case, the lag-time approach will result in a significant underestimate of the rate of each orogenic (exhumation) event. The paleo-geothermal gradient and the time of closure are the two variables most difficult to constrain in most situations.

(iii) Transport of sediments: The time between when a rock is eroded (t_e), transported and then deposited (t_d) in a sedimentary basin is usually regarded as negligible (Brandon and Vance 1992). Sediments may be transiently stored in a basin during the transportation process, which is a more important consideration for distal basins and less for proximal basins (see Chap. 7, Malusà and Garzanti 2018). For basins containing syntectonic conglomeratic deposits close to the exhuming and eroding orogen, as well as for more distal basins including large clasts transported by gravity flows (e.g. Anfinson et al. 2016), the assumption of transport time being negligible ("geologically insignificant") is usually sound.

(iv) Deposition: It is essential to know the time of deposition (t_d) (the stratigraphic age) of the detrital sample because this is used to constrain the lag-time (Cerveny et al. 1988; Garver et al. 1999). Lag-time is equal to t_c (cooling age) minus t_d, assuming transport time is negligible. Assuming or knowing the paleo-geothermal gradient in conjunction with determination of the lag-time allows the estimation of an average exhumation rate since t_c, assuming the closure temperature concept is valid for that sample. The lag-time correlates with the average exhumation rate since t_c. When the lag-time is small, the average exhumation rate is rapid. When the lag-time is large, the exhumation rate is slow (Fig. 17.1c). The lag-time plot of thermochronologic age versus stratigraphic age and the patterns of ages compared to the lines of constant lag-time can be used to determine if the rate of erosion in the orogen is increasing, decreasing or remaining the same. Decreasing lag-times indicative of increasing exhumation rates usually correlate with the constructive phase of an orogen, whereas increasing

lag-times indicative of decreasing exhumation rates correlate with the decay phase of an orogen (e.g. Brandon and Vance 1992; Garver et al. 1999; Bernet et al. 2004; Spotila 2005). The lag-time approach to constrain averaged exhumation rates will yield meaningful estimates only when applied to those sediments (in the basin) that reside above the PAZ or PRZ of the method in question. Partial annealing or partial loss will reduce the age and compromise the use of lag-time to constrain exhumation rates. In this situation, the pattern of thermochronologic age versus stratigraphic depth can be diagnostic (Fig. 17.1b), especially if used in conjunction with kinetic parameters (e.g. confined track-length distributions) or other thermal information such as vitrinite reflectance (VR) (Burnham and Sweeney 1989) or conodont alteration (CAI) (Epstein et al. 1977). As mentioned above, confined track-length distributions measured in multiple individual apatite grains in sandstones may be meaningless to constrain the thermal history because of the mixed provenance of the individual grains, the potential for sampling across considerable paleo-relief and from different exhumation terrains. In basins dominated by continentally derived conglomerates, stratigraphic age constraints are often very poor because of the lack of fossils. Magnetostratigraphy can provide significant time constraints on these basins. In our first example, the South Pyrenean Foreland Basin, the extensive magnetostratigraphy of Beamud et al. (2003, 2011) sampled the finer-grained interbedded overbank sediments, rather than the conglomerates. In addition to existing magnetostratigraphy (Burbank et al. 1992; Meigs et al. 1996; Jones et al. 2004) this approach was crucial to constraining the stratigraphic age, and hence being able to apply detrital thermochronology.

17.3 Detrital Thermochronology Using Cobbles

Using cobbles for detrital thermochronology rather than the more conventional approach of dating ∼100+ individual grains (from a sand or sandstone) has its advantages, but also limitations. One of the main strengths is that all grains from each cobble share a common thermal history. For AFT thermochronology, this means that:

- A "central age" with its smaller uncertainty than individual grain ages (e.g. Galbraith 2005) can be used.
- The confined track-length distribution (the kinetic parameter) is representative of the thermal history of that cobble, and this can be modelled using age and track-length data to constrain a best-fit T-t envelope using programmes such as HeFTy (e.g. Ketcham 2005)

and/or QTQt (Gallagher 2012). The stratigraphic age of the particular cobble provides a strong constraint for this modelling.

- The lag-time approach can be used to constrain rates of exhumation in the hinterland (all caveats discussed above in 17.3 still apply).
- Inverse modelling of the individual cobble data constrains the thermal history and episodes of rapid cooling/exhumation (if present).
- Basin analyses on deposition rates, for example, can be compared to these episodes of rapid cooling/exhumation.
- The post-depositional history of the basin associated with burial, partial annealing of fission tracks and later inversion (or re-excavation) of the basin can be constrained using the inverse thermal models.
- Cobbles are also likely to be deposited in proximal basins or in distal basins by gravity flows, so the assumption of the transport time from erosion to deposition being negligible is likely to be valid.
- Cobbles can be treated as "bedrock", and multiple thermochronologic techniques can be undertaken on each cobble, with each method constraining a different part of a common thermal history (see the Bergell-Gonfolite example below).
- Within cobbles, minerals that are less resistant to weathering (e.g. K-feldspar) and may not survive transport as a single grain can also be analysed.

Limitations of the cobble thermochronology approach include:

- Each cobble only preserves information from a single point from the source region. From where did that cobble (with respect to provenance, paleo-relief) originate? Erosion rates usually vary across the source region. In a developing orogen, there will be uplift and the creation of relief with cobbles presumably eroded over much of that relief. If there is a positive thermochronologic age-elevation relationship, as is typical in an orogen, then a selection of cobbles from a given horizon may have a range of thermochronologic ages, dependent of which part of the orogen AFT-stratigraphy (e.g. exhumed PAZ, rapidly cooled samples) is providing cobbles.
- How many cobbles from each stratigraphic horizon should be analysed to provide a representative sample and constrain the history of the source region?
- Using cobbles, especially with multiple techniques is costly and time-consuming (e.g. Falkowski et al. 2016) which limits analysing enough cobbles for complete statistical analysis of a catchment.
- Lithology—cobbles of a suitable lithology such that they contain appropriate accessory minerals are required.

However, as cobbles are often sourced in the orogen from relatively close to the basin, there is a strong possibility that the eroded remnants of that source terrane are still present and detected and thus detrital thermochronology can be compared to in-situ thermochronology collected within the orogen itself.

- Size of cobbles—Not all cobbles are large enough to provide enough apatite grains for analysis, and in this case, small cobbles may be combined to form an aggregate sample (e.g. Rahl et al. 2007) cognizant that this may produce a mixed source-signal (e.g. from different paleo-elevations). Alternatively, a sample of the (ideally sandstone) matrix could also be analysed along with the cobbles. In essence, this combines both the cobble approach and the more conventional detrital approach. When interpreting ZFT data from sandstones, it is common to assume that the minimum age (often identical to the first peak in a grain-age distribution plot) provides a proxy for fast exhumation rates at the time of deposition (see Chap. 15, Bernet 2018), with other ages (age peaks) possibly providing information on paleo-relief or on other portions of the catchment.

We focus next on an example from the south-central Pyrenees, examining cobble AFT thermochronology data from several studies (Beamud et al. 2011; Rahl et al. 2011). We go into some detail about the interpretation of cobble data from the Pyrenees syntectonic conglomeratic deposits because the cobble approach has really not been discussed in any detail in the literature and seems under-utilised. Although it should also be noted that appropriate conglomeratic deposits can be relatively rare.

17.4 The Syntectonic Conglomerates of the Southern Pyrenean Fold-and-Thrust Belt

17.4.1 Geology and Tectonic History of the Pyrenees

The geology and tectonic history of the Pyrenees are well described in a number of papers (e.g. Muñoz 1992, 2002; Coney et al. 1996; Fitzgerald et al. 1999; Beaumont et al. 2000; Sinclair et al. 2005; Beamud et al. 2011), and parts relevant to this paper are summarised below. The Pyrenean orogen formed from Late Cretaceous to Miocene times as a result of the collision between the northward moving Iberian plate and Europe (Roest and Srivastava 1991; Rosenbaum et al. 2002; Muñoz 2002). The Pyrenees are an asymmetric doubly vergent orogen that resulted from the subduction of the Iberian lower crust and lithospheric mantle under the

European plate (Fig. 17.2b). Structural style and along-strike differences were strongly controlled by the inversion of the early Cretaceous rift system at the eastern continuation of the Bay of Biscay oceanic basin (Roca et al. 2011). Before collision, the Pyrenean realm of the north-Iberian margin was a highly segmented, hyperextended margin with a narrow strip of exhumed mantle between the Iberian and European plates (Jammes et al. 2009; Lagabrielle et al. 2010; Roca et al. 2011; Tugend et al. 2014). Inversion of these inherited extensional features and distribution of the main weak horizons controlled the geometry and evolution of the orogenic wedge along the Pyrenean orogen, and thus the different stages from exhumation and erosion to deposition of eroded materials into the adjacent basins (Beaumont et al. 2000; Jammes et al. 2014). The central and eastern Pyrenees show a core of Hercynian thrust sheets known as the Axial Zone and flanked north and south by fold-and-thrust belts, in turn flanked by foreland basins, the Aquitaine to the north and the Ebro to the south (Fig. 17.2). The Axial Zone is bounded to the north by the North Pyrenean fault, which is the reactivated extensional detachment above the exhumed mantle, at the boundary between the Iberian and European plates (Lagabrielle et al. 2010). The Axial Zone is comprised of an antiformal stack of three upper crustal Hercynian thrust sheets (nappes) named the Nogueres, Orri and Rialp. Shortening across the central Pyrenees is on the order of 165 km and was accommodated by a general north to south sequence of thrusting and movement of the Nogueres (Late Cretaceous to early Eocene), Orri (middle-late Eocene) and then Rialp (Oligocene) thrust sheets (Fig. 17.2). The Hercynian basement thrust sheets include plutons of Carboniferous to Early-Permian age (Hercynian), which now form prominent massifs in the Axial Zone. These massifs are ideal for in-situ (bedrock) thermochronological analyses, and once exposed provide the basement cobbles that were transported and deposited within the evolving southern fold-and-thrust belt. The southern fold-and-thrust belt consists of three major thrust sheets activated in a forward (north–south) propagated thrust sequence; the Bóixols thrust sheet developed from the Late Cretaceous; then the Montsec thrust sheet in the Palaeocene-early Eocene; and the Serres Marginals thrust sheet in the middle Eocene–Oligocene (Fig. 17.2). Sediments in the southern Pyrenean Ebro Basin and southern fold-and-thrust belt comprise a continuous sequence of Cretaceous to upper Miocene strata. There were both marine and continental lateral equivalents in origin up until the late Eocene, when the basin was open to the Bay of Biscay, and then, sediments were continental in origin once the basin became endorheic (Riba et al. 1983; Puigdefàbregas et al. 1986; Pérez-Rivarés et al. 2004; Costa et al. 2010). As the

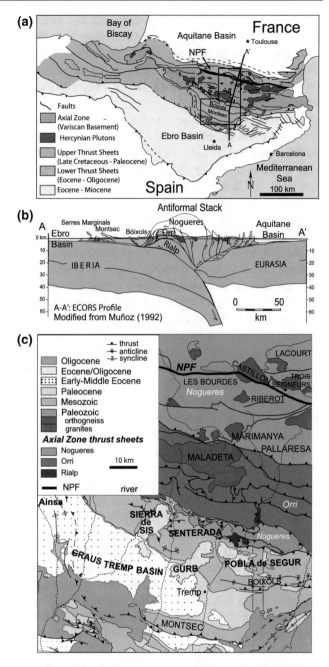

Fig. 17.2 **a** Regional tectonic setting of the Pyrenees located between the Aquitaine and Ebro Basins. The Axial Zone, Variscan basement, lies between the north-vergent Northern thrust belt and the south-vergent South Pyrenean thrust belt (modified from Coney et al. 1996). NPF = North Pyrenean Fault. Box marks the location of map in panel (**c**). **b** Cross section with thrust sheets and granitic plutons across the ECORS seismic profile (modified from Muñoz 1992). **c** Geological sketch map of the south-central Pyrenees (modified from Beamud et al. 2011). This shows the outcropping continental conglomerates of Eocene and Oligocene age sampled for cobble detrital thermochronology, and the granitic massifs where in-situ granitic samples were collected using the vertical profile approach

Axial Zone formed and was eroded, syntectonic sediments accumulated in the foreland basin and piggyback basins preserved on top of all three major fold-and-thrust sheets. Remarkably thick (km) units of syntectonic conglomerates were deposited in the southern thrust wedge during the Eocene and Oligocene.

Provenance studies have revealed pebbles were sourced from both the basement thrust sheets of the Axial Zone and the Mesozoic and Palaeogene successions (mainly carbonates) of the southern thrust sheets (e.g. Mellere 1993; Vincent 2001; Barso 2007). Magnetostratigraphy has constrained the age of these conglomeratic units as ca. 45 Ma to ca. 25 Ma (Beamud et al. 2003, 2011). These units progressively buried the southern Pyrenean thrust sheets, raising the base level of the piggyback basins and the adjacent foreland basin and burying the thrust wedge. Rapid sedimentation of these conglomerates affected the thrusting sequence depending on the depositional pattern of the sedimentary deposits (Fillion et al. 2013). The general aggradation geometry of the syntectonic conglomerates (unconformably overlying the cover thrust sheets as well as parts of the Nogueres thrust sheet), as well as the contemporaneous rapid erosion of the internal parts of the orogen, reduced the surface taper of the thrust wedge. This enhanced internal deformation and promoted break-back thrusting proceeding hindward from the thrust front on existing and new faults (Vergés and Muñoz 1990; Meigs et al. 1996; Muñoz et al. 1997; Muñoz 2002; Fillion et al. 2013).

The syn- to post-orogenic evolution of the central Pyrenees southern flank comprises some unique and somewhat unusual aspects. This is because when the South Pyrenean Foreland Basin became closed due to late Eocene–Oligocene tectonism along its margins, the syntectonic sediments eroded (mostly) from the Pyrenean orogen filled the basin and then backfilled across the actively deforming southern Pyrenean thrust belt and buried the southern flank of the orogen with up to 2–3 km of these continentally derived conglomerates (Coney et al. 1996; Muñoz et al. 1997). Following a period of tectonic quiescence after the end of Pyrenean deformation in the late Oligocene–earliest Miocene, the southern flank of the range was re-excavated as the conglomerates were flushed from the foreland basin by the Ebro River since the late Miocene. Documenting this late Miocene exhumation or re-excavation signal of the southern flank of the Pyrenees and thus testing the ideas of Coney et al. (1996) was the goal of several thermochronologic studies (Fitzgerald et al. 1999; Fillon and van der Beek 2012; Fillion et al. 2013). All these studies saw this late Miocene cooling signal using various modelling approaches, with the time of re-excavation likely beginning ~9 Ma.

This same late Miocene signal is also revealed in the thermal models for the oldest stratigraphic (most deeply buried and partly reset) conglomeratic samples from the Sierra de Sis (see below).

17.4.2 Exhumation History of the Pyrenean Orogeny Based on In-situ Thermochronology in the Hinterland

The tectonic history of the Pyrenean orogen (including the thermal and exhumation history) is associated with the inversion of the previous Early Cretaceous rift system, mainly preserved in the internal parts of the orogen and in the northern Pyrenees, and the propagation of the deformation into the Iberian crust (Muñoz 2002). The inherited structural features (Variscan and Cretaceous) affecting this crust, as well as the distribution of the Triassic salt, controlled the structural style and the geometry of the thrust system at crustal scale (Fig. 17.2b). Rocks experienced different $T - t$ histories, including thrust burial and exhumation, as this thrust system evolved during progressive shortening in the Axial Zone. Thermochronology data from within the Axial Zone (Fig. 17.3a) indicate that exhumation was asymmetric; it started in the north and progressed southward with the greatest exhumation on the southern flank of the Axial Zone (Fitzgerald et al. 1999). Information comes from a spectrum of thermochronologic methods: K-feldspar $^{40}Ar/^{39}Ar$ thermochronology including multi-diffusion domain modelling (Metcalf et al. 2009), AFT and ZFT including some inverse modelling (Yelland 1990; Morris et al. 1998; Fitzgerald et al. 1999; Metcalf et al. 2009), (U–Th)/He dating (Gibson et al. 2007; Metcalf et al. 2009) and thermo-kinematic modelling (Gibson et al. 2007; Fillon and van der Beek 2012). Exhumation, at least that constrained by low-temperature thermochronology, started in the central Pyrenean orogen by the early Eocene (~50 Ma) at a rate of ~0.2–0.3 km/Myr. At ~37–35 Ma, exhumation dramatically increased on the southern side of the orogen to rates on the order of 1–3 km/Myr. Exhumation then slowed, with the southern flank being buried by syntectonic conglomerates as the south Pyrenean basin filled and conglomeratic deposits on-lapped onto the orogen. Thrusting on the southern flank of the orogen was active up until ca. 20 Ma (Barruera thrust) followed by tectonic quiescence until ca. 9 Ma. At about that time, base level dropped in the Ebro Basin due to a connection to the Mediterranean Sea and syntectonic conglomerates filling the foreland basin and overlying the fold-and-thrust belt were largely flushed out of the system. The thermochronology adds constraints to the geologically and geophysically constrained evolution of the orogen and related basins (e.g. Muñoz 1992; Beaumont et al. 2000; Muñoz 2002) as well as confirming the ideas of Coney et al. (1996) with respect to the post-orogenic evolution and re-excavation of topography.

17.4.3 Exhumation History of the Pyrenean Orogeny Based on Detrital Thermochronology of Conglomerates in a Stratigraphic Framework

Detrital thermochronology using cobbles from the syntectonic Eocene–Oligocene conglomerates of the South Pyrenean Foreland Basin is successful because of the presence of "granitic cobbles" throughout much of the basin stratigraphy. One of the amazing peculiarities of the Pyrenees is that piggyback basins filled with syntectonic conglomerates lie very close to the interior part of the orogen and often the granitic massifs within the Axial Zone from where the granitic cobbles are sourced can be seen from the conglomeratic sampling locations. The AFT data from cobbles collected from these syntectonic conglomerates (Beamud et al. 2011; Rahl et al. 2011) provide a "classic example" of applying thermochronology to cobbles collected in a stratigraphic framework. Beamud et al. (2011) analysed 13 cobbles while Rahl et al. (2011) analysed 8 cobbles, plus 2 aggregate samples (multiple cobbles). These studies used similar approaches, namely lag-time plots and thermal modelling of the AFT data and obtained similar results, but with slight differences that serve as good examples of potential issues with this approach—as discussed below.

Lag-time plot and thermal models The lag-time plot (Fig. 17.3b) is used to visually assess the change in exhumation rate of the samples, but the timing of these events can be problematic to determine from this plot. That cobble AFT data can constrain the exhumation rate (as discussed above) relies on confined track-length distributions (that reveal the cooling history) from each cobble as well as the inverse thermal models. Best-fit $T - t$ envelopes (Fig. 17.4) constrain the timing and rate of the pre-depositional cooling events that can then be related to exhumation events. Evaluation of the thickness of the stratigraphic section as compared to the thermal range of the thermochronological method being applied along with constraints/assumption of a geothermal gradient or use of other thermal indicators such as VR also allows one to estimate when partial annealing or partial resetting is likely to have occurred within basin strata.

AFT ages (Fig. 17.3b) generally increase down-section, as would be expected from progressive unroofing of an active orogen (e.g. van der Beek et al. 2006; Rahl et al. 2007) but with variation that suggests a changing exhumation rate. In the lag-time plot, we have combined data from Beamud et al. (2011) and Rahl et al. (2011) but keep the same groups 1 through 5 that Beamud et al. used to describe their data. Mean track lengths in samples from the uppermost part of the section (group 1; stratigraphic age ∼25 Ma and group 2; stratigraphic age ∼28 Ma) down to ∼36 Ma (group 3) or ∼39 Ma (Rahl et al. 2011) indicate rapid cooling, and these samples are used to constrain the changing exhumation rate. Groups 4 and 5 with the lowest samples in the stratigraphic section have been partially reset by burial (up to temperatures of ca. 70 °C), but thermal models (Fig. 17.4) from those samples constrain both hinterland cooling events and late Miocene cooling that accompanied re-excavation of the Pyrenean fold-and-thrust belt. Note that most of these groups are defined by more than one cobble age; an important consideration given that the lowest stratigraphic sample of Rahl et al. may be significantly older than the rest of group 5, as discussed below. All but two samples of the data of Rahl et al. fall into group 5 (samples with partial annealing due to burial). One of their samples lies on the trend between group 4 and 3, and their uppermost sample labelled "R" adds to the pattern of changing exhumation rates. Sample "R" extends the trend from groups 2 to 1 to slightly older stratigraphic ages, as shown by the inverse thermal model for "R" (see Fig. 6 in Rahl et al. 2011) that is similar to other thermal models for samples in group 2.

Cooling/exhumation episodes: To constrain exhumation rates using lag-time plot, Beamud et al. (2011) used a geothermal gradient constrained independently by other workers and also compatible with VR data. Because group 5 samples have been partially annealed (shown by track-length distributions and thermal models in Fig. 17.4), they cannot be used to constrain the exhumation rate. Group 4 samples are partially annealed, but only slightly; yet it seems likely that the average exhumation rate was slow (<0.15 km/Myr) for this group. From group 3 up, calculation of exhumation rates from the lag-time plot are not affected by partial annealing. Thus while average exhumation rates of group 4 samples must have been very slow, rates increase to group 3 (∼0.3–0.4 km/Myr, underway by at least ∼49 Ma). Rates then likely decrease slightly to ∼0.25 km/Myr (group "R"), before increasing slightly to ∼0.3 km/Myr (group 2). Using

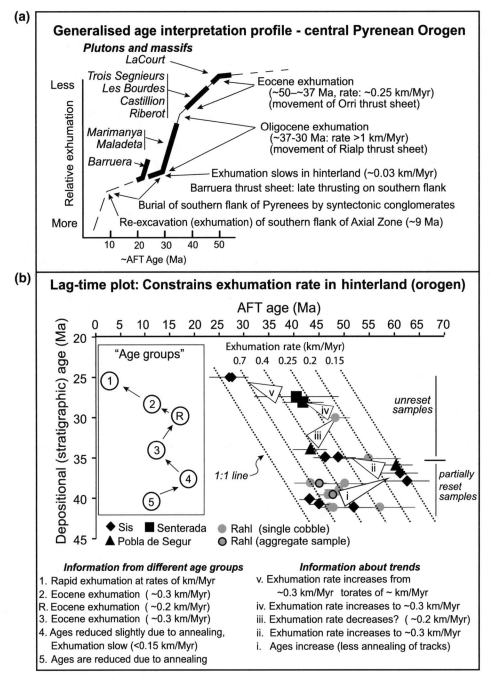

Fig. 17.3 **a** Generalised age profile interpretation for the Pyrenean orogen along the ECORS profile. This profile gives a simplified thermal and tectonic history of the central Pyrenees as revealed in thermochronology data, but it should not be interpreted as all profiles having undergone identical histories. Modified from Fitzgerald et al. (1999) with information from Sinclair et al. (2005), Gibson et al. (2007) and Fillon and van der Beek (2012). **b** Lag-time plot for samples from the Senterada, Pobla de Segur and Sierra de Sis basins (modified from Beamud et al. 2011)

information from this lag-time plot only, it can be seen that rates increased dramatically (order of km/Myr) sometime after ~40 Ma, and certainly by at least ~30 Ma (group 1) (Fig. 17.3). The calculation of exhumation rates for group 1 samples are almost certainly affected by advection in the hinterland during such rapid exhumation. Thus, the exhumation rate for group 1 samples is simply constrained as "on the order of km/Myr".

Fig. 17.4 Representative inverse thermal models (HeFTy, Ketcham 2005) from each "Group" plotted against stratigraphic position (diagram modified from Beamud et al. 2011). The models are used to constrain the time of "rapid cooling" events, either in the hinterland (before deposition) or within the South Pyrenean Foreland Basin (basin inversion) following deposition and then partial annealing. For the models, the stratigraphic ages of the sample are marked by a black square, which is a "hard" constraint for these models

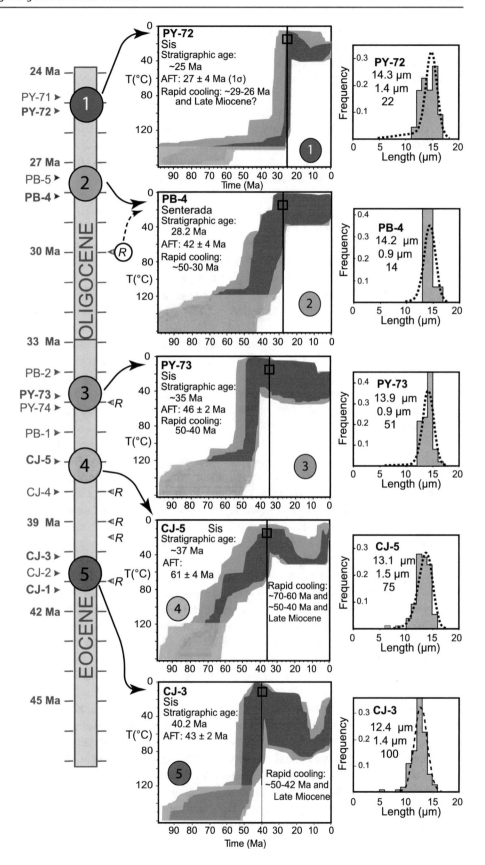

Thermal models of the cobbles (Fig. 17.4) constrain both the pre-depositional rapid cooling episodes, plus a period of late Miocene cooling following burial of the sediments in the basin. Pre-depositional cooling events within the orogen are:

(i) A 70–60 Ma event is poorly defined in group 4 thermal models, with a cooling rate of $\sim$5 °C/Myr. The exhumation rate is unconstrained from the lag-time plot as these samples are partially annealed and this cooling event is part of a complex thermal history. However, this event likely relates to relatively slow exhumation (<0.3 km/Myr).

(ii) A $\sim$50–40 Ma cooling event is evident in models from groups 2, 3, 4 and 5. The strength of the signal is dependent on the quality of age and length data from each cobble. For the Rahl et al. samples, this event is evident in all models except one. Cooling rates are $\sim$6–10 °C/Myr and exhumation rates are $\sim$0.2–0.3 km/Myr.

(iii) A $\sim$30–25 Ma rapid cooling event (group 1) has cooling rates of $\sim$30 °C/Myr and exhumation rates >1 km/Myr.

(iv) Late Miocene cooling is revealed in group 4 and 5 models, starting $\sim$<10 Ma at averaged rates of 5–10 °C/Myr.

Another way to visualise the changing thermal histories through geologic time is to plot the thermal history (rapid cooling episodes, burial and annealing) derived from the inverse models with respect to stratigraphic age (Fig. 17.5a). Rapid cooling episodes in the thermal models must be older than the depositional age, except for models with Late Miocene cooling where samples were buried, partially annealed and then later re-excavated. Thus, only cobbles from the deepest strata (groups 4 and 5) reveal the youngest cooling event. However, as a result of burial-related partial annealing, the oldest tracks recording the oldest cooling episode ($\sim$70–60 Ma) have been lost in group 5 samples, yet enough of these old tracks are still preserved in group 4 samples to reveal the older 70–60 Ma event. The dominant cooling episode (from $\sim$50 to 40 Ma) is revealed in groups 2 through 5, but these groups do not record the strong exhumation cooling beginning $\sim$35 Ma, as these samples were already exhumed by the time of this event.

Lessons learned from comparison between the Beamud et al. and Rahl et al. studies: These two studies have similar approaches and similar data, but they emphasise slightly different factors as they arrive at a fairly consistent story. Beamud et al. had a more variable exhumation history, whereas the Rahl et al. exhumation history was simpler. A difficult part with the interpretation of the detrital data is what represents a real signal and what represents noise

(dispersion in the thermochronology data). Sampling multiple cobbles at regular intervals over the greatest stratigraphic interval obviously provides more information. Beamud et al. sampled higher in the sequence, which turned out to be crucial as these youngest samples revealed the youngest pulse of orogenic exhumation. In addition, this meant Beamud et al. had a greater proportion of their samples that were not partially annealed (by burial) and hence could be interpreted in terms of the lag-time approach, whereas Rahl et al. had only two samples that were not partially annealed (Fig. 17.3b). Beamud et al. tended to have clusters of samples at each stratigraphic level, and all their AFT ages from each level had similar AFT ages, so they were more confident in their interpretation of a variable exhumation history. For example, for group 4 cobbles (stratigraphic age of $\sim$37 Ma), Beamud et al. had three cobbles in that group with similar AFT ages, and they regarded this as a real signal. In contrast, Rahl et al. had one cobble near group 4, but regarded this age as not representative as it lay off a simple trend. In another example, one of the Rahl et al. cobbles (stratigraphic age of $\sim$30 Ma, between group 2 and 3 cobbles) from a level not sampled by Beamud et al., further constrains the variable exhumation history. The combined cobble datasets also addresses issues raised in Sect. 17.3 about "the number of cobbles from a stratigraphic horizon needed to provide a representative sample of the hinterland exhumation". To illustrate this point with respect to cobble age variation at one stratigraphic level, partially reset cobbles from a stratigraphic age of $\sim$41 Ma have AFT ages that range from about 42 to 55 Ma. The oldest AFT age from this stratigraphic level lacks a younger age population of single-grain ages, which suggests this sample was likely derived from higher in the paleo-landscape than the other samples from this same level (Rahl et al. 2011). Rahl et al. (2011) also showed that AFT ages from individual cobbles are similar to aggregate samples from the stratigraphic horizons, indicating that the cobble samples are representative of the landscape exposed at the time. As suggested above, a possible method to investigate possible ambiguities with cobble data is to also undertake AFT thermochronology on the matrix to examine single-grain age variation.

Comparison of detrital cobble thermochronology to in-situ thermochronology data: When compared, there is a good agreement in both the timing and rates of exhumation events between the thermochronology data from the orogen and the detrital data (Fig. 17.5b). All pre-Miocene exhumation events have been related to erosion associated with movement of the thrust sheets (e.g. Muñoz 1992; Fitzgerald et al. 1999; Beaumont et al. 2000; Gibson et al. 2007; Metcalf et al. 2009; Rahl et al. 2011), first movement of the Nogueres thrust sheet ($\sim$70–60 Ma), then the Orri thrust sheet ($\sim$50–35 Ma) and finally the lowermost Rialp

Fig. 17.5 a Model-constrained rapid cooling events versus stratigraphic age plot for representative samples from each of the five groups. The grey boxes are the timing of the rapid cooling events constrained from HeFTy models (see Fig. 17.4) of the individual cobbles, and the blue horizontal boxes indicate which rapid cooling events are recorded in each sample's model. The orange line marks the 1:1 line where stratigraphic age = AFT age. Models will only record cooling events younger than the stratigraphic age if they are "partially reset" by burial in the basin (as shown by samples in the lower two groups 4 and 5). **b** Comparison between hinterland thermochronologic-constrained exhumation and that derived from thermochronology on South Pyrenean Foreland Basin cobbles collected within a stratigraphic framework. Data from the hinterland are mainly from in-situ samples from granitic plutons within the Axial Zone: Fitzgerald et al. (1999), Sinclair et al. (2005), Gibson et al. (2007), Metcalf et al. (2009), Fillion et al. (2013). Data from basin cobbles are summarised from Beamud et al. (2011). **c** Schematic (and simplified) figure showing the different (but similar) approaches used in the hinterland and the detrital (stratigraphic framework) approach

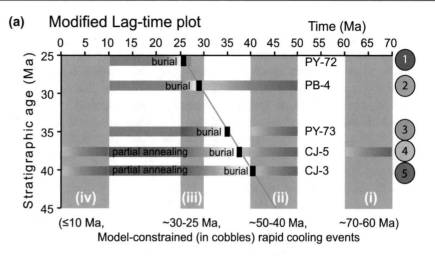

(a) Modified Lag-time plot

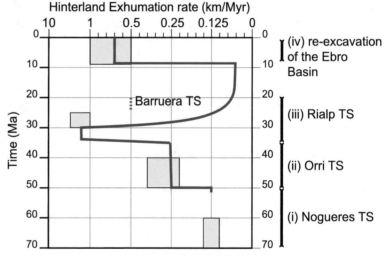

(b) Comparison between hinterland vs basin thermochronology

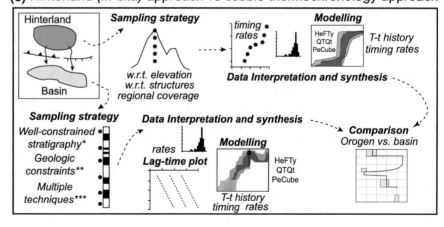

(c) Hinterland *(in-situ)* approach vs cobble thermochronology approach

thrust sheet (~35–20 Ma) that formed the antiformal stack within the Axial Zone (Fig. 17.2). As mentioned above, the poorly defined cooling/exhumation event at ~70–60 Ma has not been revealed with in-situ thermochronology data from the orogen. However, detrital zircon (U–Th)/He ages of ~80–68 Ma are found within sandstones (deposition age ~60–40 Ma) on the southern flank of the Pyrenees, although these samples were likely exhumed into the upper crust in the Late Cretaceous and then exhumed to the surface later (Filleaudeau et al. 2012). Exhumation at ~25–20 Ma related to thrusting of the Barruera thrust sheet (Gibson et al. 2007) is not revealed in the detrital data because that thrusting occurred after deposition of our youngest cobbles, and at the time of thrusting the Barruera Massif was already covered by the syntectonic conglomerates (Beamud et al. 2011). Burial of the southern fold-and-thrust belt by syntectonic conglomerates and later re-excavation since the Late Miocene (Coney et al. 1996) is revealed in the inverse thermal models (Fig. 17.4) and thermo-kinematic models (Fillon and van der Beek 2012).

17.5 The Bergell-Gonfolite Source-to-Sink System

17.5.1 Geology of the Bergell-Gonfolite Area

The Bergell-Gonfolite source-to-sink system (Fig. 17.6a) provides another example of a detrital thermochronology study using conglomerates and cobbles, from the area where the concept of blocking temperature was first defined (Jäger 1967). The Oligocene Bregaglia/Bergell pluton was intruded on top of the subducting European slab within Alpine metamorphic rocks at 30 ± 2 Ma (von Blanckenburg 1992; Oberli et al. 2004; Malusà et al. 2015). The smaller Novate granite was intruded in the same area at 25–24 Ma (Liati et al. 2000). These rocks were progressively eroded together with their country rock providing detritus to the Oligocene–Miocene turbidites of the Gonfolite Group, which is now exposed south of the Alps (Wagner et al. 1979; Giger and Hurford 1989; Spiegel et al. 2004; Malusà et al. 2011, 2016a). Widespread volcanic detritus at the base of the stratigraphic succession indicates that volcanoes were present above the Bergell pluton (Malusà et al. 2011; Anfinson et al. 2016). Estimates of unroofing depth provided by hornblende geobarometry (Davidson et al. 1996) range from ~20 km in the Bergell main body, to ~26 km in its western tail. Differential exhumation of the Bergell pluton is ascribed to N–S post-magmatic tilting, constrained to between ~25 and 16 Ma by bedrock thermochronologic data (Wagner et al. 1977, 1979; Hurford 1986; Rahn 2005; Malusà et al. 2011). The first significant detrital pulse in the Gonfolite basin marks the onset of rapid erosion in the

Bergell source area. This pulse is dated biostratigraphically at ~25 Ma (Gelati et al. 1988). This rapid sedimentation is delayed by ~5 Myr relative to the main magmatic pulse, which is instead associated to negligible erosion and starved sedimentation in the basin (Garzanti and Malusà 2008; Malusà et al. 2011). Compared to the southern Pyrenean fold-and-thrust belt, which involves magmatic rocks intruded during the Variscan orogeny, particular care is required in the Bergell-Gonfolite case to distinguish between the effects of magmatic crystallisation, and the effects of exhumational cooling during erosion and deposition of Bergell-derived cobbles in the Gonfolite basin (Chap. 8, Malusà and Fitzgerald 2018a, b).

17.5.2 Detrital Thermochronology of the Gonfolite Group and Lessons Learned from the Bergell-Gonfolite Source-to-Sink System

The diagram in Fig. 17.6b summarises all the thermochronologic ages measured in cobbles of the Gonfolite Group, including AFT, ZFT, biotite K–Ar and zircon U–Pb ages, often from the same cobbles. Ages are shown as full dots (for magmatic clasts) or empty dots (for country rock clasts) with 1σ errors indicated by horizontal bars. These ages have long been interpreted exclusively in terms of cooling during exhumation (Jäger 1967; Giger and Hurford 1989; Bernoulli et al. 1993; Fellin et al. 2005; Carrapa 2009). Because several magmatic cobbles from the basal strata of the Gonfolite Group yielded indistinguishable K–Ar, ZFT and AFT ages within error, this suggested a very fast erosion of the Bergell pluton shortly after its emplacement (Giger and Hurford 1989; Carrapa and Di Giulio 2001). However, such interpretation was in conflict with the compelling evidence of negligible erosion provided by the stratigraphic record. Previous thermochronologic interpretations exclusively considering cooling during exhumation were also unable to explain a substantial part of the thermochronologic dataset of Fig. 17.6b. In the basal strata of the Gonfolite Group (units A and B in Fig. 17.6b), ages yielded by country rock clasts are in fact systematically older than the ages yielded by magmatic clasts from the same strata, unlike expected in the case of cooling during exhumation of both country rock and encased magmatic rocks.

Malusà et al. (2011), by integrating geologic evidence from both the source and the sink areas that has become available over the last 30 years, demonstrated that the above interpretation of the Gonfolite thermochronologic record was incorrect. They showed that a pulse of shallow magmatism and hence rapid cooling was responsible for multiple thermochronologic methods yielding almost identical ages, thus

Fig. 17.6 a Sketch map of the Bergell-Gonfolite source-to-sink system. Lakes Como and Maggiore occupy Oligo–Miocene paleo-valleys funnelling detritus towards the Gonfolite basin (simplified after Malusà et al. 2011; hornblende geobarometry data after Davidson et al. 1996; biotite K–Ar ages after Villa and von Blanckenburg 1991).
b Observed age pattern in cobbles of the Gonfolite Group (modified from Malusà et al. 2011, and references therein). Full and empty dots indicate magmatic and country rock clasts, respectively (see colour coding on the top-right). Fission-track grain age distributions from detrital zircon grains are shown in boxes (after Spiegel et al. 2004). Three mineral-age units (A–C) are detected within the Gonfolite Group. Ages in unit A provide no direct information on exhumation rate

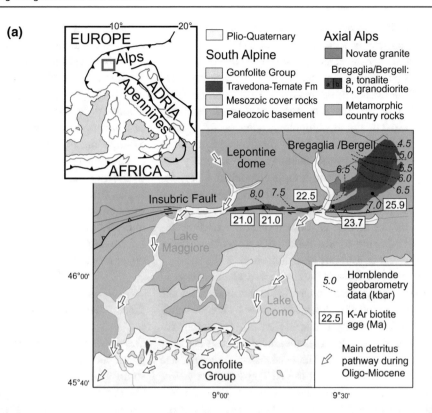

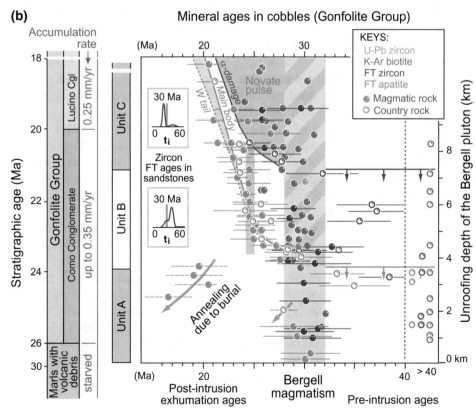

resolving the long-lasting paradox of fast erosion without detrital counterparts in adjacent foreland basins.

The cooling ages yielded by the Gonfolite clasts may either record (i) the pre-intrusion history of the country rock, (ii) the cooling history of the magmatic rock during magma crystallisation or (iii) the subsequent erosion of the magmatic complex. The thermochronologic ages recording these processes define a complex, but fully predictable detrital age pattern in sedimentary rocks, including both stationary and moving age peaks ("magmatic complex model" in Chap. 16, Malusà 2018). Ages belonging to the moving age peaks are generally set during exhumation and can provide direct constraints on the motion of rocks toward the Earth's surface. Ages belonging to the stationary age peaks, instead, provide no direct constraints on exhumation.

Trends of stationary and moving age peaks expected in the Gonfolite Group, based on available thermochronologic constraints from eroded bedrock, are shown by colour bands in Fig. 17.6b (blue for AFT, purple for ZFT, and orange for biotite K–Ar). Noteworthy, ages measured in clasts are fully consistent with these trends. The existence of a moving age peak is particularly evident for the AFT system that yields, in all of the clasts from strata younger than 24 Ma, exhumation ages that get progressively younger up-section. A broader moving age peak is also marked by ZFT ages in strata younger than 21 Ma. Within this moving age peak, ZFT ages in country rock clasts are slightly younger than in magmatic clasts, because fission tracks in zircons are generally reset at lower temperature in these older rocks, due to the greater amount of accumulated α-damage (Rahn et al. 2004). The existence of a stationary age peak is observed in Fig. 17.6b for both the main Bergell magmatic pulse and for the minor Novate magmatic pulse, which is recorded by magmatic clasts yielding ages of 25–24 Ma. Ages belonging to these peaks record magmatic crystallisation, not exhumation, whereas the oldest ages recorded by country rock clasts in units A and B record the pre-intrusion history of the country rock. The age distribution in magmatic and country rock clasts allows the recognition of three different units (A–C) in the stratigraphic succession, thus providing an example of reverse mineral-age stratigraphy (see Chap. 10, Malusà and Fitzgerald 2018b). Units A–C mark the different steps of the unroofing history of the Bergell volcano-plutonic complex and provide a reference frame for detecting tectonic repetitions by thrusting in the Gonfolite Group (Fig. 17.6a).

The magmatic complex model first proposed by Malusà et al. (2011) thus provides a coherent explanation for the whole thermochronologic dataset of the Bergell-Gonfolite system and reconciles geochronological data from cobbles with stratigraphic evidence. It also shows that a reliable interpretation of thermochronologic ages in cobbles may require the analysis of cobbles from different lithologies, and from different levels of the stratigraphic succession, in order to correctly identify stationary and moving age peaks. The analysis of cobbles exclusively from plutonic rocks, which are often preferred for thermochronologic analysis because of the higher percentage and quality of datable minerals compared to their country rocks, may lead to incorrect interpretations of magmatic cooling ages in terms of fast exhumation. Figure 17.6b also shows that the same amount of information provided by multiple methods on cobbles would not be provided by detrital ZFT analysis of sandstones (boxes in Fig. 17.6b), even if ZFT single-grain ages were coupled with U–Pb dating on the same grains. This indicates the extent that different methods provide differing information, dependant on the questions being asked and the objectives of the study.

17.6 Summary and Conclusions

Detrital thermochronology data from cobbles within conglomerates from foreland basins has the potential to constrain well the exhumation history of the adjacent orogen. Overlapping as well as complimentary information can be gained by using detrital thermochronology as well as in-situ bedrock sampling. An obvious corollary is that there is much greater confidence in the interpretation of detrital data if there is complementary hinterland data, as well as supporting geological information (e.g. basin analysis and structural restorations) with which to compare data sets. For each approach (bedrock, detrital), we stress the need to follow a well-thought out sampling strategy based on existing geological knowledge, followed by an interpretation of the data before modelling is undertaken, as this allows better evaluation of resulting best-fit $T - t$ envelopes (Fig. 17.5c). We note that detrital thermochronology may constrain older (earlier) exhumation events that are not revealed by in-situ thermochronology data as bedrock with that information may have been eroded off the orogen. In contrast, the detrital record may miss younger exhumation events, as this material has not yet eroded off the orogen. This is where multiple techniques on the same samples will prove useful.

When applying detrital thermochronology in a stratigraphic sequence, the broader the stratigraphic interval sampled means a better control on the exhumation history. AFT single-grain ages typically have large errors due to lower [U], relative to zircon, for example. An advantage to using cobbles is that all grains share a common thermal history. Using AFT ages and confined track-length measurements allows better constraints on pre-depositional cooling (exhumation) events as well as post-depositional history. Lag-time plots are useful to constrain an integrated exhumation rate, assuming that (i) the stratigraphic age of the sample is well known, (ii) ages represent a "closure age", which can be evaluated using track-length distributions and thermal models, and (iii) the time from erosion to deposition is negligible. Inverse thermal models in conjunction with the lag-time plot can be used to constrain the timing of cooling (exhumation) episodes and hence the exhumation history of the hinterland. The detrital exhumation signal can be blurred due to a number of factors: (i) partial annealing in the basin, (ii) cobbles sourced from different provenances or over significant paleo-relief resulting in cobbles at a single stratigraphic horizon that may have different ages/lengths and hence exhumation histories, (iii) samples exhumed higher into the crust during one event but not to the surface; thus, they are not eroded and deposited until the next exhumation event, (iv) samples stored in transient basins during transportation, (v) poor quality data resulting, for example from weathered samples or young samples/few tracks and/or poorly constrained thermal models.

We have summarised and synthesised AFT data from Eocene and Oligocene syntectonic conglomerates (Beamud et al. 2011; Rahl et al. 2011) and compared this data to in-situ thermochronology data from within the Pyrenean orogen (e.g. Fitzgerald et al. 1999; Gibson et al. 2007; Metcalf et al. 2009; Fillion et al. 2013). Lag-time plots in combination with inverse thermal modelling constrain two well-defined episodes of rapid cooling at ∼50–40 and ∼30–25 Ma, with a poorly defined episode at ∼70–60 Ma. These episodes relate to exhumation associated with convergence between Iberia and Europe, resultant shortening, deformation, uplift and progressive erosion of different thrust sheets. The oldest and most deeply buried samples have undergone partial annealing associated with that burial, followed by late Miocene cooling due to the re-excavation of fold-and-thrust belts by the Ebro River. In the Bergell region of the Alps, detrital thermochronology on cobbles from the Gonfolite basin indicates the onset of erosion at ∼25 Ma in the source area. A moving peak of exhumation ages that get progressively younger up-section is revealed, as is expected with progressive unroofing of the source region. Previous interpretations had rapid exhumation at ∼30 Ma, but this coincides with the main Bergell magmatism, as is shown by a stationary peak of that age in progressively younger strata, associated with older ages in cobbles of country rocks.

Acknowledgements Fitzgerald acknowledges support from grants NSF grants EAR95-06454 and EAR05-38216 that initiated this work on detrital cobble thermochronology. Fitzgerald also acknowledges Jarg Pettinga and the Erskine Program at the University of Canterbury where much of this paper was written. Malusà thanks E. Garzanti and I. M. Villa for insightful discussions on the Bergell-Gonfolite system. Muñoz acknowledges support from the SALTECRES project (CGL2014-54118-C2-1-R MINECO/FEDER, UE) as well as the Grup de Recerca de Geodinàmica i Anàlisi de Conques (2014SRG467). This paper has benefitted from discussions with Suzanne Baldwin. Thorough and thoughtful reviews by Jeff Rahl and Peter van der Beek greatly improved this paper, and we acknowledge their suggestions and comments with thanks.

References

Amidon W, Burbank D, Gehrels G (2005a) Construction of detrital mineral populations: insights from mixing of U–Pb zircon ages in Himalayan rivers. Basin Res 17:463–485

Amidon W, Burbank D, Gehrels G (2005b) U–Pb zircon ages as a sediment mixing tracer in the Nepal Himalaya. Earth Planet Sci Lett 235:244–260

Andersen T (2005) Detrital zircons as tracers of sedimentary provenance: limiting conditions from statistics and numerical simulation. Chem Geol 216:249–270

Anfinson OA, Malusà MG, Ottria G, Dafov LN, Stockli DF (2016) Tracking coarse-grained gravity flows by LASS-ICP-MS depth-profiling of detrital zircon (Aveto Formation, Adriatic foredeep, Italy). Mar Petrol Geol 77:1163–1176

Baldwin SL, Lister GS (1998) Thermochronology of the South Cyclades shear zone, Ios, Greece; effects of ductile shear in the argon partial retention zone. J Geophys Res 103:7315–7336

Baldwin SL, Harrison TM, Burke K (1986) Fission track evidence for the source of Scotland District sediments, Barbados and implications for post-Eocene tectonics of the southern Caribbean. Tectonics 5:457–468

Barso D (2007) Analisis de la procedencia de los conglomerados sinorogenicos de La Pobla de Segur (Lerida) y su relacion con la evolucion tectonica de los Pirineos centro-meridionales durante el Eoceno medio-Oligoceno. University of Barcelona, p 209

Beamud E, Garcés M, Cabrera L, Muñoz JA, Almar Y (2003) A new middle to late Eocene continental chronostratigraphy from NE Spain. Earth Planet Sci Lett 216:501–514

Beamud E, Muñoz JA, Fitzgerald PG, Baldwin SL, Garcés M, Cabrera L, Metcalf JR (2011) Magnetostratigraphy and detrital apatite fission track thermochronology in syntectonic conglomerates: constraints on the exhumation of the South-Central Pyrenees. Basin Res 23:309–331

Beaumont C, Muñoz JA, Hamilton J, Fullsack P (2000) Factors controlling the Alpine evolution of the central Pyrenees from a comparison of observations and geodynamical models. J Geophys Res 105:8121–8145

Bernet M (2018) Exhumation studies of mountain belts based on detrital fission-track analysis on sand and sandstones. In: Malusà MG, Fitzgerald PG (eds) Fission-track thermochronology and its application to geology. Springer, Berlin (Chapter 15)

Bernet M, Garver JI (2005) Fission-track analysis of detrital zircon. Rev Mineral Geochem 58(1):205–237

Bernet M, Spiegel C (2004) Introduction: detrital thermochronology. In: Bernet M, Spiegel C (eds) Detrital thermochronology—provenance analysis, exhumation, and landscape evolution of mountain belts. Geol S Am S 378:1–6

Bernet M, Zattin M, Garver JI, Brandon MT, Vance JA (2001) Steady-state exhumation of the European Alps. Geology 29:35–38

Bernet M, Brandon M, Garver J, Molitor B (2004) Fundamentals of detrital zircon fission-track analysis for provenance and exhumation studies with examples from the European Alps. Geol Soc Am Spec 378:25–36

Bernoulli D, Giger M, Müller DW, Ziegler URF (1993) Sr-isotope-stratigraphy of the Gonfolite Lombarda Group ("South-Alpine Molasse", northern Italy) and radiometric constraints for its age of deposition. Eclogae Geol Helv 86:751–767

Brandon MT (1996) Probability density plot for fission-track grain-age samples. Radiat Meas 26:663–676

Brandon MT, Vance JA (1992) Tectonic evolution of the Cenozoic Olympic subduction complex, Washington State, as deduced from fission track ages for detrital zircons. Am J Sci 292:565–636

Braun J, van der Beek P, Batt G (2006) Quantitative thermochronology: numerical methods for the interpretation of thermochronological data. Cambridge University Press, Cambridge

Brown RW (1991) Backstacking apatite fission-track "stratigraphy": a method for resolving the erosional and isostatic rebound components of tectonic uplift histories. Geology 19(1):74–77

Brügel A, Dunkl I, Frisch W, Kuhlemann J, Balogh K (2004) Geochemistry and geochronology of gneiss pebbles from foreland molasse conglomerates: geodynamic and paleogeographic implications for the Oligo-Miocene evolution of the Eastern Alps. J Geol 111:543–563

Burbank D, Puigdefabregas C, Muñoz JA (1992) The chronology of the Eocene tectonic and stratigraphic development of the eastern Pyrenean foreland basin, northeast Spain. Geol Soc Am Bull 104:1101–1120

Burnham A, Sweeney J (1989) A chemical kinetic model of vitrinite maturation and reflectance. Geochim Cosmochim Acta 53:2649–2657

Carrapa B (2009) Tracing exhumation and orogenic wedge dynamics in the European Alps with detrital thermochronology. Geology 37:1127–1130

Carrapa B, Di Giulio A (2001) The sedimentary record of the exhumation of a granitic intrusion into a collisional setting: the lower Gonfolite Group, Southern Alps, Italy. Sed Geo 139:217–228

Carrapa B, Wijbrans JR, Bertotti G (2004) Detecting provenance variations and cooling patterns within the western Alpine orogen through $^{40}Ar/^{39}Ar$ geochronology on detrital sediments: the tertiary piedmont basin, NW Italy. Geol Soc Am Spec 378:67–103

Carter A (2018) Thermochronology on sand and sandstones for stratigraphic and provenance studies. In: Malusà MG, Fitzgerald PG (eds) Fission-track thermochronology and its application to geology. Springer, Berlin (Chapter 14)

Cawood PA, Nemchin A (2001) Paleogeographic development of the East Laurentian margin: constraints from U–Pb dating of detrital zircon in the Newfoundlan Appalachians. Geol Soc Am Bull 113:1234–1246

Cerveny PF, Naeser ND, Zeitler PK, Naeser CW, Johnson NM (1988) History of uplift and relief of the Himalaya during the past 18 million years: evidence from fission-track ages of detrital zircons from sandstones of the Siwalik Group. New perspectives in basin analysis. Springer, New York, pp 43–61

Coney PJ, Muñoz JA, McClay K, Evenchick CA (1996) Syn-tectonic burial and post-tectonic exhumation of an active foreland thrust belt, southern Pyrenees, Spain. J Geol Soc 153:9–16

Costa E, Garcés M, López-Blanco M, Beamud E, Gómez-Paccard M, Larrasoaña JC (2010) Closing and continentalization of the South Pyrenean foreland basin (NE Spain): magnetochronological constraints. Basin Res 22:904–917

Davidson C, Rosenberg C, Schmid SM (1996) Symmagmatic folding of the base of the Bergell pluton, Central Alps. Tectonophysics 265(3):213–238

DeCelles PG (1988) Lithologic provenance modeling applied to the Late Cretaceous synorogenic Echo Canyon Conglomerate, Utah: a case of multiple source areas. Geology 16:1039–1043

Dodson MH (1973) Closure temperatures in cooling geochronological and petrological systems. Contrib Mineral Petr 40:259–274

Dunkl I, Székely B (2002) Component analysis with visualization of fitting—PopShare, a Windows program for data analysis. In: Goldschmidt conference abstracts 2002. Geochim Cosmochim Acta 66/A:201

Dunkl I, Frisch W, Kuhlemann J, Brügel A (2009) Pebble population dating as an additional tool for provenance studies-examples from the Eastern Alps. Geol Soc Spec Publ 324:125–140

Enkelmann E, Koons PO, Pavlis TL, Hallet B, Barker A, Elliott J, Ridgway KD (2015) Cooperation among tectonic and surface processes in the St. Elias Range, earth's highest coastal mountains. Geophysl Res Lett 42:5838–5846

Epstein AG, Epstein JB, Harris LD (1977) Conodont color alteration—an index to organic metamorphism. Boulder, Colorado, pp 1–27

Falkowski S, Enkelmann E, Drost K, Pfänder JA, Stübner K, Ehlers TA (2016) Cooling history of the St. Elias syntaxis, southeast Alaska, revealed by geo-and thermochronology of cobble-sized glacial detritus. Tectonics 8:359–378

Fellin MG, Sciunnach D, Tunesi A, Andò S, Garzanti E, Vezzoli G (2005) Provenance of detrital apatites from the upper Gonfolite Lombarda Group (Miocene, NW Italy). GeoActa 4:43–56

Filleaudeau PY, Mouthereau F, Pik R (2012) Thermo-tectonic evolution of the south-central Pyrenees from rifting to orogeny: insights from detrital zircon U/Pb and (U–Th)/He thermochronometry. Basin Res 24:401–417

Fillon C, van der Beek P (2012) Post-orogenic evolution of the southern Pyrenees: constraints from inverse thermo-kinematic modelling of low-temperature thermochronology data. Basin Res 24:418–436

Fillion C, Gautheron C, van der Beek P (2013) Oligocene-Miocene burial and exhumation of the Southern Pyrenean foreland quantified by low-temperature thermochronology. J Geol Soc 107:67–77

Fitzgerald PG, Muñoz JA, Coney PJ, Baldwin SL (1999) Asymmetric exhumation across the Pyrenean orogen: implications for the tectonic evolution of collisional orogens. Earth Planet Sci Lett 173:157–170

Fox M, Reverman R, Herman F, Fellin MG, Sternai P, Willett SD (2014) Rock uplift and erosion rate history of the Bergell intrusion from the inversion of low temperature thermochronometric data. Geochem, Geophy, Geosyst 15(4):1235–1257

Galbraith RF (2005) Statistics for fission track analysis, 1st edn. Chapman and Hall/CRC, Boca Raton, Florida

Galbraith RF, Green PF (1990) Estimating the component ages in a finite mixture. Nucl Tracks Rad Meas 17:197–206

Gallagher K (2012) Transdimensional inverse thermal history modeling for quantitative thermochronology. J Geophys Res 117(B2)

Garver JI, Brandon MT, Roden-Tice MK, Kamp PJJ (1999) Exhumation history of orogenic highlands determined by detrital fission track thermochronology. Geol Soc Spec Publ 154:283–304

Garzanti E, Malusà MG (2008) The oligocene alps: domal unroofing and drainage development during early orogenic growth. Earth Planet Sci Lett 268:487–500

Gehrels G (2014) Detrital zircon U–Pb geochronology applied to tectonics. Ann Rev Earth Pl Sci 42:127–149

Gelati R, Napolitano A, Valdisturlo A (1988) La "Gonfolite Lombarda": stratigrafia e significato nell'evoluzione del margine sudalpino. Riv It Paleont Strat 94:285–332

Gibson M, Sinclair H, Lynn G, Stuart F (2007) Late-to post-orogenic exhumation of the Central Pyrenees revealed through combined thermochronological data and modelling. Basin Res 19:323–334

Giger M, Hurford AJ (1989) Tertiary intrusives of the Central Alps: their Tertiary uplift, erosion, redeposition and burial in the south-alpine foreland. Eclogae Geol Helv 82:857–866

Gleadow AJW, Fitzgerald PG (1987) Uplift history and structure of the Transantarctic Mountains: new evidence from fission track dating of basement apatites in the Dry Valleys area, southern Victoria Land. Earth Planet Sci Lett 82(1):1–14

Gleadow AJW, Brown RW (2000) Fission track thermochronology and the long term denudational response to tectonics. In: Summerfield MA (ed) Geomorphology and global tectonics. Willey, New York, pp 57–75

Hurford AJ (1986) Cooling and uplift patterns in the Lepontine Alps South Central Switzerland and an age of vertical movement on the Insubric fault line. Contrib Mineral Petr 92:413–427

Hurford AJ, Carter A (1991) The role of fission track dating in discrimination of provenance. Geol Soc Spec Publ 57:67–78

Hurford AJ, Fitch FJ, Clarke A (1984) Resolution of the age structure of the detrital zircon populations of two Lower Cretaceous sandstones from the Weald of England by fission track dating. Geol Mag 121:269–277

Jäger E (1967) Die Bedeutung der Biotit-Alterswerte. In: Jäger E, Niggli E, Wenk E (eds) Rb-Sr Alterbestmmungen an Glimmern der Zentralalpen Beitr. Geol Kaarte Schweiz, NF, pp 28–31

Jammes S, Manatschal G, Lavier L, Masini E (2009) Tectonosedimentary evolution related to extreme crustal thinning ahead of a propagating ocean: example of the western Pyrenees. Tectonics 28: TC4012

Jammes S, Huismans R, Muñoz J (2014) Lateral variation in structural style of mountain building: controls of rheological and rift inheritance. Terra Nova 26:201–207

Jones MA, Heller PL, Roca E, Garcés M, Cabrera L (2004) Time lag of syntectonic sedimentation across an alluvial basin: theory and example from the Ebro Basin, Spain. Basin Res 16:489–506

Ketcham RA (2005) Forward and inverse modeling of low temperature thermochronometry data. Rev Mineral Geochem 58:275–314

Ketcham R (2018) Fission track annealing: from geologic observations to thermal history modeling. In: Malusà MG, Fitzgerald PG (eds) Fission-track thermochronology and its application to geology. Springer, Berlin (Chapter 3)

Ketcham RA, Donelick RA, Carlson WD (1999) Variability of apatite fission track annealing kinetics III: extrapolation to geological time scales. Am Mineral 84:1235–1255

Ketcham R, Carter A, Donelick R, Barbarand J, Hurford A (2007) Improved modeling of fission-track annealing in apatite. Am Mineralt 92:799

Lagabrielle Y, Labaume P, De Saint Blanquat M (2010) Mantle exhumation, crustal denudation, and gravity tectonics during Cretaceous rifting in the Pyrenean realm (SW Europe): insights from the geological setting of the lherzolite bodies. Tectonics 29: TC4012

Liati A, Gebauer D, Fanning M (2000) U–Pb SHRIMP dating of zircon from the Novate granite (Bergell, Central Alps): evidence for Oligocene-Miocene magmatism, Jurassic/Cretaceous continental rifting and opening of the Valais trough. Schweiz Mineral Petr Mitt 80:305–316

Lovera OM, Grove M, Harrison TM (2002) Systematic analysis of K-feldspar $^{40}Ar/^{39}Ar$ step heating results II: Relevance of laboratory argon diffusion properties to nature. Geochim Cosmochim Acta 66:1237–1255

Mahéo G, Gautheron C, Leloup PH, Fox M, Tassant-Got L, Douville E (2013) Neogene exhumation history of the Bergell massif (southeast Central Alps). Terra Nova 25:110–118

Malusà MG (2018) A guide for interpreting complex detrital age patterns in stratigraphic sequences. In: Malusà MG, Fitzgerald PG (eds) Fission-track thermochronology and its application to geology. Springer, Berlin (Chapter 16)

Malusà MG, Fitzgerald PG (2018a) From cooling to exhumation: setting the reference frame for the interpretation of thermocronologic data. In: Malusà MG, Fitzgerald PG (eds) Fission-track thermochronology and its application to geology. Springer, Berlin (Chapter 8)

Malusà MG, Fitzgerald PG (2018b) Application of thermochronology to geologic problems: bedrock and detrital approaches. In: Malusà

MG, Fitzgerald PG (eds) Fission-track thermochronology and its application to geology. Springer, Berlin (Chapter 10)

Malusà MG, Garzanti E (2018) The sedimentology of detrital thermochronology. In: Malusà MG, Fitzgerald PG (eds) Fission-track thermochronology and its application to geology. Springer, Berlin (Chapter 7)

Malusà MG, Villa IM, Vezzoli G, Garzanti E (2011) Detrital geochronology of unroofing magmatic complexes and the slow erosion of oligocene volcanoes in the Alps. Earth Planet Sci Lett 301:324–336

Malusà MG, Carter A, Limoncelli M, Villa IM, Garzanti E (2013) Bias in detrital zircon geochronology and thermochronometry. Chem Geol 359:90–107

Malusà MG, Faccenna C, Baldwin SL, Fitzgerald PG, Rossetti F, Balestrieri ML, Ellero A, Ottria G, Piromallo C (2015) Contrasting styles of (U) HP rock exhumation along the Cenozoic Adria-Europe plate boundary (Western Alps, Calabria, Corsica). Geochem Geophys Geosyst 16(6):1786–1824

Malusà MG, Anfinson OA, Dafov L, Stockli D (2016a) Tracking Adria indentation beneath the Alps by detrital zircon U–Pb geochronology: implications for the oligocene-miocene dynamics of the adriatic microplate. Geology 44:155–158

Malusà MG, Resentini A, Garzanti E (2016b) Hydraulic sorting and mineral fertility bias in detrital geochronology. Gondwana Res 31:1–19

Mancktelow NS, Grasemann B (1997) Time-dependent effects of heat advection and topography on cooling histories during erosion. Tectonophysics 270(3):167–195

Meigs AJ, Vergés J, Burbank DW (1996) Ten-million-year history of a thrust sheet. Geol Soc Am Bull 108:1608

Mellere D (1993) Thrust-generated, back-fill stacking of alluvial fan sequences, south central Pyrenees, Spain (La Pobla de Seguar conglomerates). IAS Spec Publ 20:259–276

Metcalf JR, Fitzgerald PG, Baldwin SL, Muñoz JA (2009) Thermochronology in a convergent orogen: constraints on thrust faulting and exhumation from the Maladeta Pluton in the Axial Zone of the Central Pyrenees. Earth Planet Sci Lett 287:488–503

Morris RG, Sinclair HD, Yelland AJ (1998) Exhumation of the Pyrenean orogen: implications for sediment discharge. Basin Res 10:69–85

Muñoz JA (1992) Evolution of a continental collision belt: ECORS Pyrenees crustal balanced cross-section. In: McClay K (ed) Thrust tectonics. Chapman and Hall, London, pp 235–246

Muñoz JA (2002) The Pyrenees Alpine tectonics; I, The Alpine system north of the Betic Cordillera. In: Gibbons W, Moreno T (eds) The geology of Spain. The Geological Society, London, p 649

Muñoz JA, Coney PJ, McClay KR, Evenchick CA (1997) Reply to discussion on syntectonic burial and post-tectonic exhumation of the southern Pyrenees foreland fold-thrust belt. J Geol Soc 154:361–365

Oberli F, Meier M, Berger A, Rosenberg CL, Gieré R (2004) U-Th-Pb and $^{230}Th/^{238}U$ disequilibrium isotope systematics: precise accessory mineral chronology and melt evolution tracing in the Alpine Bergell intrusion. Geochim Cosmochim Acta 68:2543–2560

Pérez-Rivarés FJ, Garcés M, Arenas C, Pardo G (2004) Magnetostratigraphy of the miocene continental deposits of the Montes de Castejón (central Ebro Basin, Spain): geochronological and paleoenvironmental implications. Geol Acta 2:221–234

Puigdefàbregas C, Muñoz JA, Marzo M (1986) Thrust belt development in the Eastern Pyrenees and related depositional sequences in the southern foreland basin. In: Allen PA, Homewood P (eds) Foreland basins. IAS Spec Publ, pp 229–246

Rahl JM, Ehlers TA, van der Pluijm BA (2007) Quantifying transient erosion of orogens with detrital thermochronology from syntectonic basin deposits. Earth Planet Sci Lett 256:147–161

Rahl JM, Haines SH, van der Pluijm BA (2011) Links between orogenic wedge deformation and erosional exhumation: evidence from illite age analysis of fault rock and detrital thermochronology of syn-tectonic conglomerates in the Spanish Pyrenees. Earth Planet Sci Lett 307:180–190

Rahn MK (2005) Apatite fission track ages from the Adula nappe: late stage exhumation and relief evolution. Schweiz Mineral Petr Mitt 85:233–245

Rahn M, Brandon MT, Batt G, Garver JI (2004) A zero-damage model for fission-track annealing in zircon. Am Mineral 89:473–484

Reiners PW (2007) Thermochronologic approaches to paleotopography. Rev Mineral Geochem 66:243–267

Reiners PW, Brandon MT (2006) Using thermochronology to understand orogenic erosion. Ann Rev Earth Planet Sci 34:419–466

Resentini A, Malusà MG (2012) Sediment budgets by detrital apatite fission-track dating (Rivers Dora Baltea and Arc, Western Alps). Geol Soc Am Spec 487:125–140

Riba O, Reguant S, Villena J (1983) Ensayo de síntesis estratigráfica y evolutiva de la cuenca terciaria del Ebro. Geol España 2:131–159

Ring U, Brandon MT, Willet SD, Lister GS (1999) Exhumation processes. Geol Soc Spec Publ 154:1–27

Roca E, Muñoz JA, Ferrer O, Ellouz N (2011) The role of the Bay of Biscay Mesozoic extensional structure in the configuration of the Pyrenean orogen: Constraints from the MARCONI deep seismic reflection survey. Tectonics 30(2)

Roest WR, Srivastava SP (1991) Kinematics of the plate boundaries between Eurasia, Iberia and Africa in the North Atlantic from the late Cretaceous to the present. Geology 19:613–616

Rosenbaum G, Lister GS, Duboz C (2002) Relative motions of Africa, Iberia and Europe during Alpine Orogeny. Tectonophysics 359:117–129

Schneider DA, Issler DR (2018) Application of low temperature thermochronology to hydrocarbon exploration. In: Malusà MG, Fitzgerald PG (eds) Fission-track thermochronology and its application to geology. Springer, Berlin (Chapter 18)

Sinclair H, Gibson M, Naylor M, Morris R (2005) Asymmetric growth of the Pyrenees revealed through measurement and modeling of orogenic fluxes. Am J Sci 305:369

Spiegel C, Kuhlemann J, Dunkl I, Frisch W (2001) Paleogeography and catchment evolution in a mobile orogenic belt: the Central Alps in Oligo-Miocene times. Tectonophysics 341:33–47

Spiegel C, Siebel W, Kuhlemann J, Frisch W (2004) Toward a comprehensive provenance analysis: a multi-method approach and its implications for the evolution of the central Alps. Geol Soc Am Spec 378:37–50

Spotila J (2005) Applications of low-temperature thermochronometry to quantification of recent exhumation in mountain belts. Rev Mineral Geochem 58:449–466

Stewart RJ, Brandon MT (2004) Detrital zircon fission-track ages for the "Hoh Formation": implications for Late Cenozoic evolution of the Cascadia subduction wedge. Geol Soc Am Bull 116:60–75

Stock JD, Montgomery DR (1996) Estimating palaeo-relief from detrital mineral age ranges. Basin Res 8:317–327

Tugend J, Manatschal G, Kusznir NJ, Masini E, Mohn G, Thinon I (2014) Formation and deformation of hyperextended rift systems: insights from rift domain mapping in the Bay of Biscay-Pyrenees. Tectonics 33:1239–1276

van der Beek P, Robert X, Mugnier JL, Bernet M, Huyghe P, Labrin E (2006) Late Miocene–recent exhumation of the central Himalaya and recycling in the foreland basin assessed by apatite fission-track thermochronology of Siwalik sediments, Nepal. Basin Res 18:413–434

Vergés J, Muñoz JA (1990) Thrust sequences in the southern central Pyrenees. Bull Soc Geol Fr 8:265–271

Vermeesch P (2004) How many grains are needed for a provenance study? Earth Planet Sci Lett 224:441–451

Vermeesch P (2012) On the visualisation of detrital age distributions. Chem Geol 312:190–194

Villa IM, von Blanckenburg F (1991) A hornblende ^{39}Ar-^{40}Ar age traverse of the Bregaglia tonalite (southeast Central Alps). Schweiz Mineral Petr Mitt 71:73–87

Vincent SJ (2001) The Sis palaeovalley: a record of proximal fluvial sedimentation and drainage basin development in response to Pyrenean mountain building. Sedimentology 48:1235–1276

von Blanckenburg F (1992) Combined high-precision chronometry and geochemical tracing using accessory minerals: applied to the Central-Alpine Bergell intrusion (central Europe). Chem Geol 100:19–40

Wagner GA, Reimer GM, Jäger E (1977) Cooling ages derived by apatite fission-track, mica Rb–Sr and K–Ar dating: the uplift and cooling history of the Central Alps. Mem Ist Geol Min Padova 30:1–27

Wagner G, Miller DS, Jäger E (1979) Fission track ages on apatite of Bergell rocks from central Alps and Bergell boulders in Oligocene sediments. Earth Planet Sci Lett 102:395–412

Wolf RA, Farley KA, Kass DM (1998) Modeling of the temperature sensitivity of the apatite (U-Th)/He thermochronometer. Chem Geol 148:105–114

Yelland A (1990) Fission track thermotectonics in the Pyrenean orogen. Nucl Tracks Rad Meas 17:293–299

Application of Low-Temperature Thermochronology to Hydrocarbon Exploration

18

David A. Schneider and Dale R. Issler

Abstract

The maturation of organic material into petroleum in a sedimentary basin is controlled by the maximum temperatures attained by the source rock and the thermal history of the basin. A cycle of continuous deposition into the basin (burial) and regional basin inversions represented by unconformities (unroofing) may complicate the simple thermal development of the basin. Applications of low-temperature thermochronology via fission-track (FT) and (U–Th)/He dating coupled with independent measurements (vitrinite reflectance, Rock-Eval) resolving the paleothermal maximum are the ideal approach to illuminate the relationship between time and temperature. In this contribution, we review the basics of low-temperature thermochronology in the context of a project workflow, from sampling to modeling, for resolving the thermal evolution of a hydrocarbon-bearing sedimentary basin. We specifically highlight the application of multi-kinetic apatite FT dating, emphasizing the usefulness of the r_{mr0} parameter for interpreting complex apatite age populations that are often present in sedimentary rocks. Still a rapidly advancing science, thermochronology can yield a rich and effective dataset when the minerals are carefully and properly characterized, particularly with regard to mineral chemistry and radiation damage.

18.1 Introduction

The timing of and degree to which a sedimentary unit has been heated after deposition is important from several practical perspectives, including the maturation of hydrocarbon source rocks (heating) and assessing the magnitude of eroded sections in basins (cooling). Differences in sedimentary facies, burial history, and tectonic processes also affect the mechanisms by which oil and gas are generated. The common model, however, is fairly simple. Following deposition of organic-rich sediments, microbial processes convert some of the organic matter into biogenic methane gas. Greater depths of burial are accompanied by increases in heat dependent on the basin's geothermal gradient. This heat causes the organic matter to gradually transform into an insoluble organic matter (kerogen). The kerogen continues its evolution as heat increases; these changes, in turn, result in petroleum compounds that are subsequently generated, yielding bitumen and petroleum. Increasing thermal maturity also causes initially complex petroleum compounds to undergo structural simplification: commencing with oil then wet gas and ending at dry gas (Fig. 18.1). The main driver for this process is temperature; therefore, resolving the thermal history of basins is a fundamental first step in evaluating the hydrocarbon resource and potential. A wide variety of paleotemperature methods may be applied to assessing the intensity of post-depositional heating in basins (e.g., Naeser and McCulloh 1989; Harris and Peters 2012). Whereas many techniques can provide a relative or qualitative measure of the magnitude of that heating, quantitative treatment is only possible using those methods for which the kinetics are understood in sufficient detail, such as vitrinite reflectance (Burnham and Sweeney 1989; Nielsen et al. 2015). Although extremely useful, such techniques provide information merely on the magnitude of the maximum paleotemperature, and little or no constraint on when it occurred. Application of thermochronologic methods (FT

D. A. Schneider (✉)
Department of Earth and Environmental Sciences, University of Ottawa, Ottawa, ON K1N 6N5, Canada
e-mail: David.Schneider@uottawa.ca

D. R. Issler
Geological Survey of Canada, Natural Resources Canada, Calgary, AB T2L 2A7, Canada

M. G. Malusà and P. G. Fitzgerald (eds.), *Fission-Track Thermochronology and its Application to Geology*,
Springer Textbooks in Earth Sciences, Geography and Environment, https://doi.org/10.1007/978-3-319-89421-8_18

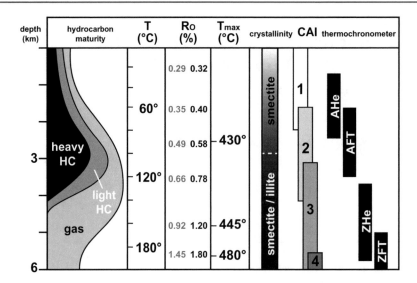

Fig. 18.1 Estimated petroleum (hydrocarbon: HC) generation zone with respect to common paleotemperature indices and thermochronometers (modified after Tissot et al. 1974, Gleadow et al. 1983). Values for each parameter are approximate because they vary with organic matter type, mineral composition and heating rate and duration. Two values are given for %Ro, which represent the models of Sweeney and Burnham (1990; gray text) and Nielsen et al. (2015; black text). Percent vitrinite reflectance (*%Ro*); Rock-Eval *Tmax*; illite crystallinity; conodont alteration index (CAI); (U–Th)/He thermochronology for apatite (AHe) and zircon (ZHe) and FT thermochronology for apatite (AFT) and zircon (ZFT)

and (U–Th)/He dating) conducted on sedimentary rocks provides the possibility to determine the timing and magnitude of the thermal maximum. This information is particularly important in hydrocarbon exploration, by resolving the timing of hydrocarbon generation in source rocks and identifying reservoirs in potential structural traps that formed after the main phase of generation.

In this chapter, we outline the tools and approaches to help determine the time–temperature histories of hydrocarbon-bearing sedimentary basins. This can often be a difficult task since the commonly used chronometers (apatite, zircon) are detrital in nature, possessing predepositional thermal histories, which may be partially preserved even after basin development. Moreover, the chemical composition of those detrital minerals will be varied, leading to mixed mineral kinetics and diffusivity, both controlling a mineral's closure temperature (T_c) and consequently cooling age. There are now a number of techniques that can be integrated to assess the thermal maturity and timing of thermal maximum of reservoir rocks and their surrounding stratigraphy. In particular, this contribution focuses on multi-kinetic apatite fission-track (AFT) thermochronology, an underutilized method that is ideally suited for resolving the thermal history of mixed apatite populations. Although this is not an exhaustive discussion, we hope that the information contained here within can assist in the design and execution of even the most challenging projects.

18.2 Thermochronologic Tools

Thermochronology is distinguished from geochronology by its ability to resolve both temporal and thermal aspects of geologic processes, and thus the timing and rates of processes. Since thermochronometers possess temperature windows over which the daughter product starts to be retained in the system, by measuring the amount of both parent nuclide and daughter product within a crystal, the time when the crystal passed through a temperature window can be calculated. This temperature window is called the partial annealing zone (PAZ) for FT dating and the partial retention zone (PRZ) for (U–Th)/He dating. Minerals such as apatite and zircon can therefore be used as thermochronometers, with their dates recording cooling rather than crystallization. The temperature ranges for PAZs and PRZs (Fig. 18.1) can vary with cooling rate, grain size, mineral chemistry, and crystal defects (e.g., Reiners and Brandon 2006).

18.2.1 Traditional FT Dating

The AFT dating method has been used extensively to resolve the low-temperature thermal histories of petroleum systems (e.g., Gleadow et al. 1983; Green et al. 1989a; Kamp and Green 1990; Burtner et al. 1994; Duddy 1997; Issler et al.

1999; Crowhurst et al. 2002; Hendriks and Andriessen 2002; Osadetz et al. 2002; Hegarty et al. 2007; Mark et al. 2008), whereas zircon FT (ZFT) dating has been applied mainly to sediment provenance studies and the cooling history of orogens due to its sensitivity at higher temperatures (e.g., Cerveny et al. 1988; Hurford 1986; Brandon and Vance 1992; Tagami et al. 1996; Bernet et al. 2004; Garver 2008; Marsellos and Garver 2010). The FT method is based on the accumulation of narrow damage trails in uranium-rich minerals, which form as a result of spontaneous nuclear fission decay of ^{238}U in nature (Price and Walker 1963; Fleischer et al. 1975). FT thermochronology is founded on the concept that tracks are partially or entirely erased at elevated temperatures, resulting in readily measured reductions in both track lengths and track densities. For apatite, kinetic models for thermally activated annealing of the fission tracks are well developed (e.g., Naeser 1979; Gleadow et al. 1986; Green et al. 1989a, b; Carlson 1990; Corrigan 1991; Crowley et al. 1991; Ketcham et al. 1999). Fluorapatite is the most common form of apatite (e.g., Pan and Fleet 2002) and is generally the least resistant to thermal annealing (see Chap. 3, Ketcham 2018). However, the annealing behavior of fission tracks, and the temperature at which all tracks are totally annealed, is influenced by apatite compositional variations and crystallographic anisotropy (Green et al. 1985, 1986, 1989b; Donelick et al. 1999; Carlson et al. 1999; Ketcham et al. 1999; Gleadow et al. 2002). It has been extensively documented that Cl content is the dominant factor with respect to compositional influences, and in fact other elements, such as Fe (Carlson et al. 1999), can also increase track retentivity. Annealing not only depends on kinetics (e.g., Green et al. 1986), but also on the duration of heating experienced by the sample (Green et al. 1989b). The degree of annealing can be determined by measuring etched horizontal (parallel to the polished mineral surface) confined track lengths within a sample (Gleadow et al. 1986; Green et al. 1989b; and Chap. 2, Kohn et al. 2018).

Empirical calibration studies of AFT data patterns demonstrate how apparent ages and track lengths vary systematically with depth and burial temperature (e.g., Green et al. 1989a, b; Dumitru 2000; Gleadow et al. 2002). For typical fluorapatite, all fission tracks are totally annealed above ~ 120 °C and partially annealed between ~ 60 and ~ 110 °C, defining the PAZ (e.g., Gleadow et al. 1986; Laslett et al. 1987; Green et al. 1989a; Fitzgerald and Gleadow 1990; Ketcham et al. 1999). This temperature range overlaps with the temperatures required for petroleum generation (Fig. 18.1), making AFT analysis well suited for such exploration purposes. Below ~ 60 °C, fission tracks in apatite anneal only at very slow rates (e.g., Fitzgerald and Gleadow 1990). Some extreme end-member fluorapatite crystals may reset at temperatures of ~ 85 °C based on the annealing experiment results in Ketcham et al. (1999), and

studies by Crowley et al. (1991) and Ketcham et al. (1999) have shown the importance of incorporating room temperature annealing into kinetic models. Unfortunately, at present there is little published literature on annealing as a function of apatite composition, as an effective lower annealing temperature would be a function of composition-dependent track retentivity. The quantitative understanding of FT annealing kinetics allows partially annealed AFT apparent ages and confined length data to be used to constrain the thermal evolution (~ 60–120 °C) of samples using stochastic inverse modeling approaches (e.g., Lutz and Omar 1991; Corrigan 1991; Gallagher 1995; Willett 1997; Ketcham et al. 2000; 2003a, b; Ketcham 2005).

Fission tracks in zircon anneal at higher temperatures than those in apatite and thus resolve higher temperature parts of the thermal history. Based on a comparison with potassium feldspar ^{40}Ar/^{39}Ar results, the ZFT T_c has been estimated to be ~ 250 °C (Foster et al. 1996; Wells et al. 2000), in good agreement with experimental data suggesting a ZFT T_c of ~ 240 °C (Zaun and Wagner 1985; Tagami 2005). Conservative estimates of the temperature bounds of the zircon PAZ are between ~ 220 and ~ 310 °C (Yamada et al. 1995; Tagami and Dumitru 1996; Tagami 2005; Yamada et al. 2007; and Chap. 3, Ketcham 2018). However, little is known about the impact of compositional variations on annealing behavior in zircon. Detailed descriptions of the fundamentals of traditional AFT and ZFT dating are readily available (e.g., Gallagher et al. 1998; Dumitru 2000; Gleadow et al. 2002; Donelick et al. 2005; Tagami 2005, and Chap. 12).

18.2.2 Multi-kinetic AFT Thermochronology

Traditional AFT dating requires that samples have a statistically uniform population of apatite grains with similar thermal annealing behavior and single grain AFT ages that pass the χ^2 test (Galbraith 1981; see also Chap. 6, Vermeesch 2018). In the absence of annealing kinetic parameter data, a fluorapatite composition may be assumed but more track-retentive AFT populations can be modeled if corresponding kinetic data are available. Problems can be encountered when trying to apply traditional FT methods to the study of detrital apatite in sedimentary basins. Such apatite can be derived from multiple source areas, with different pre-depositional cooling histories and with variable apatite chemical composition. These factors can all contribute to discordant AFT grain-age distributions when the variance of the grain-ages is greater than expected for the analytical error, causing samples to fail the χ^2 test. This is particularly common for detrital AFT samples from sedimentary basins that have had long residence times in the PAZ, which leads to differential annealing of compositionally variable grains. Under such conditions, reliable kinetic parameters are required to "unmix" and sort the apatite grains into statistical kinetic populations for thermal modeling. Difficulties in resolving AFT kinetic populations may be one of the main reasons why there are so few published studies on this subject.

AFT retentivity increases with increasing Cl content, as observed in studies of apatite from borehole samples (e.g., Green et al. 1986), except at very high Cl concentrations where retentivity appears to be less compared to the more intermediate compositions (Carlson et al. 1999; Ketcham et al. 1999; Gleadow et al. 2002; Kohn et al. 2002). Other elements such as Fe and Mn as well as OH content also affect AFT annealing (Carlson et al. 1999; Ketcham et al. 1999; Barbarand et al. 2003; Ravenhurst et al. 2003). AFT annealing kinetic parameters have been expressed in terms of the concentration of Cl or OH in atoms per formula unit (apfu) in the crystal, or the diameter of FT etch pits parallel to the apatite crystal c-axis, defined as D_{par} (Donelick 1993; Carlson et al. 1999; Donelick et al. 1999; Ketcham et al. 1999). D_{par}, which is related to composition-dependent apatite solubility, is the most commonly used proxy for annealing kinetics due to its low cost and relative ease of measurement. In general, more easily annealed F-rich apatite has smaller D_{par} values than more track-retentive Cl-rich apatite, although significant OH content can enhance apatite solubility and perturb this relation (Ketcham et al. 1999). AFT annealing experiments for apatite of variable composition suggest that heating/cooling rate-dependent total annealing temperatures can vary from ~80 to >200 °C, depending on apatite composition (Ketcham et al. 1999).

In spite of their limitations, D_{par} and Cl content have been the recommended practical kinetic parameters to use for the interpretation and modeling of mixed, multi-kinetic AFT samples (Ketcham et al. 1999; Barbarand et al. 2003). These parameters may be adequate if Cl is the main element controlling annealing but these factors are likely to fail when applied to more diverse apatite compositions, a common situation for apatite-bearing Phanerozoic sedimentary rocks of northern Canada (e.g., Issler and Grist 2008a, b, 2014; Issler 2011; Powell et al. 2017, Issler unpublished). The empirical r_{mr0} parameter (Ketcham et al. 1999) accounts for the effect of variable elemental composition on AFT thermal annealing, and it is a measure of the relative resistance to annealing for a given apatite compared to the most resistant apatite (Bamble, Norway) in the calibration dataset for the annealing experiments (see Chap. 3, Ketcham 2018). A multi-variate equation was developed to predict r_{mr0} values using a set of compositional variables determined from electron probe microanalysis (Carlson et al. 1999), and this parameter has much wider application for resolving statistical kinetic populations than the conventional kinetic parameters. A revised r_{mr0} equation was published (Ketcham et al. 2007a, b) as a result of combining and re-analyzing the experimental annealing results of

Ketcham et al. (1999) and Barbarand et al. (2003), but it is unclear that the new equation is superior to the initial one given the uncertainties in correcting for all systematic differences between analysts and experimental conditions.

For samples with discordant AFT grain-age distributions, it can be useful to display single grain AFT ages according to their relative precision on a radial plot (Galbraith 1988, 1990) and apply mixture modeling software such as Binomfit (Brandon 2002) or RadialPlotter (Vermeesch 2009) to resolve different statistical age populations. On a radial plot, the most precise AFT ages plot closest to the age axis whereas lower precision ages plot closer to the origin (see Chap. 6, Vermeesch 2018). Binomfit was developed for AFT data obtained using the conventional external detector method (Fleischer et al. 1964; Hurford and Green 1982), whereas RadialPlotter can also be used for AFT data acquired using the LA-ICP-MS method (Hasebe et al. 2004; Donelick et al. 2005; Chew and Donelick 2012). In addition, single grain AFT ages can be displayed with respect to their corresponding kinetic parameter value on a simple x–y plot so that interpreted statistical kinetic populations can be compared with the age populations derived from mixture modeling. This type of analysis can help to determine whether the different age populations within a sample are mainly the result of differential annealing of multi-kinetic apatite grains or due to variations in provenance.

Figure 18.2 shows an example radial plot of AFT age data for a drill core sample recovered from an Aklak

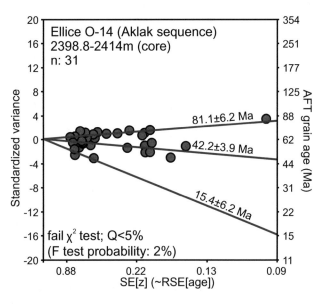

Fig. 18.2 Radial plot of single grain AFT ages for a core sample at ~2400 m depth from the Paleocene-to-early Eocene Aklak Sequence of the Ellice O-14 well from the Canadian Beaufort-Mackenzie region. The plot generated using Binomfit (Brandon 2002) and defines three age populations. The x-axis is approximately equivalent to the relative standard error on age; ages with the highest precision plot closest to the age axis on the right

Fig. 18.3 Plots of AFT data for the Ellice O-14 Aklak Sequence core sample with respect to various kinetic parameters. Two statistical kinetic populations are resolved on plots of AFT single grain-age (**a**) and track length versus r_{mr0} (effective Cl equivalent) (**b**). Kinetic populations overlap on plots of AFT age versus measured Cl content (**c**) and Dpar (**d**). Two ages have anomalously high fluorine

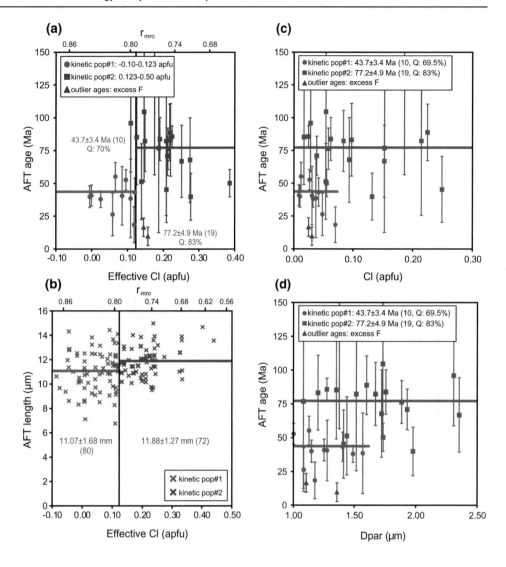

Sequence sandstone of late Paleocene-to-early Eocene age (c. 56 Ma) at 2400 m depth in the Ellice O-14 petroleum exploration well located in the Beaufort-Mackenzie Basin of northern Canada (Issler et al. 2012). This sample fails the χ^2 test, and Binomfit resolves three statistically significant age populations shown by the radial lines: 81 Ma, 42 Ma, and 15 Ma. The 81 Ma population is significantly older than the stratigraphic age of the sample, suggesting that post-depositional annealing was insufficient to erase its inherited pre-depositional thermal history. The 42 Ma and 15 Ma populations are younger than the stratigraphic age, which implies significant post-depositional annealing. The youngest AFT age population consists of two apatite grains with non-stoichiometric excess F content that appear to correlate with unusually low track retentivity.

Figure 18.3a shows a plot of single grain AFT ages for the same detrital sample with respect to the r_{mr0} parameter (Carlson et al. 1999), also expressed as an effective Cl value (apfu) using the r_{mr0}-Cl equation of Ketcham et al. (1999).

Two main kinetic populations with statistically significant pooled ages of 44 Ma and 77 Ma (similar to the Binomfit ages) using the r_{mr0} kinetic parameter (Fig. 18.3a) compared with a plot of AFT length versus r_{mr0} show that all of the shortest lengths are associated with the younger age population at low effective Cl values (Fig. 18.3b). The boundary dividing these statistical populations is placed at an effective Cl value of 0.12 apfu but some population overlap is expected; duplicate elemental analyses for selected grains indicate that effective Cl values are usually accurate to within 0.02–0.04 apfu and larger differences up to 0.24 apfu have been observed in a few cases. In contrast, these two statistical populations cannot be resolved using the conventional kinetic parameters, Cl content (Fig. 18.3c) and D_{par} (Fig. 18.3d). Effective Cl values (Fig. 18.3a) are significantly higher than measured Cl values (Fig. 18.3c) due to the variable cation and OH contents of the apatite grains. Paleogene AFT samples from the Ellice O-14 well have Fe and Na contents up to 0.08 apfu, Mg values up to 0.09 apfu,

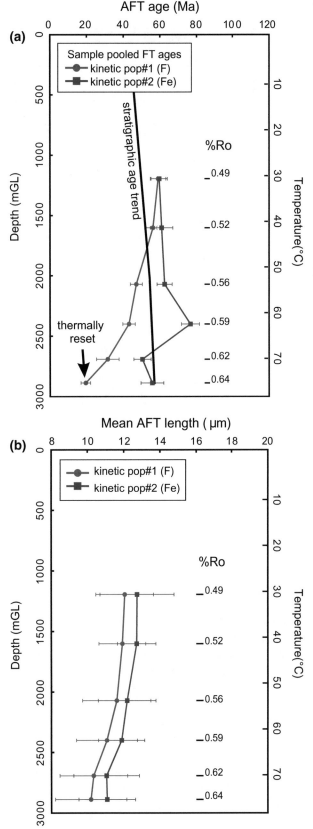

◀ **Fig. 18.4** Variation in pooled AFT age (**a**) and mean track length (**b**) versus depth for two statistical kinetic populations in six Paleogene core samples from the Ellice O-14 well. The observed pattern can be explained by differential annealing of two different kinetic populations defined by Fe- and F-rich mineral compositions of similar provenance

OH values as high as 1 apfu, and Ce values up to 0.06 apfu. The sample attained a mean random percent vitrinite reflectance (*%Ro*) level of ~0.6 (~100 °C), indicating that paleotemperatures were hot enough to cause significant annealing of fission tracks. These results strongly suggest that differential annealing of this multi-kinetic sample is the main reason for the observed discordant AFT grain-age distribution. The youngest two age-grains have low Cl and D_{par} values (Fig. 18.3c, d) but high effective Cl values (Fig. 18.3a) due to their variable cation content. These effective Cl values are inconsistent with their 15 Ma population age, which requires an unusually low resistance to annealing that may be related to their excess F content. Although anomalously high F values may be an artifact caused by F migration under the electron beam during electron probe microanalysis (Stormer et al. 1993; Stock et al. 2015), excess F grains in other samples from the region also appear to have very low track retentivity.

The procedure detailed above using Binomfit age analysis and plots of AFT parameters versus r_{mr0} was applied to five other core samples from Paleogene stratigraphic intervals in the Ellice O-14 well. Figure 18.4 shows the relationship of AFT age and mean length versus depth and present temperature for six core samples between 1000 and 3000 m depth in the well. Organic maturity values for each sample (ranging from 0.49 to 0.64 *%Ro*) were derived from detailed downhole measurements (Issler et al. 2012). Organic maturity levels indicate that maximum paleotemperatures were tens of degrees higher than present values in accordance with geological evidence for late Cenozoic exhumation at this location. Two statistical kinetic populations, a low retentivity fluorapatite population and a higher retentivity Fe-rich population, were resolved for each sample using the r_{mr0} parameter (Fig. 18.4a). The similar AFT ages for the two kinetic populations in the shallowest sample are consistent with complete thermal annealing of both AFT populations at high temperature, followed by rapid cooling of the apatite source region and subsequent minimal post-depositional annealing. The AFT ages of the two kinetic populations diverge with increasing depth and organic maturity due to differential annealing. The fluorapatite population shows a consistent decrease in AFT age with depth, becoming younger than the stratigraphic age at ~1800 m. The more retentive Fe-rich AFT population shows a more variable AFT age pattern reflecting provenance influences; AFT ages are younger than the stratigraphic age below 2600–2700 m.

Mean AFT lengths are shorter for the fluorapatite kinetic population, and both kinetic populations show parallel trends downhole (Fig. 18.4b). The fluorapatite population for the deepest sample is interpreted as being thermally reset based on the results of thermal modeling (see Sect. 18.5).

The Ellice O-14 example demonstrates the importance of collecting elemental data for constraining the annealing kinetics of multi-kinetic AFT samples. The r_{mr0} parameter provides a means for sorting the AFT data into different kinetic populations to be treated as separate thermochronometers with different annealing temperatures for thermal modeling purposes (see Sect. 18.5). The Ellice O-14 samples could not be properly interpreted using the conventional kinetic parameters, Cl content or Dpar. Two of the six AFT samples from the Ellice O-14 well passed the χ^2 test (although Binomfit analysis indicates they have two statistically significant populations), and in the absence of mineral chemistry data, they might have been treated erroneously as single uniform populations. The other four samples that failed the χ^2 test may have been deemed to be uninterpretable and rejected as unsuitable for further analysis.

18.2.3 (U–Th)/He Dating

Apatite (U–Th)/He (AHe) dating (e.g., Zeitler et al. 1987; Lippolt et al. 1994; Wolf et al. 1996, 1998) is now a well-established thermochronologic technique and is widely applied across a range of geological environments. AHe dating is based on the decay of ^{235}U, ^{238}U, and ^{232}Th (and to a lesser extent ^{147}Sm) by alpha (^{4}He nucleus) emission. Whereas a T_c of ~ 70 °C is often invoked for the AHe system, this concept is only applicable in samples that have cooled rapidly from high temperatures (Reiners and Brandon 2006). In most cases, helium diffusion in apatite with protracted thermal histories is sensitive to the effects of cooling rate, grain size, radiation damage, and grain chemistry (Stockli et al. 2002; Flowers et al. 2009; Gautheron et al. 2013). In particular, the amount of radiation damage, as assessed through the proxy "effective uranium concentration" (eU = [U] + 0.235 · [Th]) is positively correlated with increasingly retentive AHe systems (Shuster et al. 2006; Flowers et al. 2009; Gautheron et al. 2009). In effect, the AHe system may be sensitive to temperatures between 30 and 90 °C depending on the amount of radiation damage to the crystal lattice (Flowers et al. 2009). Gautheron et al. (2013) used the r_{mr0} parameter (Carlson et al. 1999; Ketcham et al. 1999) to investigate the connection between grain chemistry and radiation damage annealing in the AHe system. Their results suggest that grain chemistry, and its effects on fission track and alpha damage annealing, may influence helium diffusion in apatite and the temperatures to which the system is susceptible. This system is more sensitive at a lower range of temperatures than most isotopic thermochronometers. Assuming a mean annual surface temperature of 10 °C and a geothermal gradient of 25 °C/km, the relevant temperature range is equivalent to depths of ~ 1 to 3 km. Thus, the AHe system can be applied to investigate a variety of geologic processes in the uppermost part of the crust, making the technique particularly useful for hydrocarbon exploration in basin settings (Fig. 18.1).

The (U–Th)/He system in zircon (ZHe) is widely used as a thermochronometer because of its high U–Th concentrations, common occurrence in a wide range of lithotypes, refractory nature under elevated thermal conditions, and resistance to physical and chemical weathering (see Chap. 7, Malusà and Garzanti 2018). Several studies (e.g., Kirby et al. 2002; Reiners et al. 2004; Stockli 2005) have presented results of ZHe ages from rocks with potassium feldspar ^{40}Ar/^{39}Ar cooling models (350–150 °C; Lovera et al. 1989, 1991, 1997), confirming a ZHe T_c of ~ 170–190 °C (Reiners et al. 2004) and a PRZ ~ 130–200 °C (Reiners and Brandon 2006). Although in most cases potassium feldspar models suggested relatively rapid cooling rates through the ZHe T_c, the results showed good agreement between the two techniques in nearly all cases, suggesting that the experimentally determined helium diffusion parameters for zircon and its inferred T_c apply in natural settings. ZHe ages from some samples commonly show larger dispersion than expected from analytical precision and a single set of kinetic parameters for helium diffusion. In principle, this dispersion could have several origins, including effects arising from implantation (Spiegel et al. 2009; Gautheron et al. 2012), anisotropic diffusion (Farley 2007; Reich et al. 2007; Cherniak et al. 2009; Saadoune et al. 2009), compositional influences (i.e., zoning) on helium diffusion (Hourigan et al. 2005), and crystallographic defects. However, one of the most important known influences on helium diffusivity in zircon is radiation damage. The effects of high radiation doses on helium diffusion in zircon have been recognized for some time (e.g., Hurley 1952; Holland 1954; Nasdala et al. 2004), but only recently have these effects been quantitatively integrated with helium diffusion models (Guenthner et al. 2013, 2015; Powell et al. 2016).

18.3 Integration with Independent Data

Geoscientists employ a variety of techniques to evaluate the hydrocarbon-generating capacity of source rocks. Techniques for extracting thermal maxima that a rock has witnessed include two broad groups: measurements on organic matter (including certain fossils) and measurements on minerals. The results help ascertain how much and what kind of petroleum might have been generated, defining the source rock's potential. The history of organic thermal parameters is long, and several techniques have been developed (Fig. 18.1) (see Harris and Peters 2012 for recent reviews). Among these, the most notable are vitrinite reflectance and the Rock-Eval pyrolysis parameter T_{max}. Rock-Eval pyrolysis involves the rapid heating of rock samples in an inert atmosphere, which expels existing hydrocarbons and thermally decomposes kerogen in the rock. The temperature at which most of the kerogen pyrolyses, referred as T_{max}, has been shown to be a sensitive and predictable measure of maturity (Peters 1986). As the sample is incrementally heated, pyrolysis temperatures produce T_{max} peaks that correspond to the pyrolysis oven temperature during maximum generation of hydrocarbons. T_{max} is attained during cracking of the kerogen, which should not be confused with geologic temperatures, and it can help better understand the extent of thermal maturation of the sample.

Vitrinite reflectance relies on chemical changes in vitrinite, which is an organic maceral derived from the woody tissue of plants that is present in kerogen. Vitrinite reflectance, referred as $\%Ro$, is the percentage of incident light that is reflected off a polished particle of vitrinite; it is dependent upon the maximum paleotemperature of the sample and on the time spent at that temperature (e.g., Senftle and Landis 1991). $\%Ro$ tends to increase with increasing burial in a sedimentary basin and is commonly used to evaluate organic maturity in sedimentary rocks (e.g., Tissot and Welte 1978). Burnham and Sweeney (1989) and Sweeney and Burnham (1990) presented a kinetic model for the evolution of $\%Ro$ with temperature and time, which showed that temperature has a larger influence on $\%Ro$ than time. The EASY$\%Ro$ kinetic model of Sweeney and Burnham (1990) was based largely on pyrolysis experiments; the model has been recalibrated by Nielsen et al. (2015; basin$\%$ Ro) using well-constrained sedimentary successions. Duddy et al. (1991) illustrated that the AFT annealing kinetics (Laslett et al. 1987) are similar to the kinetics of $\%Ro$ (Burnham and Sweeney 1989) and that a given degree of annealing in apatite will be associated with the same value of $\%Ro$. For example, a $\%Ro$ value of 0.7 is roughly associated with total annealing of all fission tracks in Durango fluorapatite reference material (Duddy et al. 1994), with some differences depending on end-member heating/cooling rates. A typical standard deviation for a $\%Ro$ measurement is 10%, which depends on factors such as material recycling, number of measurements, oil staining, oxidation, disseminated pyrite. AFT and $\%Ro$ techniques complement each other because $\%Ro$ values vary with the heating history and can be used to determine maximum paleotemperatures, which is especially useful when temperatures were greater than the annealing temperature of fission tracks. In contrast, AFT data are most sensitive to cooling yet both techniques provide information on the thermal history of a sample, primarily useful for middle Paleozoic and younger clastic sediments.

Other approaches for determining thermal maturity using organic substances include the spore coloration index (SCI, Staplin 1969; Marshall 1991) and thermal alteration index (TAI, e.g., Batten 1996) that measure the color of palynomorphs. With these techniques, chemical reactions occur in response to increasing temperatures resulting in darkening of the material. A standard set of colors has been developed to categorize samples, standardized to the vitrinite reflectance (Ro) maturity scale, and it includes the conodont alteration index (CAI, Epstein et al. 1977). Field and laboratory data have demonstrated that conodonts experience a progressive and permanent sequence of eight color changes (i.e., CAI) that record maximum temperatures ranging from ~ 50 to ~ 550 °C (Epstein et al. 1977; Harris 1979). Scanning electron microscopy studies of conodonts have demonstrated that during progressive diagenesis and low-grade metamorphism, conodonts can experience an increase in the size and change in morphology (from anhedral to euhedral) of apatite crystallites. Those changes are largely restricted to the conodont surface for CAI of 1–5, but above CAI of 5, internal re-crystallization may occur (Burnett 1988; Helson 1994; Nöth 1998). Notably, using conodonts as a (U–Th)/He thermochronometer has witnessed some success (Peppe and Reiners 2007; Landeman et al. 2016; Powell et al. 2018) and key advantages of conodonts over traditional thermochronometers are that conodonts can be routinely used as geothermometers and commonly occur in limestones. Although the role of CAI on the microstructural character of conodonts is clear, how those changes impact parent isotope distributions and mobility, He diffusivity, and (U–Th)/He ages is unknown. Other less commonly used optical maturity scales include the transmittance color index (TCI, Robison et al. 2000), acritarch fluorescence (Obermajer et al. 1997) and the foraminiferal coloration index (FCI, McNeil et al. 1996; McNeil 1997; Gallagher et al. 2004; McNeil et al. 2010, 2015), which measure color changes in the organic cement of agglutinated foraminifera.

Among the approaches that rely on variable reaction rates within minerals, the mixed-layer clay illite–smectite reaction is commonly applied to helping determine the extent of diagenesis (Fig. 18.1). Hower et al. (1976) were the first to document the phenomenon of smectite reacting to illite over a range of temperatures. Illite has been reported to form in a variety of environments, from soils to deeply buried sediments. Evidence that elevated temperatures promote the authigenic precipitation of illite or the diagenetic conversion of illite–smectite mixed-layer (I/S) into illite, both termed illitization, comes from experimental work as well as geologic studies with well-documented geothermal regimes (e.g., Frey et al. 1980) or local temperature increases resulting from circulation of hydrothermal solutions at depth (e.g., Lampe et al. 2001; Meunier and Velde 2004; Timar-Geng et al. 2004). This technique has been extended to incorporate isotope geochemistry (Clauer and Lehrman 2012).

18.4 Sampling Strategy

Sampling and sample size are commonly limited by bedrock exposure, the logistics of carrying samples during traverses and of shipping samples back to the laboratory, and the nature of the material extracted during drilling (cuttings, core). When possible, bedrock samples should be fresh and collected away from ridge tops, lightning strike-prone areas, and areas devoid of any historic forest fires to avoid any effects of thermal resetting (Mitchell and Reiners 2003; see also Chap. 8 Malusà and Fitzgerald 2018). Contamination is a major concern, and the rock samples should be clean of exotic soil and sediment. For bedrock samples, blocks ~ 10 cm^3 (0.5–1.0 kg) are typically adequate for recovering an appropriate amount of material for statistically robust dating and independent thermal maturity analyses (e.g., vitrinite reflectance, Rock-Eval). Although coarse-grained material will likely yield larger accessory minerals, a grain size as small as 60 µm will produce adequate crystals (see Chap 7 Malusà and Garzanti 2018). For finer-grained sedimentary rocks, such as shales and siltstones, larger volumes of samples may be required for adequate yield. Intermediate or felsic crystalline rocks are favored for zircon recovery, whereas apatite can be present across the range of felsic to mafic lithotypes. Quartz arenites typically have a low apatite yield, whereas polygenic and lithic-rich sandstones are more likely to contain apatite; zircon will be found in both end-members. Limestones and dolostones do not possess apatite and zircon, but the carbonate sequences are commonly interbedded with sandy units that will yield these phases. Bentonites, layers of volcanic ash, are particularly attractive to help reconstruct basin histories since its zircon U–Pb crystallization age provides unequivocal stratigraphic control, and its ZHe and AHe ages, and AFT ages and lengths record the thermal evolution.

The number of samples taken from a vertical transect or drill core may be restricted by the availability of core/cuttings or outcrop, and the availability of other data. For example, %Ro data can yield maximum paleotemperatures and is a relatively low-cost method, so samples every 250 m or so would be recommended to get a reliable maturity gradient depending on the thickness of section. This type of sampling density would be helpful for conventional AHe dating as well, restricting sampling intervals according to temperature, but it could be prohibitively expensive for AFT analysis if apatite chemistry data are also collected to determine r_{mr0} values. In such a case, defining the thermal maturity of the borehole would help assist in choosing a few

strategic AFT samples that have experienced significant annealing and/or thermal resetting (typically at maturity values of 0.6–0.7 %Ro or higher). Although a series of samples from a well is preferred, a well-chosen sample from a borehole that is strongly annealed and includes multi-kinetic data can still yield a lot of useful thermal history information. Such samples behave as multiple thermochronometers with different annealing temperatures that can provide information on multiple thermal events in areas with complicated geological histories (Issler et al. 2005). The trade-off between vertical versus lateral coverage must also be considered if the study area is large.

Rock cores are the most desirable type of well sample for thermochronology studies because of minimal contamination and accurate depth control, but they are seldom collected during petroleum exploration. Well cuttings are common but only a small fraction is collected at regular depth intervals and retained after drilling so it is better to collect thermochronology samples during drilling of the well in order to obtain sufficient material. Sampling restrictions on post-drilling cuttings' samples mean that material may have to be combined over larger depth intervals influenced by changes in stratigraphy, lithology, and temperature. Furthermore, cuttings can be contaminated by drilling mud additives (e.g., bentonite) and mixing of rock material from different stratigraphic intervals through borehole caving and recirculation of cuttings during drilling. Thermochronologic data that are anomalous with respect to present temperature and observed AFT age and length versus depth trends down a borehole can indicate contamination from drilling mud additives. For example, Eocene cuttings from 4100 m (110 °C) near the base of the Taglu West H-06 well of the Canadian Beaufort-Mackenzie Basin (Issler et al. 2012) contain some apatite grains with a pooled AFT age of 608 Ma and long mean lengths of 12.9 µm. These AFT parameters are inconsistent with the observed AFT age (38 Ma) and mean length (11 µm) of the indigenous Fe-rich apatite population at this depth. Both populations have similar r_{mr0} values but the older AFT population has higher Ce, Sr, S, and Na contents indicating a different source for the apatite, most likely a mud additive. Contamination due to caving and/or cuttings recirculation can be harder to identify unless significantly different stratigraphic units are involved. Contamination of Devonian cuttings by overlying Cretaceous sediments has been documented for a well drilled in the Mackenzie Valley region south of Norman Wells (Northwest Territories, Canada; Issler and Grist 2008b). Evidence includes the presence of Cretaceous palynomorphs, organic-rich shale, and sandstone mixed with the fine-grained sandstones and siltstones of the Devonian Imperial Formation. Multi-variate statistical analysis of elemental data shows that a fraction of the apatite in the Devonian sample has the same chemical composition as the overlying Cretaceous apatite, and such information can be used to ignore the contaminant grains.

Apatite and zircon are separated from whole rock or chips by crushing using a jaw crusher alternating with sieving (see Chap. 2, Kohn et al. 2018). Care is required not to overcrush (or re-crush) the sample, which may produce broken grains. Alternatively, SELFRAG© technology allows liberation or weakening of material along natural grain boundaries using high-voltage electrical pulses. It also allows for controlled crushing without contamination. A water table may be used to provide the first level of density-based liquid separation, followed by magnetic and heavy-liquid density separations. Whereas FT dating can utilize both whole and broken grains for dating, (U–Th)/He dating is a little less forgiving, and a careful selection of inclusion and crack-free idiomorphic crystals is the best practice. This matters because the fragments will all yield ages that are different from each other and from the whole grain-age if the ^{4}He distribution within the whole grain is not homogeneous, because of partial loss due to thermal diffusion (e.g., Brown et al. 2013).

There is no general guideline for the number of crystals to use for AFT analysis because it depends on the nature of each sample. Although 50–100 grains may be an ideal case, most laboratories usually analyze 20 age-grains and try to measure 100 confined track lengths (Donelick et al. 2005; Galbraith 2005). This may be sufficient for conventional AFT analysis of single uniform apatite populations that are common for igneous rocks but more age and length measurements are needed to properly characterize multi-kinetic detrital AFT samples. Laboratories with experience in multi-kinetic AFT analysis such as some high volume commercial laboratories (Apatite to Zircon, Inc.; GeoSep Services, LLC) generally provide 40 AFT ages and 150–200 confined track length measurements per sedimentary rock sample. This is normally sufficient for typical sedimentary samples with two or three statistical kinetic populations. The FT mount can then be used for mineral chemistry analysis via LA-ICP-MS or EMPA techniques. For the (U–Th)/He methods, the thermochronology community has not agreed on a suitable number of analyses per rock required for reliable data. The number of crystals analyzed for a study is driven by the scientific question to be answered, and the time and budget allocated for the project. For crystalline rocks, 5–6 single crystal analyses per rock are becoming more common, although some investigations still only report 2–3 single crystals per rock. For detrital samples, at least 10–15 single crystals per rock should be carried out, particularly if the sample has a broad spectrum of detrital ages. Some research has been successfully undertaken to utilize broken apatite for He dating (Brown et al. 2013; Beucher et al. 2013). The challenge is balancing a representative sampling of age data that allows the researcher to assess age, chemistry, and size correlations, in order to provide the necessary data for successful modeling.

18.5 Thermal Modeling

Continuous formation and annealing of fission tracks through geologic time mean that observed AFT age and track length distributions contains a record of a sample's thermal history (see Chap. 3, Ketcham 2018). Further, advances in understanding the effect of radiation damage on temperature-dependent helium retention in apatite and zircon have shown that the (U–Th)/He age–eU dispersion of aliquots from a single sample is dependent upon the thermal history experienced by that sample (Shuster et al. 2006; Flowers et al. 2009; Guenthner et al. 2013; Powell et al. 2016). Therefore, thermal modeling of He age–eU data can be used to investigate the range of possible thermal histories that could produce the observed He age–eU distribution (e.g., Flowers et al. 2009; Guenthner et al. 2013, 2014; Powell et al. 2016). Forward thermal modeling involves the prediction of observations (AFT ages, lengths; AHe ages) given the values of the parameters that define the model which include the laboratory-based AFT annealing kinetics and/or AHe diffusion and radiation damage parameters, from a proposed time–temperature path. Forward modeling may be used to check if a time–temperature path provides a plausible explanation of measured thermochronologic data and is also a useful way to predict and understand the effect of a thermal history on AFT ages and track length distributions, or on ages and eU dispersions for He ages. Although forward models can be constrained by independent geological data, they do not provide a unique time–temperature solution, only representing plausible time–temperature candidates. Inverse modeling involves using the observed thermochronologic data (AFT ages, lengths; AHe ages) to infer the values of the model parameters that successfully simulate the data. Under most applications, the AFT annealing kinetics and AHe diffusion kinetics are fixed parameters whereas time–temperature paths are adjusted by the model so that calculated AHe ages and AFT ages and lengths match the measured thermochronologic data to within a specified amount of statistical error. Inverse modeling allows for a more thorough exploration of potential solution space, and it provides a more realistic assessment of our ability to resolve temperature histories using measured data. The present-day sample conditions and any known independent geological controls (e.g., unconformities, burial events, deposition age, organic maturity) are also applied to constrain the inversion. Commonly, a best-fit time–temperature path and a range of good- and acceptable-fit paths are found using Monte Carlo simulations (Gallagher 1995; Willett 1997; Issler et al. 2005; Ketcham 2005).

Just as a broad or bimodal distribution of FT lengths indicate that a sample has resided at temperatures within the FT PAZ, so a correlation of (U-Th)/He age with eU concentration indicates that a sample has resided at temperatures within the He PRZ. Such ages cannot simply be interpreted as the time elapsed since the sample passed through the T_c of the thermochronometer. Modeling is the only way to gain understanding of the thermal histories of slowly cooled samples, or of samples that have resided in the PAZ or PRZ for a significant time (relative to their age) and have a partially reset age. Ages calculated from forward modeling and time–temperature paths determined from inverse modeling may both be useful for understanding real He and FT ages but should be interpreted with caution. Modeling results are dependent on the kinetic parameters that constrain annealing and diffusion. Whereas both forward and inverse modeling are based on a wealth of annealing and diffusion studies (e.g., Green et al. 1986; Laslett et al. 1987; Duddy et al. 1988; Green et al. 1989b; Wolf et al. 1998; Ketcham et al. 1999, 2007a,b; Farley 2000, 2002; Barbarand et al. 2003; Reiners et al. 2004; Shuster et al. 2006; Flowers et al. 2009; Guenthner et al. 2013), model solutions are non-unique. Although the models should be used to test explicit hypotheses, models need to be applied carefully—any new hypothesis should be rooted in (good) data and any unexpected adjustment to the thermal history should be faithful to the data and independent geological constraints with some differences depending on end-member heating/cooling rates.

Early inverse models used mono-compositional AFT annealing kinetics and employed a variety of stochastic optimization and Monte Carlo search techniques to derive thermal history information from AFT data (Corrigan 1991; Lutz and Omar 1991; Gallagher 1995; Willett 1997). However, these models have been superseded by newer models that incorporate multi-kinetic AFT annealing. AFT-Solve (Ketcham et al. 2000), the precursor to the widely used HeFTy model (Ketcham 2005), was the first publically available model developed for forward and inverse thermal modeling of multi-kinetic AFT data. The newer HeFTy model (Ketcham 2005) can be used for forward and inverse modeling of multi-kinetic AFT and AHe data and is used extensively by the thermochronology community (see Chap. 3, Ketcham 2018). AFTINV (Issler 1996), originally developed as a user-friendly version of the inverse model of Willett (1997), has been extensively upgraded to deal with multi-kinetic AFT data (Issler et al. 2005). AFTINV shares many common features with AFTSolve and HeFTy, including the multi-kinetic scheme of Ketcham et al. (1999), but differs in how thermal histories are constructed and how geological constraints are applied. The QTQt model (Gallagher 2012) uses a Bayesian trans-dimensional Markov Chain Monte Carlo method to extract thermal history information from various combinations of data types (AFT,

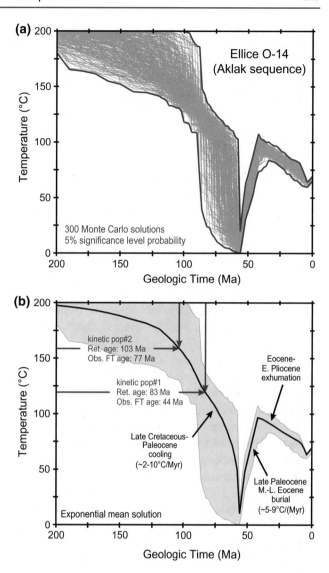

Fig. 18.5 **a** AFTINV thermal model results for the Ellice O-14 sample of Figs. 18.2 and 18.3 showing acceptable solution space defined by 300 Monte Carlo solutions. The envelopes bounding the 300 Monte Carlo solutions are not acceptable solutions. **b** The exponential mean thermal history (bold curve) provides a good fit to the AFT data and shows that there was rapid cooling of the apatite source area (constrained by the more retentive Fe-rich kinetic population) and slower, steady cooling (constrained by F-apatite population) following rapid burial. Model retention ages are theoretical ages marking the time when tracks >2 μm in length are preserved. Ret. retention; Obs. observed

U–Th/He, $^{40}Ar/^{39}Ar$) and it has forward and inverse modeling capabilities. Unlike HeFTy, this model uses the data to determine the number of model parameters (i.e., number of time–temperature points, kinetic parameters) subject to user-defined constraints. Vermeesch and Tian (2014) compare and discuss the advantages and drawbacks of the HeFTy and QTQt models. Although most modeling applications have been directed at individual samples, some models have been designed to deal with multiple samples collected along vertical profiles (Gallagher et al. 2005; Gallagher 2012; Ketcham et al. 2016). Braun et al. (2006) discuss modeling methods that allow thermochronology data to be interpreted in the context of heat transfer processes within the Earth in different tectonic settings.

18.5.1 Multi-kinetic AFT Modeling

The utility of multi-kinetic AFT modeling for a petroleum exploration well is illustrated in Fig. 18.5, which shows AFTINV thermal model results for the detrital AFT data corresponding to the Ellice O-14 core sample shown in Figs. 18.2 and 18.3. Temperature histories are generated randomly within geologically constrained a priori limits to obtain a set of statistically acceptable thermal solutions that yield calculated AFT ages and lengths that fit observed AFT data within prescribed statistical error. Initial temperature search space is defined loosely based on sample attributes and geological information (e.g., time of deposition, the presence of unconformities, thermal maturity), and users specify limits for heating and cooling rates that apply within these temperature limits. Thermal histories are generated forwards and backward by random selection of heating or cooling rate (projection angle to next point) from randomly selected initial points, subject to rate and temperature limits. Trial thermal solutions are evaluated according to the degree of misfit between calculated and observed AFT and %Ro values using the combined merit function approach described in Willett (1997) and Ketcham (2005). Track length distributions are assessed using the Kolmogorov–Smirnov (KS) statistic (e.g., Miller and Kahn 1962; Press et al. 1992), and calculated AFT ages are required to be within two standard deviations of the observed ages. A significance level probability of 0.05 provides a pass/fail test of the null hypothesis that the measured and calculated distributions are the same. The model converges when a set of statistically acceptable Monte Carlo solutions (typically 300) has accumulated (Fig. 18.5a). Convergence to 300 solutions can take thousands to millions of iterations depending on model complexity. There is no attempt to find the optimal fit to the data because the data are not optimal and will vary with the number of measurements. However, following Willett (1997), the exponential mean of the 300 solutions is taken as a representative, good-fitting solution (Fig. 18.5b).

Both kinetic populations were modeled simultaneously, and results are consistent with a common rapid exhumation and cooling history for the apatite source areas (Fig. 18.5). Calculated model AFT retention ages for the exponential mean history suggest that tracks started to be retained in the Fe-rich apatite population as temperature cooled below 160 °C at ~ 100 Ma, whereas track retention started in the fluorapatite population at temperatures below 120 °C

at ~ 80 Ma. Corresponding AFT ages for both populations are significantly younger due to partial annealing of tracks with continued cooling and subsequent reheating. Following deposition, model results show rapid Paleocene-to-mid–late Eocene heating associated with rapid burial of deltaic sediments and then slower, steady cooling associated with late Cenozoic exhumation (Fig. 18.5). The final phase of heating is poorly resolved due to the thinness of the Plio-Pleistocene strata and the rapid changes in surface temperature that occurred in the Arctic during this time period. The less annealed Fe-rich AFT kinetic population retains a record of the pre-depositional exhumation history, whereas the more annealed fluorapatite population is most sensitive to the post-depositional part of the thermal history.

Figure 18.6 shows exponential mean thermal solutions for the six Ellice O-14 AFT core samples presented in Fig. 18.4. Although each sample was modeled independently, they show similar thermal histories that are offset with respect to stratigraphic age (shift in time–temperature inflection points) and temperature related to their downhole position and thermal maturity. The overall coherence in thermal histories may in part be related to the uniform application of modeling constraints but it is also related to the remarkable consistency of the kinetic parameters for these detrital Paleocene–Eocene samples. This is demonstrated by the narrow range in effective Cl values for the fluorapatite (0–0.05 apfu) and Fe-rich apatite (0.19–0.23 apfu) kinetic populations and the consistency in calculated model retention temperatures (Fig. 18.6). The fluorapatite

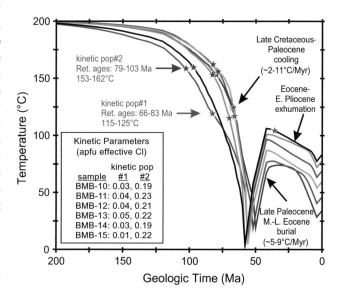

Fig. 18.6 Exponential mean thermal histories corresponding to the six Ellice O-14 core samples of Fig. 18.4. Although modeled separately, all samples yield similar thermal histories due to similar kinetic parameters and geological constraints. Note that the resolution on timing of maximum temperature decreases as the degree of AFT annealing decreases. Ret. retention

population for the deepest sample (0.64 *%Ro*) has the lowest effective Cl value (0.005 apfu), and thermal resetting of this AFT population is demonstrated by the young calculated retention age ($\sim$ 36 Ma). As expected, the deepest and most annealed AFT samples show better resolved Cenozoic thermal peaks and therefore provide the best constraints on the time and magnitude of maximum burial temperature.

The Ellice O-14 example clearly illustrates the importance of the proper interpretation and modeling of multi-kinetic AFT data. Different AFT kinetic populations can be treated as separate thermochronometers that are sensitive to different parts of the thermal history, and this can have important applications for resolving thermal histories in geologically complicated regions. Issler et al. (2005) used multi-kinetic AFT thermochronology to study the thermal history for a petroleum exploration well in the Mackenzie Corridor region of northern Canada, a frontier area with limited exploration success. Two statistical AFT populations, a fluorapatite (0.055 apfu Cl) and a higher retentivity apatite population (0.21 apfu Cl), were recovered from a well sample within the Devonian stratigraphic succession. The kinetic parameters for this sample are very similar to those of the Ellice O-14 samples, indicating that annealing temperatures differ by tens of degrees for these two AFT populations. Model results suggest that Devonian petroleum source rocks in the region reached peak temperature conditions ($\sim$ 125 °C) and generated petroleum during the Early Triassic to Middle Jurassic, prior to the development of Late Cretaceous–Cenozoic structural traps. Peak Cenozoic temperatures as recorded by the fluorapatite population ($\sim$ 100 °C) were insufficient to reactivate petroleum generation in the Devonian source rocks. Early Mesozoic heating thermally reset the fluorapatite FT population but the more retentive AFT kinetic population preserved a record of the early higher temperature event.

18.5.2 (U–Th)/He Modeling

Helium diffusion from apatite is a function of the volume fraction of radiation damage to the crystal, a quantity that varies over the lifetime of the apatite, and He diffusivity decreases with increasing damage (Shuster et al. 2006; Flowers et al. 2009; Gautheron et al. 2009; Shuster and Farley 2009). This has been interpreted as a result of preferential partitioning (trapping) of the He in damage zones, impeding diffusion. The relationship manifests as positive correlations between AHe age and effective uranium (eU), which among specimens from a sample that experienced a common time–temperature (t–T) history, is a proxy for relative extents of radiation damage. The radiation damage accumulation and annealing model (RDAAM; Flowers et al.

2009) and its variant of a damage-enhanced kinetic diffusivity model (Gautheron et al. 2009) adopt the effective FT density as a proxy for accumulated radiation damage. This proxy incorporates creation of crystal damage proportional to α-production from U and Th decay, and the elimination of that damage governed by the kinetics of FT annealing. It also has the potential to explain at least some cases in which (U–Th)/He ages are actually older than the corresponding FT ages.

Because of its lower T_c and the better documentation of apatite annealing kinetics which have been extended to the radiation damage models of apatite in the (U–Th)/He realm, the application of apatite thermochronology is quite widespread. Slightly less clear is the understanding of how radiation damage affects He diffusion in zircon. A zircon radiation model (ZRDAAM) has been applied to assess the thermal evolution of Neoproterozoic stratigraphy, which witnessed protracted (100 s m.y.) cooling through the uppermost crust (Guenthner et al. 2013; Powell et al. 2016). Single crystal ZHe ages from these older rocks possess intrasample age dispersion as great as 350 Myr, indicating that the strata have not been heated sufficiently to fully reset the ZHe system. The modeling has its most utility in samples where self-irradiation of zircon has occurred over long geologic timescales without significant annealing of damaged zones. Partially to fully reset detrital datasets can contain a tremendous amount of time–temperature information, due to the wide range in grain size, eU, and potential for variable pre-depositional histories and inherited radiation damage in the zircon population. These variables result in a broad spectrum of T_c within a single sample and in tectonic settings where strata are never buried to sufficiently high temperatures can potentially record more than the most recent thermal event. Additionally, modeling of these datasets has added value in strata where apatite is absent or too small for AHe analysis, as highly damaged zircon can record similar parts of the cooling history as the AHe system. However, several factors complicate interpretation of these datasets including the influences of pre-depositional history (e.g., inherited [4]He and radiation damage on ZHe ages and age–eU relationships). As a result of these variables, sampling from the same stratigraphic succession can yield substantially different ZHe age populations, despite having experienced the same thermal history, should the strata have different grain sizes or provenance. For these reasons, we believe that detailed thermal modeling is required to understand the probable geologic history exhibited by the datasets. We do not recommend selecting only the youngest ages from samples or averaging (U–Th)/He ages, as these methods do not acknowledge the complexity of the (U–Th)/He system and potentially exclude non-obvious, but equally probable, geologic scenarios. To this extent, using the vertical profile approach to assess exhumation rates from cooling age data may also provide an inaccurate result if the strata have not been buried to sufficient temperatures to completely reset any prior thermal history. As an alternative, analysis of more grains from individual samples and a combination of data from similar structural regions to assess regional trends in thermal history is the best approach (Guenthner et al. 2013; Powell et al. 2016). We believe that this tactic does an appropriate job of acknowledging the errors and assumptions involved in the technique and providing meaningful information on thermal history of a region.

In an example integrating multi-kinetic AFT and AHe dating, Powell et al. (2017) investigated the Upper Cretaceous Slater River Formation of the Mackenzie Plain of the Canadian Northwest Territories, which has potential as a regional source rock due to the high organic content and the presence of both oil- and gas-prone kerogen. That study examined a basal bentonite unit to understand the timing and magnitude of Late Cretaceous burial experienced by the Slater River Formation. Moreover, LA-ICP-MS and EPMA methods were employed to assess the chemistry of apatite and use these values to derive the AFT kinetic parameter r_{mr0}. The AFT ages and track lengths, respectively, range from 201.5 ± 36.9 to 47.1 ± 12.3 Ma, and 16.8 to 10.2 μm, and single crystal AHe ages are between 57.9 ± 3.5 and 42.0 ± 2.5 Ma with eU concentrations from 17.3 to 35.6 ppm. The AFT data exhibited no relationship with the kinetic parameter D_{par} and failed the χ^2 test indicating that the data do not comprise a single statistically significant population. However, when plotted against their r_{mr0} value, the data are separated into three statistically significant kinetic populations with distinct track length distributions: 154.3 ± 10.2, 89.0 ± 3.7, 53.5 ± 6.5 Ma. Inverse thermal history modeling of both the multi-kinetic AFT and AHe datasets reveal that the Slater River Formation reached maximum burial temperatures of ~ 65–80 °C between the Maastrichtian and Paleocene, indicating that at best the source rock matured to the early stages of hydrocarbon generation. The Powell et al. (2017) study emphasizes the importance of kinetic parameter choice in AFT thermochronology, as both the D_{par} measurements and the r_{mr0} values calculated without fluorine measurements are unable to explain the AFT age dispersion and track length distribution recognized in the F–OH apatite dataset. Apatite chemistry will have an effect on the helium diffusion in the AHe system as well, as it controls the temperatures at which the radiation damage anneals in apatite. Studies that integrate AFT and AHe thermochronology should consider these implications or risk thermal history model results that are inconsistent with the present understanding of annealing and diffusion kinetics in apatite.

18.6 Summary

The FT and (U–Th)/He systems have proven to be powerful but challenging tools for basin thermal history analysis, as they often have the unique potential to uncover information about passage through an important temperature window. The integration of these low-temperature thermochronometers has many potential uses in understanding the thermal and structural evolution of petroleum-bearing regions. Age-depth profiles from boreholes within a sedimentary package provide information on the timing and amount of burial and exhumation of the basin. If the cooling and exhumation are directly related to tectonics, they also constrain the timing of deformation, which in turn may have implications for hydrocarbon migration and trapping. It is also clear that, despite recent advances, important challenges remain for low-temperature thermochronologic systems. Both systems discussed commonly exhibit age dispersion beyond the analytical uncertainty of the techniques. When the samples are carefully and properly characterized, particularly with regard to mineral chemistry and radiation damage, a rich and effective dataset can emerge.

The toolbox for deciphering the thermal histories of hydrocarbon-bearing regions is growing, and advancing somewhat rapidly, particularly the numerical modeling. Several new low-temperature chronometers are being developed for application in sedimentary basins, including calcite (Copeland et al. 2007; Cros et al. 2014; Pagel et al. 2018) and fossils such as conodonts (Peppe and Reiners 2007) and crinoids (Copeland et al. 2015). These are particularly exciting since the science currently lacks reliable chronometers for carbonate-dominated basins. Moreover, technological advances in the acquisition of isotopic data now permit integrated in situ (U–Th–Sm)/He and U–Pb dating on single crystals (e.g., Evans et al. 2015; see Chap. 5 Danišík 2018) to obtain double dating of detrital minerals that will be beneficial for sediment provenance and recycling studies and in exploration applications. Defining the appropriate scientific question at the beginning of the project will help determine which techniques to implement. Although the examples outlined in this chapter primarily address conventional petroleum exploration in sedimentary basins, the thermochronology approaches can be also applied to non-conventional petroleum systems.

Acknowledgements The authors wish to thank the editors for the invitation to write this chapter, and for the constructive reviews by A. Gleadow and M. Zattin, which helped us clarify our thoughts. AFT analyses reported in this study were done by Dr. Sandy Grist at Dalhousie University. The Beaufort-Mackenzie study has been funded by a consortium of companies: Anadarko Canada Corporation, BP Canada Energy Company, Chevron Canada Limited, ConocoPhillips Canada Resources Corporation, Devon Canada Corporation, EnCana Corporation, Imperial Oil Resources Ventures Limited, MGM Energy Corporation, Petro-Canada (now Suncor), Shell Canada Limited, Shell

Exploration and Production Company, the Program of Energy Research and Development (PERD), and Natural Resources Canada. The NRCan contribution number is 20170139.

References

Barbarand J, Carter A, Wood I, Hurford T (2003) Compositional and structural control of fission track annealing in apatite. Chem Geol 198:107–137

Batten DJ (1996) Chapter 26B. Palynofacies and petroleum potential. In: Jasonius, J, McGregor DC (eds) Palynology: principles and applications. American Association of Stratigraphic Palynologists Foundation, pp 1065–1084

Bernet M, Brandon MT, Garver JI, Molitor B (2004) Fundamentals of detrital zircon fission-track analysis for provenance and exhumation studies with examples from the European Alps. Geol Soc Am 378:25–36

Beucher R, Brown R, Roper S, Persano C, Stuart F, Fitzgerald P (2013) Natural age dispersion arising from the analysis of broken crystals. Part II. Practical application to apatite (U-Th)/He thermochronometry. Geochim Cosmochim Acta 120:395–416

Brandon MT (2002) Decomposition of mixed grain age distributions using Binomfit. On Track 24:13–18

Brandon MT, Vance JA (1992) New statistical methods for analysis of FT grain age distributions with applications to detrital zircon ages from the Olympic subduction complex, western Washington State. Am J Sci 292:565–636

Braun J, van der Beek P, Batt G (eds) (2006) Quantitative thermochronology. Cambridge University Press, New York

Brown R, Beucher R, Roper S, Persano C, Stuart F, Fitzgerald P (2013) Natural age dispersion arising from the analysis of broken crystals. Part I: Theoretical basis and implications for the apatite (U–Th)/He thermochronometer. Geochim Cosmochim Acta 122:478–497

Burnham AK, Sweeney JJ (1989) A chemical kinetic model of vitrinite reflectance maturation. Geochim Cosmochim Acta 53(2):649–657

Burnett RD (1988) Physical and chemical changes in conodonts from contact-metamorphosed limestones. Irish J Earth Sci 9:79–119

Burtner RL, Nigrini A, Donelick RA (1994) Thermochronology of Lower Cretaceous source rocks in the Idaho-Wyoming thrust belt. AAPG Bull 78:1613–1636

Carlson WD (1990) Mechanisms and kinetics of apatite fission track annealing. Am Mineral 75:1120–1139

Carlson WD, Donelick RA, Ketcham RA (1999) Variability of apatite fission track annealing kinetics: I. Experimental results. Am Mineral 84:1213–1223

Cerveny PF, Naeser ND, Zeitler PK, Naeser CW, Johnson NM (1988) History of uplift and relief of the Himalaya during the past 18 million years; evidence from sandstones of the Siwalik Group. In: Kleinspehn KL, Paola C (eds) New perspectives in basin analysis. Springer, New York, pp 43–61

Cherniak D, Watson E, Thomas J (2009) Diffusion of helium in zircon and apatite. Chem Geol 268(1–2):155–166

Chew DM, Donelick RA (2012) Combined apatite fission track and U-Pb dating by LA-ICP-MS and its application in apatite provenance analysis. Mineral Ass Canada Short Course 42:219–247

Clauer N, Lehrman K (2012) Analyzing thermal histories of sedimentary basins: methods and case studies—Introduction. SEPM Special Publication 103:1–4

Copeland P, Cox K, Watson EB (2015) The potential of crinoids as (U + Th + Sm)/He thermochronometers. Earth Planet Sci Lett 422:1–10

Copeland P, Watson EB, Urizar SC, Patterson D, Lapen TJ (2007) Alpha thermochronology of carbonates. Geochim Cosmochim Acta 71:4488–4511

Corrigan J (1991) Inversion of apatite fission track data for thermal history information. J Geophys Res 96:10347–10360

Cros A, Gautheron C, Pagel M, Berthet P, Tassan-Got L, Douville E, Pinna-Jamme R, Sarda P (2014) ^{4}He behavior in calcite filling viewed by (U-Th)/He dating, ^{4}He diffusion and crystallographic studies. Geochim Cosmochim Acta 125:414–432

Crowhurst PV, Green PF, Kamp PJJ (2002) Appraisal of (U-Th)/He apatite thermochronology as a thermal history tool for hydrocarbon exploration: an example from the Taranaki Basin, New Zealand. Am Assoc Petrol Geol Bull 86:1801–1819

Crowley KD, Cameron M, Schaefer RL (1991) Experimental studies of annealing of etched fission tracks in fluorapatite. Geochim Cosmochim Acta 55:1449–1465

Danišík M (2018) Chapter 5. Integration of fission-track thermochronology with other geochronologic methods on single crystals. In: Malusà MG, Fitzgerald PG (eds) Fission-track thermochronology and its application to geology. Springer, Berlin

Donelick R (1993) A method of fission track analysis utilizing bulk chemical etching of apatite. USA Patent No. 5,267,274

Donelick RA, Ketcham RA, Carlson WD (1999) Variability of apatite fission track annealing kinetics: II. Crystallographic orientation effects. Am Mineral 84:1224–1234

Donelick RA, O'Sullivan PB, Ketcham RA (2005) Apatite fission track analysis. Rev Mineral Geochem 58:49–94

Duddy IR (1997) Focussing exploration in the Otway Basin: understanding timing of source rock maturation. Austr Petrol Prod Explor Ass J 37:178–191

Duddy I, Green P, Laslett G (1988) Thermal annealing of fission tracks in apatite: 3. Variable temperature behaviour. Chem Geol 73:25–38

Duddy I, Green P, Hegarty K, Bray R (1991) Reconstruction of thermal history in basin modelling using apatite fission track analysis: what is really possible? In: Offshore australia conference proceedings, vol 1, pp III-49–III-61

Duddy IR, Green PF, Bray RJ, Hegarty KA (1994) Recognition of the thermal effects of fluid flow in sedimentary basins. Geol Soc Spec Publ 78:325–345

Dumitru TA (2000) Fission track geochronology. In: Noller JS, Sowers JM, Lettis WR (eds) Quaternary geochronology: methods and applications. American Geophysical Union Ref Shelf 4, Washington, DC, pp 131–155

Epstein AG, Epstein JB, Harris LD (1977) Conodont color alteration—an index to organic metamorphism. USGS Prof Paper 995:1–27

Evans N, McInnes B, McDonald B, Danisik M, Becker T, Vermeesch P, Shelley M, Marillo-Sialer E, Patterson D (2015) An in situ technique for (U-Th-Sm)/He and U-Pb double dating. J Analyt Atomic Spect 30:1636–1645

Farley K (2000) Helium diffusion from apatite: general behavior as illustrated by Durango fluorapatite. J Geophys Res 105:2903–2914

Farley K (2002) (U-Th)/He dating: techniques, calibrations, and applications. Rev Mineral Geochem 47:819–844

Farley K (2007) He diffusion systematics in minerals: evidence from synthetic monazite and zircon structure phosphates. Geochim Cosmochim Acta 71:4015–4024

Fitzgerald P, Gleadow A (1990) New approaches in fission track geochronology as a tectonic tool: examples from the Transantarctic Mountains. Nucl Tracks Rad Meas 17:351–357

Fleischer RL, Price PB, Walker R (1964) Fission track ages of zircons. J Geophys Res 69:4885–4888

Fleischer RL, Price PB, Walker R (1975) Nuclear tracks in solids. University of California Press, Berkeley, p 595

Frey M, Teichmüller M, Teichmüller R, Mullis J, Künzi B, Breitschmid A, Gruner U, Schwizer B (1980) Very low-grade metamorphism in external part of the Central Alps: illite crystallinity, coal rank and fluid inclusion data. Eclogae Geol Helv 73:173–203

Flowers RM, Ketcham RA, Shuster DL, Farley KA (2009) Apatite (U-Th)/He thermochronometry using a radiation damage accumulation and annealing model. Geochim Cosmochim Acta 73:2347–2365

Foster D, Kohn B, Gleadow A (1996) Sphene and zircon fission track closure temperature revisited: empirical calibrations from $^{40}Ar/^{39}Ar$ diffusion studies on K-feldspar and biotite. In: International workshop on fission track dating 37

Galbraith RF (1981) On statistical models for fission track counts. J Math Geol 13:471–488

Galbraith RF (1988) Graphical display of estimates having different standard errors. Technometrics 30:271–281

Galbraith RF (1990) The radial plot: graphical assessment of spread in ages. Nucl Tracks Rad Meas 17:207–214

Galbraith RF (2005) Statistics for fission track analysis. Chapman & Hall/CRC, Boca Raton

Gallagher K (1995) Evolving temperature histories from apatite fission track data. Earth Planet Sci Lett 136:421–435

Gallagher K (2012) Transdimensional inverse thermal history modeling for quantitative thermochronology. J Geophys Res 117:B02408

Gallagher K, Brown R, Johnson C (1998) Fission track analysis and its applications to geological problems. Ann Rev Earth Planet Scis 26:519–572

Gallagher K, Stephenson Brown R, Holmes C, Fitzgerald P (2005) Low temperature thermochronology and modeling strategies for multiple samples 1: Vertical profiles. Earth Planet Sci Lett 237:193–208

Gallagher SJ, Duddy IR, Quilty PG, Smith AJ, Wallace MW, Holdgate GR, Boult PJ (2004) The use of Foraminiferal Colouration Index (FCI) as a thermal indicator and correlation with vitrinite reflectance in the Sherbrook Group, Otway Basin, Victoria. In: Boult PJ, Johns DR, Lang SC (eds) PESE Eastern Australasian basins symposium II. Petroleum Exploration Society of Australia, Adelaide, pp 643–653

Garver J (2008) Fission track dating. In: Gornitz V (ed) Encyclopedia of paleoclimatology and ancient environments. Encyclopedia of earth science series. Springer, Berlin, pp 247–249

Gautheron C, Tassan-Got L, Barbarand J, Pagel M (2009) Effect of alpha damage annealing on apatite (U-Th)/He thermochronology. Chem Geol 266:157–170

Gautheron C, Tassan-Got L, Ketcham RA, Dobson KJ (2012) Accounting for long alpha-particle stopping distances in (U-Th-Sm)/He geochronology: 3D modeling of diffusion, zoning, implantation, and abrasion. Geochim Cosmochim Acta 96:44–56

Gautheron C, Barbarand J, Ketcham R, Tassan-Got L, van der Beek P, Pagel M (2013) Chemical influence on α-recoil damage annealing in apatite: implications for (U-Th)/He dating. Chem Geol 351:257–267

Gleadow PF, Duddy IR, Lovering JF (1983) Fission track analysis: a new tool for the evaluation of thermal histories and hydrocarbon potential. Austr Petrol Explor Ass J 23:93–102

Gleadow AJW, Duddy IR, Green PF, Lovering JF (1986) Confined fission track lengths in apatite: a diagnostic tool for thermal history analysis. Contrib Mineral Petr 94:405–415

Gleadow AJW, Belton DX, Kohn BP, Brown RW (2002) Fission track dating of phosphate minerals and the thermochronology of apatite. Rev Mineral Geochem 48:579–630

Green PF, Duddy IR, Gleadow AJW, Tingate PR, Laslett GM (1985) Fission track annealing in apatite: track length measurements and the form of the Arrhenius plot. Nucl Tracks 10:323–328

Green PF, Duddy IR, Gleadow AJW, Tingate PR, Laslett GM (1986) Thermal annealing of fission tracks in apatite: a qualitative description. Chem Geol 59:237–253

Green PF, Duddy IR, Gleadow A, Lovering JF (1989a) Apatite fission track analysis as a paleotemperature indicator for hydrocarbon exploration. In: Naeser ND, McCulloh TH (eds) Thermal history of sedimentary basins—methods and case histories. Springer, New York, pp 181–195

Green PF, Duddy IR, Laslett GM, Hegarty KA, Gleadow AJW, Lovering JF (1989b) Thermal annealing of fission tracks in apatite 4. Quantitative modelling techniques and extension to geological timescales. Chem Geol 79:155–182

Guenthner WR, Reiners PW, Ketcham RA, Nasdala L, Geister G (2013) Helium diffusion in natural zircon: radiation damage, anisotropy, and the interpretation of zircon (U-Th)/He thermochronology. Am J Sci 313:145–198

Guenthner WR, Reiners P, DeCelles P, Kendall J (2015) Sevier belt exhumation in central Utah constrained from complex zircon (U-Th)/He data sets: radiation damage and He inheritance effects on partially reset detrital zircons. Geol Soc Am Bull 127(3–4):323–348

Harris AG (1979) Conodont color alteration, an organomineral metamorphic index and its application to Appalachian Basin geology. In: Scholle PA, Schluger PR (eds) Aspects of diagenesis, vol 26. SEPM Special Publication, pp 3–16

Harris N, Peters KE (eds) (2012) Analyzing the thermal history of sedimentary basins: methods and case studies, vol 103. SEPM Special Publication

Hasebe N, Barbarand J, Jarvis K, Carter A, Hurford AJ (2004) Apatite fission track chronometry using laser ablation ICP-MS. Chem Geol 207:135–145

Hegarty KA, Foland SA, Cook AC, Green PF, Duddy IR (2007) Direct measurement of timing: underpinning a reliable petroleum system model for the Mid-Continent Rift system. AAPG Bull 91:959–979

Helson S (1994) Micromorphological changes in Pridolian Lochkovian conodonts from low grade metamorphosed Naux Limestone (Ardennes, France). Bull Soc Belge Geol T103(1-2):205–207

Hendriks BWH, Andriessen P (2002) Pattern and timing of the post-Caledonian denudation of northern Scandinavia constrained by apatite fission track thermochronology. Geol Soc Spec Publ 196:117–137

Holland HD (1954) Radiation damage and its use in age determination. In: Faul H (ed) Nuclear geology. Wiley, New York, pp 175–179

Hower J, Eslinger EV, Hower M, Perry EA (1976) Mechanism of burial metamorphism of argillaceous sediments: 1. Mineralogical and chemical evidence. Geol Soc Am Bull 87:725–737

Hourigan JK, Reiners PW, Brandon MT (2005) U-Th zonation-dependent alpha ejection in (U-Th)/He chronometry. Geochim Cosmochim Acta 69:3349–3365

Hurford AJ (1986) Cooling and uplift patterns in the Lepontine Alps south central Switzerland and age of vertical movement on the Insubric fault line. Contrib Mineral Petr 92:413–417

Hurford AJ, Green PF (1982) A users' guide to fission track dating calibration. Earth Planet Sci Lett 59:343–354

Hurley PM (1952) Alpha ionization damage as a cause of low helium ratios. Eos Am Geophys Union 33:174–183

Issler DR (1996) An inverse model for extracting thermal histories from apatite fission track data: instructions and software for the Windows 95 environment. Geol Survey Canada Open File 2325: 84 p

Issler DR (2011) Integrated thermal history analysis of sedimentary basins using multi-kinetic apatite fission track thermochronology: examples from northern Canada. AAPG Foundation Distinguished Lecturer Series, AAPG Search and Discovery Article #90119

Issler DR, Grist AM (2008a) Integrated thermal history analysis of the Beaufort-Mackenzie basin using multi-kinetic apatite fission track thermochronology. Geochim Cosmochim Acta 72:A413

Issler DR, Grist AM (2008b) Reanalysis and reinterpretation of apatite fission track data from the central Mackenzie Valley, NWT, northern Canada: implications for kinetic parameter determination and thermal modeling. In: Proceedings from the 11th international conference on thermochronometry, pp 130–132

Issler DR, Grist AM (2014) Apatite fission track thermal history analysis of the Beaufort-Mackenzie Basin, Arctic Canada: a natural laboratory for testing multi-kinetic thermal annealing models. In: Proceedings from the 14th international conference on thermochronometry, pp 125–126

Issler DR, Willett SD, Beaumont C, Donelick RA, Grist AM (1999) Paleotemperature history of two transects across the Western Canada Sedimentary Basin: constraints from apatite fission track analysis. Bull Can Petrol Geol 47:475–486

Issler DR, Grist AM, Stasiuk LD (2005) Post-early Devonian thermal constraints on hydrocarbon source rock maturation in the Keele Tectonic Zone, Tulita area, NWT, Canada, from multi-kinetic apatite fission track thermochronology, vitrinite reflectance and shale compaction. Bull Can Petrol Geol 53:405–431

Issler DR, Reyes J, Chen Z, Hu K, Negulic E, Grist A, Stasiuk L, Goodarzi F (2012) Thermal history analysis of the Beaufort-Mackenzie Basin, Arctic Canada. In: SEPM Bob F. Perkins research conference vol 32, pp 609–641

Kamp PJJ, Green PF (1990) Thermal and tectonic history of selected Taranaki Basin (New Zealand) wells assessed by apatite fission track analysis. AAPG Bull 74:1401–1419

Ketcham RA (2003a) Effects of allowable complexity and multiple chronometers on thermal history inversion. Geochim Cosmochim Acta 67:A213

Ketcham RA (2003b) Observations on the relationship between crystallographic orientation and biasing in apatite fission track measurements. Am Mineral 88:817–829

Ketcham RA (2005) Forward and inverse modeling of low-temperature thermochronometry data. Rev Mineral Geochem 58:275–314

Ketcham R (2018) Chapter 3. Fission track annealing: from geologic observations to thermal history modeling. In: Malusà MG, Fitzgerald PG (eds) Fission-track thermochronology and its application to geology. Springer, Berlin

Ketcham RA, Donelick RA, Carlson WD (1999) Variability of apatite fission track annealing kinetics: III. Extrapolation to geological time scales. Am Mineral 84:1235–1255

Ketcham RA, Donelick RA, Donelick MB (2000) AFTSolve: A program for multi-kinetic modeling of apatite fission track data. Geol Mat Res 2:1–32

Ketcham RA, Carter A, Donelick RA, Barbarand J, Hurford AJ (2007a) Improved modeling of fission track annealing in apatite. Am Mineral 92:799–810

Ketcham RA, Carter A, Donelick RA, Barbarand J, Hurford AJ (2007b) Improved measurement of fission track annealing in apatite using c-axis projection. Am Mineral 92:789–798

Ketcham RA, Mora A, Parra M (2016) Deciphering exhumation and burial history with multi-sample down-well thermochronometric inverse modeling. Basin Res. https://doi.org/10.1111/bre.12207

Kirby E, Reiners PW, Krol M, Hodges K, Farley KA, Whipple K, Yiping L, Tang W, Chen Z (2002) Late Cenozoic uplift and landscape evolution along the eastern margin of the Tibetan plateau: Inferences from [40]Ar/[39]Ar and U-Th-He thermochronology. Tectonics. https://doi.org/10.1029/2000TC001246

Kohn BP, Foster DA, Farley KA (2002) Low temperature thermochronology of apatite with exceptional compositional variations: the Stillwater Complex, Montana revisited. Fission Track Analysis Workshop: Theory and Applications, El Puerto de Santa Maria, Spain. Geotemas 4:103–105

Kohn B, Chung L, Gleadow A (2018) Chapter 2. Fission-track analysis: field collection, sample preparation and data acquisition. In: Malusà

MG, Fitzgerald PG (eds) Fission-track thermochronology and its application to geology. Springer, Berlin

Lampe C, Person M, Nöth S, Ricken W (2001) Episodic fluid flow within continental rift basins: some insights from field data and mathematical models of the Rhinegraben. Geofluids 1:42–52

Landman R, Flowers R, Rosenau N, Powell J (2016) Conodont (U-Th)/He thermochronology: a case study from the Illinois Basin. Earth Planet Sci Lett 456:55–65

Laslett GM, Green PF, Duddy IR, Gleadow AJW (1987) Thermal annealing of fission tracks in apatite. 2. A quantitative analysis. Chem Geol 65:1–13

Lippolt HJ, Leitz M, Wernicke RS, Hagedorn B (1994) (U + Th)/He dating of apatite: experience with samples from different geochemical environments. Chem Geol 112:179–191

Lovera OM, Richter FM. Harrison M (1989) The [40]Ar/[39]Ar thermochronology for slowly cooled samples having a distribution of Diffusion Domain Sizes. J Geophys Res 94(B12):17,917–17,935

Lovera OM, Richter FM. Harrison M (1991) Diffusion domains determined by [39]Ar released during step heating. J Geophys Res 96:2057–2069

Lovera OM, Grove M, Harrison TM, Mahon KI (1997) Systematic analysis of K-feldspar [40]Ar/[39]Ar step heating results I: relevance of laboratory argon diffusion properties to nature. Geochim Cosmochim Acta 61:3171–3192

Lutz TM, Omar G (1991) An inverse method of modeling thermal histories from apatite fission track data. Earth Planet Sci Lett 104:181–195

Malusà MG, Fitzgerald PG (2018) Chapter 8. From cooling to exhumation: setting the reference frame for the interpretation of thermocronologic data. In: Malusà MG, Fitzgerald PG (eds) Fission-track thermochronology and its application to geology. Springer, Berlin

Malusà MG, Garzanti E (2018) Chapter 7. The sedimentology of detrital thermochronology. In: Malusà MG, Fitzgerald PG (eds) Fission-track thermochronology and its application to geology. Springer, Berlin

Mark D, Green PF, Parnell J, Kelly SP, Lee MR, Sherlock SC (2008) Late Palaeozoic hydrocarbon migration through the Clair field, West of Shetland, UK Atlantic margin. Geochim Cosmochim Acta 72:2510–2533

Marsellos A, Garver J (2010) Radiation damage and uranium concentration in zircon as assessed by Raman spectroscopy and neutron irradiation. Am Mineral 95:1192–1202

Marshall J (1991) Quantitative spore colour. J Geol Soc 148:223–233

McNeil DH (1997) Diagenetic regimes and the foraminiferal record in the Beaufort-Mackenzie Basin and adjacent cratonic areas. Ann Soc Geol Pol 67:271–286

McNeil DH, Issler DR, Snowdon LR (1996) Colour alteration, thermal maturity, and burial diagenesis in fossil foraminifers. Geol Surv Canada Bull 499:34 p

McNeil DH, Dietrich JR., Issler DR, Grasby SE, Stasiuk LD, Dixon J (2010) A new method for recognizing subsurface hydrocarbon seepage and migration using altered foraminifera from a gas chimney in the Beaufort-Mackenzie Basin. In: Wood L (ed) Shale tectonics. AAPG Mem 93: 197–210

McNeil DH, Schulze HG, Matys E, Bosak T (2015) Raman spectroscopic analysis of carbonaceous matter and silica in the test walls of recent and fossil agglutinated foraminifera. AAPG Bull 99:1081–1097

Meunier A, Velde B (2004) Illite: Origins, evolution and metamorphism. Springer, Berlin, p 288

Miller RL, Kahn JS (1962) Statistical analysis in the geological sciences. Wiley, New York, p 483

Mitchell SG, Reiners PW (2003) Influence of wildfires on apatite and zircon (U-Th)/He ages. Geology 31:1025–1028

Naeser CW (1979) Thermal history of sedimentary basins in fission track dating of subsurface rocks. SEPM Special Publication 26:109–112

Naeser ND, McCulloh TH (eds) (1989) Thermal history of sedimentary basins: methods and case histories. Springer, Berlin

Nasdala L, Reiners P, Garver J, Kennedy AK, Stern RA, Balan E, Wirth R (2004) Incomplete retention of radiation damage in zircon from Sri Lanka. Am Mineral 89:219–231

Nielsen SB, Clausen OR, McGregor E (2015) basin%Ro: a vitrinite reflectance model derived from basin and laboratory data. Basin Res 29 S1:515–536

Nöth S (1998) Conodont color (CAI) versus microcrystalline and textural changes in Upper Triassic conodonts from Northwest Germany. Facies 38:165–173

Obermajer M, Stasiuk LD, Fowler MG, Osadetz KG (1997) Acritarch fluorescence as a new thermal maturity indicator. AAPG Bull 81:1561

Osadetz KG, Kohn BP, Feinstein S, O'Sullivan PB (2002) Thermal history of Canadian Williston basin from apatite fission track thermochronology—implications for petroleum systems and geodynamic history. Tectonophysics 349:221–249

Pagel M, Bonifacie M, Schneider DA, Gautheron C, Brigaud B, Calmels D, Cros A, St-Bezar B, Landrein P, Davis D, Chaduteau C (2018) A big step in paleohydrological and diagenetic reconstructions in calcite veins and breccia of a sedimentary basin by combining d^{47} temperature, $d^{18}O_{water}$ and U-Pb age. Chem Geol 481:1–17

Pan Y, Fleet ME (2002) Compositions of the apatite-group minerals: substitution mechanisms and controlling factors. Rev Mineral Geochem 48:13–49

Peppe DJ, Reiners PW (2007) Conodont (U-Th)/He thermochronology: Initial results, potential, and problems. Earth Planet Sci Lett 258:569–580

Peters KE (1986) Guidelines for evaluating petroleum source rock using programmed pyrolysis. AAPG Bull 70:318–329

Powell J, Schneider DA, Stockli D, Fallas K (2016) Zircon (U-Th)/He thermochronology of Neoproterozoic strata from the Mackenzie Mountains, Canada: Implications for the Phanerozoic exhumation and deformation history of the northern Canadian Cordillera. Tectonics 35:35.1–35.27

Powell J, Schneider DA, Issler D (2017) Assessing source rock thermal history through multi-kinetic apatite fission track and (U-Th)/He thermochronology. Basin Res 30 S1:497–512

Powell J, Schneider DA, Desrochers A, Flowers, RM, Metcalf J, Gaidies F, Stockli DF (2018) Low-temperature thermochronology of Anticosti Island: A case study on the application of conodont (U-Th)/He thermochronology to carbonate basin analysis. Marine and Petroleum Geology. https://doi.org/10.1016/j.marpetgeo.2018.05.018

Press WH, Teukolsky SA, Vetterling, WT, Flannery BP (1992) Numerical recipes in FORTRAN: the art of scientific computing, 2nd edn. Cambridge University Press, 963 p

Price PB, Walker RM (1963) Fossil tracks of charged particles in mica and the age of minerals. J Geophys Res 68:4847–4862

Ravenhurst CE, Roden-Tice MK, Miller DS (2003) Thermal annealing of fission tracks in fluorapatite, chlorapatite, manganapatite, and Durango apatite: experimental results. Can J Earth Sci 40:995–1007

Reich M, Ewing RC, Ehlers TA, Becker U (2007) Low-temperature anisotropic diffusion of helium in zircon: Implications for zircon (U-Th)/He thermochronometry. Geochimica Cosmochimica Acta 71:3119–3130

Reiners P, Brandon M (2006) Using thermochronology to understand orogenic erosion. Annu Rev Earth Planet Sci 34:419–466

Reiners P, Spell T, Nicolescu S, Zanetti K (2004) Zircon (U-Th)/He thermochronometry: He diffusion and comparisons with $^{40}Ar/^{39}Ar$ dating. Geochim Cosmochim Acta 68:1857–1887

Robison CR, Van Gijzel P, Darnell LM (2000) The transmittance color index of amorphous organic matter: a thermal maturity indicator for petroleum source rocks. Int J Coal Geol 43:83–103

Saadoune I, Purton JA, de Leeuw NH (2009) He incorporation and diffusion pathways in pure and defective zircon $ZrSiO_4$: a density functional theory study. Chem Geol 258:182–196

Senftle JT, Landis CR (1991) Vitrinite reflectance as a tool to assess thermal maturity. In: Merrill RK (ed) Source and migration processes and evaluation techniques. AAPG treatise of petroleum geology, Handbook of petroleum geology, pp 119–125

Shuster DL, Farley K (2009) The influence of artificial radiation damage and thermal annealing on helium diffusion kinetics in apatite. Geochim Cosmochim Acta 73:183–196

Shuster DL, Flowers R, Farley K (2006) The influence of natural radiation damage on helium diffusion kinetics in apatite. Earth Planet Sci Lett 249:148–161

Spiegel C, Kohn B, Belton D, Berner Z, Gleadow A (2009) Apatite (U-Th-Sm)/He thermochronology of rapidly cooled samples: the effect of He implantation. Earth Planet Sci Lett 285:105–114

Staplin FL (1969) Sedimentary organic matter, organic metamorphism, and oil and gas occurrence. Can Petrol Geol Bull 17:47–66

Stock MJ, Humphreys MCS, Smith VC, Johnson RD, Pyle DM (2015) New constraints on electron-beam induced halogen migration in apatite. Am Mineral 100:281–293

Stockli DF (2005) Application of low-temperature thermochronometry to extensional tectonic settings. Rev Mineral Geochem 58:411–448

Stockli DF, Surpless BE, Dumitru TA, Farley KA (2002) Thermochronological constraints on the timing and magnitude of Miocene and Pliocene extension in the central Wassuk Range, western Nevada. Tectonics 21:4

Stormer JC Jr, Pierson ML, Tacker RC (1993) Variation of F and Cl X-ray intensity due to anisotropic diffusion in apatite during electron microprobe analysis. Am Mineral 78:641–648

Sweeney JJ, Burnham AK (1990) Evaluation of a simple model of vitrinite reflectance based on chemical kinetics. AAPG Bull 74:1559–1570

Tagami T (2005) Zircon fission track thermochronology and applications to fault studies. Rev Mineral Geochem 58:95–122

Tagami T, Dumitru TA (1996) Provenance and thermal history of the Franciscan accretionary complex: constraints from zircon fission track thermochronology. J Geophys Res 101:11,353–11,364

Tagami T, Carter A, Hurford AJ (1996) Natural long-term annealing of the zircon fission track system in Vienna Basin deep borehole samples: constraints upon the partial annealing zone and closure temperature. Chem Geol 130:147–157

Timar-Geng Z, Fügenschuh B, Schaltegger U, Wetzel A (2004) The impact of the Jurassic hydrothermal activity on zircon fission track data from the southern Upper Rhine Graben area. Schweiz Mineral Petr Mitt 84:257–269

Tissot BP, Welte DH (1978) Petroleum formation and occurrence: a new approach to oil and gas exploration. Springer, Berlin

Tissot B, Durand B, Espitalie J, Combaz A (1974) Influence of nature and diagenesis of organic matter in formation of petroleum. AAPG Bull 58:499–506

Vermeesch P (2009) RadialPlotter: a Java application for fission track, luminescence and other radial plots. Radiat Meas 44:409–410

Vermeesch P (2018) Chapter 6. Statistics for fission-track thermochronology. In: Malusà MG, Fitzgerald PG (eds) Fission-track thermochronology and its application to geology. Springer, Berlin

Vermeesch P, Tian Y (2014) Thermal history modelling: HeFTy vs QTQt. Earth-Sci Rev 139:279–290

Wells ML, Snee LW, Blythe AE (2000) Dating of major normal fault systems using thermochronology: an example from the Raft River detachment, Basin and Range, western United States. J Geophys Res 105:16,303–16,327

Willett SD (1997) Inverse modeling of annealing of fission tracks in apatite 1: a controlled random search method. Am J Sci 297:939–969

Wolf RA, Farley KA, Silver LT (1996) Helium diffusion and low temperature thermochronometry of apatite. Geochim Cosmochim Acta 60:4231–4240

Wolf RA, Farley KA, Kass DM (1998) Modeling of the temperature sensitivity of the apatite (U-Th)/He thermochronometer. Chem Geol 148:105–114

Yamada R, Tagami T, Nishimura S, Ito H (1995) Annealing kinetics of fission tracks in zircon: an experimental study. Chem Geol 122:249–258

Yamada R, Murakami M, Tagami T (2007) Statistical modeling of annealing kinetics of fission tracks in zircon; reassessment of laboratory experiments. Chem Geol 236:75–91

Zaun PE, Wagner GA (1985) Fission track stability in zircons under geological conditions. Nucl Tracks 10:303–307

Zeitler PK, Herczeg AL, McDougall I, Honda M (1987) U-Th-He dating of apatite: a potential thermochronometer. Geochim Cosmochim Acta 51:2865–2868

The Application of Low-Temperature Thermochronology to the Geomorphology of Orogenic Systems

19

Taylor F. Schildgen and Peter A. van der Beek

Abstract

The geomorphologic evolution of orogens has been a subject of revived interest and accelerated development over the past few decades, thanks to both the increasing availability of high-resolution data and computing power and the realisation that orogenic topography plays a central role in coupling deep-earth and surface processes. Low-temperature thermochronology takes a central place in this revived interest, as it allows us to link quantitative geomorphology to the spatial and temporal patterns of exhumation. In particular, rock cooling rates over million-year timescales derived from thermochronological data have been used to reconstruct rock exhumation histories, to detect km-scale relief changes, and to document lateral shifts in relief. In this chapter, we review how classic approaches of determining exhumation histories have contributed to our understanding of landscape evolution, and we highlight novel approaches to quantifying relief changes that have been developed over the last decade. We discuss how patterns of exhumation in laterally accreting orogens are recorded by low-temperature thermochronology, and how such data can be applied to infer temporal variations in exhumation rates, providing indirect constraints on topographic development. We subsequently review recent studies aimed at quantifying relief development and modification associated with river incision, glacial modifications of landscapes, and shifts in the position of range divides. We also point out how interpretations of some datasets are non-unique, emphasizing the importance of understanding the full range of processes that may influence landscape morphology and how each may affect spatial patterns of thermochronologic ages.

19.1 Introduction

Understanding the development of topography not only helps to reconstruct the geodynamic and surface processes responsible for landscape development, but it is also one of the most important requirements for correctly interpreting stratigraphic records, speciation patterns, and regional to global paleoclimatic changes (Ruddiman 1997; Crowley and Burke 1998). Despite its clear importance, some of the most common techniques for reconstructing paleotopography, involving reconstructing changes in paleotemperature from paleobotany or changes in stable-isotope ratios, have limited resolution, and the latter in particular require detailed knowledge of air circulation patterns, isotope lapse rates in space and time, continental (evapo-) transpiration, and vapor recycling (Mulch 2016). Moreover, although these approaches can help reveal the existence of high topography, they typically do not provide information on the distribution of elevation across the landscape (i.e., the relief), such as may be created by rivers or glaciers. Low-temperature thermochronology complements and may offer some advantages over these approaches. Although it does not provide direct constraints on paleo-elevations, low-temperature thermochronology can be used to (1) resolve changes in rock cooling rates over million-year timescales, which can be interpreted in terms of rock exhumation and may provide indirect constraints on topographic development (e.g., Montgomery and Brandon 2002); (2) detect km-scale relief changes; and (3) document lateral shifts in relief.

One of the most common applications of low-temperature thermochronology in the realm of geomorphology has been to test for changes in erosion/exhumation rates, which are typically linked to changes in climate, uplift rates, and/or

T. F. Schildgen (✉)
Helmholtz Zentrum Potsdam, GeoForschungsZentrum (GFZ), Potsdam, Germany
e-mail: tschild@gfz-potsdam.de

P. A. van der Beek
Institut des Sciences de la Terre (ISTerre), Université Grenoble Alpes, Grenoble, France

© Springer International Publishing AG, part of Springer Nature 2019
M. G. Malusà and P. G. Fitzgerald (eds.), *Fission-Track Thermochronology and its Application to Geology*,
Springer Textbooks in Earth Sciences, Geography and Environment, https://doi.org/10.1007/978-3-319-89421-8_19

topography. But changes in exhumation rates are typically linked to changes in topographic relief, making it difficult to separate the two. Although the influence of changing relief on low-temperature thermochronometers has been appreciated for decades (e.g., Stüwe et al. 1994; Mancktelow and Grasemann 1997), it is only within the past decade that a number of novel approaches to quantify changes in landscape relief have emerged. These relief changes include those associated with river incision, glacial modifications of landscapes, and shifts in the position of range divides. Because thermochronometers are typically limited to spatial resolution on the km scale and temporal resolution on the million-year scale, thermochronological data is most relevant for discerning long-term changes in relief that occur in orogenic or post-orogenic settings.

In this chapter, we review how classic approaches of determining exhumation histories have contributed to our understanding of landscape evolution, and we highlight novel approaches to quantifying relief changes. In each case, a range of different sampling schemes have provided useful information (Fig. 19.1): bedrock samples have been collected in transects that traverse topography either in very steep (Fig. 19.1b, c) or in sub-horizontal (Fig. 19.1d) transects, over distances that are either localised or span the width of an orogen. Detrital samples (Fig. 19.1e) have been collected from modern river sediments throughout the landscape and from within dated sedimentary stratigraphic sections. Previous reviews (e.g., Braun 2005; Spotila 2005; Braun et al. 2006; Reiners and Brandon 2006) have laid out the groundwork for many of these techniques; here, we will focus on studies that have taken advantage of the record of past thermal structure and erosion patterns contained in low-temperature thermochronometers to study landscape development in orogenic systems.

19.2 Exhumation Histories from Low-Temperature Thermochronology

On the scale of a compressional orogenic system, the tectonic (accretionary) influx of crustal material into the system leads to thickening of the crust, isostatic uplift of the surface, and increasing topographic relief. Because erosion rates tend to increase with steeper slopes (Ahnert 1970), topographic relief is predicted to increase until the flux of material leaving the system through erosion matches the tectonic influx (Jamieson and Beaumont 1988). An important refinement to this concept in orogenic settings was the notion of "threshold hillslopes," which are the strength-limited maximum slopes that can be achieved despite further increases in erosion rates related to landslide frequency (Larsen and Montgomery 2012). This concept, first illustrated by the nonlinear

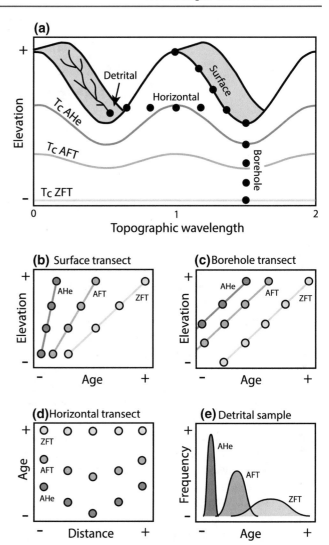

Fig. 19.1 Sampling schemes for thermochronologic data. **a** Illustration of various surface- and subsurface sampling schemes. Solid black circles are sample sites. T_c refers to closure temperature of the associated thermochronologic system. AHe: apatite (U–Th)/He; AFT: apatite fission track; ZFT: zircon fission track. **b** Age-elevation plot of a steep elevation surface transect. For the thermal structure shown in (**a**), the slope of the ZFT data provides the correct exhumation rate, while those of the AFT and AHe data overestimate the exhumation rate (see Sect. 19.2.2.2). **c** Age-elevation plot of borehole samples, in which each thermochronologic system provides the correct exhumation rate if the thermal structure is steady over time. **d** Age-distance plot of samples from a horizontal transect, which reflect the inverted shape of the T_c isotherm at the time that the samples crossed it. **e** Age-frequency plot of a detrital sample, illustrating the distribution of ages that result from material derived from a range of elevations within the source area; note that the width of the PDF is comparable to the range of ages shown in (**b**). Parts (**a–d**) modified from Braun et al. (2012), reproduced with permission from Elsevier

relationship between mean slope and erosion rates derived from thermochronological data (Burbank et al. 1996; Montgomery and Brandon 2002), has since been supported with catchment-mean erosion rates derived from cosmogenic nuclides (e.g., Binnie et al. 2007; Ouimet et al. 2009).

If the accretionary flux remains constant, topography is likely to be relatively stable as well, with steeper slopes enabling faster erosion in areas of faster rock uplift (Willett and Brandon 2002). Under these conditions, exhumation steady state, reflecting the rates and pathways through which rocks are exhumed to the surface, should also be reached (Willett and Brandon 2002) (Fig. 19.2). Hence, topographic and exhumational steady state are presumably closely linked. In the examples that follow, we will illustrate how exhumation histories derived from thermochronology data have been used to infer either steady-state topography or changes in uplift rates and the evolution of topography.

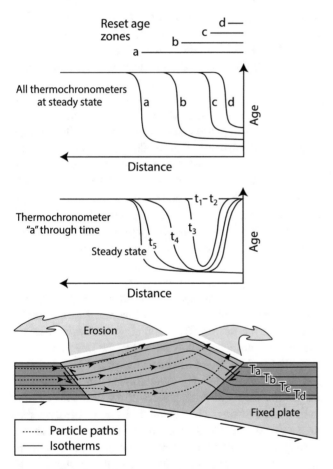

Fig. 19.2 Brandon 2002, reproduced with permission from the Geological Society of America). Closure temperature "*T*" for each thermochronometer (with subscripts a through d) is illustrated in the cartoon below. Exhumation pathways that bring samples below their respective closure isotherms result in reset ages at the surface, which are much younger than those of samples that did not pass below their closure isotherm. Note that thermochronometers with lower closure temperatures show a wider distribution of reset ages across the orogen, with reset zones of higher-temperature thermochronometers nested within

19.2.1   Exhumation Patterns in Laterally Accreting Orogens

The spatial distribution of thermochronometric ages across an orogen can help reveal whether or not exhumational steady state has been reached (Batt and Braun 1999; Willett and Brandon 2002). Material that passes through a compressional orogenic system follows a trajectory that is controlled by the direction of accretion and the pattern of erosion at the surface (Fig. 19.2). Material that enters the system through the accretionary flux will be heated to a temperature that depends on the depth of its trajectory before being cooled as it approaches the surface. Across a laterally accreting orogen, material that reached greatest depths tends to occur near the center, or is offset toward the retro side of the orogen (opposite from the accreting side) (Fig. 19.2). This characteristic of the material pathways implies that when an orogen reaches exhumational steady state (unchanging rates and patterns of exhumation), different thermochronometers will show distinct patterns of reset ages, with higher-temperature thermochronometers only reset in the region revealing deepest exhumation pathways, and lower-temperature thermochronometers showing wider zones of reset ages. Overall, this should create a pattern of higher-temperature reset age zones successively nested within lower-temperature reset age zones (Fig. 19.2).

Several studies have taken advantage of this characteristic of compressional orogenic systems to test for exhumational steady state. Batt and Braun (1999) used the pattern of nested ages in the southern Alps of New Zealand together with thermal modeling (see also Chap. 13, Baldwin et al. 2018) to illustrate the west-to-east transport of material across the orogen, as well as to show that the system is close to or at exhumational steady state. A nested pattern of ages across the Cascadia wedge in NW Washington state (Brandon et al. 1998) was argued to illustrate the existence of flux steady state, but not yet exhumational steady state, because the reset age zones have not yet reached the northeastern margin of the orogen (Batt et al. 2001). In Taiwan as well, the pattern of reset zircon (ZFT) and apatite (AFT) fission-track ages published by Liu et al. (2001) and Willett et al. (2003) was used to argue that the central portion of the mountain range is in exhumational steady state. However, because the collision propagates to the south, steady state has not yet been reached for the entire range, leading to a narrowing and eventual disappearance of the steady-state zone to the south (Willett and Brandon 2002; Willett et al. 2003).

The importance of lateral rock advection for creating asymmetric topography and controlling exhumation pathways has been demonstrated on smaller spatial scales as well, including an individual fault-bend fold (e.g., Miller et al. 2007). Across the Himalayan front, Whipp et al. (2007) showed how lateral rock advection with spatial variations in erosion rates was necessary to explain complex patterns of AFT data. Since then, numerous studies have used thermochronologic data from the Himalaya to reconstruct exhumation pathways across individual structures and/or ramps with duplex structures (e.g., Robert et al. 2009, 2011; Herman et al. 2010; Landry et al. 2016; van der Beek et al. 2016).

19.2.2 Determining Exhumation Histories

Although obtaining a wide spatial distribution of cooling ages is effective for evaluating the exhumational "state" of an orogen, useful information can also be derived from more spatially targeted sets of samples. In the following section, we review various approaches to deriving exhumation histories based on (i) multiple thermochronometers and shorter-term erosion rate data, (ii) age-elevation transects, and (iii) detrital thermochronology (see also Chap. 9, Fitzgerald and Malusà 2018 and Chap. 10, Malusà and Fitzgerald 2018b). It is important to keep in mind that exhumation rates derived from thermochronology are not measured directly (see Chap. 8, Malusà and Fitzgerald 2018a), but rather are inferred from cooling ages and an assumed crustal thermal structure. As such, we discuss several of the complications that can make it difficult to derive accurate exhumation rates without the aid of thermal modeling.

19.2.2.1 Exhumation Histories from Multiple Thermochronometers

As argued in Sect. 19.2, topographic and exhumational steady states are often presumed to be linked (Willett and Brandon 2002). Hence, exhumation histories derived from multiple thermochronometers have commonly been used to test for topographic steady state. Studying the Smoky Mountains (USA), Matmon et al. (2003) reconstructed an exhumation history not only with low-temperature thermochronometry, but also with cosmogenic nuclides and sediment yields to demonstrate similar exhumation rates over timescales of 10^8–10^2 years. This apparent longevity of the topography of the range was suggested to result from a thickened crustal root that has responded isostatically to erosion for nearly 200 million years (Matmon et al. 2003). However, even in the case of persistently steady exhumation in post-orogenic settings, changes in topography may occur. Within the Dabie Shan of eastern China, Reiners et al. (2003b) used AFT, ZFT and apatite (U–Th)/He data (AHe) to infer similar exhumation rates over the past ~115 million years, but the data were best predicted by assuming a decay of topography over time. Braun and Robert (2005) refined the estimates of relief loss to between a factor of 2.5–4.5 since the cessation of tectonic activity.

In many other cases, exhumation histories spanning millions of years reveal changes through time. Some of the earliest studies of exhumation histories based on multiple thermochronometers were used to infer uplift and the development of topography in the European Alps (e.g., Wagner et al. 1977; Hurford 1986; Hurford et al. 1991) and in the northwest Himalaya (e.g., Zeitler et al. 1982). However, linking an exhumation history to topographic development is not always straightforward. As we will illustrate below, even in the case of similar derived exhumation histories, differing interpretations of paleotopography can emerge.

Several studies have used exhumation histories from multiple thermochronometers to infer multi-km-uplift of the Tibetan plateau. In eastern Tibet, Kirby et al. (2002) used biotite $^{40}Ar/^{39}Ar$ thermochronology, multiple diffusion-domain modeling of alkali feldspar ^{40}Ar release spectra, and (U–Th)/He thermochronology of zircon (ZHe) and apatite to construct cooling histories of two different study areas, one within the Tibetan plateau interior, and another at the eastern plateau margin adjacent to the Sichuan Basin. The plateau margin site revealed very slow cooling from Jurassic to the late Miocene or early Pliocene, followed by rapid cooling. The plateau interior site, in contrast, showed relatively slow cooling throughout the same period, with only a small increase in cooling rates starting sometime within the middle Tertiary. Kirby et al. (2002) suggested that the rapid cooling along the plateau margin site was induced by the development of relief in that area, implying that the high topography adjacent to the Sichuan Basin was created in the late Miocene or early Pliocene.

A similar approach was used by van der Beek et al. (2009) in the northwest Himalaya to help determine the age and origin of high elevation, low-relief surfaces characterising that part of the orogen. By combining AFT with AHe and ZHe thermochronology, they found that the Deosai plateau, a low-relief surface at ~4 km elevation east of Nanga Parbat, had experienced relatively slow cooling since at least the middle Eocene (~35 Ma), just 15–20 Myr after the onset of India–Asia collision. They therefore inferred that the region marks one of several remnants of a plateau region that had already been uplifted by middle Eocene time. Rohrmann et al. (2012) extended the multi-thermochronometer approach into central Tibet, where they found moderate to rapid cooling from Cretaceous to Eocene time, followed by relatively slow cooling since ~45 Ma. They suggested that the early phase of rapid cooling was related to crustal shortening, thickening, and erosion associated with India-Asia collision, whereas the subsequent extended phase of slow cooling represents the establishment of a high elevation, low-relief plateau in the Eocene, in line with the interpretations of van der Beek et al. (2009).

Similar to the studies by van der Beek et al. (2009) and Rohrmann et al. (2012), Hetzel et al. (2011) found evidence for rapid cooling in south-central Tibet between ~70 and 55 Ma followed by a phase of very slow cooling since ~50 Ma based on AFT, ZHe, and AHe data. However, unlike the other interpretations, Hetzel et al. (2011) suggested that the slow exhumation since ~50 Ma represented a phase of peneplanation from laterally migrating rivers at low elevations, and that uplift must have occurred sometime later, subsequent to India-Eurasia collision, without inducing any faster exhumation. Highlighting the differing assumptions that led to these very different interpretations, Rohrmann et al. (2012) argued that low-relief surfaces can form at high elevation, calling into question the suggestions of Hetzel et al. (2011). The contrasting interpretations illustrate the potential difficulty of inferring uplift and topographic development from cooling (exhumation) histories alone (see also Chap. 8, Malusà and Fitzgerald 2018a).

19.2.2.2 Exhumation Histories from Age-Elevation Relationships

In regions of high topographic relief, steep elevation transects of bedrock samples can be used to derive exhumation rates over a range of time defined by the age distribution of the samples, with the slope of the line on an age-elevation plot (for a single thermochronometer) reflecting, to a first order, the exhumation rate. Wagner and Reimer (1972) and Wagner et al. (1977) were the first to propose and apply this approach to determine exhumation histories at several locations within the European Alps. Assumptions inherent to this approach include: (1) the sample's elevation is an accurate proxy for distance from the closure-temperature (T_c) isotherm; (2) there are minimal differences in the erosion rates across the horizontal distance of the samples; and (3) the isotherm's position has not changed over the timescale of exhumation (Reiners and Brandon 2006). The first assumption is most problematic for low-temperature systems, because their associated T_c isotherms more closely mimic topography compared to high-temperature systems (Stüwe et al. 1994, Fig. 19.1a). Braun (2002a) noted that in these cases, exhumation rates estimated directly from age-elevation relationships will be overestimated, as the change in elevation from one sample to the next is greater than the change in the distance traveled from the T_c isotherm (Fig. 19.1a, b). The second assumption may be valid in cases where the horizontal distance between samples is minimal (e.g., Braun 2002a; Valla et al. 2010), as will be discussed in more detail in Sect. 19.3.1.2. The third assumption may be reasonable for areas that have undergone exhumation at constant rates.

Generally, the onset of faster exhumation will produce a region over which slopes of the age-elevation relationship change, because thermochronometers tend to exhibit a zone over which daughter products (or fission tracks) are only partially retained (or annealed) (Gleadow and Fitzgerald 1987; Baldwin and Lister 1998; Wolf et al. 1998). In the case of AHe thermochronology, this "partial retention zone" (PRZ) occurs between ~40–80 °C for typical cooling rates, with the exact bounds dependent on the chemical composition of the crystal, the cooling rate, and any accumulated radiation damage that affects He diffusion (Reiners and Brandon 2006). In the case of AFT thermochronology, tracks that are created will be instantaneously annealed at temperatures above 110 °C, and slowly annealed within the "partial annealing zone" (PAZ) between ~60 and 110 °C (Reiners and Brandon 2006), the exact temperatures again dependent on apatite composition and cooling rate. Hence, even in the case of a sudden increase in exhumation rates, age-elevation relationships tend to show a curved or double-kinked zone, with the zone of intermediate or changing slope representing the width of the exhumed PAZ/PRZ (see Chap. 9, Fitzgerald and Malusà 2018 for details).

Another complication to directly inferring exhumation rates from age-elevation relationships arises because isotherms tend to be advected upward with the onset of faster exhumation; hence, the onset of faster cooling (which involves rocks crossing isotherms) will be delayed. Moore and England (2001) showed that for a step-wise increase in exhumation rates, the cooling ages will initially reflect a gradual increase in exhumation rates. Reiners and Brandon (2006) estimated the time required for each isotherm to reach a steady position following the onset of faster exhumation: for an increase in exhumation rate from 0 to 1 km/Myr, the T_c isotherm for He diffusion in apatite slows to 10% of its initial (advected) upward velocity after ~2.4 Myr, while it takes ~7.5 Myr for the T_c isotherm of Ar in muscovite to slow to 10% of its initial velocity.

Changes in relief can also affect age-elevation relationships. Using synthetic data extracted from the finite-element thermal-kinematic model Pecube (Braun 2003), Braun (2002a) illustrated how increasing relief leads to a shallower slope of the plot, whereas decreasing relief leads to a steepening. In extreme cases, decreases in relief may even result in an inverted slope, with ages decreasing with increasing elevation (see Chap. 9, Fitzgerald and Malusà 2018). Although low-temperature thermochronometers are most sensitive to changes in relief, Braun (2002a) illustrated that age-elevation plots of all thermochronometers with T_c below 300 °C are affected by it. A complication that follows from these findings is that a change in slope in an age-elevation relationship may result from a change in relief, a change in exhumation rates, or both. In a synthetic modeling study, Valla et al. (2010) investigated how well changes in exhumation rates and landscape relief could be quantified from age-elevation relationships. They illustrated how multiple thermochronometers can be most effective in resolving both, but in the case of AHe and AFT data (ages and track lengths), the rate of relief growth must be 2–3 times higher than the background exhumation rate in order to be quantified and precisely resolved. This limitation was illustrated with field data from La Meije Peak in the western Alps, where a combination of ZFT, AFT and AHe data was effective in constraining temporal changes in exhumation rates but not the relief history, due to the relatively high background denudation rates (van der Beek et al. 2010).

These complications related to the PRZ/PAZ, thermal response times, and changes in relief can make it very difficult to infer exhumation histories directly from age-elevation relationships. It is only since the development of 1D thermal models (Brandon et al. 1998; Ehlers et al. 2005; Reiners and Brandon 2006), spectral analyses (Braun 2002b), 3D thermal-kinematic models (Braun 2003), and linear inversion approaches (Fox et al. 2014), which can take into account many or all of these effects, that more accurate reconstructions of exhumation histories have been possible. However, in a densely sampled transect of Denali in Alaska, Fitzgerald et al. (1995) illustrated the power of combining AFT ages and track-length distributions to derive an accurate exhumation history without thermal modeling (see Chap. 9, Fitzgerald and Malusà 2018). Because fission tracks are slowly annealed (shortened) between $\sim$60 and 110 °C (in the PAZ), but rapidly (instantaneously) annealed at temperatures above 110 °C, the pattern of track-length distributions can be used to reconstruct the position of the exhumed base of the fossil PAZ (110 °C isotherm). Indeed, in addition to finding a sharp steepening in slope of the age-elevation relationship below 4500 m elevation (starting at $\sim$6 Ma), Fitzgerald et al. (1995) found a change from a wide distribution of track lengths above 4500 m (13.2 $\pm$ 2.4 μm), indicating relatively slow cooling and annealing through the PAZ, to a narrow distribution of long track lengths below 4500 m (14.6 $\pm$ 1.4 μm), indicating rapid cooling through the PAZ with minimal annealing. Hence, they identified the sharp break in slope and change in track-length distributions as the exhumed base of the fossil PAZ, which marked the 110 °C isotherm prior to the onset of rapid uplift. By reconstructing the current depth of the 110 °C isotherm based on an assumed geothermal gradient and using independent (sedimentary) constraints on the paleotopography, Fitzgerald et al. (1995) were able to infer the total rock uplift, the amount of surface uplift, as well as the amount of exhumation that occurred since the start of rapid exhumation.

19.2.2.3 Detrital Sediment Lag-Times and Age Distributions

Detrital material, either from modern river networks or from sedimentary rocks, can also be used to constrain exhumation rates across the contributing source area. One common application is to convert the difference between cooling ages and depositional ages (the "lag-time") into an exhumation rate (e.g., Brandon and Vance 1992; Brandon et al. 1998; Garver et al. 1999). Assumptions underlying this approach are discussed in Chap. 10 (Malusà and Fitzgerald 2018b). Within a single detrital sample, however, there is a range of ages, which record spatial variations in exhumation rates across the contributing area and/or topographic relief, which is typically characterized by increasing ages at higher elevations even for uniform exhumation rates (Garver et al. 1999, Fig. 19.1e). As such, a maximum exhumation rate may be determined for the youngest age population within a sample, based on the T_c isotherm depth beneath the lowest elevation point within the contributing area. In cases where multiple samples from a stratigraphic section are available, it is possible to evaluate how the lag-times, or exhumation rates, evolve through time (see Chap. 15, Bernet 2018). These patterns have been linked to the topographic evolution of orogens, with decreasing lag-times representing orogenic growth, stable lag-times representing steady state, and increasing lag-times representing topographic decay (e.g., Bernet and Garver 2005).

In an early application to the European Alps, Bernet et al. (2001) found that the ZFT grain-age distributions from a stratigraphic section spanning the last 15 Myr showed constant lag-times, suggesting steady-state exhumation since at least 15 Ma. What remained unclear, however, was how well the approach could resolve exhumation-rate changes that might have occurred. In a more recent study from the Western Alps, Glotzbach et al. (2011b) found constant AFT lag-times since 10 Ma. From their sensitivity analysis based on 3D thermal modeling with Pecube, they illustrated that the data should have been capable of resolving a twofold increase in exhumation rates at 5 Ma, but not an increase as recent as 1 Ma.

In an alternative approach, Brewer et al. (2003) and Ruhl and Hodges (2005) used the age distribution from a modern detrital sample and measurements of modern catchment relief to determine a catchment-averaged erosion rate in the Nepalese Himalaya. In essence, they use the detrital data to predict a vertical distribution of cooling ages. Both studies point out the range of assumptions inherent to the approach, including vertical exhumation pathways, steady and uniform erosion over the T_c interval represented by the detrital ages, and no change in catchment relief compared to the present. This approach is also affected by the variable mineral fertility in different source rocks, which is the variable propensity of different parent rocks to yield detrital grains of specific minerals when exposed to erosion (Malusà et al. 2016; see also Chap. 7, Malusà and Garzanti 2018). Brewer et al. (2003) used a modeling approach to explore what average erosion rate gives an age distribution most similar to the catchment hypsometry and found that good matches could be obtained for a slowly eroding catchment. Poorer results were obtained from a more rapidly eroding catchment, which they suggested was a consequence of higher uncertainties in the younger ages, the steeper age-elevation relationship (leading to less spread in expected ages), and non-uniform erosion rates. Ruhl and Hodges (2005) simply divided the range of modern elevations by the range of ages to determine an average erosion rate, but compared the age distribution to the shape of catchment hypsometries as a first-order test for the steady-state assumptions. Out of their three studied catchments, only one showed similarity between detrital age distribution and catchment hypsometry, suggesting steady, uniform erosion over the age range of the samples (between 11 and 2.5 Ma). For the others, Ruhl and Hodges (2005) suggested that a single detrital sample was unlikely to fully characterize the temporally and spatially transient erosion of the source area.

More recent studies have incorporated significant improvements in these approaches. For example, Brewer and Burbank (2006) used a kinematic and thermal model to better predict bedrock cooling ages along the Himalayan front in central Nepal, where lateral rock advection is an important component of exhumation pathways. Furthermore, Avdeev et al. (2011) illustrated how Bayesian statistics and the Markov chain Monte Carlo algorithm could be used to invert for the timing of erosion-rate changes. When applying this approach to single-grain AFT and AHe ages from large rivers along the southeast margin of the Tibetan plateau, Duvall et al. (2012) found that all the rivers recorded an increase in exhumation rate between 11 and 4 Ma.

19.3 Relief Development and Modification in Landscapes

Although testing for changes in exhumation rates in an active orogen is one way to indirectly constrain paleotopography or paleorelief, a number of novel approaches in thermochronology have provided more direct constraints on the timing and magnitude of relief change. Most of these approaches take advantage of the tendency of near-surface isotherms to mimic the shape of topography, with strong changes in the shape of the isotherms resulting from changes in surface morphology (Fig. 19.1) (Stüwe et al. 1994; Mancktelow and Grasemann 1997). Because these effects are strongest for isotherms that are closest to the surface, most of these studies employ systems with very low T_c such as AHe or ^{4}He/^{3}He thermochronometry. In the subsections that follow, we explore a number of different approaches that have been used to investigate changes in landscape relief in various contexts: fluvial landscapes, glacial landscapes, and along drainage divides. As these examples show, (very) low-temperature thermochronology data have the potential

to constrain models of landscape evolution when collected using a targeted sampling strategy. However, careful consideration of all possible scenarios as well as the detailed kinetic behavior of the samples is required to discriminate among competing models.

19.3.1 Relief Changes in Fluvial Landscapes

The timing and magnitude of km-scale river-valley incision can provide critical information on the influence of climatic or tectonic forcing on a landscape, or may alternatively record a major change in a river network, such as a large capture event. In the following examples, we highlight various approaches that have successfully constrained the relief history of fluvial landscapes, individual valleys, or individual reaches of a river channel.

19.3.1.1 Landscape Relief Derived from Sub-horizontal Transects

One of the earliest applications of a single thermochronometer to constrain both the magnitude and age of topographic relief within a fluvially sculpted landscape is the work by House et al. (1997, 1998, 2001) in the Sierra Nevada, California. In their approach, samples were collected at a similar elevation along the full length of the range (similar to the sampling approach illustrated in Fig. 19.1c), with the presumption that variations in age would reflect deflection of the T_c isotherm at the time the samples passed through it, and hence, paleorelief. AHe ages across the northern sector of the Sierra Nevada (the Kings and San Joaquin river valleys) ranged from $\sim$40 to $\sim$70 Ma and, interestingly, showed a pattern of ages that varied inversely with the long-wavelength topography of the range: older ages occurred across the broad valleys and younger ages occurred around the peaks. Accordingly, House et al. (1997, 1998, 2001) inferred that km-scale topographic relief must have existed at $\sim$70–40 Ma.

While the existence of significant paleorelief was well demonstrated with the AHe data of House et al. (1998, 2001), later work has modified some details of the original interpretations. Braun (2002b) found from a spectral analysis of the available data that relief has decreased by at least $\sim$50% since the end of the 70–80 Ma Laramide Orogeny, although he noted that the data points were not ideally spaced for an accurate reconstruction of relief. In contrast, Clark et al. (2005a) suggested a paleo-range crest elevation of only $\sim$1.5 km during the Late Cretaceous (i.e., less than today's $\sim$4 km crest elevation), which they justified by noting that river profiles, as well as increased river incision rates between 2.7 and 1.4 Ma (Stock et al. 2004), indicate two periods of renewed uplift and relief generation since $\sim$32 Ma. In a more detailed interpretation of the thermochronology data using Pecube thermal-kinematic modeling, McPhillips and Brandon (2012) suggested an alternative interpretation that potentially resolves the debate: Their results indicate that relief and elevations were likely high in the Late Cretaceous, decreased throughout the Paleogene, and then increased again during the Neogene. This work nicely illustrates that while direct investigations into relief development may provide better constraints on paleotopography, thermal modeling is often needed to explore the full range of possible interpretations.

19.3.1.2 Canyon Incision

On a more local scale, numerous approaches have been used to determine the incision history of an individual canyon, including dating volcanic flows and travertine deposits (e.g., Pederson et al. 2002; Thouret et al. 2007; Karlstrom et al. 2007; Montero-López et al. 2014), cave fluvial sediments (e.g., Stock et al. 2004; Haeuselmann et al. 2007), and carbonate deposits (e.g., Polyak et al. 2008), or by investigating changes in sedimentology (e.g., Blackwelder 1934; Lucchitta 1972; Wernicke 2011). Volcanic flows are useful geomorphic markers, but their ages provide only a minimum constraint on the time that a canyon was carved to the depth at which the flow is preserved; the relief may have been carved earlier, and the flow itself may not have reached the lowest elevations of the canyon. Cave deposits stranded along the banks of an incising river are mainly limited by the requirement that one must have a clear understanding of the hydrology (and paleo-hydrology) of the area to accurately relate the end of carbonate or fluvial-sediment deposition to river incision (e.g., Polyak et al. 2008; Karlstrom et al. 2008). Finally, sedimentology can be very effective for revealing changes in provenance and potentially changes in river flow regime, which could record capture events or incision in some cases (e.g., Wernicke 2011), but does not allow for direct reconstructions of incision magnitudes or rates.

Low-temperature thermochronology has provided an effective alternative (or complementary) approach in several areas. In principle, km-scale valley incision will locally depress near-surface isotherms, which can be recorded as localised, rapid cooling from low-temperature thermochronology (Schildgen et al. 2007, Fig. 19.3a). The earliest application of this approach was in Eastern Tibet, where rivers have incised more than 2 km beneath a low-relief regional surface (Fig. 19.3b). Clark et al. (2005b) compiled AHe data from several short elevation transects and interpreted the incision history after plotting the cooling ages versus depth beneath the regional surface. This approach is preferable to plotting all the samples on an age-elevation plot, because considering their broad spatial distribution, the shape of the T_c isotherm on a regional scale is best represented by the long-wavelength topography, i.e., the regional low-relief surface. Distance from the low-relief surface

therefore provides a better proxy for distance from the T_c isotherm compared to elevation, which inherently assumes that the isotherm is horizontal. Building on this work, AHe and ZHe data from the Dadu, Yalong, and upper Yangtze rivers along the Eastern Tibet margin were shown by Ouimet et al. (2010) to all reveal relatively rapid incision since between ∼ 10 and 15 Ma, with variations in the onset suggested to be related to local upper-crustal deformation patterns superimposed on long-wavelength, epeirogenic uplift (Fig. 19.3b). Yang et al. (2016) found similar pulses of incision at or after ∼ 6 Ma in the Mekong and Salween rivers farther to the southwest (Fig. 19.3b). They suggested that spatiotemporal variations in erosion histories, and particularly the northward temporal progression of maximum erosion rates along the Salween river, could be related to deformation associated with the northward migration of the corner of the Indian continent.

A similar approach was taken in southern Peru to constrain the incision history of the Cotahuasi-Ocoña canyon across the western margin of the central Andean plateau (Schildgen et al. 2007, 2009). In this case, plotting samples on a plot of age versus depth below a preexisting low-relief surface was critical, because the samples were collected over a very limited elevation range along a valley bottom, but over a large range of depths beneath the regional low-relief surface (Schildgen et al. 2007). Indeed, samples that were collected later from elevation transects up the canyon walls revealed a similar pattern as the valley-bottom transect on an age-depth plot (Schildgen et al. 2009). Three-dimensional thermal-kinematic modeling pointed to an onset of incision between ∼ 8.5 and 11 Ma (Schildgen et al. 2009).

The Grand Canyon of the Colorado River has been another main target of river-incision studies based on thermochronology. Flowers et al. (2008) used AHe data from both plateau-surface and canyon-interior samples to try to distinguish between regional unroofing and canyon incision events. At the eastern end of the Grand Canyon, cooling ages ranging from ∼ 25 to 20 Ma argued for a Late Cenozoic (post-6 Ma) phase of canyon incision below the modern plateau surface (Flowers et al. 2008). However, Early Cenozoic thermal histories from samples separated by 1500 m of elevation and stratigraphic position in the same area are very similar. This finding implies that there must have been km-scale relief at that time, most likely carved into units that have since been removed through regional unroofing; otherwise, the deeper samples would have experienced higher temperatures compared to the shallower samples (Flowers et al. 2008). Later analyses using apatite ^{4}He/^{3}He thermochronology on canyon samples corroborated this multi-phase cooling history of the eastern Grand Canyon, which contrasts with the evidence for a single-phase, Laramide cooling history of most (70–80%) of the western Grand Canyon (Flowers and Farley 2012). Earlier AFT

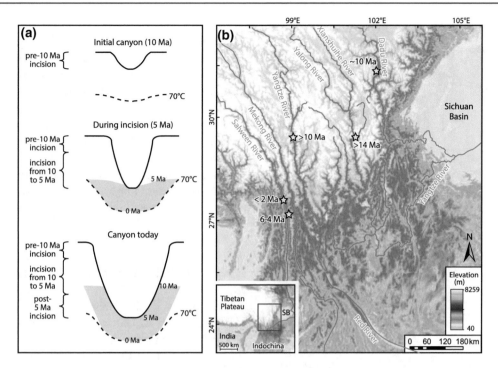

Fig. 19.3 Thermochronologic approach to constraining canyon incision history. **a** Schematic illustration of changes in topography, position of the T_c isotherm of the AHe system (70 °C), and development of a rapidly cooled zone beneath a canyon (shaded). Cartoon illustrates a scenario in which moderate relief exists prior to 10 Ma, the onset of rapid incision is at 10 Ma, and continues to the present (modified from Schildgen et al. 2007, reproduced with permission from the Geological Society of America). **b** Timing of the onset of rapid incision for rivers at various sites throughout eastern Tibet, based on thermochronometric data and modeling from Clark et al. (2005b), Ouimet et al. (2010), and Yang et al. (2016). Extent of (**b**) is outlined with red box in the inset map. SB: Sichuan Basin

studies from both the western outlet of the Grand Canyon (Fitzgerald et al. 2009) and along the eastern end of the canyon near Grand Canyon Village (Dumitru et al. 1994) also found evidence for rapid Laramide cooling. Nonetheless, the proposed Laramide incision of most of the western Grand Canyon was met with controversy. Fox and Shuster (2014) pointed out that with the uncertainties concerning how burial reheating affects the diffusion kinetics of He in apatite, it was not possible to use the data of Flowers and Farley (2012) to distinguish between the different incision scenarios. Others had argued that most of the incision in the western canyon occurred much later: Evidence pointing to post-6 Ma incision included dated volcanic flows, other thermochronology data, and the lack of Grand Canyon sediments found at its western outlet prior to 6 Ma (Karlstrom et al. 2008). But it is important to keep in mind that the high incision rates inferred from volcanic flows and speleothems represent maximum incision rates, and the lack of sediments to the west prior to 6 Ma could be explained by drainage reversal (from east- to west-flowing) at ∼6 Ma (Wernicke 2011). While the idea of some Laramide paleo-relief in the region appears to be well established, the canyon may only have become fully integrated and comparable to the modern system in the last 5–6 Ma (Karlstrom et al. 2014). Although the interpretations appear to be converging

on a story of geomorphologic evolution that is consistent with the available data, the controversy surrounding this problem illustrates the importance of appreciating the limits to each approach of constraining the evolution of landscape relief.

19.3.1.3 Knickpoint Migration

When attempting to interpret river-incision data in the context of external forcing, it is important to consider potential lags in the onset of incision. In a simple case of a regional tilting of the landscape, the entire length of a river may start to incise at a similar time, close to the timing of tilting (Fig. 19.4a Scenario 1) (e.g., Braun et al. 2014). However, following a uniform increase in uplift rates or a drop in base level, a river may not immediately start incising along its full length. Instead, incision will initiate at its downstream end and propagate upward through time, with a knickpoint separating the lower, incised reach from the upper, relict reach (Whipple and Tucker 1999, Fig. 19.4a Scenario 2). Bedrock river terraces can be useful for reconstructing river incision (e.g., Burbank et al. 1996) and potentially knickpoint migration (Harkins et al. 2007), but when considering km-scale incision waves or incision histories spanning millions of years, low-temperature thermochronology may provide more insights.

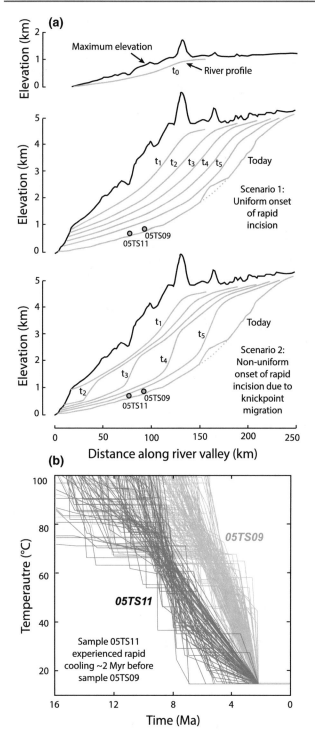

Fig. 19.4 Knickpoint propagation in the Cotahuasi Canyon (southwest Peru) from apatite $^4He/^3He$ thermochronometry (Schildgen et al. 2010, reproduced with permission from Elsevier). **a** Schematic illustration of how the river profile may evolve through time following surface uplift, either through a uniform onset of incision following monoclinal warping (Scenario 1) or upstream propagation of a knickpoint following block uplift (Scenario 2). **b** Temperature–time paths that provide good fits to $^4He/^3He$ data (see Schildgen et al. 2010 for details) illustrate the time-transgressive nature of the onset of rapid cooling for different samples collected along the valley bottom, indicating that Scenario 2 is more likely

Apatite $^4He/^3He$ thermochronometry is a promising technique for this particular application, due to its ability to resolve cooling histories at and below the He T_c (Shuster and Farley 2004, 2005). Schildgen et al. (2010) derived cooling histories from inverse modeling of apatite $^4He/^3He$ data for four different samples along the length of the canyon, in order to decipher the onset of rapid cooling in each sample. A comparison of the time-temperature paths illustrated the time-transgressive nature of the onset of rapid cooling; downstream samples showed a 1–2 Myr earlier onset of cooling compared to upstream samples, which likely represents the upstream migration of a major knickpoint (Schildgen et al. 2010, Fig. 19.4b).

19.3.2 Relief Changes in Glacial Landscapes

Glacial erosion of river valleys represents a special case of relief development. Due to the localised erosion along the lower flanks of the valley during this type of relief modification (from V- to U-shaped valley morphology, Fig. 19.5a), both spatial and temporal changes in erosion rates would be expected. However, these changes may only be resolvable with the lowest-temperature thermochronometers.

In the extensively glaciated Coast Mountains of British Columbia, Shuster et al. (2005) used apatite $^4He/^3He$ thermochronometry to resolve the onset of incision-related cooling in a steep transect of samples along a valley wall. Not only did all of the samples reveal increased cooling rates at ~1.8 Ma, but the highest sample revealed an onset of cooling that slightly predated the lower-elevation samples, illustrating the progressive nature of valley deepening. Furthermore, because a sample on the east side of the valley revealed a later onset of cooling compared to a sample at similar elevation on the west side, Shuster et al. (2005) inferred that valley widening had proceeded toward the east. In the same region, Ehlers et al. (2006) used AHe data to illustrate that the increase in exhumation appears to have extended over a wide region, with an apparent shift in the position of a major topographic divide due to widespread glacial erosion that overtopped and eroded ridgelines.

In a slightly different approach in the western Alps, Glotzbach et al. (2011a) applied inverse numerical thermal-kinematic modeling to a dense set of AFT and AHe data across and through (via a tunnel) the Mont Blanc massif. From initial 3D thermo-kinematic modeling, in which they assumed no change in relief, they found that following a period of relatively slow exhumation, the data were most consistent with a phase of rapid exhumation starting at ~1.7 Ma. However, that scenario did not effectively predict ages from the tunnel samples. A second scenario, which was consistent with both the tunnel and the

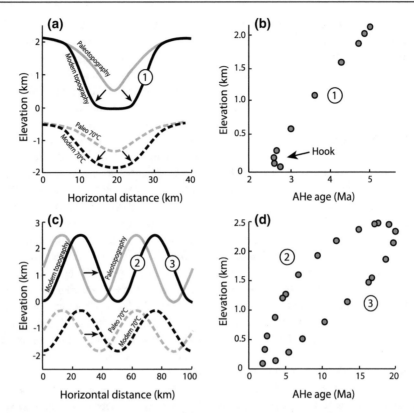

Fig. 19.5 Age-elevation transects in areas of changing relief modified from Olen et al. (2012). **a** Schematic illustration of changes in topography and position of the AHe T_c isotherm associated with increasing valley width, particularly a change from a V- to U-shaped valley. **b** Associated age-elevation transect along slope 1. **c** Illustration of changes in topography and position of the T_c isotherm associated with a lateral shift in topography, such as would be associated with asymmetric precipitation. **d** Associated age-elevation transect from the wetter (2) and drier (3) sides of the range

surface samples, included an increase in relief due to valley deepening at ∼0.9 Ma, most likely related to Alpine glaciations. AHe and $^4He/^3He$ data from the Rhône Valley in Switzerland combined with thermal-kinematic modeling revealed a similar pattern with even greater detail, indicating ∼1–1.5 km deepening of the glacial U-shaped valley starting at ∼1 Ma (Valla et al. 2011), representing an approximate doubling of valley relief since that time. The ∼1 Ma timing of glacial valley incision in the Alps is consistent with independent sedimentary and cosmogenic-isotope data (Muttoni et al. 2003; Haeuselmann et al. 2007) and suggests that glacial erosion only became an efficient relief-transforming agent after the mid-Pleistocene climatic transition.

In some cases, valley widening might be detectable from age-elevation transects alone. Within the Coast Ranges of British Columbia, low-temperature thermochronometers have revealed steepened (Densmore et al. 2007) or even hooked (Olen et al. 2012) patterns on age-elevation plots, with age minima occurring near the base of the valley due to focused erosion at the lower flanks (Fig. 19.5a, b). In Fig. 19.5a, this pattern can be understood by comparing the

"modern topography" to the "Paleo 70 °C" isotherm. The shortest distance between the isotherm and the modern topography, and hence, the youngest cooling ages, are found near the edges of the flattened valley bottom. Densmore et al. (2007) found that both temporal and spatial variations in erosion rates were needed to explain the pattern of thermochronometer ages in their elevation profile, but, in the absence of landscape evolution modeling, they could only qualitatively constrain the changes in relief that have occurred.

In an AHe and $^4He/^3He$ data set from Fiordland, New Zealand, Shuster et al. (2011) found a series of age-elevation relationships that display distinctly hooked patterns, i.e., showing a change from a positive slope higher on the valley walls to a negative slope near the base (e.g., Figure 19.5b). Using Pecube to model these data, they found that best fits to their data were achieved by models that included headward migration of km-scale topographic steps, interpreted to be facilitated by sliding-rate-dependent erosion at the base of the glaciers that carved out the landscape. Another possibility, which was not considered, is that valley widening produced or at least contributed to the distinct age-elevation patterns. Although additional modeling is required to test if both scenarios are plausible, this example highlights the potential for non-unique interpretations of thermochronological data, particularly in the case of changing landscape relief.

19.3.3 Range-Divide Migration

In areas of asymmetric precipitation across a mountain range, a situation commonly encountered where the range strikes perpendicular to oncoming winds, both numerical (Beaumont et al. 1992; Roe et al. 2003; Anders et al. 2008) and analog (Bonnet 2009) models predict a progressive migration of the drainage divide toward the drier side of the range. In the case of spatially uniform uplift, steepening of the dry-side slopes and lowering of the wet-side slopes over the course of divide migration may eventually lead to a balance in erosion rates on both sides of the range, resulting in steady, asymmetric topography (Roe et al. 2003). Where a horizontal component to rock advection occurs, drainage divides will also tend to migrate in the direction of tectonic motion (Beaumont et al. 1992; Willett 1999; Willett et al. 2001) (Fig. 19.2). Exhumational and topographic steady state may eventually be reached, with faster tectonic deformation and exhumation occurring on the wetter, or

retrowedge (non-accreting), side (Willett 1999; Willett et al. 2001). In the former case (uniform uplift), topographic steady state will be characterized by symmetric cooling ages across the range, whereas in the latter case (with a horizontal component to uplift), topographic steady state will be characterized by asymmetric ages across the range.

However, if asymmetric ages across the range are found, the interpretation may be non-unique. Rather than reflecting a coupling between tectonic deformation and climate/erosion under steady-state topographic conditions, an asymmetric pattern of ages could reflect spatially uniform uplift with a migrating drainage divide. As Olen et al. (2012) illustrated in their modeling study, under spatially uniform uplift and asymmetric erosion, the retreating (wet) side, which erodes into the landscape faster during divide migration, is expected to yield younger ages compared to samples at a similar elevations on the advancing (drier) and more slowly eroding side (Fig. 19.5c, d). Moreover, the pattern of ages on the wet side (transect "a" in Fig. 19.5d) could mistakenly be interpreted to show an increase in exhumation rates due to the steepening of the age-elevation relationship at lower elevations.

Within the Washington Cascades (Northwestern USA, Fig. 19.6a), Reiners et al. (2003a) suggested that a spatial coincidence between peak orographic precipitation and fastest exhumation from AHe data on the western flank of the

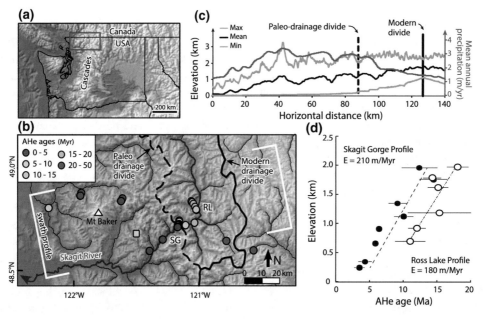

Fig. 19.6 Thermochronometric data from the Washington cascades (USA) collected over a range of mean annual precipitation values. **a** Overview of Cascades; red box outlines region shown in (**b**). **b** Sample locations colored by AHe age on digital elevation model. Samples from the Skagit Gorge (SG) elevation profile are from a wetter region on the western side of the paleo-drainage divide compared to samples from the Ross Lake (RL) profile, which is on the eastern side

of the paleo-divide. **c** Swath profile (location shown with white brackets in (**a**) shows mean, minimum, and maximum topography along the swath, and mean annual precipitation. Note the consistently younger ages on the wetter (west) side of the paleo-drainage divide, both in map view and in the age-elevation plot **d** of the two sample profiles. *E* is the exhumation rate. (**b–d**) modified from Simon-Labric et al. (2014), reproduced with permission from the Geological Society of America

range pointed to a strong coupling between climate and erosion. They noted that if topography is at or near steady state, the data would argue also for a coupling among climate, erosion, and tectonic deformation. However, denudation rates derived from cosmogenic [10]Be, which average over the past several millennia, are about four times higher than the million-year timescales, implying that the topography is not in steady state (Moon et al. 2011). A study by Simon-Labric et al. (2014) also emphasised the transient nature of topography in the area: the existence of wind gaps, drainage basins oriented parallel to the crest of the divide, and peaks in topography that are offset from the drainage divide point to a relatively recent reorganisation of the drainage system. Also, Simon-Labric et al. (2014) highlighted how AHe data reveal a pattern consistent with progressive divide migration: age-elevation plots show similar slopes, but ages are consistently younger on the wetter side at a given elevation compared to the drier side (Fig. 19.6b–d). Their study illustrates the importance of independent geomorphic interpretations of the landscape in cases where the interpretation of thermochronology data may be non-unique.

19.4 Summary

Thermochronology has proven to be a valuable tool in geomorphological studies for the past several decades. With the advent of thermal-kinematic modeling and high-resolution low-temperature thermochronology systems, we have moved from indirectly inferring paleotopography and relief development toward more precise and quantitative constraints on landscape morphology. Interpretations of thermochronological data, including age-elevation relationships, can be greatly improved with the incorporation of thermal-kinematic modeling, which can help to disentangle the influence of partial retention/annealing zones, transient motion of isotherms following a change in exhumation rate, and potentially also changes in relief versus changes in exhumation rates, on age patterns. The development of very low-temperature thermochronometers such as AHe and apatite ^{4}He/^{3}He is vital in providing the resolution necessary to quantify changes in relief and effectively distinguish them from changes in exhumation rates.

Detrital thermochronology has proven to be effective for characterising erosion rates, particularly when focusing on interpreting lag-times (the time lapse between the cooling age and depositional age). However, with respect to detecting relief changes, this approach has been hindered by insufficient characterization of complex (e.g., variable lithology, non-uniform erosion rates) source areas. As such, it seems to hold less potential for studying long-term changes in landscape relief, particularly for large, heterogenous catchments.

Finally, thermochronology data alone may be insufficient to resolve all types of changes in landscape relief. In the case of both valley widening and range-divide migration, steepened patterns at the base of age-elevation transects will result. Hence, a single transect of thermochronological data may be insufficient to distinguish valley widening or range-divide migration from an increase in regional exhumation rates. In such examples of major changes in landscape relief, multiple sets of sample transects coupled with geomorphological field observations will be crucial for deciphering the details of landscape evolution.

Acknowledgements We thank Alison Duvall and Scott Miller for constructive reviews, and Marco G. Malusà and Paul G. Fitzgerald for editorial handling.

References

Ahnert F (1970) Functional relationships between denudation, relief, and uplift in large mid-latitude drainage basins. Am J Sci 268:243–263

Anders AM, Roe GH, Montgomery DR, Hallet B (2008) Influence of precipitation phase on the form of mountain ranges. Geology 36:479–482

Avdeev B, Niemi NA, Clark MK (2011) Doing more with less: Bayesian estimation of erosion models with detrital thermochronologic data. Earth Planet Sci Lett 305(3–4):385–395

Baldwin SL, Lister GS (1998) Thermochronology of the South Cyclades Shear Zone, Ios, Greece: effects of ductile shear in the argon partial retention zone. J Geophys Res 103:7315–7336

Baldwin SL, Fitzgerald PG, Malusà MG (2018) Chapter 13. Crustal exhumation of plutonic and metamorphic rocks: constraints from fission-track thermochronology. In: Malusà MG, Fitzgerald PG (eds) Fission-track thermochronology and its application to geology. Springer

Batt GE, Braun J (1999) The tectonic evolution of the Southern Alps, New Zealand: insights from fully thermally coupled dynamical modeling. Geophys J Int 136:403–420

Batt GE, Brandon MT, Farley KA, Roden-Tice M (2001) Tectonic synthesis of the Olympic Mountains segment of the Cascadia wedge using 2-D thermal and kinematic modeling of isotopic ages. J Geophys Res 106(B11):26731–26746

Beaumont C, Fullsack P, Hamilton J (1992) Erosional control of active compressional orogens. In: McClay KR (ed) Thrust tectonics. Chapman and Hall, London, pp 1–18

Bernet M (2018) Chapter 15. Exhumation studies of mountain belts based on detrital fission-track analysis on sand and sandstones. In: Malusà MG, Fitzgerald PG (eds) Fission-track thermochronology and its application to geology. Springer

Bernet M, Garver JI (2005) Fission-track analysis of detrital zircon. Rev Mineral Geochem 58(1):205–237

Bernet M, Zattin M, Garver JI, Brandon MT, Vance JA (2001) Steady-state exhumation of the European Alps. Geology 29:35–38

Binnie SA, Phillips WM, Summerfield MA, Fifield LK (2007) Tectonic uplift, threshold hillslopes, and denudation rates in a developing mountain range. Geology 35(8):743–746

Blackwelder E (1934) Origin of the Colorado river. Geol Soc Am Bull 45:551–566

Bonnet S (2009) Shrinking and splitting of drainage basins in orogenic landscapes from the migration of the main drainage divide. Nature Geosci 2:766–771

Brandon MT, Vance JA (1992) Fission-track ages of detrital zircon grains: implications for the tectonic evolution of the Cenozoic Olympic subduction complex. Am J Sci 292:565–636

Brandon MT, Roden-Tice MK, Garver JI (1998) Late Cenozoic exhumation of the Cascadia accretionary wedge in the Olympic mountains, NW Washington state. Geol Soc Am Bull 110:985–1009

Braun J (2002a) Quantifying the effect of recent relief changes on age-elevation relationships. Earth Planet Sci Lett 200:331–343

Braun J (2002b) Estimating exhumation rate and relief evolution by spectral analysis of age-elevation datasets. Terra Nova 14:210–214

Braun J (2003) Pecube: a new finite-element code to solve the 3D heat transport equation including the effects of a time-varying, finite amplitude surface topography. Computers Geosci 29:787–794

Braun J (2005) Quantitative constraints on the rate of landform evolution derived from low-temperature themrochronology. Rev Mineral Geochem 58:351–374

Braun J, Robert X (2005) Constraints on the rate of post-orogenic erosional decay from low-temperature thermochronologic data: application to the Dabie Shan, China. Earth Surf Proc Land 30 (9):1203–1225

Braun J, van der Beek P, Batt G (2006) Quantitative thermochronology: numerical methods for the interpretation of thermochronological data. Cambridge University Press, Cambridge, p 258

Braun J, Guillocheau F, Robin C, Baby G, Jelsma H (2014) Rapid erosion of the southern African plateau as it climbs over a mantle superswell. J Geophys Res 119(7):6093–6112

Brewer ID, Burbank DW (2006) Thermal and kinematic modeling of bedrock and detrital cooling ages in the central Himalaya. J Geophys Res 111(B9). https://doi.org/10.1029/2004jb003304

Brewer ID, Burbank DW, Hodges KV (2003) Modelling detrital cooling-age populations: insights from two Himalayan catchments. Basin Res 15(3):305–320

Burbank DW, Leland J, Fielding E, Anderson RS, Brozovic N, Reid MR, Duncan C (1996) Bedrock incision, rock uplift and threshold hillslopes in the northwestern Himalayas. Nature 379:505–510

Clark MK, Maheo G, Saleeby J, Farley KA (2005a) The non-equilibrium landscape of the southern Sierra Nevada, California. GSA Today 15(9):4–10

Clark MK, House MA, Royden LH, Whipple KX, Burchfiel BC, Zhang X, Tang W (2005b) Late Cenozoic uplift of southeastern Tibet. Geology 33(6):525–528

Crowley TJ, Burke KC (eds) (1998) Tectonic boundary conditions for climate reconstructions. Oxford University Press, Oxford, p 285

Densmore MS, Ehlers TA, Woodsworth GJ (2007) Effect of Alpine glaciations on thermochronometer age-elevation profiles. Geophys Res Lett 34:L02502. https://doi.org/10.1029/2006GL028317

Dumitru TA, Duddy IR, Green PF (1994) Mesozoic-Cenozoic burial, uplift, and erosion history of the west-central Colorado plateau. Geology 22(6):499–502

Duvall AR, Clark MK, Avdeev B, Farley KA, Chen Z (2012) Widespread late Cenozoic increase in erosion rates across the interior of eastern Tibet constrained by detrital low-temperature thermochronometry. Tectonics 31(3). https://doi.org/10.1029/2011tc002969

Ehlers T, Chaudhri T, Kumar S, Fuller C, Willett S, Ketcham RA, Brandon MT, Belton DX, Kohn BP, Gleadow AJW, Dunai TJ, Fu FQ (2005) Computational tools for low-temperature thermochronometer interpretation. Rev Mineral Geochem 58(1):589–622

Ehlers TA, Farley KA, Rusmore ME, Woodsworth GJ (2006) Apatite (U–Th)/He signal of large-magnitude accelerated glacial erosion, southwest British Columbia. Geology 34(9):765–768

Fitzgerald PG, Malusà MG (2018) Chapter 9. Concept of the exhumed partial annealing (retention) zone and age-elevation profiles in thermochronology. In: Malusà MG, Fitzgerald PG (eds) Fission-track thermochronology and its application to geology. Springer

Fitzgerald PG, Sorkhabi RB, Redfield TF, Stump E (1995) Uplift and denudation of the central Alaska range: a case study in the use of apatite fission track thermochronology to determine absolute uplift parameters. J Geophys Res 100(B10):20175–20191

Fitzgerald PG, Duebendorfer EM, Faulds JE, O'Sullivan P (2009) South Virgin-White Hills detachment fault system of SE Nevada and NW Arizona: applying apatite fission track thermochronology to constrain the tectonic evolution of a major continental detachment fault. Tectonics 28(2). https://doi.org/10.1029/2007tc002194

Flowers RM, Farley KA (2012) Apatite ^{4}He/^{3}He and (U–Th)/He evidence for an ancient Grand Canyon. Science 338(6114):1616–1619

Flowers RM, Wernicke BP, Farley KA (2008) Unroofing, incision and uplift history of the southwestern Colorado plateau from (U–Th)/He apatite thermochronometry. Geol Soc Am Bull 120:571–587

Fox M, Shuster D (2014) The influence of burial heating on the (U–Th)/He system in apatite: Grand Canyon case study. Earth Planet Sci Lett 397:174–183

Fox M, Herman F, Willett SD, May DA (2014) A linear inversion method to infer exhumation rates in space and time from thermochronometric data. Earth Surf Dyn 2:47–65

Garver JI, Brandon MT, Roden-Tice M, Kamp PJJ (1999) Exhumation history of orogenic highlands determined by detrital fission-track thermochronology. Geol Soc Spec Publ 154:283–304

Gleadow AJW, Fitzgerald PG (1987) Uplift history and structure of the Transantarctic mountains: new evidence from fission track dating of basement apatites in the Dry Valley area, southern Victoria Land. Earth Planet Sci Lett 82:1–14

Glotzbach C, van der Beek PA, Spiegel C (2011a) Episodic exhumation and relief growth in the Mont Blanc massif, western Alps from numerical modeling of thermochronology data. Earth Planet Sci Lett 304:417–430

Glotzbach C, Bernet M, van der Beek P (2011b) Detrital thermochronology records changing source areas and steady exhumation in the western european alps. Geology 39(3):239–242

Haeuselmann P, Granger DE, Jeannin PY (2007) Abrupt glacial valley incision at 0.8 Ma dated from cave deposits in Switzerland. Geology 35(2):143–146

Harkins N, Kirby E, Heimsath A, Robinson R, Reiser U (2007) Transient fluvial incision in the headwaters of the Yellow river, northeastern Tibet, China. J Geophys Res 112(F3). https://doi.org/10.1029/2006jf000570

Herman F, Copeland P, Avouac J-P, Bollinger L, Mahéo G, Le Fort P, Rai S, Foster D, Pêcher A, Stüwe K, Henry P (2010) Exhumation, crustal deformation, and thermal structure of the Nepal Himalaya derived from the inversion of thermochronological and thermobarometric data and modeling of the topography. J Geophys Res 115(B6). https://doi.org/10.1029/2008jb006126

Hetzel R, Dunkl I, Haider V, Strobl M, von Eynatten H, Ding L, Frei D (2011) Peneplain formation in southern Tibet predates the India-Asia collision and plateau uplift. Geology 39(10):983–986

House MA, Wernicke BP, Farley KA, Dumitru TA (1997) Cenozoic thermal evolution of the central Sierra Nevada, California, from (U–Th)/He thermochronometry. Earth Planet Sci Lett 151:167–179

House MA, Wernicke BP, Farley KA (1998) Dating topography of the Sierra Nevada, California, using apatite (U–Th)/He ages. Nature 396:66–69. https://doi.org/10.1038/23926

House MA, Wernicke BP, Farley KA (2001) Paleogeomorphology of the Sierra Nevada, California, from (U–Th)/He ages in apatite. Am J Sci 301:77–102

Hurford AJ (1986) Cooling and uplift patterns in the Lepontine alps, south central Switzerland, and an age of vertical movement on the insubric fault line. Contrib Mineral Petrol 92:413–427

Hurford AJ, Hunziker JC, Stöckhert B (1991) Constraints on the late thermotectonic evolution of the western Alps: evidence for episodic rapid uplift. Tectonics 10(4):758–769

Jamieson RA, Beaumont C (1988) Orogeny and metamorphism: a model for deformation and *P-T-t* paths with applications to the central and southern Appalachians. Tectonics 7:417–445

Karlstrom K, Crow R, McIntosh W, Peters L, Pederson J, Raucci J, Crossey LJ, Umhoefer P, Dunbar N (2007) $^{40}Ar/^{39}Ar$ and field studies of Quaternary basalts in Grand Canyon and model for carving Grand Canyon: quantifying the interaction of river incision and normal faulting across the western edge of the Colorado plateau. Geol Soc Am Bull 119:1283–1312

Karlstrom KE, Crow R, Crossey LJ, Coblentz D, van Wijk J (2008) Model for tectonically driven incision of the younger than 6 Ma Grand Canyon. Geology 31:835–838

Karlstrom KE, Lee JP, Kelley SA, Crow RS, Crossey LJ, Young RA, Lazear G, Beard LS, Ricketts JW, Fox M, Shuster DL (2014) Formation of the Grand Canyon 5 to 6 million years ago through integration of older palaeocanyons. Nature Geosci 7(3):239–244

Kirby E, Reiners PW, Krol MA, Whipple KX, Hodges KV, Farley KA, Tang W, Chen Z (2002) Late Cenozoic evolution of the eastern margin of the Tibetan plateau: inferences from $^{40}Ar/^{39}Ar$ and (U–Th)/He thermochronology. Tectonics 21(1):1001

Landry KR, Coutand I, Whipp DM Jr, Grujic D, Hourigan JK (2016) Late Neogene tectonically driven crustal exhumation of the Sikkim Himalaya: insights from inversion of multithermochronologic data. Tectonics 35(3):833–859

Larsen IJ, Montgomery DR (2012) Landslide erosion coupled to tectonics and river incision. Nat Geosci 5:468–473

Liu T-K, Hsieh S, Chen Y-G, Chen W-S (2001) Thermo-kinematic evolution of the Taiwan oblique-collision mountain belt as revealed by zircon fission track dating. Earth Planet Sci Lett 186:45–56

Lucchitta I (1972) Early history of the Colorado river in the basin and range province. Geol Soc Am Bull 83:1933–1948

Malusà MG, Fitzgerald PG (2018) Chapter 8. From cooling to exhumation: setting the reference frame for the interpretation of thermochronologic data. In: Malusà MG, Fitzgerald PG (eds) Fission-track thermochronology and its application to geology. Springer

Malusà MG, Fitzgerald PG (2018) Chapter 10. Application of thermochronology to geologic problems: bedrock and detrital approaches. In: Malusà MG, Fitzgerald PG (eds) Fission-track thermochronology and its application to geology. Springer

Malusà MG, Garzanti E (2018) Chapter 7. The sedimentology of detrital thermochronology. In: Malusà MG, Fitzgerald PG (eds) Fission-track thermochronology and its application to geology. Springer

Malusà MG, Resentini A, Garzanti E (2016) Hydraulic sorting and mineral fertility bias in detrital geochronology. Gondwana Res 31:1–19

Mancktelow NS, Grasemann B (1997) Time-dependent effects of heat advection and topography on cooling histories during erosion. Tectonophysics 270:167–195

Matmon A, Bierman PR, Larsen J, Southworth S, Pavich M, Caffee M (2003) Temporally and spatially uniform rates of erosion in the southern Appalachian great smoky mountains. Geology 31(2):155–158

McPhillips D, Brandon MT (2012) Topographic evolution of the Sierra Nevada measured directly by inversion of low-temperature thermochronology. Am J Sci 312:90–116

Miller SR, Slingerland RL, Kirby E (2007) Characteristics of steady state fluvial topography above fault-bend folds. J Geophys Res 112 (F4). https://doi.org/10.1029/2007jf000772

Montero-López C, Strecker MR, Schildgen TF, Hongn F, Guzmán S, Bookhagen B, Sudo M (2014) Local high relief at the southern margin of the Andean plateau by 9 Ma: evidence from ignimbritic valley fills and river incision. Terra Nova 26(6):454–460

Montgomery DR, Brandon MT (2002) Topographic controls on erosion rates in tectonically active mountain ranges. Earth Planet Sci Lett 201(3–4):481–489

Moon S, Chamberlain CP, Blisniuk K, Levine N, Rood DH, Hilley GE (2011) Climatic control of denudation in the deglaciated landscape of the Washington cascades. Nat Geosci 4:469–473

Moore MA, England PC (2001) On the inference of denudation rates from cooling ages of minerals. Earth Planet Sci Lett 185:265–284

Mulch A (2016) Stable isotope paleoaltimetry and the evolution of landscapes and life. Earth Planet Sci Lett 433(1):180–191

Muttoni G, Carcano C, Garzanti E, Ghielmi M (2003) Onset of major Pleistocene glaciations in the Alps. Geology 31(11):989–992

Olen SM, Ehlers TA, Densmore MS (2012) Limits to reconstructing paleotopography from thermochronometer data. J Geophys Res 117:F01024

Ouimet W, Whipple KX, Granger DE (2009) Beyond threshold hillslopes: channel adjustment to base-level fall in tectonically active mountain ranges. Geology 37(7):579–582

Ouimet W, Whipple K, Royden L, Reiners P, Hodges K, Pringle M (2010) Regional incision of the eastern margin of the Tibetan plateau. Lithosphere 2(1):50–63

Pederson J, Karlstrom KE, Sharp W, McIntosh W (2002) Differential incision of the Grand Canyon related to Quaternary faulting—constraints from U-series and Ar/Ar dating. Geology 30(8):739–742

Polyak V, Hill C, Asmerom Y (2008) Age and evolution of the Grand Canyon revealed by U-Pb dating of water table-type speleothems. Science 321:1377–1380

Reiners PW, Brandon MT (2006) Using thermochronology to understand orogenic erosion. Annu Rev Earth Planet Sci 34:419–466

Reiners PW, Ehlers TA, Mitchell SG, Montgomery DR (2003a) Coupled spatial variations in precipitation and long-term erosion rates across the Washington cascades. Nature 426:645–647

Reiners PW, Zhou Z, Ehlers TA, Xu C, Brandon MT, Donelick RA, Nicolescu S (2003b) Post-orogenic evolution of the Dabie Shan, eastern China, from (U–Th)/He and fission-track thermochronology. Am J Sci 303(6):489–518

Robert X, van der Beek P, Braun J, Perry C, Dubille M, Mugnier J-L (2009) Assessing Quaternary reactivation of the main central thrust zone (central Nepal Himalaya): new thermochronologic data and numerical modeling. Geology 37(8):731–734

Robert X, van der Beek P, Braun J, Perry C, Mugnier J-L (2011) Control of detachment geometry on lateral variations in exhumation rates in the Himalaya: insights from low-temperature thermochronology and numerical modeling. J Geophys Res 116(B5)

Roe GH, Montgomery DR, Hallet B (2003) Orographic precipitation and the relief of mountain ranges. J Geophys Res 108:2315

Rohrmann A, Kapp P, Carrapa B, Reiners PW, Guynn J, Ding J, Heizler M (2012) Thermochronologic evidence for plateau formation in central Tibet by 45 Ma. Geology 40(2):187–190

Ruddiman WF (ed) (1997) Tectonic uplift and climate change. Plenum, New York, p 535

Ruhl KW, Hodges KV (2005) The use of detrital mineral cooling ages to evaluate steady state assumptions in active orogens: an example from the central Nepalese Himalaya. Tectonics 24(4):TC4015

Schildgen TF, Hodges KV, Whipple KX, Reiners PW, Pringle MS (2007) Uplift of the western margin of the Andean plateau revealed from canyon incision history, southern Peru. Geology 35:523–526

Schildgen TF, Ehlers TA, Whipp DM Jr, van Soest MC, Whipple KX, Hodges KV (2009) Quantifying canyon incision and Andean plateau surface uplift, southwest Peru: a thermochronometer and numerical modeling approach. J Geophys Res 114:F04014

Schildgen TF, Balco G, Shuster DL (2010) Canyon incision and knickpoint propagation recorded by apatite $^4He/^3He$ thermochronometry. Earth Planet Sci Lett 293:377–387

Shuster DL, Farley KA (2004) $^{4}He/^{3}He$ thermochronometry. Earth Planet Sci Lett 217:1–17

Shuster DL, Farley KA (2005) $^{4}He/^{3}He$ thermochronometry: theory, practice, and potential complications. Rev Mineral Geochem 58:181–203

Shuster DL, Ehlers TA, Rusmore ME, Farley KA (2005) Rapid glacial erosion at 1.8 Ma revealed by $^{4}He/^{3}He$ thermochronometry. Science 310:1668–1670

Shuster DL, Cuffey KM, Sanders JW, Balco G (2011) Thermochronometry reveals heaward propagation of erosion in an alpine landscape. Science 332:84–88

Simon-Labric T, Brocard GY, Teyssier C, van der Beek P, Reiners PW, Shuster DL, Murray KE, Whitney DL (2014) Low-temperature thermochronologic signature of range-divide migration and breaching in the north cascades. Lithosphere 6(6):473–482

Spotila JA (2005) Applications of low-temperature thermochronometry to quantification of recent exhumation in mountain belts. Rev Mineral Geochem 58:449–466

Stock GM, Anderson RS, Finkel RC (2004) Pace of landscape evolution in the Sierra Nevada, California, revealed by cosmogenic dating of cave sediments. Geology 32:193–196

Stüwe L, White R, Brown X (1994) The influence of eroding topography on steady-state isotherms; application to fission-track analysis. Earth Planet Sci Lett 124:63–74

Thouret J-C, Wörner G, Gunnell Y, Singer B, Zhang X, Souriot T (2007) Geochronologic and stratigraphic constraints on canyon incision and Miocene uplift of the central Andes in Peru. Earth Planet Sci Lett 263(3–4):151–166

Valla PG, Herman F, van der Beek PA, Braun J (2010) Inversion of thermochronological age-elevation profiles to extract independent estimates of denudation and relief history—I: theory and conceptual model. Earth Planet Sci Lett 295:511–522

Valla PG, Shuster DL, van der Beek PA (2011) Significant increase in relief of the European Alps during mid-Pleistocene glaciations. Nat Geosci 4:688–692

van der Beek P, van Melle J, Guillot S, Pêcher A, Reiners PW, Nicolescu S, Latif M (2009) Eocene Tibetan plateau remnants preserved in the northwest Himalaya. Nat Geosci 2:364–368

van der Beek PA, Valla P, Herman F, Braun J, Persano C, Dobson KJ, Labrin E (2010) Inversion of thermochronological age-elevation profiles to extract independent estimates of denudation and relief history—II: application to the French Western Alps. Earth Planet Sci Lett 296:9–22

van der Beek P, Litty C, Baudin M, Mercier J, Robert X, Hardwick E (2016) Contrasting tectonically driven exhumation and incision patterns, western versus central Nepal Himalaya. Geology 44 (4):327–330

Wagner GA, Reimer GM (1972) Fission track tectonics: the tectonic interpretation of fission track ages. Earth Planet Sci Lett 14:263–268

Wagner GA, Reimer GM, Jäger E (1977) Cooling ages derived by apatite fission track, mica Rb-Sr, and K-Ar dating: the uplift and cooling history of the central Alps. Mem Inst Geol Mineral Univ Padova 30:1–27

Wernicke B (2011) The California river and its role in carving Grand Canyon. Geol Soc Am Bull 123:1288–1316

Whipp DM Jr, Ehlers TA, Blythe AE, Huntington KW, Hodges KV, Burbank DW (2007) Plio-Quaternary exhumation history of the central Nepalese Himalaya: 2. thermokinematic and thermochronometer age prediction model. Tectonics 26(3)

Whipple KX, Tucker GE (1999) Dynamics of the stream-power river incision model: implications for height limits of mountain ranges, landscape response timescales, and research needs. J Geophys Res 104:17661–17674

Willett SD (1999) Orogeny and orography: the effects of erosion on the structure of mountain belts. J Geophys Res 104:28957–28981

Willett SD, Brandon MT (2002) On steady states in mountain belts. Geology 30(2):175–178

Willett SD, Slingerland R, Hovius N (2001) Uplift, shortening, and steady state topography in active mountain belts. Am J Sci 301:455–485

Willett SD, Fisher D, Fuller C, En-Chao Y, Chia-Yu L (2003) Erosion rates and orogenic-wedge kinematics in Taiwan inferred from fission-track thermochronology. Geology 31(11):945–948

Wolf RA, Farley KA, Kass DM (1998) Modeling of the temperature sensitivity of the apatite (U–Th)/He thermochronometer. Chem Geol 148:105–114

Yang R, Fellin MG, Herman F, Willett SD, Wang W, Maden C (2016) Spatial and temporal pattern of erosion in the three rivers region, southeastern Tibet. Earth Planet Sci Lett 433:10–20

Zeitler PK, Johnson NM, Naeser CW, Tahirkheli RAK (1982) Fission-track evidence for Quaternary uplift of the Nanga Parbat region, Pakistan. Nature 298:255–257

Fission-Track Thermochronology Applied to the Evolution of Passive Continental Margins

Mark Wildman, Nathan Cogné and Romain Beucher

Abstract

Passive continental margins (PCMs) form at divergent plate boundaries in response to continental breakup and subsequent formation of new oceanic basins. The onshore topography of PCMs is a key component to understand the evolution of extensional settings. The classic nomenclature of PCMs is derived from early investigations that suggested apparent tectonic stability after the initial phase of rifting and breakup. However, geological and geomorphic diversity of PCMs requires more complex models of rift and post-rift evolution. Fission-track (FT) thermochronology provides appropriate tools to decipher the long-term development of PCM topography and better resolve the spatial and temporal relationships between continental erosion and sediment accumulation in adjacent offshore basins. FT datasets have revealed complex spatial and temporal denudation histories across some PCMs and have shown that several kilometres of material may be removed from the onshore margin following rifting. Combining these data with geological and geomorphological observations, and with predictions from numerical modelling, suggests that PCMs may have experienced significant post-rift activity. Case histories illustrated in this chapter include the PCM of southeastern Africa and the conjugate PCMs of the North and South Atlantic.

20.1 Introduction

Fragmentation of continents involves a sequence of intra-continental rifting, breakup and sea-floor spreading, with the fringes of new continents referred to as passive continental margins (PCMs) (Fig. 20.1, Péron-Pinvidic et al. 2013). A large number of PCMs are associated with major hydrocarbon provinces and/or important mineral deposits (e.g. Campos Basin, Gabon and Angola shelves, Niger and Mississippi deltas) and have been widely documented by the oil industry (Katz and Mello 2000; Groves and Bierlin 2007). Kilometre-thick sedimentary basins adjacent to PCMs imply a direct link between the onshore and offshore domains (Whittaker et al. 2013). The onshore topography and landscape morphology control the production and transfer of material into the basins and have a critical role in the development of the entire margin. However, due to margin uplift and the destructive nature of erosion, the onshore stratigraphic record of the early stages of PCM development is generally not preserved.

Low-temperature thermochronology techniques, particularly apatite fission-track (AFT) and (U–Th)/He (AHe) analyses, provide constraints on the thermal history of rocks as they are exhumed towards the Earth's surface (e.g. Gallagher et al. 1998; Gleadow et al. 2002). These techniques have been successfully applied to constrain denudation histories across PCMs, complementing the information derived from the offshore sedimentary record, and providing independent constraints for the validation of conceptual and numerical models of PCM evolution (see Braun 2018, for review). This chapter presents an overview of the modern perspectives and debates surrounding PCM evolution and explores the framework in which FT analysis and other low-temperature thermochronology methods can be used to shed light on the structural complexity and geomorphological evolution of PCMs.

M. Wildman (✉) · N. Cogné
Géosciences Rennes Université de Rennes 1, 35000 Rennes, France
e-mail: mark.wildman@univ-rennes1.fr

R. Beucher
School of Earth Sciences, University of Melbourne, Victoria, 3010, Australia

M. G. Malusà and P. G. Fitzgerald (eds.), *Fission-Track Thermochronology and its Application to Geology*, Springer Textbooks in Earth Sciences, Geography and Environment, https://doi.org/10.1007/978-3-319-89421-8_20

Fig. 20.1 Global elevation map showing sediment thickness of offshore sediments (Divins 2003) and location of topographic profiles across several passive continental margin settings. Coastal strips (CS) and escarpment(s) (E) separating the coastal strip from the continental interior are also indicated. Dashed boxes indicate areas for maps presented in Figs. 20.5 and 20.6

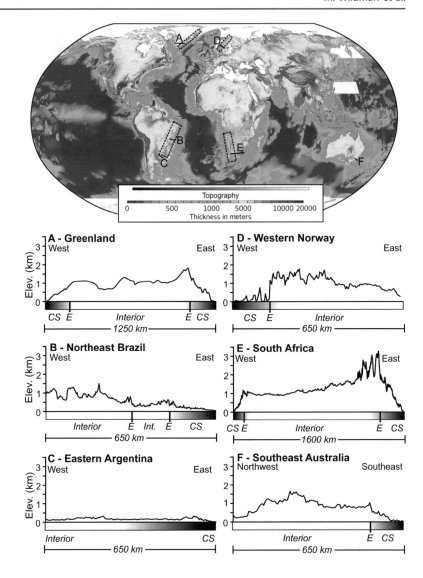

20.2 Geodynamics and Geomorphology of PCMs

20.2.1 Topography of PCMs

PCMs form through stretching and thinning of the lithosphere in response to an extensional stress field (see Watts 2012 for a review). The onshore regions of PCMs display a wide range of landscape morphologies (Fig. 20.1). High-elevation passive margins (e.g. southeastern Brazil, southwestern Africa, western India) are characterised by a low-lying coastal strip, extending ∼50–200 km landward, separated from an elevated plateau by a major, seaward-facing escarpment (e.g. Summerfield 1991). The coastal strip and elevated plateau may be characterised by a low-to-moderate relief, often with a gentle upwarping of the plateau towards the escarpment lip (Gilchrist and Summerfield 1990). The escarpment (or escarpment zone) is characterised by high relief and steep hillslopes (Brown et al. 2002; van der Beek et al. 2002; Persano et al. 2005). Low-elevation PCMs typically exhibit low-relief coastal plains where elevation gain is modest (few hundred metres) over several hundreds of kilometres away from the coast (e.g. central southern Australia, eastern Argentina and central western Africa). The creation and preservation of elevated topography at PCMs are controlled by pre- and post-breakup thermal, tectonic and surface processes. It is the relative importance of these processes over time that determines the evolution of the onshore region.

20.2.2 Styles of Lithospheric Breakup

Geodynamic models of rifting and lithospheric breakup have advanced from models assuming pure shear and uniform extension through the entire lithosphere (McKenzie 1978), to more complex models involving simple shear (Wernicke

1985; Lister et al. 1986) and a major role played by lithospheric rheology and mantle convection (e.g. Braun and Beaumont 1989; Keen and Boutilier 1995; Kusznir and Karner 2007; Huismans and Beaumont 2011). A consequence of depth-dependant extension controlled by a layered lithospheric rheology is that asymmetric patterns of uplift and subsidence, magmatism and deformation are observed across conjugate margins (Lemoine et al. 1986; Péron-Pinvidic and Manatschal 2009; Malusà et al. 2015). The isostatic response to lithospheric necking and deformation in the upper and lower crust can cause subsidence (for shallow crustal necking) or regional flexural uplift (for subcrustal necking) (Braun and Beaumont 1989). Flexural uplift of footwall blocks may also occur during the syn-rift phase because of tectonic unloading (Kusznir et al. 1991).

20.2.3 Conceptual Models of PCM Evolution

While geodynamic models have validated several possible mechanisms for creating uplift at rift flanks, one of the challenges has been producing and maintaining high-elevation topography several hundreds of kilometres away from the main rift zone, as observed in many PCMs worldwide (Weissel and Karner 1989; Gilchrist and Summerfield 1990). This issue has been addressed by surface process models that investigate the influence of lithology, climate and isostasy on the evolution and preservation of high-elevation rift flank escarpments over 10–100 Myr timescales (e.g. Kooi and Beaumont 1994; Gilchrist et al. 1994; Tucker and Slingerland 1994; van der Beek et al. 2002; Sacek et al. 2012).

Early work suggested that the step-like topography of many PCMs, characterised by low-relief surfaces and steep escarpments (e.g. southern Africa and southeastern Australia), was the result of parallel escarpment retreat. This process would involve river incision to a base level following some form of regional uplift, to generate broad, low-angle, concave surfaces (King 1962). However, surface process models have highlighted the importance of isostatic rebound in causing gradual surface uplift in concert with erosion. Numerical models combined with thermochronology-derived denudation estimates have shown that models involving a downwarped pre-rift basal unconformity (case (i) in Fig. 20.2) and escarpment retreats into an upland plateau were not likely scenarios (e.g. Ollier and Pain 1997; Seidl et al. 1996). The apparent downwarped morphology across some rift shoulders (e.g. southern Oman) has been attributed to lithospheric flexure during post-rift erosion and deposition offshore (Gunnell et al. 2007). Alternatives to the downwarp scenario require the initial escarpment to form through long-wavelength upwarping or at a syn-rift normal fault. The flexural isostatic response to

initial rifting and later to onshore denudation and offshore sedimentation causes upwarping of the plateau inland of the escarpment drainage divide, while fluvial erosion and the existing drainage network control the evolution, morphology and persistence of marginal escarpments over time (Braun 2018). The escarpment retreat model (case (ii) in Fig. 20.2) predicts that escarpments originally at the coast propagate towards their inland present-day location by fluvial erosion until the inland catchment areas are captured (Kooi and Beaumont 1994; Weissel and Seidl 1998; Braun 2018). The plateau downwearing (or degradation) scenario (case (iii) in Fig. 20.2) considers an existing pre-rift drainage divide on an elevated landscape and predicts rapid erosion of the area between the coastline and the inner divide leading to the formation of an escarpment at the position of the divide (Gallagher and Brown 1999). In this scenario, the escarpment is established close to its present-day position relatively quickly after breakup (tens of millions years) and retreats slowly thereafter (Kooi and Beaumont 1994; van der Beek et al. 2002; Cockburn et al. 2000; Persano et al. 2005; Braun 2018).

20.2.4 The Role of Deep Processes and Tectonic Inheritance

A major control on the post-rift development of rift flank topography and adjacent sedimentary basins is the coupling between surface processes and flexure of the lithosphere in response to onshore denudational unloading and offshore sediment loading (Burov and Cloetingh 1997; Rouby et al. 2013). The coupling between mantle convection, plate stresses and the strength of the lithosphere may drive tectonic uplift over short (10–100 km) and long (100–1000 km) wavelengths (Cloetingh et al. 2008), with transient uplift phases possibly controlled by small-scale mantle convection (Moucha et al. 2008; Sacek 2017).

The reactivation of pre-existing structures is very important for the localisation of the main rift zone and may also influence post-rift deformation (Ziegler and Cloetingh 2004). Redfield and Osmundsen (2013) propose that geometric relationships exist between gradients in crustal thickness, drainage networks, elevated topography and brittle structures across a margin. Redistribution of sediment through erosion and deposition across thinned crustal sections can lead to flexural uplift of the margin and may induce lateral stresses triggering fault reactivation and footwall uplift (e.g. Redfield et al. 2005; Redfield and Osmundsen 2013). Changes in intraplate stress triggering deformation may occur due to changes in regional plate movements (e.g. Torsvik et al. 2009; Pérez-Díaz and Eagles 2014) and by lithospheric resistance to plate rotation (e.g. Bird et al. 2006) triggering fault reactivation. Uplift over hot upwelling

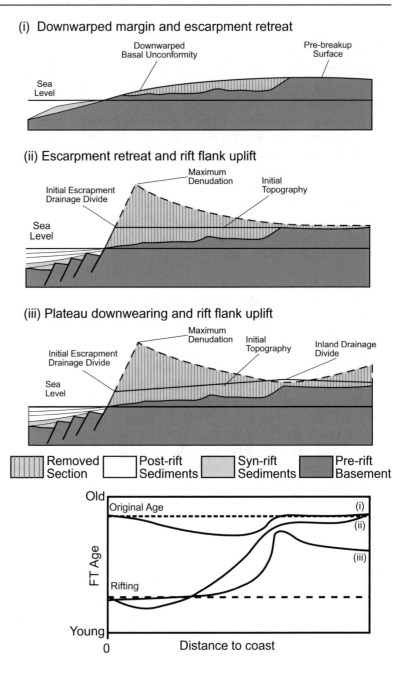

mantle plumes can introduce additional stresses that can potentially propagate through pre-existing lithospheric structures (e.g. Burov and Gerya 2014; Koptev et al. 2017).

20.2.5 Dynamic Geomorphology Models Versus Polycyclic Landscape Evolution Models

In the context of creating and preserving PCM topography, models of polycyclic landscape evolution (see Green et al. 2013; Lidmar-Bergström et al. 2013 for reviews) provide an alternative to dynamic geomorphology models where erosion and the flexural isostatic response to unloading can explain the major topographic features observed at PCMs (Bishop 2007). Polycyclic landscape evolution models typically involve cycles of uplift and erosion to form regional peneplains—low-relief erosion surfaces worn to a distinct and common base level (e.g. sea level) (Phillips 2002). However, the tectonic implications that follow from this concept have been fervently debated (e.g. Phillips 2002; van der Beek et al. 2002; Nielsen et al. 2009; Green et al. 2013; Lidmar-Bergström et al. 2013). The polycyclic landscape evolution approach involves correlating regional planation surfaces based on their form, weathering profile, spatial relationship with other planation surfaces and additional geological constraints (Green et al. 2013). Correlating these surfaces assumes that they once formed a single planation surface at low elevations. If these surfaces are subsequently uplifted and dissected by incising rivers, the remnant surfaces (i.e. their present-day form) can be correlated based on their topographic concordance or similar weathering profiles (e.g. Partridge and Maud 1987; Ollier and Pain 1997; Burke and Gunnell 2008; Japsen et al. 2012a; Green et al. 2013; Lidmar-Bergström et al. 2013). However, the use of erosion surfaces to infer cyclic episodes of uplift and erosion is fraught with uncertainties concerning, for example, the initial correlation of remnant surfaces (Summerfield 1985; van der Beek et al. 2002; Burke and Gunnell 2008), and the more fundamental issue of whether it is plausible that an evolving landscape is permitted the time and tectonic stability to become a peneplain (e.g. Phillips 2002). Alternative explanations for the formation of low-relief surfaces have been proposed, such as etchplanation, lateral stream erosion, local drainage variations controlled by lithology and/or structure and glacial erosion in glaciated regions (Pavich 1989; Summerfield 1991; Mitchell and Montgomery 2006; Steer et al. 2012; Gunnell and Harbor 2010; Yang et al. 2015).

20.3 The Application of FT Thermochronology to PCMs

20.3.1 Sampling Strategies for FT Analysis

Numerical models of PCM landscape evolution imply major erosion along the rift flanks and less erosion further inland. The total amount of erosion and the rate and style of escarpment retreat depends on the pre-rift topography, amount of flexural isostatic rift flank uplift, with additional controls on the topographic evolution exerted by lithology and climate (Gilchrist et al. 1994; Braun 2018). Transects comprising of AFT age and mean track length (MTL) data, running inland perpendicular to the coastline through the escarpment to the continental interior, and combined with coast-parallel sampling, should provide information on the regional thermal evolution of a PCM (Braun and van der Beek 2004). Results of FT analysis can be interpreted alongside predictions of erosion made by numerical landscape evolution models to better understand the large-scale processes that produce the first-order topographic features of a PCM.

In addition to a regional sampling approach, the growing evidence for post-rift phases of margin uplift and/or localised fault reactivation invites further study of PCMs using lower temperature techniques and sampling of more localised geological and structural complexity. In high relief regions, elevation profiles on hill slopes and tilted fault

blocks are highly informative on the timing and rate of exhumation, whereas high spatial resolution sampling across known structures may reveal if reactivation during the syn- or post-rift has been important. Sampling should also seek to exploit well-dated geological and geomorphic markers that may provide independent constraints for thermal history modelling.

20.3.2 Regional AFT Transects

Samples collected from regional transects, which have been exhumed from different palaeotemperatures, will have experienced different degrees of thermal annealing. In this case, AFT ages and MTLs form a distinct concave up (or 'boomerang') pattern (Green 1986; Gallagher et al. 1998; see also Chap. 10, Malusà and Fitzgerald 2018b). These 'boomerang' style relationships have been observed along the rifted margins of southeastern Australia and northeastern Africa, and along the Red Sea (Fig. 20.3). On the right side of these plots (Fig. 20.3), older AFT ages have relatively long MTLs corresponding to samples that have resided at temperatures colder than 60 °C after an initial (older) cooling event. At these lower temperatures, fission tracks have not been annealed after initial (earlier) cooling and retain thermal information corresponding to this earlier event. AFT ages and MTLs decrease in the central part of the plot, reflecting samples that have resided at higher temperatures and have experienced annealing in the partial annealing zone (PAZ) prior to a later (young) cooling event. Towards the left (younger) side of the plot, MTLs become progressively longer as AFT ages become younger. Where MTLs are very long ($\sim$15 µm), the AFT ages approximate the time of the onset of rapid cooling (Fig. 20.3).

The downwarp, scarp retreat and downwearing models all imply limited erosion landward of the escarpment (Fig. 20.2). AFT ages in this inland region are significantly older than the age of rifting and reflect the preservation of a pre-rift cooling event (e.g. Gallagher and Brown 1999). If an interior drainage divide exists, some denudation of the plateau may occur; however, the magnitude of denudation is typically too low to exhume rocks with completely reset AFT ages following rifting. Differences in AFT age because of the different style in PCM evolution are best observed seaward of the escarpment (Fig. 20.2). In the downwarp scenario (case (i) in Fig. 20.2), a pre-rift basal unconformity is eroded through backwearing. This produces a pattern of old AFT ages at the coastline, which young inland towards the base of the present-day escarpment. Unless the pre-rift elevation of the plateau was high (>2 km), magnitudes of erosion predicted by this model may not be sufficient to completely exhume any annealed samples (van der Beek et al. 1995). In this case, AFT ages across the coastal strip

Fig. 20.3 (i) Plot of AFT age against mean track length (MTL) showing boomerang shapes with a clear peak of ages with long (>13 μm) MTLs at 80–100 Ma (for southeastern Australia) and <40 Ma (for northeastern Africa). (ii) AFT age versus distance to coast showing that younger ages are found closer to the coast and older, pre-rift, ages are found inland. Datasets from Moore et al. (1986) (grey circles); Persano et al. (2002, 2005) (white circles); Omar and Steckler (1995) (white diamonds); Abbate et al. (2002) and Ghebreab et al. (2002) (dark grey diamonds); Balestrieri et al. (2009) (light grey diamonds)

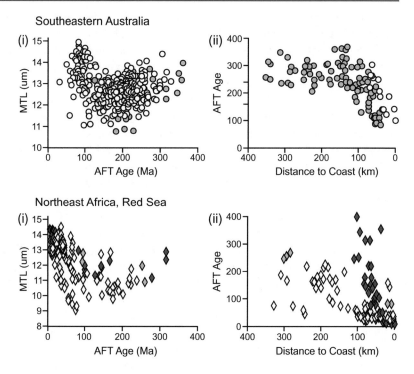

seaward of the escarpment would typically be older than the age of rifting (Fig. 20.2), and no boomerang pattern will be expected. In the escarpment retreat and plateau downwearing scenario (cases (ii) and (iii) in Fig. 20.2), denudation is highest at the coast and can be several kilometres depending on the initial escarpment height and isostatic response to erosion. For escarpment retreat, as the total amount of denudation decreases to a minimum above the present-day escarpment, AFT ages gradually increase. If an inland drainage divide is present, AFT ages on the coastal strip are approximately syn-rift and uniform due to the magnitude and pattern of denudation seaward of the drainage divide (Fig. 20.2) (Gallagher et al. 1998).

20.3.3 Vertical Profiles and Borehole Samples

Sampling vertical profiles and borehole profiles can provide greater insight into the cooling history of a region. Obtaining AFT data from samples from a vertical section that have resided at different palaeotemperatures prior to cooling (either from a borehole or elevation profile) will reveal distinct trends of AFT age and MTL with depth/elevation (see Chap. 9, Fitzgerald and Malusà 2018). By modelling these data, the thermal history may be better resolved and unjustified complexity, introduced by modelling samples individually, possibly removed (Gallagher et al. 2005). Moreover, vertical profile and deep borehole sampling can provide site-specific constraints on the palaeogeothermal gradient of the upper crust (Gallagher 2012; see also Chap. 8, Malusà and

Fitzgerald 2018a), which can be used to convert thermal histories to denudation histories more reliably.

20.3.4 Integration of AFT with Other Methods

Apatite (U–Th)/He Apatite (U–Th)/He (AHe) thermochronology has become a popular technique to obtain tighter constraints on magnitudes of erosion lower than the sensitivity of AFT data are able to achieve (see Chap. 5, Danišík 2018). While the theoretically lower closure temperature (T_c) of the AHe system makes its use a logical choice, several factors can influence the T_c of an individual apatite grain (Brown et al. 2013). At long-lived and relatively slowly cooled settings such as PCMs and plate interiors, this can lead to dispersed single-grain ages from a sample. Although these data can be challenging to interpret and reproduce during thermal history modelling, the natural dispersion in many datasets may also contain additional thermal history information (Fitzgerald et al. 2006; Hansen and Reiners 2006; Ksienzyk et al. 2014; Wildman et al. 2016).

Cosmogenic Nuclides Recent erosion across many PCMs has been constrained by cosmogenic nuclide dating (Fleming et al. 1999; Bierman and Caffee 2001; Cockburn et al. 2000), which have generally predicted moderate to low rates of erosion and scarp retreat. While these data have been used to support the conclusions drawn from AFT and AHe studies, confidence in the extrapolation of erosion rates derived from cosmogenic nuclide analysis into the geological past is limited, depending largely on how long the present climatic and tectonic conditions have prevailed.

Weathering Geochronology Dating techniques to constrain the timing of weathering profile formation, such as $^{40}Ar/^{39}Ar$ on K–Mn oxides, have provided both an age of the preservation of landscapes and estimates of erosion based on the dissection of these weathering profiles (de Oliveira Carmo and Vasconcelos 2006; Beauvais et al. 2016; Bonnet et al. 2016). These constraints can be incorporated alongside thermochronology and cosmogenic nuclide data to achieve a more complete record of the erosional history of the margin (e.g. Vasconcelos et al. 2008).

20.4 Application to the Southeastern African Margin

The prominent morphology of the Drakensberg Escarpment in southeastern Africa is a classic example of a high-elevation PCM, and the region has proved itself to be an excellent natural case study for long-term landscape evolution. The geomorphology is characterised by a low-relief, low-lying coastal plain and a low-relief, elevated inland plateau separated by a ∼200-km-wide region of high relief and increasing maximum elevation with distance from the coastline (i.e. Lesotho Highlands) (Fig. 20.4). The escarpment is clearly defined by a rapid increase in elevation

Fig. 20.4 a–c Coast perpendicular transect across southeastern Africa showing AFT ages (a), mean track lengths (MTL) (b) and topography (c). **d** AFT data versus depth for two boreholes (LA and SW), see location in (c) (after Brown et al. 2002)

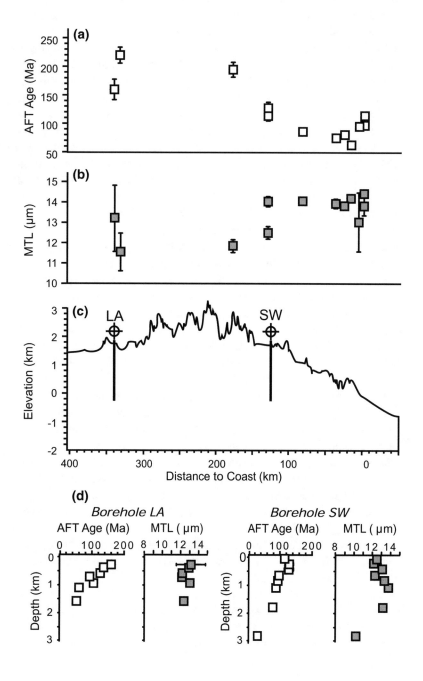

of around 1 km over a distance of 10 km, and the continental drainage divide coincides with the location of the escarpment summit (Brown et al. 2002). The coastal geology is comprised of Palaeozoic basement rocks while the escarpment and plateau preserve a sequence of Permian–Triassic sedimentary rocks capped by Lower Jurassic basalts.

Initially, the formation of the Drakensberg Escarpment was attributed to parallel escarpment retreat across a downwarped monocline, which formed during continental breakup (e.g. King 1962; Ollier and Marker 1985; Partridge and Maud 1987). More recently, AFT data were collected from outcrop samples along a transect through the southeastern African PCM and from two boreholes: one located inland of the escarpment and one located seaward of the escarpment (Brown et al. 2002). The aim of the study was to test different models for the evolution of PCM topography. The AFT ages of outcrop samples are ∼95–115 Ma at the coastline, become slightly younger on the coastal strip, and get progressively older with elevation at the escarpment, reaching ages of ∼200 Ma on the plateau (Fig. 20.4a). MTLs are generally longer on the coastal strip (13.5–14.5 μm) and become shorter further inland (11.5–12.5 μm) (Fig. 20.2b). These outcrop data imply that a minimum of 4.5 km of denudation occurred since 130 Ma along the coastal plain (Brown et al. 2002). Borehole data show a relationship of decreasing AFT age with depth (Fig. 20.4d); however, the relationship of MTL versus depth is more complicated because of annealing experienced by each sample due to their different palaeotemperatures. Thermal modelling of these samples predicts that a second pulse of denudation in the Late Cretaceous affected both the region immediately seaward of the escarpment and the hinterland (Brown et al. 2002).

A companion paper (van der Beek et al. 2002) to the study of Brown et al. (2002) investigated the role of surface processes and isostatic rebound in producing an erosion history consistent with the observed FT data and cosmogenic nuclide analysis (Fleming et al. 1999). The surface process modelling results support a scenario where an initially elevated plateau experiences rapid degradation forming an escarpment at an inland drainage divide tens of millions of years after breakup, which experienced minimal retreat (∼20–30 km) since then. The spatial and temporal patterns of post-breakup denudation predicted along the southeastern African margin provide strong evidence against a downwarp model. However, the surface process models produced by van der Beek et al. (2002) predict constant rates of denudation throughout the Cretaceous without a Late Cretaceous pulse in denudation as suggested by thermal modelling of borehole data. Apatite FT borehole data from southern Africa (Tinker et al. 2008a) also record this Late Cretaceous denudation event suggesting that regional uplift

and denudation may have been more important in the post-rift phase than processes typically associated with 'classical' escarpment evolution models.

20.5 The Conjugate South Atlantic Passive Margins

An extensive AFT dataset now exists along the eastern margin of South America and the western margin of southern Africa (Fig. 20.5). This dataset provides insights into the timing and style of topographic development on both sides of the Atlantic since the Mesozoic. The breakup of western Gondwana and the opening of the South Atlantic occurred in the Late Jurassic and propagated northwards with complete separation of South America and Africa occurring by the mid–Late Cretaceous (Macdonald et al. 2003; Heine et al. 2013). Along both margins, the AFT ages seaward of the escarpment are markedly younger than the age of rifting. A general observation that can be made from Fig. 20.5 is that younger AFT ages are typically associated with longer MTLs and older ages with shorter MTLs. In detail, these data reveal a more complex pattern of crustal exhumation than is predicted by conceptual models, which is linked to different modes of post-rift evolution along the conjugate margins.

20.5.1 The Atlantic Margin of South America

In southeastern Brazil, the first regional AFT dataset (Gallagher et al. 1994) qualitatively agreed with theoretical landscape evolution models of scarp retreat following rifting (see Fig. 20.2). The low-lying coastal strip exhibits ages younger than rifting, whereas the elevated inland plateau shows older ages. However, the geomorphology, characterised by two different escarpments (Serra do Mar and Serra da Mantiqueira), geological evidence of post-rift activity and more recent thermochronological data suggest such a simple evolution since rifting is unlikely.

In addition to the large AFT database that now exists in southeastern Brazil (Fig. 20.5), ZFT and U–Th/He dating on both apatite and zircon provide more detailed thermal history constraints and are supportive of post-rift reactivation. Along the southeastern Brazilian margin, modelled thermal histories imply that at least three phases of rapid cooling occurred during the Late Cretaceous, Palaeogene and Neogene (Tello-Saenz et al. 2003; Hiruma et al. 2010; Cogné et al. 2011, 2012; Franco-Magalhaes et al. 2014). Although an inferred Neogene cooling pulse may be an artefact of modelling (Dempster and Persano 2006; Redfield 2010) there are several lines of evidence that support a phase of Neogene cooling in Brazil. First, cooling is coeval with a

Fig. 20.5 Interpolation maps of AFT age (**a**) and mean track lengths (**b**) for the southeast American margin and the conjugate margin of southwestern Africa. Dashed black lines represent major structural features and tectonic boundaries: 1—Borborema Plateau, 2—Pernambuco Shear Zone, 3—Recôncavo-Tucano-Jatoba Basin, 4—San Francisco Craton, 5—Mantiqueira Range, 6—Cabo Frio Lineament, 7—Serro do Mar; A—Damara Belt, B—Kaapvaal Craton, C—Doringberg–Hartbees Fault Zone, D—Cango Fault, E—Cape Fold Belt, F—South African Plateau. Data sources for SE America: Gallagher et al. (1994, 1995), Amaral et al. (1997), Harman et al. (1998), Tello Saenz et al. (2003), Turner et al. (2008), Morais-Neto et al. (2009), Cogne et al. (2011, 2012), Karl et al. (2013), Franco-Magalhaes et al. (2014), Jelinek et al. (2014), de Oliveira et al. (2016), Kollenz et al. (2016). Data sources for SW Africa: Haack (1983), de Wit (1988), Brown et al. (1990, 2014), Raab et al. (2002, 2005), Jackson et al. (2005), Tinker et al. (2008a), Kounov et al. (2009, 2013), Wildman et al. (2015, 2016, 2017), Green et al. (2017), Green and Machado (2017)

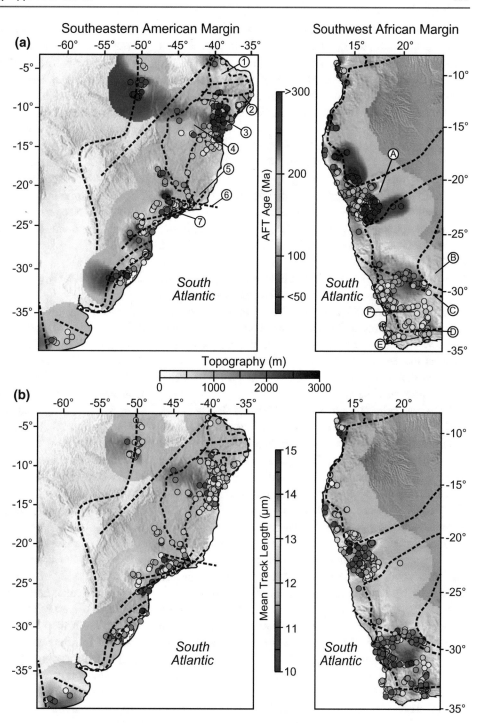

period of increased sedimentary flux into the offshore basins (Assine et al. 2008; Contreras et al. 2010). Second, structural evidence argues for transpressional deformation in the onshore Cenozoic basins (Cobbold et al. 2001; Riccomini et al. 2004; Cogné et al. 2012) and offshore Campos Basin (Fetter 2009). Finally, Modenesi-Gauttieri et al. (2011) suggest climate changes along with Neogene uplift produced weathering profiles in the Serra da Mantiqueira. The inferred Cretaceous cooling episode is coeval with the emplacement of alkaline bodies (Riccomini et al. 2005) and a period of rapid sedimentation offshore (Assine et al. 2008). Palaeocene cooling is localised on the area around the Cenozoic basins and is coeval with their formation (Cogné et al. 2013). Overall, in the Serro do Mar region of southeastern Brazil the main post-breakup cooling events are linked to the reactivation of old shear zones that run parallel to the coast. South of 25°S (see Fig. 20.5), similar cooling episodes are identified (Karl et al. 2013; De Oliveira et al. 2016) and are also coeval with main pulses in sedimentation into offshore basins. Karl et al. (2013) identified three distinct blocks that are separated by fracture zones inherited from the rift and are reactivated during the Palaeogene and Neogene in a similar manner to the reactivation of the Cabo Frio lineament during the Late Cretaceous (Riccomini et al. 2005; Cobbold et al. 2001).

In contrast to southeastern Brazil, northeastern Argentina is a relatively low-elevation PCM with low-elevation (<1000 m) topography running perpendicular to the coast (Fig. 20.1). Here, AFT ages are older than rifting and there is no indication of post-rift reactivation (Fig. 20.5). Instead, the geomorphological characteristics of the margin better reflect the Gondwana tectonic history (Kollenz et al. 2016). On the other hand, the margin of Brazil north of Rio de Janeiro also exhibits low coastal plains and elevated inland regions. Harman et al. (1998) showed that some of their samples, north of the São Francisco Craton, have ages younger than rifting and could be as young as Late Cretaceous near the Pernambuco Shear Zone. While most of their samples indicate an episode of cooling during continental breakup, the young ages indicate reactivation of old structures during the Late Cretaceous.

More recently, Turner et al. (2008) and Japsen et al. (2012b) studied the Recôncavo-Tucano-Jatobá Rift, collecting samples from basement rocks and sedimentary rocks in the rift basin (Fig. 20.5). Through thermochronology and geomorphological analysis, they conclude that three phases of reactivation-driven cooling occurred during the Late Cretaceous, Eocene and Miocene. The Late Cretaceous (Campanian) and Eocene uplift phases led to the erosion of sediments deposited since the Early Cretaceous breakup, with the consequent formation of a peneplain. The peneplain was then covered by sediments that were in turn eroded during the Miocene uplift phase. This erosive period generated a second low-lying peneplain. Data from the eastern margin (Morais-Neto et al. 2009; Jelinek et al. 2014) also provide evidence for two cooling episodes during the Late Cretaceous and Neogene in the Mantiqueira Range and the Borborema Plateau. However, while Jelinek et al. (2014) recognise the phases of uplift, they argue against the intervening episodes of sedimentation. Indeed, their data do not require heating episodes and show variability among the same alleged palaeosurface. As in southeastern Brazil, the main phases of cooling are coeval with phases of sedimentation in the offshore basins (Jelinek et al. 2014).

At a regional scale, thermochronological data suggest the Brazilian margin experienced at least two major episodes of accelerated post-rift cooling during the Late Cretaceous and Neogene, respectively, which was more significant near inherited structures than in the interior cratons. A possible third phase is locally inferred, such as in the Paraiba do Sul Valley (Cogné et al. 2012) and in the Recôncavo-Tucano-

Jatobá Rift system (Japsen et al. 2012b). These three phases at the scale of the margin agree with the sedimentary record in the offshore basins. The synchronicity of these cooling episodes with the main phases of building of the Andes led several authors (e.g. Cogné et al. 2012; Japsen et al. 2012b) to link cooling episodes to structural reactivation under transmitted compressive stress. However, as described by Jelinek et al. (2014), the ability to transmit this deformation to the PCMs is largely dependent on lithospheric rheology and may have required a thermal anomaly in the mantle. According to Jelinek et al. (2014) and Morais-Neto et al. (2009), the Neogene denudation episode was likely due to a change to a more erosive climate.

20.5.2 The Atlantic Margin of Southern Africa

Samples along the western and southern margins of South Africa yield AFT ages that fall within the mid-Triassic to Late Cretaceous (~230–65 Ma) and do not show any significant correlation with MTL, which are mainly between 12 and 15 μm. One of the most intriguing features of the AFT data across southwestern Africa is that post-rift ages and long MTLs extend well inland of the escarpment (~500–600 km) (Fig. 20.5). Despite samples from different studies yielding similar AFT data, different approaches to modelling and interpreting thermal histories have led to different conclusions on the timing, magnitude and distribution of cooling across the margin. However, a consistent conclusion is that multiple cooling events have taken place since the beginning of the Mesozoic.

Along the western South African margin, Kounov et al. (2009) attribute cooling from 160 to 138 Ma to thermal relaxation after Karoo magmatism at 180 Ma, but acknowledge that Early Cretaceous rift-related tectonics may also have played a role. Wildman et al. (2015, 2016) show that the onset of cooling between 150 and 130 Ma can be observed in samples from along the coastal strip and landward of the escarpment, suggesting that cooling was a result of exhumation driven by erosion of rift-related topography. Green et al. (2017) suggest that cooling during this time along the southern margin could be explained by differential block uplift (and subsequent differential erosion) during the rifting and separation of Africa. However, Tinker et al. (2008a) predict cooling at 140–120 Ma attributed to denudation triggered by regional uplift above a buoyant mantle.

Post-rift cooling episodes during the mid–Late Cretaceous have been described by all thermochronology studies across South Africa. However, the spatial and temporal nature of mid–Late Cretaceous cooling is still debated. Thermal histories derived from AFT data presented by Kounov et al. (2009) and Wildman et al. (2015) and AFT and AHe data from Wildman et al. (2016) suggest that the southwestern African margin experienced heterogeneous patterns of crustal exhumation during 110–90 Ma related to post-rift reactivation of basement structures. Green et al. (2017) also propose that mid-Cretaceous (∼ 120–100 Ma) fault reactivation of the Cango Fault along the Cape Fold Belt caused greater exhumation north of the fault. On the South African Plateau, AFT borehole data (Tinker et al. 2008a), joint AFT and AHe data from outcrop samples (Kounov et al. 2013; Wildman et al. 2016, 2017) and AHe from Cretaceous kimberlite intrusions (Stanley et al. 2013, 2015) and the eastern Kaapvaal Craton (Flowers and Schoene 2010) show the regional extent of mid-Cretaceous cooling and suggest that several kilometres of material have been removed from the plateau. Regional uplift due to increased mantle buoyancy (Tinker et al. 2008a), isostatic or 'dynamic' mantle-sourced uplift (Kounov et al. 2013) or thermochemical changes in the lithosphere (Stanley et al. 2015) has been proposed for triggering denudation. However, Wildman et al. (2017) suggest that block uplift along the Doringberg-Hartbees fault zone at the craton–mobile belt boundary may have also occurred in response to regional horizontal stresses. The younger ages from this area likely record cooling subsequent to heating associated with the Etendeka magmatic episode, as samples from this region were taken from the Etendeka basalts or closely underlying lithologies. The older ages were obtained from Archaean basement rocks and likely record partially reset 'mixed ages' after reburial since the Carboniferous and subsequent exhumation in the Late Cretaceous. The spatial relationship in the AFT data observed in northern and south-central Namibia is such that the position of the present-day escarpment broadly defines the boundary between young post-rift ages (seaward of the escarpment) and pre-Gondwana breakup ages (on the plateau) (Fig. 20.5a). This could be due to samples in the coastal zone having been exhumed from greater depths (i.e. due to greater erosion) than the interior following rifting or due to samples in the coastal zone being reset at ∼ 130 Ma by the emplacement of the Etendeka lavas (Raab et al. 2002). However, in the Damara Belt in central Namibia, a NE–SW trending zone of Late Cretaceous (∼ 60–80 Ma) AFT ages, with long MTLs, extends 400 km inland and implies rapid cooling of this block. This cooling has been attributed to reactivation of NE–SW shear zones along the Damara Belt (Raab et al. 2002; Brown et al. 2014).

An alternative model for the Late Cretaceous and Cenozoic evolution of southern Africa was proposed by Green et al. (2017). Their interpretation of AFT data and thermal histories from samples collected along the Cape Fold Belt is that post-rift subsidence and burial, potentially extending far enough inland to cover the southern Karoo basin, preceded regional exhumation during the Late Cretaceous (85–75 Ma)

and Late Cenozoic (∼ 30–20 Ma). The total amount of cover that was deposited and removed over this time is suggested to be between ∼ 1.2 and 2.2 km. They also suggest that sedimentary basins along the southwestern Cape and elevated escarpment topography were buried beneath an overburden of ∼ 1 km during the Oligocene. They argue that instead of the Great Escarpment being a remnant feature related to continental breakup, it is produced by a period of enhanced denudation in response to Late Cenozoic regional uplift (e.g. Burke and Gunnel 2008). Although Cenozoic denudation of ∼ 1 km is consistent with results from thermal models from other studies (e.g. Tinker et al. 2008a; Stanley et al. 2013; Wildman et al. 2015), these models predict that the material was removed over the entire Cenozoic at low rates, consistent with regional estimates of erosion over the last few million years predicted from cosmogenic nuclide studies (Kounov et al. 2007; Scharf et al. 2013; Decker et al. 2011).

The sedimentary record in the western and southern offshore basins shows that the peak of accumulation occurs during the mid–Late Cretaceous whereas the Cenozoic is characterised by much lower sediment volumes (Guillocheau et al. 2012; Tinker et al. 2008b) supporting the importance of a mid–Late Cretaceous denudation phase. Denudation during the mid–Late Cretaceous has been attributed to enhanced erosion in response to uplift of the entire subcontinent. Dynamic uplift, or tilting, of the South African Plateau due to the movement of the continent over a large buoyant 'superplume' in the deep mantle has frequently been cited as a possible cause (e.g. Lithgow-Bertelloni and Silver 1998; Gurnis et al. 2000; Braun et al. 2014). However, the contribution to uplift from dynamic vertical forces from the deep mantle relative to isostatic responses due to chemical and density contrasts in the lithospheric mantle is still debated (Molnar et al. 2015; Stanley et al. 2015; Artemieva and Vinnik 2016). Moreover, horizontal plate movements are suggested to be linked to post-rift uplift (Colli et al. 2014), with Moore et al. (2009) suggesting that flexural uplift is related to plate reorganisations and tensional stresses in the lithosphere. Irrespective of such interpretations, the heterogeneous pattern of cooling across southwestern Africa, predicted from thermochronology data, implies a complex tectonic history with an interaction between regional uplift mechanisms and structural reactivation at basement structures along the margins and at major intraplate structural zones.

20.5.3 Analogies and Differences Across the South Atlantic

Tectonic and magmatic asymmetries are observed on the opposite sides of the South Atlantic mid-ocean ridge, with structural and magmatic changes being described both along the southeastern American and southwestern African margins (Blaich et al. 2013). A complex asymmetric rifting history may be due to tectonic inheritance, rift migration and plume–rift interaction (Becker et al. 2014; Brune et al. 2014) and may lead to different patterns of uplift and subsidence on the conjugate margins. Determining and comparing the magnitude and spatial patterns of erosion of syn-rift topography along the southeastern American and southwestern African margins are challenging, due to the overprint of post-rift tectonic events and localised structural reactivation. During the breakup and separation of South America and Africa, plate kinematics change the intracontinental stress field possibly inducing intraplate deformation (Moulin et al. 2010; Pérez-Díaz and Eagles 2014). In addition to stresses related to divergent plate motions, South America and Africa are subject to distinct regional tectonic and mantle processes during their post-rift phase and their influence on the tectonic and geomorphic development of the margins is still debated. In southeast America, far-field stresses propagating from the subduction zone beneath the Andes may have driven post-rift margin uplift, whereas in South Africa, a mantle 'superplume' may have caused long-wavelength uplift of the entire subcontinent. Despite different regional mechanisms acting on each continent, thermochronology data have revealed spatially distinct post-rift cooling events and suggest a major control on the cooling pattern exerted by local tectonic structures.

20.6 The Conjugate Passive Margins of the North Atlantic

The North Atlantic PCMs are the result of a prolonged, multi-phase history of rifting beginning in the Permo–Triassic (Péron-Pinvidic et al. 2013). Rifting culminated in the Palaeogene with continental breakup leading to the formation of new oceanic crust and the opening of the North Atlantic. Despite extensive investigation, the origin of high-elevation topography along the western Scandinavian and the eastern Greenland margins is still debated. While the AFT data from different studies are broadly similar, their modelling and interpretation have been used to support quite different hypotheses on the effective age of the present-day topography. On the one end, the topography is interpreted to be the result of a relatively young Neogene event involving uplift and subsequent erosion; on the other end, the formation of the topography is interpreted as an essentially old (Caledonian) event, with the present-day morphology reflecting continuous slow erosion of the Caledonides and associated isostatic rebound.

20.6.1 The Scandinavian Margin of the North Atlantic

The earliest set of AFT data (Rohrman et al. 1994, 1995) from the western Scandinavian PCM (Fig. 20.6) was interpreted in terms of two phases of rapid denudation related to uplift of the margin. The first Triassic–Jurassic event was ascribed to erosion of the margin driven by a lowering of base levels and uplift of the rift flanks. The latter Neogene event was proposed to have occurred in response to a doming-style uplift caused by mantle convection. The occurrence of uplift and erosion in the Oligocene to Neogene has long been suggested by unconformities in offshore sediments, but the mechanisms and timing of uplift are still debated (Japsen and Chalmers 2000; Mosar 2003). An episodic evolution of the North Atlantic margins has been advocated by combining AFT data (Fig. 20.6) with geomorphological observations of concordant erosion surfaces and extrapolation of offshore unconformities onshore (Green et al. 2011; Japsen et al. 2012a). Thermal histories derived using this approach imply multiple cooling–heating episodes involving erosion of a landscape to form peneplains at sea level, followed by burial and then uplift.

The debate surrounding the tectonic and geomorphic significance of low-relief surfaces has been particularly important for the evolution of the western Scandinavian margin (e.g. Schemer et al. 2017). A major criticism concerns the lack of age constraints on surfaces and whether they were eroded to sea level or not. In Norway, the absence of local stratigraphic age constraints on surfaces due to the lack of dateable sediment overlying them has resulted in age information being extrapolated from widely distributed stratigraphic sequences (Nielsen et al. 2009). The more fundamental issue on whether a landscape is afforded the tectonic and climatic stability to be eroded to a regional planation surface (e.g. Phillips 2002) is questioned by the other mechanisms that may have produced these surfaces. For example, erosion in heavily glaciated regions has been proposed to limit topography (Mitchell and Montgomery 2006; Steer et al. 2012; Egholm et al. 2009) and drainage reorganisation may have resulted in low-relief surfaces forming at different times and elevations (Gunnell and Harbor 2010; Yang et al. 2015). Correlating poorly dated surfaces is also challenging across PCMs, like western Scandinavia, that may have experienced regional tectonic and isostatic uplift, localised fault reactivation and small-scale mantle-driven uplift (Redfield et al. 2005; Praeg et al. 2005; Osmundsen and Redfield 2011). Although the alternative mechanisms for producing elevated low-relief surfaces are debated in the context of the Scandinavian margin (e.g. Green et al. 2013) and the evolution of

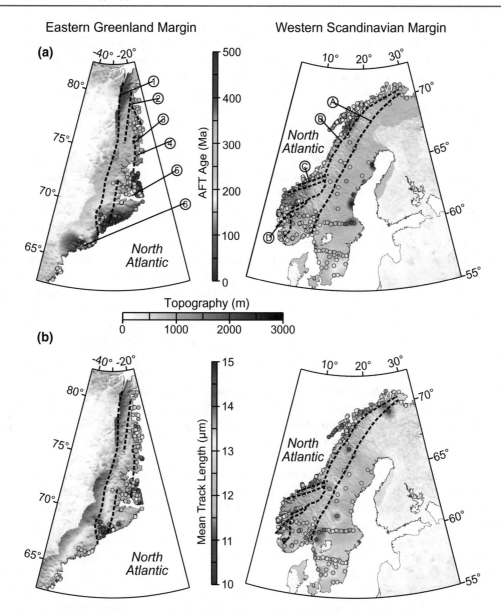

Fig. 20.6 Interpolation maps of AFT age (**a**) and mean track lengths (**b**) for the east Greenland margin and the conjugate margin of western Scandinavia. Dashed black lines represent major structural features and tectonic boundaries: 1—Caledonian Deformation Front, 2—Storstrommen Shear Zone, 3—Western Fault Zone, 4—SW Clavering Ø, 5—Scoresby Sund, 6—Kangertittivatsiaq; A—Caledonian Deformation Front, B—Inner Boundary Fault, C—Møre Trøndelag Fault Complex, D—Hardangerfjord Shear Zone, E—Fennoscandian Shield. Data sources for east Greenland: Gleadow and Brooks (1979), Hansen (1992), Thomson et al. (1999), Johnson and Gallagher (2000), Mathiesen et al. (2000), Hansen et al. (2001), Hansen and Brooks (2002), Pedersen et al. (2012), Japsen et al. (2014). Data sources for western Scandinavia: van den Haute (1977), Andriessen and Bos (1986), Andriessen (1990), Gronlie et al. (1994), Rohrman et al. (1994, 1995), Hansen et al. (1996), Cederbom et al. (2000), Cederbom (2001), Hendriks and Andriessen (2002), Hendriks (2003), Huigen and Andriessen (2004), Redfield et al. (2004, 2005), Hendriks et al. (2010), Davids et al. (2013), Ksienzyk et al. (2014)

topography globally (e.g. Whipple et al. 2017), the complex coupling between tectonics, climate and structural inheritance in the crust seems to imply that the stability required to produce regional peneplains is rarely obtained. However, understanding how these low-relief surfaces have formed may yield insight into the overall development of PCM topography (Schermer et al. 2017).

An alternative interpretive model of Scandinavian topography, proposed by Nielsen et al. (2009), suggests that southern Norway has been elevated since the Caledonian orogeny, with isostatic rebound in response to slow erosion that would have maintained topography since then. Modelling of AFT data in their study, without the addition of assumed geological constraints, leads to much simpler thermal histories than those mentioned by other studies (e.g. Japsen et al. 2012a) and implies protracted cooling through the PAZ consistent with slow exhumation. Despite this more conservative approach to modelling the data, significant details were still resolved such as slow exhumation that started earlier, in the Late Palaeozoic, on the western islands and along the western coast than inland, and the occurrence of a younger (Cenozoic) cooling event required to bring the samples from within the PAZ to the Earth's surface. The former event is suggested to represent enhanced erosional denudation of the margin caused by lowering of base levels during rifting (Dunlap and Fossen 1998), which progressively propagated inland. The timing and magnitude of the latter event are poorly constrained due to both the temperature sensitivity of the AFT system and the regional, multi-sample approach used by Nielsen et al. (2009).

On a more local scale, detailed sampling across the major structural features along western Norway has revealed localised vertical movements and fault reactivations. Transects perpendicular to the strike of the Møre-Trøndelag Fault Complex led to a reinterpretation of the two-stage model of margin development proposed by Rohrman et al. (1995). Instead of Neogene domal uplift, Redfield et al. (2005) report offset AFT ages across major faults and advocate that reactivation during the Late Cretaceous and Cenozoic caused different exhumation histories on individual fault blocks. The mechanisms proposed to drive fault reactivation are linked to stresses induced by flexure of the lithosphere in response to erosional unloading of the margin and loading in adjacent basins (Osmundsen and Redfield 2011). More recent AFT and AHe data from southwestern Norway (Ksienzyk et al. 2014) have similarly reported offset ages across faults and support the importance of fault reactivation in the development of PCM topography. The role of reactivation either during the syn-rift or post-rift phase would distort and complicated the pattern of ages across the margin predicted for traditional models (Fig. 20.2). Offshore loading (by sediment deposition) and onshore unloading (by erosion and glacial retreat) may lead to flexure of the lithosphere and generate compressional and extensional stress regimes, respectively, in the upper part of the lithosphere (Osmundsen and Redfield 2011). The location of reactivated structures across the Norwegian PCM is linked to the gradient of the thinning crust offshore.

20.6.2   The Eastern Greenland Margin of the North Atlantic

Long-lived preservation of Caledonian topography has also been suggested to explain the topography of the eastern Greenland PCM (Pedersen et al. 2012), challenging previous proposals for major Neogene uplift (e.g. Japsen and Chalmers 2000). AFT ages presented by Pedersen et al. (2012) from northeast Greenland (75–80°N) range from 191 to 358 Ma and do not show characteristic AFT age or MTL trends with distance from the coast expected for a 'classic' passive margin (Fig. 20.6). Pedersen et al. (2012) conclude that their data can be explained by slow exhumation since the collapse of the Caledonian orogen in Permo–Triassic times. The implications of this model is that the present-day topography, which is relatively subdued, is a remnant of the original Caledonian topography and has been maintained due to the isostatic response to unloading. More rapid cooling related to rifting is suggested to have occurred until ~250 Ma due to denudation in response to rift flank uplift and base-level changes. In contrast, Japsen et al. (2013) argue that northeastern Greenland was buried following the Caledonian orogeny and basement rocks previously at the surface were buried under ~1–2 km of sediments. Uplift and erosion then formed a low-lying low-relief surface during the mid-Jurassic, which was subsequently buried under a further ~1–2 km of sediments prior to exhumation since post-mid-Jurassic times. The AFT data and inverse geodynamic modelling approach adopted by Pedersen et al. (2012) imply that reheating in the Mesozoic and Cenozoic was not significant across northeastern Greenland, and that any sediment cover would have caused insufficient burial to reheat samples above 60 °C. As is the case across other margins described previously, the nature of burial (reheating) events across the eastern Greenland margin and the ability of low-temperature thermochronology data to constrain this is the source of much debate. Much of the conflict between these two end-member scenarios described above stems from fundamentally different interpretations of the preserved post-rift sediments across the onshore margin. Whereas Japsen et al. (2013) suggest that these sediments are a remnant of a much more extensive sedimentary cover, Pedersen et al. (2012) advocate that these sediments were deposited in small fault basins. This minor onshore faulting would be related to post-Caledonian continental rifting, which would be largely confined to the offshore domain (Osmundsen and Redfield 2011).

AFT ages younger than those reported by Pedersen et al. (2012) from northeastern Greenland (i.e. post-Triassic) are recorded along the central and southeastern Greenland margins between 72 and 65°N (Fig. 20.6) (Johnson and Gallagher 2000; Bernard et al. 2016). The data generally show an increase in AFT age with elevation and distance from the coast. However, the post-rift tectonic, thermal and glacial history of the margin is complex, and care is required when relating topographic evolution to the thermal history predicted by AFT data. AFT ages from the interior hinterland indicate that the region has resided at temperatures less than 100–120 °C since the Early Mesozoic

(Fig. 20.6). AFT data from an elevation profile from the sides of a glacial valley at SW Clavering Ø show increasing age (151–226 Ma) with increasing elevation (up to 1000 m) (Johnson and Gallagher 2000). AFT ages and track lengths from these samples were jointly inverted to obtain a thermal history that predicts two Mesozoic cooling episodes in the Early Jurassic and mid-Cretaceous, respectively. However, the Cretaceous thermal history is poorly resolved in this study and other FT studies have suggested that cooling through the Cretaceous was slow (Hansen et al. 2001; Hansen and Brooks 2002) or involved reheating (Mathiesen et al. 2000). The exhumation and burial history are partially resolved by the accumulation of sediment in adjacent rift basins and an erosional unconformity beneath Palaeocene lavas. The emplacement of these lavas is in part responsible for the lack of preservation of the pre-Cenozoic thermal history. Extensive basaltic volcanism characterises the final stage of rifting and the onset of sea-floor spreading in the North Atlantic during the Palaeocene–Eocene (Skogseid et al. 2000). Palaeocene and younger AFT ages indicate that apatite from these samples was buried beneath the thick lava pile and partially reset. A component of this heating may also have been due to minor and short-lived thermal effects from the conductive heat transfer from the basalt (Gallagher et al. 1994; Hansen et al. 2001; Bernard et al. 2016) or from hot fluid flow (Hansen et al. 2001). AFT ages from the margin predict Neogene cooling reflecting erosion of the basalts (Hansen and Brooks 2002; Bernard et al. 2016). AHe ages from Kangertittivatsiaq (Fig. 20.6) were interpreted in the context of their correlation with grain size and elevation and were tentatively interpreted as recording 1.5 km of glacial incision in the Neogene (Hansen and Reiners 2006). Hansen and Reiners (2006) describe an apparent incompatibility between the thermal history predicted by AFT data and that predicted by the AHe data. However, joint modelling of AFT and AHe data from sub-vertical profiles further north between 68 and 76°N reveals a 30 Ma cooling event attributed to the initiation of glacial erosion (Bernard et al. 2016).

Japsen et al. (2014) propose that AFT data across the southeast Greenland margin are consistent with three phases of post-rift uplift and erosion across the margin. The first inferred uplift event occurred during the Late Eocene followed by complete erosion of the margin topography to produce a regional planation surface near sea level. This surface was then uplifted in the Late Miocene and incised to produce a lower planation surface at sea level. The third uplift phase, during the Pliocene, led to incision of valleys and fjords below the lower planation surface. This interpretation is based on mapping what are interpreted to be coeval planar surfaces to construct the remnants of the three

planation surfaces, and inferring tectonic significance to their respective elevations. These three uplift episodes would be linked to a combination of mantle convection and changes in plate motion. This link has not been tested quantitatively and cannot be constrained from AFT data and surface morphology alone.

20.6.3 Polycyclic Versus Monotonic Cooling of North Atlantic PCMs: The Role of Thermochronology

Although AFT analysis has been widely applied along the Scandinavian margin to develop models of landscape evolution, considerable debate remains, often due to the non-uniqueness of thermal history models derived from thermochronology data. Furthermore, interpretations and speculations concerning the nature of physical mechanisms to explain the thermal histories can become distant from the original AFT data. Contrasting views on the assumed geological history of the North Atlantic PCMs can also be propagated back into the interpretation of thermochronology data in the form of constraints imposed on thermal history models. As multiple time–temperature paths may be able to replicate in models, the observed AFT data, the choice of geological constraints, annealing model and statistical criteria for fitting the data are critically important (see Chap. 3, Ketcham 2018).

On the one hand, imposing multiple constraints to honour a complex geological history involving cycles of burial and exhumation and searching for palaeothermal maxima constrained by AFT data will lead to complex thermal histories involving several reheating and cooling episodes (Green et al. 2011; Japsen et al. 2014). On the other hand, imposing no (or limited) constraints due to the lack of well-dated geological information can result in monotonic cooling histories (e.g. Nielsen et al. 2009) that may still adequately explain the data. In the latter approach, the data and their uncertainty ultimately control the inferred complexity in the thermal history. Consequently, in the absence of well-justified geological constraints, more detail will only be resolved by using more robust FT data (consisting of single-grain ages, track lengths and compositional information), joint modelling of multiple samples, (e.g. from borehole/elevation profiles), multiple thermochronometers (e.g. AHe), as well as appropriate sampling techniques. For a more complete overview of the debate of these opposing models on the development of the North Atlantic PCMs and the role of AFT thermochronology within this debate, see discussion papers by Lidmar-Bergström and Bonow (2009), Nielsen et al. (2009), Pedersen et al. (2012), Japsen et al. (2013), and the review by Green et al. (2013).

20.7 Conclusion

Over the last few decades, efforts have been made to integrate tectonic and geomorphological processes to achieve a more holistic understanding of PCM evolution. Research from field observations and numerical and analog models have revealed important insights into the tectonic processes involved during the pre-, syn- and post-rift phases and have shown how margin topography can be sculpted and preserved over time. By providing unique insights into the thermal history of rocks, FT thermochronology has had a pivotal role in constraining the timing and rate of erosion associated with these processes.

Many studies have shown 'classical' AFT trends across PCMs, where data from the coastal strip record cooling in response to erosion due to rifting, and data from the continental interior reflect pre-breakup processes. However, a growing body of AFT data, and more recently AHe data, points to significant post-rift erosion across several PCMs, and spatially localised cooling events, related to structural reactivation and landscape development. The driving mechanisms for post-rift erosion remain debated and are likely to reflect the surface response to deformation driven by a complex interaction between mantle convection, in-plate stresses and lithospheric properties.

Obtaining high-quality FT data from regional outcrop samples remain a highly informative approach to obtaining thermal history information across a PCM setting. However, it is apparent that regional FT datasets are often not explained by a single conceptual model, or even a few 'end-member' models, of PCM evolution. Therefore, consideration should be given to the different structural, geomorphological and lithological factors that can control or perturb a particular style of landscape evolution. To further probe the thermal, tectonic and geomorphic history, sampling strategies to extend the spatial resolution (e.g. vertical/borehole profiles, high-density sampling of structural zones) and/or temporal resolution (e.g. AHe dating, cosmogenic nuclide analysis) should also be considered. Carefully selected sampling strategies, perhaps coupled with preliminary modelling studies to identify preferred regions for discriminating hypotheses, will likely be more informative than a random sampling approach, particularly in more well-studied regions. Thermochronological data can provide fairly direct information on the thermal history of rocks, although the inferred thermal history depends on the modelling approach and assumptions made during thermal modelling. As inferences on tectonic and geodynamic processes are typically made based on the results of thermal modelling, the approach used should be carefully considered and described. In this way, thermochronology data can be integrated with other geochronology and geological data to achieve a more comprehensive understanding of the physical processes and mechanisms occurring at PCMs.

Acknowledgements We would like to thank Roderick Brown, David Chew, Kerry Gallagher, Cristina Persano and Fin Stuart for sharing their knowledge and experience in low-temperature thermochronology analysis, thermal history modelling and PCM evolution. We thank Kerry Gallagher for additional constructive comments on this work, and Peter van der Beek and Marco G. Malusà for their constructive and detailed reviews. We are grateful to Bart Hendriks for sharing databases of AFT data, to Lauren Wildman for collating data from additional sources, and to Andrea Licciardi for producing the maps for Figs. 20.5 and 20.6.

References

Abbate E, Balestrieri ML, Bigazzi G (2002) Morphostructural development of the Eritrean rift flank (southern Red Sea) inferred from apatite fission track analysis. J Geophys Res B: Solid Earth 107:B11

Amaral G, Born H, Hadler JCN et al (1997) Fission track analysis of apatites from Sao Francisco craton and Mesozoic alkaline-carbonatite complexes from central and southeastern Brazil. J S Am Earth Sci 10(3–4):285–294

Andriessen PA, Bos ARJAN (1986) Post-caledonian thermal evolution and crustal uplift in the Eidfjord area, western Norway. Nor Geol Tidsskr 66(3):243–250

Andriessen PAM (1990) Anomalous fission track apatite ages of the Precambrian basement in the Hunnedalen region, south-western Norway. Int J Rad Appl Instrum D 17(3):285–291

Artemieva IM, Vinnik LP (2016) Density structure of the cratonic mantle in southern Africa: 1 implications for dynamic topography. Gondwana Res 39:204–216

Assine ML, Corrêa FS, Chang HK (2008) Migração de depocentros na Bacia de Santos: importância na exploração de hidrocarbonetos. Rev Brasil Geoci 38:111–127

Balestrieri ML, Abbate E, Bigazzi G et al (2009) Thermochronological data from Sudan in the frame of the denudational history of the Nubian Red Sea margin. Earth Surf Proc Land 34(9):1279–1290

Beauvais A, Bonnet NJ, Chardon D et al (2016) Very long-term stability of passive margin escarpment constrained by [40]Ar/[39]Ar dating of K–Mn oxides. Geology 44(4):299–302

Becker K, Franke D, Trumbull R et al (2014) Asymmetry of high-velocity lower crust on the South Atlantic rifted margins and implications for the interplay of magmatism and tectonics in continental breakup. Solid Earth 5(2):1011

Bernard T, Steer P, Gallagher K et al (2016) Evidence for Eocene–Oligocene glaciation in the landscape of the east Greenland margin. Geology 44(11):895–898

Bierman PR, Caffee M (2001) Slow rates of rock surface erosion and sediment production across the Namib desert and escarpment, southern Africa. Am J Sci 301(4–5):326–358

Bird P, Ben-Avraham Z, Schubert, G et al (2006) Patterns of stress and strain rate in southern Africa. J Geophys Res B: Solid Earth 111 (B08402)

Bishop P (2007) Long-term landscape evolution: linking tectonics and surface processes. Earth Surf Proc Land 32(3):329–365

Blaich OA, Faleide JI, Tsikalas F et al (2013) Crustal-scale architecture and segmentation of the South Atlantic volcanic margin. Geol Soc London Spec Publ 369(1):167–183

Bonnet NJ, Beauvais A, Arnaud N et al (2016) Cenozoic lateritic weathering and erosion history of Peninsular India from [40]Ar/[39]Ar dating of supergene K–Mn oxides. Chem Geo 446:33–53

Braun J (2018) A review of numerical modeling studies of passive margin escarpments leading to a new analytical expression for the rate of escarpment migration velocity. Gondwana Res 53:209–224

Braun J, Beaumont C (1989) A physical explanation of the relation between flank uplifts and the breakup unconformity at rifted continental margins. Geology 17(8):760–764

Braun J, Guillocheau F, Robin C et al (2014) Rapid erosion of the southern African plateau as it climbs over a mantle superswell. J Geophys Res B: Solid Earth 119(7):6093–6112

Braun J, van der Beek P (2004) Evolution of passive margin escarpments: what can we learn from low temperature thermochronology? J Geophys Res: Earth Sur 109(F4)

Brown R, Summerfield M, Gleadow A et al (2014) Intracontinental deformation in southern Africa during the late Cretaceous. J Afr Earth Sci 100:20–41

Brown RW, Beucher R, Roper S et al (2013) Natural age dispersion arising from the analysis of broken crystals. Part I: theoretical basis and implications for the apatite (U–Th)/He thermochronometer. Geochim Cosmochim Acta 122:478–497

Brown RW, Rust DJ, Summerfield MA et al (1990) An early Cretaceous phase of accelerated erosion on the south-western margin of Africa: evidence from apatite fission track analysis and the offshore sedimentary record. Int J Rad Appl Instrum D 17(3):339–350

Brown RW, Summerfield MA, Gleadow AJW (2002) Denudational history along a transect across the Drakensberg Escarpment of southern Africa derived from apatite fission track thermochronology. J Geophys Res Solid Earth 107(B12)

Brune S, Heine C, Pérez-Gussinyé M et al (2014) Rift migration explains continental margin asymmetry and crustal hyper-extension. Nat Comm 5:4014

Burke K, Gunnell Y (2008) The African erosion surface: a continental-scale synthesis of geomorphology, tectonics, and environmental change over the past 180 million years. Geol Soc Am Mem 201:1–66

Burov E, Cloetingh SAPL (1997) Erosion and rift dynamics: new thermomechanical aspects of post-rift evolution of extensional basins. Earth Planet Sci Lett 150(1–2):7–26

Burov E, Gerya T (2014) Asymmetric three-dimensional topography over mantle plumes. Nature 513(7516):85–89

Cederbom C (2001) Phanerozoic, pre-Cretaceous thermotectonic events in southern Sweden revealed by fission track thermochronology. Earth Planet Sci Lett 188(1):199–209

Cederbom C, Larson SÅ, Tullborg EL et al (2000) Fission track thermochronology applied to Phanerozoic thermotectonic events in central and southern Sweden. Tectonophysics 316(1):153–167

Cloetingh S, Beekman F, Ziegler PA et al (2008) Post-rift compressional reactivation potential of passive margins and extensional basins. Geol Soc London Spec Publ 306(1):27–70

Cobbold PR, Meisling KE, Mount VS (2001) Reactivation of an obliquely rifted margin, Campos and Santos basins, southeastern Brazil. AAPG bull 85(11):1925–1944

Cockburn HAP, Brown RW, Summerfield MA et al (2000) Quantifying passive margin denudation and landscape development using a combined fission-track thermochronology and cosmogenic isotope analysis approach. Earth Planet Sci Lett 179(3):429–435

Cogné N, Gallagher K, Cobbold PR (2011) Post-rift reactivation of the onshore margin of southeast Brazil: evidence from apatite (U–Th)/He and fission-track data. Earth Planet Sci Lett 309(1):118–130

Cogné N, Cobbold PR, Riccomini C et al (2013) Tectonic setting of the Taubaté Basin (southeastern Brazil): insights from regional seismic profiles and outcrop data. J South Amer Earth Sci 42:194–204

Cogné, N, Gallagher K, Cobbold PR et al (2012) Post-breakup tectonics in southeast Brazil from thermochronological data and combined inverse-forward thermal history modeling. J Geophys Res Solid Earth 117(B11)

Colli L, Stotz I, Bunge HP et al (2014) Rapid South Atlantic spreading changes and coeval vertical motion in surrounding continents: evidence for temporal changes of pressure-driven upper mantle flow. Tectonics 33(7):1304–1321

Contreras J, Zühlke R, Bowman S et al (2010) Seismic stratigraphy and subsidence analysis of the southern Brazilian margin (Campos, Santos and Pelotas basins). Mar Petr Geol 27(9):1952–1980

Danišík M (2018) Chapter 5. Integration of fission-track thermochronology with other geo-chronologic methods on single crystals. In: Malusà MG, Fitzgerald PG (eds) Fission-track thermochronology and its application to geology. Springer

Davids C, Wemmer K, Zwingmann H et al (2013) K–Ar illite and apatite fission track constraints on brittle faulting and the evolution of the northern Norwegian passive margin. Tectonophysics 608:196–211

de Oliveira Carmo I, Vasconcelos PM (2006) $^{40}Ar/^{39}Ar$ geochronology constraints on late Miocene weathering rates in Minas Gerais, Brazil. Earth Planet Sci Lett 241(1):80–94

de Oliveira CHE, Jelinek AR, Chemale F et al (2016) Evidence of post-Gondwana breakup in southern Brazilian shield: insights from apatite and zircon fission track thermochronology. Tectonophysics 666:173–187

de Wit MCJ (1988) Aspects of the geomorphology of the north-western Cape, South Africa. In: Dardis GF, Moon BP (eds) Geomorphological studies in southern Africa. CRC Press, Rotterdam, pp 57–69

Decker JE, Niedermann S, de Wit MJ (2011) Soil erosion rates in South Africa compared with cosmogenic 3 He-based rates of soil production. S Afr J Geol 114(3–4):475–488

Dempster TJ, Persano C (2006) Low-temperature thermochronology: resolving geotherm shapes or denudation histories? Geology 34:73–76

Divins DL (2003) Total sediment thickness of the world's oceans & marginal seas. NOAA National Geophysical Data Center, Boulder CO

Dunlap WJ, Fossen H (1998) Early Paleozoic orogenic collapse, tectonic stability, and late Paleozoic continental rifting revealed through thermochronology of K-feldspars, southern Norway. Tectonics 17(4):604–620

Egholm DL, Nielsen SB, Pedersen VK et al (2009) Glacial effects limiting mountain height. Nature 460(7257):884–887

Fetter M (2009) The role of basement tectonic reactivation on the structural evolution of Campos basin, offshore Brazil: evidence from 3D seismic analysis and section restoration. Mar Petr Geol 26:873–886

Fitzgerald PG, Malusà MG (2018) Chapter 9. Concept of the exhumed partial annealing (retention) zone and age-elevation profiles in thermochronology. In: Malusà MG, Fitzgerald PG (eds) Fission-track thermochronology and its application to geology. Springer

Fitzgerald PG, Baldwin SL, Webb LE, O'Sullivan PB (2006) Interpretation of (U–Th)/He single grain ages from slowly cooled crustal terranes: a case study from the Transantarctic Mountains of southern Victoria Land. Chem Geol 225(1–2):91–120

Fleming A, Summerfield MA, Stone JO et al (1999) Denudation rates for the southern Drakensberg escarpment, SE Africa, derived from in-situ-produced cosmogenic 36C1: initial results. J Geol Soc 156(2):209–212

Flowers RM, Schoene B (2010) (U–Th)/He thermochronometry constraints on unroofing of the eastern Kaapvaal craton and significance for uplift of the southern African plateau. Geology 38(9):827–830

Franco-Magalhaes AOB, Cuglieri MAA, Hackspacher PC et al (2014) Long-term landscape evolution and post-rift reactivation in the southeastern Brazilian passive continental margin: Taubaté basin. Int J Earth Sci 103(2):441–453

Gallagher K, Brown R (1999) Denudation and uplift at passive margins: the record on the Atlantic margin of southern Africa. Phil Trans R Soc London Ser A 357(1753):835–859

Gallagher K, Brown R, Johnson C (1998) Fission track analysis and its applications to geological problems. Annu Rev Earth Planet Sci 26 (1):519–572

Gallagher K, Hawkesworth CJ, Mantovani MSM (1994) The denudation history of the onshore continental margin of SE Brazil inferred from apatite fission track data. J Geophys Res Solid Earth 99 (B9):18117–18145

Gallagher K, Hawkesworth CJ, Mantovani MSM (1995) Denudation, fission track analysis and the long-term evolution of passive margin topography: application to the southeast Brazilian margin. J South Amer Earth Sci 8(1):65–77

Gallagher K. (2012) Transdimensional inverse thermal history modeling for quantitative thermochronology. J Geophys Res Solid Earth 117(B2)

Gallagher K, Stephenson J, Brown R et al (2005) Low temperature thermochronology and modeling strategies for multiple samples 1: vertical profiles. Earth Planet Sci Lett 237(1):193–208

Ghebreab W, Carter A, Hurford AJ et al (2002) Constraints for timing of extensional tectonics in the western margin of the Red Sea in Eritrea. Earth Planet Sci Lett 200(1):107–119

Gilchrist AR, Kooi H, Beaumont C (1994) Post-Gondwana geomorphic evolution of southwestern Africa: implications for the controls on landscape development from observations and numerical experiments. J Geophys Res Solid Earth 99(B6):12211–12228

Gilchrist AR, Summerfield MA (1990) Differential denudation and flexural isostasy in formation of rifted-margin upwarps. Nature 346 (6286):739–742

Gleadow AJW, Belton DX, Kohn BP et al (2002) Fission track dating of phosphate minerals and the thermochronology of apatite. Rev Mineral Geochem 48(1):579–630

Gleadow AJW, Brooks CK (1979) Fission track dating, thermal histories and tectonics of igneous intrusions in east Greenland. Contrib Mineral Petrol 71(1):45–60

Green PF (1986) On the thermo-tectonic evolution of northern England: evidence from fission track analysis. Geol Mag 153:493–506

Green PF, Duddy IR, Japsen P et al (2017) Post-breakup burial and exhumation of the southern margin of Africa. Basin Res 29(1):96–127

Green PF, Japsen P, Chalmers JA (2011) Thermochronology, erosion surfaces and missing section in west Greenland. J Geol Soc 168 (4):817–830

Green PF, Lidmar-Bergström K, Japsen P et al (2013) Stratigraphic landscape analysis, thermochronology and the episodic development of elevated, passive continental margins. Geol Surv Den Green Bull (30)

Green PF, Machado V (2017) Pre-rift and synrift exhumation, post-rift subsidence and exhumation of the onshore Namibe margin of Angola revealed from apatite fission track analysis. Geol Soc London Spec Publ 438(1):99–118

Grønlie A, Naeser CW, Naeser ND et al (1994) Fission-track and K–Ar dating of tectonic activity in a transect across the Møre-Trøndelag Fault Zone, central Norway. Nor Geol Tidsskr 74(1):24–34

Groves DI, Bierlein FP (2007) Geodynamic settings of mineral deposit systems. J Geol Soc 164:19–30

Guillocheau F, Rouby D, Robin C et al (2012) Quantification and causes of the terrigeneous sediment budget at the scale of a continental margin: a new method applied to the Namibia–South Africa margin. Basin Res 24(1):3–30

Gunnell Y, Carter A, Petit C et al (2007) Post-rift seaward downwarping at passive margins: new insights from southern Oman using stratigraphy to constrain apatite fission-track and (U–Th)/He dating. Geology 35(7):647–650

Gunnell Y, Harbor DJ (2010) Butte detachment: how pre-rift geological structure and drainage integration drive escarpment evolution at rifted continental margins. Earth Surf Proc Land 35(12):1373–1385

Gurnis M, Mitrovica JX, Ritsema J et al (2000) Constraining mantle density structure using geological evidence of surface uplift rates: the case of the African superplume. Geochem Geophys Geosyst 1 (7)

Haack U (1983) Reconstruction of the cooling history of the Damara orogen by correlation of radiometric ages with geography and altitude. In: Martin H, Eder FW (eds) Intracontinental fold belts: case studies in the Variscan belt of Europe and the Damara belt in Namibia. Springer, Berlin, pp 873–884

Hansen K (1992) Post-orogenic tectonic and thermal history of a rifted continental margin: the scoresby Sund area, east Greenland. Tectonophysics 216(3):309–326

Hansen K, Bergman SC, Henk B (2001) The Jameson land basin (east Greenland): a fission track study of the tectonic and thermal evolution in the Cenozoic North Atlantic spreading regime. Tectonophysics 331(3):307–339

Hansen K, Brooks CK (2002) The evolution of the east Greenland margin as revealed from fission-track studies. Tectonophysics 349 (1):93–111

Hansen K, Pedersen SVEND, Fougt H et al (1996) Post-Sveconorwegian exhumation and cooling history of the Evje area, southern Setesdal, central south Norway. Nor Geol Unders 431:49–58

Hansen K, Reiners PW (2006) Low temperature thermochronology of the southern east Greenland continental margin: evidence from apatite (U–Th)/He and fission track analysis and implications for intermethod calibration. Lithos 92(1):117–136

Harman R, Gallagher K, Brown R et al (1998) Accelerated denudation and tectonic/geomorphic reactivation of the cratons of northeastern Brazil during the late Cretaceous. J Geophys Res Solid Earth 103 (B11):27091–27105

Heine C, Zoethout J, Müller RD (2013) Kinematics of the South Atlantic rift. Solid Earth 4(2)

Hendrik BWH (2003) Cooling and denudation of the Norwegian and Barents sea margins, northern Scandinavia. Constrained by apatite fission track and (U–Th/He) thermochronology. Netherlands Research School of Sedimentary Geology (NSG), Amsterdam

Hendriks BW, Andriessen PA (2002) Pattern and timing of the post-Caledonian denudation of northern Scandinavia constrained by apatite fission-track thermochronology. Geol Soc London Spec Publ 196(1):117–137

Hendriks BWH, Osmundsen PT, Redfield TF (2010) Normal faulting and block tilting in Lofoten and Vesterålen constrained by apatite fission track data. Tectonophysics 485(1):154–163

Hiruma ST, Riccomini C, Modenesi-Gauttieri MC et al (2010) Denudation history of the Bocaina plateau, Serra do Mar, southeastern Brazil: relationships to Gondwana breakup and passive margin development. Gondwana Res 18(4):674–687

Huigen Y, Andriessen P (2004) Thermal effects of Caledonian foreland basin formation, based on fission track analyses applied on basement rocks in central Sweden. Phys Chem Earth (A/B/C) 29 (10):683–694

Huismans R, Beaumont C (2011) Depth-dependent extension, two-stage breakup and cratonic underplating at rifted margins. Nature 473(7345):74–78

Jackson MPA, Hudec MR, Hegarty KA (2005) The great west African tertiary coastal uplift: fact or fiction? a perspective from the Angolan divergent margin. Tectonics 24(6)

Japsen P, Bonow JM, Green PF et al (2012a) Episodic burial and exhumation in NE Brazil after opening of the South Atlantic. Geol Soc Am Bull 124(5–6):800–816

Japsen P, Chalmers JA (2000) Neogene uplift and tectonics around the north Atlantic: overview. Global Planet Change 24(3):165–173

Japsen P, Green PF, Bonow JM et al (2014) From volcanic plains to glaciated peaks: burial, uplift and exhumation history of southern east Greenland after opening of the NE Atlantic. Global Planet Change 116:91–114

Japsen P, Green PF, Chalmers JA (2013) The mountains of north-east Greenland are not remnants of the Caledonian topography. A comment on Pedersen et al (2012): tectonophysics vol 530–531, pp 318–330. Tectonophysics 589:234–238

Japsen P, Chalmers JA, Green PF et al (2012b) Elevated, passive continental margins: not rift shoulders, but expressions of episodic, post-rift burial and exhumation. Global Planet Change 90:73–86

Jelinek AR, Chemale F, van der Beek PA et al (2014) Denudation history and landscape evolution of the northern east-Brazilian continental margin from apatite fission-track thermochronology. J South Amer Earth Sci 54:158–181

Johnson C, Gallagher K (2000) A preliminary Mesozoic and Cenozoic denudation history of the north east Greenland onshore margin. Global Planet Change 24(3):261–274

Karl M, Glasmacher UA, Kollenz S et al (2013) Evolution of the South Atlantic passive continental margin in southern Brazil derived from zircon and apatite (U–Th–Sm)/He and fission-track data. Tectonophysics 604:224–244

Katz BJ, Mello MR (2000) Petroleum systems of South Atlantic marginal basins. AAPG Mem 73

Keen CE, Boutilier RR (1995) Lithosphere-asthenosphere interactions below rifts. In: Banda E, Torné M, Talwani M (eds) Rifted ocean-continent boundaries. Springer, Netherlands, pp 17–30

Ketcham R (2018) Chapter 3. Fission track annealing: from geologic observations to thermal history modeling. In: Malusà MG, Fitzgerald PG (eds) Fission-track thermochronology and its application to geology. Springer

King LC (1962) Morphology of the Earth. Oliver and Boyd, Edinburgh

Kollenz S, Glasmacher UA, Rossello EA et al (2016) Thermochronological constraints on the Cambrian to recent geological evolution of the Argentina passive continental margin. Tectonophysics 716:182–203

Kooi H, Beaumont C (1994) Escarpment evolution on high-elevation rifted margins: insights derived from a surface processes model that combines diffusion, advection, and reaction. J Geophys Res Solid Earth 99(B6):12191–12209

Koptev A, Cloetingh S, Burov E et al (2017) Long-distance impact of Iceland plume on Norway's rifted margin. Sci Rep 7(1):10408

Kounov A, Niedermann S, de Wit MJ et al (2007) Present denudation rates at selected sections of the South African escarpment and the elevated continental interior based on cosmogenic ^{3}He and ^{21}Ne. S Afr J Geol 110(2–3):235–248

Kounov A, Viola G, de Wit M et al (2009) Denudation along the Atlantic passive margin: new insights from apatite fission-track analysis on the western coast of South Africa. Geol Soc London Spec Publ 324(1):287–306

Kounov A, Viola G, Dunkl I et al (2013) Southern African perspectives on the long-term morpho-tectonic evolution of cratonic interiors. Tectonophysics 601:177–191

Ksienzyk AK, Dunkl I, Jacobs J et al (2014) From orogen to passive margin: constraints from fission track and (U–Th)/He analyses on Mesozoic uplift and fault reactivation in SW Norway. Geol Soc London Spec Publ 390:390–27

Kusznir NJ, Karner GD (2007) Continental lithospheric thinning and breakup in response to upwelling divergent mantle flow: application to the Woodlark, Newfoundland and Iberia margins. Geol Soc London Spec Publ 282(1):389–419

Kusznir NJ, Marsden G, Egan SS (1991) A flexural-cantilever simple-shear/pure-shear model of continental lithosphere extension: applications to the Jeanne d'Arc Basin, Grand Banks and Viking Graben, North Sea. Geol Soc London Spec Publ 56(1):41–60

Lemoine M, Bas T, Arnaud-Vanneau A et al (1986) The continental margin of the Mesozoic Tethys in the western Alps. Mar Petr Geol 3:179–199

Lidmar-Bergström K, Bonow JM, Japsen P (2013) Stratigraphic landscape analysis and geomorphological paradigms: Scandinavia as an example of phanerozoic uplift and subsidence. Glob Planet Change 100:153–171

Lidmar-Bergström K, Bonow JM (2009) Hypotheses and observations on the origin of the landscape of southern Norway—a comment regarding the isostasy-climate-erosion hypothesis by Nielsen et al 2008. J Geodyn 48(2):95–100

Lister GS, Etheridge MA, Symonds PA (1986) Detachment faulting and the evolution of passive continental margins. Geology 14(3):246–250

Lithgow-Bertelloni C, Silver PG (1998) Dynamic topography, plate driving forces and the African superswell. Nature 395(6699):269–272

Macdonald D, Gomez-Perez I, Franzese J et al (2003) Mesozoic breakup of SW Gondwana: implications for regional hydrocarbon potential of the southern South Atlantic. Mar Petr Geol 20(3):287–308

Malusà MG, Fitzgerald PG (2018) Chapter 8. From cooling to exhumation: setting the reference frame for the interpretation of thermochronologic data. In: Malusà MG, Fitzgerald PG (eds) Fission-track thermochronology and its application to geology. Springer

Malusà MG, Fitzgerald PG (2018) Chapter 10. Application of thermochronology to geologic problems: bedrock and detrital approaches. In: Malusà MG, Fitzgerald PG (eds) Fission-track thermochronology and its application to geology. Springer

Malusà MG, Faccenna C, Baldwin SL et al (2015) Contrasting styles of (U) HP rock exhumation along the Cenozoic Adria-Europe plate boundary (western Alps, Calabria, Corsica). Geochem Geophys Geosyst 16(6):1786–1824

Mathiesen A, Bidstrup T, Christiansen FG (2000) Denudation and uplift history of the Jameson land basin, east Greenland—constrained from maturity and apatite fission track data. Global Planet Change 24(3):275–301

McKenzie D (1978) Some remarks on the development of sedimentary basins. Earth Planet Sci Lett 40(1):25–32

Mitchell SG, Montgomery DR (2006) Influence of a glacial buzzsaw on the height and morphology of the cascade range in central Washington state, USA. Quat Res 65(1):96–107

Modenesi-Gauttieri MC, de Toledo MCM, Hiruma ST et al (2011) Deep weathering and landscape evolution in a tropical plateau. CATENA 85:221–230

Molnar P, England PC, Jones CH (2015) Mantle dynamics, isostasy, and the support of high terrain. J Geophys Res B: Solid Earth 120(3):1932–1957

Moore ME, Gleadow AJ, Lovering JF (1986) Thermal evolution of rifted continental margins: new evidence from fission tracks in basement apatites from southeastern Australia. Earth Planet Sci Lett 78(2–3):255–270

Moore A, Blenkinsop T, Cotterill FW (2009) Southern African topography and erosion history: plumes or plate tectonics? Terra Nova 21(4):310–315

Morais-Neto JM, Hegarty KA, Karner GD et al (2009) Timing and mechanisms for the generation and modification of the anomalous topography of the Borborema province, northeastern Brazil. Mar Petr Geol 26(7):1070–1086

Mosar J (2003) Scandinavia's north Atlantic passive margin. J Geophys Res Solid Earth 108(B8)

Moucha R, Forte AM, Mitrovica JX et al (2008) Dynamic topography and long-term sea-level variations: there is no such thing as a stable continental platform. Earth Planet Sci Lett 271(1):101–108

Moulin M, Aslanian D, Unternehr P (2010) A new starting point for the south and equatorial Atlantic ocean. Earth Sci Rev 98(1):1–37

Nielsen SB, Gallagher K, Leighton C et al (2009) The evolution of western Scandinavian topography: a review of neogene uplift versus the ICE (isostasy–climate–erosion) hypothesis. J Geodyn 47(2):72–95

Ollier CD, Marker ME (1985) The great escarpment of southern Africa. Z Geomorph Suppl 54:37–56

Ollier CD, Pain CF (1997) Equating the basal unconformity with the palaeoplain: a model for passive margins. Geomorphology 19(1–2):1–15

Omar GI, Steckler MS (1995) Fission track evidence on the initial rifting of the Red sea: two pulses, no propagation. Science 270 (5240):1341

Osmundsen PT, Redfield TF (2011) Crustal taper and topography at passive continental margins. Terra Nova 23(6):349–361

Partridge TC, Maud RR (1987) Geomorphic evolution of southern Africa since the mesozoic. S Afr J Geol 90(2):179–208

Pavich MJ (1989) Regolith residence time and the concept of surface age of the Piedmont "peneplain". Geomorphology 2(1–3):181–196

Pedersen VK, Nielsen SB, Gallagher K (2012) The post-orogenic evolution of the northeast Greenland Caledonides constrained from apatite fission track analysis and inverse geodynamic modelling. Tectonophysics 530:318–330

Pérez-Díaz L, Eagles G (2014) Constraining South Atlantic growth with seafloor spreading data. Tectonics 33(9):1848–1873

Péron-Pinvidic G, Manatschal G (2009) The final rifting evolution at deep magma-poor passive margins from Iberia-Newfoundland: a new point of view. Int J Earth Sci 98(7):1581–1597

Péron-Pinvidic G, Manatschal G, Osmundsen PT (2013) Structural comparison of archetypal Atlantic rifted margins: a review of observations and concepts. Mar Petr Geol 43:21–47

Persano C, Stuart FM, Bishop P et al (2005) Deciphering continental breakup in eastern Australia using low-temperature thermochronometers. J Geophys Res Solid Earth 110(B12)

Persano C, Stuart FM, Bishop P et al (2002) Apatite (U–Th)/He age constraints on the development of the Great Escarpment on the southeastern Australian passive margin. Earth Planet Sci Lett 200 (1):79–90

Phillips JD (2002) Erosion, isostatic response, and the missing peneplains. Geomorphology 45(3):225–241

Praeg D, Stoker MS, Shannon PM et al (2005) Episodic Cenozoic tectonism and the development of the NW European 'passive' continental margin. Mar Petr Geol 22(9):1007–1030

Raab MJ, Brown RW, Gallagher K et al (2002) Late Cretaceous reactivation of major crustal shear zones in northern Namibia: constraints from apatite fission track analysis. Tectonophysics 349 (1):75–92

Raab MJ, Brown RW, Gallagher K et al (2005) Denudational and thermal history of the early Cretaceous Brandberg and Okenyenya igneous complexes on Namibia's Atlantic passive margin. Tectonics 24(3)

Redfield TF (2010) On apatite fission track dating and the tertiary evolution of west Greenland topography. J Geol Soc 167:261–271

Redfield TF, Braathen A, Gabrielsen RH et al (2005) Late Mesozoic to early Cenozoic components of vertical separation across the Møre-Trøndelag fault complex, Norway. Tectonophysics 395 (3):233–249

Redfield TF, Osmundsen PT (2013) The long-term topographic response of a continent adjacent to a hyperextended margin: a case study from Scandinavia. Geol Soc Am Bull 125:184–200

Redfield TF, Torsvik TH, Andriessen PAM et al (2004) Mesozoic and Cenozoic tectonics of the Møre Trøndelag fault complex, central Norway: constraints from new apatite fission track data. Phys Chem Earth 29(10):673–682

Riccomini C, Sant'Anna LG, Ferrari AL (2004) Evolução geológica do Rift Continental do Sudeste do Brasil. In: Mantesso Neto V, Bartorelli A, Carneiro CDR et al (eds) Geologia do continente Sul-Americano: evolução da Obra de Fernando Flávio Marques de Almeida. Edições Beca, São Paulo, pp 383–405

Riccomini C, Velázquez VF, Gomes CB (2005) Tectonic controls of the Mesozoic and Cenozoic alkaline magmatism in central-southeastern Brazilian Platform. In: Comin-Chiaramonti P, Gomes CB (eds) Mesozoic to Cenozoic alkaline magmatism in the Brazilian platform, vol 123. EdUSP, Sao Paulo, pp 31–56

Rohrman M, van der Beek P, Andriessen P (1994) Syn-rift thermal structure and post-rift evolution of the Oslo rift (southeast Norway): new constraints from fission track thermochronology. Earth Planet Sci Lett 127(1–4):39–54

Rohrman M, van der Beek P, Andriessen P et al (1995) Meso-Cenozoic morphotectonic evolution of southern Norway: Neogene domal uplift inferred from apatite fission track thermochronology. Tectonics 14(3):704–718

Rouby D, Braun J, Robin C et al (2013) Long-term stratigraphic evolution of Atlantic-type passive margins: a numerical approach of interactions between surface processes, flexural isostasy and 3D thermal subsidence. Tectonophysics 604:83–103

Sacek V (2017) Post-rift influence of small-scale convection on the landscape evolution at divergent continental margins. Earth Planet Sci Lett 459:48–57. https://doi.org/10.1016/j.epsl.2016.11.026

Sacek V, Braun J, van der Beek P (2012) The influence of rifting on escarpment migration on high elevation passive continental margins. J Geophys Res Solid Earth 117(B4)

Scharf TE, Codilean AT, de Wit M et al (2013) Strong rocks sustain ancient postorogenic topography in southern Africa. Geology 41 (3):331–334

Schermer ER, Redfield TF, Indrevær K (2017) Geomorphology and topography of relict surfaces: the influence of inherited crustal structure in the northern Scandinavian mountains. J Geol Soc 174 (1):93–109

Seidl MA, Weissel JK, Pratson LF (1996) The kinematics and pattern of escarpment retreat across the rifted continental margin of SE Australia. Basin Res 8(3):301–316

Skogseid J, Planke S, Faleide JI et al (2000) NE Atlantic continental rifting and volcanic margin formation. Geol Soc London Spec Publ 167(1):295–326

Stanley JR, Flowers RM, Bell DR (2013) Kimberlite (U–Th)/He dating links surface erosion with lithospheric heating, thinning, and metasomatism in the southern African plateau. Geology 41 (12):1243–1246

Stanley JR, Flowers RM, Bell DR (2015) Erosion patterns and mantle sources of topographic change across the southern African plateau derived from the shallow and deep records of kimberlites. Geochem Geophys Geosyst 16(9):3235–3256

Steer P, Huismans RS, Valla PG et al (2012) Bimodal Plio-Quaternary glacial erosion of fjords and low-relief surfaces in Scandinavia. Nat Geosci 5(9):635–639

Summerfield MA (1985) Plate tectonics and landscape development on the African continent. In: Morisawa M, Hack JT (eds) Tectonic geomorphology, vol 15. Allen and Unwin, Boston, pp 27–51

Summerfield MA (1991) Sub-aerial denudation of passive margins: regional elevation versus local relief models. Earth Planet Sci Lett 102(3–4):460–469

Tello-Saenz C, Hackspacher PC, Neto JH et al (2003) Recognition of Cretaceous, Paleocene, and Neogene tectonic reactivation through apatite fission-track analysis in Precambrian areas of southeast Brazil: association with the opening of the South Atlantic ocean. J South Amer Earth Sci 15(7):765–774

Thomson K, Green PF, Whitham AG et al (1999) New constraints on the thermal history of north-east Greenland from apatite fission-track analysis. Geol Soc Am Bull 111:1054–1068

Tinker J, de Wit M, Brown R (2008a) Mesozoic exhumation of the southern Cape, South Africa, quantified using apatite fission track thermochronology. Tectonophysics 455(1):77–93

Tinker J, de Wit M, Brown R (2008b) Linking source and sink: evaluating the balance between onshore erosion and offshore sediment accumulation since Gondwana break-up, South Africa. Tectonophysics 455(1):94–103

Torsvik TH, Rousse S, Labails C et al (2009) A new scheme for the opening of the South Atlantic ocean and the dissection of an Aptian salt basin. Geophys J Int 177(3):1315–1333

Tucker GE, Slingerland RL (1994) Erosional dynamics, flexural isostasy, and long-lived escarpments: a numerical modeling study. J Geophys Res Solid Earth 99(B6):12229–12243

Turner JP, Green PF, Holford SP et al (2008) Thermal history of the Rio Muni (West Africa)–NE Brazil margins during continental breakup. Earth Planet Sci Lett 270(3):354–367

van den Haute P (1977) Apatite fission track dating of Precambrian intrusive rocks from the southern Rogaland (south-western Norway). Bull Belg Ver Geologie 86:97–110

van der Beek P, Andriessen P, Cloetingh S (1995) Morphotectonic evolution of rifted continental margins: inferences from a coupled tectonic-surface processes. Tectonics 14(2):406–421

van der Beek P, Summerfield MA, Braun J et al (2002) Modeling post-breakup landscape development and denudational history across the southeast African (Drakensberg Escarpment) margin. J Geophys Res Solid Earth 107(B12)

Vasconcelos PM, Knesel KM, Cohen BE et al (2008) Geochronology of the Australian Cenozoic: a history of tectonic and igneous activity, weathering, erosion, and sedimentation. Aust J Earth Sci 55(6–7):865–914

Watts AB (2012) Models for the evolution of passive margins. In: Roberts DG, Bally AW (eds) Regional geology and tectonics: Phanerozoic rift systems and sedimentary basins. Elsevier, Amsterdam, pp 32–57

Weissel JK, Karner GD (1989) Flexural uplift of rift flanks due to mechanical unloading of the lithosphere during extension. J Geophys Res Solid Earth 94(B10):13919–13950

Weissel JK, Seidl MA (1998) Inland propagation of erosional escarpments and river profile evolution across the southeast Australian passive continental margin. In: Tinkler KJ, Wohl EE (eds) Rivers over rock: fluvial processes in bedrock channels. American Geophysical Union, Washington DC, pp 189–206

Wernicke B (1985) Uniform-sense normal simple shear of the continental lithosphere. Can J Earth Sci 22(1):108–125

Whipple KX, DiBiase RA, Ouimet WB et al (2017) Preservation or piracy: diagnosing low-relief, high-elevation surface formation mechanisms. Geology 45(1):91–94

Whittaker JM, Goncharov A, Williams SE et al (2013) Global sediment thickness data set updated for the Australian-Antarctic southern ocean. Geochem Geophys Geosyst 14(8):3297–3305

Wildman M, Brown R, Persano C et al (2017) Contrasting Mesozoic evolution across the boundary between on and off craton regions of the South African plateau inferred from apatite fission track and (U-Th-Sm)/He thermochronology. J Geophys Res Solid Earth 122(2):1517–1547

Wildman M, Brown R, Watkins R et al (2015) Post breakup tectonic inversion across the southwestern cape of South Africa: new insights from apatite and zircon fission track thermochronometry. Tectonophysics 654:30–55

Wildman M, Brown R, Beucher R et al (2016) The chronology and tectonic style of landscape evolution along the elevated Atlantic continental margin of South Africa resolved by joint apatite fission track and (U-Th-Sm)/He thermochronology. Tectonics 35(3):511–545

Yang R, Willett SD, Goren L (2015) In situ low-relief landscape formation as a result of river network disruption. Nature 520 (7548):526–529

Ziegler PA, Cloetingh S (2004) Dynamic processes controlling evolution of rifted basins. Earth Sci Rev 64(1):1–50

Application of Low-Temperature Thermochronology to Craton Evolution

21

Barry Kohn and Andrew Gleadow

Abstract

The view that cratons are tectonically and geomorphologically inert continental fragments is at odds with a growing body of evidence partly based on low-temperature thermochronology (LTT) studies. These suggest that large areas of cratons may have undergone discrete episodes of regional-scale Neoproterozoic and/or Phanerozoic heating, and cooling from modestly elevated paleotemperatures. Cooling is often attributed to the km-scale erosion of overlying low-conductivity sediments, rather than to removal of large sections of crystalline basement. Independent evidence for sedimentary burial includes: preservation of outliers, the sedimentary record in intracratonic basins, and sedimentary xenoliths entrained within kimberlites periodically emplaced into cratons. Further, stratigraphic and isotopic data from basinal sediments proximal to some cratons carry a record of the detritus removed, which can be linked temporally to cooling episodes in their inferred cratonic source areas. Differences in denudation rates reported from cratonic basement reconstructed from LTT data (long-term) and cosmogenic isotope and chemical weathering studies (short-term) reflect the strong contrast in erodibility potential between cover sediments since removed and the preserved crystalline rocks. Underlying processes involved in cratonic heating and cooling may include one of, or a complex interplay between: proximity to sediment sources from elevated orogens forming extensive foreland basins, structural deformation transmitted by far-field horizontal stresses from active plate boundaries, and the development of dynamic topography driven by vertical mantle stresses. Dynamic topography may also explain elevation changes observed in some cratons, where no clear deformation is apparent. LTT studies from classic cratons in Fennoscandia, Western

Australia, Southern Africa, and Canada are reviewed, with emphasis on different aspects of their more recent evolution.

21.1 Introduction

Cratons (Greek *kratos* "strength") are the scattered remnants of the most ancient regions of the terrestrial crust and provide a direct record of the Earth's early history. They have conventionally been defined as areas of continental crust that have attained and maintained long-term tectonic and geomorphologic stability as relatively inert crustal fragments (e.g., Fairbridge and Finkl 1980; de Wit et al. 1992; Lenardic and Moresi 1999), by and large resistant to internal deformation, with tectonic reworking being confined to their margins (e.g., Lenardic et al. 2000). The term craton is commonly applied to stable segments of Archean (>2.5 Ga) crust, but there is no strict age implication in the craton definition, as many of these terranes may only have attained final amalgamation and stability during the Proterozoic (Bleeker 2003).

Archean cratons (Fig. 21.1) form <15% of continental area (Bowring and Williams 1999) and are mostly characterized by relatively subdued topography (<1000 m relief). Their stability is thought to be partly a function of their thicker lithosphere (high effective elastic thickness >100 km), characterized by a relatively cool, but compositionally buoyant keel of upper mantle (e.g., O'Neill et al. 2008). Apart from some tectonic reworking at their margins, cratonization has often been viewed as the end point in the evolution of continental lithosphere, beyond which it enters into a relatively quiescent state.

Some lines of evidence, however, attest to more recent periodic disturbances of cratonic environments. These include: scattered concentrations of low-level seismicity, particularly at cratonic margins (e.g., Mooney et al. 2012); the presence of Phanerozoic sedimentary sequences,

B. Kohn (✉) · A. Gleadow
School of Earth Sciences, University of Melbourne, Melbourne, VIC 3010, Australia
e-mail: b.kohn@unimelb.edu.au

© Springer International Publishing AG, part of Springer Nature 2019
M. G. Malusà and P. G. Fitzgerald (eds.), *Fission-Track Thermochronology and its Application to Geology*,
Springer Textbooks in Earth Sciences, Geography and Environment, https://doi.org/10.1007/978-3-319-89421-8_21

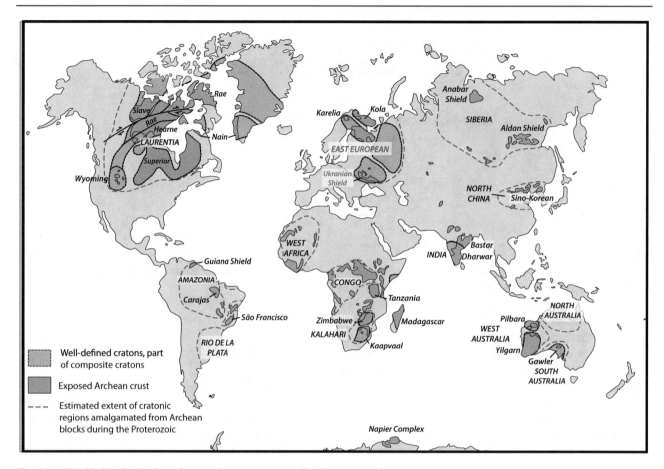

Fig. 21.1 Worldwide distribution of exposed Archean crust, > 2.5 Ga. Names of Archean cratons and shield area are labeled in lower case. Outlines of composite cratons that were assembled during the Proterozoic are labeled in upper case (modified after Bleeker 2003)

including intracratonic basins preserving a record of regional subsidence and uplift (e.g., Sloss and Speed 1974, Allen and Armitage 2012); sporadic magmatic episodes (e.g. kimberlites, Heaman et al. 2004); far-field deformation transmitted to cratons over hundreds of km from collisional plate interactions (e.g., Pinet 2016); and evidence for fluid movement and paleoclimate on heat flow (e.g., Popov et al. 1999; Mottaghy et al. 2005). However, knowledge of the degree to which cratons have participated in the geodynamic and morphological changes to continents in the last few hundreds of millions of years of the Earth's history is rather fragmentary. This incomplete understanding has been strongly influenced by the paucity of records that underpin many conventional geological investigations, such as a suitably preserved stratigraphic record, as well as the lack of any structural controls.

It is in this context that low-temperature thermochronology (LTT) has played an increasingly important role in helping to decipher a more recent subtle record of possible geodynamic and/or climatic interactions that may have shaped the upper levels of the cratonic lithosphere. A growing body of data, mainly based on apatite fission-track and (U–Th)/He studies from different cratons, has revealed that episodes of regional-scale Phanerozoic heating and/or cooling from modestly elevated paleotemperatures are the characteristic of many cratons (e.g., Brown 1992; Harman et al. 1998; Kohn et al. 2002; Lorencak et al. 2004; Danišík et al. 2008; Flowers and Schoene 2010; Ault et al. 2013; Japsen et al. 2016; Kasanzu 2017). These findings are at odds with long-held views of craton stability and have triggered further investigations aimed at a more comprehensive understanding of the later history of craton evolution. In part, this is because LTT studies may inform on the preservation of some of the world's most valuable mineral resources hosted by cratons and provide fundamental information on their potential as storage sites for radioactive waste. Further, such studies often elucidate unroofing histories, which may be particularly relevant to sediment supply into intracratonic and adjacent basins, with important implications for hydrocarbon prospectivity.

21.2 Low-Temperature Thermochronology (LTT)

LTT involves the use of radiometric dating methods characterized by temperature-sensitive daughter products that accumulate in minerals and are retained over temperatures ranging from ~40 to 300 °C. Within the upper crustal environment, temperature can often be used as a proxy for depth, so that reconstructed cooling histories may reveal a record of rock movement toward the surface (see Chap. 8, Malusà and Fitzgerald 2018). The unique ability to detect and quantify the signature of such movements or thermal perturbations in the near-surface environment, which are largely invisible to other analytical techniques, has been the basis for the application of LTT to a broad range of interdisciplinary problems in the earth sciences (e.g., Gleadow et al. 2002a; Farley 2002).

21.2.1 Methods

Two principal methods commonly employed for LTT studies are fission-track (FT) analysis and (U–Th)/He dating, and the most common minerals investigated are apatite, zircon, and titanite. The fundamental principles underpinning these techniques, interpretation of data, and their applications are outlined in several works; for FT see Chap. 2, Kohn et al. (2018) and references therein, and for (U–Th)/He see, e.g., Farley (2002) and Reiners (2005).

Briefly, fission tracks are formed continuously, but may be annealed (i.e., track lengths are shortened) over specific temperature ranges. If this process occurs gradually, then the temperature range over which annealing occurs is termed as the partial annealing zone (PAZ). For apatite fission track (AFT), which is the best understood and most commonly used mineral system, the temperature range over which partial annealing occurs typically ranges between ~60 and 110 °C for a heating time of ~10 Myr, but this may also vary with chemical composition. For zircon fission track (ZFT), the PAZ varies between ~180 and 350 °C for a ~10 Myr heating time (e.g., Tagami 2005), but this range varies with degree of radiation damage, while for titanite (TFT) it falls between ~265 and 310 °C for the same heating time (Coyle and Wagner 1998). Above the higher temperature limits of these zones, any tracks formed are totally annealed, so no physical record of the daughter product remains and ages are reset to zero.

In recent years, an important advance in LTT has been the resurgence and modern development of (U–Th)/He thermochronometry (e.g., Zeitler et al. 1987, Farley 2002). In the (U–Th)/He system, the partial retention zone (PRZ) is a comparable concept to the PAZ. But in this case, the PRZ is related to the retention of ^{4}He, which can be readily lost progressively by volume diffusion over a range of temperatures. The apatite (U–Th–Sm)/He system provides information to lower temperatures, (PRZ is typically ~35–85 °C over relatively short heating times, e.g., Farley and Stockli 2002), than those accessible to the AFT system. For zircon (U–Th)/He (ZHe) and titanite (THe), the typical temperature range of the PRZ is ~130–200 °C (Reiners 2005; Wolfe and Stockli 2010) and ~100–180 °C (Stockli and Farley 2004), respectively. Above the higher temperature limits of the PRZ, no helium is retained and ages are totally reset. The temperature range of daughter product retention sensitivity in apatite, zircon, and titanite in the FT and (U–Th)/He thermochronometry systems is therefore ideal for providing complementary thermal history information relevant to the uppermost ~10 km or so of the continental crust.

Where samples have cooled relatively rapidly through the PAZ or PRZ, the concept of closure temperature (T_c) can be used (Dodson 1973). While this concept is very useful for comparing the results of different thermochronology systems, it is based on the assumption that since passing through the T_c samples have cooled progressively through the PAZ or PRZ. However, this assumption does not always hold in cratonic settings, where samples may have resided for lengthy periods in these zones, providing complications for data interpretation (e.g. Reiners 2005; Guenthner et al. 2013). Further, the T_c (and temperature range of daughter product retention zones) may vary as a function of different factors such as radiation damage, mineral chemistry, grain volume, and cooling rate (e.g., Gleadow et al. 2002a; Reiners 2005; Shuster et al. 2006; Guenthner et al. 2013, see also Chap. 9, Fitzgerald and Malusà 2018).

21.2.2 Methodological Issues

21.2.2.1 Apatite

In principle, the AHe system should provide complementary data to AFT analysis and, on the basis of their predicted relative decay-product retention properties, in most geological scenarios AFT ages would be expected to yield older ages. However, AHe age data may yield results that are difficult to replicate and which are older than the seemingly more consistent AFT data, and this discrepancy may become more pronounced in cratonic samples (e.g., Lorencak 2003; Belton et al. 2004; Green and Duddy 2006; Kohn et al. 2009).

Several factors can result in seemingly anomalous or dispersed intra-sample AHe ages (e.g., see Table 3 in Wildman et al. 2016). Of particular interest for cratonic environments is the finding that accumulating α-radiation damage progressively increases the ^{4}He retentivity of apatites (the trapping model), thereby increasing the effective T_c and apparent ages obtained from slowly cooled rocks (e.g., Shuster et al. 2006). This phenomenon results in a range of apparent ages correlating with both ^{4}He concentration and effective U concentration *(eU)* expressed as [U ppm + 0.235 Th ppm], which weights the decay of the parents for their α productivity. Enhanced ^{4}He retention in response to the accumulation of α-radiation damage, often leads to positive correlations between grain age and *eU*, such that grains with greater accumulated radiation damage may have a higher T_c than lower *eU*, less damaged grains. The amplification of this effect and that of α-damage annealing on ^{4}He retentivity, especially under conditions of slow cooling or reheating of basement rocks as might be expected in cratonic environments (e.g., Gautheron et al. 2009), is also consistent with the empirical observations between coexisting apatite FT and He ages reported by Green and Duddy (2006) and Kohn et al. (2009).

21.2.2.2 Zircon

For both the ZFT and ZHe systems, α-radiation damage is a key factor in evaluating the possibility of dating zircons from cratonic terranes. Radiation damage is a function of U and Th isotope content, accumulation time and the degree of annealing and internal disorder repair as a function of the thermal history. Because of their old age and relatively high parent isotope concentration, the accumulated radiation damage in cratonic zircons may progressively transform the structurally ordered crystalline state, which eventually becomes metamict (amorphous or non-crystalline), while preserving the external crystalline form. If grains are too damaged, then due to the loss of parent isotopes and/or daughter products, they may not be useful for geo-thermochronological measurement. This has fundamental implications for the retention of both fission tracks (e.g., Kasuya and Naeser 1988; Tagami 2005) and He (e.g. Reiners 2005; Guenthner et al. 2013). ZFT ages within a sample may reveal a wide spectrum of different ages or age populations (especially in detrital samples) reflecting different provenances and/or degrees of radiation damage, and this requires careful interpretation (e.g., Bernet and Garver 2005, see also Chap. 16, Malusà 2018). For technical aspects related to ZFT dating of old and/or high U zircons commonly found in cratonic rocks, see Sect. 2.7 in Chap. 2, Kohn et al. (2018).

With respect to the profound influence of radiation damage on the ZHe system, for zircons that have experienced low to moderate α-radiation damage, ages may correlate positively with *eU* and yield some information of direct geo-thermochronological significance. In contrast, more highly damaged grains display distinctly negative *eU*-age correlations (Reiners 2005; Guenthner et al. 2013). This dispersed age spectrum is more pronounced in samples that have experienced a prolonged thermal history. However, an approach incorporating zircon radiation damage and an annealing model (ZRDAAM which follows zircon FT annealing kinetics) of Guenthner et al. (2013) calculates diffusion kinetics of individual zircons measured for (U–Th)/He analysis. This is achieved by integrating the *eU* over the time since the zircon cooled below ZFT partial annealing temperatures, allowing for the quantification of thermal histories from ZHe data sets in cratonic environments (e.g., Orme et al. 2016; Guenthner et al. 2017). A notable feature of some *eU*-age plots is that some of the more damaged zircon grain ages may form a quasi-plateau or pediment without significant age dispersion across a wide range of *eU* values (e.g., Orme et al. 2016). This suggests relatively rapid cooling across a range of relatively low closure temperatures, accompanied by "concurrent" closure of He loss and damage accumulation. The ages defining these relatively invariant *eU*-age plots may also be related to co-existing AFT and AHe data, and in these cases suggest that the ZHe T_c is only slightly above or within the range of the AFT PAZ and even the AHe PRZ (e.g., Johnson et al. 2017; Mackintosh et al. 2017). Recent work has further underlined complexities in ZHe systematics, particularly in slowly cooled rocks (e.g., Danišík et al. 2017; Anderson et al. 2017; Mackintosh et al. 2017).

21.2.3 Sampling

Cratonic terranes comprise vast tracts of mainly igneous and metamorphic rocks. Preferred lithologies for LTT studies and some sampling strategies are outlined in Chap. 2, Kohn et al. (2018). It is important to sample across major structures or lithospheric boundaries and from deep drillholes (where available), for the latter also collecting any available downhole temperature data. It can also be useful to sample rocks with contrasting petrology in close proximity, e.g., granites and gabbros. Minerals in these rocks, e.g., apatite, often exhibit different chemical composition and daughter product retentivity properties, which may allow a more detailed thermal history to emerge. It is always useful to collect non-basement samples from which stratigraphic/geochronological or organic maturity data may provide important geological constraints in thermal history models (see section below). This may include outliers of sedimentary cover or weathered regolith, in which case both this material and underlying basement should be sampled. Geo-thermochronological data from samples of post-cratonization magmatism (including estimates of emplacement depth) or impact structures can also serve as useful time-marker inputs for thermal history modeling. Where kimberlites are present, primary material and also any mantle or sedimentary xenoliths entrained within them should be collected, if possible. LTT data together with thermal maturation studies on the latter may yield useful information for estimating burial depth prior to erosion of overlying sediments, providing further important thermal history controls.

21.2.4 Thermal History Modeling

Forward and inverse thermal history modeling is commonly used to extract geological interpretations that best match the acquired LTT data. The two main software programs used by the thermochronology community are HeFTy (Ketcham 2005) and QTQt (Gallagher 2012—see also Chap. 3, Ketcham 2018, Chap. 6, Vermeesch 2018). Vermeesch and Tian (2014) evaluated the two approaches and their assessment has generated some spirited debate (e.g., Gallagher and Ketcham 2018; Vermeesch and Tian 2018 and references therein). Recent LTT craton studies have used both protocols (together with geological constraints) for generating thermal history simulations, e.g., HeFTy (Flowers et al. 2012; Ault et al. 2013; Guenthner et al. 2017, all using mainly He data) or QTQt (Kasanzu 2017; Mackintosh et al. 2017, using both AFT and He data).

Due to complexities in He systematics, wherever possible, multiple thermochronometers should be used to provide additional constraints for generating thermal history models, particularly in ancient, slowly cooled rocks (e.g., Anderson et al. 2017). In this context, the study of Mackintosh et al. (2017) from the Zimbabwe Craton is instructive in presenting a strategy for generating thermal history models for large, complex LTT data sets (AFT, AHe, and ZHe) in such a setting. QTQt was chosen for that study as (U–Th–Sm)/He data from broken apatite grains (with one or two crystal terminations preserved) could be incorporated using the fragmentation model of Brown et al. (2013). Furthermore, as the AHe and ZHe ages often showed large intra-sample dispersion, a resampling protocol could be applied using a scaling factor to input some of the undefined uncertainty for less precise data, while still honoring the observed ages (Gallagher 2012). Because our present understanding of the AFT system is arguably more clearly established, each sample was first modeled with only the AFT data. Using AFT as a baseline, modeling of AFT with AHe data was then carried out and finally the corresponding ZHe data was added, so that a comparison could be made of all model outputs to ensure reliability. An example of the model output from the Zimbabwe Craton using this sequential approach together with inputs and available geological constraints is shown in Fig. 21.2.

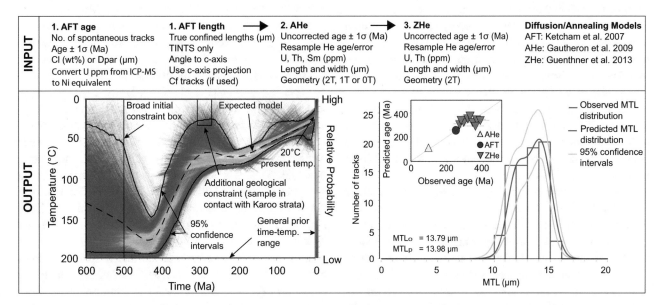

Fig. 21.2 Example of a QTQt thermal history inversion of AFT, AHe and ZHe data from a granite gneiss, northern Zimbabwe Craton. Upper panel: input single grain LTT data and relevant diffusion/annealing models used to generate the thermal history—numbers 1–3 and arrows indicate order in which data was used to create the inversion. Note 2T, 1T, and 0T refer to number of crystal terminations in analysed grain. Lower left panel: model output of accepted thermal histories together with constraints used, displayed as a color map indicating the posterior distribution of the model histories. The expected model (weighted mean of the posterior distribution) is regarded as the most representative thermal history with the 95% confidence intervals shown. Lower right panel: assessment of data fit using plot of observed ages against predicted ages (with resampled He error values). Comparison of data fit of the observed (o) and predicted (p) mean track length (MTL) values and distribution together with associated uncertainties (modified after Mackintosh 2017)

21.3 Evidence for a Dynamic Phanerozoic Cratonic Lithosphere

Petrological studies, particularly of suites of xenoliths and xenocrysts, indicate that some cratonic lithospheric mantle "keels" has undergone episodic and irreversible post-Archean changes in composition (e.g., O'Neill et al. 2008). Such "re-fertilization" of cratonic lithosphere by fluids related to subduction or magmatic processes results in thinner, hotter, and chemically metasomatized lithosphere, causing changes in the density (reducing cratonic root viscosities) and geothermal gradient, e.g., the Sino-Korean Craton (Xu 2001; Wenker and Beaumont 2017). This may lead to a reduction in cratonic mantle lithosphere strength, making it more susceptible to respond to plate boundary forces, with significant thermal and tectonic consequences that should be detected in the LTT record from the near-surface environment.

Kimberlites, commonly found intruding cratons, have been studied extensively because they are a major source for diamonds and also occasionally include lower crustal and subcontinental lithosphere mantle xenocrysts and xenoliths (e.g., Heaman et al. 2004; Stanley et al. 2015). These provide a rich resource for understanding mantle evolution, and a range of tectonic and geodynamic processes has been invoked to explain their time–space relationships. Kimberlite diatremes from several cratons contain sedimentary xenoliths, indicating that sediments covered the host craton at the time of their emplacement, but in most cases this sedimentary cover has since been eroded away (e.g., Stasiuk et al. 2006). Ault et al. (2015) reported a pattern of synchroneity between phases of sedimentary burial largely based on modeling of LTT data, and gaps in the kimberlite record for some Canadian and Australian cratons. The implication being that there may be a preservational bias of kimberlites intruded during burial phases.

Interactions between horizontal tectonic plate motions over the Earth's surface may transmit far-field in-plane stresses to cratonic areas resulting in crustal folding, reactivation of ancient structures, uplifting of basement blocks, and inversion of sedimentary basins (e.g., Pinet 2016). Closer to orogenic fronts, extensive foreland basins may form across adjacent cratons as a result of lithospheric flexure due to tectonic loading resulting in extensive km-scale sedimentary piles (e.g., DeCelles and Giles 1996) that have since been largely eroded away (e.g., Huigen and Andriessen 2004). Intracratonic and epicratonic basins consisting of thick sections of sedimentary infill characterized by continental and shallow marine paleoenvironments have long histories of relatively slow subsidence. Classic Phanerozoic examples include the Williston, Michigan, and Hudson Bay basins of North America, and these show irregular subsidence histories marked by periods of exhumation (e.g., Crowley 1991; Osadetz et al. 2002). Several mechanisms may be responsible for these histories including: thermal relaxation following slow lithospheric stretching or emplacement of warmer mantle, magmatic upwelling, reactivation of ancient rift structures or basins underlying intracratonic basins, mechanical coupling of intracratonic basins with adjacent foreland basins, or mantle convection and dynamic topography (e.g., Allen and Armitage 2012).

Phanerozoic strata on continental platforms reveal a more dynamic history of vertical motions than previously assumed and have been used to infer the epeirogenic movements of continents (e.g., Bond 1979). Since the late 1980s mounting evidence has accumulated that viscous stresses created by deep mantle flow and changing mantle flow patterns can have a profound influence on the history of vertical motions of continental interiors inducing surface deformation leading to dynamic topography (e.g., Mitrovica et al. 1989; Gurnis 1993; Lithgow-Bertelloni and Silver 1998; Braun 2010). For example, the influence of mantle circulation above shallow-dipping subduction zones and associated vertical displacement of the Earth's surface generated by such flow has been proposed to explain the distribution and tilting of seemingly enigmatic Phanerozoic sedimentary sequences across the North American craton (e.g., Mitrovica et al. 1989; Burgess et al. 1997). Dynamic topography may also help explain why some stratigraphic sequences are temporally out of phase with the pattern of continental marine transgressions predicted by worldwide eustatic sea-level changes (e.g., Moucha et al. 2008). However, strata deposited in long-wavelength depressions (reflecting the length scale of mantle processes of several hundred to thousands of km) with amplitudes generally ≤ 1 km on continents formed by dynamic topography have a low preservation potential. This is a result of the ephemeral nature of the topography; therefore, the record of sedimentation related to subduction-related dynamic topography is often most likely represented by unconformities (e.g., Burgess et al. 1997) and is unlikely to be detected by conventional methods.

LTT data from cratons indicate that despite their longevity, they are mostly characterized by relatively young ages (<500 Ma) (e.g. Brown et al. 1994; Harman et al. 1998; Feinstein et al. 2009; Flowers and Schoene 2010; Ault et al. 2013; Kasanzu et al. 2016). However, there are some exceptions where ages up to ~1 Ga have been reported, e.g., North American shield (e.g., Crowley et al. 1986; Flowers et al. 2006; Kohlmann 2010) and Fennoscandian shield (e.g. Hendriks et al. 2007; Kohn et al. 2009; Guenthner et al. 2017).

As noted above, thermal histories reconstructed from apatite LTT data suggest that many cratons record episodes of Phanerozoic heating and cooling. Due to the extremely low average heat flow reported for most Archean cratons (mean 41 ± 12 mW/m^2—Rudnick et al. 1998), the erosion purely of crystalline basement would imply the removal of unrealistic thicknesses (e.g., Brown et al. 1994). However, the thermal history patterns observed in many cratonic areas have often been linked to burial and unroofing of overlying strata, rather than to the removal of large sections of crystalline shield rocks (e.g., Harman et al. 1998; Kohn et al. 2002; Ault et al. 2013).

Estimation of the cover thickness, however, depends on the robustness of the techniques used, the paleogeothermal gradient that prevailed at the time of cooling and the thermal conductivity of any removed cover material (e.g., Braun et al. 2016; Luszczak et al. 2017). Further, Braun et al. (2013) demonstrated that fluvial erosion of dynamic topography of long wavelength (~ 1000 km width) and amplitude (~ 1 km) may attain sufficient depth to expose rocks in which the apatite FT and (U–Th–Sm)/He systems have been totally reset.

In some cases, a sedimentary cover on a craton formed as part of an extensive foreland basin from an adjacent orogenic belt (e.g., Huigen and Andriessen 2004; Kohn et al. 2005). Sedimentary sequences still preserved around the margins of, and occasionally on some cratons, together with stratigraphic and isotopic evidence from proximal onshore and offshore sediments, carry a record of the detritus removed, which can be linked temporally to cooling episodes in their source areas (e.g., Kohn et al. 2002; Patchett et al. 2004; Kasanzu et al. 2016). However, long-term denudation rates (10^7–10^8 yr) derived from AFT studies on cratons (e.g., Kohn et al. 2002) appear to be at odds with geomorphic arguments (e.g., Fairbridge and Finkl Jr 1980; Gale 1992), and the very low erosion rates derived from cosmogenic isotope dating (e.g., Bierman and Caffee 2002; Belton et al. 2004) and chemical weathering studies (e.g., Edmond et al. 1995). It has therefore been argued that such studies based on substantially shorter time scales are more consistent with the traditional view of the relative tectonic and thermal stability of cratons. However, for AFT (and He) studies it is important to consider that the rate of basement cooling being measured is usually related to removal of the more erodible overlying strata rather than a considerable section of cratonic basement.

Fewer zircon and titanite LTT data have been reported from ancient terranes, some of which have been reviewed by Enkelmann and Garver (2016). However, these ages are also mostly younger than the crystallization or metamorphic ages of their host rocks. ZHe ages of ~ 1.58–1.7 Ga have been reported from the western Canadian shield (Flowers et al. 2006) and approaching 1 Ga from Archean Wyoming Province rocks in western North America (Orme et al. 2016) and the Fennoscandian shield (Guenthner et al. 2017). The

oldest ZFT and TFT ages reported from the Fennoscandian shield range between ~ 560–930 Ma (Rohrman 1995; Larson et al. 1999) and for THe ages up to ~ 1.2 Ga from the Kaapvaal Craton (Baughman et al. 2017) and ~ 940 Ma from the Fennoscandian shield (Guenthner et al. 2017).

21.4 Case Histories

LTT data aimed at evaluating the more recent thermal history of cratons and other ancient terranes have been reported in numerous studies. It is not the intention here to review all of these but rather to describe examples from some of the more widely studied archetypal cratons, illustrating how LTT data have been used to reconstruct different aspects of their more recent thermal history evolution.

21.4.1 Fennoscandian Shield

The Fennoscandian shield (also known as the Baltic shield) includes the exposed Precambrian crystalline basement forming the northwestern terrane of the East European Craton (Fig. 21.1). The Phanerozoic Caledonian Orogen marks the western boundary, while the Phanerozoic cover of the Russian platform defines the eastern outcrop boundary (Fig. 21.3a). The shield can be divided geochronologically into three main crustal domains, each related to successive orogenic events during Archean to Mesoproterozoic time (Gaál and Gorbatschev 1987). These contributed to continental growth and show a younging trend to the southwest (Fig. 21.3a). Following its amalgamation, the Fennoscandian crust was substantially reworked during the Sveconorwegian Grenvillian (~ 1.25–0.9 Ga) and Caledonian (~ 0.43–0.39 Ga) orogenies (Gorbatschev and Bogdanova 1993; Gee and Sturt 1985, respectively).

TFT and ZFT ages from crystalline basement in southern Sweden and southern Finland range between ~ 560 and ~ 930 Ma (averaging ~ 850 Ma) (Larson et al. 1999). The ages are interpreted as reflecting partial annealing due to sedimentary burial under a foreland basin developed from the Sveconorwegian Orogen (Fig. 21.3a). The thickness of the sediment pile is estimated to have been at least 8 km, at ~ 100 km east of the preserved orogen, and covered large areas of Sweden and Finland (Larson et al. 1999). Most of these strata were eroded away and by the end of the Proterozoic large areas of the present surface were exposed or lay close to the present-day surface (e.g., Lidmar-Bergström 1996). ZFT and ZHe and THe data reported by Rohrman (1995) and Guenthner et al. (2017), respectively, from deep borehole samples (Figs. 21.3a and 21.4) are also consistent with this conclusion. The resulting sub-Cambrian peneplain (Fig. 21.3a) was subsequently

Fig. 21.3 **a** Simplified geological overview of Fennoscandian shield (modified after Tullborg et al. 1995). Sub-Cambrian peneplain CDF = Caledonide deformation front, OG = Oslo Graben, GoB = Gulf of Bothnia. Deep drillhole sites for low-temperature thermochronology studies numbered with relevant references. 1 = Gravberg-1 (Rohrman 1995), 2 = Kola deep borehole (Rohrman 1995; Kohn et al. 2004—see also Fig. 21.4), 3 = Oskarshamm KLX01 and KLX02, Laxemar region (Söderlund et al. 2005; Guenthner et al. 2017). **b** Schematic cross section along line A-B-C of the Caledonide Foreland basin (after Larson et al. 1999)

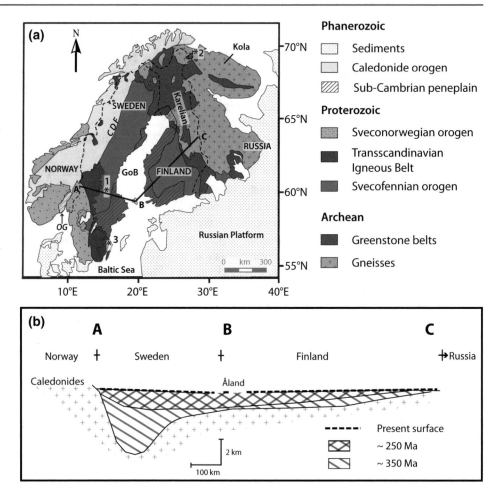

covered by a few hundred meters of Cambrian to Silurian ($\sim$ 540–420 Ma) platform sediments (Lidmar-Bergström 1996). On the Fennoscandian shield, apart from a Cambro–Ordovician sequence in the Gulf of Bothnia, only scattered remnants of this sequence are preserved, mainly in southern Finland. The peneplain forms an important geological constraint for thermal history modeling because the present-day surface must have been either exposed or at a near-surface level by the end of the Proterozoic and that over much of the Phanerozoic a sedimentary cover has preserved it.

The Late Siluran—Early Devonian Caledonian orogeny is marked by continent–continent collision between $\sim$430 and $\sim$390 Ma (e.g., Gee and Sturt 1985). The resulting mountain range is thought to have rivaled the modern day Himalayas in scale (Streule et al. 2010). Fault-controlled basins along the west coast of Norway are filled with up to tens of km of Upper Silurian–Devonian conglomerates, and the offshore Oslo Graben area (Fig. 21.3a) preserves some 4–6 km of Cambro–Silurian sediments in down-faulted blocks (e.g., Steel et al. 1985). However, Silurian or Devonian strata are largely absent from onshore Sweden or Finland. Nevertheless, it has been proposed that the orogeny caused flexure of the Fennoscandian shield and, following

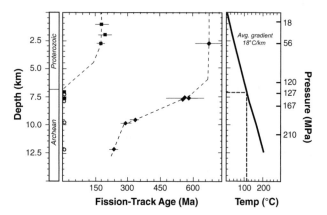

Fig. 21.4 Apatite and zircon fission-track ages (left- and right-hand curves, respectively), geological section, geothermal gradient and in situ effective pressure for the superdeep Kola SG3 borehole. Solid symbols = data from Rohrman (1995), open circles = data from Kohn et al. (2004). See text for further details

gravitational collapse of the orogen resulting from uplift and erosion, an extensive foreland basin was developed. This basin extended for several hundreds of km, thinning to the east, and was progressively redeposited over southeastern Scandinavia (see Fig. 21.3b). Geological arguments, including evidence from different AFT studies, for the existence of such a foreland basin across central and southern Sweden are summarized by Larson et al. (2006). This also includes evidence for Phanerozoic radiogenic Pb loss from zircons in crystalline basement rocks across the Swedish segment of the shield to the Åland Islands, southwestern Finland (Larson and Tullborg 1998). AFT data from transects across Sweden, extending from the Caledonian deformation front eastwards to the Gulf of Bothnia (Fig. 21.2a), are compatible with a history of burial beneath an asymmetrical Caledonian foreland basin, which is thought to have attained a thickness of at least $\sim$3.5 km near the deformation front, thinning to $\sim$2.5 km near the Gulf of Bothnia coast (e.g., Zeck et al. 1988; Larson et al. 1999; Cederbom et al. 2000; Huigen and Andriessen 2004; Japsen et al. 2016, see also Fig. 21.3b).

In southern Finland, AFT ages ranging between $\sim$310 and 1030 Ma have been reported by Hendriks et al. (2007) and Kohn et al. (2009). Ages are younger in the south and southwest but are mostly >600 Ma and include some of the oldest AFT ages reported from anywhere. The range of age differences, sometimes over relatively small distances, cannot be attributed solely to variations in sedimentary overburden. Rather they may be due to a combination of different factors, such as variations in apatite chemistry, especially for gabbros, in which apatites usually display higher Cl content coupled with the oldest AFT ages (Kohn et al. 2009). Further, variations in heat flow, rock conductivity or the relatively low heat production in gabbros may also lead to significant differences in thermal histories (e.g., Kukkonen 1998; Łuszczak et al. 2017). In this respect, the variable and generally high heat flow within a modern foreland basin overlying Precambrian basement (Western Canada Sedimentary Basin) is instructive (Weides and Majorowicz 2014).

Thermal history models for southern Finland indicate a major phase of Late Proterozoic cooling, which was attributed to intra-plate stress propagation from the Sveconorwegian orogeny (Murrell and Andriessen 2004). Track length measurements reported mostly suggest different degrees of partial annealing. Larson et al. (1999) also presented AFT data from three $\sim$1000 m deep drill holes across southern Finland and estimated that sedimentary detritus derived from the Scandanavian Caledonides had buried most of Finland by 0.5–1.5 km. These estimates are in accord with time–temperature models presented by Lorencak (2003), Murrell and Andriessen (2004), and Kohn et al. (2009). The lack of substantial Caledonide burial in southern Finland (see also images of AFT age and track length patterns across the Fennoscandian shield by Hendriks et al. 2007) suggests that samples in this area may still retain older Neoproterozoic AFT ages largely related to removal of a former thick Sveconorgwegian foreland section, which had totally reset

zircon and titanite FT and He ages to the southwest in eastern Sweden.

Hendriks and Redfield (2005, 2006) challenged the notion of deep burial of the eastern Fennoscandian shield under a regionally extensive Caledonian foreland basin, especially in Finland. Rather they proposed that a hitherto overlooked process of α-radiation-enhanced annealing (REA) of fission tracks, resulting principally from the decay of ^{238}U, operates at low-temperatures (<60 °C) in areas of prolonged thermotectonic stability such as southern Finland. The effect of this postulated REA in addition to temperature sensitivity, implied that modeled AFT data in cratonic settings (and in U- and/or Th-rich apatites in general) using conventional thermal annealing models may lead to an overestimation of paleotemperatures, and hence the amount of section removed.

The case for REA was partly based on inverse correlations between AFT grain age and U content from southern and central Finland samples. Donelick et al. (2006) also reported a range of inverse relationships (weak to moderately strong) between AFT age and α-emitter actinides within samples from Finland and Canada, with the older grains from Finland showing the strongest negative correlation. Kohn et al. (2009), however, using new and old AFT data from southern and central Finland, as well as from the southern Canadian and Western Australian shields were unable to reproduce the consistent inverse correlations presented by Hendriks and Redfield. Furthermore, despite the large range of U content, sometimes varying three to sixfold between intra-sample grains (with a limited range of Cl content), AFT ages in southern Finland yield chi-square values >5%, which are consistent with the data being treated as a single age population.

Further, Hendriks and Redfield (2005) also argued that REA could explain the observation that, contrary to predicted expectations, a significant number of AHe ages from the Fennoscandian shield are older than their coexisting AFT ages. However, many of the AHe data reported from southern and central Finland samples (e.g., Murrell 2003; Lorencak 2003; Kohn et al. 2009) show age dispersion and poor reproducibility. In cratons, effects such as slow cooling, lengthy HePRZ residence, annealing of α-damage due to burial, variable eU and wildfires can enhance intra-sample He age variation between single grains. Several of these factors (see Sect. 21.2.2.1) were only recognized following the publication of Hendriks and Redfield (2005). This makes it difficult to directly compare at face value the relationship between an AFT age with a number of coexisting dispersed single grain AHe ages (see Sect. 21.2.4). The applicability of the postulated REA to AFT studies has raised considerable debate (e.g., Green and Duddy 2006; Larson et al. 2006; Hendriks and Redfield 2006; Donelick et al. 2006; Kohn et al. 2009) and is still somewhat controversial.

The variation of fission-track parameters in deep borehole samples has been of critical importance in extrapolating laboratory-annealing experiments to geological environments. In this respect, FT studies from the superdeep Kola borehole (SG3) located in the northeastern Fennoscandian shield (Fig. 21.3a), drilled to a depth of 12,262 m (the deepest drillhole on Earth), have also provided important information on geological annealing constraints (Rohrman 1995) (see Fig. 21.4). The bottom hole temperature of the well is ~216 °C although the geothermal gradient is variable, particularly in the upper ~2 km, due to movement of fluids and post-glacial isostatic adjustment leading to pressure reduction in deep-seated faults (Popov et al. 1999). ZFT data from the well suggest a ~200–310 °C ZFT PAZ for a heating time of ~100 Myr. AFT data from samples at depths from ~7060 to 12,150 m essentially show a zero age. This suggests that total track annealing, approximating the base of the AFT PAZ, occurs at temperatures of ~110–115 °C (Rohrman 1995; Kohn et al. 2004) or possibly at a slightly lower temperature at a higher level where no samples were available, which is typical of that reported from wells elsewhere (e.g., Gleadow et al. 2002a). At this temperature in the well, the depth equates to a pressure of ~127 MPa (Morrow et al. 1994), indicating that no significant pressure effect on AFT annealing is evident. Worthy of note here is that in almost all natural geological environments a pressure range up to 127 MPa covers the entire range of the AFT PAZ. Taken together, FT data in the well are interpreted as evidence for rapid denudation at ~600 Ma, coincident with the formation of the sub-Cambrian peneplain, as described from further south, as well as substantial cooling commencing between ~180 and ~250 Ma (Rohrman 1995), consistent with later re-exhumation of this surface.

21.4.2 Western Australia Shield

The Western Australian shield is an amalgamation of two Archean cratons (Pilbara ~60,000 km^2 and Yilgarn ~657,000 km^2) and several Proterozoic basins and terranes, surrounded by Phanerozoic basins (Fig. 21.5). Overlying Proterozoic basins obscure the suture between the two cratons, which consist predominantly of Archean granite–greenstone belts, comprising an assemblage of several smaller terranes with distinct geological histories and geochemical features (e.g., Myers 1993). The cratons contain some of the oldest well-preserved rocks and fossil stromatolites on Earth, and host a rich endowment of precious and base metal deposits, and major banded-iron formations. At Jack Hills in the Narryer Gneiss Terrane (Fig. 21.5), zircons in a metamorphosed quartz pebble conglomerate have been dated at >4 Ga (e.g., Harrison 2009).

The Phanerozoic cooling history of the Western Australian Shield has been investigated using AFT thermochronology (Gleadow et al. 2002b; Kohn et al. 2002; Weber 2002; Weber et al. 2005). AFT ages are up to ~450 Ma (toward the eastern craton boundary), but mostly range between ~200 and 375 Ma, with mean confined horizontal track lengths varying between 11.5 and 14.3 µm. Remarkably, these AFT parameters are similar in range over many areas of the cratons, as well as in most of the intervening Proterozoic basement (e.g., Gascoyne complex) and basins (e.g., Hamersley Basin), the Mesoproterozoic Albany-Fraser Orogen and the Neoproterozoic–Cambrian Leeuwin complex (see Fig. 21.5). Thermal history models from the southwestern Yilgarn suggest the onset of a regional cooling episode from temperatures >110 °C in the Late Carboniferous–Early Permian, which continued into the Late Jurassic–Early Cretaceous (Kohn et al. 2002; Weber et al. 2005). AHe and ZHe data acquired from the southwestern Yilgarn Craton and adjacent Proterozoic basement mostly fall within a similar age range to that reported for AFT data (Lu 2016).

The Yilgarn Craton is currently largely devoid of sedimentary cover. However, remnants of Permian and Cretaceous strata are preserved within the fault-bounded Collie, Wilga, and Boyup Basins located within the Yilgarn Craton ~150–180 km SSE of Perth (Fig. 21.5). The largest Collie Basin contains ~1.4 km of Permian tillite, shale, and coal measures (Le Blanc Smith 1993). Vitrinite reflectance data ($R_0 = 0.41$–0.61) and the stratigraphy suggest that maximum burial temperature for the coal was up to ~95–100 °C. According to Le Blanc Smith (1993), these "basins" are in-faulted remnants of a formerly much more extensive cover sequence, several kilometers of which was removed in the Permian, with a later erosion cycle removing further section between the Permian and Early Cretaceous.

The Perth Basin, located west of the Yilgarn Craton (Fig. 21.5), hosts up to 15 km of strata in major depocenters, ranging in age from Ordovician to Quaternary, but most is Permian to Lower Cretaceous (Baillie et al 1994). The sediments show a predominance of Mesoproterozoic detrital U–Pb zircon ages, with relatively few Archean grains (e.g., Cawood and Nemchin 2000), suggesting that despite its close proximity, the extensive Yilgarn Craton itself was not a major source for basin sediments. Further, U–Pb zircon ages in the Permian Collie Basin coal measures and Lower Triassic rocks of the northern Perth Basin, peak at ~1200 Ma, suggesting a possible Albany-Fraser Province source to the south (Veevers et al. 2005). Other possible sources for sedimentary cover over the craton are from the relatively poorly exposed Mesoproterozoic–Cambrian Pinjarra orogen (e.g., Fitzsimons 2003), lying to the west of the Darling Fault (Fig. 21.5) Cooling recorded on the craton can therefore most likely be attributed to the removal of a sedimentary

Fig. 21.5 Simplified geological map of the West Australian shield, showing cratons and regions covered by previous low-temperature thermochronology studies. A—Weber (2002), Kohn et al. (2002), Gleadow et al. (2002b), Weber et al. (2005)—all AFT and B—Kohn et al. (2002), Gleadow et al. (2002b) and Lu (2016). JH = Jack Hills

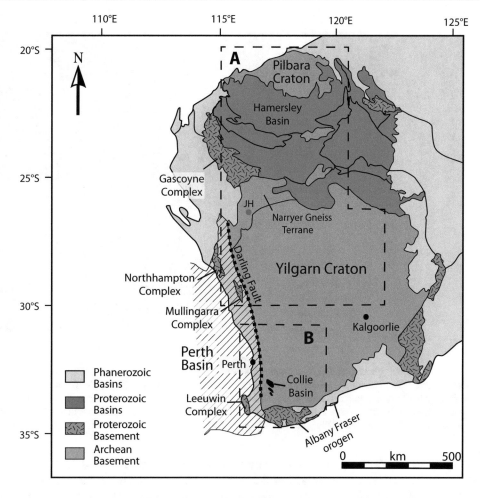

cover rather than major denudation from the underlying craton itself. Killick (1998) calculated the volume of sediment deposited in basins marginal to the Western shield (Yilgarn and Pilbara Cratons and intervening Proterozoic basins) between Early Ordovician to Late Cretaceous, after which clastic sedimentation effectively ceased. This led to an estimate that ∼4.1 km of material had been removed from the West Australian shield since the onset of basin development and sedimentation, at an average denudation rate of 8.87 m/Myr. This estimate along with those presented by Kohn et al. (2002) are all significantly higher than the rates of 0.1–0.2 and 0–2 m/Myr previously reported on the basis of geomorphological considerations for the Yilgarn Craton by Fairbridge and Finkl (1980) and Gale (1992), respectively, although such low rates may well be typical for the post-Cretaceous denudation of the craton basement.

Present-day heat-flow data from the Western Australia shield range from ∼30 to ∼50 mW/m² but measurements are both sparse and heterogeneously distributed (e.g., Cull 1982). Nevertheless, despite the uncertainties in paleogeothermal gradients, the vitrinite reflectance and LTT data suggest that a substantial thickness of upper Paleozoic–Mesozoic sediment extended across much of the craton and

that the Collie and adjacent basins are preserved remnants of that accumulation.

The trigger for onset of late Paleozoic–Mesozoic cooling across the southwestern Western Australian shield is not clear, but could be related to far-field effects linked to compressional events throughout Gondwana due to collisions on the Panthalassa margin (e.g., Harris 1994). It is also noted that a continental ice sheet covered much of the Western Australian shield in Late Carboniferous–Early Permian time. Glacial traces and deposits are found over a wide area in all Western Australia Paleozoic basins including those peripheral to the shield (Mory et al. 2008), as well as in the Collie Basin within the Yilgarn itself. The isostatic response to the waxing and waning of a thick ice sheet over a large area of the shield and its implications for denudation has not been fully evaluated.

21.4.3   Kalahari Craton

The Kaapvaal and Zimbabwe Cratons sutured by the Limpopo mobile belt underlying southeastern Africa, form part of the Kalahari Craton and include some of the best

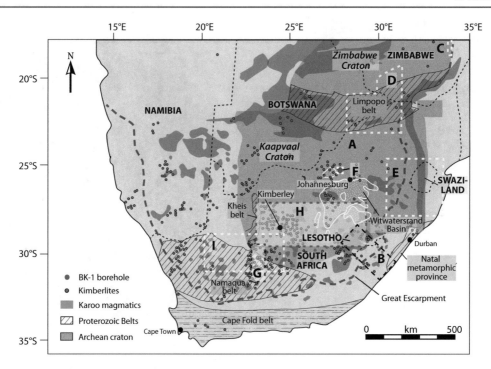

Fig. 21.6 Simplified geological map of southern Africa showing Kalahari Craton area, kimberlites, and Karoo magmatics, as well as the Great Escarpment. Regions outlined refer to areas covered by previous low-temperature thermochronology studies, which also included samples from the Archean cratons. These are: A—Brown (1992), B— Brown et al. (2002), C—Belton (2006), D—Belton (2006), Belton and Raab (2010), E—Flowers and Schoene (2010), F—Gallagher et al. (2005), Beucher et al. (2013), G—Stanley et al. (2013), H—Stanley et al. (2015), I—Wildman et al. (2017). Not outlined is a LTT study of the Zimbabwe Craton by Mackintosh et al. (2017)

preserved early Archean crystalline nuclei anywhere (Fig. 21.6). The cratons, forming part of the southern African Plateau, are anomalous in that they are elevated (averaging 1.0–1.15 km, but increasing up to 2.5 km in Lesotho) by more than 0.5–0.7 km compared to all other cratonic areas in the world (apart from the Tanzania Craton) (Artemieva and Vinnik 2016). Further, the southeastern Africa cratons are bordered by the "Great Escarpment" (Fig. 21.6), located about 50–250 km inland from coast and separating the elevated interior from the more highly denuded coastal plain (see Chap. 20, Wildman et al. 2018). The southern Africa Plateau is located above a significant low-seismic-velocity mantle anomaly, commonly called the "African superswell," which is considered to be a probable source of dynamic buoyancy (e.g., Lithgow-Bertelloni and Silver 1998). The timing and cause of the southern African topographic anomaly have been the subject of considerable scientific interest and debate (e.g., Burke and Gunnell 2008; Moore et al. 2009; Zhang et al. 2012; Stanley et al. 2015; Artemieva and Vinnik 2016).

Archean lithologies of the Kaapvaal and Zimbabwe Cratons are dominantly granite–gneiss–greenstone. On the Kaapvaal Craton, most Archean rocks outcrop in the northeast and east, while there is considerably more exposure across the Zimbabwe Craton. Late Archean to Proterozoic tectonic activity across the southern African

Plateau is mainly confined to mobile belts and terranes adjacent to the cratons. Upper Carboniferous to Lower Jurassic Karoo Supergroup strata, Jurassic volcanic rocks of the Karoo and Cretaceous Etendeka Large Igneous Provinces (LIPs) and Upper Cretaceous-Cenozoic sediments of the Kalahari Supergroup dominate the Phanerozoic record of southern Africa. Karoo Basin sedimentation, attaining thicknesses of several kms, records an almost continuous sequence from marine glacials to terrestrial deposition, over much of the Kalahari Craton (Catuneanu et al. 2005). Karoo deposition terminated with the eruption of Lower Jurassic Karoo LIP lavas, which are coeval with the widespread emplacement of the Ferrar LIP extending across Antarctica (see Chap. 13, Baldwin et al. 2018) to Tasmania. To the west, the Lower Cretaceous off-craton Etendeka LIP volcanics were emplaced coevally with eruption of the Paraná flood basalts in South America. Both these eruptive phases are linked to incipient Gondwana breakup and are also associated with kimberlite magmatism on the craton.

Numerous kimberlites and related alkaline magmas were emplaced in southern Africa (Fig. 21.6) between the Proterozoic to Cenozoic (Jelsma et al. 2009). The most prominent Phanerozoic kimberlites were erupted during two episodes, each marked by a distinctive composition; Group II between ∼200 and ∼110 Ma (peaking between ∼140 and ∼120 Ma) and Group I erupted between ~100 and ~80 Ma (Jelsma et al. 2009). Xeno-lithic clasts down-rafted at the time of eruption into the outcropping diatreme facies of many South African kim-berlites from overlying strata, but since eroded away, pro-vide valuable insights into the stratigraphy at the time of kimberlite emplacement (Stanley et al. 2015). Crustal xenoliths suggest that Karoo sediments and flood basalts covered much of southern Kaapvaal Craton when the Group II pipes erupted, but that the basalts had been removed from the southwestern craton by the time the Group I pipes were emplaced (Hanson et al. 2009).

LTT studies in southern Africa have largely focused on sampling across the Great Escarpment and passive conti-nental margin of South Africa and Namibia, with the mainly AFT studies focusing on the evolution and retreat of the passive margin during and following Gondwana breakup (e.g., Brown et al. 2002; Kounov et al. 2013; Wildman et al. 2016; Green et al. 2017—see also Chap. 20, Wildman et al. 2018). Only a few studies, however, have specifically investigated the thermotectonic history of the cratonic inte-rior regions.

The earliest AFT data reported from the Kaapvaal Craton (northeastern region) are strongly correlated with elevation (Brown 1992). Ages range from ~90 Ma at ~500 m to ~450 Ma at elevations of ~2000 m (with all ages >300 Ma above elevations of 1200 m). Mean track lengths decrease from lengths of ~14 μm to a minimum of ~11.0 ± 0.5 μm at elevations of 1250 m before increasing to lengths of 12–13 μm at elevations ~2000 m. This pattern, together with a TFT age from the adjacent Natal metamorphic Province (Fig. 21.6), suggests that apa-tites were totally overprinted prior to and/or during the Pan-African orogeny. Further, younger ages, located toward the craton margin, cooled from at least ~110 ± 10 °C since the Late Cretaceous, with older ages from the craton interior preserving an older cooling record from lower temperatures. The Late Cretaceous cooling episode, similar in age range to the younger cooling episode above, is attributed to uplift and km-scale denudation of the overlying relatively low-conductivity Karoo Basin sediments and volcanics. Late Cretaceous cooling is also recorded in AFT data from the Ladybrand (LA/168) borehole, located at the NW termination of transect "B" in Fig. 21.6, penetrating Karoo Basin sediments and underlying Silurian quartzites (Brown et al. 2002, Chap. 20, Wildman et al. 2018). AHe data from the eastern Kaapvaal Craton and across the eastern escarpment also record significant Cretaceous cooling, clustering around ~100 Ma with little evidence for sub-stantial Cenozoic cooling (Flowers and Schoene 2010). AFT and AHe data from the BK-1 (Bierkraal) drillhole at an elevation of ~1500 m (Fig. 21.6) in the Paleoproterozoic Bushveld Complex intruding the Kaapvaal Craton indicate protracted residence within the AFT PAZ between ~400

to ~100 Ma, followed by relatively rapid Late Cretaceous cooling (Gallagher et al. 2005; Beucher et al. 2013). AFT and AHe data from the southwestern Kaapvaal Craton also suggest that the region resided at near-surface tempera-tures <60 °C since at least ~300 Ma (Wildman et al. 2017). The study also suggests that the craton margin and sur-rounding Proterozoic mobile belt record two discrete cooling episodes during the Early and Late Cretaceous with only minor (<1 km) denudation during the Cenozoic.

Evidence for discrete Early and Late Cretaceous cooling phases are also recorded by AFT data from the Limpopo Belt and southern Zimbabwe Craton periphery (Belton and Raab 2010) and from two vertical profiles in the Eastern Highlands of the craton (Belton 2006—see area C in Fig. 21.6). Samples at elevations >1000 m from one of the vertical profiles, as well as from a few samples in peripheral areas of the southern and northern Zimbabwe Craton (area not shown in Fig. 21.6) also record a protracted cooling history through the PAZ from Late Carboniferous time and are considered to have resided near the surface since the Late Jurassic (Noble 1997; Belton 2006). However, a more comprehensive LTT study of the Zimbabwe Craton by Mackintosh et al. (2017) revealed two cooling episodes inferred to be a denudational response to surface uplift. The most significant being related to stress transmission related to Pan-African orogenesis during Gondwana amalgamation. A second minor episode, commencing in the Paleogene, is recorded by ZHe data only (which are younger than their coexisting AFT and AHe ages) and removed km-scale Karoo sedimentary cover.

Stanley et al. (2013, 2015) and Wildman et al. (2017) proposed that the spatially variable denudation pattern of southern Africa based on LTT data may be a response to both horizontal plate tectonic stresses at the craton margin, amplified by changes in plate motions in the mid-late Cre-taceous, and a lithospheric thermal anomaly at that time. Vertical mantle stresses created by a buoyant mantle upwelling may have driven long-wavelength uplift of the strong cratonic interiors, while weakened lithosphere around the cratonic margins possibly experienced short wavelength deformation. Thus, the evolution of the southern African landscape may have resulted from the complex interplay between these horizontal and vertical forces.

21.4.4 Western Canadian Shield

The Canadian shield was formed by a collage of Archean plates and accreted juvenile arc terranes and sedimentary basins that were progressively amalgamated during the Proterozoic (Fig. 21.7). Most accretion and growth occurred during the Paleoproterozoic Trans-Hudson orogeny (e.g., Schneider et al. 2007). The shield constitutes most of the

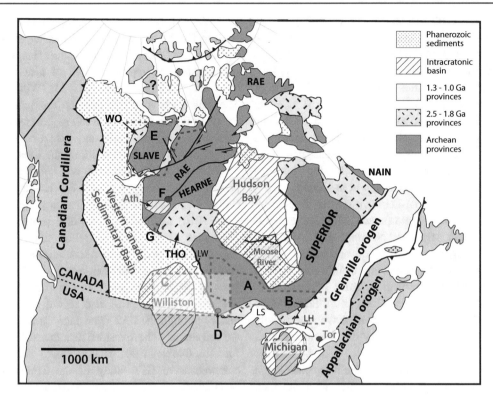

Fig. 21.7 Simplified geological map of the Canadian shield showing Archean cratons and some intracratonic basins. Regions outlined refer to areas covered by previous LTT studies, which have also included samples from Archean cratons. These are: A—Crowley (1991) [mainly Michigan Basin basement and crystalline basement to the north to the Moose River Basin and northeast into the Grenville Province], Lorencak (2003), Kohn et al. (2005), B—Sudbury deep drillhole, Lorencak et al. (2004), C—Basement from Williston Basin and adjacent crystalline rocks—Crowley et al. (1985), Crowley and Kuhlman (1988), Kohn et al. (1995), Osadetz et al. (2002), D—URL Pinawa deep drillhole, Feinstein et al. (2009), E—Arne (1991), Kohlmann et al. (2007), Kohlmann (2010), Ault et al. (2009, 2013), F—Flowers et al. (2006), Flowers (2009), G—Flowers et al. (2012). Not outlined is the area covered by Pinet et al. (2016), which includes the intracratonic Hudson Bay Basin basement and adjacent shield outcrop samples to the west. WO = Wopmay Orogen, Ath = Athabasca Basin, TH = Trans-Hudson Orogen, LW = Lake Winnipeg, LS = Lake Superior, LH = Lake Huron, Tor = Toronto

ancient geological core of North America (also known as Laurentia) and forms the greatest area of exposed Archean rocks anywhere (Fig. 21.1) and the Slave Craton hosts the oldest dated rocks (Acasta Gneiss) on Earth. LTT studies of the shield have mainly focused on the southern Superior Craton (e.g., Kohn et al. 2005), and the western shield, particularly on the Slave Craton, as well as on crystalline basement rocks in the intracratonic Michigan and Hudson Bay basins and the epicratonic Williston Basin (see references cited in Fig. 21.7). The discussion here will focus on the western Canadian shield.

The most intensively studied area in the western Canadian shield for LTT is the Slave Craton (area E, Fig. 21.7). AFT data from Paleozoic sediments and crystalline basement outcrops in the SW Slave Craton area suggest heating following deposition of Devonian carbonates (Arne 1991). Heating was attributed to burial, with samples attaining maximum paleotemperatures of ~85–100 °C during the Late Cretaceous (note this approximation was based on direct interpretation of the data) followed by an

estimated ~2.0–2.5 km of uplift and denudation. These paleotemperatures are consistent with previous studies on organic maturity and conodont alteration indices in the Devonian carbonate, as well as fluid inclusion data in Pb–Zn orebodies hosted by the carbonates adjacent to the craton margin (Arne 1991).

Kohlmann et al. (2007), Kohlmann (2010) reported AFT data from a broad coverage of basement and kimberlites across the Slave Craton. Basement ages range from ~130 to 855 Ma, with older ages located in the north and interior of the craton, with younger ages being more prevalent toward the western margin. However, there is some age scatter, even between samples in close proximity and in those cases the older age apatites almost invariably show significantly higher Cl content. Thermal history models across the craton show a distinct Devonian to Pennsylvanian heating episode, with maximum paleotemperatures sufficient to have partially or totally annealed fission tracks in all samples. Cooling is recorded from the latest Paleozoic with samples reaching near-surface temperatures in Triassic to Jurassic time. In some samples, models suggest a phase of slight reheating in Cretaceous time, but as samples barely attained temperatures of 60 °C, this was considered to be difficult to confirm unequivocally.

AHe data from the Slave Craton yield mean ages ranging between 382 and 210 Ma, with ages younging across the craton from east to west (Ault et al. 2009, 2013). When combined with geologic and stratigraphic constraints, particularly from kimberlites of varying age, thermal history simulations are broadly similar to those derived from AFT data, indicating that apatites experienced complete He loss in middle to late Paleozoic time. This was followed by removal of most Paleozoic strata by the Late Jurassic and then a lesser phase of reheating during which maximum paleotemperatures ranged between ~50 and ~70 °C. Younger ages from the Wopmay Orogen ranged between 212 and 231 Ma (Ault et al. 2009), although statistically indistinguishable from the westernmost Slave Craton results, these may be due to higher heat flow or a slightly younger timing for cooling in the western portion of the terranes studied.

ZHe and AHe data from basement adjacent to the NE Athabasca Basin to the south of the Slave Craton (area F in Fig. 21.7) yield relatively old ages ~1.73–1.58 Ga and ~0.95–0.12 Ga, respectively (Flowers et al. 2006, Flowers 2009). AFT ages range from ~1.02 to ~0.63 Ga (all with Dpar values >1.87, suggesting relatively high Cl content). Time–temperature LTT models are consistent with Phanerozoic reheating, indicating partial to total He loss at peak temperatures between ~62 and ~95 °C. A similar thermal history pattern is evident from AFT and AHe data from basement lithologies to the southwest in area G (Fig. 21.7), but with slightly lower peak temperatures (Flowers et al. 2012).

A broadly consistent Phanerozoic thermal history pattern has emerged from LTT studies on the western Canadian shield. Peak temperatures were achieved at the time of a major phase of mid-Paleozoic–early Mesozoic heating, during which the AHe system was totally reset, with the AFT system partially to totally reset to varying degrees. Regionally, heating was highest in the Slave Craton–Wopmay Orogen and decreased southwards. These data are interpreted as indicating that beginning at ~450 Ma most of the shield was flooded and blanketed by a Paleozoic marine sedimentary cover. The Silurian to Devonian Caledonian–Franklynian orogeny (Andresen et al. 2007) to the north of the Slave Craton in Greenland and the Canadian Arctic islands has been linked to the formation of these blanketing strata. Based on a Sm–Nd isotopic study of Canadian clastic lithologies, Patchett et al. (2004) suggested that this orogen provided the source of the dominant Devonian marine shales deposited on the Canadian shield. Flowers et al. (2012), however, noted that the burial and unroofing history observed over an extensive area of the western Canadian shield was out-of-phase with that expected from eustatic sea-level chronologies and noted that vertical continental displacements may have been an important control on the inferred depositional and erosional history. They proposed that a process that can induce long-wavelength (over >1000 km) elevation change in a continental interior region without significant crustal deformation, such as dynamic topography, could have been a feasible mechanism to explain the vertical motions. Hence, significant Paleozoic burial may have been due to subsidence during cold mantle downwelling at the time of Pangea assembly, followed by unroofing due to low amplitude vertical motion induced by warm mantle upwelling during Pangea breakup (see also Zhang et al. 2012). As such it was suggested that such characteristics might largely eliminate the need for plate margin tectonism as an underlying mechanism for the heating–cooling history observed.

Although the western Canadian shield is currently largely devoid of Phanerozoic strata, reconstructions indicate that the thickness of Ordovician and Devonian strata varied between 1 and >4 km (Ault et al. 2009; Kohlmann 2010). Further evidence for wide Paleozoic sedimentary cover over the shield includes: early Paleozoic limestone xenoliths entrained in the Late Ordovician Cross diatreme, as well as in four other kimberlites in the southwestern Slave (Pell 1997; Heaman et al. 2004), Middle Devonian fossiliferous limestone clasts in the Middle Jurassic Jericho kimberlite pipe in the north-central Slave (Cookenboo et al. 1998), outliers of deformed Ordovician dolomite in the Trans-Hudson Orogen (Elliott 1996), the presence of a thick section of lower Paleozoic strata preserved in the northern Western Canada Sedimentary Basin to the west of the Wopmay Orogen (e.g. Hamblin 1990) and a ~2.5 km thick Upper Ordovician–Upper Devonian section in the intracratonic Hudson Bay Basin (Pinet et al. 2013).

A second phase of milder reheating in Cretaceous time, barely discernable in the AFT data, is best established by AHe data in the Slave Craton. It is possibly related to 1.6 km of sedimentary burial during the development of a foreland basin (still partially preserved in the West Canada Sedimentary Basin) related to evolution of the Canadian Cordillera to the west. Erosion of these strata was followed by terrestrial deposition in the Eocene, which was later removed (Ault et al. 2013). Cretaceous–early Tertiary marine mudstone and shale clasts on which organic maturity measurements have been made in the Cretaceous to Eocene Lac de Gras kimberlites in the central Slave (e.g., Stasiuk et al. 2006) provide evidence that sedimentary cover at this time extended at least 300 km further east than the preserved limit of the foreland basin (Ault et al. 2009). Further south, evidence for a Laramide (Cordilleran) foreland basin is represented by the upper Zuni sequence (Upper Cretaceous–Paleogene) in the Williston Basin (e.g. Osadetz et al. 2002—area C in Fig. 21.7).

21.5 Conclusions

Despite their longevity, most cratons yield AFT and AHe ages <500 Ma, but some may be as old as ~1 Ga, with occasional older zircon and titanite LTT ages. Older age apatites often display higher Cl content or Dpar values (AFT) or high eU (AHe). Thermal histories reconstructed from craton apatite LTT data often show Neoproterozoic and/or Phanerozoic episodes of heating and cooling. These have mostly been related to widespread km-scale sedimentary burial followed by unroofing, which has largely erased this sedimentary record. Evidence for burial includes: preservation of sedimentary outliers on cratons or correlatives in adjacent basins, studies on sediments and basement rocks (in drillholes) in epicratonic or intracratonic basins, and sedimentary xenoliths entrained within kimberlites. Further, stratigraphic and isotopic provenance data, (e.g., U–Pb zircon detrital grain ages) from onshore or offshore basinal sediments proximal to some cratons may carry a record of the detritus removed, whose accumulation can be linked temporally to cooling episodes in their cratonic source areas.

Cratonic environments are characterized by low heat flow and this has led to seemingly high estimates for the thickness of overburden required to totally reset or even partially reset the AFT and AHe systems. Evaluation of the thickness of cover removed, however, depends on the robustness of the techniques used, the paleogeothermal gradient that prevailed at the time of cooling and the thermal conductivity of any earlier cover removed.

Long-term average denudation rates from cratons based on LTT data appear to be at odds with geomorphic arguments and rates derived from cosmogenic isotope dating and chemical weathering studies. However, there is an inherent bias in extrapolating such short-term rates into deep geological time, because they are measured on currently exposed basement, whereas histories simulated from models are often relevant to rates determined during the removal of sedimentary overburden in much earlier periods. Thus, the strong contrast in erosion potential between basement and sediment needs to be taken into account, as well as the difference in timing of the denudation being measured.

Fundamental mechanisms related to cratonic heating and cooling episodes reconstructed from LTT data are complex and in any region more than one factor may be at play. These include: adjacent orogens supplying vast volumes of sediment into widespread foreland basins; horizontal far-field transmission of stresses from active plate boundaries resulting in brittle deformation, often along inherited structures; and the effects of dynamic topography produced by vertical mantle stresses over long wavelengths. Dynamic topography also provides an explanation for differences in degrees of elevation changes observed in some cratons, particularly where no clear surface deformation is apparent.

Recent developments in understanding the relationship between radiation damage and He diffusion in the zircon and titanite (U–Th)/He systems, add considerable value to FT data for unraveling thermal history paths in cratonic environments over a wider temperature range than previously possible. Future studies could also be undertaken in conjunction with $^{40}Ar/^{39}Ar$ K-feldspar and U–Pb apatite data to provide greater intersystem calibration, together with further exploration of the interpretation of lower-intercept ages on U–Pb zircon concordia.

Acknowledgements We are grateful to many past PhD students and researchers of the Melbourne thermochronology research group, including David Belton, Rod Brown, Fabian Kohlmann, Matevz Lorencak, Song Lu, Vhairi Macintosh, Wayne Noble, Paul O'Sullivan, Himansu Sahu and Ursula Weber, and other external researchers including Richard Everitt, Shimon Feinstein, Becky Flowers, Kerry Gallagher, Ilmo Kukkonen, Sven åke Larson, Kirk Osadetz, and Peter Sorjonen-Ward for their contributions toward low-temperature thermochronology studies in different cratonic settings and discussion around this topic. Low-temperature thermochronology craton studies at the University of Melbourne have been supported by funding from: the Australian Research Council (ARC), the Australian Institute of Nuclear Science and Engineering (AINSE), the Australian Geodynamics Cooperative Research Center, the Geological Survey of Canada, the Office for Energy Research and Development (Canada), and the AuScope program of the National Collaborative Research Infrastructure Strategy (NCRIS). Danielle Majer-Kielbaska assisted with drafting of some figures. We appreciate the thoughtful reviews provided by Mark Wildman, Ulrich Glasmacher, and Paul Fitzgerald.

References

Allen PA, Armitage JJ (2012) Cratonic basins. In: Busby C, Azor A (eds) Tectonics of sedimentary basins: recent advances. Blackwell Publishing Ltd., pp 602–620

Anderson AJ, Hodges KV, van Soest MC (2017) Empirical constraints on the effects of radiation damage on helium diffusion in zircon. Geochim Cosmochim Acta 218:308–322

Andresen A, Rehnstrom EF, Holte M (2007) Evidence for simultaneous contraction and extension at different crustal levels during the Caledonian orogeny in NE Greenland. J Geol Soc London 164:869–880

Arne DC (1991) Regional thermal history of the Pine Point area, Northwest Territories, Canada, from apatite fission-track analysis. Econ Geol 86:428–435

Artemieva IM, Vinnik LP (2016) Density structure of the cratonic mantle in southern Africa: 1. Implications for dynamic topography. Gondwana Res 39:204–216

Ault AK, Flowers RM, Bowring SA (2009) Phanerozoic burial and unroofing history of the western Slave craton and Wopmay orogen from apatite (U–Th)/He thermochronometry. Earth Planet Sci Lett 284:1–11

Ault AK, Flowers RM, Bowring SA (2013) Phanerozoic surface history of the Slave craton. Tectonics 32:1066–1083

Ault AK, Flowers RM, Bowring SA (2015) Synchroneity of cratonic burial phases and gaps in the kimberlite record: episodic magmatism or preservational bias? Earth Planet Sci Lett 410:97–104

Baillie PW, Powell CM, Li ZX, Ryall AM (1994) The tectonic framework of Western Australia's Neoproterozoic to recent sedimentary basins. In: Proceedings of the petroleum exploration society of Australia symposium, Perth, pp 45–62

Baldwin SL, Fitzgerald PG, Malusà MG (2018) Crustal exhumation of plutonic and metamorphic rocks: constraints from fission-track thermochronology (Chapter 13). In: Malusà MG, Fitzgerald PG (eds) Fission-track thermochronology and its application to geology. Springer

Baughman JS, Flowers RM, Metcalf JR, Dhansay T (2017) Influence of radiation damage on titanite He diffusion kinetics. Geochim Cosmochim Acta 205:50–64

Belton DX, Brown RW, Kohn BP, Fink D, Farley KA (2004) Quantitative resolution of the debate over antiquity of the central Australian landscape: implications for the tectonic and geomorphic stability of cratonic interiors. Earth Planet Sci Lett 219:21–34

Belton DX (2006) The low-temperature thermochronology of cratonic terranes. PhD thesis, The University of Melbourne

Belton DX, Raab MJ (2010) Cretaceous reactivation and intensified erosion in the Archean-Proterozoic Limpopo Belt, demonstrated by apatite fission track thermochronology. Tectonophysics 480:99–108

Bernet M, Garver JI (2005) Fission-track analysis of detrital zircon. Rev Mineral Geochem 58:205–237

Beucher R, Brown RW, Roper S, Stuart F, Persano C (2013) Natural age dispersion arising from the analysis of broken crystals. Part II. Practical application to apatite (U-Th)/He thermochronometry. Geochim Cosmochim Acta 120:395–416

Bierman PR, Caffee M (2002) Cosmogenic exposure and erosion history of Australian bedrock landforms. Geol Soc Am Bull 114:787–803

Bleeker W (2003) The late Archean record: a puzzle in ca. 35 pieces. Lithos 71:99–134

Bond GC (1979) Evidence for some uplifts of large magnitude in continental platforms. Tectonophysics 61:285–305

Bowring SA, Williams IS (1999) Priscoan (4.00–4.03 Ga) orthogneisses from northwestern Canada. Contrib Mineral Petrol 134:3–16

Braun J (2010) The many surface expressions of mantle dynamics. Nature Geosci 3:825–833

Braun J, Robert X, Simon-Labric T (2013) Eroding dynamic topography. Geophys Res Lett 40:1494–1499

Braun J, Stippich C, Ulrich A. Glasmacher UA (2016). The effect of variability in rock thermal conductivity on exhumation rate estimates from thermochronological data. Tectonophysics 690:288–297

Brown RW (1992) A fission track thermochronology study of the tectonic and geomorphic development of the sub-aerial continental margins of southern Africa. Ph.D. thesis, La Trobe University

Brown RW, Summerfield MA, Gleadow AJW (1994) Apatite fission track analysis: Its potential for the estimation of denudation rates and implications for models of long term landscape development. In: Kirkby MJ (ed) Process models theoretical geomorphology. Wiley, Chichester, pp 23–53

Brown RW, Summerfield MA, Gleadow AJW (2002) Denudational history along a transect across the Drakensberg Escarpment of southern Africa derived from apatite fission track thermochronology. J Geophys Res 107: ETG 10-1–ETG 10-18

Brown RW, Beucher R, Roper S, Persano C, Stuart F, Fitzgerald P (2013) Natural age dispersion arising from the analysis of broken crystals. Part I: theoretical basis and implications for the apatite (U–Th)/He thermochronometer. Geochim Cosmochim Acta 122:478–497

Burgess PM, Gurnis M, Moresi L (1997) Formation of sequences in the cratonic interior of North America by interaction between mantle, eustatic, and stratigraphic processes. GSA Bull 108:1515–1535

Burke K, Gunnell Y (2008) The African erosion surface: a continental scale synthesis of geomorphology, tectonics, and environmental change over the past 180 million years. GSA Memoir 201, 66 p

Catuneanu O, Wopfner H, Eriksson PG, Cairncross B, Rubidge B, Smith RMH, Hancox PJ (2005) The Karoo basins of south-central Africa. J African Earth Sci 43:211–253

Cawood P, Nemchin A (2000) Provenance record of a rift basin: U/Pb ages of detrital zircons from the Perth basin, Western Australia. Sed Geol 134:209–234

Cederbom C, Larson SÅ, Tullborg E-L, Stiberg J-P (2000) Fission track thermochronology applied to Phanerozoic thermotectonic events in central and southern Sweden. Tectonophysics 316:153–167

Cookenboo HO, Orchard MJ, Doaud DK (1998) Remnants of Paleozoic cover on the Archean Canadian Shield: limestone xenoliths from kimberlite in the central Slave craton. Geology 26:391–394

Coyle DA, Wagner GA (1998) Positioning the titanite fission-track partial annealing zone. Chem Geol 149:117–125

Crowley KD (1991) Thermal history of Michigan Basin and southern Canadian Shield from apatite fission track analysis. J Geophys Res 96:697–711

Crowley KD, Kuhlman SL (1988) Apatite thermochronometry of western Canadian Shield: implications for the origin of Williston Basin. Geophys Res Lett 15:221–224

Crowley KD, Ahern JL, Naeser CW (1985) Origin and epeirogenic history of the Williston Basin: evidence from fission-track analysis of apatite. Geology 13:620–623

Crowley KD, Naeser CW, Babel CA (1986) Tectonic significance of Precambrian apatite fission-track ages from the midcontinent United States. Earth Planet Sci Lett 79:329–336

Cull JP (1982) An appraisal of Australian heat-flow data. BMR Aust Geol Geophys 7:11–21

Danišík M, Sachsenhofer RF, Privalov VA, Panova EA, Frisch W, Spiegel C (2008) Low-temperature thermal evolution of the Azov Massif (Ukrainian Shield–Ukraine)—implications for interpreting (U–Th)/He and fission track ages from cratons. Tectonophysics 456:171–179

Danišík M, McInnes BIA, Kirkland CL, McDonald BJ, Evans NJ, Becker T (2017) Seeing is believing: visualization of He distribution in zircon and implications for thermal history reconstruction on single crystals. Sci Adv 3:e1601121

DeCelles PG, Giles KA (1996) Foreland basin systems. Basin Res 8:105–123

De Wit MJ, Roering C, Hart RJ, Armstrong RA et al (1992) Formation of an Archaean continent. Nature 357:553–562

Dodson MH (1973) Closure temperature in cooling geochronological and petrological systems. Contrib Mineral Petrol 259–274

Donelick RA, O'Sullivan PB, Ketcham RA, Hendriks BWH, Redfield TF (2006) Relative U and Th concentrations from LA-ICP-MS for apatite fission-track grain-age dating. Geochim Cosmochim Acta 66:A143. https://doi.org/10.1016/j.gca.2006.06.1595 (Abstract)

Edmond JM, Palmer MR, Measures CI, Grant B, Stallard RF (1995) The fluvial geochemisty and denudation rate of the Guyana Shield in Venezuela, Colombia, and Brazil. Geochem Cosmochim Acta 59:3301–3325

Enkelmann E, Garver J (2016) Low temperature thermochronology applied to ancient settings. J Geodyn 93:17–30

Elliott CG (1996) Phanerozoic deformation in the "stable" craton, Manitoba, Canada. Geology 24:909–912

Fairbridge RW, Finkl CW Jr (1980) Cratonic erosional unconformities and peneplains. J. Geology 88:69–86

Farley KA (2002) (U-Th)/He dating: techniques, calibrations and applications. Rev Mineral Geochem 47:819–843

Farley KA, Stockli DF (2002) (U-Th)/He dating of phosphates: apatite, monazite and xenotime. Rev Mineral Geochem 48:559–577

Feinstein S, Kohn BP, Osadetz KG, Everitt R, O'Sullivan P (2009) Variable Phanerozoic thermal history in the Southern Canadian Shield: evidence from an apatite fission track profile at the underground research laboratory (URL), Manitoba. Tectonophysics 475:190–199

Fitzgerald PG, Malusà MG (2018) Concept of the exhumed partial annealing (retention) zone and age-elevation profiles in thermochronology (Chapter 9). In: Malusà MG, Fitzgerald PG (eds) Fission-track thermochronology and its application to geology. Springer

Fitzsimons ICW (2003) Proterozoic basement provinces of southern and southwestern Australia, and their correlation with Antarctica. Geol Soc London Spec Publ 206:93–130

Flowers RM (2009) Exploiting radiation damage control on apatite (U-Th)/He dates in cratonic regions. Earth Planet Sci Lett 277:148–155

Flowers RM, Schoene B (2010) (U-Th)/He thermochronometry constraints on unroofing of the eastern Kaapvaal craton and significance for uplift of the southern African Plateau. Geology 38:827–830

Flowers RM, Bowring SA, Reiners PW (2006) Low long-term erosion rates and extreme continental stability documented by ancient (U–Th)/He ages. Geology 34:925–928

Flowers RM, Ault AK, Kelley SA, Zhang N, Zhong S (2012) Epeirogeny or eustasy? Paleozoic-Mesozoic vertical motion of the North American continental interior from thermochronometry and implications for mantle dynamics. Earth Planet Sci Lett 317–318:436–445

Gaál G, Gorbatschev R (1987) An outline of the Precambrian evolution of the Baltic Shield. Precambr Res 35:15–52

Gale SJ (1992) Long-term landscape evolution in Australia. Earth Surf Proc Land 17:323–343

Gallagher K, Stephenson J, Brown R, Holmes C, Fitzgerald P (2005) Low temperature thermochronology and modeling strategies for multiple samples 1: vertical profiles. Earth Planet Sci Lett 237:193–208

Gallagher K (2012) Transdimensional inverse thermal history modeling for quantitative thermochronology. J Geophys Res 117:1–16

Gallagher K, Ketcham R (2018) Comment on "Thermal history modelling: HeFTy vs QTQt" by Vermeesch and Tian. Earth Sci Rev 176:387–394

Gautheron C, Tassan-Got L, Barbarand J, Pagel M (2009) Effect of alpha-damage annealing on apatite (U–Th)/He thermochronology. Chem Geol 266:157–170

Gee DG, Sturt BA (eds) (1985) Caledonide orogen—Scandinavia and related areas: parts 1 and 2. Wiley, Chichester

Gleadow AJW, Belton DX, Kohn BP, Brown RW (2002a) Fission track dating of phosphate minerals and the thermochronology of apatite. Rev Mineral Geochem 48:579–630

Gleadow AJW, Kohn BP, Brown RW, O'Sullivan PB, Raza A (2002b) Fission track thermotectonic imaging of the Australian continent. Tectonophysics 349:5–21

Green PF, Duddy IR (2006) Interpretation of apatite (U–Th)/He ages and fission track ages from cratons. Earth Planet Sci Lett 244:541–547

Green PF, Duddy IR, Japsen P, Bonow JM, Malan JA (2017) Post-breakup burial and exhumation of the southern margin of Africa. Basin Res 29:96–127

Gorbatschev R, Bogdanova S (1993) Frontiers in the Baltic Shield. Precambr Res 64:3–22

Guenthner WR, Reiners PW, Ketcham RA, Nasdala L, Giester G (2013) Helium diffusion in natural zircon: radiation damage anisotropy, and the interpretation of zircon (U-Th)/He thermochronology. Am J Sci 313:145–198

Guenthner WR, Reiners PW, Drake H, Tillberg M (2017) Zircon, titanite and apatite (U-Th)/He ages and age-eU correlations from the Fennoscandian Shield, southern Sweden. Tectonics 36:1–21

Gurnis M (1993) Phanerozoic marine inundation of continents driven by dynamic topography above subducting slabs. Nature 364:589–593

Hanson EK, Moore JM, Bordy EM, Marsh JS, Howarth G, Robey JVA (2009) Cretaceous erosion in Central South Africa: evidence from upper-crustal xenoliths in kimberlite diatremes. S Afr J Geol 112:125–140

Hamblin AP (1990) Petroleum potential of the Cambrian Mount clarke formation (Tedji Lake Play), Colville Hills area, Northwest Territories. Geol Surv Canada Open File 2309

Harman R, Gallagher K, Brown RW, Raza A, Bizzi L (1998) Accelerated denudation and tectonic/geomorphic reactivation of the cratons of northeastern Brazil during the Late Cretaceous. J Geophys Res 103:27091–27105

Harris L (1994) Structural and tectonic synthesis for the Perth Basin, Western Australia. J Petroleum Geol 17:129–156

Harrison TM (2009) The hadean crust: evidence from >4 Ga zircons. Annu Rev Earth Planet Sci 37:479–505

Heaman LM, Kjarsgaard BA, Creaser RA (2004) The temporal evolution of North American kimberlites. Lithos 76:377–397

Hendriks BWH, Redfield TF (2005) Apatite fission track and (U–Th)/He data from Fennoscandia: an example of underestimation of fission track annealing in apatite. Earth Planet Sci Lett 236:443–458

Hendriks BWH, Redfield TF (2006) Reply to: Comment on "Apatite fission track and (U–Th)/He data from Fennoscandia: an example of underestimation of fission track annealing in apatite" by BWH Hendriks and TF Redfield. Earth Planet Sci Lett 248:568–576

Hendriks BWH, Andriessen P, Huigen Y, Leighton C, Redfield T, Murrell G, Gallagher K, Nielsen SB (2007) A fission track data compilation for Fennoscandia. Norw J Geol 87:143–155

Huigen Y, Andriessen PAM (2004) Thermal effects of Caledonian foreland basin formation, based on fission track analyses applied on basement rocks in central Sweden. Phys Chem Earth 29:683–694

Japsen P, Green PF, Bonow JM, Erlström M (2016) Episodic burial and exhumation of the southern Baltic Shield: epeirogenic uplifts during and after break-up of Pangaea. Gondwana Res 35:357–377

Jelsma H, Barnett W, Richards S, Lister G (2009) Tectonic setting of kimberlites. Lithos 112S:155–165

Johnson JE, Flowers RM, Baird GB, Mahan KH (2017) "Inverted" zircon and apatite (U–Th)/He dates from the front range, Colorado: high-damage zircon as a low-temperature (<50 °C) thermochronometer. Earth Planet Sci Lett 466:80–90

Kasanzu CH, Linol B, de Wit MJ, Brown R, Persano C, Stuart FM (2016) From source to sink in central Gondwana: coeval exhumation of the Precambrian basement rocks of Tanzania and sediment accumulation in the adjacent Congo Basin. Tectonics 35:2034–2051

Kasanzu CH (2017) Apatite fission track and (U–Th)/He thermochronology from the Archean Tanzania Craton: contributions to cooling histories of Tanzanian basement rocks. Geosci Front 8:999–1007

Kasuya M, Naeser CW (1988) The effect of α-damage on fission-track annealing in zircon. Nuclear Tracks Radiat Meas 14:477–480

Ketcham RA (2005) Forward and inverse modeling of low-temperature thermochronometry data. Rev Mineral Geochem 58:275–314

Ketcham R (2018) Fission-track annealing: from geologic observations to thermal modeling (Chapter 3). In: Malusà MG, Fitzgerald PG (eds) Fission-track thermochronology and its application to geology. Springer

Killick MF (1998) Phanerozoic denudation of the Western Shield of Western Australia. Geol Soc Aust Abstr 49:248

Kohlmann F (2010) Insights into the nanoscale formation of fission tracks in solids and a low temperature thermochronology study of the Archaean Slave Province, Northwest Territories, Canada. Ph.D. thesis, The University of Melbourne

Kohlmann F, Kohn BP, Gleadow AJW, Osadetz KG (2007) Low temperature thermochronology Phanerozoic kimberlites and Archaean basement, Slave Province Canada. Geochim Cosmochim Acta 71(15):A505

Kohn BP, Osadetz KG, Bezys RK (1995) Apatite fission-track dating of two crater structures in the Canadian Williston Basin. Bull Can Pet Geol 43:54–64

Kohn BP, Gleadow AJW, Brown RW, Gallagher K, O'Sullivan PB, Foster DA (2002) Shaping the Australian crust over the last 300 million years: insights from fission track thermotectonic and denudation studies of key terranes. Aust J Earth Sci 49:697–717

Kohn BP, Gleadow AJW, Raza A, Mavrogenes J, Raab MJ, Belton, DX, Kukkonen IT (2004) Revisiting apatite under pressure: further experimental data on fission track annealing and deep borehole measurements. In: Abstracts, 10th international conference on fission track dating and thermochronology conference, Amsterdam, p 54

Kohn BP, Gleadow AJW, Brown RW, Gallagher K, Lorencak M, Noble WP (2005) Visualising thermotectonic and denudation histories using apatite fission track thermochronology. Rev Mineral Geochem 58:527–565

Kohn BP, Lorencak M, Gleadow AJW, Kohlmann F, Raza A, Osadetz KG, Sorjonen-Ward P (2009) Low temperature thermochronology of the eastern Fennoscandia Shield and radiation enhanced apatite fission track annealing revisited. Geol Soc London Spec Publ 324:193–216

Kohn B, Chung L, Gleadow A (2018) Chapter 2. Fission-track analysis: field collection. sample preparation and data acquisition. In: Malusà MG, Fitzgerald PG (eds) Fission-track thermochronology and its application to geology. Springer

Kounov A, Viola G, Dunkl I, Frimmel HE (2013) Southern African perspectives on the long-term morpho-tectonic evolution of cratonic interiors. Tectonophysics 601:177–191

Kukkonen IT (1998) Temperature and heat flow density in a thick cratonic lithosphere: the Sveka transect, central Fennoscandian Shield. J Geodyn 26:111–136

Larson SÅ, Tullborg E-L (1998) Why Baltic Shield zircons yield late Palaeozoic, lower-intercept ages on U-Pb concordia. Geology 26:919–922

Larson SÅ, Tullborg E-L, Cederbom C, Stiberg J-P (1999) Sveconor-wegian and Caledonian foreland basins in the Baltic Shield revealed by fission track thermochronology. Terra Nova 11:210–215

Larson SÅ, Cederbom C, Tullborg E-L, Stiberg J-P (2006) Comment on "Apatite fission track and (U–Th)/He data from Fennoscandia: an example of underestimation of fission track annealing in apatite" by Hendriks and Redfield [Earth Planet Sci Lett 236 (443–458)]. Earth Planet Sci Lett 248:561–568

Le Blanc Smith G (1993) Geology and Permian coal resources of the Collie Basin, Western Australia, Report 38. Geological Survey of Western Australia, 86 p

Lenardic A, Moresi L (1999) Some thoughts on the stability of cratonic lithosphere: effects of buoyancy and viscosity. J Geophys Res 104:12747–12758

Lenardic A, Moresi L, Mühlhaus H (2000) The role of mobile belts for the longevity of deep cratonic lithosphere: the crumple zone model. Geophys Res Lett 1235–1238

Lidmar-Bergström K (1996) Long-term morphotectonic evolution in Sweden. Geomorphology 16:33–59

Lithgow-Bertelloni C, Silver PG (1998) Dynamic topography, plate driving forces and the African superswell. Nature 395:269–272

Lorencak M (2003) Low temperature thermochronology of the Canadian and Fennoscandian Shields: integration of apatite fission track and (U-Th)/He methods. Ph.D. thesis, The University of Melbourne

Lorencak M, Kohn BP, Osadetz KG, Gleadow AJW (2004) Combined apatite fission track and (U-Th)/He thermochronometry in a slowly cooled terrane: results from a 3440 m deep drill hole in the southern Canadian Shield. Earth Planet Sci Lett 227:87–104

Lu S (2016) The thermotectonic evolution of the south-west Yilgarn Craton, Western Australia. Ph.D. thesis, The University of Melbourne

Łuszczak K, Persano C, Braun J, Stuart FM (2017) How crustal thermal properties influence the amount of denudation derived from low-temperature thermochronometry. Geology 45:779–782

Mackintosh V (2017) Thermochronological insights into the morpho-logical evolution of Zimbabwe, southern Africa. Ph.D. thesis, The University of Melbourne

Mackintosh V, Kohn B, Gleadow A, Tian Y (2017) Phanerozoic morphotectonic evolution of the Zimbabwe Craton: unexpected outcomes from a multiple low-temperature thermochronology study. Tectonics 36:2044–2067

Malusà MG (2018) A guide for interpreting complex detrital age patterns in stratigraphic sequences (Chapter 16). In: Malusà MG, Fitzgerald PG (eds) Fission-track thermochronology and its appli-cation to geology. Springer

Malusà MG, Fitzgerald PG (2018) From cooling to exhumation: setting the reference frame for the interpretation of thermochronologic data (Chapter 8). In: Malusà MG, Fitzgerald PG (eds) Fission-track thermochronology and its application to geology. Springer

Mitrovica JX, Beaumont C, Jarvis GT (1989) Tilting of continental interiors by the dynamical effects of subduction. Tectonics 8:1079–1094

Mooney WD, Ritsema J, Hwang YK (2012) Crustal seismicity and the earthquake catalog maximum moment magnitude (M_{cmax}) in stable continental regions (SCRs): correlation with the seismic velocity of the lithosphere. Earth Planet Sci Lett 357–358:78–83

Moore A, Blenkinsop T, Cotterill F (2009) Southern African topog-raphy and erosion history: plumes or plate tectonics? Terra Nova 21:310–315

Morrow C, Lockner D, Hickman S, Rusanov M, Röckel T (1994) Effects of lithology and depth on the permeability of core samples from the Kola and KTB drill holes. J Geophys Res 99:7263–7274

Mory A, Redfern J, Martin J (2008) A review of Permian-Carboniferous glacial deposits in Western Australia. Geol Soc Am Spec Pap 441:29–40

Mottaghy D, Schellschmidt TR, Popov YA, Clauser C, Kukkonen IT, Nover G, Milanovsky S, Romushkevich RA (2005) New heat flow data from the immediate vicinity of the Kola super-deep borehole: vertical variation in heat flow confirmed and attributed to advection. Tectonophysics 401:119–142

Moucha R, Forte AM, Mitrovica JX, Rowley DB, Quéré S (2008) Dynamic topography and long-term sea-level variations: there may be no such thing as a stable continental platform. Earth Planet Sci Lett 271:101–108

Murrell GR (2003) The long-term thermal evolution of Central Fennoscandia. Ph.D. thesis, Vrjie Universiteit, Amsterdam

Murrell GR, Andriessen PAM (2004) Unravelling a long-term multi-event thermal record in the cratonic interior of southern Finland through apatite fission track thermochronology. Phys Chem Earth 29:695–706

Myers JS (1993) Precambrian history of the west Australian craton and adjacent orogens. Ann Rev Earth Planet Sci 21:453–485

Noble PW (1997) Post Pan African tectonic evolution of Eastern Africa: an apatite fission track study. Ph.D. thesis, La Trobe University

O'Neill CJ, Lenardic A, Griffin WL, O'Reilly SY (2008) Dynamics of cratons in an evolving mantle. Lithos 102:12–24

Orme DA, Guenthner WR, Laskowski AK, Reiners PW (2016) Long-term tectonothermal history of Laramide basement from zircon–He age-eU correlations. Earth Planet Sci Lett 453:119–130

Osadetz KG, Kohn BP, Feinstein S, O'Sullivan PB (2002) Williston basin thermal history from apatite fission track thermochronology—implications for petroleum systems and geodynamic history. Tectonophysics 349:221–249

Patchett PJ, Embry AF, Ross GM, Beauchamp B, Harrison JC, Mayr U, Isachsen CE, Rosenberg EJ, Spence GO (2004) Sedimentary cover of the Canadian shield through Mesozoic time reflected by Nd isotopic and geochemical results for the Sverdrup Basin, Arctic Canada. J Geol 112:39–57

Pell JA (1997) Kimberlites in the Slave craton, Northwest Territories, Canada. Geosci Can 24:77–90

Pinet N (2016) Far-field effects of Appalachian orogenesis: a view from the craton. Geology 44:83–86

Pinet N, Kohn BP, Lavoie D (2016) The ups and downs of the Canadian Shield: 1-preliminary results of apatite fission track analysis from Hudson Bay region. Geol Surv Canada Open File 8110

Pinet N, Lavoie D, Dietrich J, Hu K, Keating P (2013) Architecture and subsidence history of the Hudson Bay intracratonic basin. Earth Sci Rev 125:1–23

Popov YA, Pevzner SL, Pimenov VP, Romushkevich RA (1999) New geothermal data from the Kola superdeep well SG-3. Tectonophysics 306:345–366

Reiners PW (2005) Zircon (U-Th)/He thermochronometry. Rev Min Geochem 58:151–179

Rohrman MHEJ (1995) Thermal evolution of the Fennoscandian region from fission track thermochronology: an integrated approach. Ph.D. thesis, Vrije Universiteit, Amsterdam

Rudnick RL, McDonough WF, O'Connell RJ (1998) Thermal structure, thickness and composition of continental lithosphere. Chem Geol 145:395–411

Schneider DA, Heizler MT, Bickford ME, Wortman GL, Condie KC, Perilli S (2007) Timing constraints of orogeny to cratonization: thermochronology of the Paleoproterozoic Trans-Hudson orogen, Manitoba and Sakatchwan, Canada. Precambr Res 153:65–95

Shuster DL, Flowers RM, Farley KA (2006) The influence of natural radiation damage on helium diffusion kinetics in apatite. Earth Planet Sci Lett 249:148–161

Sloss LL, Speed RC (1974) Relationships of cratonic and continental-margin tectonic episodes. Soc Econ Paleontol Mineral Spec Publ 22:98–119

Söderlund P, Juez-Larré J, Page LM, Dunai TJ (2005) Extending the time range of apatite (U–Th)/He thermochronometry in slowly cooled terranes: Palaeozoic to Cenozoic exhumation history of southeast Sweden. Earth Planet Sci Lett 239:266–275

Stanley JR, Flowers RM, Bell DR (2013) Kimberlite (U-Th)/He dating links surface erosion with lithospheric heating, thinning, and metasomatism in the southern African Plateau. Geology 41:1243–1246

Stanley JR, Flowers RM, Bell DR (2015) Erosion patterns and mantle sources of topographic change across the southern African Plateau

derived from the shallow and deep records of kimberlites. Geochem Geophys Geosyst 16:3235–3256

Stasiuk LD, Sweet AR, Issler DR (2006) Reconstruction of burial history of eroded Mesozoic strata using kimberlite shale xenoliths, volcaniclastic and crater facies, Northwest Territories, Canada. Int J Coal Geol 65:129–145

Steel R, Siedlecka A, Roberts D (1985) The old red sandstone basins of Norway and their deformation: a review. In: Gee DG, Sturt BA (eds) The Caledonide orogen: Scandinavia and related areas. Wiley, Chichester, pp 293–315

Stockli DF, Farley KA (2004) Empirical constraints of the titanite (U–Th)/He partial retention zone from the KTB drillhole. Chem Geol 207:223–236

Streule MJ, Strachan RA, Searle MP, Law RD (2010) Comparing Tibet-Himalayan and Caledonian crustal architecture, evolution and mountain building processes. Geol Soc London Spec Publ 335:207–232

Tagami T (2005) Zircon fission-track thermochronology and applications to fault studies. In: Reiners P, Ehlers T (eds) Low-temperature thermochronology. Rev Min Geochem 58:95–122

Tullborg E-L, Larson SÅ, Björklund L, Samuelsson LJ, Stigh J (1995) Thermal evidence of Caledonide foreland, molasse sedimentation in Fennoscandia: Stockholm. Swedish Nuclear and Waste Management Company, Technical report 95–18, 38 p

Veevers J, Saeed A, Belousova E, Griffin W (2005) U-Pb ages and source composition by Hf isotope and trace-element analysis of detrital zircons in Permian sandstone and modern sand from southwestern Australia and a review of the paleogeographical and denudational history of the Yilgarn Craton. Earth Sci Rev 68:245–279

Vermeesch P, Tian Y (2014) Thermal history modelling: HeFTy vs QTQt. Earth Sci Rev 139:279–290

Vermeesch P, Tian Y (2018) Reply to comment on "Thermal history modelling: HeFTy vs. QTQt" by Gallagher and Ketcham. Earth Sci Rev 176:395–396

Vermeesch P (2018) Statistics for fission-track thermochronology (Chapter 6). In: Malusà MG, Fitzgerald PG (eds) Fission-track thermochronology and its application to geology. Springer

Weber UD (2002) The thermotectonic evolution of the northern Precambrian Shield, Western Australia. Ph.D. thesis, The University of Melbourne

Weber UD, Kohn BP, Gleadow AJW, Nelson DR (2005) Low temperature Phanerozoic history of the northern Yilgarn Craton, Western Australia. Tectonophysics 400:127–151

Weides S, Majorowicz J (2014) Implications of spatial variability in heat flow for geothermal resource evaluation in large foreland basins: the case of the Western Canada Sedimentary Basin. Energies 7:2573–2594

Wenker S, Beaumont C (2017) Can metasomatic weakening result in the rifting of cratons? Tectonophysics (in press)

Wildman M, Brown R. Beucher R, Persano, C, Stuart, F Gallagher K, Schwanethal J, Carter A (2016) The chronology and tectonic style of landscape evolution along the elevated Atlantic continental margin of South Africa resolved by joint apatite fission track and (U-Th-Sm)/He thermochronology. Tectonics 35:511–545

Wildman M, Brown R, Persano C, Beucher R, Stuart FM, Mackintosh V, Gallagher K, Schwanethal J, Carter A (2017) Contrasting Mesozoic evolution across the boundary between on and off craton regions of the South African plateau inferred from apatite fission track and (U-Th-Sm)/He thermochronology. J Geophy Res-Solid Earth 122:1–31

Wildman M, Beucher R, Cogne N (2018) Fission-track thermochronology applied to the evolutionof passive continental margins (Chapter

20). In: Malusà MG, Fitzgerald PG (eds) Fission-track thermochronology and its application to geology. Springer

Wolfe MR, Stockli DF (2010) Zircon (U-Th)/He thermochronometry in
the KTB drill hole, Germany, and its implications for bulk He
diffusion kinetics in zircon. Earth Planet Sci Lett 295:69–82

Xu YG (2001) Thermo-tectonic destruction of the Archean lithospheric
keel beneath the Sino-Korean craton in China: evidence, timing and
mechanism. Phys Chem Earth Part A, Solid Earth Geodesy 26:747–
757

Zeck HP, Andriessen PAM, Hansen K, Jensen PK, Rasmussen BL
(1988) Paleozoic paleo-cover of the southern part of the Fennoscandian Shield—fission track constraints. Tectonophysics 149:61–66

Zeitler PK, Herczeg AL, McDougall I, Honda M (1987) U-Th-He
dating of apatite: a potential thermochronometer. Geochim Cosmochim Acta 51:2865–2868

Zhang N, Zhong S, Flowers R (2012) Predicting and testing continental
vertical motion histories since the Paleozoic. Earth Planet Sci Lett
317–318:426–435

Printed in the United States
By Bookmasters